Jörg Knäblein (Ed.)

Modern Biopharmaceuticals

Modern Biopharmaceuticals

Volume 3

Design, Development and Optimization

Edited by
Jörg Knäblein

WILEY-VCH Verlag GmbH & Co. KGaA

Editor

Dr. Jörg Knäblein
Head Microbiological Chemistry
Schering AG
Müllerstraße 178
13342 Berlin
Germany

Library of Congress Card No.:
applied for

British Library Cataloguing-in-Publication Data
A catalogue record for this book is available from the British Library

**Bibliographic information published by
Die Deutsche Bibliothek**
Die Deutsche Bibliothek lists this publication in the Deutsche Nationalbibliografie; detailed bibliographic data is available in the Internet at http://dnb.ddb.de.

Printed in the Federal Republic of Germany
Printed on acid-free paper

Cover   Tim Fonseca, www.fonsecatim.com
Typsetting   K+V Fotosatz GmbH, Beerfelden
Printing   betz-druck GmbH, Darmstadt
Bookbinding   J. Schäffer GmbH, Grünstadt

ISBN-13   978-3-527-31184-2
ISBN-10   3-527-31184-X

Contents

Volume 1

Modern Biopharmaceuticals. Edited by J. Knäblein
Copyright © 2005 WILEY-VCH Verlag GmbH & Co. KGaA, Weinheim
ISBN: 3-527-31184-X

Volume 4

Part IV
Production of Biopharmaceuticals

The Industry's Workhorses – Mammalian Expression Systems

1
Manufacture of Recombinant Biopharmaceutical Proteins by Cultivated Mammalian Cells in Bioreactors

Florian M. Wurm

Abstract

The first recombinant protein therapeutic made in cultivated mammalian cells obtained market approval in 1986. This event made the use of Chinese hamster ovary (CHO) cells in large-scale bioreactors known to a wider public. These cells are now the dominating host system for recombinant protein production, as more than 60% of all new target proteins in the clinical pipelines of pharmaceutical and biotechnology companies are being produced in hamster-derived cells. This chapter will cover aspects of gene transfer, cell line development and process development for mammalian protein expression systems using CHO cells as the main example, but also making reference to other mammalian cells that are used for the large-scale production of therapeutic proteins. Most importantly, the scientific and technological insights that resulted in the rapid and surprising yield improvements from such processes, bringing the volumetric productivity of mammalian cell culture processes into the gram per liter range, will be discussed. Not only is this level of productivity equal to that of microbial systems, but the recombinant proteins in mammalian cells have all the necessary secondary modifications that only a higher eukaryote can execute. The regulatory framework for the use of mammalian host systems will also be discussed, as the perceived risks of transmission of adventitious agents to patients have resulted in stringent rules to which all manufacturers must adhere.

Abbreviations

ADCC	antibody-dependent cellular cytotoxicity
BHK	baby hamster kidney
BSE	bovine spongiform encephalopathy
cGMP	current Good Manufacturing Practice
CHO	Chinese hamster ovary
CIP	clean in place

Modern Biopharmaceuticals. Edited by J. Knäblein
Copyright © 2005 WILEY-VCH Verlag GmbH & Co. KGaA, Weinheim
ISBN: 3-527-31184-X

DHFR	dihydrofolate reductase
FISH	fluorescence *in situ* hybridization
GHT	glycine, hypoxanthine, and thymidine
GS	glutamate synthetase
IND	investigational new drug
MSCB	Master Seed Cell Bank
MSX	methionine sulphoximine
MTX	methotrexate
MVM	minute virus of the mouse
MWCB	Manufacturer's Working Cell Bank
PCV	packed cell volume
Pip	pristinamycin-induced protein
PLA	Process Licence Applications
PrP_{sc}	proteinaceous-infectious particles, scrapie
S/MARs	scaffold or matrix attachment regions
SEAP	secreted alkaline phosphatase
SIP	sterilize in place
tPA	tissue plasminogen activator
UCOS	ubiquitous chromatin opening elements

1.1

Introduction

A small number of immortalized mammalian cells have become host systems for production of kilogram (ton) per year quantities of complex recombinant proteins for clinical applications. In 2004, these cells will produce about 60–70% of all recombinant biopharmaceutical proteins. The most popular cells are Chinese hamster ovary (CHO), Mouse myeloma-derived NS0 (NS0), Baby hamster kidney (BHK) and human retina-derived immortalized PER.C6 (PER.C6) cells (see also Part IV, Chapter 3). One compelling reason why mammalian cells are now so popular is the exceptional productivity reported in a few cases. Sophisticated processes with highly optimized cell lines can provide grams per liter of recombinant antibodies or chimeric immunoglobulin-fusion proteins, in extended batch cultures of volumes of up to 12 000 L. The know-how and technology behind large-scale processes for mammalian cells have evolved over 20 years and have resulted in a more than 100-fold improvement in volumetric productivity (Table 1.1).

Hundreds of proteins are presently being produced in mammalian cells for clinical evaluation. The decision to employ more complex *in vitro* cultures of recombinant mammalian cells has been driven by the need to obtain proteins with complex biochemical structures and resulting superior biological activity, reflecting their native structure and function. Biological activity and the pharmacokinetic characteristics of recombinant proteins frequently depend on a number of complex protein modifications (i.e., proper folding, disulfide bridge formation, oligomerization, proteolytic processing, phosphorylation and the addition of specific and complex

Table 1.1 Volumetric productivities from cell culture processes developed for the production of recombinant proteins for clinical use: observations over a 20-year period

	Volumetric productivity titers of recombinant proteins in cell culture media		
Period	1982–1985	1992–1995	2002–2004
Titer range	5–50 mg L^{-1}	50–500 mg L^{-1}	500–5000 mg L^{-1}

carbohydrate groups). These and other protein processing steps can be executed with high efficiency by mammalian cells. Thus, *Escherichia coli* and fungi – the production hosts of first choice during the early years of applied DNA technology – are now playing only a minor role for large-scale expression of proteins for clinical use.

The terms "expression" and "production" used herein refer to secreted proteins only. To my knowledge, all biopharmaceutical proteins from mammalian cells were developed from gene constructs that allow secretion of the desired protein into the culture medium. This is one of the most intriguing advantages of protein production in mammalian cell culture. The cell is truly used as a machine that converts a given DNA construct into a protein product while assuring that the product is secreted and thus easily separated from the majority of the cell's contents. Fig. 1.1 shows an SDS-PAGE analysis of a full-length recombinant antibody in the supernatant from a CHO culture or in an *E. coli* lysate. For the latter, a weak set of IgG-specific bands and a large number of bands corresponding to bacterial proteins are observed. In contrast, the antibody product in the supernatant of CHO cells represents a much higher percentage of the total protein.

The use of CHO and NS0 cells for the large-scale production of recombinant proteins has been facilitated by the fact that both cell types grow well in single-cell suspension culture with the highest cell densities from an extended batch process reported to be greater than 10^7 cells mL^{-1}. This is about five times higher than was seen 20 years ago. Optimized cell lines can achieve secretion of 50 pg per cell per day of recombinant protein. The record on volumetric titer from recombinant mam-

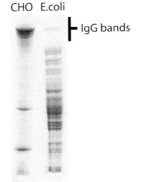

CHO E.coli

IgG bands

Fig. 1.1 SDS-Polyacrylamide gel electrophoresis of samples of supernatants and lysates, respectively, of a Chinese hamster ovary process and of an *E. coli* fermentation producing full-length human antibodies. (Image provided by Genentech Inc., courtesy Drs. Brad Snedecor and Lynne Krummen)

malian cells (CHO) for a secreted antibody made in an extended batch process stands at 4.6 g L^{-1} [1].

CHO cells, originally derived from the ovary of a female Chinese hamster, emerged as a spontaneously immortalized line, in the hands of Kao and Puck more than 40 years ago [2, 3]. The cells have been the object of many studies since then, and are very well characterized with respect to a variety of aspects, including karyotype, chromosome structures, gene mapping, general culture conditions, cell physiology and media requirements. For the purpose of genetic and physiological studies on the activity of dihydrofolate reductase (DHFR) and its genomic alleles, during the late 1970s, Urlaub and Chasin [4] mutagenized the original cell line with the help of radiation and chemicals. The goal was to create a mutant cell line lacking functional activity of both alleles of the DHFR gene locus. The resulting cell line CHO-DUX (sometimes also referred to as CHO-DUKX-B11), though not initially intended for this purpose, proved to be an ideal substrate for a number of pioneering gene transfer experiments [5–7].

CHO cells were chosen as the host substrate for the first recombinant protein from mammalian cells, human tissue plasminogen activator (Activase®; tPA) [8], which achieved market approval in the US and other global markets in 1986/1987. Subsequently, other companies chose this production system because the approval barriers for a second product from the same host were considered easier to overcome, especially with respect to the regulatory process in the USA. CHO cells are unique in many ways and exhibit a full set of advantages, ranging from ease of introduction of exogenous DNA to capacity for growth and high-level productivity at large scales. An excellent selection of articles concerning the scientific history, general and cellular biology, cytogenetics and molecular biology of Chinese hamsters and CHO cells can be found in a comprehensive compendium edited by Gottesman [9]. For two decades now not only CHO cells but also NS0 cells have spearheaded the development of animal cell technology in an unprecedented way, and are now the basis of accessory industries that have developed around the clinical manufacturers. Numerous companies provide complex media formulations that boost the growth and productivity of these cells for large-scale operations, while eliminating undefined mixtures of growth factors and nutritional components, such as fetal bovine serum or animal tissue-derived peptones.

Any host system different from CHO will be subject to the same regulatory scrutiny. For obvious reasons, due to the evolutionary relatedness of all mammalian species, regulatory concerns for the transmission of unknown disease-causing principles is higher when utilizing a hamster-derived cell line than when using a microbial host. Safety concerns were recently raised to a new level because of the transmission of a bovine prion disease to humans, causing variant Creutzfeld–Jakob disease in hundreds of consumers. This article attempts to summarize arguments, issues, advantages, questions and ongoing research for the industrial production of high value proteins derived from mammalian cells. Section 1.2 provides an introduction to the principles of expression of recombinant proteins from mammalian cells. Sections 1.3, 1.4, and 1.5 address, in a more profound way, the molecular and cellular biology of gene transfer and gene amplification of recombinant DNA in mammalian cells, with the emphasis being on the genetic stability of recombinant cells. Section 1.6 discusses process issues for scale-up and manufacturing, and Section 1.7 is an extensive discussion on regulatory aspects of these processes. The importance of "regulatory" issues should not be underestimated, because most of the money and time invested in the development of a manufacturing scheme based on mammalian cells will go to addressing the safety, consistency and quality of the product.

Issues raised and discussed herein are not meant to be comprehensive. However, it is hoped that the most critical points impacting developmental efforts for protein production with mammalian cells in culture will be addressed.

1.2
Vectors, Transfections, and Cell Line Generation

1.2.1
Calcium Phosphate DNA Coprecipitates for Transfection

Graham and van der Eb [10] showed more than 30 years ago that exposing cells to micro-precipitates of DNA and calcium phosphate allowed the transfer of DNA into cultivated mammalian cells. This simple method was termed calcium phosphate transfection, and a number of modified versions have become widely used. With optimized transfection conditions, up to 100% of cells take DNA into their cytoplasm. However, in addition to the transit across the cells' plasma membrane, the transport of DNA from the cytoplasm to the nucleus seems to be a significant barrier. Axel and collaborators first showed stable integration of transfected DNA into chromosomes of mammalian cells [11, 12]. The number of emerging colonies upon transfection by the classical calcium phosphate technique is low: usually between 0.05% and 1% of the transfected cells give rise to recombinant colonies [13]. Recent improvements of crucial physico-chemical parameters of the calcium phosphate transfection methodology have increased the frequency of stable recombinant cells to about 5–10% of the transfected population. In addition, the efficiency of transient transfection has increased to levels of 50% or higher [14–16]. Calcium phosphate transfection is still a frequently used method for the generation of recombinant CHO cells. Other methods, to be discussed below, are also appropriate.

With calcium phosphate, and also with other methods, a large excess of DNA molecules over the number of cells is nor-

mally used, in the region of 100 000 or more per cell. This is probably necessary, as much of the DNA will not reach the nucleus, due to degradation. However, it has been shown that association of the DNA with calcium phosphate protects against nuclease attack during transport to the nucleus. Eventually, these complexes become dissociated and nuclear endo- and exonucleases will have access to the "naked" DNA. Moreover, it has been shown that supercoiled plasmid DNA molecules will be converted into relaxed (single strand cut) and linear molecules (double strand cut) within the nucleus after 1–2 hours [17]. This is an essential step for integration into the linear DNA backbone of a chromosome.

The mechanism by which the DNA is transported across the nuclear membrane (nuclear pores are not large enough for diffusive transport) and finally to the site of chromosomal integration is not yet known. There is evidence however that disruption of the nuclear membrane during mitotic activity of cells is an important aspect of overcoming this barrier [18]. On the receiving side of the process of integration, at least one strand of the chromosomal DNA needs to be opened while the linear plasmid molecule(s) is in sufficiently close proximity to integrate. Nuclear ligases could then mediate covalent linkages. Little is known about the mechanism affecting the site specificity of integration. It is assumed that genome DNA replication and repair could facilitate entry of exogenous DNA into chromosomal DNA.

The site of integration is a critical factor since it influences the transcription rate of the integrated DNA. The current assumption is that exogenous DNA will integrate randomly within the genome. This is also true if the transfection cocktail contains

DNA with homology to sequence segments of the genome. Gene targeting experiments in mammalian cells is inefficient; sequence-specific integration is a rare event. Only 1 in 1000 to 1 in 10 000 events result in targeted integration [19, 20]. With respect to actual sites of integration in CHO cells there are few data available. If integration in general is random, only a small proportion of recombinant cell lines should contain the transferred DNA within transcriptionally active regions of chromosomes, since very little DNA of any higher eukaryote is transcribed at any time. However, the frequency in which exogenous DNA integrates into transcriptionally inactive regions of the genome cannot be determined since cells that do not overcome the selection step cannot be analyzed.

We performed a small study using fluorescence *in situ* hybridization (FISH) to determine the integration sites in 12 clonal cell lines following calcium phosphate transfection [21]. Interestingly, we found only a single integration site in each cell line. There was no preference for a specific chromosomal region, but there may have been a slight preference for larger chromosomes (1 to 3). This bias, however, is most likely due to the fact that these three chromosomes represent a large fraction of the genomic DNA.

Nuclear enzymes such as endo- and exonucleases, but also ligases [22] and possibly recombinogenic enzymes [23, 24] acting on the population of plasmid molecules within the nucleus, are responsible for the modifications of transfected DNA that eventually becomes integrated into the genome. These modifications may not only degrade many plasmid molecules, but nuclear ligase activity is also responsible for the creation of larger DNA complexes containing numerous copies of the plasmid DNA. These DNA molecules seem to be created before integration into the genome. They provide the physical basis a genetic link between the selection marker and the gene(s) of interest in those transfections that utilize separate vectors. In my laboratory, co-transfections are being executed routinely, and we have found a high degree of covalent linkage of all individual plasmid molecules when analyzing the integration sites of stable cell lines [25]. One should be aware that co-integration of multiple plasmids is probably a general phenomenon in eukaryotic cells. Chen and coworkers reported the generation of transgenic rice plants receiving and expressing 13 different plasmids out of 14 that were used in the DNA cocktail. Analysis by Mendelian genetic approaches revealed integration into one locus [26]. It should be noted here that DNA transfer methods other than calcium phosphate might deliver different quantities of DNA to the nucleus. This may have profound effects on the structure and copy number of integrated DNA molecules (see Sect. 1.4).

Among individual cell lines from a single transfection there is usually extensive heterogeneity in productivity of the recombinant protein. The expression levels from mammalian cell clones generally have a very wide range, sometimes exceeding two orders of magnitude [27, 28]. Numerous laboratories have verified this observation with CHO, NS0 [29] or PerC6 cells [30]. As a consequence, the identification of high producer cell lines is a tedious and labor-intensive exercise, and requires the screening of hundreds of individual cell lines. It is generally necessary to invest between 2 and 4 months of laboratory work into this task. Only then can the upper range of sustainable expression for a recombinant protein be assessed.

1.2.2
Other DNA Transfer Methods
for Mammalian Cells

In addition to calcium phosphate transfection, methods for DNA transfer into mammalian cells by electroporation [31, 32] and by transfection mediated through cationic lipids, liposome [33–35], biolistics [36] and polymers [37, 38] have been developed. Most of these techniques have been reported to mediate higher transfection efficiencies as compared to calcium phosphate-mediated DNA transfer. Such claims must be regarded with some caution, as all DNA transfer techniques established so far suffer from high variability due to technical difficulties. Other factors to cause major variations in transfection efficiency are the type of cells used and the condition of the cells prior to transfection.

Since the mechanism for the transfer of DNA to cells can vary widely from one method to another, one suspects that very different consequences may result within the cell – events over which the experimentation has no or little control. For example, the DNA transfer method may affect the plasmid copy number average in individual clones, depending on the amount of DNA transferred to the nucleus. Electroporation may result in a lower copy number of integrated DNA than calcium phosphate-based transfection. In view of the applied selection procedures for the generation of recombinant cell lines, very different structures of integrated plasmid DNA may result.

Another factor to be taken into consideration for the integration of plasmid DNA is the physical form of the transfected DNA. Since linearization of plasmid DNA is a prerequisite for integration, the form of the plasmid prior to transfection will affect the integrated form. When circular plasmids are used, as was done in the past, linearization is dependent on cellular enzyme activity. Since eukaryotic DNA degrading enzymes act on random sequences, a significant number of molecules will be linearized within the DHFR sequence, jeopardizing the functionality of the plasmid. Therefore, opening the circular plasmid molecule by restriction enzyme digestion with appropriate enzymes prior to transfection is recommended for improved transfection efficiencies in protocols that aim at integration of plasmid DNA.

We have seen in our laboratory that linearization of plasmids improves the efficiency of stable transfections [15]. Linear plasmid molecules are also used routinely to transfer the DNA into a specific (homologous) DNA sequence of the host genome (gene targeting). In these experiments, linearization is executed in cutting the homologous DNA fragment approximately in half. The molecules for DNA transfer carry homologous DNA sequences at its ends [39, 40]. We have used such an approach in our laboratory by employing (defective) retroviral DNA sequences derived from the CHO genome in expression vector cocktails. They have improved significantly the frequency of high producer cell lines upon transfection [27, 28]. Several conference presentations have hinted at the use of gene targeting DNA, though no specific data have been published in this respect. This approach requires well-designed vectors in combination with plasmids providing enzymes favoring targeted integration such as Bacteriophage P1 Cre recombinase, lambda phage integrase, or yeast Flp recombinase. These proteins are capable of exchanging long stretches of DNA, that are bordered by regions of homology, between the genome and the vector DNA [41–43].

1.2.3
Vectors for Expression of Recombinant Proteins

Mammalian expression vectors are generally made with constitutive promoters. A strong promoter–enhancer cassette, usually of viral origin, drives the expression of a cloned recombinant gene [5]. Recently non-viral [44] or chimeric viral/non-viral promoters [45] have also been used. These gene-of-interest constructs can be transfected together with separate vectors that confer resistance to a selection agent such as DHFR. Frequently, the selection gene and the gene-of-interest are inserted into the same vector. The expectation is that a better genetic link between the two genes would be provided this way. This precaution is not necessary due to the abundant ligase activity in mammalian nuclei (see above) [46]. In order to increase the chance of obtaining high-level producer cell lines, the selective gene can be driven from a weak promoter. Although this approach is expected to reduce the efficiency of stable transfection, cells that survive selection would be expected to produce more product. Another approach is to use a 1:5 ratio between selection plasmid and the gene-of-interest plasmid. This strategy assumes co-integration of several plasmid molecules, one of which would mediate selectivity. Polycistronic vectors have also been proposed for obtaining high expressors [47]. Unfortunately, the different strategies discussed above have never been formally compared.

More recently, inducible promoters have been developed for mammalian expression vectors because of their potential use in gene therapy where constitutive expression would frequently not be desired. One of the beneficial aspects of induction would be the separation of culture phases for large-scale production whereby gene constructs beneficial for rapid growth would be "switched on" during expansion of cell populations, and then "switched off" when not required. Other gene constructs for protein expression would be induced during the final production phase (see also the section on host cell engineering) [48, 49]. Inducible promoters can also be used for some protein products that confer toxicity when expressed from a constitutive promoter in mammalian cells [50, 51].

Another important aspect for high-level expression is the structure of the mRNA produced by the integrated vector DNA. Intron-free cDNA constructs are not ideal in mammalian cells to obtain efficient cytoplasmic transport of the mRNA. Most expression vectors now include at least one intron sequence that is usually located between the promoter/enhancer and the cDNA coding sequence [45].

Transgene expression in animal cells or in animals is rapidly silenced in many cases, probably under the influence of surrounding endogenous condensed chromatin (heterochromatin). This gene silencing correlates with histone hypoacetylation, methylation of lysine 9 of histone H3, and an increase in CpG methylation in the promoter region [52]. Heterochromatin is different in structural organization from euchromatin (transcriptionally active), and the border between the two chromatin types has been suggested to be marked by sequence elements such as scaffold or matrix attachment regions (S/MARs) [53]. These elements and ubiquitous chromatin opening elements (UCOS) [54] have attracted considerable attention since they are thought to increase and maintain high-level production of recombinant proteins. S/MAR elements act to partition silent regions of the chromosomes from domains permissive to gene expression [55, 56]. They also recruit factors such as histone acetyltransferases that reconfigure chromatin locally to adopt a structure that is more permissive to gene expression [57, 58].

When inserted into expression vectors or when co-transfected on separate plasmids, these elements act to significantly increase transgene expression [59]. These elements could therefore decrease the screening times for the identification of suitable cell lines.

Recently, the genetic code of MARs has been broken down to a collection of short genetic sequences that can be recognized using bioinformatics. Even more potent MAR elements have been unraveled from genomic sequences and are currently entering the recombinant protein field (N. Mermod, personal communication).

A very recent publication speaks of yet another class of sequence elements that may mediate and maintain high-level productivity in mammalian cells upon gene transfer. Highly conserved anti-repressor elements of 1500–2000 base pairs have been identified and cloned from human genomic libraries and inserted into expression vectors. Some of these elements have shown to allow the generation of CHO cell clones producing secreted alkaline phosphatase (SEAP) at 50–80 pg per cell per day [60].

Another approach towards inhibition of silencing is to block deacetylation of histones. Acetylated histones are considered the primary hallmark of active chromatin [61]. In recent investigations by Hacker and colleagues, it was shown that increased expression from silenced transgenes in recombinant CHO cell lines could be mediated by transient expression of Gam-1, an avian adenovirus protein that is known to block deacetylase activity [62]. It should be mentioned in this context that the successful application of butyrate in large-scale manufacturing processes for

the induction of increased specific productivity of mammalian cells [63] is thought to be based, at least in part, on the inhibition of histone deacetylation [64, 65].

Unfortunately, a reliable and comprehensive analysis of the various options for vector design has never been carried out, in part because this would be a very difficult and time-consuming task. Most comparisons of promoter/enhancer elements have been made with model proteins and in transient transfections since expression from individual clonal cell lines or from populations from stable transfections range dramatically from one experiment to another, and are also dependent on the transfection method.

In prokaryotic expression systems, a conversion of mammalian codon use to bacterial codon use is an accepted and widely used concept. In general, highly expressed genes exhibit a codon bias towards more abundant tRNAs. However, few data are found in the literature on codon use when human genes are over-expressed in mammalian host cells. It is questionable that original exon sequences for all the desired proteins of interest will always provide a codon utilization that is compatible with high-level expression. In some cases, codon optimization has been shown to increase transgene expression dramatically in mammalian cells. A review of these studies has been provided by Makrides [66].

1.3
Host Cell Engineering

The use of serum and other undefined and complex media additives for the growth of mammalian cells may result in problems of reproducibility in the process. To avoid batches with negative impact on cells, testing procedures must be incorporated before use of these additives. Serum and animal-derived growth factors cannot be heat-sterilized prior to use – a typical process stage for the elimination or reduction of infectious agents. In general therefore, chemically well-defined materials are preferred as substrates for mammalian cell culture processes. Variable and undefined additives to culture media are expected to be removed and to be replaced eventually by chemically defined components that have little or no variations in quality and are of higher purity.

Serum serves as a source of growth factors, and one way to avoid this is to have them produced by the host cell line. Such engineered hosts can subsequently be employed as a "platform" for "generic" production processes. The hope is to generate hosts that would achieve superior growth rates, that survive death-inducing insults in extended batch processes, for example when media components are exhausted, and that have higher productivity. Oncogenes, cell cycle genes (cyclines), hormone genes (insulin-like growth factor) and anti-apoptotic genes [67–73] have been individually transferred into the cellular genome resulting in novel and possibly superior production hosts [74]. More recently, due to observed limitations in protein folding and secondary modification in cell lines, chaperone genes and genes encoding enzymes for glycosylation have been transfected into mammalian cells, and this has resulted in superior protein quality and quantity in bioreactors [75]. Insights into the genomic organization and function of mammals will dramatically increase in the years to come. It can be expected that this information will provide leads towards a more efficient use of mammalian cells for protein production. Most likely, targeted knock-out mutations as well as designed enhancements of metabolic pathways for

efficient nutrient use, will make mammalian cells even more useful for production purposes.

Targeted transgene expression control in mammalian cells is another exciting new opportunity for host cell engineering. Tetracycline has been used in the Tet-on, Tet-off system, developed by Bujard and colleagues [76]. In order selectively to use independent gene control of two different gene activities in the same cells, Fussenegger's group developed a repressible as well as an inducible system based on the repressor Pip (pristinamycin-induced protein) [77]. Such systems allow control over growth and productivity. Rapid cell mass expansion would be a first goal for the generation of biomass, followed by the growth arrest and boost of high-level productivity [78, 79]. Host cell engineering for metabolic benefits and improved productivity has already been shown in a hybridoma cell line by introduction of the glutamine synthetase gene, resulting in independence of cells from glutamine addition to the medium and in a reduction of the waste product ammonium [80].

Another highly promising aspect of host cell engineering concerns the improvement in post-translational protein modification and processing. A number of therapeutic antibodies produced in CHO cells have been successful products. Yet, the efficacy of these antibodies can probably be improved by enhancing the potency of their natural immune effector functions. In particular, the affinity of the interaction between the antibody Fc region and Fc-gamma receptor appears to be crucial for *in vivo* biological activity [81]. These molecular interactions are affected by the presence of carbohydrates at conserved sites in the antibody Fc region [82]. Engineering the Fc oligosaccharides can be explored as a means to enhance Fc-gamma receptor binding and the associated immune effector functions. Umaña and co-workers were the first to demonstrate that recombinant DNA-based technology could be used to manipulate the apparatus of cells for secondary modification, thus generating antibodies with a modified glycosylation pattern and an associated increased immune effector function. These authors have developed stable over-expression of a-1,4-N-acetylglucosaminyltransferase-III in recombinant antibody-producing CHO cells in order to generate IgGs with high levels of bisected, non-fucosylated oligosaccharides in the Fc region, and to obtain large increases (over two orders of magnitude) in antibody-dependent cellular cytotoxicity (ADCC) [75].

1.4
Gene Transfer and Gene Amplification in Mammalian Cells

The underlying principles for use of mammalian cells as recipients of protein encoding DNA vectors and for the most popular way to improve productivity, by experimentally induced gene amplification, are described in this section. It is well accepted that these principles are applicable to any mammalian systems used in the industry, albeit with the appropriate modifications in methods.

A mutant CHO cell line, lacking DHFR activity can be cultivated in the presence of glycine, hypoxanthine, and thymidine (GHT). When these cells are transfected with a functional DHFR gene, cells that have acquired the gene can be selected and expanded in media lacking GHT. A second expression cassette for a product of interest (for example a protein with therapeutic value) can be included on the DHFR plasmid or on a separate vector(s) (Fig. 1.2). It should be noted that co-transfection of several plasmids is possible due to an apparently unlimited capacity for the uptake of foreign DNA by mammalian cells. Surprisingly, integration of individually co-transfected DNAs at the same site in the genome seems to be the rule in mammalian cells [25]. In spite of this experience, which has been verified in numerous cell lines created by the co-transfection of individual vectors, an interesting approach was proposed recently to tightly link the expression of DHFR with expres-

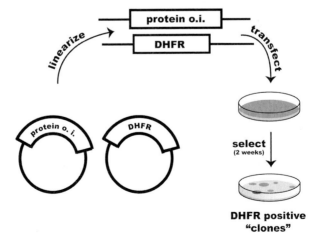

Fig. 1.2 Generation of stable cell lines using DHFR-minus CHO cells. The example here uses co-transfection of restriction enzyme linearized plasmids. Clones appear 2–3 weeks after exposure to selective environmental conditions – here, culture of cells in media lacking hypoxanthine and thymidine.

sion of an antibody. DNA segments containing the coding sequences of each a half of a DHFR-protein were integrated into the two vectors containing cassettes for a heavy and light chain, respectively, of an antibody gene. The functional assembly of the two halves of the DHFR-protein was assured by the addition of leucine-zipper sequences to the respective DHFR-protein segments [83]. While the average expression level derived from these clones is reported to be good and in the range of 10–25 pg per cell and day, the screening of a large number of cell lines will be always required in order assure satisfying productivities. Typical screening efforts will evaluate 100 to 500 individually established cell lines, preferably from several independent transfections.

While DHFR is still the most frequent selection approach with CHO cells, other selection systems can be used (e.g., antibiotics such as neomycin, hygromycin or puromycin), as well as fluorescence proteins [84]. An important consideration for the choice of a selection agent is the degree of selectivity (stringency). The more stringent the selective agent used, the smaller will be the number of obtainable clones. However, a more stringent agent will select for colonies of cells that express the resistance marker gene at higher levels and, frequently, also the desired gene of interest. The DHFR system is – even when the gene is driven for example by a relatively strong SV-40 promotor – a rather stringent selection system and will produce, after transfection of cells, fewer clones than would selection with the antibiotics neomycin [85].

With both DHFR selection and the glutamate synthetase (GS) system, expression of both the selection gene and the gene of interest can be augmented by exposing recombinant cells to drugs that block the activity of the product of the selection gene (see also Part IV, Chapter 4). For DHFR, the drug methotrexate (MTX) has been used successfully in a large number of cases [86–88]. MTX is a folate derivative that blocks DHFR activity completely and irreversibly. Usually, after 2–3 weeks of exposure to MTX, a majority of cells die while a few survive that are resistant to MTX toxicity due to elevated expression of DHFR. Essentially any given level of MTX is overcome by a small number of cells that produce more DHFR than would be inhibited by the given intracellular quantity of MTX. It was found that these cells frequently contain chromosomally integrated plasmid sequences in a higher copy number than observed in cells before exposure to MTX. Stepwise treatment with elevated concentrations of MTX can be repeated several times and may result in the isolation of cells that contain dramatically increased copy numbers of the transferred genes. The phenomenon of MTX-mediated gene amplification had been observed before the use of recombinant DNA technology, most notably in cancer patients [89]. CHO cell lines containing several hundred to a few thousand copies of transfected plasmid DNA have been established [50, 90]. In most cases, the amplified segments contain the gene of interest, but large segments of 100 to 10000 kilobases of the surrounding region have also been amplified in the process [91, 92]. Most "amplified" cells produce more product than the unamplified host cells did previously. However, the improvement of specific productivity (up to 10- to 20-fold) is highly variable when studying individual clones [93], and also varies from product to product [94].

The principle for MTX-driven amplification also applies to other immortalized cell lines [95]. However, DHFR gene transfer followed by amplification in MTX works best with cells that lack a functional endogenous DHFR gene. A popular approach with NS0 cells utilizes GS as a selective gene. This system relies on the fact that NS0 cells express very low levels of GS. These cells require either an exogenous source of glutamine or an exogenous GS gene in order to survive in the absence of glutamine. A specific and irreversible inhibition of GS can be mediated by the addition of methionine sulphoximine (MSX) to the culture medium. At a concentration of 10 to 100 μM MSX, resistant clones can be identified in selected NS0 cell populations that have amplified the transgene complex containing the GS gene and the desired gene(s) of interest [96, 97]. The GS system can also be applied to cells such as CHO that have a normal level of glutamate synthetase. In this case, the starting concentration of MSX needs to be higher than that used for the selection of recombinant NS0 clones in order to block the endogenous GS and to select for clones that over-express the exogenous GS gene [98]. Unfortunately, the literature does not provide any information on the cytogenetics of gene amplification in the GS system.

With recent publication of the genome sequences of man, mouse and rat, we have learned that mammalian genomes are exceptionally dynamic due to the presence of repetitive sequences, remnants of retroviral genomes and transposable elements. This phenomenon can be termed "sequence mobility". Mammalian cells – particularly immortalized cells – have an even more intrinsic genomic fluidity. This becomes evident when studying chromosome numbers in metaphase spreads [99]. It appears that immortalized cell populations will diverge, even when established as clonal cell lines, in the number and structure of their chromosomes within very short time frames (weeks to months). Due to sequence mobility and chromosomal instability immortalized mammalian cells are ideal substrates for experimentally induced gene amplification.

1.4.1
Cytogenetics of CHO Cell Lines, Genetic and Production Stability

Subpopulations and clones of MTX-treated CHO cells may contain the transfected DNA sequences at very high copy numbers [90, 100]. Studies concerning genetic features of these amplified DNA sequences within the CHO genome have been ambiguous. It is still controversial today, whether continued presence of a selective agent (i.e., MTX) in long-term cultures of recombinant CHO cells is required for production stability. Whereas some studies suggest that MTX is required for the stable production of recombinant proteins [101], others indicate that continuous cultivation of clonal cell lines in MTX might not be necessary [102]. Cytogenetic studies, using FISH [103] were performed in my own laboratory in the late 1980s with clonal and non-clonal recombinant CHO cell lines, and showed that continuous exposure to MTX at the same concentration, under which the cell populations were initially established, promotes genetic *instability* at the chromosomal level [104, 105]. Cell lines selected at micromolar concentrations of MTX showed elongated chromosomal structures that hybridize to probes representing transfected DNA. Some of the chromosomes that contained a large number of tightly arranged bands of fluorescence differed dramatically, most notably in length, from the "normal" CHO chromosome. Frequently, we found chromosomes with transgenic DNA up to the very end of one arm, indicating the absence of a telomere. If in fact these chromosomes do not have functional telomeres, then this raises questions about the stability of the amplified DNA during continuous subcultivation.

In summary, when cultivating these cell lines in the absence of MTX, unique and characteristic integrations were found in 95–99% of metaphase spreads. We continued to cultivate these cell lines for extended periods in the absence of MTX, and occasionally performed FISH analyses. We found that the chromosomal structures described above were stable within the observation period (a minimum of 60 days, and in one case of 160 days). Cytogenetically, these observations indicate a high degree of genetic stability of chromosomally amplified sequences in the absence of MTX. Equivalent observations as those presented above have been made recently by Kim and Lee [106].

1.4.2
Transgene Structure
and Locus Determination

"Southern" hybridization of genomic DNA is a useful tool to determine the molecular structure of integrated plasmid DNA. Using suitable restriction enzymes, this technique will provide information on the integrity of the transgene. Estimates of the copy number of the chromosomally integrated DNA can be established when using in the same experiment known quantities of plasmid DNA restricted with the same enzyme as that used for the genomic DNA. Southern hybridization may also provide information on the question of whether one or more than one integration locus exists for the plasmid sequences. However, conclusions on multiplicity of integration must be made with a

degree of caution. Aberrations from the expected signals can be due to post-integration rearrangement in a fraction of the cell population or in the initial co-integration of a few copies of the plasmid sequences that had been subjected to nuclease attack, resulting in the deletion of the restriction enzyme site used for the analysis.

All Southern hybridizations are based on DNA extracted from thousands of individual cells. Even if the cell lines are based on a "cloning" step, one must be aware that none of the cell lines is clonal in the most narrow sense: Genetic variations occur very rapidly in immortalized cells, due to their inherent chromosomal instability.

FISH can be used to gather knowledge about the degree of chromosomal amplification (not to be confused with copy number estimates) and the chromosomal location of the recombinant DNA. In order to provide some useful information, FISH studies must be supplemented by a statistical analysis of identified integration sites (and structures observed). They should also take into consideration the time point of analysis with respect to the total time of cultivation of the cells. A reasonable value may be gained from a FISH study performed shortly after cells have been thawed from a "bank" of cells stored in liquid nitrogen. Depending on the culture conditions (with or without MTX or serum) and the length of cultivation time, the results of FISH analyses may vary considerably. Despite this problem, in the studies discussed above we were able to determine the identity of one recombinant cell line from another by identifying a chromosomal marker containing hybridizing DNA which was present in a large fraction of the individual cells of the populations studied. The identifying chromosomal markers containing recombinant sequences were termed "master integrations" which we found to be the genetically stable entities in the cell lines.

1.5
Production Principles for Mammalian Cells: Anchorage-dependent Cultures and Suspension Cultures

Process scientists and their managers must decide which type of production system to choose for the product in question. Chiefly, the anticipated scale of operation for the manufacturing process drives the choice. A number of very successful recombinant products from mammalian cells such as erythropoeitin (Epogen®) are given to tens of thousands of patients. Epogen® is a protein hormone developed by Amgen for the treatment of dialysis patients with chronic renal failure. The single dose necessary for treatment is relatively small (about 100 µg/patient). In contrast, treatment with another protein therapeutic, the recombinant antibody Herceptin®, developed by Genentech, requires multiple doses over weeks and months with a maintenance dose of about 150 mg per patient. Herceptin® is a humanized IgG directed against the Her2 receptor that is over-expressed in a percentage of breast cancer patients (see also Part I, Chapter 5). The number of patients treated per annum is approximately the same for these two products. It is clear that a 1000- to 10 000-fold difference in the annual quantities of Epogen® and Herceptin® needed will require entirely different decisions on the scale and mode of operation when developing the manufacturing processes for these two products.

Processes for recombinant proteins from mammalian cells can be established on the basis of two cellular growth modes: adherent and suspension cultures. CHO and other hosts such as BHK and HEK293 cells can be grown in either mode. NS0 cells that were derived from a mineral oil-induced plasmacytoma in mice will only grow in suspension, and will not firmly attach to a surface that is exposed to mixing induced shear force.

1.5.1
The Rollerbottle Process

In the case of erythropoeitin and a few other hormone-type protein products, a process based on rollerbottles appears to be sufficient to supply the market with product. In this case, adherent cells are cultivated on the inner surface of a cylindrical bottle having a volume of 1, 2 or 3 L. A typical 2-L bottle provides an inner surface of 850 cm^2, but there are variations of these bottles that provide extended areas for attachment of cells. A simple and reproducible process can be established with minimal initial investment in equipment using such rollerbottles. Provided that there are sufficient human resources available, this process can be easily scaled up since the number of rollerbottles handled in parallel determines scale.

Cells thawed from a cell bank can be expanded by subcultivation into a fixed number of rollerbottles. The standard 2-L rollerbottle is usually filled with 300–500 mL of medium. The remaining volume provides the necessary oxygen, while the closed bottles are slowly rolled at about 1 r.p.m. in an incubator at 37 °C. A sufficiently large number of rollerbottles containing an adequate cell population represents the starting point for several production cycles. For scale-up, the cells from a single confluent rollerbottle can seed up to 20 rollerbottles. From freshly seeded rollerbottles to confluent rollerbottles requires 3–6 days, depending on seeding density, growth rate, and composition of the medium. Since adherence of cells to the inner surface of the rollerbottle is required, serum is frequently used in such a process at a concentration of 1–10%, providing necessary attachment factors to the cells. Adherence can also be assured in media lacking serum if fibronectin and other cell attachment factors obtained from animal sources are added to the culture medium. Within two to three subcultivations, starting from a seed vial obtained from the working cell bank, a sufficiently large number of inoculated rollerbottles can be generated that can constitute a production phase. A part of the cell mass generated in the last subcultivation cycle can be used for the generation of seed culture for the subsequent production cycle.

Media for the production phase are usually richer in nutrient content than the seed train medium in order to maintain viability and productivity of the cells for a minimum of 1–2 weeks. For facilitating recovery and purification of the product and for cost reduction, serum is not used for the production phase. The attachment of cells to the surface of the rollerbottle will not be compromised by such a modification in the medium. However, gentle handling is required or the sheets of cells will detach. Upon incubation of the confluent cells with the enriched, serum-free medium, the secreted product will be harvested, leaving the adherent cells inside the bottle. Sometimes, a refeeding with fresh medium for a second production cycle is possible, on the condition that the product of the first and the second harvest will be similar in composition and quality. Since a standard 2-L rollerbottle will contain about 300 mL of medium for harvest, 1000 rollerbottles will provide from 300 L to 600 L of supernatant. Over a one-year period, a manufacturing process based on this schedule will deliver 15 000–30 000 L of

cell-free culture medium containing the product of interest. Product concentrations in the 50 to 200 mg L^{-1} ranges are possible, thus providing the protein in the kilogram range annually. Such a process is labor-intensive, requires the repeated use of trypsin for detachment of cells, and is at considerable risk of contamination with adventitious agents through handling.

Epogen® (erythropoietin) has been developed on the basis of a rollerbottle process. Today's Epogen® process is essentially a robot-based manufacturing procedure whereby all the critical handling steps – including the seeding of cells, filling of bottles with media and harvesting of cell culture fluids – are executed within air-filtered environments and without human interaction.

A variation of the above process involves stirred tanks or hollow-fiber bioreactors for growth of the seed culture. The growth of CHO cells in both the suspension and adherent modes allows streamlining the rollerbottle production process. The seed culture for the rollerbottle production phase can be generated in spinner flasks or in bioreactors. The advantage of such a process is that fewer subcultivations are needed to generate sufficient cell mass. It also reduces the risk of contamination by adventitious agents. Hollow-fiber bioreactors can also be used in which very high cell densities can be achieved through the continuous perfusion of the reactor with fresh medium. Several reactor volumes of fresh medium can be perfused through such a system, resulting eventually in cell densities approaching tissue-like character. These rather compact systems provide sufficient cells to seed a very large number of production rollerbottles. Again, the goal of such an approach is to reduce human interaction and the risk of contamination.

1.5.2
Adherent Cell Culture in Bioreactors

Van Wezel [107] proposed the use of water-suspended polymer spheres termed "microcarriers" for the culture of adherent cells in stirred-tank bioreactors. The purpose of growing cells in stirred bioreactors instead of on fixed surfaces is to allow for easier scale-up and increased homogeneity in supply of nutrients in media, but also in supply of oxygen and carbon dioxide exchange. Several processes have been developed in the human and animal vaccine industry using the microcarrier concept, always with cells that have a high anchorage dependency [108]. These cells serve as substrates for the multiplication of viruses such a measles, polio or mumps [109]. CHO cells are being used for the production of several human recombinant proteins, most notably at Serono, on microcarriers in stirred bioreactors [110]. These processes date from the early phase of recombinant mammalian cell culture technology and require the use of serum, at least for parts of the process. For scale-up, cells are seeded at a density of about one to five cells per bead, and these will subsequently grow to confluency on the beads. Widely used microcarriers are Cytodex™1 and Cytodex 3, both marketed by Amersham Biosciences. Cytodex 1 is based on a cross-linked dextran matrix which is substituted with positively charged *N,N*-diethylaminoethyl groups. The charged groups are distributed through the microcarrier matrix. Cytodex 3 consists of a thin layer of denatured collagen chemically coupled to a matrix of cross-linked dextran. The denatured collagen layer is susceptible to digestion by a variety of proteases including trypsin and collagenase, allowing for removal of cells from the microcarriers while maintaining maximum cell viability. Once cells have been detached, additional carriers can be added, while both cells and carriers are gravity settled in the bioreactor. The microcarrier approach has certain advantages with respect to harvesting of product from cell culture fluids, but also for perfusion processes where fresh medium is added to a culture while spent medium is withdrawn. Spier and Kadouri have reviewed the evolution of commercial production processes based on anchorage-dependent cultivation [111].

Processes without the use of microcarriers are however less cumbersome, since the transfer of cells from one scale to the next and thus reseeding of fresh carriers is tedious and complicates processes beyond need, especially when the preferred host cells for recombinant protein production can now easily be cultivated without any matrix.

Microcarriers will remain important in the field of vaccine production, since several viral products are dependent on strictly adherent cell lines and in tissue engineering. For the latter, different cell types are needed to reconstruct multi-layered organs and tissues. Macroporous carriers and matrices can be used to generate structured cellular complexes that are molded into functional organ/tissue systems (see also Part I, Chapter 15).

1.5.3
Stirred-tank Bioreactor Processes

Increased worldwide needs for recombinant biopharmaceutical proteins drive major investments into the construction of new bioreactor facilities. In addition to those companies that produce their own protein pharmaceuticals in large-scale manufacturing plants, a few contract manufacturers offered in 2004 a bioreactor capacity of about 130 000 L. These contract manufacturers serve an increasingly competitive market. Projections state a capacity shortfall of about 400 000 L for the year 2006 [112]. The clinical and commercial success that recombinant proteins have had during the past 10 years has clearly stimulated many newcomers in the field to try to develop similar clinical targets, thereby creating a demand for bioreactor capacity which exceeds current availability. Most of the successful antibody or antibody-like proteins are given to patients in rather large doses (hundreds of milligrams to grams per patient) and thus require very large facilities for manufacturing. Likewise, new markets are generated after the approval of a given product that widen the application of a biopharmaceutical. Off-label use increases the product demand even further.

Suspension culture of mammalian cells is the most popular approach for large-scale manufacturing [113]. This approach using CHO cells and a few other cell lines now dominates the domain of mass production of recombinant protein products. With the exception of blood-derived cells, most of the other cells used in the industry were of fibroblast or epitheloid character and were therefore initially anchorage-dependent. In the early 1980s, the adaptation of CHO cells to suspension culture was a tedious process, mostly because of the lack of media formulations that facilitate suspension growth. Today, multiple factors seem to have made the transition from adherent to suspension culture much easier. For example, cell culture media have been developed which support the growth of cells in suspension better than earlier formulations based on DMEM and Ham's F12. Also, the selection of cell populations in media with reduced serum and calcium concentrations has resulted in cell lines that support the transition from adherent to suspension growth more readily. Some scientific reports have claimed facilitated serum-free suspension growth due to genetic modification [114]. However, these advances must be regarded with caution as non-modified cells do readily grow now in optimized suspension media. When handled correctly and when using appropriate media formulation, seeding densities and stirring conditions, the transition of CHO cells from adherent to suspension cultures, even without genetic modification, can be executed in a few weeks.

In a "simple" bioreactor-based process, the scale-up to very large volumes can occur rather rapidly. This is usually executed by diluting the entire volume of one bioreactor into 5–20 volumes of fresh medium held prewarmed in a larger reactor (Fig. 1.3). Within 10–15 days, a suspension culture at the 50-L scale can be used to inoculate a 10 000 L reactor. It is a major goal of process development work to optimize media for the production phase. Such a medium needs to support good growth initially in order to achieve the highest cell density possible, and then it needs to provide the nutritional basis and physiological balance to maintain viability and productivity for extended periods. The periods for production (6–14 days) usually exceed in time the typical subcultivation periods of 3–5 days. While the termination (i.e., harvest) of such a culture is driven mainly by plant capacity and volumetric productivity, the other important issue to consider is the quality of the derived product. The continuously changing composition of the culture medium during the production phase can affect the quality of earlier synthesized product through degradative activities mediated by cell-released enzymes. Also, a diminishing supply of nutrients as energy providers or as build-

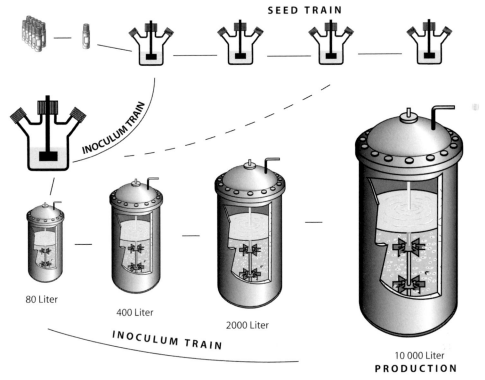

SEED TRAIN

INOCULUM TRAIN

80 Liter

400 Liter

2000 Liter

INOCULUM TRAIN

10 000 Liter
PRODUCTION

Fig. 1.3 Diagram of a simple batch (or extended-batch) process with suspension cells. Cells are obtained from a Master- or Working Cell bank (MCB/WCB) and inoculated into spinners for a defined subcultivation period (every 3–4 days, usually for up to 100 days or more). For maintenance purposes of the culture, the cells in the spinner are referred to as the "seed train". Cells from spinners (1–5 L volume, filling volume up to 40%) are used to inoculate bioreactors at increasing scales of operation, until the final volume for production is obtained. Cells in vessels with increasing volume are referred to as the "inoculum train". The final and largest vessel is used for production purposes.

ing blocks for the synthesized product will most likely change the molecular composition of recombinant proteins. The most probable alteration of protein being made early or late in the production process would involve the structure and extent of glycosylation. This topic is discussed in more detail elsewhere in this book (see Part IV, Chapter 7).

1.5.4
Batch and Extended-batch Perfusion

Batch and extended-batch processes have achieved unprecedented productivity. These are the results of many months – if not years – of work that went into the development of the manufacturing process. This development work is summarized in a very simplified way in the Fig. 1.4, starting from gene transfer to cells and ending with the establishment of a well characterized masterbank. Scientists from Genentech reported in March, 2004 on volumetric titers of more than $4\,g\,L^{-1}$ of a secreted antibody-product in the supernatant of CHO cells in large-scale bioreactors. A single production run, executed at a volume of 10 000 L can therefore produce more than 30 kg of purified product (assuming a recovery yield of about 70%). Repeating such a production run successfully 20 times each year would provide 600 kg of product. A handful of companies have invested heavily in large-scale production facilities, with several having up to six parallel trains for scale-up. The largest mammalian bioreactor system is presently being constructed by Roche in Basle, with a bioreactor volume of 25 000 L. Just 20 years ago, $5\,mg\,L^{-1}$ from CHO cells was considered sufficient to justify investments into the development of a CHO platform for recombinant protein production. Clearly, the success of several antibody and antibody-fusion products in clinic, for the treatment of cancer and diseases such as rheumatoid arthritis, has driven huge investments in order to assure market supply. Therapeutic antibodies are projected to obtain six to eight market approvals per year and to reach sales in the USA of $20 billion by 2010.

The "batch process" is considered the most simple – and thus most robust – production process for stirred bioreactors. The term "batch" is connected to the very last phase of the process, the phase during which accumulated product is maintained in a final production vessel. Since all manufacturing cell lines used so far drive the expression of the product gene from constitutive promoters, product will be synthesized during earlier phases of the process, but not harvested. One popular approach is to define the entire process from the thawing of cells from a bank to the production vessel as three separate phases. These are the "seed train", the "inoculum train", and the final production phase (see Fig. 1.3). In the step preceding "production", the cells in a smaller bioreactor are cultivated to maximal cell density and then transferred along with the exhausted growth medium into the production reactor. The timing of cell culture subcultivation and the target density of inoculation of the subsequent culture step are the subjects of process development questions and must be determined on a case-by-case basis. The production process begins when cells and fresh medium are mixed in the reactor, and it ends at a predetermined time-point when the synthesis of recombinant protein diminishes due to exhaustion of nutritional components in the medium and/or accumulation of toxic end products of cellular metabolism. Usually, with CHO and NS0 cells the production phase lasts for between 7 and 14 days after inoculation of the reactor, depending on the susceptibility of the proteins to degradative enzymes, as well as a number of other process-related factors. The advantage of such a process is obvious. Provided that the inoculating cell mass can be generated reproducibly, the resulting production process will show a high degree of similarity with respect to cell growth, viability, and quantity and quality of the product being synthesized.

The issue of reproducibility of process parameters and of achievable product quantity and quality is of highest significance as this will ultimately be evaluated by the regulatory agencies. For Investigational New Drug applications (IND) and Process Licence Applications (PLA), rather specific requirements must be met with respect to the minimal number of product batches analyzed. For INDs, no less than three product "runs" are recommended. Shorter batch processes (5–7 days) have the advantage of generating more data within a given time frame. This can be a very important cost factor, since the time period necessary to acquire and evaluate necessary data from the new process will eventually affect the overall time necessary for entering the market.

On the other hand, there are a number of arguments that would sway process development decisions to another direction. The option to prolong the synthetic activities of cells in the production vessel would capitalize on the "process investment" which allowed generating the necessary cell mass for the production vessel. Increasingly therefore, extended-batch or perfused-batch cultures are used. With a longer-lasting production phase in cell culture, feeding additional medium components becomes necessary. Clearly, extending the process for a considerable period of time (e.g., from 8 to 14 days or longer) only makes sense when the return for this "investment" in labor and in occupation of the production facility result in a sufficiently high increase in product concentration within the vessel. There are various ways that medium and medium components can be added to a culture that had been initiated a few days earlier in the same tank. This might be done by feeding (batch wise) highly concentrated mixtures of essential amino acids and other medium components, thereby not significantly affecting the volume in the tank. Alternatively, a culture can be started in the production vessel at half or so of the working volume, after which standard concentration medium can be pumped slowly and continuously or batch-wise into the tank until the final working volume has been reached. The choice of either mode – or combinations thereof – is in the hand of the process development scientist, who must evaluate carefully any advantages and disadvantages. No matter what principle will be used for extending the production phase, the ultimate overall result will always be a tank that contains – in one batch – the entire protein population for subsequent recovery and purification.

Continuously perfused production processes represent an entirely different philosophy for manufacturing. Here, the goal is to achieve the highest cell concentrations possible – within smaller tanks that hold the cells. Sometimes the reasoning used is that up-front investment for manufacturing equipment is reduced and product quantities can be quite high from such processes. The much-improved knowledge base in technology and in the physiology of mammalian cells in culture have made this more complicated approach to manufacturing attractive. Perfused cultures can be maintained for many weeks and months, with product harvests occurring repeatedly throughout that period. A protein of high interest to the pharmaceutical industry for several decades – the antihemophilic Factor VIII (see also Part II, Chapter 3) – is reliably being manufactured using perfusion technology with BHK cells. The glycoprotein, which is probably the largest secreted single peptide chain protein ever produced in bioreactors, is harvested continuously through Bayer's cell retention technology that allows cells to be returned to the bioreactor. This process, when run for up to 6 months, improves the yields of fragile proteins that would be degraded if left in the fermentor for the typical time used in fed-batch processes. While the production of Cognate® Factor VIII [115] has pioneered the use of perfusion technology for recombinant proteins with mammalian cells in a non-CHO cell, other products from CHO cells have also been approved using this technology [116].

Finally, many monoclonal antibodies have been produced for some years at the laboratory scale in perfusion systems using hollow-fiber technology. Hence, the transfer of this technology to production scale for pharmaceutical manufacture has, for some time, been obvious [117, 118]. A diagnostic monoclonal antibody for imaging in patients with prostate cancer (ProstaScint^{TM}) was the first FDA-approved product to have been produced by a hollow-fiber perfusion process [119]. The application of this and other perfusion technologies using immobilization of mammalian cells on macroporous carriers producing recombinant antibodies or other proteins is likely to be yet another option for future manufacturing processes [120].

1.6
Large-scale Transient Expression

Mammalian cells have been used as production hosts in some of the first pioneering experiments for the design of novel proteins. One of these proteins – for example, the first chimeric therapeutic candidate, the fusion of the CD4-receptor and a human immunoglobulin – was first designed as mammalian gene construct and then generated by transient expression in HEK-293 cells [121]. Another designed human therapeutic – the thrombolytic biopharmaceutical protein drug TNKase®, a mutagenized tissue-plasminogen activator – was developed based on hundreds of TPA-variants that have been expressed initially by transient expression [122]. While these investigations in the early 1990s were carried out on a small scale (i.e., 1–10 mL cell culture and microgram quantities of protein), large-scale transient expression from mammalian cells is a new technology addressing an urgent need in biotechnology for the rapid production of recombinant proteins in the milligram to gram range. With better tech-

nologies for the reliable growth of mammalian cells, and with better nucleic acid transfer systems, the opportunity arose to explore transient expression in mammalian cells beyond the laboratory scale. In addition, many companies in the field of somatic gene therapy, using artificial or modified virus vectors, depend on transient DNA transfer to mammalian cells as one of the key manufacturing steps for their products [123]. Other than with stable expression, vector DNA is not required to integrate into the chromosome DNA of the host cell, but remains shortly (transiently) in the nuclear environment where at least some transgene DNA is utilized as templates for transcription into mRNA [124]. The highly improved DNA transfer systems developed over the past 10 years (see also Part VI, Chapter 6) allow to supply frequently 50% or more of cells in a population with sufficient DNA. The most popular large-scale transient expression systems are based on non-viral DNA delivery and utilize calcium phosphate [125] and PEI [126, 127] as vehicles, and the preparation of these vehicles with DNA has been modified for use with stirred single cell suspensions in bioreactors. Calcium phosphate and PEI are both cheap components – an important consideration for scale-up. Several groups have reported the scale-up of transient expression to bioreactors of 10 to 100 L [128, 129], mainly for the production of research materials used in pre-clinical research. The yields from these exploratory experiments are in the range of 1 to 50 mg L^{-1} for antibodies [130], and referred in one report to the expression of recombinant protein at 100 mg L^{-1} from 100-L scale operations with transiently transfected CHO cells [131]. These yields are clearly far below those observed with highly optimized production processes that have proven their robustness and reproducibility in large-scale operations at the 1000 or 10 000 L scale. Why then the need to engage into the development of an alternative technology?

The reason is speed. At only days after the availability of an expression vector, milligrams to hundreds of milligrams of a recombinant protein can be delivered into the hands of the researcher. Vectors for transient expression do not require a selection marker – the goal is to deliver DNA to a maximal number of cells in the population. Several vectors can be transfected simultaneously into cells and will be expressed simultaneously. With calcium phosphate as a vehicle, it was shown that approximately 20 000 plasmid molecules per cell can be delivered [132]. After a few days, the copy number of plasmid molecules will decline in the nucleus and the production of mRNA ceases. Depending on the protein at the time point of the highest accumulated yield, the product is harvested and cells are discarded. A new production can be re-started at any time when sufficient fresh cells can be provided and a new DNA-vehicle preparation is ready for transfection.

Large-scale transfection requires significant quantities of DNA. With both calcium phosphate and PEI, approximately 1–2 mg of plasmid DNA are usually needed per liter of suspension culture. Media and culture conditions for large-scale transient transfection are under further development, as are the vehicle preparation techniques. With calcium phosphate as a vehicle, a small concentration (1–2%) of fetal bovine serum may be required for high transfection efficiency. Here as well, it is a goal to generate processes that are low or free of undefined components.

It remains to be seen whether transient expression technologies will eventually be used under conditions for clinical production and thus provide eventually products for human medical use.

1.7
Regulatory Issues

All mammalian cells used for the large-scale production of recombinant proteins are considered "immortalized", as they can be grown continuously for an indefinite period if correct culture conditions are provided. This is an exceptional characteristic for animal-derived cells, since the tissues and organs of animals are constructed of cells with a defined lifespan. The limited lifespan of cells in animals was detected first by Hayflick [133], and is linked – among other reasons – to a declining telomerase activity on chromosomes of somatic cells, but not in germ cells.

The climate for permission by regulatory agencies, particularly by the FDA in the United States to use immortalized CHO cells for the production of recombinant proteins was not favorable in the early 1980s. Discussions about risks associated with the use of mammalian cells were controversial and had been initiated more than two decades earlier [134] when a first generation of "classical" biological products (i.e., vaccines and the natural interferons) were developed on the basis of primary monkey kidney cells, human diploid cells and, later, transformed mammalian cells.

The manufacturers of recombinant proteins for clinical applications and regulatory agencies were in agreement that it was extremely important to minimize eventual risks associated with the use of recombinant mammalian cell hosts. Risks were seen in "tumour" principles, carried by the DNA of the host and in adventitious agents (viruses, mycoplasma, etc.) that could infect the host cell lines and thus eventually be transmitted to patients receiving products from those hosts. Also, the consistency and quality of the recombinant proteins were discussed in the context of risk assessment and risk control. The result of a long series of scientific discussions in journals and at conferences held over a decade was that stringent controls, regulations and monitoring procedures were enforced as a prerequisite for manufacture of proteins from such cells [135]. A balance had to be found between the almost assured clinical benefit of some of those first recombinant products and the perceived risks associated with the unavoidable necessity to produce them in "tumour" cells.

The first product – recombinant tissue plasminogen activator (Activase®-rtPA) – proved to be a good candidate to achieve such a balance, since the benefit – the saving of lives of heart-attack patients – outweighed by far the anticipated risks. However, approval was achieved only after a large amount of data were provided to the regulatory agencies which showed: 1) that consistently only a minute quantity of CHO DNA (<10 pg per dose, later relaxed to <100 pg per dose) was present in the final product; 2) that the product itself could be produced with a high degree of reproducibility; and 3) that it was produced with a purity not achieved before in any biological derived from mammalian cell culture.

1.7.1
Bacterial and Fungal Contamination

The prevention of bacterial or fungal infections in cell culture and recovery systems can be assured, to a high degree of confidence, by the use of a piping and vessel system which maintains absolute containment of the sterile medium fluids. The equipment used must be of a nature to allow cleaning and sterilizing by Clean in place (CIP) and Sterilize in place (SIP) procedures (usually high-quality stainless steel). Most cell culture processes require complex media containing amino acids, vitamins, protein hormones and fetal bovine serum. Some of the components of mammalian cell culture media cannot be autoclaved, and thus sterility (freedom from viruses and microbial organisms) cannot be assured to a 100% confidence level. To exclude the introduction of bacterial and fungal contamination through raw materials, prior testing and, in addition, filtration through membranes of 0.2 μm or even 0.1 μm pore size into pre-sterilized containers is employed. It is, of course, well understood that most (small) viruses and prions cannot be excluded through filtration procedures.

Rigorous testing of the Master Seed Cell Bank (MSCB) and the Manufacturers Working Cell Bank (MWCB) (for a review, see Ref. [136]), which is accomplished by analyzing cells of a number of representative cryovials, assures that the production cell line itself is not contaminated with viruses, bacteria, mycoplasma, and fungi. Sterility testing of the cell line must be carried out in appropriate media lacking antibiotic or anti-fungal compounds, for obvious reasons. Virus testing is performed in suitable cell systems that are validated for each of the individual virus species. Mycoplasmas, which represent the smallest living cells, are frequent contaminants of cells derived from patients and from animal sources. They can remain undetected in cell culture for extended periods of time, and are therefore more threatening to cultures for large-scale processes than typical bacteria that multiply rapidly.

1.7.2
Prions

The use of sera or other products derived from bovine sources in culture media represents a potential risk of transfer of the causative agent for bovine spongiform encephalopathy (BSE) to patients. Therefore, regulatory agencies request detailed information on the origin and processing of products derived from bovine sources. For example, sera obtained from countries in which BSE was diagnosed, even in a small number of animals, are considered unacceptable by regulatory agencies. Companies have, in most cases anticipating these regulations, assured their supply of bovine-derived process materials from countries such as New Zealand or Australia, where BSE has not been reported so far. The US had been considered a BSE-free country until recently (2003) when a single cow was diagnosed with BSE. In view of new insights into the molecular biology of prion diseases, one must consider now that these agents are more widely present in nature than previously thought. Disease risk perception rose in Europe, in the US and elsewhere, and has initiated public safety discussions and even the implementation of stringent process regulations. The use of components of animal origin in media including the use of amino acids purified from animal sources is considered increasingly unacceptable. The high degree of concern regarding BSE is based on findings that: 1) a small compound PrP_{sc} (proteinaceous-infectious particles, scrapie) is likely to be responsible for the disease; 2) transfer of the bovine disease to human populations as a variant Creutzfeld-Jakob disease has occurred in hundreds of cases; 3) detection of the causative agent is only possible with rather sophisticated techniques, and then only in tissues that are typically highly affected; and 4) the inactivation of infectivity of PrP_{sc} is difficult. Even autoclaving procedures (121 °C, 20 min) do not completely eliminate infectivity. Reviews on prion biology were published by Prusiner (1997) [137] and Aguzzi et al. (2004) [138].

1.7.3
Viral Contaminants

Why were *hamster* cells chosen as a host for making *human* recombinant proteins in the early 1980s? A strong argument for favoring non-human cell lines over human cell lines for the production of proteins is the fact that certain life-threatening human viruses cannot be propagated at all, or multiply only poorly, in non-human cell lines. CHO cells do not support the replication of pathogenic viruses such as polio, herpes, hepatitis B, HIV, measles, adenoviruses, rubella, and influenza. Thus, the risk of a viral adventitious agent of being involuntarily carried along with the product of interest can be considered extremely low. Wiebe et al. tested a total of 44 human pathogenic viruses for replication in CHO cells and found only seven (reo 1,2,3, mumps, and parainfluenza 1,2,3) that were able to infect these cells [139]. Exclusion of these virus species and others that can be propagated on CHO cells, such as the parvovirus MVM (Minute Virus of the Mouse), can be assured to a high degree of certainty through testing. Tests can be performed with all materials which enter the manufacturing process and which would support the "viability" of the virus in question. Tests are also obligatory with fluids which contain the product of interest.

Sterility filtration of fluids containing a variety of raw materials, some of which may have been exposed to viruses, does not prevent the introduction of viral contaminants into the process. Only recently have membranes with pore sizes small enough to exclude passage of virus particles become available for industrial-scale operations. However, these membranes cannot be introduced into existing processes, without complex consequences on regulatory issues. Therefore, testing still appears to be the most efficient method to exclude viruses which may reside within the host cell line itself, or which could be introduced via the biologically derived raw materials required for cell culture process. Specialized service companies have developed, in close collaboration with the pharmaceutical client companies, batteries of validated test procedures. They utilize cell culture systems in which supernatants or lysates of the production cells are co-cultivated with the corresponding virus-sensitive substrates. Since cells in MSCBs and in MWCBs are of the highest importance for the cell culture production process, samples from these are the first to be considered for the rather expensive and time-consuming testing exercise.

A complementary approach to virus safety is the design of virus kill and removal steps of the protein recovery process. These include the physical and chemical principles of separating (theoretical) viral contaminants from the product, or inactivating them. Again, appropriate testing procedures and the demonstration of inactivation and removal of model viruses, as discussed by Wiebe et al. [139] is considered a major provision for the safety of recombinant products from hamster cells.

While the argument remains a strong one – non-human host cells for human protein drugs for biosafety reasons – it should be noted that recently (2002) a human cell line was approved for the production of a recombinant protein. A human embryo kidney cell line, transformed by a shared adeno-virus DNA (Human Embryo Kidney 293-cells), was used to produce Activated Protein C (E. Lilly).

1.7.4
Product Consistency, Quality, and Purity

Within the past two decades, a rich collection of methods has become available for the analysis of purified proteins. In addition, most of these methods have been optimized and fine-tuned to very high sensitivities and resolution. When employed as routine analytical procedures during the manufacturing process, they are able to assure high quality and consistency of protein products [140–145]. Nonetheless, the manufacturer of a biopharmaceutical protein has one major concern: Will the essential characteristics of a product that has demonstrated its efficacy in clinical trials remain the same when produced over many years in a defined manufacturing process? It seems surprising, but due to the large size and complexity of proteins under study for clinical use today, their structure and function within the human body may not be fully understood by the manufacturer after completion of clinical trials. This is especially true for the newer generation of pharmaceutical proteins that are larger in size, and often contain multiple polypeptides and/or specialized domains with secondary modifications. Subtle changes – which sometimes are difficult to detect due to inherent heterogeneity in protein populations – may result in a loss or modification of activity and could pose risks to the patient. In order to reduce this possibility, batteries of in-process controls and tests are an inherent part of the production of clinical biopharmaceutical proteins. In the following section, sensitive analytical techniques are outlined and discussed. The objective is to: 1) prevent the occurrence of even small changes in the procedures for production of the product; and 2) enable the detection and exclusion from the final product variants differing in a major way from that tested in clinical trials.

The first level of protection against inadvertent changes of the product rests in effective management of the production process over time. The challenge is that of any mature industry: to produce large amounts of material at competitive cost, while ensuring that product consistency and quality are maintained. Clearly, manufacturing teams and their supervisors undertake serious efforts to reproduce the manufacturing process to the utmost detail in every production run. Defining and describing each of the various steps in the form of detailed protocols achieve this. Almost every aspect of the procedure is documented, these documents establishing, in the form of cGMP (current Good Manufacturing Practice) protocols, the basis for the overall procedure. A "sign-off" procedure by supervisors represents an integral part of this procedure, assuring that the operating personnel for the manufacturing process are in fact controlling, assessing and executing it according to the established protocols. At critical points of the overall process, the signatures are prerequisites to allow the progression of the process to proceed to the consecutive steps.

The time period of cell line and process development, leading to the establishment of the manufacturing process, is important for the definition of critical check points. During this period, knowledge about parameters and steps is acquired that can result in product changes. Once certain limits of variations of process conditions have been identified (within which no change was observed), the cGMP protocol is drafted and finalized. Specific events – defined in precise terms as part of the manufacturing protocols – can trigger a more elaborate investigation. Supervisors and

managers can even order an interruption of the manufacturing process. In extreme cases, crude product batches are withheld from further processing and are discarded.

Quality control (QC) is an integral part of the manufacturing process for recombinant products (see also Part VII, Chapter 1). A comprehensive approach, utilizing independent validated techniques, is applied to assess the quality and identity of the product from various angles (for a review, see also Ref. [146]). It is the goal of QC efforts to assure that products made over years of manufacturing will meet the stated specifications in terms of identity, quantity, activity, and purity.

In principle, it is no longer difficult to produce large quantities of highly purified recombinant proteins, especially when proteins are secreted into the medium. However, methods to produce recombinant proteins are still part of a young technology, since its basis is the manipulation of genetic material in the laboratory. Those manipulations involve the creation of plasmid vectors and their transfer into mammalian cells cultivated *in vitro*. A major concern has been the fidelity (amino acid sequence identity) of the final product, particularly in view of the high degree of ignorance with respect to gene transfer mechanisms in higher eukaryotes. It also appears that transfected DNA may have a somewhat elevated propensity for mutation during or following transfer into mammalian cells [147].

Based on a history of experience with this technology for more than 20 years, it can be stated that this "young" technology is very reliable. It has been suggested that rigorous and extensive nucleic acid-based tests – most notably a complete sequence assessment of the integrated DNA (or transcribed RNA from recombinant cells) – should be performed [148, 149]. How-

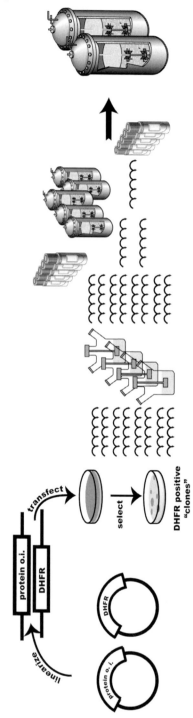

Fig. 1.4 Outline of process development efforts for a large-scale manufacturing process on the basis of mammalian cells. Usually, many candidate cell lines must be evaluated for productivity and long-term stability from different transfections. The wavy lines indicate subcultivations. Over time, the majority of these candidate cell lines are eliminated to obtain a final candidate line for establishing a Master Seed Cell Bank that leads into cGMP manufacturing. In order to secure the supply of cells for experimental work, small cell banks are generated along the way.

ever, it appears that these complex, expensive and time-consuming tests are not necessary. As pointed out above, powerful and reliable new methods have emerged in analytical protein chemistry, and these have increased the capacity to characterize purified protein preparations to a high degree of sensitivity and resolution. These methods represent efficient tools to assess identity, including the amino acid sequence, quantity, potency, and purity of the product immediately prior to administration to the patient [150].

This is not to say that mutants may not occasionally emerge. A 1000-L reactor contains usually more than 10^{12} cells. Mutations will occur at a frequency similar to that in mammalian genomes (1 bp change in 10^9 bp for each generation) [151]. Of course, such mutation will be "scattered" over the entire genome, and it is highly unlikely that individual and specific variants of a given protein product will emerge in the population of protein molecules. This is of course different if a mutant plasmid DNA molecule was integrated into the genome of the host cell at the time of gene transfer.

A telling example for the power of in-process controls and the associated biochemical assays to identify variants in a population of molecules is given by Sliwkowski et al. [152]. This describes the detection of an amino-acid exchange mutant of a recombinant monoclonal antibody (anti-Her2). This variant was detected early during process development efforts when using a MTX-amplified clone of CHO cells. The variant represented about 10% of the total population of antibody molecules. The origin of this mutant remains mysterious, but it seems it was the result of an early event during the development of the cell line.

1.8
Concluding Remarks

The technology to use mammalian cells for recombinant biopharmaceutical protein production is, surprisingly, still in its infancy. Much must be done to establish production processes in a more straightforward way, and also to make them more productive. CHO, NS0, BHK, and PER.C6 cells have been developed that express, in highly optimized manufacturing processes, several grams per liter of secreted proteins, usually antibodies or antibody-fusion proteins. Thus, recent claims [97] that cells of lymphoid origin (e.g., NS0 cells) are especially equipped for the secretion of proteins and therefore are preferential high producers must be questioned. It appears that immortalized mammalian cells – of whatever origin – have tremendous plasticity, both for the uptake of foreign DNA, allowing high level protein synthesis under bioreactor conditions. Even with the newly obtained yields in highly optimized processes of several grams per liter, one should not feel that the end of the opportunity for further improvement has been reached. Indeed, yields of 10–20 g L^{-1} and higher product concentrations should be possible in the near future, particularly if one considers the fact that batch and extended-batch processes obtain, at best, cell densities of about 10^6 mL^{-1}. These cell numbers correspond to about 2% biomass with respect to the total volume in the bioreactor (2% packed cell volume (PCV)). Highly developed microbial processes achieve 20–30% PCV.

A data explosion is occurring presently in biology and in biomedical research. Knowledge gained from genomics will also lead us to a much better understanding of the biochemistry and physiology of mammalian cells. As a result, there is every reason to take a highly very optimistic view that mammalian cells will continue to be preferred as hosts for recombinant protein production.

References

1 Andersen, D. (2004) Scientific report at a cell culture conference in Cancun.

2 Kao, F.-T., Puck, T.T. (1968) Genetics of somatic mammalian cells VII. Induction and isolation of nutritional mutants in Chinese hamster cells. Proc. Natl. Acad. Sci. USA 60:1275–1281.

3 Puck, T.T. (1985) Development of the Chinese hamster ovary (CHO) cell for use in somatic cell genetics. In: M.M. Gottesman (Ed) Molecular Cell Genetics, John Wiley & Sons, New York, pp. 37–64.

4 Urlaub, G. and Chasin, L.A. (1980) Isolation of Chinese hamster cell mutants deficient in dihydrofolate reductase activity. Proc. Natl. Acad. Sci. USA 77:4216–4220.

5 Ringold, G., Dieckmann, B., and Lee, F. (1981) Co-expression and amplification of dihydrofolate reductase cDNA and the *Escherichia coli* XGPRT gene in Chinese hamster ovary cells. J. Mol. Appl. Genet. 1:165–175.

6 Kaufman, R.J. and Sharp, P. (1982) Amplification and Expression of Sequences Cotransfected with a Modular Dihydrofolate Reductase Complementary DNA Gene. J. Mol. Biol. 159:601–621.

7 Kaufman, R.J., Wasley, L.C., Spiliotes, A.J., Gossels, S.D., Latt, S.A., Larsen, G.R., and Kay, R.M. (1985) Coamplification and coexpression of human tissue-type plasminogen activator and murine dihydrofolate reductase in Chinese hamster ovary cells. Mol. Cell. Biol. 5:1750–1759.

8 Pennica, D., Holmes, W.E., Kohr, W.J., Harkins, R.N., Vehar, G.A., Ward, C.A., Bennett, W.F., Yelverton, E., Seeburg, P.H., Heyneker, H.L., Goeddel, D.V., and Collen, D. (1983) Cloning and expression of human tissue-type plasminogen activator cDNA in *E. coli*. Nature 301:214–221.

9 Gottesman, M.M. (1985) Molecular Genetics. John Wiley & Sons, New York.

10 Graham, F.L. and van der Eb, A.J. (1973) A new technique for the assay of infectivity of human adenovirus 5 DNA. Virology 52:456–467.

11 Wigler, M., Silverstein, S., Lee, L.-S., Pellicer, A., Cheng, Y.-C., and Axel, R. (1977) Transfer of purified Herpes Virus thymidine kinase gene to cultured cells. Cell 11:223–232.

12 Pellicer, A., Wigler, M., and Axel, R. (1978) The transfer and stable integration of the HSV thymidine kinase gene into mouse cells. Cell 14:133–141.

13 Robins, D.M., Ripley, S., Henderson, A., and Axel, R. (1981) Transforming DNA integrates into the host chromosome. Cell 23:29–39.

14 Chen, C. and Okayama, H. (1987) High efficiency transformation of mammalian cells by plasmid DNA. Mol. Cell. Biol. 7, 8:2745–2752.

15 Jordan, M., Schallhorn, A. and Wurm, F.M. (1996) Transfecting mammalian cells: optimization of critical parameters affecting calcium-phosphate precipitate formation. Nucleic Acids Res. 24, 4:596–601.

16 Jordan M., Köhne, C. and Wurm, F.M. (1997) Principles of a scaleable transient transfection and expression technology for mammalian cells. In: M.J. Carrondo, B. Griffiths and J.P. Moreira (Eds) Animal Cell Technology – from Vaccines to Genetic Medicine, pp. 47–50. Kluwer Academic Publishers.

17 Finn, G.K., Kurz, B.W., Cheng, R.Z. and Shmookler, R.J. (1989) Homologous plasmid recombination is elevated in immortally transformed cells. Mol. Cell. Biol. 9:4009–4017.

18 Grosjean, F., Batard, P., Jordan, M. and Wurm, F.M. (2002) S-phase synchronized CHO cells show elevated transfection efficiency and expression using CaPi. Cytotechnology 38, 1/2:57–62.

19 Mansour, S.L., Thomas, K.R., and Capecchi, M.R. (1988) Disruption of the proto-oncogene int-2 in mouse embryo-derived stem cells: a general strategy for targeting mutations to non-selectable genes. Nature 336:348–352.

20 Zheng, H. and Wilson, J.H. (1990) Gene targeting in normal and amplified cell lines. Nature 344:170–173.

21 Wurm, F.M., Johnson, A., Ryll, T., Köhne, C., Scherthan, H., Glaab, F., Lie, Y.S., Petropoulos, C.J. and Arathoon, W.R. (1996) Gene transfer and amplification in CHO cells – efficient methods for maximising specific productivity and assessment of genetic consequences. Ann. N.Y. Acad. Sci. 782:70–78.

22 Perucho, M., Hanahan, D., and Wigler, M. (1980) Genetic and physical linkage of exogenous sequences in transformed cells. Cell 22:309–317.

23 Finn, G. K., Kurz, B. W., Cheng, R. Z., and Shmookler Reis, R. J. (1989) Homologous Plasmid Recombination is Elevated in Immortally Transformed Cells. Mol. Cell. Biol. 9, 9:4009–4017.

24 Folger, K. R., Wong, E. A., Wahl, G., and Capecchi, M. R. (1982) Patterns of integration of DNA microinjected into cultured mammalian cells: evidence for homologous recombination between injected plasmid molecules. Mol. Cell. Biol. 2, 11:1372–1387.

25 Jordan, M. and Wurm, F. M. (2003) Co-transfer of multiple plasmids/viruses as an attractive method to introduce several genes in mammalian cells. In: S. C. Makrides (Ed.) Gene Transfer and Expression in Mammalian Cells. New Comprehensive Biochemistry, Volume 38, pp. 337–357.

26 Chen, L., Marmey, P., Taylor, N. J., Brizard, J.-P., Espinoza, C., D'Cruz, P., Huet, H., Zhang, S., Kochko, A., de Beachy, R. N., and Fauquet, C.M. (1998) Expression and inheritance of multiple transgenes in rice plants. Nature Biotechnol. 16:1060–1064.

27 Wurm, F. M. (2004) Production of recombinant protein therapeutics in cultivated mammalian cells. Nature Biotechnology 22, 11:1393–1398.

28 Wurm, F. M., Johnson, A., Lie, Y. S., Etcheverry M. T., and Anderson, K. P. (1992) Host Cell Derived Retroviral Sequences Enhance Transfection and Expression Efficiency in CHO Cells. In: Spier, R. E., Griffiths, J. B., Mac-Donald, C. (Eds) Animal Cell Technology: Developments, Processes and Products. Butterworth-Heinemann, Oxford, UK, pp. 35–41.

29 Barnes, L. M., Bentley, C. M., and Dickson, A. J. (2004) Molecular definition of predictive indicators of stable protein expression in recombinant NS0 myeloma cells. Biotechnol. Bioeng. 85:115–121.

30 Jones, D., Kroos, N., Anema, R., von Montfort, B., Vooys, A., van der Kraats, S., van der Helm, E., Smits, S., Schouten, J., Brouwer, K., Lagerwerf, F., van Berkel, P., Opstelten, D.-J., Logtenberg, T., and Bout, A. (2003) High level expression of recombinant IgG in the human cell line PER.C6. Biotechnol. Prog. 19:163–168.

31 Chu, G., Hayakawa, H., and Berg, P. (1987) Electroporation for the efficient transfection of mammalian cells with DNA. Nucleic Acids Res. 15:1311–1326.

32 Barsoum, J. (1990) Introduction of stable high-copy-number DNA into Chinese hamster ovary cells by electroporation. DNA Cell Biol. 9, 4:293–300.

33 Felgner, P. L. and Ringold, G. M. (1989) Cationic liposome-mediated transfection. Nature 337:387–388.

34 Behr, J.-P., Demeneix, B., Loeffler, J.-P., and Perez-Mutul, J. (1989) Efficient gene transfer into mammalian primary endocrine cells with lipopolyamine-coated DNA. Proc. Natl. Acad. Sci. USA 86:6983–6986.

35 Muller, S. R., Sullivan, P. D., Clegg, D. O., and Feinstein, S.C. (1990) Efficient transfection and expression of heterologous genes in PC12 cells. DNA Cell Biol. 9, 3:221–229.

36 O'Brien, J. and Lummis, S. C. R. (2004) Biolistic and diolistic transfection: using the gene gun to deliver DNA and lipophilic dyes into mammalian cells. Methods 33:121–125.

37 Brokx, R. and Gariepy, J. (2004) Peptide- and polymer-based gene delivery vehicles. Methods Mol. Med. 90:139–160.

38 Putnam, D., Gentry, C. A., Pack, D. W., and Langer, R. (2001) Polymer-based gene delivery with low cytotoxicity by a unique balance of side-chain termini. Proc. Natl. Acad. Sci. USA 98(3):1200–1205.

39 Jasin, M. and Berg, P. (1988) Homologous integration in mammalian cells without target gene selection. Genes Dev. 2:1353–1363.

40 Smithies, O., Gregg, R. G., Boggs, S. S., Koralewski, M. A., and Kucherlapati, R. S. (1985) Insertion of DNA sequences into the human chromosomal beta-globin locus by homologous recombination. Nature 317:230–234.

41 Bode, J., Benham, C., Ernst, E., Knopp, A., Marschalek, R., Strick, R., and Strissel, P. (2000) Fatal connections: when DNA ends meet on the nuclear matrix. J. Cell Biochem. 35:3–22.

42 Bode, J., Götze, S., Ernst, E., Hüsemann, Y., Baer, A., Seibler, J., and Mielke, C. (2002) Architecture and utilization of highly expressed genomic sites. In: S. C. Makrides (Ed) Gene transfer and expression in mammalian cells. Elsevier Science B.V., pp. 551–571.

43 Wilson, T.J. and Kola, I. (2001) The LoxP/CRE system and genome modification. Methods Mol. Biol. 158:83–94.

44 Gopalkrishnan, R.V., Christiansen, K.A., Goldstein, N.I., DePinho, R.A., and Fisher, P.B. (1999) Use of the human EF-1alpha promoter for expression can significantly increase success in establishing stable cell lines with consistent expression: a study using the tetracycline-inducible system in human cancer cells. Nucleic Acids Res. 27(24):4775–4782.

45 Kim, S.Y., Lee, J.H., Shin, H.S., Kang, H.J., and Kim, Y.S. (2002) The human elongation factor 1 alpha (EF-1 alpha) first intron highly enhances expression of foreign genes from the murine cytomegalovirus promoter. J. Biotechnol. 93:183–187.

46 Wilson, J.H., Berget, P.B., and Pipas, J.M. (1982) Somatic cells efficiently join unrelated DNA segments end to end. Mol. Cell. Biol. 2:1258–1269.

47 Balland, A., Faure, T., Carvallo, D., Cordier, P., Ulrich, P., Fournet, B., de la Salle, H., and Lecocq, J.P. (1988) Characterisation of two differently processed forms of human recombinant factor IX synthesised in CHO cells transformed with a polycistronic vector. Eur. J. Biochem. 172(3):565–572.

48 Meents, H., Enenkel, B., Werner, R.G. and Fussenegger, M. (2002) p 27Kip1-mediated controlled proliferation technology increases constitutive sICAM production in CHO-DUKX adapted for growth in suspension and serum-free media. Biotechnol. Bioeng. 79, 6:619–627.

49 Kaufmann, H. and Fussenegger, M. (2003) Metabolic engineering of mammalian cells for higher protein yield. In: S.C. Makrides (Ed) Gene transfer and expression in mammalian cells. Elsevier Science B.V., pp. 547–569.

50 Wurm, F.M., Gwinn, K.A., and Kingston, R.E. (1986) Inducible overexpression of the mouse c-myc protein in mammalian cells. Proc. Natl. Acad. Sci. USA 83:5414–5418.

51 Weber, W., Fux, C., Daoud-El Baba, M., Keller, B., Weber, C.C., Kramer, B.P., Heinzen, C., Aubel, D., Bailey, J.E., and Fussenegger, M. (2002) Macrolide-based transgene control in mammalian cells and mice. Nature Biotechnol. 20:901–907.

52 Richards, E.J. and Elgin, S.C. (2002) Epigenetic codes for heterochromatin formation

and silencing: rounding up of the usual suspects. Cell 108:489–500.

53 Girod, P.-A. and Mermod, N. (2003) Use of scaffold/matrix-attachment regions for protein production. In: S.C. Makrides (Ed) Gene transfer and expression in mammalian cells. Elsevier Science B.V., pp. 359–379.

54 Antoniou, M., Harland, L., Mustoe, T., Williams, S., Holdstock, J., Yague, E., Mulcahy, T., Griffiths, M., Edwards, S., Ioannou, P.A., Mountain, A., and Crombie, R. (2003) Transgenes encompassing dual-promoter CpG islands from the human TBP and HNRPA2B1 loci are resistant to heterochromatin-mediated silencing. Genomics 82(3):269–279.

55 Stief, A., Winter, D.M., Stratling, W.H., and Sippel, A.E. (1989) A nuclear attachment element mediates elevated and position-independent gene activity. Nature 341:343–345.

56 Phi-Van, L., von Kries, J.P., Ostertag, W., and Stratling, W.H. (1990) The chicken lysozyme 5' matrix attachment region increases transcription from a heterologous promoter in heterologous cells and dampens position effects on the expression of transfected genes. Mol. Cell. Biol. 10:2302–2307.

57 Martens, J.H., Verlaan, M., Kalkhoven, E., Dorsman, J.C., and Zantema, A. (2002) Scaffold/matrix attachment region elements interact with a p300-scaffold attachment factor A complex and are bound by acetylated nucleosomes. Mol. Cell. Biol. 22:2598–2606.

58 Fernandez, L.A., Winkler, M., and Grosschell, R. (2001) Matrix attachment region-dependent function of the immunoglobulin mu enhancer involves histone acetylation at a distance without changes in enhancer occupancy. Mol. Cell. Biol. 21:196–208.

59 Zahn-Zabal, M., Kobr, M., Imhof, M.O., Chatellard, P., de Jesus, M., Wurm, F., and Mermod, N. (2001) Development of stable cell lines for production or regulated expression using matrix attachment regions. J. Biotechnol. 87:29–42.

60 Kwaks, T.H.J., Barnett, P., Hemrika, W., Siersma, T., Sewalt, R.G.A.B., Satijn, D.P.E., Brons, J.F., van Blokland, R., Kwakman, P., Kruckeberg, A.L., Kelder, A. and Otte, A.P. (2003) Identification of anti-repressor elements that confer high and stable protein production in mammalian cells. Nature Biotechnol. 21:553–558.

61 Mutskov, V. and Felsenfeld, G. (2004) Silencing of transgene transcription precedes methylation of promoter DNA and histone H3 lysine. EMBO J. 23:138–149.

62 Hacker, D. L., Derow E. and Wurm, F. M. (2005) The CELO Adenovirus Gam1 Protein Enhances Transient and Stable Recombinant Protein Expression in Chinese Hamster Ovary Cells. J. Biotechnology 117:21–29.

63 Gorman, C. M., Howard, B. H., and Reeves, R. (1983) Expression of recombinant plasmids in mammalian cells is enhanced by sodium butyrate. Nucleic Acids Res. 11:7631–7648.

64 Riggs, M. G., Whittaker, R. G., Nuemann, J. R., and Ingram, V. M. (1977) n-Butyrate causes histone modification in HeLa and Friend erythroleukaemia cells. Nature 268:462–464.

65 Cuisset, L., Tichonicky, L., and Delpech, M. (1998) A protein phosphatase is involved in the inhibition of histone diacetylation by sodium butyrate. Biochem. Biophys. Res. Commun. 246:760–764.

66 Makrides, S. C. (1999) Components of vectors for gene transfer and expression in mammalian cells. Protein Express. Purif. 17:183–202.

67 Shirahata, S., Teruya, K., Seki, K., Mori, T., Ohashi, H., Tachibana, H., and Murakami, H. (1992) Oncogene-activated production of recombinant proteins for animal cells. In: Spier, R.E., Griffiths, J.B., MacDonald, C. (Eds) Animal Cell Technology: Developments, Processes and Products. Butterworth-Heinemann, Oxford, pp. 54–59.

68 Lee, K. H., Guerini-Sburlati, A., Renner, W. A., and Bailey, J. E. (1997) Deregulated expression of cloned transcription factor E2F-1 in Chinese Hamster Ovary cells shifts protein patterns and activates growth in protein-free medium. In: Carrondo, M.J.T. et al. (Eds), Animal Cell Technology – from vaccines to genetic medicine. Kluwer Academic Publishers, pp. 621–623.

69 Ishaqe, A. and Al-Rubeai, M. (2002) Role of vitamins in determining apoptosis and extent of suppression by bcl-2 during hybridoma cell culture. Apoptosis 7:231–239.

70 Pak, S. C. O., Hunt, S. M. N., Bridges, M. W., Sleigh, M. J., and Gray, P. P. (1996) Super CHO – a cell line capable of autocrine growth under fully defined protein-free conditions. Cytotechnology 22:139–146.

71 Simpson, N. H., Singh, R. P., Emery, A. and Al-Rubeai, M. (1999) Bcl-2 overexpression reduces growth rate and prolongs G1 phase in continuous chemostat cultures of hybridoma cells. Biotechnol. Bioeng. 64:174–186.

72 Sunstrom, N.-A., S., Gay, R. D., Wong, D. C., Kitchen, N. A., Deboer, L., Gray, P. P. (2000) Insulin-like growth factor-1 and transferrin mediate growth and survival of Chinese hamster ovary cells. Biotechnol. Prog. 16:698–702.

73 Renner, W. A., Lee, K. H., Hstzimanikatis, V., Bailey, J. E., and Eppenberger (1995) Recombinant cyclin E expression activates proliferation and obviates surface attachment of Chinese hamster ovary (CHO) cells in protein-free medium. Biotechnol. Bioeng. 47:476–482.

74 Arden, N., Nivitchanyong, T., and Betenbaugh, M. J. (2004) Cell engineering blocks cell stress and improves biotherapeutic production. Bioprocess. J. 2:23–28.

75 Umaña, P., Jean-Mairet, J., Moudry, R., Amstutz, H., and Bailey, J. E. (1999) Engineered glycoforms of an antineuroblastoma IgG1 with optimized antibody-dependent cellular cytotoxic activity. Nature Biotechnol. 17:176–180.

76 Gossen, M. and Bujard, H. (1992) Tight control of gene expression in mammalian cells by tetracycline-responsive promoters. Proc. Natl. Acad. Sci. USA 89:5547–5551.

77 Fussenegger, M., Morris, R. P., Fux, C., Rimann, M., von Stockar, B., Thompson, C.J., and Bailey, J. E. (2000) Streptogramin-based gene regulation systems for mammalian cells. Nature Biotechnol. 18:1203–1208.

78 Simpson, N. H., Singh, R. P., Perani, A., Goldenzon, C., and Al-Rubeai, M. (1999) In hybridoma cultures, deprivation of any single amino acid leads to apoptotic death, which is suppressed by expression of the bcl-2 gene. Biotechnol. Bioeng. 59:90–98.

79 Mazur, X., Fussenegger, M., Renner, W. A., and Bailey, J. E. (1998) Higher productivity of growth-arrested Chinese hamster ovary cells expressing the cyclin-dependent kinase inhibitor p27. Biotechnol. Prog. 14:705–713.

80 Birch, J. R., Boraston, R. C., Metcalfe, H., Brown, M. E., Bebbington, C. R. and Field, R. P. (1994) Selecting and designing cell lines for improved physiological characteristics. Cytotechnology 15:11–16.

81 Clynes, R. A., Towers, T. L., Presta, L. G., and Ravetch, J. V. (2000) Inhibitory Fc receptors modulate *in vivo* cytotoxicity against tumor targets. Nature Med. 6:443–446.

82 Wright, A. and Morrison, S. L. (1997) Effect of glycosylation on antibody function: implications for genetic engineering. Trends Biotechnol. 15:26–32.

83 Bianchi, A. A. and McGrew, J. T. (2003) High-level expression of full-length antibodies using trans-complementing expression vectors. Biotechnol. Bioeng. 84:439–444.

84 Hunt, L., M. De Jesus, M. Jordan, and F. M. Wurm (1999) GFP expressing mammalian cells for fast, accurate, non-invasive cell growth assessment in the kinetic mode. Biotechnol. Bioeng. 65:201–205.

85 Wurm, F. M., unpublished observation.

86 Griffiths, B. and Wurm, F. (2002) Mammalian cell culture. Encycl. Phys. Sci. Technology 9:31–47.

87 Wurm, F. M., Johnson, A., Ryll, T., Köhne, C., Scherthan, H., Glaab, F., Lie, Y. S., Petropoulos, C. J., and Arathoon, W. R. (1996) Gene transfer and amplification in CHO cells. Ann. N. Y. Acad. Sci. 782:70–78.

88 Gandor, C., Leist, C., Fiechter, A., and Asselbergs, F. A. (1995) Amplification and expression of recombinant genes in serum-independent Chinese hamster ovary cells. FEBS Lett. 377:290–294.

89 Biedler, J. L. and Spengler, B. A. (1976) A novel chromosome abnormality in human neuroblastoma and antifolate-resistance Chinese hamster cell lines in culture. J. Natl. Cancer Inst. 57, 3:683–689.

90 Wurm, F. M., Gwinn, K. A., and Kingston, R. E. (1986) Inducible overexpression of the mouse c-myc protein in mammalian cells. Proc. Natl. Acad. Sci. USA 83:5414–5418.

91 Coquelle, A., Pipiras, E., Toledo, F., Buttin, G., and Debatisse, M. (1997) Expression of fragile sites triggers intrachromosomal mammalian gene amplification and sets boundaries to amplicons. Cell 89:215–225.

92 Stark, G. R., Debatisse, M., Giulotto, E., and Wahl, G. M. (1989) Recent progress in understanding mechanisms of mammalian DNA amplification. Cell 57:901–908.

93 Zettlmeissl, G., Wirth, M., Hauser, H. J., and Küpper, H. A. (1988) Isolation of overproducing recombinant mammalian cells by a fast and simple selection procedure. Gene 73:419–426.

94 Wurm, F. M., unpublished observation.

95 Hendricks, M. B., Luchette, C. A., and Banker, M. J. (1989) Enhanced Expression of an immunoglobulin based vector in myeloma cells mediated by coamplification with a mutant dihydrofolate reductase gene. Biotechnology 7:1271–1274.

96 Bebbington, C. R., Renner, G., Thomson, S., King, D., Abrams, D., and Yaranton, G. T. (1992) High level expression of a recombinant antibody from myeloma cells using a glutamine synthetase gene as an amplifiable selectable marker. Bio/Technology 10:169–175.

97 Barnes, L. M., Bentley, C. M. and Dickson, A. J. (2000) Advances in animal cell recombinant protein production: GS-NS0 expression system. Cytotechnology 32:109–123.

98 Bebbington, C. R. (1991) Expression of antibody genes in non lymphoid mammalian cells. Methods: A companion to Methods in Enzymology 2:136–145.

99 Hsu, T. C. (1961) Chromosomal evolution in cell populations. Int. Rev. Cytol. 12:69–121.

100 Nunberg, J. H., Kaufman, R. J., Schimke, R. T., Urlaub, G., and Chasin, L. A. (1978) Amplified dihydrofolate reductase genes are localized to a homogeneously staining region of a single chromosome in a methotrexate-resistant Chinese hamster ovary cell line. Proc. Natl. Acad. Sci. USA 75:5553–5556.

101 Weidle, U. H., Buckel, P. and Wienberg, J. (1988) Amplified expression constructs for human tissue-type plasminogen activator in Chinese hamster ovary cells: instability in the absence of selective pressure. Gene 66:193–203.

102 Vitek, J. A. (1987) Similarity in dynamics of single and double minute chromosomes incidence and number of chromosomal aberrations during long-term treatment of a human cell line with methotrexate. Neoplasma 34, 6:665–670.

103 Pinkel, D., Straume, T., and Gray, J. W. (1986) Cytogenetic analysis using quantitative, high sensitivity fluorescence hybridization. Proc. Natl. Acad. Sci. USA 83:2934–2938.

104 Pallavicini, M. G., DeTeresa, P. S., Rosette, C., Gray, J. W., and Wurm, F. M. (1990) Effects of Methotrexate (MTX) on Transfected DNA Stability in Mammalian Cells. Mol. Cell. Biol. 10:401–404.

105 Wurm, F. M., Pallavicini, M. G., and Arathoon, R. (1992) Integration and stability of CHO amplicons containing plasmid sequences. Dev. Biol. Stand. 76:69–82.

106 Kim, S.J. and Lee G.M. (1999) Cytogenetic analysis of chimeric antibody-producing CHO cells in the course of dihydrofolate reductase-mediated gene amplification and their stability in the absence of selective pressure. Biotechnol. Bioeng. 64, 6:741–749.

107 Van Wezel, A.L. and van der Velden-de Groot, C.A.M. (1978) Large scale cultivation of animal cells in microcarrier culture. Process Biochem. 13:6.

108 Van Wezel, A.L., van der Velden-de Groot, C.A.M., de Haan, H.H., van den Heuvel, N., and Schasfoort, R. (1984) Large scale animal cell cultivation for production of cellular biologicals. Dev. Biol. Stand. 60:229–236.

109 Berry, J.M., Barnabe, N., Coombs, K.M., and Butler, M. (1999) Production of reovirus type-1 and type-3 from Vero cells grown on solid and macroporous microcarriers. Biotechnol. Bioeng. 62(1):12–19.

110 Loumaye, E., Dreano, M., Galazka, A., Holes, C., Ham, L., Munafo, A., Eshkol, A., Giudice, E., De Luca, E., Sirna, A., Antonetti, F., Giartosio, C-E., Scaglia, L., Kelton, C., Campbell, R., Chappel, S., Duthu, B., Cymbalista, S., and Lepage, P. (1998) Recombinant follicle stimulating hormone: development of the first biotechnology product for the treatment of infertility. Hum. Reprod. Update 4, 6:862–881.

111 Spier, R. and Kadouri, A. (1997) The evolution of processes for the commercial exploitation of anchorage-dependent animal cells. Enzyme Microb. Technol. 21:2–8.

112 Wood Mackenzie, Horizons, Pharmaceuticals Issue 6 (January 2003).

113 Birch, J.R. and Arathoon, W.R. (1990) Suspension culture of mammalian cells. In: A.S. Lubiniecki (Ed) Large Scale Mammalian Cell Culture Technology. Dekker Inc. N.Y., pp. 251–270.

114 Afandi, V. and Al-Rubeai M. (2003) Stable transfection of CHO cells with the c-myc gene results in increased proliferation rates, reduces serum dependency and induces anchorage independence. Cytotechnology 41:1–10.

115 Bödecker, B.G.D., Newcomb, R., Yuan, P., Braufman, A., and Kelsey, W. (1994) Production of recombinant Factor VIII from perfusion cultures: I. Large Scale Fermentation. In: Spier, R.E., Griffiths, J.B., Berthold, W.

(Eds), Animal Cell Technology, Products of Today, Prospects for Tomorrow, pp. 580–590.

116 Annual Report 1995, ARES-SERONO: Follicle Stimulating Hormone.

117 van Wedel, R.J. (1987) Mass culture of mouse and human hybridoma cells in hollow-fiber culture. In: S. Seaver (Ed) Commercial Production of Monoclonal Antibodies. Dekker, New York, pp. 159–173.

118 Davies, J.M., Lavender, C.M., Bowes, K.J., Hanak, J.A.J., Combridge B.S., and Kingsland, S.L. (1995) Human therapeutic monoclonal Anti-D antibody produced in long-term hollow-fibre culture. In: Beuvery, E.C. et al. (Eds) Animal Cell Technology: Developments towards the 21st century. Kluwer Academic Publishers, pp. 149–153.

119 Heimbuch, S. (1996) Press Release of Cellex Biosciences: Cellex Biosciences perfusion equipment used for 1st injectable product produced in hollow fiber licensed by FDA. October 30, 1996.

120 Preissmann, A., Bux, R., Schorn, P., and Noe, W. (1995) Comparative study for the propagation of anchorage-dependent cells using different forms of macroporous carriers. In: Beuvery, E.C. et al. (Eds) Animal Cell Technology: Developments towards the 21st century. Kluwer Academic Publishers, pp. 841–845.

121 Byrn, R.A., Mordenti, J., Lucas, C., Smith, D., Marsters, S.A., Johnson, J.S., Chamow, S.M., Wurm, F.M., Gregory, T., Groopman, J.E. and Capon D.J. (1990) Biological Properties of a CD4 Immunoadhesin. Nature 344, 6267:667–670.

122 Bennett, W.F., Paoni, N., Botstein, D., Jones, A.J.S., Keyt, B., Presta, L., Wurm, F.M. and Zoller M. (1991) Functional Properties of a Collection of Charged-to-Alanine Substitution Variants of Tissue-Type Plasminogen Activator. J. Biol. Chem. 266, 8:5191–5201.

123 Ostrove, J.M., Iyer, P., Marshall, J., and Vacante, D. (1998) Comparison of manufacturing techniques for Adenovirus production. In: Merten, O.W., et al. (Eds) New Developments and New Applications in Animal Cell Technology. Kluwer Academic Publishers, pp. 515–521.

124 Gorman, C.M., Gies, D.R., McCray, G. (1990) Transient Production of Proteins Using an Adenovirus Transformed Cell Line. DNA Protein Eng. Technol. 2:1–28.

125 Meissner, P., Kulangara, A., Pick, H., Chatellard, P. and Wurm, F. M. (2001) Transient Gene Expression: Recombinant Protein Production with Suspension-Adapted HEK-293EBNA Cells. Biotechnol. Bioeng. 75, 2:197–203.

126 Boussif, O., Lezoualc'h, F., Zanta, M., Mergny, M., Scheman, D., Demeneix, B., and Behr, J.-P. (1995) A versatile vector for gene and oligonucleotide transfer into cells in culture and *in vivo*: Polyethyleneimine. Proc. Natl. Acad. Sci. USA 92:7297–7301.

127 Derouazi, M., Girard, P., Van Tilborgh, F., Muller, N., Bertschinger, M. and Wurm, F. M. (2004) Serum-free large scale transient transfection of CHO cells. Biotechnol. Bioeng. 84, 7:537–545.

128 Schlaeger, E. J., Legendre, J. Y., Trzeciak, A., Kitas, E. A., Christensen, K., Deuschle, U., and Supersaxo, A. (1998) Transient transfection in mammalian cells. In: Merten, O. W. et al. (Eds) New Developments and New Applications in Animal Cell Technology. Kluwer Academic Publishers. pp. 105–112.

129 Girard, P., Derouazi, M., Baumgartner, G., Bourgeois, M., Jordan, M., Jacko, B. and Wurm, F. M. (2002) 100 Liter-transient transfection. Cytotechnology 38:15–21.

130 Wurm, F. M. and Bernard, A. (2001) Transient gene expression from mammalian cells – a new chapter in animal cell technology? Cytotechnology 35:155–156.

131 Reilly, D. (Genentech), conference report 2004.

132 Batard, P., Jordan, M., Chatellard, P. and Wurm, F. M. (2001) Transfer of high copy number plasmid into mammalian cells by calcium phosphate transfection. Gene 270:61–68.

133 Hayflick, L. and Moorhead, P. S. (1961) The serial cultivation of human diploid cell strains. Exp. Cell. Res. 25:585–621.

134 Hayflick, L. (1997) SV 40 and Human Cancer, letter. Science 276:336–337.

135 Food and Drug Administration, Center for Biologics Evaluation and Research (1987) Points to Consider in the Characterization of Cell Lines Used to Produce Biologicals, revised 1993.

136 Wiebe, M. E. and May, L. H. (1990) Cell banking. In: Lubiniecki, A. S. (Ed) Large Scale Mammalian Cell Culture Technology. Dekker, New York, pp. 147–160.

137 Prusiner, S. B. (1997) Prion diseases and the BSE crisis. Science 278:245–251.

138 Aguzzi, A. and Polymenidou, M. (2004) Mammalian Prion Biology: One century of evolving concepts. Cell 116:313–327.

139 Wiebe, M. E., Becker, F., Lazar, R., May, L., Casto, B., Semense, M., Fautz, C., Garnick, R., Miller, C., Masover, G., Bergman, D., and Lubiniecki, A. S. (1989) A multifaceted approach to assure that recombinant tPA is free of adventitious virus. In: Spier, R. E., Griffiths, J. B., Berthold, W. (Eds) Advances in Animal Cell Biology and Technology for Bioprocesses. Croy, pp. 68–71.

140 O'Connor, J. V., Keck, R. G., Harris, R. J., and Field, M. J. (1994) In: Brown, F., Lubiniecki A. S. (Eds) Genetic Stability and Recombinant Product Consistency. Dev. Biol. Stand. Karger, Basel, pp. 165–173.

141 Garnick, R. L. (1992) Peptide mapping for detecting variants in protein products. Dev. Biol. Stand. 76:117–130.

142 Lu, H. S., Tsai, L. B., Kenney, W. C. and Lai, P.-H. (1988) Identification of unusual replacement of methionine by norleucine in recombinant interleukin-2 produced by *E. coli*. Biochem. Biophys. Res. Commun. 156:807–813.

143 Chloupek, R. C., Harris, R. J., Leonard, C. K., Keck, R. G., Keyt, B. A., Spellman, M. W., Jones, A. J. S., and Hancock, W. S. (1989) Study of the primary structure of recombinant tissue plasminogen activator by reversed-phase high-performance liquid chromatographic tryptic mapping. J. Chromatogr. 463:375–396.

144 O'Connor, J. V. (1993) The use of peptide mapping for the detection of heterogeneity in recombinant DNA-derived proteins. Biologicals 21:111–117.

145 Stephenson, R. C. and Clarke, S. (1989) Succinide formation from aspartyl and asparaginyl peptides as a model for the spontaneous degradation of proteins. J. Biol. Chem. 264:6164–6170.

146 Anicetti, V. and Hancock, W. S. (1994) Analytical Considerations in the Development of Protein Purification Processes. In: R. Harrison (Ed) Protein Purification Process Engineering. Marcel Dekker, Inc., New York.

147 Calos, M., Lebkowski, J. S., and Botchan, M. R. (1983) High mutation frequency in

DNA transfected into mammalian cells. Proc. Natl. Acad. Sci. USA 80:3015–3019.

148 Galibert, F. (1990) Stability of a gene recombinant: what does it mean and how to check for it. Biologicals 18:221–224.

149 Galibert, F. (1994) Assessing genetic stability at the nucleic acid level. In: Brown, F., Lubiniecki, A.S. (Eds) Genetic Stability and Recombinant Product Consistency. Dev. Biol. Stand. Karger, Basel, pp. 27–30.

150 Wurm, F.M., Petropoulos, C.J. and O'Connor, J.V. (1996) Manufacture of proteins based on recombinant CHO cells: Assessment of genetic issues and assurance of consistency and quality. In: E.R. Schmidt, Th.

Hankeln (Eds) Transgenic Organisms and Biosafety – Horizontal Gene Transfer, Stability of DNA, and Expression of Transgenes. Springer, New York, pp. 283–304.

151 Drake, J.W. (1991) Spontaneous mutation. Annu. Rev. Genet. 25:125–146.

152 Harris, R.J., Murnane, A.A., Utter, S.L., Wagner, K.L., Cox, E.T., Polastri, G.D., Helder, J.C., and Sliwskowski, M.B. (1993) Assessing genetic heterogeneity in production cell lines: Detection by peptide mapping of a low level Tyr to Gln sequence variant in a recombinant antibody. Bio/Technology, pp. 1293–1297.

2
Alternative Strategies and New Cell Lines for High-level Production of Biopharmaceuticals

Thomas Rose, Karsten Winkler, Elisabeth Brundke, Ingo Jordan and Volker Sandig

Abstract

Complex glycosylated biopharmaceutical proteins are typically produced in mammalian cells, and the majority originate from Chinese hamster ovary (CHO) cells and mouse NS0 cells. The development of mammalian super-producer cells from these starter cell lines is an unpredictable and time-consuming effort, requiring the identification of rare clones which combine integration of the expression unit into a highly active genomic locus with superior folding, processing and secretion capabilities. Fine tuning the selection and vector, which includes new cellular promoters, allows us to reproducibly generate productive clone pools of CHO cells suitable for immediate production of test material and improves identification of superior clones. Alternatively, the fast and reliable generation of clones is achieved by site-specific cassette exchange based on heterospecific *flp* sites. We have expanded the strategy to use the strong IgH locus of the G-line, a human/mouse heterohybridoma: replacement of the endogenous human IgM heavy chain gene provides the environment for efficient transcription, secretion and a mostly human glycosylation pattern for Ig fusion proteins. As a new platform alternative to CHO and NS0, which supports the production of fully human proteins, we evaluate human designer cell lines of various tissues created directly from primary cells.

Abbreviations

BHK	baby hamster kidney
CHO	Chinese hamster ovary
CMV	cytomegalovirus
DHFR	dihydrofolate reductase
GS	glutamine synthetase
IRES	internal ribosome entry site
LCR	locus control region
MSX	methionine sulfoximine
MTX	methotrexate
PCR	polymerase chain reaction

2.1
Mammalian Cells as a Workhorse to Produce Protein-based Biopharmaceuticals

The majority of biopharmaceutical proteins are complex glycoproteins. Among them, monoclonal antibodies have experienced tremendous growth over recent years with some products reaching blockbuster status (see also Part V, Chapters 1 and 2). They are followed by cytokines and fusion proteins – truncated re-

Modern Biopharmaceuticals. Edited by J. Knäblein
Copyright © 2005 WILEY-VCH Verlag GmbH & Co. KGaA, Weinheim
ISBN: 3-527-31184-X

ceptors or ligands equipped with additional effector domains (see also Part V, Chapters 6 and 7). Replacement therapies using recombinant versions of human glycoproteins represent the major treatment option for many monogenic genetic diseases (see also the Introduction to this book). All these proteins contain multiple domains, and have substantial requirements for folding and post-translational processing. Their function is often dependent on, or at least modulated by, carbohydrate structures. The glycosylation pattern is a crucial factor for correct protein folding, intracellular trafficking and secretion, as well as for *in vivo* clearance rate, immunogenicity, proteolytic stability and full biological activity of the recombinant glycoprotein (see also Part IV, Chapter 7) [1–4]. Moreover, therapeutic glycoproteins may be rendered antigenic upon exposure of epitopes that are normally masked by oligosaccharides (see also Part VI, Chapter 3). Whereas lower eukaryotic systems such as yeast can cope with some aspects of folding, proteolytic processing and phosphorylation (see also Part IV, Chapter 13), only mammalian cells perform carboxylation, isoprenylation, and add the expected N- and O-linked sugars (see also Part IV, Chapter 12).

This capability comes at a high price: mammalian cell lines are substantially more demanding with respect to media and fermentor design (see also Part IV, Chapter 1). Lower cell densities and product yields per cell result in comparatively low volumetric productivity. Under these conditions manufacturing costs become a substantial parameter affecting the success or failure of a biopharmaceutical. In addition, slow replication of mammalian cells (duplication time 36–48 h) compared to prokaryotes and lower eukaryotes increases the time required for establishment of pro-

ducer lines and generation of clinical material. However, modern expression vectors and cell lines as well as improved culture media and process designs have raised yields from below 100 mg L^{-1} to 5 g L^{-1} for individual antibodies (see also Part IV, Chapter 16). Although already at a very high degree of complexity, careful analysis of the existing technology, rational design of cell lines and modulation of biochemical pathways is expected to boost this number even further. New approaches capable of improving yields or shortening time lines are of great importance. This chapter will summarize general strategies in mammalian cell line development, highlight the most essential factors, and provide a more detailed description of alternative approaches exploring new unconventional cell substrates and locus-specific gene targeting.

2.2
The Cell Line of Choice

Any mammalian cell line has the basic machinery to express and secrete recombinant protein, and huge numbers of cell lines with suitable growth properties are available from various tissues and species. The small number of cell lines industrially used for manufacturing is, therefore, surprising. Two hamster cell lines, the Chinese hamster ovary cell line (CHO) and the baby hamster kidney cell line (BHK), and two genetically related mouse cell lines, the myeloma NS0 derived from BALB/c mice, and the hybridoma SP2-0, a fusion of the myeloma with B cells from the same mouse strain, supply most of the mammalian cell-based biopharmaceuticals, whether marketed or still under development. Once commonly accepted as producers, a large body of information about these cell lines has accumulated and allowed us to build improvements on top of sophisticated existing technology, further increasing the acceptance of the respective cell lines. Moreover, clinical studies and marketed products have provided substantial safety information about CHO and NS0 cell lines, resulting in a higher level of acceptance by regulatory agencies such as the FDA (see also Part VII, Chapters 4 and 5).

The production cell lines were selected mainly for their growth properties: they are propagated in synthetic or chemically defined media with a doubling time of 24–36 h. However, originating from natural tumors (plasmacytoma, NS0) or embryonic tissue (CHO), these cells have lost most differentiated features. This also includes loss of the highly specialized expression and secretion apparatus of differentiated cells.

In contrast, in living organisms, most of the secretory proteins are provided by terminally differentiated resting cell types equipped with a unique set of transcription factors to activate specific promoters and induce complex adaptations in the endoplasmic reticulum and Golgi apparatus. Examples are plasma cells, secretory cells of the pituitary, pancreatic island cells and hepatocytes. Special pluripotent precursor cells or stem cells (see also Part I, Chapters 11 and 12) are required to maintain homeostasis. Proliferation and efficient production of secretory proteins seem to be mutually exclusive. This conflict may be specifically addressed in new or engineered producer cell lines. One example is the separation of growth and production in a biphasic process: the cell is engineered to express a protein inducing differentiation or blocking cell cycle in a drug-regulated fashion. The cyclin kinase inhibitor p27 which prevents phosphoryla-

tion of Rb causing arrest in the G_1 phase of the cell cycle may serve as an example.

Taking the current selection of producer cells into account, it may seem that the mammalian species of origin does not have any impact on the quality of the product. However, much care is taken that human biopharmaceuticals contain human coding sequences. Whereas the first antibodies applied in clinical trials were derived from mouse genes and created a severe human anti-mouse antibody response [5], today's antibody therapeutics are mainly constituted of human sequences (see also Part V, Chapter 2). They originate from phage-displayed antibody libraries expressing variable domains of human origin or from transgenic mice in which IgG genes are replaced with their human counterparts [6–8]. While this improves the pharmacological features substantially, even these advanced biopharmaceuticals may induce an immunologic response or suffer more rapid clearance. Post-translational modifications of human or humanized immunoglobulins produced on cells of nonhuman origin may contribute to this phenomenon. Although mammalian cells in general provide complex N- and O-linked glycosylation (sugars attached to asparagine and threonine residues of the polypeptide chain), the specific pattern depends on the tissue type and species of origin as well as on cell culture conditions [9–11] (see also Part IV, Chapter 1).

For instance, proteins produced in mouse cells carry glycans containing Gal α1–3Gal residues, which are missing in human cells [12]. A high titer of anti-Gal α1–3Gal antibodies in humans [13] causes a rapid clearance of proteins carrying this residue in their glycans. Antibodies produced in CHO cells which lack Gal α1–3Gal residues still require high dosages. Therefore, it is likely that other post-translational modifications are involved in a specific human immune response against antibodies with human primary sequence, but produced in CHO cells.

There are even indications that not just species, but individual tissues, provide a specific glycosylation pattern with functional implications. For instance, brain-derived glycoproteins are reported to contain a higher degree of fucosylation and high amounts of bisecting *N*-acetylglucosamine [14], whereas in blood-derived glycoproteins a high rate of terminal sialic acid is evident which is likely to be required to protect the protein from clearance via the hepatic asialoglycoprotein receptor.

Human cells or cells with human glycosylation machinery should minimize these problems. However, for many years the regulatory hurdles for human cells have been even stronger than those for rodent cells. The lack of a species barrier allowing easier transfer of adventitious agents was considered as a major limitation. On the other hand, it can be argued that infection with human pathogenic agents is likely to result in a full-blown pathogenic effect in human cells that is easy to detect, whereas the agent may be dormant in rodent cells. For all new cell lines, whether of animal or human origin, the risk of transmission of prion-based diseases is addressed with strict documentation requirements and the lack of contact with any potentially infected bovine material (see also Part I, Chapter 6). So far, only one such cell line, PER.C6, a transformed human retinoblast, has entered the market (see also Part IV, Chapter 3).

2.3
Pushing Expression Levels – Impact of Vector Design and Cell Clone Selection

During the 1980s, multiple strong promoters and enhancers were described, and functional models for the relationship between the core promoters and upstream elements were proposed [15, 16]. Most of these promoters are of viral origin (from human or mouse cytomegalovirus, SV40 or Rous sarcoma retrovirus). Their core promoter activity is dominated by a TATA box 20–30 bp upstream of the start site, which directs accurate transcription initiation via binding of a protein called TBP (TATA-binding protein), recruitment of associated factors and formation of the polymerase II pre-initiation complex. The core promoter was found to be functionally separated from the enhancer, a collection of transcription factor-binding sites acting independent of position and orientation, and mediating promoter strength via removal of nucleosomal repression. Despite the 10- to 50-fold different promoter activity in transient assays (expression measured 2–3 days post-introduction of recombinant DNA), stable producer clones containing the strongest promoter [human cytomegalovirus (hCMV) IE] have no clear advantage over clones derived with other viral promoters. Moreover, expression levels vary greatly between individual clones containing the same vector and in many clones expression declines with prolonged propagation. One explanation for this observation is that viral promoters integrated into the host genome preferentially become inactivated by DNA methylation [17] or progressive deacetylation of histones H3 and H4 [18–20]. Both processes are linked: DNA methylation induces deacetylation of histones making the region inaccessible to transcription factors and extensive acetylation is able to prevent methylation at promoter sites [21]. The hCMV IE promoter, one of the most active and frequently used promoters in cell line establishment, is affected so strongly that only very few stable CHO clones maintain expression at a medium or higher level. Specific sequences such as the chicken HS4 insulator adjacent to the promoter/enhancer can protect from both methylation and histone deacetylation [22]. The search for stable and highly expressing clones after random integration of the vector, which makes cell line generation so tedious and time consuming, simply identifies rare genomic sites with functions similar to those mentioned above. Once found, time-efficient approaches can be established by using these same advantageous locations for other transgenes. This makes homologous recombination and integration by site-specific recombinases so attractive in cell line design. Typically, a reporter gene (such as β-galactosidase) linked to a site for recombinases such as *flp* or *cre* is used to identify a preferable locus for integration (see also Part III, Chapter 2). During a secondary transfection in the presence of recombinase, the gene of interest is inserted at the predetermined position and the test gene is inactivated. In this chapter we describe a *flp* recombinase-based exchange system applied to a selected locus in CHO and to the highly active immunoglobulin locus of a human mouse hetero-hybridoma.

Alternatively, sequences proposed to stabilize or increase expression may be inserted into the vector. Multiple such elements have been described such as ubiquitous chromatin opening (UCOE) element or the EASE element [23]; US Patent 6,312,951). Comparable to matrix attachment regions, insulators or locus control regions (LCR), these elements act in cis

(upon the same DNA molecule) in stably transformed cell lines by rendering the DNA accessible to transcription independent of the site of integration and/or by protecting CpG islands in the proximity of promoters from methylation. In contrast to LCR regions, however, the elements act in a tissue-independent manner. It is no surprise that the effect of these elements was demonstrated and is most pronounced with inactivation-sensitive promoters such as hCMV promoter.

We and others have isolated regions from cellular genes that are strong promoters and enhancers and in addition transfer the property of locus-independent expression and prevent transgene deactivation. As an example for this strategy, 12 kb of upstream and 3 kb of downstream areas of the hamster EF1 α gene have provided stable expression levels exceeding those of the hCMV promoter by at least an order of magnitude [24]. In contrast, a 1.3-kb region of the human EF1 α promoter enables only moderate expression levels.

In addition to the transgene cassette, expression vectors typically contain selection marker genes. They primarily serve to eliminate untransfected and transiently transfected cells after transfection, and help to generate a clone pool from which high producers can be selected. However, they may also be used to substantially enrich the fraction of high producers. It is believed that a transcriptional link between the marker and the gene of interest is required to achieve this goal. For this strategy, both genes are placed on a bicistronic message and driven by a single strong promoter. While the gene of interest positioned close to the cap site at the 5′-end of the message is expressed in a cap-dependent manner, expression of the marker in the second position is ensured by an internal ribosome entry site (IRES) often taken from picorna viruses (encephalomyocarditis virus or poliovirus). Despite the presence of the IRES element, expression of the marker gene is impaired. We have found that this reduced marker expression is most critical to rich selectivity for high producers with increasing drug concentrations. We have achieved the same effect by expressing both genes as separate transcripts located in close proximity. Marker gene expression in IRES-based constructs strongly depends on the nature of the gene of interest in the first position [25]. This complicates selection as appropriate drug concentrations have to be determined for each new protein. In contrast, generic selection strategies can be applied when separate transcription units are used.

The nature of the marker itself is crucial to the efficacy of the selection process. One class represented by neomycin phosphotransferase (*npt*), hygromycin B-phosphotransferase (*hpt*) or blasticidin deaminase (*bda*) and puromycin N-acetyl-transferase (*pac*) encodes enzymes to inactivate drugs blocking protein biosynthesis; the other – auxotrophic markers such as glutamine synthetase (GS) (see also Part IV, Chapter 4) and dihydrofolate reductase (DHFR) (see also Part IV, Chapter 1) – encodes metabolic enzymes which eliminate specific nutritional requirements. Auxotrophic markers require target cell lines deficient in the respective genes like CHO *dhfr⁻* cell line clones (DUXB11 and DG44) [26, 27] from which the gene has been mutated or deleted, or myeloma cells possessing very little GS activity per se [28]. Drug inhibition of the enzyme [methionine sulfoximine (MSX) and methotrexate (MTX) for GS and DHFR, respectively] in multiple steps induces amplification of the marker gene and the colocalising transgene [28, 29]. This time-consuming process has provided most of the earlier production cell lines. Used at a single drug concentration, selection with MSX or MTX eliminates low producing clones. The combination of markers from both classes and the use of stable cellular promoters has allowed us to generate CHO clone pools reaching up to $14\,\mathrm{pg\,cell^{-1}\,day^{-1}}$ of a recombinant glycoprotein. This strategy even competes with the specific targeting approaches described below.

2.4
A Single CHO High-producer Clone for Multiple Products

The investment into a producer cell is substantial and increasing when a cell line enters later stage phases of clinical development. Defining and fine tuning media and processes often takes more than a year, and requires a larger team of process engineers. The time is well spent because volumetric productivity can often be increased by an order of magnitude (see also Part IV, Chapter 1). This effort focuses on a particular producer clone rather than a starting cell line. With every new product candidate introduced into a given starting cell line the investment has to be repeated from scratch. It would be intriguing to exchange one protein for another and keep most of the features (e.g., high expression level) of the particular clone. As discussed above, the genomic locus harboring the foreign gene substantially contributes to productivity. In addition, the preferred clone has adapted to efficient protein folding, glycosylation and secretion and has escaped the unfolded protein response [30, 31], a protective biochemical pathway induced by stressful overexpression of proteins. This selection therefore cannot be carried out with an empty producer cell line.

We have explored *flp* recombinase-mediated exchange of one transgene for another to approach this issue. As a test gene we used the adipostatic hormone leptin linked to an IgG4 Fc-domain (hobFc) which represents the large group of Fc fusion proteins. The gene was cloned downstream of a human CMV promoter linked to the first intron of EF1*a* gene. An SV40-driven blasticidin gene was used for selection. The two expression units were flanked by target recognition sites (*frt*) for *flp* recombinase (see also Part III, Chapter 2). To favor gene exchange over excision we have used mutated *frt* sites differing in there core sequence F3 and F5 [32]: these sites efficiently recombine with identical *frt* sites, but fail to interact with each other. A third heterospecific site (wild-type *frt*) was inserted between the leptin gene and its promoter to allow exchange of the gene only. We positioned a promoterless ATG-deficient *neo* gene outside of the replacement cassette. The exchange vector was equipped with a minimal promoter as well as an in-frame ATG to activate the *neo* gene allowing selection for correct exchange (Fig. 2.1). As proof for this strategy we introduced a promoterless *gfp* gene and the signals for activation of the *neo* gene into a clone pool harboring the target vector (Fig. 2.2) and found that *gfp* expression was activated in all *neo* resistant cells. To select a superior starter clone the target vector was introduced into a pre-selected CHO DUXB11 clone by electroporation, the transfection method providing the highest degree of single-copy integration events [33, 34]. Screening of 1500 clones yielded three clones with productivities between 6 and 10 pg cell^{-1} day^{-1}. While leptin–Fc could be reproducibly exchanged by the gene of interest in several of the clones, we unexpectedly observed variations in the expression level among clones originating from a single individual recipient cell. Some, but not all, variations were reflected by different RNA levels (Fig. 2.3). This heterogeneous expression can be attributed to the perturbance of the architecture of a pre-

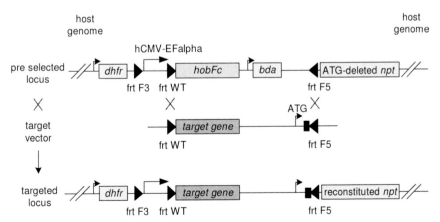

Fig. 2.1 Gene replacement at a predefined locus. The test gene leptin–Fc (hobFc) and the primary selection marker blasticidin deaminase (*bda*) residing between heterospecific *frt* sites (frtF3, frtwt and frtF5) are exchanged for the target gene. The minimal promoter and an in-frame ATG present in the targeting vector activate the neomycin resistance gene (*npt*) used to select for correct recombinants.

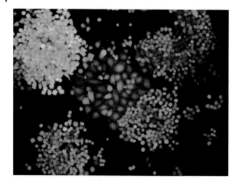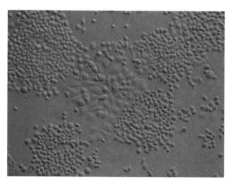

Fig. 2.2 Highly efficient gene targeting in CHO cells. A promoterless *gfp* gene is activated in all cell clones surviving G418 selection after *flp*-mediated recombination in a pool of clones carrying the targeting vector.

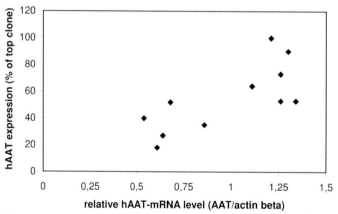

Fig. 2.3 Heterogeneity of expression from a single genomic locus. A leptin–Fc-expressing starter clone was targeted with a promoterless α_1-anti-trypsin gene. Individual daughter clones were analyzed for daily cellular productivity at a defined cell density (10^6 mL^{-1}). Levels of α_1-antitirypsin RNA relative to endogenous hamster β-actin RNA were determined in a SYBR Green-based real-time PCR assay. Variable productivity dependent and independent of RNA levels is observed.

viously stable locus caused by epigenomic phenomena. Moreover, the nature of the transgene itself influenced the level of expression as well as the degree of variation. While the system was not able to completely replace clone screening, it still provides a substantial advantage for the evaluation of multiple product candidates: medium or high producers are obtained with high reproducibility within 5 weeks after screening of 10 or 20 clones. Morphologic features and growth parameters of secondary clones usually are inherited from the primary clones. Therefore, the search for a process-friendly starter clone may render this *flp* system even more valuable. For some proteins superior productivity has been already achieved: CHO clones secreting up to 40 pg cell^{-1} day^{-1} were generated with the *flp* strategy for a human proteoglycan.

2.5
The G-line: Use of the Immunoglobulin Locus of a Human/Mouse Heterohybridoma for Heterologous Gene Expression

A well-characterized natural cell line with long-term, high-level protein secretion and a known locus responsible for this expression may provide a template for more reliable heterologous gene expression as an alternative to identification of superior genomic loci in common cell lines after random integration of a target vector and large-scale screening. We have explored a human mouse heterohybridoma, which expressed a human IgM antibody in a stable configuration for over 2 years, as a potential protein producer cell line.

The heterohybridoma CB03 was created by fusing human B lymphocytes from a patient with chronic thrombocytopenia, obtained by therapeutic splenectomy, with the mouse myeloma line P3X63Ag8. CB03 secretes human autoantibodies of the IgMλ type which react with human platelets, and double- and single-stranded DNA [35]. The hybridoma was shown to secrete the antibody in a stable manner at a rate of 45 pg cell^{-1} day^{-1} over a period of 2 years. This is in striking contrast to the majority of heterohybridomas, which tend to quickly lose human chromosomes, resulting in unstable immunoglobulin expression. Moreover, expression was preserved when the cell line was cultivated in high-density fermentation systems of the CellPharm family (Unisyn Technologies) and remained stable in five independent fermentor runs which had a mean duration of 66 days. Expression did not decrease below 30 pg cell^{-1} day^{-1} when the medium was exchanged for a protein-free medium in a continuous fermentation run.

This suggests that the immunoglobulin loci of CB03 are highly accessible and stable, and are, therefore, well suited to drive a heterologous transgene. To assess the possibility to target these loci, thereby not only introducing a transgene cassette, but also abolishing IgM expression, the cell line was submitted to spectral karyotype analysis to find out whether the respective loci are present as single copies. The typical CB03 cell contains 69–94 chromosomes with a dominance of mouse chromosomes. Via hybridization with specifically labeled human chromosome libraries, eight complete human chromosomes [4, 5, 7, 10, 14, 17, 18 and 22] and fragments of others [4, 8–11, 14 and 16], each linked to a mouse chromosome, were identified (Fig. 2.4). Since a complete and a partial copy of chromosome 14 (the chromosome harboring the IgH genes) were found, *in-situ* hybridization with an IgH probe was performed and a single copy of the human IgH region was identified. With a single chromosome 22 present, a single copy of the Igλ locus was expected as well.

Using the known cDNA sequence and the IMGT database, we have identified V1–2, D1, J6 and μ as the elements participating in constitution of the heavy chain gene. A sequence map of the rearranged IgH locus of CB03 was built including sequences located upstream of the V$_H$ promoter region of V1–2. Using polymerase chain reaction (PCR) primers located 2000 bp upstream of the transcription start point for V1–2 and within J$_H$6, the predicted structure was confirmed. A typical targeting vector for the IgH locus was constructed using a proofreading PCR system.

This vector consists of a short flank (1930 bp), which represents sequences upstream of the V$_H$ promoter, the promoter itself, the transcription initiation point and the RNA leader sequence without the start codon and a long flank (7400 bp) ranging from J$_H$6 to C$_H$1 spanning the entire C$_\mu$ intron.

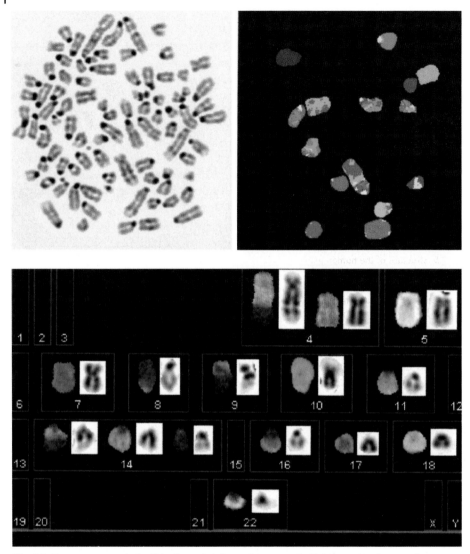

Fig. 2.4 Chromosome analysis of the heterohybridoma CB03. GTG banding (upper left), spectral karyotype analysis (upper right) and identification of human chromosomes by hybridization with specifically labeled human chromosome libraries (lower panel). Eight complete human chromosomes [4, 5, 7, 10, 14, 17, 18 and 22] were identified. In addition, fragments of human chromosomes 4, 8–11, 14, and 16 were found, each linked to a mouse chromosome

As for the CHO approach, the targeting vector was equipped with the blasticidin gene for selection and hobFc as the reporter gene. Either the CMV/EF1 fusion promoter or the endogenous V_H promoter were used to drive the reporter. The expression units were again flanked by heterospecific *flp* sites to allow for secondary exchange (Fig. 2.5).

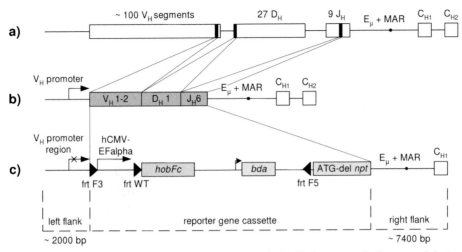

Fig. 2.5 Structure of the human germ line (a), and rearranged (b) and targeted (c) IgH I locus of the G-line. The targeting vector used in homologous recombination contains the leptin–Fc reporter gene (hobFc) and the blasticidin deaminase gene flanked by heterospecific *frt* sites and the inactive neomycin phosphotransferase gene. The cassette is flanked by regions homologous to the IgH locus at either side.

The composition of the λ gene locus was investigated and a targeting construct was designed using a similar approach. The $V_{\lambda}3$–19 gene (upstream) and J_C2 (downstream) are the components forming the active λ gene of CB03. The flanks were limited to 4000 (V_{λ}) and 4500 bp (J_C2) in order to exclude highly repetitive sequences located further upstream and downstream. Hygromycin and a_1-antitrypsin were used as selection marker and reporter, respectively. An independent gene replacement system similar to that of the heavy chain, but based on alternative *frt* sites and an inactive histidinol resistance marker, was included.

In order to screen clone numbers large enough to detect homologous recombinants, we developed direct cell staining techniques for secreted IgM and IgG (Fc) using Texas Red- and AMCA-labeled antibodies. By optimizing the concentration and incubation time, we were able to form antibody precipitates in situ in the absence of methylcellulose or agarose, which are usually employed to limit product diffusion (Fig. 2.6). This procedure allowed not only the identification of high producers, but also the quick isolation of these clones by micro-capillary picking. From approximately 800 clones, 32 with intense IgG staining were identified. Of these clones, 14 showed no staining for IgM, which was confirmed by Western blot for a total of 11 clones. PCR tests using primer pairs located outside the targeting vector and in the transgene region confirmed a homologous targeting event in the clones analyzed. Interestingly, and in confirmation for our strategy, expression remained stable over 3 months in clones originating from homologous recombination, whereas several clones resulting from random integration lost expression. We observed the highest expression levels of 25 pg cell^{-1} day^{-1} for clones containing the hCMV/EF1a hybrid promo-

 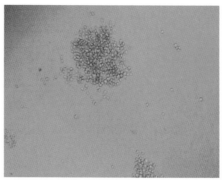

Fig. 2.6 Identification of recombinant clones by direct immunostaining. Colonies were incubated with Texas Red-labeled antibodies against IgG at concentrations allowing the formation of immune precipitates at colonies expressing leptin–Fc (left, upper large colony) in contrast to colonies without expression (bottom, small colony). Phase contrast (×4) of the same area for comparison (right).

ter. The resulting cell line was named "G-line".

Based on the presence of the set of human chromosomes in the heterohybridoma, a glycosylation pattern (Fig. 2.7) different from that of mouse myelomas such as NS0 was expected. Thus, we analyzed the oligosaccharide structure of hobFc generated in a roller bottle process. The single N-linked oligosaccharide chain located in the Fc region was sialylated at 37%, a rate close to average sialylation on antibodies in human blood. Sialic acids were mainly N-acetylneuraminic acid, typical for human cells. Only 2% were represented by N-glycolylneuraminic acid, the immunogenic form dominating in mouse myeloma cells. $a1$–3 Gal structures, not made in human cells and recognized by pre-existing antibodies, were only found in 1.3% of the glycans. This suggests that the G-line indeed executes glycosylation in a way that better resembles the pattern of human compared to mouse cell lines. Moreover, we observed an unexpectedly low degree of core fucose (Tab. 2.1). The inhibition of fucosylation, a secondary effect from artificial introduction of β–$1,4$-N-acetylglucosa-minyltransferase III into CHO cells, is known to enhance Fc effector functions such as antibody-dependent cellular cytotoxicity [36]. For some applications this may enhance potency of Fc-fusion proteins or antibodies generated from the G-line.

Based on the presence of *frt* sites and selection systems in the targeting vectors, the G-line allows the simple introduction of secondary target genes via recombinase mediated cassette exchange. Multiple glycoproteins have been introduced into the IgH locus. For a_1-antitrypsin (*aat*) introduced into the IgH locus expression levels reached 9 pg cell^{-1} day^{-1}. In general, expression levels of secondary transgenes were comparable to those achieved in CHO cells. Absolute levels as well as the degree of homogeneity are transgene dependent. The G-line seems particularly well suited for Fc-fusion proteins which are more difficult to produce in other systems.

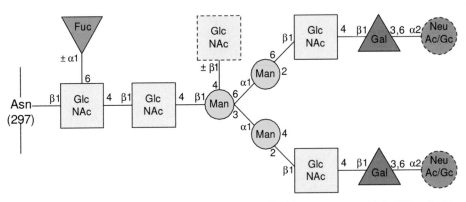

Fig. 2.7 Typical structure of the complete bian-tennary N-linked glycan of an antibody. GlcNAc, N-acetyl-D-glucosamine; Fuc, L-fucose; Man, D-mannose; Neu Ac/Gc, N-acetyl-D-neuraminic acid or N-glycolyl-D-neuraminic acid. In CHO cells, bisecting GlcNAc (mannose position 4) is absent and neuraminic acids are missing in a high percentage of the glycans.

Table 2.1 Differences in glycan structures between production cell lines.

Feature	Impact	G-line	CHO	NS0	Human
Sialylation	Proteolytic sensitivity clearance rate	37%	Variable	Variable	35–40%
N-acetylneuraminic acid	Human glycostructure	98%	High	Low	100%
N-glycolylneuraminic acid	Immunogenic	2%	Low	>50%	No
2–6 linkage	Unknown	No	No	No	Variable
α1–3Gal	Pre-existing antibodies (increased clearance rate)	1.3%	Variable	High	No
Bisecting N-acetyl-D-glucosamine	Inhibits core fucosylation	ND	No	No	10%
Lack of core fucosylation	Increased antibody-dependent cellular cytotoxicity and Fcγ-binding	60%	5%	10–50%	5%
G_0 structures	Increased dimerization; G_2 increases complement-dependent cytotoxicity	4.3%	Variable	Variable	Low

2.6
Human Designer Cell Lines

The quest for genuine human protein processing is best addressed with human cell lines.

Natural tumor cells growing well *in vitro* appear to be the first choice. The strongest regulatory arguments against these cell lines have been a poorly documented history, propagation in mice during establishment and risk of transferring oncogenes to recipients. These hurdles can be overcome when cell lines are made directly from primary cells from a well-documented source using a set of weak and well-known oncogenes. Allowing a better risk assessment, the use of such designer cell lines is permitted for the manufacture of live-attenuated viral vaccines or viral vectors. Some of the criteria also apply to cell lines intended for glycoprotein manufacturing. Whereas tumorigenicity and the transfer of oncogenes are much better controlled through minimal levels of contaminating DNA in protein preparations, a well-documented cell history is just as critical. Although the designer cell approach provides a theoretical solution, there are substantial practical obstacles. Human primary cells have been quite refractory to transformation: at least four different barriers protect against tumor formation.

- Blocking of cell cycle progression by Rb.
- Apoptosis induction via p53 caused by deregulation of the Rb pathway.
- Growth factor dependence for cell cycle progression and apoptosis prevention.
- Telomere shortening and crisis in the absence of telomerase [37] (see also Part 1, Chapter 1).

While it became feasible to transfect primary cells at reasonable and in some cases high efficiencies, stable integration of foreign DNA is extremely inefficient compared to established cell lines. Often several of the immortalizing/transforming genes have to be introduced simultaneously to prevent the immediate induction of apoptosis.

To mirror the process of oncogenic transformation *in vitro*, typically three to four genes are applied: SV40 large T antigen to inactivate Rb and p53 pathways, the catalytic subunit of the telomerase enzyme hTERT to maintain telomere length, and v-*ras* to abolish the strong dependency on external growth factors. This process has been reviewed as a major milestone in cancer research [37]. However, these factors are not suitable for designer cell lines. Transforming genes of SV40 and virally activated *ras* are considered strong and dangerous oncogenes for recipients of therapeutic preparations. The same applies to other genes frequently used in research projects, such as E6 and E7 of human papilloma virus type 16 or 18, viruses highly associated with malignant cervical tumors.

One exemplary set of proteins suitable for immortalization originates from adenoviruses (see also Part I, Chapter 6 and Part IV, Chapter 3). The adenoviral E1A (12S and 13S) proteins bind Rb and family members, and act as a transcriptional modulator, while the E1B protein 55k converts p53 from a transcriptional activator to a repressor. The second E1B protein (19 k) is a homolog to Bcl-2 and blocks apoptosis interacting with *bax* and *bad* genes. These proteins have an extremely high safety profile: the E1 proteins of group C adenoviruses have never been associated with human tumors despite an almost complete exposure of the human population to the virus causing the common cold (over 90% of the human population in Europe has neutralizing antibodies). Moreover, the E1A genes possess tumor-inhibiting, apoptosis-inducing features when introduced in the absence of E1B.

This desirable feature makes immortalization using this strategy more challenging compared to introduction of SV40 large T antigen. Only few cell lines have been generated. The first cell line, HEK293 [38], was made by transfecting sheared adenovirus DNA (see also Part IV, Chapter 12). It originates from embryonic kidney, but is believed to be of neuronal origin [39]. The 911 [40] and PER.C6 [41] carry a defined adenovirus E1 fragment linked to heterologous promoter and poly(A) signals (see also Part IV, Chapter 3).

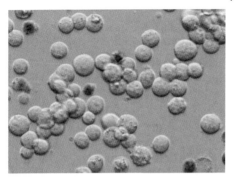

Fig. 2.8 Neuronal cell line NCA1 growing in serum-free medium in suspension. Hoffmann modulation contrast, ×20. Dead cells are stained blue due to Trypan blue uptake.

Both are based on embryonic retinoblasts, a cell type highly susceptible to immortalization. A third published cell type is an E1-immortalized amniocyte [42].

We have developed human cell lines from several tissues using multiple vectors and approaches: new cell lines have been generated by transfection with immortalizing cellular or viral genes followed by continuous passage, subcloning and adaptation to serum-free conditions. These procedures have been carried out in designated laboratories, separated from other cell culture activities.

Although we were able to identify and expand clones from other tissues as well, neuronal precursor cells were most susceptible to transformation, yielding up to six clones from a single transfection of 2×10^6 cells. By continuous cultivation for 8 months, cells derived from some of the cell clones stabilized and became more homogenous in size and cell morphology. Doubling time in fetal calf serum-containing culture was reduced from above 72 down to 40 h when cells were kept at higher densities. Adaptation to serum-free growth in suspension, an ultimate requirement for any new cell line applicable to production of biopharmaceuticals, was suc-

cessful only for a fraction of clones and served as a criterion for further development. In contrast to CHO cells for which a stepwise reduction of serum concentration and transfer to serum-free medium (so called weaning) is recommended [43], we succeeded when cells were transferred directly to appropriate serum-free media (Fig. 2.8) (see also Part II, Chapter 3).

In order to assess their capacity to produce and process recombinant proteins we introduced the a_1-antitrypsin gene driven by a mouse CMV/EF1-a hybrid promoter together with selection markers into one of the cell clones derived from neuronal precursors. Twelve individual clones were isolated and specific cellular productivity was determined. Several of the clones secreted more than 75 pg cell^{-1} day^{-1}, at least 6-fold more than the best CHO or G-line based producers isolated for this protein so far. Titers of 0.5 g L^{-1} accumulated over 17 days of stationary culture in T-flasks.

2.7
Summary and Conclusion

Mammalian cells allow the production of complex biopharmaceutical proteins. Advances in the generation of stable recombinant clones, media formulation, and process and fermentor design have significantly increased yields over the past two decades. Exciting networks across the proteome and transcriptome are elucidated that show the enormous potential still hidden in mammalian cells. The near future will definitely show further improvement with cell lines specifically designed and metabolically engineered for industrial-scale production of biopharmaceuticals.

References

1 Berg DT, Burck PJ, Berg DH, Grinnell BW. 1993. Kringle glycosylation in a modified human tissue plasminogen activator improves functional properties. *Blood 81*, 1312–1322.

2 Helenius A. 1994. How N-linked oligosaccharides affect glycoprotein folding in the endoplasmic reticulum. *Mol Biol Cell 5*, 253–265.

3 Fiedler K, Simons K. 1995. The role of N-glycans in the secretory pathway. *Cell 81*, 309–312.

4 Sareneva T, Pirhonen J, Cantell K, Julkunen I. 1995. N-glycosylation of human interferon-gamma: glycans at Asn-25 are critical for protease resistance. *Biochem J 308*, 9–14.

5 Shawler DL, Bartholomew RM, Smith LM, Dillman RO. 1985. Human immune response to multiple injections of murine monoclonal IgG. *J Immunol 135*, 1530–1535.

6 McCafferty J, Griffiths AD, Winter G, Chiswell DJ. 1990. Phage antibodies: filamentous phage displaying antibody variable domains. *Nature 348*, 552–554.

7 Bruggemann M, Spicer C, Buluwela L, Rosewell I, Barton S, Surani MA, Rabbitts TH. 1991. Human antibody production in transgenic mice: expression from 100 kb of the human IgH locus. *Eur J Immunol 21*, 1323–1326.

8 Lonberg N, Taylor LD, Harding FA, Trounstine M, Higgins KM, Schramm SR, Kuo CC,

Mashayekh R, Wymore K, McCabe JG. 1994. Antigen-specific human antibodies from mice comprising four distinct genetic modifications. *Nature 368*, 856–859.

9 Goochee CF, Monica T. 1990. Environmental effects on protein glycosylation. *Biotechnology 8*, 421–427.

10 Hahn TJ, Goochee CF. 1992. Growth-associated glycosylation of transferrin secreted by HepG2 cells. *J Biol Chem 267*, 23982–23987.

11 Wright A, Morrison SL. 1997. Effect of glycosylation on antibody function: implications for genetic engineering. *Trends Biotechnol 15*, 26–32.

12 Borrebaeck CA. 1999. Human monoclonal antibodies: the emperor's new clothes? *Nat Biotechnol 17*, 621.

13 Galili U, Anaraki F, Thall A, Hill-Black C, Radic M. 1993. One percent of human circulating B lymphocytes are capable of producing the natural anti-Gal antibody. *Blood 82*, 2485–2493.

14 Hoffmann A, Nimtz M, Getzlaff R, Conradt HS. 1995. "Brain-type" N-glycosylation of asialo-transferrin from human cerebrospinal fluid. *FEBS Lett 359*, 164–168.

15 Mueller-Storm HP, Sogo JM, Schaffner W. 1989. An enhancer stimulates transcription in *trans* when attached to the promoter via a protein bridge. *Cell 58*, 767–777.

16 Kermekchiev M, Pettersson M, Matthias P, Schaffner W. 1991. Every enhancer works with every promoter for all the combinations tested: could new regulatory pathways evolve by enhancer shuffling? *Gene Expr 1*, 71–81.

17 Chen C, Yang MC, Yang TP. 2001. Evidence that silencing of the HPRT promoter by DNA methylation is mediated by critical CpG sites. *J Biol Chem 276*, 320–328.

18 Wolffe AP, Pruss D. 1996. Targeting chromatin disruption: transcription regulators that acetylate histones. *Cell 84*, 817–819.

19 Grunstein M. 1997. Histone acetylation in chromatin structure and transcription. *Nature 389*, 349–352.

20 Wade PA, Pruss D, Wolffe AP. 1997. Histone acetylation: chromatin in action. *Trends Biochem Sci 22*, 128–132.

21 Cervoni N, Szyf M. 2001. Demethylase activity is directed by histone acetylation. *J Biol Chem 276*, 40778–40787.

22 Mutskov VJ, Farrell CM, Wade PA, Wolffe AP, Felsenfeld G. 2002. The barrier function of an insulator couples high histone acetylation lev-

els with specific protection of promoter DNA from methylation. *Genes Dev 16*, 1540–1554.

23 Antoniou M, Harland L, Mustoe T, Williams S, Holdstock J, Yague E, Mulcahy T, Griffiths M, Edwards S, Ioannou PA, Mountain A, Crombie R. **2003**. Transgenes encompassing dual-promoter CpG islands from the human TBP and HNRPA2B1 loci are resistant to heterochromatin-mediated silencing. *Genomics 82*, 269–279.

24 Running Deer J, Allison DS. **2004**. High-level expression of proteins in mammalian cells using transcription regulatory sequences from the Chinese hamster EF-1 gene. *Biotechnol Prog 20*, 880–889.

25 Mizuguchi H, Xu Z, Ishii-Watabe A, Uchida E, Hayakawa T. **2000**. IRES-dependent second gene expression is significantly lower than cap-dependent first gene expression in a bicistronic vector. *Mol Ther 1*, 376–382.

26 Urlaub G, Chasin LA. **1980**. Isolation of Chinese hamster cell mutants deficient in dihydrofolate reductase activity. *Proc Natl Acad Sci USA 77*, 4216–4220.

27 Urlaub G, Mitchell PJ, Kas E, Chasin LA, Funanage VL, Myoda TT, Hamlin J. **1986**. Effect of gamma rays at the dihydrofolate reductase locus: deletions and inversions. *Somat Cell Mol Genet 12*, 555–566.

28 Bebbington CR, Renner G, Thomson S, King D, Abrams D, Yarranton GT. **1992**. High-level expression of a recombinant antibody from myeloma cells using a glutamine synthetase gene as an amplifiable selectable marker. *Biotechnology 10*, 169–175.

29 Kaufman RJ, Sharp PA. **1982**. Amplification and expression of sequences cotransfected with a modular dihydrofolate reductase complimentary DNA gene. *J Mol Biol 159*, 601–621.

30 Harding HP, Calfon M, Urano F, Novoa I, Ron D. **2002**. Transcriptional and translational control in the Mammalian unfolded protein response. *Annu Rev Cell Dev Biol 18*, 575–599.

31 Cudna RE, Dickson AJ. **2003**. Endoplasmic reticulum signaling as a determinant of recombinant protein expression. *Biotechnol Bioeng 81*, 56–65.

32 Schlake T, Bode J. **1994**. Use of mutated FLP recognition target (FRT) sites for the exchange of expression cassettes at defined chromosomal loci. *Biochemistry 33*, 12746–12751.

33 Mielke C, Maass K, Tummler M, Bode J. **1996**. Anatomy of highly expressing chromo-somal sites targeted by retroviral vectors. *Biochemistry 35*, 2239–2252.

34 Baer A, Schubeler D, Bode J. **2000**. Transcriptional properties of genomic transgene integration sites marked by electroporation or retroviral infection. *Biochemistry 39*, 7041–7049.

35 Jahn S, Niemann B, Winkler T, Kalden JR, von Baehr R. **1994**. Expansion of a B-lymphocyte clone producing IgM auto-antibodies encoded by a somatically mutated V_HI gene in the spleen of an autoimmune patient. *Rheumatol Int 13*, 187–196.

36 Sburlati AR, Umana P, Prati EG, Bailey JE. **1998**. Synthesis of bisected glycoforms of recombinant IFN-beta by overexpression of beta-1,4-N-acetylglucosaminyltransferase III in Chinese hamster ovary cells. *Biotechnol Prog 14*, 189–192.

37 Weitzman JB, Yaniv M. **1999**. Rebuilding the road to cancer. *Nature 400*, 401–402.

38 Graham FL, Smiley J, Russell WC, Nairn R. **1977**. Characteristics of a human cell line transformed by DNA from human adenovirus type 5. *J Gen Virol 36*, 59–74.

39 Shaw G, Morse S, Ararat M, Graham FL. **2002**. Preferential transformation of human neuronal cells by human adenoviruses and the origin of HEK 293 cells. *FASEB J 16*, 869–871.

40 Fallaux FJ, Kranenburg O, Cramer SJ, Houweling A, Van Ormondt H, Hoeben RC, Van Der Eb AJ. **1996**. Characterization of 911, a new helper cell line for the titration and propagation of early region 1-deleted adenoviral vectors. *Hum Gene Ther 7*, 215–222.

41 Fallaux FJ, Bout A, van der Velde I, van den Wollenberg DJ, Hehir KM, Keegan J, Auger C, Cramer SJ, van Ormondt H, van der Eb AJ, Valerio D, Hoeben RC. **1998**. New helper cells and matched early region 1-deleted adenovirus vectors prevent generation of replication-competent adenoviruses. *Hum Gene Ther 9*, 1909–1917.

42 Schiedner G, Hertel S, Kochanek S. **2000**. Efficient transformation of primary human amniocytes by E1 functions of Ad5, generation of new cell lines for adenoviral vector production. *Hum Gene Ther 11*, 2105–2116.

43 Kim EJ, Kim NS, Lee GM. **1999**. Development of a serum-free medium for dihydrofolate reductase-deficient Chinese hamster ovary cells (DG44) using a statistical design: beneficial effect of weaning of cells. *In Vitro Cell Dev Biol Anim 35*, 178–182.

3
PER.C6® Cells for the Manufacture of Biopharmaceutical Proteins

Chris Yallop, John Crowley, Johanne Cote, Kirsten Hegmans-Brouwer, Fija Lagerwerf,
Rodney Gagne, Jose Coco Martin, Nico Oosterhuis, Dirk-Jan Opstelten, and Abraham Bout

Abstract

The PER.C6® human cell line was generated by immortalizing retina cells with the E1 genes of human adenovirus type 5. Master and Working cell banks were laid down and characterized in detail. Initially, the cell-line was used for the efficient and safe manufacture of recombinant adenoviral vectors for use in gene therapy and as vaccines. In total, six adenoviral vectors manufactured on PER.C6 are currently in clinical trials in the US and in Europe, of which one is used as a vaccine. In addition, PER.C6 is used for the manufacture of classic vaccines such as the influenza virus and West-Nile virus vaccines. The latest application of PER.C6 is in the field of protein production. A monoclonal antibody manufacture process has been developed to determine the growth and metabolic properties of PER.C6 and to investigate the yield and quality of the produced proteins. This chapter details the history of the PER.C6 cell line, the generation of antibody-producing PER.C6 cells, and the performance of these cells in production processes. In general, PER.C6 can be easily adapted to serum-free medium and can grow to very high cell concentrations in fed-batch ($>10^7$ cells mL^{-1}) and, in particular, continuous perfusion ($>10^8$ cells mL^{-1}). Specific productivity can be maintained at these high cell concentrations, resulting in high product yields. In addition, the high cell densities have no impact on product quality. Such cell densities are novel in the industry and will have a significant impact on the cost of manufacturing biopharmaceutical proteins, in particular those that are difficult to manufacture.

Abbreviations

APAC	analytical protein A chromatography
CBER	Center for Biologics Evaluation and Research
CHO	Chinese hamster ovary
cIEF	capillary isoelectric focusing
CMV	cytomegalovirus
CSPR	cell specific perfusion rate
DCW	dry cell weights
DHFR	dehydrofolate reductase
DMEM	Dulbecco's modified eagle medium
E1	transcription unit
ELISA	enzyme linked immuno sorbent assay
FBS	fetal bovine serum
G418	geneticin
HER	human embryonic retina

Modern Biopharmaceuticals. Edited by J. Knäblein
Copyright © 2005 WILEY-VCH Verlag GmbH & Co. KGaA, Weinheim
ISBN: 3-527-31184-X

HPAEC-PAD	high performance anion exchange chromatography with pulsed amperometric detection
HPLC	high performance liquid chromatography
HP-SEC	high-performance size exclusion chromatography
IEF	Isoelectric focusing
LC	liquid chromatography
LC-MS	liquid chromatography coupled mass spectroscopy
MALDI	matrix assisted laser desorption ionization
MALDI-TOF	matrix assisted laser desorption ionization-time of flight
MCB	Master Cell Bank
MDM2	morse double mint 2
MS	mass spectrometry
MW	molecular weight
N/D	not detected
NS0	hybrid cells
PGK	phosphoglycerate kinase
PNG	peptide:N-glycosidase
PrPsc	prior specific protein (scrapy)
SDS-PAGE	sodium dodecyl sulfate polyacrylamide gel electrophoresis
SP2	hybrid cells
UVA 280	ultaviolet absorption at a wavelength of 280 nm
VCD	viable cell density
VPR	volumetric production rate

3.1
Introduction

The PER.C6® human cell line was generated by the immortalization of primary retina cells with E1 sequences of human adenovirus serotype 5. The cell line was initially developed for the safe production of pharmaceutical grade recombinant human adenoviral vectors. Such vectors are currently used for vaccine and gene therapy purposes. In addition, the cell line is exploited for the manufacture of classical vaccines including influenza and West Nile Virus vaccines.

More recently, PER.C6 cells were evaluated for the production of therapeutic proteins, the global market of which has grown rapidly over the past five years, with an average annual growth rate of approximately 21% and sales reaching approximately \$41 bn in 2003 (AS Insights 2003; Reuters Business Insight 2003). Furthermore, with approximately one-third of all pipeline candidates currently in clinical development, this growth looks set to continue. One group of therapeutic proteins – the monoclonal antibodies – has shown particularly rapid growth in recent years, increasing from approximately 1% of therapeutic protein sales in 1995 to 14% in 2001. There are currently 15 approved antibodies on the market and many more in the late stages of clinical development.

The production of therapeutic proteins is commonly performed using mammalian cell lines, most commonly Chinese hamster ovary (CHO), but including also NS0 and SP2/0 cell lines. Mammalian cell lines are currently responsible for more than 60% of all licensed products. Their importance is due to an ability to perform the correct complex post-translational modifications required by many therapeutic proteins for their physiological activity. However, a drawback of mammalian cells is that yields are typically low compared to bacterial and yeast systems, while development and manufacturing costs are high. A major goal of process development groups over recent years has therefore been the increase of product yields (from both upstream and downstream process improvements) and the reduction of development and manufacturing costs and timelines (see also Part IV, Chapter 1).

Due to the high doses required for many antibody therapies, high product yields are particularly desirable. Yields of 1–2 g L^{-1} are current industry standard targets. Yields above 2 g L^{-1} have also been achieved, but at present these do not generally result in significant cost savings due to current limitations in downstream processing. However, as progress is made in this area, demands for higher yields can be expected. At the same time, there is also a drive to reduce development costs and timelines. One goal is to reduce the time to clinic by reducing the time taken to generate production cell lines and to generate the material for pre- and early clinical phase studies. Current timelines may vary slightly depending on the individual situation, cell line, antibody, etc., but typically range from 14 to 16 months for cDNA to production of clinical trial material. However, these may be expected to decrease in the coming years. The more efficient use of process development resources is another major driver, particularly for projects up to early clinical phase studies where the risk of failure is highest and where it may be necessary to run a number of projects in parallel.

An approach adopted by many has been the development of platform technologies. The aims of such a platform include for example, the provision of technology required to generate cell lines with high cell-specific productivity, to ensure the selection of production cell lines that perform well in the desired production process, to develop cost-effective production media, and to provide high-yielding, efficient and cost-effective production and purification processes suitable for large-scale manufacture.

Of particular importance is the development of generic processes. By developing processes that are generic, timelines can be shortened and development costs reduced for each new cell line that is generated. For example, the development of generic production and purification processes removes the need to perform lengthy and costly process development for each new cell line, thus reducing costs, timelines and allowing multiple projects to proceed simultaneously. They may also act as a basis for development of the final manufacturing process, thus minimizing investment in process development for Phase III and beyond. Moreover, the inclusion of a generic production process in the cell line generation program allows the selection of cell lines that perform optimally in the desired final production process.

It is the aim of Crucell and DSM Biologics to establish the human PER.C6 cell line as a platform for the production of therapeutic proteins, with particular emphasis on monoclonal antibodies. The approach taken has been to develop an integrated production platform that combines the rapid generation of high-yielding production cell lines with high-yielding generic production (batch, fed-batch and perfusion) and purification processes and a metabolically characterized host cell line. Data generated from the metabolic characterization of PER.C6 cell lines was used to design generic, high-yielding batch, fed-batch and perfusion production processes, matched to the metabolic requirements of the cells. Cell lines are evaluated as early as possible in the desired production process so that lead clones are selected that match and will perform optimally in the desired production process.

The investigations described in this chapter provide an overview of clone generation, fed-batch and perfusion process development, as well as detailing the history of the PER.C6 cell line, and how it has been characterized. These studies have been conducted in an alliance between Crucell Holland N.V. (Leiden, The Netherlands) and DSM Biologics (Groningen, The Netherlands and Montreal, Canada).

3.2
Generation of PER.C6 Cells

3.2.1
Immortalization of Cells by E1 Proteins of Human Adenovirus

Human adenoviruses are associated only with mild disease in healthy humans [1].

Adenoviruses have a DNA genome of approximately 36 kb that encodes proteins of the virus capsid, and proteins that dedicate the cell to replicate the viral genome and synthesize viral proteins. Among the latter are the so-called E1 gene products of adenovirus. It has long been known that the isolated E1 genes can immortalize primary human cells [2]. This property of E1 genes of adenovirus was used to generate the PER.C6 cell line. The E1 region of adenovirus 5 consists of two transcriptional units, E1A and E1B. The E1A transcription unit encodes two proteins, which are generated by alternative splicing. The proteins

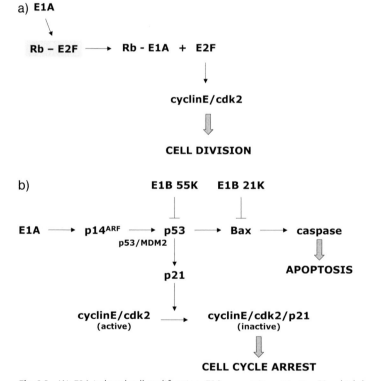

Fig. 3.1. (A) E1A-induced cell proliferation. E1A proteins release E2F from Rb, which subsequently induces CyclinE/cdk2 gene expression and pushes cells into S-phase. (B) E1A-induced cell cycle arrest and apoptosis counteracted by E1B. E1A pro-teins activate p53, which leads to cell cycle arrest and apoptosis. These effects are counteracted by E1B 55K, which binds directly to p53, and E1B 21K, which inactivates cytochrome-c.

are acidic in nature, are 243 and 289 amino acids long, respectively, and are located in the nucleus of the cell. A detailed description of the E1 regions of adenovirus and the function of E1A proteins of adenovirus type 5 is available on (http://www.geocities.com/jmymryk.geo/).

The E1B region generates one RNA, which is translated into two proteins, with molecular weights of 21 and 55 kDa.

For efficient immortalization and transformation of primary cells, both the E1A and E1B regions are required, although it has been described that the E1A region by itself can immortalize rodent cells [3] and occasionally human cells [4], with very low efficiency. Expression of E1A alone usually results in the induction of programmed cell death (apoptosis), which can be prevented by co-expression of E1B [5].

E1A proteins affect major cellular processes such as cell cycle control, differentiation, apoptosis, and transformation. The immortalization of primary cells occurs by binding of E1A proteins to the tumor suppressor protein pRB, p107 and p130, as well as to the co-activator p300 [6]. These proteins have in common that they can form complexes with E2F transcription factor proteins, leading to inactivation of the E2F factors. The binding of E1A leads to the release of E2F transcription factors from the complexes, which results in activation of cellular genes that have E2F binding sites in their promoters (Fig. 3.1A). Amongst these genes is cyclinE/cdk2 that stimulates the cell to enter the cell cycle. However, the strong proliferation signal of E1A causes the cell to activate p53 (Fig. 3.1B). p53 is complexed to MDM2, which renders it inactive. E1A mediates the induction of p14ARF protein expression, which inhibits the activity of MDM2, thereby causing release of p53 [7]. P53 is an activator or transcription of genes that cause cell cycle arrest and apoptosis of the cells. So E1A stimulates cells to proliferate but also induces a stress response in the cell, leading to growth arrest or apoptosis. The stress response is counteracted by the E1B proteins (Fig. 3.1B). E1B 55K, which is located in the nucleus, forms a complex with p53 and thereby inactivates it. E1B 21K interferes with the apoptotic effects of p53-induced Bax protein in the cell [8]. Bax is a pro-apoptotic protein, that causes release of cytochrome-c from mitochondria, which in turn causes caspase-mediated apoptosis. E1B 21K is a homologue of the cellular anti-apoptotic protein Bcl2 [8]. It inhibits Bax-induced release of cytochrome-c from mitochondria, thereby preventing apoptosis.

3.2.2
Generation of PER.C6 Cells

The DNA construct pIG.E1A.E1B (Fig. 3.2) was used for making PER.C6 cells. The E1 genes are driven by the human phosphoglycerate kinase (PGK) promoter, which is a known house-keeping promoter [9] and the poly(A) sequences are derived from the Hepatitis B surface antigen gene [10, 11].

The primary cells selected for transfection with the E1 construct were human embryonic retina (HER) cells, which can be immortalized relatively easily by E1 of human Ad5 [4, 12, 13] and Ad12 [14].

459   **3510**

| PGK | E1A – E1B | p(A) |

Fig. 3.2 DNA construct used to generate PER.C6 cells. In this construct, the E1A gene is driven by the human phosphoglycerate kinase (PGK) promoter. Transcription is terminated by the hepatitis B virus surface antigen poly(A) sequences.

Primary HER cells have a limited life span, and can be cultured only a few passages, after which the cells senesce. Transfection of HER cells with E1 constructs results in transformation and immortalization of the cells, reflected by focus formation in the cultures. This is easily recognized by both macroscopic and microscopic examination of the cultures. Such foci can be isolated and cultured further. In this way, PER.C6 cells were isolated after transfection of primary HER cells with pIG.E1A.E1B [15, 16]. The cells were apparently immortalized also, without passing through a crisis phase.

Transformation and immortalization of primary cells with E1 sequences of adenovirus guarantees: 1) a stable expression of E1 proteins, as the cells need E1 expression for growth; and 2) that no external selection marker is needed to distinguish E1 expressing from non-expressing cells.

PER.C6 cells stably express the E1 proteins. In particular, the 21K and 55K E1 proteins that counteract apoptosis and p53-mediated cell cycle arrest, respectively are expressed to high levels as compared to, for example, HEK293 cells [15]. We assume that this makes the cells relatively insensitive to apoptosis, and may be one of the factors that make the PER.C6 cells grow to high cell densities and support production of a wide variety of proteins, without further manipulation of the cells.

At passage number 29, a research Master Cell Bank was laid down, which was extensively characterized and tested for safety (including sterility testings). Research cell banks were made at passage numbers 33 and 36.

The characterization and safety testing of the cell banks has been described extensively elsewhere [16]. In brief, the identity, sterility, viral safety, absence of PrPsc protein, tumorigenicity and genetic characterization, including chromosome analysis, has been performed.

A description of the history of the cell line – as well as study protocols and reports of all safety studies carried with the cell line – has been filed as a Biologics Master File at CBER.

3.3
PER.C6 Cells for the Manufacture of Recombinant Proteins

The first step in the manufacturing train is the generation and selection of a high-producing cell line. This has been performed

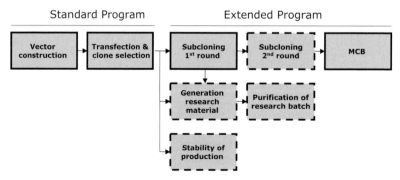

Fig. 3.3 Overview of the generation of PER.C6 clones producing high levels of monoclonal antibodies.

in total more than 15 times, mainly for IgG1 (κ and λ light chains, and for both allelic variants of IgG1 heavy chains [17]). In addition, IgM and IgA, as well as F(ab)$_2$ fragments have been expressed. A brief description of the selection of high-producing cell lines (summarized in Fig. 3.3), as well as adaptation to serum-free suspension conditions is presented, followed by a summary of the generic fed-batch and perfusion processes that have been developed for PER.C6 cells producing recombinant protein.

3.3.1
Vector Construction and Transfection

The first step in the production of monoclonal antibodies is generation of the expression construct. For antibody generation, the antibody construct depicted in Fig. 3.4 has mostly been used. Here, expression of both the light chain and the heavy chain genes is driven by a cytomegalovirus (CMV) promoter that has been modified to

obtain high levels of gene expression in PER.C6 cells. Adherent PER.C6 cells, cultured in medium containing fetal bovine serum (FBS) are transfected with this construct. Cells that contain a stably integrated construct are selected using G418 (Geneticin). G418-resistant colonies are transferred to 96-well plates.

3.3.2
Primary and Secondary Screens

A total of 300–400 clones is isolated and transferred to 96-well plates and cultured in DMEM supplemented with 10% FBS. After 5–10 days, culture supernatants are sampled and screened for the presence of IgG either by Protein A HPLC or by ELISA. Production titers from two independent screening rounds are used to rank the transfectants, and the top 20–30 are selected and expanded for cryopreservation and further evaluation. Selection pressure is maintained until cryopreservation, after

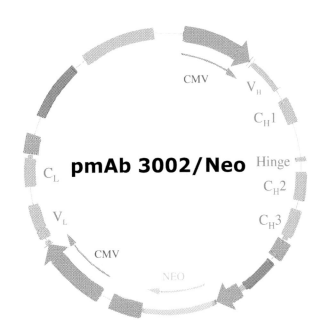

Fig. 3.4 Expression plasmid encoding IgG heavy and light chains used for transfection into PER.C6 cells.

which G418 is removed from the culture medium.

The highest ranked cell lines selected from the primary screens are then screened in 6-well plates using DMEM plus 10% FBS. Cells are seeded at 0.5×10^6 cells per well in duplicate and incubated for 4 days at 37 °C and 10% CO_2. Culture supernatants are then harvested and the IgG concentration determined by Protein A HPLC or ELISA. Final cell concentrations are measured and the data used to calculate cell-specific production rates. The cell lines with the highest cell specific production rate are selected for adaptation to serum-free conditions. Fig. 3.5 illustrates the results of a typical secondary screen performed for an internal clone generation program at Crucell. Fig. 3.5 A shows the final antibody concentration, and Fig. 3.5 B the cell-specific productivity. There is usually a good correlation between volumetric and cell-specific productivity; that is, cells with a high volumetric productivity show a correspondingly high specific productivity, and vice versa. Occasionally, a cell line with a high specific

A

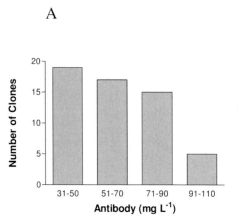

B

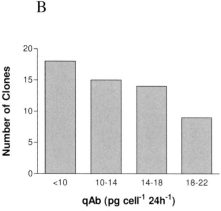

Fig. 3.5 Results of a secondary screen of 57 clones from a cell-line generation program conducted at Crucell. The screen was performed in 6-well plates using DMEM + 10% FBS. Cells were seeded at 0.5×10^6 mL^{-1} and the supernatant was harvested at day 4. (A) volumetric productivity; (B) specific productivity.

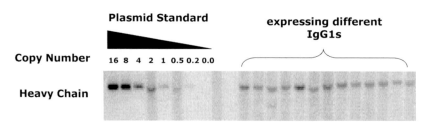

Fig. 3.6 Southern blot indicating the copy number of DNA encoding the light chain in different IgG expressing PER.C6 clones. Plasmid copies are measured to a standard comprising a known amount of plasmid DNA in a background of human chromosomal DNA.

A   B

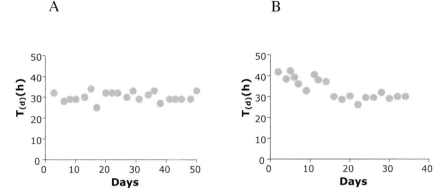

Fig. 3.7 Typical adaptation profiles for two different antibody-producing PER.C6 cell lines (A and B). Cells cultured in the presence of 10% FBS are transferred directly to serum-free media in Erlenmeyer shake flasks. Cells are passaged and population doubling time is calculated.

productivity will show a low volumetric production due to poor growth, or vice versa. In this example, 20 clones were selected for adaptation to serum-free conditions.

Selected PER.C6 cell lines possess low copy numbers, typically below 10 copies per cell as measured by Southern analysis (Fig. 3.6).

3.3.3
Selection Serum-free

Adherent cells are trypsinized and re-suspended directly in shake flasks containing serum-free medium. Cells are cultured every 2–4 days and incubated at 37 °C, 5% CO_2 and 100 r.p.m. The adaptation period for PER.C6 cell lines using this strategy is typically quite short, up to a maximum of 15–20 days. Fig. 3.7 shows two typically observed adaptation profiles for antibody-producing cell lines. Once adaptation is complete and a stable doubling time is observed, cell lines are evaluated in the desired production process, typically batch or fed-batch. Growth, production and metabolite profiles are characterized. The product is purified and analyzed by SDS-PAGE (no

major contaminating and/or unexpected bands), IEF (conform previous produced material), HPLC-SEC (>90% monomer) and glycan analysis (correct galactosylation). A selection of one to three lead cell lines is then made based on process performance (growth rate, productivity, metabolic profile) and product quality. Fig. 3.8 shows the final antibody concentrations from batch culture for the 20 selected clones. In this example, seven clones showed yields above 0.5 g L^{-1} and were selected for further evaluation in fed-batch.

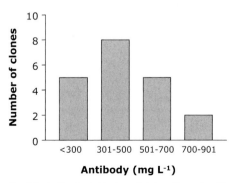

Fig. 3.8 Antibody yields from batch cultures for 20 antibody-producing cell lines adapted to serum-free media.

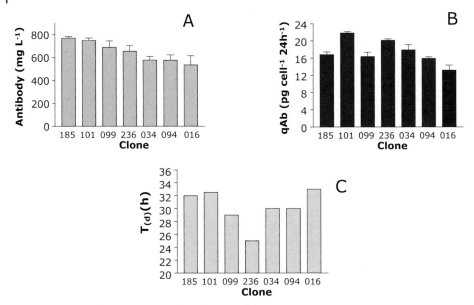

Fig. 3.9 Overview of the seven antibody-producing cell lines yielding more than 500 mg L^{-1} in batch culture (A), specific productivity ranging from 12 to 22 pg per cell per day (B), and population doubling time ranging from 24 to 34 hours (C).

Fig. 3.9 A shows the final antibody concentration, Fig. 3.9 B specific productivity, and Fig. 3.9 C the doubling time for each selected cell line.

Once final selection has been made, cryopreserved cell stocks are prepared in serum-free medium.

3.3.4
Sub-cloning

Cell lines that are carried forward as potential production lines are sub-cloned. Cells are plated at an average of 0.3 cells per well in 96-well plates, and out-growing colonies screened, expanded, frozen, and tested as described for the initial clones.

3.3.5
Cell-line Generation Timelines

The aim of cell-line generation is rapidly to select high-yielding cell lines that perform optimally in the desired production process. The process from transfection to final selection of the lead clone (including evaluation in batch or fed-batch) takes 6–7 months (see Fig. 3.10). The aim is to move as quickly as possible to serum-free conditions and to make the final selection based on performance in one of the generic production processes, whether batch, fed-batch, or perfusion. The inclusion of such generic production processes in the selection program not only ensures that the lead cell line that will perform optimally in a production environment, but also reduces the amount of process development work required for each new cell line. An

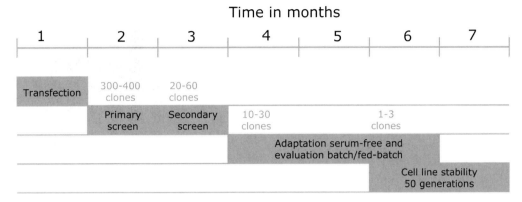

Fig. 3.10 Timelines for the generation of stable antibody-producing PER.C6 cell lines.

additional feature of the PER.C6 cell line that reduces the timeline for cell-line generation is the easy and rapid adaptation to serum-free conditions (see also Fig. 3.7), which typically requires less than 3 weeks by a direct adaptation in shake flask. Finally, expression in PER.C6 cells does not involve amplification of gene copy number, as for example in CHO DHFR⁻ cell lines. As a result, recombinant PER.C6 cell lines can be relatively quickly selected and evaluated in the required production system, without the time normally needed for amplification.

3.4
Fed-batch Process Development

A generic fed-batch process has been developed for the production of monoclonal antibodies in PER.C6 cells. The process typically results in a 3- to 4-fold increase in antibody yields compared to the batch process, with yields of 1–3.5 g L^{-1} after 16–18 days. The feed strategy is based on the metabolic requirements of the PER.C6 cell line. Metabolic characterization of several antibody-producing cell lines identified nutrients and medium components which

are important for the maintenance of growth and productivity. These were assembled in a nutrient concentrate consisting of glucose, phosphate and amino acids and a component concentrate consisting of vitamins, lipids, trace elements, salts and growth factors.

The feed strategy involves the addition of these nutrients based on cell-specific requirements in order to supply the nutrients only as required by the culture and to limit overflow metabolism or the build-up of nutrients or metabolites that may result in reduced process performance (antibody yields) and product quality [18–25].

In addition to a controlled feed strategy, physico-chemical process parameters have been optimized for process efficiency. For example, the growth rate of PER.C6 cells is optimal at pH 7.3 (Table 3.1). The cell-specific rates of nutrient utilization are highest at that pH (Table 3.1) however, with values for glucose, glutamine and phosphate for example up to two- or three-fold higher than at pH 6.9. This increased rate of nutrient utilization at pH 7.3 does not result in higher maximum cell yields or cell-specific productivities, and can thus be regarded as metabolically less efficient. It also has a significant influence on the

Table 3.1 Summary of metabolic and growth data for antibody-producing PER.C6 cell line (antibody A) in batch culture at different culture pH values.

	pH 6.9	pH 7.3	No low limit pH control
qGlc	0.6	1.6	0.7
qGln	0.17	0.32	0.18
qPhos	0.05	0.12	0.07
Nv (max.)	9.8	10.2	10.5
Avge T(d) day 1–4	38	28	31
qAb	14–16	14–16	14–16

design and efficiency of a fed-batch process, as a feed strategy at pH 7.3 would involve the addition of two to three times the nutrient concentrations as for a process at pH 6.9. This would give increased osmolality and result in reduced process performance. The problem with operating a process at pH 6.9 is the sub-optimal growth rate compared with pH 7.3, which results in a longer process. This was overcome by controlling the starting pH of cultures to 7.3, but then operating without a low limit pH control. In PER.C6 cell cultures this resulted in a pH "drift" down to approximately 6.9 during growth, which led to a culture that showed optimal growth rates and nutrient utilization profiles. Operating the process with such a pH drift also reduces lactate accumulation. PER.C6 cells possess a lactate transport system that is a proton symport system and thus is dependent on a low extracellu-

A   B

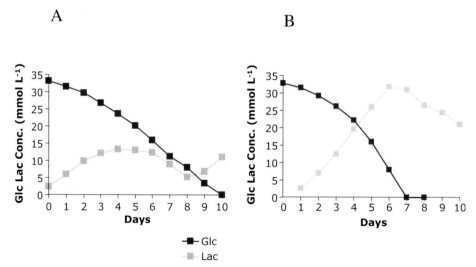

Fig. 3.11 Glucose utilization and lactate production profiles for: (A) a batch culture operated with no low limit pH control (initial culture pH 7.3); and (B) a batch culture operated with pH control at 7.3.

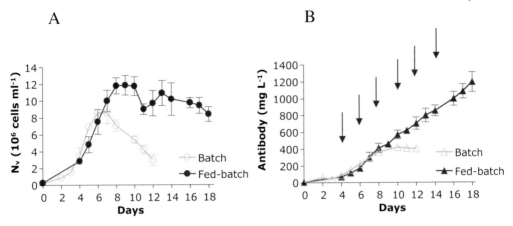

Fig. 3.12 (A) Cell and (B) antibody profiles for batch (open symbols) and fed-batch cultures for a PER.C6 cell-line expressing antibody A. The data represent an average of eight 2-L bioreactor runs.

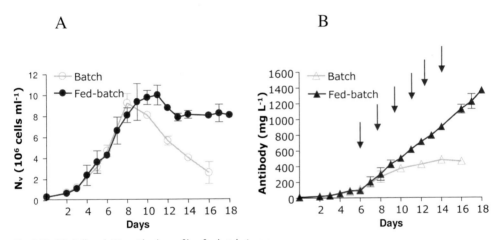

Fig. 3.13 (A) Cell and (B) antibody profiles for batch (open symbols) and fed-batch cultures for a PER.C6 cell-line expressing antibody B. The data represent an average of three 2-L bioreactor runs.

lar pH. When cultures are operated with no low limit pH control therefore, there is a period of lactate release and the pH decreases. As this occurs, lactate transport starts and extracellular lactate concentrations plateau and begin to decrease. However, if pH is maintained at 7.3, no lactate transport is observed and lactate accumulates in the culture (Fig. 3.11).

A typical feed strategy involves the addition of four to six bolus feed additions at regular intervals during a 16- to 18-day process. Similar growth and production profiles are observed for all antibody-pro-

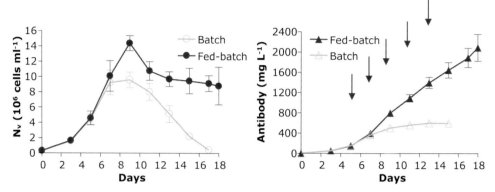

Fig. 3.14 (A) Cell and (B) antibody profiles for batch (open symbols) and fed-batch cultures for a PER.C6 cell-line expressing antibody C. The data represent an average of 18 shake flask runs.

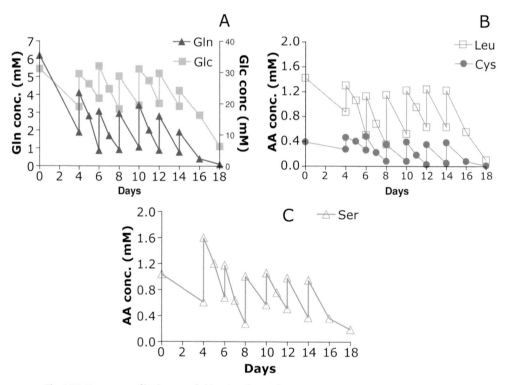

Fig. 3.15 Nutrient profile during a fed-batch culture of a PER.C6 cell-line expressing antibody A. The data show the concentration of (A) glutamine (closed triangles), glucose (closed squares); (B) leucine (open squares), cystine (closed circles); and (C) serine (open triangles).

Table 3.2 Summary of results obtained with four PER.C6 cell lines expressing different IgG1

	Batch [g L^{-1}]	Fed-batch [g L^{-1}]	Process length [days]	qAb [pg/cell/day]
Antibody 1	0.4	1.3	18	12–15
Antibody 2	0.5	1.4	18	10–12
Antibody 3	0.6	2.1	18	15–18
Antibody 4	0.6	1.8	16	16–19

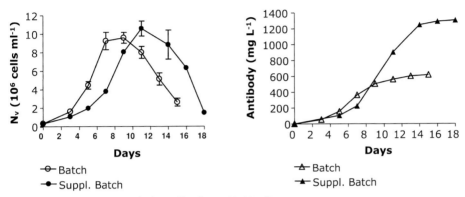

Fig. 3.16 (A) Cell and (B) antibody profiles for a PER.C6 cell-line expressing antibody C in batch and supplemented batch cultures. Supplementation of the batch was made prior to inoculation, with 50% of the feed added to a typical fed-batch culture. The data represent the average of four shake flasks.

ducing cell lines that have been evaluated in the fed-batch process. Figs. 3.12, 3.13 and 3.14 show the growth and production profiles of three different antibody-producing cell lines (cell lines A, B, and C, respectively) in batch and fed-batch. Table 3.2 shows a summary of the same three cell lines (A, B, and C) and a fourth (cell line D) evaluated in batch and fed-batch, and includes the average specific productivity. Fig. 3.15 illustrates selected nutrient profiles during the fed-batch for one of the cell lines (cell line A), showing that concentrations of the added nutrients remain stable during the fed-batch and that the feed is accurately matched to the metabolic requirements.

3.4.1
Supplemented Batch Process

Metabolic data from the fed-batch development was used to develop a supplemented batch process involving the addition of up to three of the feeds added in the fed-batch process, to the culture medium prior to inoculation of the cells. Final antibody yields are not as high as for the fed-batch process, typically an increase of 2-fold over batch yields compared to 3- to 4-fold increases for the fed-batch. However, the process offers a relatively simple way of obtaining increased antibody yields. Fig. 3.16 shows a supplemented batch culture for clone C where the final antibody concentration reached 1.3 g L^{-1}.

3.5

Operation of PER.C6 Cells in Continuous Perfusion

Perfusion presents a number of advantages over other modes of cultivation, such as increased volumetric productivity and rapid removal of easily inactivated products from the culture environment [26]. These homogeneous perfusion systems can be operated under conditions of total cell retention, or with the removal of part of the biomass through culture bleeding [27]. Total cell retention facilitates kinetics studies and, as demonstrated in other studies [28], prevents unnecessary cell division, allowing for cells to produce product at higher rates [29].

A perfusion process has been developed for the production of monoclonal antibodies in PER.C6 cells. The process typically results in a more than 30-fold increase in antibody yield compared to the batch process, and a 10-fold increase compared to the fed-batch process. As for the development of the fed-batch process, the perfusion strategy is based on the metabolic and physico-chemical requirements of the PER.C6 cell line.

3.5.1

Initial Assessment

The first feature investigated was the impact of cell perfusion rate (CSPR) on cell culture performance using a spinfilter as the cell retention device. Fig. 3.17 shows the viable cell concentration of a PER.C6 cell line expressing antibody A operated under the same conditions with two different CSPR implemented at the 7-L scale.

It was observed that a doubling of the CSPR resulted in a 65% increase in the upper cell density. The fact that such an increase was achieved suggested that significant improvements in cell density (and related productivity) could be obtained by modification of the CSPR. By day 18, both of these cultures were prematurely ended due to clogging of the spinfilter. Upon inspection of the spinfilter material, a cake of cells was observed to have gathered, which suggested that the high cell densities achieved with this culture are not compatible with a standard spinfilter operation as a cellular retention device.

A complete metabolic characterization of these cultures was performed (data not

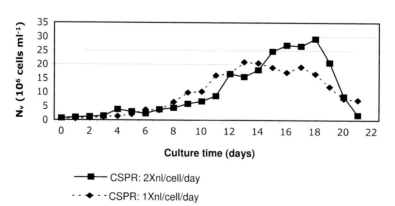

Fig. 3.17 Cell concentration ($\times 10^6$ cells mL^{-1}) versus culture time (days) for PER.C6 cell line expressing antibody A operated at two different cell perfusion rates (CSPR).

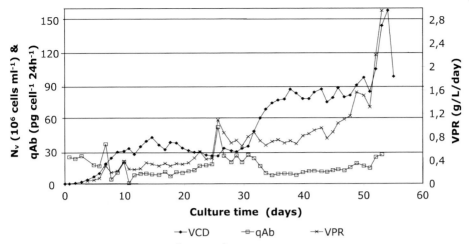

Fig. 3.18 Viable cell density (VCD; $\times 10^6$ cells mL^{-1}), cell-specific production rate (qAb; pg per cell per day) and volumetric production rate (VPR; g L^{-1} per day) versus culture time (days) for PER.C6 cell line expressing antibody A in a continuous perfusion.

shown); post analysis of a modified perfusion strategy was assessed.

3.5.2
Stage 1 Development

Fig. 3.18 details the cell profile achieved during round 1 development of a continuous perfusion process using a PER.C6 cell line expressing antibody A. The viable cell concentration reached approximately 30×10^6 mL^{-1} by day 12 and remained constant at this level for a further 20 days. At this point, a series of actions were taken which resulted in the cell concentration increasing to approximately 80×10^6 mL^{-1} by day 38. Conditions were then kept constant for a 14-day period during which a constant cell concentration was maintained. On day 52, the same process change as operated at day 32 was implemented which led to a further increase in the viable cell concentration to 155×10^6 mL^{-1} by day 53. During the next 24 hours, the cell retention device be-

came unusable due to the high cell concentrations, and this led to termination of the run. The cell viability throughout the process was greater than 95%.

The specific production profile (q$_{Ab}$) of the cells was stable for the majority of the process, with an average of 15 pg per cell per day. Higher values at the end of the process were due to cell retention device failure. High volumetric production rates (VPR) were achieved during this process, with an average VPR of 0.76 g L^{-1} per day calculated over the entire process, 0.9 g L^{-1} per day at approximately 80×10^6 mL^{-1} and an upper VPR of 3 g L^{-1} per day achieved at 150×10^6 mL^{-1}.

3.5.3
Stage 2 Development

Round 2 development consisted of identifying the critical features necessary to achieve the extreme cell densities. Fig. 3.19 shows the viable cell concentration ($\times 10^6$ cells

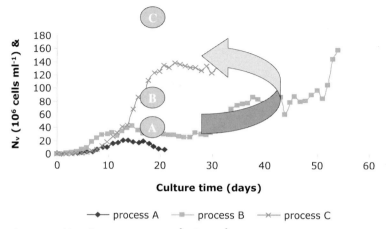

Fig. 3.19 Viable cell concentration ($\times 10^6$ cells mL^{-1}) versus culture time (days) of the optimized modified process compared to the initial experiments.

mL^{-1}) versus culture time (days) achieved using the round 2 modified process.

The progression gained from each of the stages can be visualized in Fig. 3.19. It should be noted that biphasic growth profile observed in process B is reduced to a monophasic growth profile by process C. In addition, there is significant reduction in the process time required to achieve cell concentrations of 100×10^6 cell mL^{-1}. Process C requires approximately 16 days to achieve cell concentrations of 100×10^6 cells mL^{-1}, while process B required approximately 40 days to reach 80×10^6 cells mL^{-1}.

In summary, and to the present authors' best knowledge, the cell concentrations achieved with the PER.C6 cell line (see Fig. 3.20) are the highest reported value to date for a mammalian cell line. These results propel the overall productivity for a PER.C6 cell line to the highest reported values for an antibody-producing process.

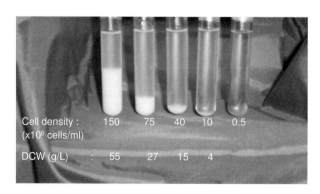

Fig. 3.20 PER.C6 cell concentration with corresponding dry cell weights (DCW) (g L^{-1}) obtained during a perfusion mode of operation.

3.6
Characterization of Antibodies Produced by PER.C6 Cells

One of the cornerstones in the development of biopharmaceuticals is the availability of reliable, sensitive and accurate analytical methods to characterize recombinant products [30]. Typically, these methods are divided into several categories: product quantity, identity, purity, and potency testing [31]. Product characterization methods can be general for any recombinant protein or specific to a particular drug substance. In addition, new analytical technologies and modifications to existing technologies should be used whenever the methods can add valuable data to process development and lead to a better understanding of the consequences of process changes to ensure the safety and efficacy of the product in patients. These methods should be set up during the process development phase of a biopharmaceutical product. This section summarizes the analytical results obtained in parallel with the development of processes for the production of monoclonal antibodies in the PER.C6 cell line. General analytical methods that are used to release materials from production such as sterility testing or in-process testing such as process-related impurities and indicators are beyond the scope of the present review, or are shown as part of process development.

Quantity is an important product characteristic, and is usually the first method to be implemented during process development. There are a number of physicochemical tests available to measure the antibody content of cell culture samples, including ELISA, UVA$_{280}$ and affinity chromatography [32]. Typically, ELISA methods to assay IgG content are very sensitive and used to quantify low levels of crude biological products such as those

seen in clone selection or in microscale processes; a large number of these methods are commercially available [33]. Several ELISA methods have already been successfully implemented to quantify antibody titers produced in the PER.C6 cell line.

Our method of choice to quantify recombinant antibodies produced in PER.C6 cells is that of analytical protein A chromatography (APAC). This method is automated, rapid (total run time of 3.8 min), precise (standard deviation <5%), and sensitive (quantitation limit 25 µg mL^{-1}), and generally outperforms ELISA methods, except for sensitivity.

There are a number of features of monoclonal antibodies that are identified as critical quality attributes. Much of the discussion between biotechnology companies and regulatory agencies centers on the choice of the appropriate methods to demonstrate product consistency for lot release. In the present project, since the novelty of the technology is the use of human cells in the production of recombinant antibodies, the focus of the characterization studies is general characteristics such as glycosylation and protein profiles [34]. In particular, the exhibition of the predictive nature and consistency of glycan and protein heterogeneity is important in order to validate the PER.C6 cell line as an attractive expression system for the manufacture of antibodies.

It was necessary to purify the monoclonal antibodies that were produced in PER.C6 cells in order to examine the structural features of these products. To accomplish this, a two-step purification method was executed whereby cell culture material was loaded onto a semi-preparative protein A column and the antibodies were eluted at low pH [35]. The eluted antibodies were injected directly on a desalting column and then concentrated in order to prepare the samples for characterization analyses. The purification of the antibodies produced in PER.C6 cells was straightforward, and recoveries exceeded 75%. The samples were very pure after the affinity step, as shown by SDS–PAGE analysis and high-performance size exclusion chromatography (HP-SEC). The level of higher molecular weight impurities in the samples was very low (typically < 1.5%). Once purified, samples were subjected to a number of identity assays including gel and capillary electrophoresis and glycosylation analysis.

3.6.1
SDS-PAGE

The conventional slab gel technique SDS-PAGE is an important technique for the routine identity and purity analysis of biotechnology products [36]. A typical SDS-PAGE analysis, run under reducing conditions with Coomassie blue staining, is shown in Fig. 3.21 A. The heavy and light chains of the purified antibody samples are clearly seen in all of the loaded samples. Molecular weight (MW) markers that encompass the bands of interest are also loaded onto the gel. No differences have been detected between cultures run in batch, fed-batch or perfusion modes.

3.6.2
Isoelectric Focusing

Isoelectric focusing (IEF) is another routine slab gel technique for the identification of biotechnology products [37]. The isoelectric points (pI) of the produced protein and its variants can be monitored. The presence of several charge variants is a common feature of recombinant antibodies and can be the result of, for example, deamidation or differences in processing of C-terminal lysine residues [38, 39]. The consistency of the charge heterogeneity can be monitored by using IEF. A typical IEF analysis is shown in Fig. 3.21 B. The pI values of the protein isoforms can be identified for all of the samples. The IEF patterns for antibodies produced in batch, fed-batch and perfusion modes with PER.C6 cells are very similar.

Unlike its conventional slab gel counterpart, capillary isoelectric focusing (cIEF) is automated, precise and quantitative [40]. A typical cIEF profile of a purified human IgG1 that was produced in PER.C6® cells is shown in Fig. 3.22. In this case, five isoforms can be identified and quantified. The major isoform has a pI value of 8, which is typical for human IgG1.

During the process development of monoclonal antibodies production in PER.C6

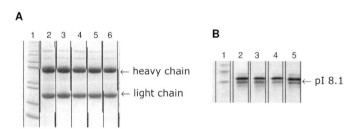

Fig. 3.21 (A) SDS-PAGE analysis under reducing conditions of human IgG1 produced in PER.C6 cells. Lane 1: MW markers; lanes 2 and 4: IgG1 produced in batch process; lanes 3, 5, and 6: IgG1 produced in fed-batch process. (B) IEF gel of human IgG1 produced in PER.C6 cells. Lane 1: pI markers; lanes 2 and 4: IgG1 produced in fed-batch process; lanes 3 and 5: IgG1 produced in batch process.

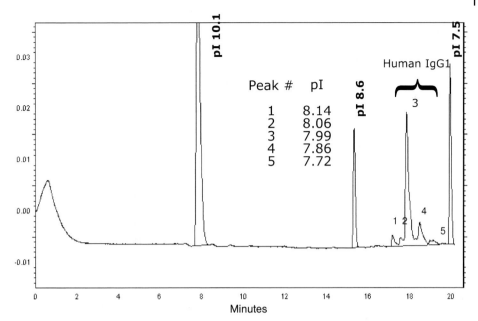

Fig. 3.22 A typical capillary isoelectric focusing profile of purified human IgG1 from PER.C6 cells. The peaks at pI values of 10.1, 8.6, and 7.5 are pI markers.

Table 3.3 Typical pI values of the five isoforms for purified human IgG1 samples produced in batch, fed-batch, and perfusion modes.

Sample	Peak 1 [%]	Peak 2 [%]	Peak 3 [%]	Peak 4 [%]	Peak 5 [%]
Batch – end	3	6	59	24	8
Fed-batch – mid	N/D	4	66	21	9
Fed-batch – end	4	7	67	17	5
Perfusion – early	N/D	4	87	6	3
Perfusion – mid	N/D	N/D	87	10	3
Perfusion – late	N/D	N/D	87	10	3

Relative peak areas were obtained by the integration of the major isoforms in the cIEF electropherograms of the samples.
N/D=not detected.

cells, several in-process samples were taken from the production vessel, purified, and subjected to cIEF analysis. Typical results of these analyses are shown in Table 3.3. Excellent quality control is seen in both fed-batch and perfusion modes, since the cIEF profiles of samples taken at the mid and end-points of the runs were very similar. Inter-assay variation is ± 3% for each isoform of the same sample. Minor differences in cIEF profiles can be seen for samples taken from the three modes of manufacturing. Presumably, the charge heterogeneity is due to differences in deamidation levels caused by the different production conditions.

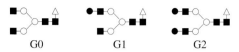

G0   G1   G2

- ■ N-acetylglucosamine
- △ Fucose
- ○ Mannose
- ● Galactose

Fig. 3.23 Structures of the three most common glycans found in human IgG1 molecules, G0, G1, and G2.

3.6.3
Glycan Analyses

All eukaryotic recombinant expression systems produce therapeutic proteins that are glycosylated in their native state for *in vivo* activity [41]. The expressed glycoprotein contains glycosylation variants, called glycoforms, which are the subject of a number of recent reviews [42–45]. They are also discussed in detail in this book (Part IV, Chapter 1, 2, 7; and Part VI, Chapter 2). Since glycosylation of biological molecules is achieved through a complex, post-translational pathway involving several enzymes, carbohydrate structures are very sensitive to even subtle differences in the environment in which they are formed. The main factors that influence the oligosaccharide profile of glycoproteins are cell line, cell culture medium, bioreactor parameters, harvest time, and manufacturing site changes.

In the present case, all IgG1 molecules contain a conserved N-glycosylation site at asparagine 297 in the constant region of the molecule. In human serum, the major

glycoforms of IgG1 contain a biantennary structure with a core fucose [46]. The three most common glycans found in IgG1 molecules, G0, G1, and G2, are shown in Fig. 3.23.

Purified antibody samples from development runs were subjected to PNGase F treatment to release the glycans, and these were analyzed by either MALDI-MS or HPAEC-PAD to determine the glycan structures.

IgG1 produced by PER.C6® cells in batch culture show a similar galactosylation profile to human serum IgG [34], with approximately 30% G0, 50% G1, and 20% G2 (Fig. 3.24). This can be compared to CHO-produced antibodies, which are typically

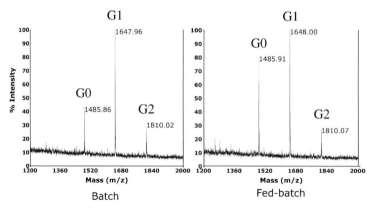

Fig. 3.24 MALDI-TOF traces of glycans of antibody B produced in batch and fed-batch.

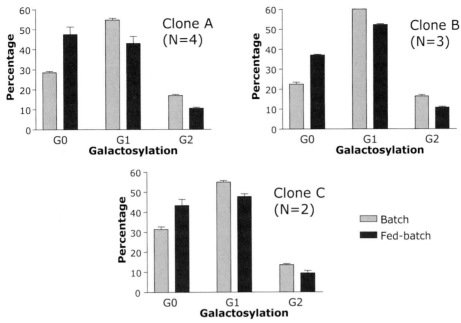

Fig. 3.25 Galactosylation profile of three different monoclonal antibodies produced in batch and fed-batch processes. The profiles were made by MALDI-TOF analyses.

produced predominantly in the G0 form. For example, Hills et al. [47] reported a galactosylation profile for an antibody produced in NS0 and CHO cells of approximately 60% G0, 35% G1, and 5% G2 for the NS0 cell line, and 63% G0, 33% G1, and 4% G2 for the CHO cell line. A small decrease in galactosylation is typically observed in the fed-batch-produced antibody, with the percentage of G0 glycoforms in-

creased from 30 to 45%, and the percentage of G1 and G2 glycoforms decreased from 50% and 20% to 40% and 15%, respectively. Fig. 3.25 shows the galactosylation profile of three different antibodies produced in batch and fed-batch culture from three different PER.C6 cell lines. This reduction in galactosylation is likely due to the different culture conditions between the two processes, such as an increase in process length and in final

Table 3.4 Glycan profile by HPAEC-PAD chromatography of human IgG1 produced in PER.C6 cells in the three manufacturing modes. The normal ranges of the three major isoforms are presented

Sample	G0 [%]	G1 [%]	G2 [%]
Batch	18–27	55–59	15–25
Fed-batch	18–27	55–59	15–25
Perfusion	11–20	53–57	20–33

ammonia concentrations (ammonia concentrations typically reach 10–12 mM by the end of the fed-batch compared to 3–4 mM for the batch process). The effects of ammonia on glycosylation have been well reported [48–50], and appear to be due to the activity of ammonia as a weak base, increasing the pH of the lumen of the Golgi body. In general, the distribution of the G0, G1, and G2 isoforms varies only slightly between cell culture runs. Inter-assay variation is ±0.5% for each isoform of the same sample. Only slight differences have been observed between the cell lines analyzed to date in either batch and fed-batch modes. Little or none bisecting *N*-acetylglucosamine or sialic acid is present, and no evidence of structures that may be immunogenic in humans, such as high-mannose or hybrid structures has been detected. These structures have been reported for glycoproteins produced in non-human cell lines, such as NS0 (a mouse myeloma cell line) [51]. An interesting observation is that the recombinant antibody is generally more galactosylated in perfusion mode (Table 3.4), where the cell counts are the highest ($\sim 10^8$ cells mL^{-1}).

3.6.4
Peptide Mapping

Peptide mapping using liquid chromatography (LC) coupled with either UV detection or mass spectrometry (MS) is a powerful technique to study the primary structure of the antibody, and to further investigate post-translational or chemical modifications. The peptide map is a chromatographic finger-print which is obtained after reduction/alkylation and subsequent proteolytic digestion of the antibody (Fig. 3.26). Thus obtained UV-patterns are used routinely to screen for structural integrity after process changes or for quality control of production lots [52]. LC-MS is applied to confirm the amino acid sequence and to detect and identify modifications.

The primary sequences of PER.C6 cell-derived IgG1 and IgG4 antibodies were confirmed in LC-MS peptide maps, and no changes compared to the sequence expected from DNA-transcription were detected. The presence of the typical glycan structures (see Fig. 3.23) could be confirmed, and it was demonstrated that non-glycosylated heavy chains were not present. Modifications of the N- and C-terminus of the heavy chains were observed in both IgG1 and IgG4 antibodies. In all cases, the N-terminal glutamine residue was converted by cycliza-

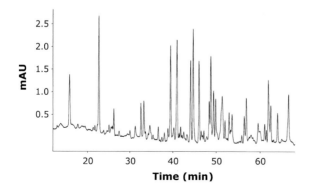

Fig. 3.26 Peptide map of a PER.C6-derived IgG1 using liquid chromatography coupled with UV detection.

tion into a pyroglutamate residue, which is a common chemical modification of antibodies. The C-terminal modification is caused by removal of the C-terminal lysine residue from the heavy chain. This is due to the activity of carboxypeptidases, and is a frequently observed characteristic of proteins produced in mammalian cell culture [53]. Both the pyroglutamate conversion and removal of the lysine residue can contribute to charge heterogeneity of the product. However, in the described antibodies the conversions were 100%, resulting in homogeneous N- and C-termini. The charge heterogeneity observed in the IEF-analyses is most likely caused by deamidation of asparagine residues. The exact deamidation sites and the percentage of deamidated forms could be established with LC-MS.

3.6.5
Summary Biochemical Analyses

A number of analytical methods have been used to characterize monoclonal antibodies produced in PER.C6 cells. It has been shown that the antibodies can be quantified and purified easily, and several quality attributes can be maintained in batch, fed-batch, and perfusion production modes. The protein and carbohydrate structures of the antibodies are completely human in nature, and excellent control of these attributes during process development has been observed.

In line with this, the bio-activity of PER.C6 cell-produced monoclonal antibodies was shown to be equal or better than antibodies produced in CHO or murine cell lines (results not shown).

3.7
Conclusion

The PER.C6 cell line was generated from retina-derived primary human cells, which were immortalized by insertion of the adenovirus E1 gene. In comparison to CHO and NS0 cells, for example, the PER.C6 cell line has been used for only a few years in protein production. Nonetheless, it is already an attractive expression platform that exhibits many favorable characteristics for the production of IgG and other proteins. These include the very high cell numbers and hence high yields that are obtained, the rapid generation of high-expressing clones that match a generic and robust fed-batch process, and the high quality of the antibodies produced. These issues are discussed below.

3.7.1
Rapid Generation of High-producing Clones

The transfection of PER.C6 cells with expression plasmids is very efficient, as is the subsequent generation of stable cell lines (an overview of the process is shown in Fig. 3.3). Importantly, high expression levels of recombinant antibodies are observed in the absence of gene amplification, giving a considerable time advantage over cell lines which require amplification for efficient protein expression. High-expressing PER.C6 cell lines contain between two and 10 copies of antibody genes per genome, compared to hundreds of copies in amplified cell lines. A high gene copy number is associated with instability of expression over time. In contrast, expression from PER.C6 cell lines producing antibodies is very stable over several months.

Because the amplification of gene copy number is not required in PER.C6 cells, and thanks to the rapid and easy adaptation to serum-free and animal component-free media, the timelines for cell line generation are minimal – 6–7 months from transfection to final lead clone selection for antibody-expressing PER.C6 cell lines. These cell lines are selected based on their performance in generic batch or fed-batch processes, thereby ensuring the selection of cell lines that perform optimally in the desired production process and reducing the investment in process development required for the rapid generation of material for pre-clinical and Phase I clinical studies.

3.7.2
High Numbers of Viable Cells

The generic production processes are based on the metabolic and physico-chemical requirements of PER.C6 cells. The specific productivity of antibody-producing cell lines analyzed to date ranges typically from 10 to 20 pg per cell per day. This does not compare with highly amplified CHO cell lines, for example, but it does result in high batch (0.4–0.8 g L^{-1}), fed-batch (1.3–2.2 g L^{-1}) and perfusion (up to 1 g L^{-1} per day) due to the high cell numbers obtained in these processes, which is on a par with levels obtained in industry with CHO and other cell lines. The reason for the high viable cell numbers may be related to the high-level expression of E1B proteins, which are known to be anti-apoptotic.

3.7.3
Consistent Product Quality of the Antibodies

Antibodies produced on PER.C6 cells show consistent product quality, as measured by IEF, SDS–PAGE and glycan analysis in batch, fed-batch, and perfusion processes. Typical glycan profiles for antibodies produced in PER.C6 cells are similar to human serum IgG, with a ratio of G0:G1:G2 of approximately 30%:55%:15% (see Fig. 3.25). Galactosylation is slightly reduced in the fed-batch, with an increase in G0 to 40–45% and a decrease in G1 and G2 to 45–50% and 10%, respectively. This decrease is probably due to the influence of ammonia, which reaches a higher concentration in the fed-batch.

To date, the profiles obtained are similar for all antibodies produced in PER.C6 cells, independent of the production levels obtained. Thus, the outcome of a production process becomes highly predictable.

By contrast, the majority of CHO-derived IgGs contain low levels of galactose, which may diminish the antibody's ability to initiate effector functions. In addition, in CHO cells sialic acid is added only via an a(2-3) linkage, whereas the sialic acid linkage in human serum may be a(2-6) or a(2-3). NS0 cells exhibit similar characteristics, but may also add an extra galactose to an existing terminal galactose via an a(1-3) linkage. Humans lack the enzyme that adds this structure, and such a Gal a(1-3) structure is highly immunogenic in humans: indeed, it is estimated that 1% of circulating Ig is directed against this moiety. Glycans with high-mannose structures and hybrid structures have also been observed on IgGs produced in CHO and NS0 cells. However, no such structures have been identified in antibodies produced in PER.C6 cells.

3.7.4
Future Prospects

As yet, the PER.C6 cell line has been used for protein production for only a relatively short period of time, but the high cell densities obtained, the generic fed-batch process and consistent product indicate that the cell line has vast potential. Unequalled high cell densities ($>150 \cdot 10^6$ cells mL^{-1}) obtained in continuous perfusion can be used to manufacture very large amounts of proteins in a small-scale reactor. In addition, this procedure may be used to produce unstable proteins in an efficient manner.

The process also demonstrates that the maximum cell densities obtained in fed-batch can be improved significantly, and hence further increase the yields. The ultimate aim is a production platform on which large quantities of high-quality protein are produced at low cost, thereby allowing more people to benefit from effective, but expensive, biopharmaceuticals.

References

1 Horwitz, M.S., Adenoviridae and their replication, in: B.N. Fields and D.M. Knipe (Eds) Virology, 1990, Raven Press, Ltd, New York, pp. 1679–1740.

2 Graham, F.L., J. Smiley, W.C. Russell, and R. Nairn, Characteristics of a human cell line transformed by DNA from adenovirus type 5. J. Gen. Virol., 1977, 36, 59–72.

3 Houweling, A., P.v.d. Elsen, and A.v.d. Eb, Partial transformation of primary rat cells by the left-most 4.5% of adenovirus 5 DNA. Virology, 1980, 105, 537–550.

4 Gallimore, P.H., R.J.A. Grand, and P.J. Byrd, Transformation of human embryo retinoblasts with simian virus 40, adenovirus and ras oncogenes. Anticancer Res., 1986, 6, 499–567.

5 Teodoro, J. and P. Branton, Regulation of apoptosis by viral gene products. J. Virol., 1997, 71, 1739–1746.

6 Sang, N., J. Caro, and A. Giordano, Adenoviral E1A: everlasting tool, versatile applications, continuous contributions and new hypotheses. Front. Biosci., 2002, 4, 407–413.

7 Zhang, Y. and Y. Xiong, Mutations in human ARF exon 2 disrupt its nucleolar localization and impair its ability to block nuclear export of MDM2 and p53. Mol. Cell, 1999, 3(5), 579–591.

8 Cuconati, A. and E. White, Viral homologues of BCL-2: role of apoptosis in the regulation of virus infection. Genes Dev., 2002, 16, 2465–2478.

9 Singer-Sam, J., D.H. Keith, K. Tani, R.L. Simmer, L. Shively, S. Lindsay, A. Yoshida, and A.D. Riggs, Sequence of the promoter region of the gene for X-linked 3-phosphoglycerate kinase. Gene, 1984, 32, 409–417.

10 Valerio, D., M.G.C. Duyvesteyn, B.M.M. Dekker, G. Weeda, T.M. Bervens, L.v.d. Voorn, H.v. Ormondt, and A.J.v.d. Eb, Adenosine deaminase: characterization and expression of a gene with a remarkable promoter. EMBO J., 1985, 4, 437–443.

11 Simonsen, C. and A. Levinson, Analysis of processing and polyadenylation signals of the hepatitis B Virus surface antigen gene by using simian virus 40-Hepatitis B Virus chimeric plasmids. Mol. Cell Biol., 1983, 3, 2250–2258.

12 Vaessen, R.T.M.J., A. Houweling, A. Israel, P. Kourilsky, and A.J.v.d. Eb, Adenovirus E1A-mediated regulation of class I MHC expression. EMBO J., 1986, 5, 335–341.

13 Fallaux, F.J., O. Kranenburg, S.J. Cramer, A. Houweling, H.v. Ormondt, R.C. Hoeben, and A.J.v.d. Eb, Characterization of 911: a new helper cell line for the titration and propagation of early-region-1-deleted adenoviral vectors. Hum. Gene Ther., 1996, 7, 215–222.

14 Byrd, P., K. Brown, and P. Gallimore, Malignant transformation of human embryo retinoblasts by cloned adenovirus 12 DNA. Nature, 1982, 298, 69–71.

15 Fallaux, F.J., A. Bout, I.v.d. Velde, D.J.M.v.d. Wollenberg, K. Hehir, J. Keegan, C. Auger, S.J. Cramer, H.v. Ormondt, A.J.v.d. Eb, D. Valerio, and R.C. Hoeben, New helper cells and matched early region 1-deleted adenovirus vectors prevent generation of replication-competent adenoviruses. Hum. Gene Ther., 1998, 9, 1909–1917.

16 Nichols, W.W., R. Lardenoie, B.J. Ledwith, K. Brouwer, S. Manam, R. Vogels, D. Kaslow, D. Zuidgeest, A.J. Bett, J. Chen, M. Van der Kaaden, S.M. Galloway, R.B. Hill, S.V. Machotka, C.A. Anderson, J.B. Lewis, D. Martinez, J. Lebron, C. Russo, D. Valerio, and A. Bout, Propagation of adenoviral vectors: use of PER.C6 cells, in: D.T. Curiel and J.T. Douglas (Eds) Adenoviral vectors for gene therapy, 2002, Elsevier. San Diego, pp. 129–167.

17 Paterson, T., J. Innes, L. McMillan, I. Downing, and M.C. Carter, Variation in IgG1 heavy chain allotype does not contribute to differences in biological activity of two human anti-

Rhesus (D) monoclonal antibodies. Immuno-technology, 1998, 4, 37–47.

18 Ljunggren, J. and L. Haggstrom, Glutamine limited fed-batch culture reduces the overflow metabolism of amino acids in myeloma cells. Cytotechnology, 1992, 8, 45–56.

19 Ljunggren, J. and L. Haggstrom, Catabolic control of hybridoma cells by glucose and glutamine limited fed-batch cultures. Biotechnol. Bioeng., 1994, 44, 808–818.

20 Xie, L.Z. and D.I.C. Wang, Fed-batch cultivation of animal cells using different medium design concepts and feeding strategies. Biotechnol. Bioeng., 1994, 43, 1175–1189.

21 Xie, L.Z. and D.I.C. Wang, Stoichiometric analysis of animal cell growth and its application in medium design. Biotechnol. Bioeng., 1994, 43, 1164–1174.

22 Sanfeliu, A., C. Paredes, J.J. Cairo, and F. Godia, Identification of key patterns in the metabolism of hybridoma cells in culture. Enzyme Microb. Technol., 1997, 21, 421–428.

23 Zhou, W.C., C.C. Chen, B. Buckland, and J. Aunins, Fed-batch culture of recombinant NS0 myeloma cells with high monoclonal antibody production. Biotechnol. Bioeng., 1997, 55, 783–792.

24 Zhou, W.C., J. Rehm, A. Europa, and W.S. Hu, Alteration of mammalian metabolism by dynamic nutrient feeding. Cytotechnology, 1997, 24, 99–108.

25 Europa, A.F., A. Gambhir, P.C. Fu, and W.S. Hu, Multiple steady states with distinct cellular metabolism in continuous culture of mammalian cells. Biotechnol. Bioeng., 2000, 67, 25–34.

26 Mercille, S., M. Johnson, S. Lanthier, A.A. Kamen, and B. Massie, Understanding factors that limit the productivity of suspension based perfusion cultures operated at high medium renewal rates. Biotechnol. Bioeng., 2000, 67(4), 435–447.

27 Banik, G.G. and C.A. Heath, An investigation of cell density effects on hybridoma metabolism in a homogeneous reactor. BioProcess Eng., 1994, 11, 229–237.

28 Brosie, D.d.l., M. Nosieux, R. Lemieux, and B. Massie, Long-term perfusion culture of hybridoma: a 'grow or die' cell cycle system. Biotechnol. Bioeng., 1991, 38, 781–787.

29 Richieri, R.A., L.S. Williams, and P.C. Chou, Cell cycle dependency of monoclonal antibody production in asynchronous serum-free hybri-

doma cultures. Cytotechnology, 1991, 5, 972–976.

30 Schenerman, M.A., B.R. Sunday, S. Kozlowski, K. Webber, H. Gazzano-Santoro, and A. Mire-Sluis, CMC Strategy Forum Report: Analysis and structure characterization of monoclonal antibodies. BioProcess Int., 2004, February, 42–52.

31 International Conference on Harmonisation Guidance on Specifications: Test procedures and acceptance criteria for biotechnological/biological products. Fed. Reg., 1996, 64, 2733.

32 Krips, D.M., R.O. Sitrin, and C.N. Oliver, A very rapid 2 min protein A HPLC assay for monoclonal antibodies. FASEB J., 1991, 5, A465.

33 Brown, M.A., L.M. Stenberg, U. Persson, and J. Stenflo, Identification and purification of vitamin K-dependent proteins and peptides with monoclonal antibodies specific for gamma-carboxyglutamyl (Gla) residues. J. Biol. Chem., 2000, 275, 19795–19802.

34 Jones, D., N. Kroos, R. Anema, B. van Montfort, A. Vooys, S. van der Kraats, E. van der Helm, S. Smits, J. Schouten, K. Bouwer, F. Lagerwerf, P. van Berkel, D.-J. Opstelten, T. Logtenberg, and B. Bout, High-level expression of recombinant IgG in the human cell line PER.C6. Biotechnol. Prog., 2003, 19, 163–168.

35 Biedermann, K., Sabater, M., Sorensen, J., Fiedler, H., and Emborg, C, Quantitative binding studies of a monoclonal antibody to immobilized protein-A. Bioseparation, 1991, 2, 309–314.

36 Shapiro, A., E. Vinuela, and J. Maizel, Molecular weight estimation of polypeptide chains by electrophoresis in SDS polyacrylamide gels. Biochem. Biophys. Res. Commun., 1967, 28, 815.

37 Righetti, P.G., Isoelectric Focussing: theory, methodology and applications. 1983, Elsevier Biomedical Press, Amsterdam.

38 Harris, R.J., B. Kabakoff, F.D. Macchi, F.J. Shen, M. Kwong, J.D. Andya, S.J. Shire, N. Bjork, K. Totpal, and A.B. Chen, Identification of multiple sources of charge heterogeneity in a recombinant antibody. J. Chromatogr. B, 2001, 752, 233–245.

39 Perkins, M., R. Theiler, S. Lunte, and M. Jeschke, Determination of the origin of charge heterogeneity in a murine monoclonal antibody. Pharm. Res., 2000, 17, 1110–1117.

40 Wehr, T., R. Rodriguez-Diaz, and M. Zhu, Recent advances in capillary isoelectric focussing. Chromatographia Suppl., 2001, 53, S47–S58.

41 Jenkins, N., R. B. Parekh, and D. C. James, Getting the glycosylation right: Implications for the biotechnology industry. Nature Biotechnol., 1996, 14, 975–981.

42 Cumming, D. A., Glycosylation of recombinant protein therapeutics: Control and functional implications. Glycobiology, 1991, 1, 115–130.

43 Wright, A. and S. Morrison, Effect of glycosylation on antibody function: implications for genetic engineering. Trends Biotechnol., 1997, 15, 26–32.

44 Jefferis, R., Glycosylation of Human IgG antibodies: Relevance to therapeutic applications. BioPharmacology, 2001, September, 19–27.

45 Raju, T. S., Glycosylation variations with expression systems and their impact on biological activity of therapeutic immunoglobulins. BioProcess Int., 2003, April, 44–53.

46 Raju, T. S., J. B. Briggs, S. M. Borge, and A. J. S. Jones, Species-specific variation in glycosylation of IgG: Evidence for the species-specific sialylation and branch-specific galactosylation and importance for engineering recombinant glycoprotein therapeutics. Glycobiology, 2000, 10, 477–486.

47 Hills, A. E., A. K. Patel, P. N. Boyd, and D. C. James, Control of therapeutic antibody glycosylation, in: A. Bernard, et al. (Ed) Animal Cell Technology: Products from cells, cells as products, 1999, Kluwer Academic Press, Dordrecht, The Netherlands, pp. 255–257.

48 Borys, M. C., D. H. Linzer, and E. T. Papoutsakis, Ammonia affects the glycosylation patterns of recombinant mouse placental lactogen-I by Chinese hamster ovary cells in pH dependent manner. Biotechnol. Bioeng., 1994, 43, 505–514.

49 Andersen, D. C. and C. Goochee, The effect of ammonia on the O-linked glycosylation of granulocyte colony stimulating factor by Chinese hamster ovary cells. Biotechnol. Bioeng., 1995, 47, 96–105.

50 Gawlitzek, M., U. Valley, and R. Wagner, Ammonium ion and glucosamine dependent increases of oligosaccharide complexity in recombinant glycoproteins secreted from cultivated BHK-21 cells. Biotechnol. Bioeng., 1998, 57, 518–528.

51 Hills, A. E., A. Patel, P. Boyd, and D. C. James, Metabolic control of recombinant monoclonal antibody N-glycosylation in GS-NS0 cells. Biotechnol. Bioeng., 2001, 75, 239–251.

52 Bongers, J., J. J. Cummings, M. B. Ebert, M. M. Federici, L. Gledhill, D. Gulati, G. M. Hilliard, B. H. Jones, K. R. Lee, J. Modzanowski, M. Naimoli, and S. Burman, Validation of a peptide mapping method for a therapeutic antibody: what could we possibly learn about a method we have run 100 times? J. Pharm. Biomed. Anal., 2000, 21, 1099–1128.

53 Harris, R. J., Processing of C-terminal lysine and arginine residues of proteins isolated from mammalian cell culture. J. Chromatogr. A., 1995, 705, 129–134.

4

Use of the Glutamine Synthetase (GS) Expression System for the Rapid Development of Highly Productive Mammalian Cell Processes

John R. Birch, David O. Mainwaring, and Andrew J. Racher

Abstract

Mammalian cell culture is becoming increasingly important for the production of high-volume biopharmaceutical proteins. This is driving improvements in process efficiency. This chapter provides examples of improvements in both the creation of cell lines and in cell culture optimization, focusing particularly on experience with the glutamine synthetase (GS) expression system.

Abbreviations

ACF&PF	animal component-free and protein-free
ADCC	antibody dependent cellular cytoxicity
BHK	baby hamster kidney
cdk	cyclin-dependent kinase
CHO	Chinese hamster ovary
DHFR	dihydrofolate reductase
FACS	fluorescence-activated cell sorter
GS	glutamine synthetase
hCMV	human cytomegalovirus
IVC	time integral of the viable cell concentration
MIE	major intermediate early
MSX	methionine sulfoximine
NS0	non-secreting murine myeloma
OUR	oxygen uptake rate
SV40	simian virus 40
TIMP	tissue inhibitor of metalloproteinases
UTR	untranslated region

4.1
Introduction

Mammalian cell culture is an established technology for the manufacture of biopharmaceuticals. In recent years, there has been a significant increase in the number of proteins – particularly monoclonal antibodies – which are used in relatively large volumes: this has been a significant driver for improvements in manufacturing processes. Substantial progress has been made, both in the optimization of upstream processes and in the design of expression systems, for the creation of highly productive cell lines. A high-yielding biopharmaceutical protein manufacturing process is the result of using a number of approaches that affect the cell line *per se*, the cell culture process, product recovery, and purification activities. Improved cell lines are the result of increasing the efficiency of gene expression and protein secretion, together with the use of stringent selection protocols to isolate the rare high producers. The optimization of cell culture processes by, for example, improving media and by developing advanced feeding strategies that support high space-time yields of viable biomass has increased substantially the product concentrations achieved in the bioreactor [1–3] (see also Part IV, Chapter 1). Antibody concentrations of 5.1 and 4.3 g L^{-1} have been reported in fed-batch culture of GS-NS0 and GS-CHO respectively [4, 5] (see also Part IV, Chapter 16). This chapter reviews an expression technology based on the use of the glutamine synthetase gene and the integration of this technology into the development of high-yielding, large-scale manufacturing processes for biopharmaceuticals.

4.2
Cell Line Construction and Selection

4.2.1
Choice of Cell Line

The enzyme glutamine synthetase (GS) catalyzes the formation of glutamine from glutamic acid and ammonia, driven by the hydrolysis of ATP. Glutamine has multiple roles in cell metabolism, particularly as an energy source, protein constituent and as a nitrogen donor in purine and pyrimidine synthesis. Cell lines that do not produce GS have an absolute requirement for glutamine and do not grow in glutamine-free culture media. This provides the basis for using the enzyme as a selectable marker in gene expression vectors. The cloning of the GS gene from Chinese hamster ovary (CHO) cells was described by Sanders and Wilson [6]. The utility of GS as a selectable marker is increased by the availability of an efficient inhibitor of GS, methionine sulfoximine (MSX), which can be used to improve the stringency of selection, to select for gene amplification and to inhibit enzyme activity

in those cell lines which produce endogenous GS. Most myeloma and hybridoma cells have an absolute requirement for glutamine. In contrast, many other cell types such as BHK-21, L-cells and the widely used CHO do not require glutamine, provided that glutamic acid is present in the culture medium. In these cases GS can still be used as a selectable marker, but it is necessary to use a specific inhibitor of GS (e.g., MSX) to inhibit the endogenous enzyme. The toxicity of MSX at very low concentrations (3 µM) for wild-type CHO cells has been demonstrated [6]. Since the enzyme is used as a dominant selectable marker, it is not necessary to create relevant mutant host cells.

GS expression vectors designed for use in mouse NS0 cells [1] and CHO cells [7, 8] have been described. The NS0 cell line was chosen because it has an absolute requirement for glutamine – in contrast to other lymphoid cells, which gave a relatively high frequency of glutamine-independent variants. Other factors contributing to the choice of this cell line included the ease with which it could be grown in serum-free suspension culture and, given its B-cell lineage, an expectation that it has the machinery for efficient antibody secretion.

In principle, selection based on GS can be used with a wide variety of cell lines, and in practice the most commonly used are NS0 and CHO. Whilst GS-NS0 has been used most often for antibody production, GS-CHO has been used to express a large range of proteins in CHO cells. In addition to antibodies (e.g., [9]), enzymes (e.g., [10]), interleukins [11] and membrane-bound proteins (e.g., [12]) are examples of the range of proteins produced using the GS expression system. For many proteins – and particularly many non-antibody products – the glycosylation properties of CHO are preferable to those found in NS0. It has been found, for example, that NS0 has a limited ability to add sialic acid to glycoproteins, and this can have a significant effect on the clearance rate of the protein *in vivo*. Flesher et al. [13] compared the *in vivo* clearance profile of a soluble form of the membrane receptor CTLA4 produced in CHO and NS0 cells. A correlation was observed between the quantity of N-acetylneuraminic acid in the product and *in vivo* clearance rates. Product made in NS0 cells had no detectable N-acetylneuraminic acid and exhibited an accelerated clearance rate. Baker et al. [14] compared the glycosylation of recombinant tissue inhibitor of metalloproteinases (TIMP) made in GS-CHO and GS-NS0; these authors found significant differences, and in particular a high proportion (30%) of the NS0 glycans terminated in α-1,3-linked galactose. In addition a high proportion of the sialic acid in NS0 material was in the form of N-glycolylneuraminic acid as opposed to N-acetylneuraminic acid. Lifely et al. [15] compared the glycosylation of the monoclonal antibody Campath®-1H made in CHO, GS-NS0 and rat Y0 myeloma cells. The glycan profiles of CHO and NS0 were similar, although NS0 antibody was underglycosylated to a significant extent. The Y0-derived antibody had fucosylated and non-fucosylated glycans containing a bisecting GlcNAc, and was observed to have enhanced ADCC activity.

4.2.2
Expression Vector Design

The ability to express the product at a high level is the critical issue for any manufacturing process using recombinant cell lines. Consequently, expression vectors have been developed that, through a combination of suitable promoters and favorable RNA processing signals, can achieve high levels of transcription from the genes of interest.

Strong promoters used to drive expression of the genes of interest are generally of viral origin or from highly expressed genes in a mammalian cell. A number of different viral promoters have been evaluated for use with the GS expression vectors [8]. These authors screened different promoters using a test system based on the transient transfection of CHO-K1 cells with the gene for TIMP. The efficiency of the transcription units was in the order hCMV > SV40 early > hybrid Moloney murine leukemia virus-SV40 promoter > SV40 late. Expression of the TIMP gene driven by the hCMV promoter produced five to ten times more TIMP than the SV40 early promoter. The hCMV promoter fragment chosen consists of the complete enhancer-promoter and 5'-UTR (untranslated region) of the major intermediate early (MIE) gene. This promoter, unlike the SV40 early promoter, is highly efficient in most cell types including lymphocytes (see Ref. [1]). Various vector constructions were evaluated in NS0 cells, and the system chosen is shown in Fig. 4.1. Product gene expression is driven by the hCMV-MIE promoter. A polylinker site is incorporated downstream of the hCMV sequence to allow incorporation of the product gene, and an SV40 polyadenylation site is situated downstream of the polylinker. For the expression of monoclonal antibodies, both the heavy and light chain genes are contained in the same vector, with the heavy chain gene downstream of the light chain gene. The GS gene, which is upstream of the product genes, is driven by the SV40 early promoter.

4.2.3
Increasing Transcription

Several options exist to increase transcription. In early expression systems this was generally by gene amplification. Gene amplification is usually achieved by construct-

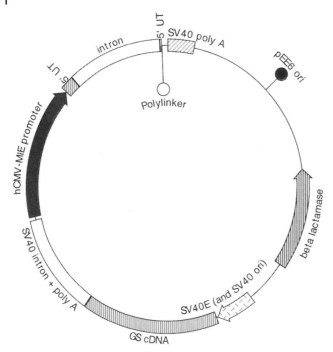

Fig. 4.1 Glutamine synthetase (GS) expression vector.

ing the expression vector so that the genes of interest are linked to an amplifiable gene. Although high-yielding GS-NS0 and GS-CHO cell lines can be isolated without amplification – particularly if attention is paid to the stringency of selection – several groups have used amplification, especially for CHO cells.

For the GS expression system, amplification is achieved by applying increasing concentrations of the GS inhibitor, MSX. Laubach et al. [10], for instance, comment that amplification (by growth in the presence of 400 μM MSX) of two GS-CHO cell lines making inducible nitric oxide synthase resulted in a 3- to 4-fold increase in productivity. The amplification of a GS-CHO cell line making a monoclonal antibody increased the antibody concentration from 110 to 250 mg L^{-1} [16].

Bebbington et al. [1] found that, for a GS-NS0 cell line, amplification using MSX was accompanied by an increase in copy number of the vector from one to four. Similar results have been described by other groups [17, 18]. This is much lower than the levels observed with the dihydrofolate reductase (DHFR)-CHO system [17]. In CHO cell lines the copy numbers of DHFR-linked genes can be as high as 1000, whereas in NS0 clones the copy number of GS-linked genes rarely exceeds 20. For a GS-CHO cell line, amplification with 200 μM MSX increased the copy number from 5 to 200 [16].

Peakman et al. [17] studied the impact of amplification upon the productivity of both GS-CHO and GS-NS0 cell lines. For GS-NS0 cell lines expressing a recombinant antibody, the copy number of either the GS or immunoglobulin heavy chain genes may be the same in different clones, but it does not follow that mRNA levels will be the same. Although amplification of DHFR-CHO and GS-NS0 cell lines may result in markedly different copy numbers, the two cell lines may still express approximately the same amount of antibody due to mRNA levels being virtually identical. Therefore, for cell lines generated using the GS vector system, amplification of copy number is not as critical for generating high-producing clones as with the DHFR expression system. With the GS expression system, the position of integration of the transfected DNA is the important factor in determining whether the cell line will ultimately be a high producer.

The high copy numbers of the expression vector seen upon amplification – especially with the DHFR expression system – may increase the cell-specific productivity, but it can also have a detrimental effect upon other properties of the cell. Amplification of the desired gene will frequently result in poor growth performance of the resulting cell population and may alter cellular metabolism. These effects have been seen in both GS-NS0 and DHFR-CHO cell lines [17, 19]. Gu et al. [20] suggested that the poor growth seen upon amplification of the DHFR gene is not due to increased expression of a recombinant protein; rather, it is a consequence of the higher metabolic burden imposed upon the cell by the increase in specific DHFR activity. The problem of poor growth (low values for the maximum viable cell concentration and time integral of the viable cell concentration) can be counteracted to some extent by using the combination of good growth characteristics and a high specific production rate as selection criteria. Amplification and the resulting variation in copy number can also alter the inherent stability of expression, and often requires the continued presence of the amplification agent [16, 21]. If the selective agent is required in the production bioreactor, it will be necessary to demonstrate that the purification process removes this compound from the bulk drug product.

Expression from the hCMV-MIE promoter in CHO-K1 cells can also be enhanced by expression of the adenovirus 5 E1A transactivator or a mutant of E1A that has lost the oncogenic transformation function [22]. After optimization of E1A expression, since over-expression led to inhibition of cell growth, it was possible to raise expression levels of TIMP in non-amplified GS cell lines to levels which were previously only achievable after vector amplification. CHO cell lines constitutively expressing the transactivator were created. The transactivator increased the specific production rate of recombinant protein product in both a transient expression system, and in stable cell lines – either by incorporating it into a producer cell line or by transfecting the product gene into a host cell line that already contained E1A. Vector amplification did not produce higher-producing cell lines in E1A-containing cells. Likewise, the introduction of E1A into an amplified GS cell line making TIMP failed to enhance productivity.

4.2.4
Selection of High-producers without Amplification

Amplification of the number of copies of expression vector can lead to an increase in productivity of the cell line, but this is not the only method for generating high-yielding cell lines. The GS system [1] and some variants of the DHFR system [23] do not rely upon amplification to achieve high productivities. Instead, these systems rely upon insertion of the antibody construct into a transcriptionally active region to achieve high productivities. Specific production rates for non-amplified antibody-producing GS cell lines of about 65 pg per cell·day, and an antibody concentration of more than $1.8 \, g \, L^{-1}$, can be obtained in fed-batch bioreactors [24].

Currently, the generation of high-yielding cell lines is typically achieved by screening large numbers of transfectants, or a combination of amplification plus screening. There are problems with these approaches due to the time needed to obtain a cell line with an acceptable productivity, the potential instability of amplified cell lines, and the deleterious effect of amplification upon other characteristics of cell line. The reason it is difficult to obtain high-yielding cell lines is that large regions of the genome are organized into heterochromatin, which is believed to be transcriptionally inactive. The level of mRNA expression from a vector that integrates into the heterochromatin will be low. Since there are only a few loci within the genome capable of expressing the selectable marker gene and the linked gene(s) of interest gene at high levels, it follows that the probability of integration into such a transcriptionally active locus is low. Thus, large numbers of transfectants normally have to be screened to isolate those few clones where the vector has integrated into transcriptionally active loci, with concomitant high product expression levels.

Several approaches have been developed to reduce the time needed to obtain high-yielding cell lines. These approaches exploit the importance of the chromosomal locus in determining the level of gene expression to increase the proportion of transfectants with the expression vector integrated into a transcriptionally active locus by up to 10-fold. Since these approaches generate a clone with only one or at most a few copies of the expression vector [1, 25], the problems associated with amplification are eliminated.

One approach is to use site-specific recombination of the gene(s) of interest into a known transcriptionally active locus. Expression vectors can be constructed that contain a specific targeting sequence that will direct the vector to integrate by homologous recombination into a particular active site. Such a sequence has been identified in the immunoglobulin locus of NS0 cells [26]. Vectors containing this sequence are targeted to the immunoglobulin locus in more than 50% of high-producing GS-NS0 clones.

A corollary of this approach is to take the sequences flanking the transcriptionally active locus and incorporate them into the expression vector. Thus, the vector should create a favorable environment for expression independent of its integration site in the genome. Vectors based on ubiquitous chromatin opening elements [27] or the flanking sequences of the Chinese hamster elongation factor-1α gene [28] have been described.

An alternative approach is to transfect the cells with a conventional expression vector (i.e., randomly to integrate the expression vector into the genome), but then bias the selection method so that only transfectants where the vector integrated into a transcriptionally active site are progressed. This can be done by using a selection system that only allows transfectants producing sufficient levels of the selectable marker gene product to proliferate. Expression systems using a selectable marker gene with either the weak SV40 promoter [1] or an impaired Kozak sequence upstream of the marker gene [23] are included in this class of selection system. Linkage of the antibody construct to the selectable marker gene results in the over-production of antibody, as both genes are integrated into a transcriptionally active locus. The choice of selection conditions is extremely important for the success of this approach. The data in Table 4.1 show that increasing the selection stringency for GS cell lines (through increased MSX concentration) reduces the number of stable transfectants but, by optimizing the transfection conditions, it is possible to maintain the number of transfectants generated. A higher selection stringency resulted in shift in the position of the median antibody concentration (from about 45 mg L^{-1} to about 90 mg L^{-1}) and interquartile range (from 25 to 70 mg L^{-1}, to about 60 to 130 mg L^{-1}). This approach shows that it is possible to increase the average productivity without restricting the number of transfectants.

The function of the expression vectors described in the previous sections is to generate cell lines with high specific production rates of the protein of interest. However, a transfectant with a high specific production rate does not necessarily result in a cell line that performs well in the production process. Hence, a sufficient number of cell lines need to be generated to allow for the attrition in numbers when screening for other desired characteristics.

By definition, transfectants with the highest productivities are rare: this is shown in Fig. 4.2. The figure shows the probability of finding a transfectant that produced antibody at a defined concentration. The probability of finding a primary transfectant producing 150 mg L^{-1} is about 0.0005. The majority of transfectants (90%) produced less than 90 mg L^{-1} antibody, and only 1.5% produced more than 150 mg L^{-1}. The issue is therefore, how can the hit rate for finding highly productive cell lines or the number of hits be increased?

Finding these rare events requires the combination of a number of approaches. The simplest approach is to screen more

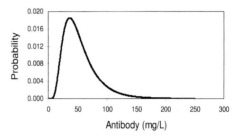

Fig. 4.2 Productivity distribution of antibody concentrations for primary GS-CHO transfectants. Ninety-two primary GS-CHO transfectant colonies were transferred from 96-well to 24-well plates and grown for 14 days: the mean concentration at harvest was 48 mg L^{-1}. A log-normal probability density function was fitted to the antibody concentration data.

Table 4.1 Influence of transfection and selection conditions upon the yield of stable antibody-producing GS-CHO transfectants

Electroporation condition	Selection condition MSX [μM]	Number of stable transfectants per 5×10^6 cells electroporated
250 V, 400 μF	25	68
	50	32
275 V, 650 μF	25	124
	50	57
300 V, 900 μF	25	197
	50	70

transfectants, but how many? Simulations of screening experiments using the 150 mg L^{-1} cut-off suggest that, over the long term, at least 500 transfectants would have to be screened to avoid any individual screening experiment having no transfectants over 150 mg L^{-1} (A.J.R., unpublished results). To obtain tens of transfectants above this cut-off, then several thousand transfectants should be screened.

Conventional methods for the screening of cell lines after cloning are labor-intensive, and this limits the number of cell lines that can be screened. Increasingly, robotics are being used to automate the liquid handling and cell transfer stages, but this does not address the need to screen large numbers of transfectants to identify sufficient high producers to screen against the additional growth criteria that contribute to high productivity in a manufacturing process. Flow cytometry can be used to identify cells making high levels of the target product, while fluorescence-activated cell sorting (FACS) can be used to collect cells aseptically with the desired characteristics from large heterogeneous populations. Cells can be sorted into large populations ("bulk sorting"), from which cell lines can be isolated by conventional cloning methods, or by single cell sorting.

A number of FACS-based approaches have been reported for the isolation of cell lines secreting high levels of antibody. These include encapsulating the secreting cells in a biotinylated agarose droplet, which captures the secreted antibody [29], trapping the secreted protein in the membrane [30], or using a matrix constructed on the cell surface to trap the secreted antibody [31]. Holmes and Al-Rubeai [32] used a surface capture methodology to isolate clones with higher specific production rates from the GS-NS0 cell line 6A1(100)3. On average, the sorted clones had a specific production

rate which was 25% higher than the original GS-NS0 population. Racher [33] described a modification of this surface capture method that uses Protein A immobilized on the cell surface as a capture method for monoclonal antibodies.

The identification of high producers is the first step in isolating high-producing cell lines. The next step is to screen the pool of high producers against criteria that fit the cell line to the manufacturing process. Fig. 4.3 shows a schematic for a cell line selection program for GS cell lines. High-producing transfectants are identified and expanded through static into suspension culture. Once acceptable and reproducible growth is achieved, the cell lines are adapted to animal component and protein-free (ACF&PF) medium. Initially, transfectants are screened against productivity criteria: once in suspension culture, the cell lines are screened against additional criteria.

Typically, several criteria are used to select the production cell line. The criteria include: a high specific production rate; growth characteristics such as the magnitude of the time integral of the viable cell concentration (IVC) and maximum cell concentration; product concentration at harvest; cell line stability; and product quality. The importance of screening prior to cell line selection in a system that has relevance to the manufacturing process was demonstrated by Brand et al. [34]. These workers found there to be poor correlation between productivity of recombinant myelomas in static culture (cloning plates and flasks) and agitated suspension culture.

A key feature of any selection scheme is that it is important either to undertake the screening in an acceptable model of the manufacturing process, or to know the predictive power of the screen. In Fig. 4.4, the cell lines are evaluated in suspension culture using the same media, feeds and sub-

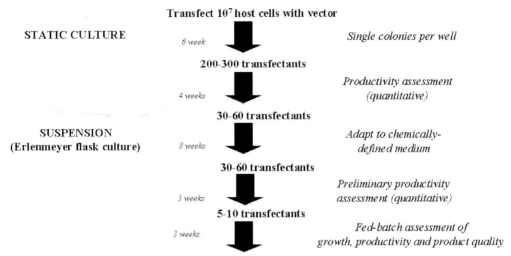

Transfect 10^7 host cells with vector

STATIC CULTURE

6 week

Single colonies per well

200–300 transfectants

4 weeks

Productivity assessment (quantitative)

30–60 transfectants

SUSPENSION
(Erlenmeyer flask culture)

8 weeks

Adapt to chemically-defined medium

30–60 transfectants

3 weeks

Preliminary productivity assessment (quantitative)

5–10 transfectants

3 weeks

Fed-batch assessment of growth, productivity and product quality

Select 3 cell lines for further analysis, including cell line stability study

Fig. 4.3 Schematic of a cell line selection programme for glutamine synthetase (GS) cell lines. High-producing transfectants are identified and expanded through static into suspension culture, and then adapted to chemically defined medium. For the suspension phase, the cell lines are grown in Erlenmeyer flasks. Initially, transfectants are screened against productivity criteria: once in suspension culture, the cell lines are screened against productivity, growth and product quality criteria.

culture regimes as used in the manufacturing process. Using this approach, it is possible routinely to obtain cell lines producing more than 1 g L^{-1}. Fig. 4.5 shows the predictive power of the screening process outlined in Fig. 4.3. The data in Fig. 4.5 are obtained from the cell lines eventually chosen for the manufacture of seven randomly cho-

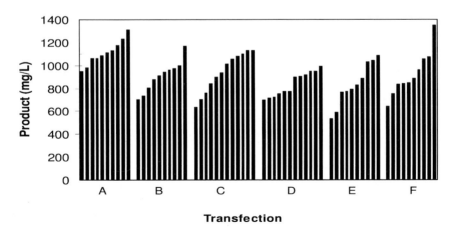

Fig. 4.4 Data are from six cell line construction programs. The data are the antibody concentrations achieved by panels of 10 GS-NS0 cell lines during the fed-batch assessment phase of the programme outlined in Fig. 4.3.

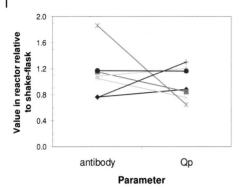

Fig. 4.5 Relationship between productivity characteristics of the lead cell lines making seven different antibodies when evaluated during the fed-batch assessment phase of Fig. 4.3 and in the production bioreactor. The data are either the antibody concentration or the specific production rate of the bioreactor culture (Qp) normalized to the values obtained in the fed-batch assessment phase that uses Erlenmeyer flasks.

sen antibodies. In general, the values obtained in bioreactors are 0.8- to 1.2-fold the values obtained in the Erlenmeyer flask assessment, although there are exceptions. The Erlenmeyer flask fed-batch model used in the assessment to select GS cell lines for the GMP manufacture of antibodies appears to have high predictive power.

In summary, although the transfectants with the highest specific production rates are (by definition) rare, the current approaches are successful in making productive cell lines. By using a combination of expression vectors with strong promoters and a stringent selection system it is possible to construct and then select high-producing, non-amplified, cell lines. Selection against productivity criteria should not be the only consideration. Multiple selection criteria should be used in an acceptable scale-down model of the manufacturing process to select cell lines that fit the manufacturing cell culture process.

4.3
Cell Line Stability

An important issue in creating cell lines is to maintain stability over the number of generations required for manufacturing – in practice, several tens of generations from a cell bank – for a fed-batch process at 10- to 20 000-L scale. Stability will be influenced by factors such as copy number and site of integration of the foreign gene(s). One would expect non-amplified lines with low copy number to be more stable than amplified lines with high copy number. In general it is possible to isolate stable, non-amplified lines which do not require MSX to maintain productivity. In the case of amplified cell lines MSX may be required. Hassell et al. [16] monitored the stability of GS-NS0 and GS-CHO cells making antibodies. Two amplified GS-NS0 cell lines making different antibodies were stable in shake-flask culture in both the presence and absence of MSX. In contrast, an amplified CHO cell line was stable in the presence of MSX but not in its absence. Other groups have also reported on the need to maintain MSX selection in amplified GS-CHO cells. Cosgrove et al. [35] used MSX to maintain stability of an CHO cell line producing insulin receptor ectodomain; MSX was added to culture media until the final production step. Similarly, Guerini et al. [12] found that expression of an ATPase in an amplified CHO cell line was stable for 6 months in the presence of MSX, but decreased substantially over 30 to 40 generations in its absence.

The stability of cell lines will be dependent not only on the characteristics of the cell line and gene inserts, but also on culture conditions. Bird et al. [36] presented evidence that stability of a GS-NS0 cell line was reduced under conditions where glutamine accumulated – for example, when cultures were grown in hollow-fiber devices. The authors found that the presence of 60 µM glutamine was sufficient to cause instability, presumably overcoming the selection pressure of the glutamine-free medium.

The mechanisms underlying instability in recombinant cell lines are poorly understood. Barnes et al. [37, 38] determined that there may be molecular features of transfectants that predicate instability. These authors studied a series of GS-NS0 cell lines making an anti-CD38 monoclonal antibody. Although copy number remained constant in these cell lines, there was a loss in expression of mRNA during prolonged culture. This did not result in loss of productivity in all of the cell lines. It seems that productivity was not influenced provided that levels of antibody mRNA remained above a critical threshold value.

4.4
Cell Engineering to Increase Productivity

4.4.1
Delaying Apoptosis

An important tenet for achieving highly productive processes is the achievement within the bioreactor of a high viable cell concentration and its subsequent maintenance for an extended period. The latter requires the death rate to be minimized. This section describes cell engineering approaches evaluated with GS cell lines to minimize the death rate.

Al-Rubeai et al. [39] showed that the major cause of cell death in animal cell culture is through the induction of apoptosis (programmed cell death) pathways by chronic, rather than acute, insults. As apoptosis can be induced by a variety of insults and is mediated by several pathways, diverse environmental and genetic strategies to limit cell death have been proposed (for a review, see Ref. [40]). There are a number of perceived advantages from increasing cell robustness by engineering apoptosis resistance. These include increased space-time yields for viable biomass with a concomitant increase in product concentrations, enhanced survival in nutrient limited conditions, and more efficient clarification of the feedstock prior to downstream processing.

Since apoptosis can be induced by nutrient deprivation, one approach to limit its extent is to prevent nutrient limitation. The use of fed-batch operations can delay the onset of apoptosis in GS-NS0 cell lines and substantially reduce its extent [3, 41], thereby increasing the IVC. However, the use of a fed-batch process does not completely eliminate apoptosis. Another approach to increasing process productivity is to engineer resistance to apoptosis into the cell lines.

As activation of the apoptotic pathways results in destruction of the cell, the pathways must be tightly regulated. The best understood regulatory mechanism involves the Bcl-2 family of proteins. Some members of the Bcl-2 family stimulate apoptosis (e.g., Bax, Bak and Bid), whilst others have an anti-apoptotic function (e.g., Bcl-2 and Bcl-x_L). Bcl-2 family members have been postulated to inhibit apoptosis by a number of mechanisms [42, 43].

The anti-apoptotic properties of Bcl-2 family members have been used to protect industrially important cell lines, including GS-NS0 and GS-CHO cells [44, 45], from insults typically experienced during cell culture operations. Interestingly, although Tey et al. [45] report that over-expression of Bcl-2 protects a GS-NS0 cell line against apoptosis, Murray et al. [46] reported no benefit in another GS-NS0 cell line. These workers found that this NS0 cell line expressed Bax and Bcl-x_L. Given that Bcl-x_L is a sequence and functional homologue of Bcl-2, they postulated that Bcl-2 is redundant in the NS0 cell background [46]. These authors postulate further that cell lines such as NS0 express only a subset of genes important in apoptosis. Modulation of death characteristics in such cells will have to take account of the expression profile of such genes and their regulatory interactions.

A number of authors also report an increase in product concentration achieved in cultures of the Bcl-2 over-expressing cell lines compared to the control cell line [45, 47, 48]. A fed-batch culture of the Bcl-2 over-expressing GS-NS0 cell line 6A1-bcl2 made more antibody than the parental cell line 6A1(100)3: the antibody concentration increased from about 26 mg L^{-1} at harvest to about 38 mg L^{-1} [45]. Again, there are also reports of no benefit [44]: the antibody concentration achieved by a GS-CHO over-expressing Bcl-2 was similar to that of the control cell line at about 40 mg L^{-1}, although differences in growth kinetics were seen.

At least one alternative to over-expressing Bcl-2 family proteins has also been evaluated in NS0 cell lines. Studies in an antibody-producing GS-NS0 cell line using the specific inhibitor Z-VAD-fmk, which targets a range of caspases, showed that although the extent of apoptosis was reduced there was no benefit to productivity [49].

The data from studies of over-expressing Bcl-2 in GS-NS0 cell lines are contradictory, whilst no improvement in antibody concentration was seen with the use of caspase inhibitors. These observations, coupled with the complexity of the circuits controlling apoptosis, suggest that apoptosis will have to be modulated at several sites simultaneously if a substantial increase in product concentration is to be achieved. However, this will increase the metabolic load upon the cell.

Most of the studies of Bcl-2 over-expressing cell lines are limited in that, although they used industrially important cell lines, the systems used were only simple models of modern biopharmaceutical manufacturing processes. Characteristics of modern commercial cell culture processes include the use of serum- or protein-free media and feeding strategies that support high viable cell concentrations (ca. 10^7 mL^{-1}) for extended periods (more than 240 h). In contrast, the media used in the reports described above often contained serum and were not highly developed. The development of media, feeds and processes may have eliminated or postponed the appearance of the insults that trigger apoptosis. For example, we [50] have evaluated the Bcl-2 over-expressing GS-NS0 cell line 6A1-bcl2 [45] in a scale-down model (10 L) of a state-of-the-art fed-batch process used to manufacture therapeutic antibodies at the 5000-L scale. Again, as reported by Tey et al. [45], over-expression of Bcl-2 resulted in a substantial increase in the space-time yield of viable biomass and protects against apoptosis. However, unlike the results of Tey et al. [45] with the same cell lines, no improvement in antibody concentration was seen, with both cell lines producing 500 to 700 mg L^{-1} antibody.

4.4.2
Manipulating the Cell Cycle to Increase Productivity

Studies with hybridomas [51, 52] showed an inverse correlation between growth rate and specific antibody production rate. Methods to achieve growth arrest whilst maintaining high viability therefore have potential for improving specific production rates. Thus, an ideal production process would involve a period of rapid cell growth to a high viable cell concentration, with the cells in a physiological state capable of maintaining a high specific production rate but with a low death rate. This phase is then preserved by induction of a sustained growth arrest. It is hypothesized (e.g. [53]) that the cell diverts metabolism from growth-associated processes to maintenance processes, which include the synthesis of constitutively expressed recombinant proteins. This section describes approaches evaluated with GS cell lines to arrest growth.

The cell cycle and cell proliferation are controlled by the activity of cyclin-dependent kinases (cdks) (for a review, see Ref. [54]). The cdks are activated by association with cyclin regulatory subunits and phosphorylation, and inhibited by binding of inhibitors such as $p21^{CIP1}$ and $p27^{KIP1}$. The inhibitor $p21^{CIP1}$ inhibits cdk2, which is known to have a role in the G_1/S transition: over-expression of $p21^{CIP1}$ in a variety of cell lines results in G_1-phase cell cycle arrest.

Al-Rubeai and co-workers have investigated the effect of expressing $p21^{CIP1}$ in both GS-CHO and GS-NS0 cell lines [55, 56]. In one study, an antibody-producing GS-CHO cell line was engineered to express inducibly the $p21^{CIP1}$ cdk inhibitor [56]. Upon induction, cell growth was arrested and the specific production rate increased, the largest increase being from about 60 pg to about 250 pg per cell·day. However, the induced cells actually produced less antibody than the non-induced cells, most likely because the loss in viable biomass outweighed the increase in specific production rate. If $p21^{CIP1}$ was induced at higher cell concentrations (above ca. 5×10^5 mL^{-1}), cell death was observed. In-

duction of apoptosis in growth-arrested cells is a possibility, and it has previously been shown for the parental GS-CHO cell line that growth arrest induced apoptosis that could be protected against by over-expression of Bcl-2 [44].

Over-expression of both a cdk inhibitor and an anti-apoptosis protein has been evaluated with a GS-NS0 cell line. When cell line 6A1(100)3 was engineered to express $p21^{CIP1}$ from an inducible promoter, the specific production rate increased by up to 1.5- to 4.5-fold to 35 to 45 pg per cell·day [55]. However from the data presented, it can be inferred that, overall, there was no increase in volumetric productivity. The GS-NS0 cell line 6A1(100)3 has also been engineered to constitutively express a mutant Bcl-2 with $p21^{CIP1}$ under the control of an inducible promoter [53]. Again, an increase in specific production was seen from about 10 pg per cell·day, which is similar to the parent 6A1(100)3, to about 50 pg per cell·day. Examination of the growth curve data for batch cultures of these cell lines again suggests that the loss of viable biomass outweighs the increase in specific production rate so that there was no benefit to volumetric productivity. Interestingly, the choice of Bcl-2 gene used to transfect the GS-NS0 parent had a profound affect upon the degree of protection against apoptosis. Previous studies [44] used the wild-type protein and saw an increase in IVC compared to the non-transfected parent, 6A1(100)3. When a mutant Bcl-2 protein that lacks any cycle activity was introduced into cell line 6A1(100)3, no increase in the IVC was observed [53].

The "rational design" approaches to improving the phenotype are based upon the direct manipulation of the transcriptome through control of specific genes. The problem with such approaches is that the

"...profile of an ideal cell depends on a multitude of genes that are rather poorly understood, mostly unknown, and broadly distributed throughout the genome" [57]. Although we may be able to manipulate genes (e.g., *Bcl-2* or *p21^{CIP1}*) that are known to have a major role in regulating complex pathways, the impact of these changes upon other complex pathways – for example, the synthesis and secretion of a recombinant antibody – cannot be predicted. This can be seen when the impact of over-expression of Bcl-2 is examined. Improvements in volumetric productivity were seen for some cell lines in some cultures systems, but not others. Thus, there still appears to be a role for the "classical" strain improvement methods where cells with desired phenotypes are isolated from a mutagenized population.

Cell cycle mutants – especially temperature-sensitive (*ts*) ones – are a good source of cell lines in which progression through the cell cycle can be reversibly arrested. Typically, these cell lines have the potential to maintain high viability for extended periods. Jenkins and Hovey [58] isolated *ts*-mutants from CHO-K1 and engineered these mutants to express TIMP using a GS expression vector. Optimization of temperature control was investigated by repeatedly exposing the culture to the non-permissive temperature (39 °C), with recovery at the permissive temperature (34 °C). The concentration of TIMP increased from 200 to 300 mg L^{-1} due to a 3-fold increase in specific production rate to 3.4 pg per cell·day, with growth arrest and no loss of culture viability.

4.4.3
Summary

In summary, a number of cell engineering approaches have been evaluated with GS cell lines to improve volumetric productivity. These approaches have included uncoupling cell growth from productivity and increasing the space-time yield of viable biomass by decreasing the death rate. Some of these approaches resulted in an increase in the specific production rate, but no increase in the volumetric production rate was seen: the latter is the key parameter for a commercial manufacturing organization. A few of these approaches have been evaluated in state-of-the-art manufacturing processes for biopharmaceutical proteins, where different results to those obtained in laboratory studies were obtained. The reasons for this are not clear, but they may be due to the elimination of apoptosis triggers during process optimization.

4.5
Selection of Useful Cell Sub-populations

In addition to metabolic engineering, it is sometimes possible to isolate useful subpopulations of cell lines.

A limitation in the use of CHO cell lines for producing biopharmaceutical proteins has been the long time it can take to adapt such cell lines to single cell suspension culture in serum- or protein-free media. A variant of the CHO-K1 cell line that grows spontaneously in protein-free suspension culture has been described for use with the GS system [59]. The isolation of natural variants has also been exploited to isolate an NS0 clone which no longer requires cholesterol [60]. This nutrient is insoluble and its addition to protein-free media is not straightforward.

4.6
Process Development

4.6.1
Media

In recent years there has been a drive to remove serum, serum proteins and other animal-derived materials from cell culture media, motivated in large part by concerns regarding the potential introduction of adventitious agents. The removal of complex additions such as proteins offers other advantages; particularly cost reduction and easier purification of product. In addition, chemical definition of the medium greatly assists process optimization.

Serum and serum proteins have diverse functions which are now reasonably well understood for the industrially important cell lines, and which can generally be substituted by non-protein alternatives. Mammalian cells typically require a source of fatty acids, which were historically supplied by serum. To supply these, serum-free media usually contain plasma lipoprotein fractions, free fatty acids complexed to serum albumin or fatty acid/phospholipid microemulsions [61]. A high-density lipoprotein serum-fraction in medium containing bovine serum albumin was used by Seamans et al. [62] to replace serum in cultures of a recombinant antibody-producing GS-NS0 cell line. Further, they found that the serum-fraction could be replaced with a commercially available non-proteinaceous lipid emulsion and a pluronic F-68/cholesterol emulsion. This gave equivalent growth and productivity (100 mg L^{-1}).

The requirement for cholesterol supplementation for the serum-free culture of NS0 cells is thought to be a function of their ancestry as they are derived from the NS-1 cell line. The NS-1 cell line is deficient in 3-ketosteroid reductase activity, which is responsible for the conversion of lathosterol to cholesterol, and leads to a requirement for cholesterol [63]. However, the requirement for cholesterol can be circumvented. Birch et al. [60] successfully isolated cholesterol-independent variants of the NS0 host. They achieved this by dilution cloning in a medium that was free of both serum and cholesterol. One of these variants was able to grow in protein-free chemically defined medium, without the addition of any lipids, and with a population doubling time equivalent to the parental cell line. In contrast, without cholesterol supplementation the original NS0 cell line died within 24 hours. Keen and Hale [64] adapted an antibody-producing GS-NS0 cell line to grow in the absence of cholesterol. This removed the final animal-derived raw material from their medium, which was further improved by elevating the concentration of glutamate, asparagine, ribonucleosides, and choline chloride.

Iron delivery to cells in culture needs careful consideration: transferrin has been used successfully for many years, human transferrin being more effective than bovine [65]. However, this is still an animal-derived raw material and thus undesirable for biopharmaceutical manufacturing. Some cells do not need an iron carrier and can be supplied with soluble iron compounds such as ferric ammonic citrate [66]. In other cases an iron carrier may be required such as the synthetic lipophilic iron carrier tropolone [67, 68].

4.6.2
Glutamine-free Media

Aside from its use as a selectable marker, there are physiological advantages in introducing GS to remove the glutamine dependence of cells. Glutamine is relatively unstable in culture media and degrades to release ammonia, which can accumulate to inhibitory levels. Several studies have described the metabolic engineering of hybridoma cell lines with GS to achieve glutamine prototrophy [60, 69, 70]. Birch et al. [60] demonstrated that a hybridoma transfected with GS had increased antibody productivity when grown in the absence of glutamine.

4.6.3
Culture Conditions

It is usual to control pH, dissolved oxygen, and temperature in bioreactors. Small changes in pH can have dramatic effects on process performance. Wayte et al. [71] compared the effect of pH on a GS-NS0 and a hybridoma in shake-flask culture. Using this approach, they found that the specific growth rate was relatively constant over the range pH 7.05 to 7.4, but that below this pH culture growth was significantly inhibited. In fed-batch bioreactor cultures the response of different cell lines to culture pH was variable. One hybridoma had an increased IVC and a lower specific production rate at low culture pH, whilst a second hybridoma showed an increased IVC and no change in specific production rate. However, in both cases, decreasing the culture pH from 7.2 to 7.1 caused an increase in the harvest antibody concentration. The GS-NS0 cell line examined in this study was less sensitive to culture pH than the hybridomas, and larger changes in culture pH were needed to affect the culture. At pH 7.1, both the IVC and the specific production rate were increased compared to pH 7.4, resulting in an increase in antibody concentration from 119 mg L^{-1} to 194 mg L^{-1}.

Osman et al. [72] investigated the effect of pH shifts and perturbations in cultures of the antibody-producing GS-NS0 cell line,

6A1(100)3, cultured in serum containing batch culture. Cells growing at pH 7.3 were able to continue growing after a shift in culture pH in the range of pH 7.0 to 8.0. A shift in culture pH of greater than 0.2 pH units caused a transient increase in the proportion of apoptotic cell in the culture, but the cultures were able to recover from this. However, cultures were not able to recover if the pH was decreased below 7.0 or increased above 8.0. The culture pH affected both growth and metabolism. The antibody concentration was highest at pH 7.0 as a result of increased IVC, whilst the specific production rate was constant over the relatively wide pH range of 6.5 to 8.0. The authors also investigated the effect of transient shifts in culture pH which could potentially occur as a result of zoning in large-scale reactors as a result of, for example, alkali addition to poorly mixed areas. Transient shifts had to be quite large to have an affect on growth. Increases of culture pH to above 8.5 for longer than 10 minutes induced a lag and caused a reduction in the maximum viable cell concentration. Similar effects were seen at low pH (below pH 6.5), but the perturbation needed to be for several hours.

Using a GS-NS0 producing an IgG$_1$ in protein-free fed-batch culture, Moran et al. [73] investigated the effect of a range of parameters on the growth rate, specific production rate, IVC and antibody concentration at harvest. In contrast to the results of Wayte et al. [71] and Osman et al. [72], there was no statistically significant effect on any of these parameters within the range of culture pH from 7.1 to 7.5. More importantly, there was no detectable change in the distribution of glycoforms of the antibody. It is not clear why there are such differences but it may depend upon the particular cell line as well as the process.

4.6.4
Fed-batch Cultures

The early mammalian cell processes were typically batch. As culture media and processes have developed over the years, advances in feeding strategies for fed-batch processes have increased productivity to several grams per liter for both GS-NS0 [4] and GS-CHO cell lines [5].

One approach to implementing a fed-batch strategy is to feed cultures with medium concentrates. This can offer a rapid approach to increasing productivity, and can also be relatively simple to implement [17, 74]. Using GS-NS0 cells producing an antibody, Bibila et al. [17] fed cultures with 10× basal medium concentrates (Iscove's Modified Dulbecco's medium) to increase productivity. Sodium chloride, potassium chloride and sodium bicarbonate were omitted from the medium concentrates in order to minimize the increases in osmolarity caused by feeding. In their system, feeding basal medium concentrates did not result in an increase in the maximum viable cell concentration or the IVC. However, the final antibody concentration was increased 1.9-fold as a result of an increase in the specific production rate.

A further refinement of the fed-batch method is to feed the supplements added to the medium in addition to the basal medium concentrates [17]. This approach was shown to be more effective than concentrates alone, and led to increases in the maximum viable cell concentration (1.7- to 2-fold), the IVC (2.3- to 3.3-fold) and the specific production rate (2-fold). These effects combined to produce an up to a 7-fold increase in antibody concentration. Further increases might be expected by feeding more nutrients, though above a certain volume of additions a decrease in process performance was observed. This was thought to be the result of increases in osmolarity caused by the medium components. Information on metabolism gained from the medium concentrate experiments was then used to develop an optimized fed-batch process. The strategy chosen for this was to maintain nutrient homeostasis, where the amino acid concentrations were maintained at their original concentrations and the culture was supplemented with glucose, lipids and proteins. Further development of the fed-batch process required "significant process development time and effort". However, this led to product concentration at harvest of 1.8 and 1.2 g L^{-1}.

The effect of medium osmolarity on the growth of GS-NS0 cells was investigated by Bibila et al. [17]. Cell growth was reduced when the osmolarity was increased to 400 mOsm and completely inhibited above 500 mOsm. The specific production rate increased as the osmolarity was increased from the baseline of 270 mOsm to 300 and 400 mOsm. However, as a direct result of reduced growth, the cultures at 400 mOsm reached a lower product concentration than the controls. Zhou et al. [2] noted that increases in osmolarity below 450 mOsm had little impact upon productivity, but above this level there was a rapid increase in the specific production rate. However, growth cessation occurred at this elevated osmolarity.

Zhou et al. [2] refined the nutrient homeostasis approach further by feeding cultures based on the IVC, with the aim of keeping nutrient concentrations around their original concentrations. However, this assumes that the consumption and yields of these nutrients are constant throughout the culture, which may not be correct. On-line measurement of the oxygen uptake rate (OUR) was used to infer nutrient depletion. Rapid decreases in OUR were observed that could be reversed by addition of amino

(a)

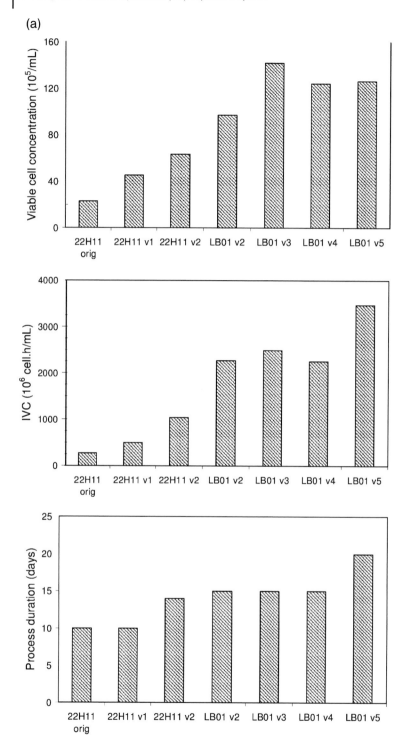

(b)

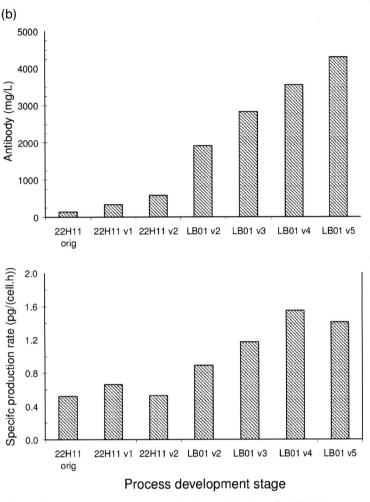

Fig. 4.6 Changes in process parameters during optimization of a GS-CHO process producing an IgG$_4$ antibody using chemically defined animal-component free media in 10-L laboratory-scale airlift bioreactors: (a) growth parameters; (b) productivity parameters.

acids. This did not result in an increased cell concentration, indicating that another nutrient was limiting or that some factor had accumulated to a growth-inhibitory level. The addition of an increased amount of a cholesterol complex in conjunction with the amino acid feed was able to restore growth and increased the product concentration to 2.7 g L^{-1}. Despite responsive feeding based on the OUR, it was not possible to maintain growth indefinitely. It was a reduction in the cell death rate that resulted in a prolonged culture lifetime. During the decline phase there was a much slower linear decrease in the viable cell concentration and OUR, rather than the rapid decreases in OUR observed previously. The authors suggested that this indicated that cell death might be caused by environmental conditions such as high osmolarity rather than nutrient limitation. One drawback noted was that although no base was used for pH control, because the amino acid solution had a pH of 9.5, feeding ultimately resulted in a rise in the CO_2 concentration in the reactor as a result of maintaining the culture pH set point. As nutrient metabolism changes through the different growth phases, they proposed a two-feed strategy where one feed is used to extend the cell growth phase, after which a different feed is used to prolong culture longevity.

deZengotita et al. [3], using the GS-NS0 cell line described by Zhou et al. [2], found that feeding phosphate prolonged the cell growth phase and delayed the onset of apoptosis, resulting in a doubling of the maximum viable cell concentration. An increased IVC resulted in an increase in the product concentration at harvest from 0.5 to 1.3 g L^{-1}. This also delayed the metabolic shift from lactate production to lactate consumption.

Sauer et al. [74] discussed the need for high-yielding generic fed-batch processes to decrease the amount of development time prior to manufacturing. They used a similar approach to that of Bibila et al. [17], starting with partial media concentrates and initially controlled feed additions based on glucose concentration. For an Sp2/0 cell line producing an antibody, this led to a 3-fold increase in product concentration, from 70 to 220 mg L^{-1}. By changing the glucose concentration at which feeds were added, it was possible to demonstrate the effect of underfeeding, comparable to batch culture, and overfeeding: both conditions showed a reduction in the final antibody concentration. That the feeding regime was robust was demonstrated by the range of glucose concentration over which the process could operate without adversely affecting process performance. To test the general applicability of the process it was tested on a panel of cell lines. In each case, compared to batch culture, feeding increased both the exponential growth phase

Table 4.2 Oligosaccharide profiles determined by MALDI-TOF MS for a GS-NS0 IgG$_4$ antibody during process optimization in chemically defined animal component-free fed-batch culture

Structure	Relative Peak Intensity [%]				
	0.37 g L^{-1}	0.48 g L^{-1}	0.75 g L^{-1}	1.0 g L^{-1}	1.4 g L^{-1} [a]
G2F+2 (α-Gal)	4.0	3.5	3.3	2.4	3.1
G2F+(α-Gal)	8.4	6.8	7.6	6.0	6.0
G2F	39.1	43.2	41.8	41.9	40.4
G1F	34.7	32.1	33.7	32.7	37.8
G0F	9.1	9.9	10.5	12.6	12.7
G1F-GN	1.5	1.6	1.2	1.7	ND
G0	0.7	0.5	0.8	0.6	ND
G0F-GN	1.5	1.5	1.2	1.4	ND
Man-5	1.0	0.9	0.0	0.9	ND

a) Analyzed in separate assay to other samples.
  ND = Not detected.

and the culture duration. Most of the increase in product concentration was a result of an increased IVC rather than increased specific production rate. Between the different cell lines there were marked differences in the specific glucose consumption rate, up to a factor of 4-fold, whilst the apparent yield of lactate on glucose was relatively unchanged. Interestingly, there was an inverse correlation between specific glucose utilization rate and IVC.

Similar improvements in productivity were obtained by Dempsey et al. [68], who performed repeated rounds of nutrient supplementation and analysis to develop nutrient supplements for their GS-NS0 cultures. They tested these supplements on three cell lines producing different antibodies, and attained a 10-fold increase over the original product concentrations.

Shaw et al. [24] showed that the chemically defined animal component-free process they developed using the GS-NS0 cell line 6A1(100)3 was applicable to other cell lines. Using a different cell line that was making 1 g L^{-1} in a serum-free process, with no optimization for this second cell line, an antibody concentration of 1.8 g L^{-1} was attained. This has subsequently been confirmed with other cell lines producing above 1 g L^{-1} (unpublished results).

Process optimization using our model GS-CHO cell line (22H11) was achieved using multiple rounds of fermentations in chemically defined media (Fig. 4.6). The initial optimization was performed by changing the base medium, and the feeds were modified using the approach of spent medium analysis and re-supplementation. This increased the yield from 139 mg L^{-1} to 585 mg L^{-1} – a 4-fold increase in productivity. This optimized process was then used as the starting point for a new, non-amplified GS-CHO cell line (LB01), and this resulted in a 14-fold increase over the

original process, to 1917 mg L^{-1}. Further process optimization was then performed using the new cell line. The progress of the optimization is shown in Fig. 4.6. For iterations 4 and 5 (compare LB01 v4 and LB01 v5 in Fig. 4.6), the pH control was optimized, which resulted in a further improvement in productivity to 4301 mg L^{-1}, a 31-fold increase over the original process. It is apparent from the data shown in Fig. 4.6 that it is possible to improve productivity by optimising several parameters, namely specific production rate, IVC, and maximum viable cell concentration.

4.6.5
Process Optimization and Product Quality

One of the concerns with increasing the product concentration is that the product quality characteristics are maintained. We have monitored the product quality of a GS-NS0 cell line throughout the optimization process. Through successive rounds of optimization involving changes in the composition of the feeds, culture pH and extending culture duration, the product concentration from the GS-NS0 process was increased from 0.37 to 1.4 g L^{-1}. There were no major changes observed in the oligosaccharide profiles during this optimization process (Table 4.2). It cannot however be assumed that changes will not occur, and it is essential to monitor product quality during process development. For example, we found an increased proportion of an aglycosyl variant of an antibody produced in GS-NS0 during process optimization. This was shown to be a result of glucose becoming limiting under revised feeding conditions.

4.7
Summary

Significant progress has been made in recent years in the development of high-yielding processes for the production of biopharmaceuticals. Highly efficient non-amplified gene expression systems such as that based on glutamine synthetase, in combination with new approaches to screening, have provided highly productive cell lines. We can expect to see further improvements to cell lines resulting from deliberate engineering of desirable characteristics. Improved understanding of cell physiology using modern "omics" tools will contribute significantly to these efforts, and we are already seeing the first indications of this [75]. In parallel with these developments in the design of cell lines, we have also seen impressive progress in the optimization of culture processes, particularly through the use of sophisticated feeding strategies for fed-batch culture. For recombinant antibodies it is now possible regularly to achieve yields in excess of 1 g L^{-1} in completely chemically defined media, and it is probable that yields for modern biopharmaceuticals of at least 10 g L^{-1} will be achieved in the foreseeable future.

References

1 C. R. Bebbington, G. Renner, S. Thomson, D. King, D. Abrams, G. T. Yarranton, *Bio/Technol.* **1992**, 10, 169–175.

2 W. Zhou, C-C. Chen, B. C. Buckland, J. G. Aunins, *Biotechnol. Bioeng.* **1997**, 55, 783–792.

3 V. M. deZengotita, W. M. Miller, J. G. Aunins, W. Zhou, *Biotechnol. Bioeng.* **2000**, 69, 566–576.

4 Varma, A. E. Schmelzer, S. White, M. Patel, C. DiGirolamo, S. Case. *The Waterside Conference, 8th Annual Conference, San Francisco (Abst).* **2003**.

5 D. O. Mainwaring, *227th American Chemical Society Meeting, Anaheim (Abst).* **2004**.

6 P. G. Sanders, R. H. Wilson, *EMBO J.* **1984**, 3, 65–71.

7 C. R. Bebbington, *Methods: A Companion to Methods in Enzymology.* **1991**, 2, 136–145.

8 M. I. Cockett, C. R. Bebbington, G. T. Yarranton, *Bio/Technology* **1990**, 8, 662–667.

9 D. R. Burton, J. Pyati, R. Koduri, S. J. Sharp, G. B. Thornton, P. W. H. I. Parren, L. S. W. Sawyer, R. M. Hendry, N. Dunlop, P. L. Nata, M. Lanacchia, E. Garratty, E. R. Stiehm, Y. J. Bryson, Y. Cao, J. P. Moore, D. D. Ho, C. F. Barbas, *Science* **1994**, 266, 1024–1027.

10 V. E. Laubach, E. P. Garvey, P. Sherman, *Biochem. Biophys. Res. Commun.* **1996**, 218, 802–807.

11 J. McKnight, B. J. Classon, *Immunology* **1992**, 75, 286–292.

12 D. Guerini, S. Schroder, D. Foletti, E. Carafoli, *J. Biol. Chem.* **1995**, 270, 14643–14650.

13 R. Flesher, J. Marzowski, W.-C. Wang, H. V. Raff, *Biotechnol. Bioeng.* **1995**, 46, 399–407.

14 K. N. Baker, M. H. Rendall, A. E. Hills, M. Hoare, R. B. Freedman, D. C. James, *Biotechnol. Bioeng.* **2001**, 73, 188–202.

15 M. R. Lifely, C. Hale, S. Boyce, M. J. Keen, J. Phillips, *Glycobiology* **1995**, 5, 813–822.

16 T. E. Hassell, H. N. Brand, G. L. Renner, A. J. Westlake, R. P. Field, In: *Animal Cell Technology: Developments, Processes and Products* (R. E. Spier, J. B. Griffiths, C. MacDonald, Eds.), Butterworth-Heinemann, UK, pp. 42–47, **1992**.

17 T. A. Bibila, C. Ranucci, K. Glazomitsky, B. C. Buckland, J. G. Aunins, *Ann. N. Y. Acad. Sci.* **1994**, 745, 277–284.

18 T. C. Peakman, J. Worden, R. H. Harris, H. Cooper, J. Tite, M. J. Page, D. R. Gewert, M. Bartholomew, J. S. Crowe, S. Brett, *Hum. Antibod. Hybrid.* **1994**, 5, 65–74.

19 G. J. Pendse, S. Karkare, J. E. Bailey, *Biotechnol. Bioeng.* **1992**, 40, 119–129.

20 M. B. Gu, P. Todd, D. S. Kompala, *Cytotechnology.* **1996**, 18, 159–166.

21 U. H. Weidle, P. Buckel, J. Wienberg, *Gene* **1988**, 66, 193–203.

22 M. I. Cockett, C. R. Bebbington, G. T. Yarranton, *Nucleic Acids Res.* **1991**, 19, 319–325.

23 M. E. Reff, Internl. Pat. WO 94/11523, **1994**.

24 M. Shaw, D. O. Mainwaring, T. S. Root, E. E. Allen, *Cell Culture Engineering XIV, Cancún (Abst).* **2004**.

25 R. S. Barnett, K. L. Limoli, T. B. Huynh, E. A. Ople, M. E. Reff, In: *Antibody Expression and Engineering* (H. Y. Wang, Ed.), American Chemical Society, USA, pp. 27–40, **1995**.

26 G. F. Hollis, G. E. Mark, Internl. Pat. WO95/17516, **1995**.

27 T. Benton, T. Chen, M. McEntee, B. Fox, D. King, R. Crombie, T. C. Thomas, C. R. Bebbington, *Cytotechnology* **2002**, 38, 43–46.

28 J. Running Deer, D. S. Allison, *Biotechnol. Prog.* **2004**, 20, 880–889.

29 K. T. Powell, J. C. Weaver, *Bio/Technology* **1990**, 8, 333–337.

30 S. C. G. Brezinsky, G. G. Chiang, A. Szilvasi, S. Mohan, R. I. Shapiro, A. MacLean, W. Sisk, G. Thill, *J. Immunol. Methods* **2003**, 277, 41–155.

31 R. Manz, M. Assenmacher, E. Pflüger, S. Miltenyi, A. Radbruch, *Proc. Natl. Acad. Sci. USA* **1995**, 92, 1921–1925.

32 P. Holmes, M. Al-Rubeai, *J. Immunol. Methods* **1999**, 230, 141–147.

33 A. J. Racher, *Third Annual Biological Production Forum, Edinburgh (Abst.).* **2004**.

34 H. N. Brand, S. J. Froud, H. K. Metcalfe, A. O. Onadipe, A. Shaw, A. J. Westlake, In: *Animal Cell Technology: Products for Today, Prospects for Tomorrow* (R. E. Spier, J. B. Griffiths, W. Berthold, Eds.), Butterworth-Heinemann Ltd, UK, pp. 55–60, **1994**.

35 L. Cosgrove, G. O. Lovrecz, A. Verkuylen, L. Cavaleri, L. A. Black, J. D. Bentley, G. J. Howlett, P. P. Gray, C. W. Ward, N. M. McKern, *Prot. Purificat.* **1995**, 6, 789–798.

36 P. Bird, E. Bolam, L. Castell, O. Obeid, N. Darton, G. Hale, In: *New Developments and New Applications in Animal Cell Technology* (O.-W. Merten, P. Perrin, J. B. Griffiths, Eds.), Kluwer Academic Publishers, The Netherlands, pp. 43–49, **1998**.

37 L. M. Barnes, C. M. Bentley, A. J. Dickson, *Biotechnol. Bioeng.* **2001**, 73, 261–270.

38 L. M. Barnes, C. M. Bentley, A. J. Dickson, *Biotechnol. Bioeng.* **2003**, 85, 115–121.

39 M. Al-Rubeai, D. Mills, A. N. Emery, *Cytotechnology* **1990**, 4, 13–28.

40 N. Arden, M. J. Betenbaugh, *Trends Biotechnol.* **2004**, 22, 174–180.

41 D. J. DiStefano, G. E. Mark, D. K. Robinson, *Biotechnol. Lett.* **1996**, 18, 107–1072.

42 H. Zou, W. J. Henzel, X. Liu, A. Lutschg, X. Wang, *Cell.* **1997**, 90, 405–413.

43 D. G. Kirsch, A. Doseff, B. N. Chau, D.-S. Lim, N. C. de Souza-Pinto, M. B. Kastan, Y. A. La-

zebnik, J. M. Hardwick, *J. Biol. Chem.* **1999**, 274, 21155–21161.

44 B. T. Tey, R. P. Singh, L. Piredda, M. Piacentini, M. Al-Rubeai, *Biotechnol. Bioeng.* **2000**, 68, 31–43.

45 B. T. Tey, R. P. Singh, L. Piredda, M. Piacentini, M. Al-Rubeai, *J. Biotechnol.* **2000**, 79, 147–159.

46 K. Murray, C.-E. Ang, K. Gull, J. A. Hickman, A. J. Dickson, *Biotechnol. Bioeng.* **1996**, 51, 298–304.

47 S. Terada, T. Komatsu, T. Fujita, A. Terakawa, T. Nagamune, S. Takayama, J. C. Reed, E. Suzuki, *Cytotechnology* **1999**, 31, 141–149.

48 N. S. Kim, G. M. Lee, *Biotechnol. Bioeng.* **2001**, 71, 184–193.

49 S. L. McKenna, T. G. Cotter, *Biotechnol. Bioeng.* **2000**, 67, 165–176.

50 Perani, J. Kenworthy, I. Alam, G. Gilchrist, H. K. Metcalfe, J. R. Birch, A. J. Racher, *Second European Meeting on Cell Engineering*, Costa Brava, Spain (Abst), **2001**.

51 W. M. Miller, H. W. Blanch, C. R. Wile, *Biotechnol. Bioeng.* **1988**, 32, 947–965.

52 E. Suzuki, D. F. Ollis, *Biotechnol. Prog.* **1990**, 6, 231–236.

53 N. Ibarra, S. Watanabe, J.-X. Bi, J. Shuttleworth, M. Al-Rubeai, *Biotechnol. Prog.* **2003**, 19, 224–228.

54 M. Fussenegger, J. E. Bailey, *Biotechnol. Prog.* **1998**, 14, 807–833.

55 S. Watanabe, J. Shuttleworth, M. Al-Rubeai, *Biotechnol. Bioeng.* **2002**, 77, 1–7.

56 J.-X. Bi, J. Shuttleworth, M. Al-Rubeai, *Biotechnol. Bioeng.* **2004**, 85:741–749.

57 G. Stephanopoulos, *Nature Biotechnol.* **2002**, 20, 666–668.

58 N. Jenkins, A. Hovey, In: *Animal Cell Technology: Basic and Applied Aspects. Vol. 5.* Kluwer Academic Publishers, The Netherlands, pp. 267–272, **1993**.

59 M. H. Rendall, A. Maxwell, D. Tatham, P. Khan, R. D. Gay, R. C. Kallmeier, J. R. T. Wayte, A. J. Racher, In: Animal Cell Technology Meets Genomics (F. Gòdia, M. Fussenegger, Eds.) Springer, The Netherlands, pp. 701–704, **2005**.

60 J. R. Birch, R. C. Boraston, H. K. Metcalfe, M. E. Brown, C. R. Bebbington, R. P. Field, *Cytotechnology.* **1994**, 15, 11–16.

61 S. Schmidt, In: *Production of Biologicals from Animal Cells in Culture* (R. E. Spier, J. B. Grif-

fiths, B. Meigner, Eds.) Butterworth-Heinemann, UK, pp. 67–69, **1991**.

62 T.C. Seamans, S.L. Gould, D.J. DiStefano, M. Silberklang, D.K. Robinson, *Ann. N.Y. Acad. Sci.* **1994**, 745, 240–243.

63 J.D. Sato, T. Kawamoto, T. Okamoto, *J. Exp. Med.* **1987**, 165, 1761–1766.

64 M.J. Keen, C. Hale *Cytotechnology* **1996**, 18, 207–217.

65 T.O. Messmer, *Biochim. Biophys. Acta* **1973**, 320, 663–670.

66 M. Titeux, U. Testa, F. Louache, P. Thomopoulos, H. Rochant, J. Breton-Gorius, *J. Cell Physiol.* **1984**, 121, 251–256.

67 H.K. Metcalfe, R.P. Field, S.J. Froud, In: *Production of Biologicals from Animal Cells in Culture* (R.E. Spier, J.B. Griffiths, B. Meigner, Eds.), Butterworth-Heinemann, UK, pp. 88–90, **1994**.

68 J. Dempsey, S. Ruddock, M. Osborne, A. Ridley, S. Sturt, R.P. Field, *Biotechnol. Prog.* **2003**, 19, 175–178.

69 S.L. Bell, C.R. Bebbington, M.F. Scott, J.N. Wardell, R.E. Spier, M.E. Bushell, P.G. Sanders, *Enzyme Microb. Technol.* **1995**, 17, 98–106.

70 C. Paredes, E. Prats, J.aJ. Cairó, F. Azorín, L. Cornudella, F. Gòdia, *Cytotechnology* **1999**, 30, 85–93.

71 J.R.T. Wayte, R.C. Boraston, H. Bland, J. Varley, M.E. Brown, *Genet. Engineer. Biotechnol.* **1997**, 17, 125–132.

72 J.J. Osman, J.R. Birch, J. Varley, *Biotechnol. Bioeng.* **2001**, 75, 63–73.

73 E.B. Moran, S.T. McGowan, J.M. McGuire, J.E. Frankland, I.A. Oyebade, W. Waller, L.C. Archer, L.O. Morris, J. Pandya, S.R. Nathan, L. Smith, M.L. Cadette, J.T. Michalowski, *Biotechnol. Bioeng.* **2000**, 69, 242–255.

74 P.W. Sauer, J.E. Burky, M.C. Wesson, H.D. Sternard, L. Qu, *Biotechnol. Bioeng.* **2000**, 67, 585–597.

75 C.M. Smales, D.M. Dinnis, S.H. Stansfield, D. Alete, E.A. Sage, J.R. Birch, A.J. Racher, C.T. Marshall, D.C. James, *Biotechnol. Bioeng.* **2004**, 88, 474–488.

Vivat, Crescat, Floreat –
A Ripe and Blooming Market for Transgenic Animals and Plants

5
Biopharmaceuticals Derived from Transgenic Plants and Animals

Julio Baez

Abstract

Technical advances made during the past 20 years have enabled the genetic transformation and regeneration of transgenic plants and animals for the tissue-specific accumulation of recombinant human proteins. These transgenic systems provide production technology for biopharmaceuticals requiring complex multi-subunit assembly (e.g., vaccines and secretory antibodies) and for proteins that can not be efficiently synthesized by current commercial blood fractionation microbial mammalian cell culture systems. The manufacture of biotherapeutics in transgenic animals and plants grown using conventional agronomic and farming practices also offers the opportunity to produce practically unlimited supplies of life-saving products at low cost. In addition, the production of biotherapeutics using transgenic systems (e.g., milk) offers the highest accumulation level of heterologous protein accumulation obtained from a recombinant production system. Transgenic plants allow the production of bio- pharmaceuticals free of potential animal-derived contaminants and pathogens such as prions in a matrix that can be used for oral delivery, without additional purification and with no requirement for refrigeration. Seeds provide a stable matrix for handling and storing biopharmaceuticals for years after harvest decoupling downstream processing from biosynthesis. In addition to these attractive advantages, the implementation of transgenic systems for biopharmaceutical production offers the opportunity to improve agricultural efficiency and profitability whilst reinforcing the public perception of biotechnology as an important tool to enhance both agriculture and healthcare. Transgenic systems can deliver innovative biotherapeutics to treat cancer, infectious diseases, inflammation, organ rejection, skin conditions, genetic deficiencies, and respiratory ailments. These biopharmaceuticals will be both affordable and accessible to broad segments of the population and developing regions of the world that currently do not have access to these treatments. This chapter will focus on biopharmaceuticals derived from transgenic animals and plants that are currently commercialized, or that have human clinical experience. Exploring the properties and performance of these transgenic-derived biopharmaceuticals used as diagnostics, protein replacement therapy, cancer therapeutics, immunoprophylactics, as anti-infectives, nutraceuticals, excipients, and in medical devices will provide understanding of the status and of the potential for transgenic-based production systems.

Abbreviations

AAT	alpha-1-antitrypsin
AIDS	aquired immunodeficiency syndrome
AT-III	antithrombin III
BSSL	bile salt-stimulated lipase
DMF	drug master file
EpCAM	epithelial cellular-adhesion molecule
HAE	hereditary angioedema
hBChE	human butyrylcholinesterase
HIV	human immunodeficiency
HSA	human serum albumin
IND	investigational new drug
LTB	labile toxin B
MPS-I	mucopolysaccharidosis
rhAAT	recombinant human alpha-1-antitrypsin
rhAGLU	recombinant human alpha-glucosidase
rhC1I	recombinant human C1 inhibitor
rhFIB	recombinant human fibrinogen
rhLF	recombinant human lactoferrin
rhLZ	recombinant human lysozyme
TMV	tobacco mosaic virus
USDA	US Department of Agriculture
USDA/ APHIS	US Department of Agriculture/Animal and Plants Health Inspection Service

5.1
Introduction

During the past 20 years, the application of recombinant DNA technology to healthcare has enabled the introduction of more than 140 biopharmaceutical products providing innovative diagnostic, preventive, and therapeutic treatment for cancer, cardiovascular disease, diabetes, sepsis, infectious diseases, inflammation, organ rejection, skin ailments, autoimmune conditions, respiratory ailments, genetic deficiencies, and asthma [1]. About 370 biopharmaceuticals are currently undergoing clinical trials, targeting more than 200 diseases [1]. Recombinant human proteins used as biotherapeutics are derived from mammalian cell culture and microbial fermentation. The application of these recombinant production technologies during the past 25 years has delivered many innovative products that have provided new opportunities for growth to the pharmaceutical industry, while making available life-saving diagnostic, prophylactic, and therapeutic approaches to health providers and to patients. Many of these recombinant human products have replaced biologics prepared from animal/human tissues, whilst others have been made available for the first time, as they could not be recovered from natural sources. Microbial and mammalian culture-based recombinant production technologies, supplemented with insect cell culture and solid-phase protein synthesis, have also provided valuable reagents for the discovery, development, and analysis of proteins and non-protein-based new chemical entities used as drugs [2].

In parallel to these efforts to develop recombinant systems for the production of innovative drugs, the same recombinant DNA technology has been applied to improvements in agricultural systems for the successful commercialization of transgenic crop plants and farm animal recombinant products. Taken together, this has resulted in enhanced agronomic performance and productivity. The first wave of agricultural biotechnology products – herbicide/pest-tolerant plants and bovine growth hormone – are currently providing higher profits to farmers and agricultural companies whilst minimizing the negative impact of agricultural activities on the environment. Farmers have quickly embraced recombinant technology when it is available, as illustrated by its successful implementation in the US. By 2004 – just eight years after the introduction of the first commercial pest-resistant crops – 45% of the corn, 85% of soy and 76% cotton fields planted in the US will have genetically enhanced plants [3]. A second wave of genetically enhanced crop plants are on the horizon, these being designed to deliver improved shelf-life and quality food products with higher concentrations of designes health-enhancing oils, proteins, and vitamins. Alternatively, crop plants have been designed for improved agronomic performance to further enhance the value of agricultural biotechnology to society.

The use of transgenic animals and plants as factories for valuable pharmaceutical and industrial products represents the third wave of agricultural biotechnology products derived from genetically enhanced organisms. Recombinant DNA technology allows the accumulation of recombinant human proteins in all tissues of a transgenic organism, or selectively in a particular tissue. Biopharmaceutical accumulation can be directed into conventional agricultural products such as milk, eggs, foliage, fruits, stems, and seeds that are normally harvested from farm animals and crop plants. Directed accumulation of

biopharmaceuticals into these familiar agricultural products facilitates the implementation of transgenic production technology using established agricultural practices. The use of recombinant DNA technology to generate transgenic-derived biopharmaceuticals is a continuation to the historical use of animal and plant tissues and derived products to provide valuable health-enhancing agents. Since the beginning of medicine, human/animal blood, animal tissues, and plants had been the source of many oral, topical, and injectable therapeutics, such as Factor VIII for hemophiliacs, serum albumin used as plasma expander, porcine insulin for diabetes treatment, egg viral vaccines for immunization, therapeutic polyclonal antibodies, steroids, and plant morphine for pain treatment. Of the 100 most frequently prescribed US drugs, about one-fifth are obtained directly from plants; representing products such as birth control pills (Mexican yam), digitalis (foxglove), and recent anticancer therapeutics such as taxol from the Pacific yew tree. Transgenic animals and plants are simply providing innovative ways to enhance the use of animal and plant tissues and derived products to provide new biopharmaceuticals.

The initial implementation of recombinant DNA technology in agriculture to deliver improved agronomic performance to crops and to animals had limited direct impact upon the efficiency and profitability of food production, and did not deliver value-added products to consumers. This created the perception in the food industry and consumers that agricultural biotechnology has no direct value to them. These groups instead focused on the perceived high risk associated with the implementation of agricultural biotechnology, and ignored its potential to improve nutrition and healthcare. This perception of low value/high risk has created significant controversy related to the implementation of products derived from agricultural biotechnology. Valid technical concerns related with the unknown long-term impact of genetic modifications on food safety and the environment combined with questions about the actual cost benefits to the farming community, food industry, consumers and governments providing farming subsidies resulted in the generation of significant opposition to the implementation of agricultural biotechnology, especially outside the US. Extensive testing has been conducted to certify the safety to both consumers and the environment of products derived from genetically enhanced plants and animals. However, no long-term exposure data are available, and this has created the demand by many organizations to slow down the implementation of genetically enhanced food and to provide proper labeling for these products. The use of crop plants and farm animals for the production of biopharmaceuticals will have to be implemented in this controversial environment. Those involved in developing the technology, the food industry, and the regulatory agencies are working together to ensure that the production of biopharmaceuticals using transgenic animals and plants can be implemented without contaminating the food supply or the environment [4].

The first successful therapeutic protein made in a transgenic system was human tissue plasminogen activator regulated by a milk-directed promoter for accumulation in mouse milk [5]. Human growth hormone, which was one of the first proteins produced using recombinant microbial systems in the early 1980s, became the first human protein expressed in plants (tobacco cells) in 1986 [6]. Since then, over 200 biotherapeutics of diverse origin,

structure, and function such as antibodies, enzymes, antigens for vaccines, and hormones have been successfully expressed in tissues capable of being regenerated into transgenic animals and plants. Today, transgenic-derived recombinant human proteins are commercially available for non-human use in research, processing, and as diagnostics. There is no biotherapeutic derived from a transgenic plant or animal approved for human therapeutic use. One product, goat milk-derived injectable recombinant human antithrombin, was submitted for European approval (see Part IV, Chapter 11). Transgenic production strategies, material used for extracting the recombinant product based on these strategies, host organisms, commercial or academic institutions developing these strategies, and product examples are listed in Table 5.1. For transgenic animals, production is conducted in specially built barns designed and managed exclusively for recombinant protein manufacturing. Using transgenic plants, there are two basic strategies for manufacturing plant-derived biopharmaceuticals: 1) production in open fields, greenhouses or underground; and 2) production in bioreactors containing transgenic aquatic plants, cells, or tissues in suspension such as roots, mechanically chopped foliage, or germinated seeds. Biopharmaceuticals are recovered from milk, semen, whole organisms, blood, and eggs using transgenic animals; from foliage (transgenic or using viral infection), tubers, stems, fruits, and seeds using plants. Most farming animals and commercial crops have been used for the production of biopharmaceuticals by over 120 institutions (commercial and academic; see Table 5.1), though several of the commercial establishments listed are no longer operational. Table 5.1 also lists the large number of recombinant proteins

Table 5.1 Production strategies, material used for extraction, host organisms, institutions, and products from transgenic systems.

Production strategy	Material used for extraction	Host organism	Institutions	Product examples	Reference
Transgenic animals housed in specialized barns	Milk	Mouse Rabbit Cow Goat Sheep Pig	Astra, Sweden INRA, France	Human extracellular superoxide dismutase	36
			BioProtein Technologies	Antibodies, vaccines, human C1 inhibitor, erythropoetin, superoxide dismutase	37, 38
			Gala Biotechnology	Not available	39
			GTC Biotherapeutics	Human antithrombin, Human serum albumin, HIV vaccine, malaria vaccine, Monoclonal antibodies, Peptides, Fusion proteins, Beta-interferon, Interferon alpha, Glutamic acid decarboxylase, Human growth hormone, Insulin, Tissue plasminogen activator	9, 40–49
			Infigen	human collagen type I, Human fibrinogen, alpha glucosidase, gelatin	50, 51
			Institut für Tierzucht Tierverhalten, Neustadt, Germany & Fraunhofer, Hannover, Germany	Human factor VIII	52
			Korea Institute of Science and Technology, Taejon	Human granulocyte colony-stimulating factor	53
			Nexia	human butyrylcholinesterase, spider silk	54, 55
			Pharming	C1 esterase inhibitor, fibrinogen, collagen I & II, Lactoferrin, Factor VII, Factor IX	56–58
			PPL Therapeutics	Bile human gastric Lipase, Fibrinogen, thrombin, Factor VII, Factor IX, alpha-antitrypsin, calcitonin (salmon), collagen, superoxide dismutase, Glucagon lipopeptide, Human serum albumin, Protein C	59–63

Table 5.1 (continued)

Production strategy	Material used for extraction	Host organism	Institutions	Product examples	Reference
			Virginia Tech, Blacksburg, VA American Red Cross	Human protein C	64, 65
			Virtanen Institute, University of Kuopio Finland	Human granulocyte-macrophage colony-stimulating factor, human erythropoietin	66
	Blood	Cow Rabbit	Hematech	Polyclonal antibodies	67, 68
			Therapeutic Human Polyclonal Inc.	Polyclonal antibodies	69
	Semen	Pig	TGN Biotech	Human follicle-stimulating hormone	70
	Urine	Mice	Catholic University of Korea, Seoul	Human granulocyte-macrophage colony-stimulating factor	71
			NYU/USDA/U. Vermont	Human growth hormone	72, 73
	Whole animal	Caterpillars Shrimp	Advanced Bionutrition	Not available	74
			Chesapeake PERL	Not available	75
	Organs/Cells	Pig Cow	Advanced Cell Technology	Cell transplantation	76
			Nextran (Baxter)	Organ xenotransplantation	77
	Eggs	Chicken	Avigenics	Interferon, antibodies	78, 79
			BioAgri	alpha-1 antitrypsin, biogeneric	80–82
			GeneWorks Inc.	Human growth factor, antibodies	83
			GenWay Biotech	Antibodies	84
			Origen Therapeutics	Not available	85
			TransGenRx	Proinsulin	86
			TransXenoGen	Anti-Neoplastic Urinary Protein, Insulin, Human Serum Albumin	87
			Viragen	Vaccines	88
			Vivalis	Vaccines	89, 90

Table 5.1 (continued)

Production strategy	Material used for extraction	Host organism	Institutions	Product examples	Reference
Open or contained growth in greenhouses or underground cultivation	Foliage	Alfalfa Potato Tobacco Melon *Brassica carinata* *Brassica napus* Lettuce Sunflower Turnip	Agriculture and Agri-Food Canada	IL-10	91
			Battelle, Pacific Northwest National Laboratories	hEGF, Factor VIII, IX, XIII, Thrombin	92–98
			Boyce Thompson Institute/Texas A&M/ Axis Genetics	hepatitis B surface antigen, enterotoxigenic *E. coli* fusion protein, Norwalk virus antigen	99–103
			Center for Genetic Engineering and Biotechnology, Havana, Cuba	sFv Anti-Hepatitis B virus surface antigen, coat protein potato leaf roll virus	104, 105
			CICV, INTA- Buenos Aires, Argentina INIA, Madrid, Spain	Structural protein VP1 of foot-and-mouth disease virus, spike protein from swine-transmissible gastroenteritis coronavirus	106–110
			Chlorogen	Plastid accumulation of vaccines, proinsulin, antibodies, human serum albumin	111–114
			Chonbuk National University, Jeonju Korea	Plastid accumulation of B subunit of *E. coli* enterotoxin	115
			Cobeto	Human Intrinsic Factor, human transcobalamin	116
			Copenhagen University, Denmark	Monoclonal antibody	117
			ENEA, Rome, Italy	Antibodies	118, 119
			EpiCyte/Scripps Research Institute	Antibodies, secretory antibodies	120–124
			ERA Plantech	Calcitonin	125
			Farmacule	Not available	126, 127
			Fraunhofer	Antibodies, vaccines	128–130
			Friederich Miescher-Institut, Basel, Switzerland	Human interferon	131
			Gent University	Antibodies	132–137
			Hebrew University of Jerusalem	Interferon beta	138, 139
			Hokkaido University, Japan	Human interferon-alpha2b, IL 8, Human tumor necrosis factor	140, 141

Table 5.1 (continued)

Production strategy	Material used for extraction	Host organism	Institutions	Product examples	Reference
			Icon	Interferon-α, β, somatotropin Restriction enzyme Single-chain antibodies Monoclonal antibodies Antigens, Glucocerebrosidase, Thaumatin. Albumin DNAse, RNAse inhibitor, Insulin	142–145
			Institute of Plant Genetics and Cultivated Plant Research in Gatersleben	Human papillomavirus (HPV), Type 16 virus-like particles, spider silk, single chain antibodies	146–151
			Jefferson Medical College, Philadelphia	Antibodies	438
			KIST, S. Korea	IL-6	152
			Kyoto University, Japan	Erythropoietin	153, 154
			Medicago	IgGs, Thrombin, Aprotinin, tPa, Superoxide dismutase, Protease inhibitor, Collagen fragment, Enzyme for CO_2 solution, hemoglobin	155–158
			Meristem	Gastric lipase, Human serum albumin, Lactoferrin, collagen, MAbs, hemoglobin, Beta-interferon	159–168
			Mogen International	Human serum albumin	169
			Monsanto/ Agracetus	Monoclonal antibodies, single chain antibodies, Human growth hormone (plastid), collagen	170–177
			Monash University Victoria, Australia	Measles virus hemagglutinin protein	178–180
			National Institute of Agrobiological Tsukuba and Ibaraki, Japan	Lactoferrin, lactoalbumin, human epidermal growth factor	181–183
			Nexgen/Guardian	Not available	184
			North Carolina State University	Canine oral papillomavirus protein	185
			Phylogix	Lectin-based proteins	186, 187
			Planet Biotechnology	Secretory antibodies CaroRX, RhinoRX, DoxoRX	188–194

Table 5.1 (continued)

Production strategy	Material used for extraction	Host organism	Institutions	Product examples	Reference
			Plantigen	GAD (glutamic acid decarboxylase) and cytokines, Interleukin-10, Interleukin-4, MHC (major histocompatibility complex) and cytokines	91, 195–199
			Roswell Park Institute, Buffalo, New York	Potato vaccine booster with injected hepatitis B vaccine	200
			St. George London John Innes Centre	Vaccines, secretory antibodies	201
			Università degli Studi di Verona, Italy	Diabetes-inducing auto-antigen glutamic acid decarboxylase	202
			University of Guelph, Canada	Porcine epidermal growth factor, Swine Viral Epitope Fusion	203, 204
			University of Kentucky	Engineered Antimicrobial Peptides	205, 206
			University of Milan, Italy	E. coli toxin B subunit tuberculosis antigen	207–209
			University of Western Ontario	Diabetes-inducing auto-antigen glutamic acid decarboxylase	196
			UTS Biotech, Rome, Italy	Human papillomavirus 16 E7 protein	210, 211
			Wageningen University, The Netherlands	Antibody subunits, glycosylation research	212–216
			York University, Toronto	HIV antigen	217
	Foliage infected by recombinant Brassica virus	Tobacco	Agrenvec	Not available	218–220
			CNR, Turin, Italy	Single-chain Fv antibody fragment	221
			Fraunhofer	Vaccines, antibodies	222–230
			Icon Genetics	Interferon-α, β, Somatotropin, Restriction enzyme, Single-chain antibodies, Monoclonal antibodies, Antigens	142, 144, 233
			Large Scale Biology	Antigen from cancer cells as personalized cancer vaccines, Aprotinin Alpha-galactosidase, Hematopoietic factors, lysosomal acid lipase	234–238, 239

Table 5.1 (continued)

Production strategy	Material used for extraction	Host organism	Institutions	Product examples	Reference
	Tuber	Potato Carrot	Arizona State University/Boyce Thompson I.	Human papillomavirus like particles, Norwalk virus antigen	101, 240
			Battelle, Pacific Northwest National Laboratories	hEGF, Factor VIII, IX, XIII, Thrombin	92, 94–98, 241
			Institute of Agrobio-technology CSIC, Pamplona, Spain	Human serum albumin	242
			Institute of Plant Genetics and Cultivated Plant Research in Gatersleben	Spider silk, viral particles	146, 147
			Loma Linda University	Lactoferrin, diabetes-inducing autoantigen fusion proteins of insulin and glutamic acid decarboxylase to cholera toxin	243–247
			MPB Cologne	scFv antibodies	248, 249
			New Zealand's Crop and Food Research	Atrial natriuretic factor	250
			Novoplant	Oral-delivered MAbs	251
			Planton (Kiel)	Human antimicrobial proteins	252
	Stems	Sugarcane Rubber Tree	Rubber Research Institute of Malaysia	Human Serum Albumin	253–255
			Texas A&M/Procane	Collagen	256
	Fruit	Banana Tomato Melon	Arizona State Boyce Thompson I.	*E. coli* endotoxin fusion protein	102, 257
			University of Colorado, Boulder	Respiratory syncytial virus fusion protein	258
			University of Delhi, New Delhi, India	Cholera toxin B subunit	259, 260
			ViroGene	Vaccines	261

Table 5.1 (continued)

Production strategy	Material used for extraction	Host organism	Institutions	Product examples	Reference
	Seed	Barley	Cropdesign	Not available	262
		Phaseolus v.	Dow	Antibodies, peptides	263
		Corn	Epicyte	Anti-herpes and anti-sperm secretory antibodies for topical gels	120, 121, 194, 264, 265
		Rice			
		Sawflower-derived Oil	Fraunhofer	Antibody fragments	266, 267
		Brassica napus derived oil	Gent University	Enkephalins	268, 269
			Helsinki University	Gelatin	270
		Peas	Institute of Plant Genetics and Cultivated Plant Research in Gatersleben	Antibodies	148
		Soybean			
		Tobacco			
			Iowa State University	E. coli enterotoxin B subunit, porcine alpha-lactalbumin	271, 272
			Lethbridge Research Centre, Alberta, Canada	Bovine virus protein	273
			Maltagene	Not available	274
			Meristem	Gastric Lipase, Human Serum Albumin, Lactoferrin	159–161, 163–165, 275
			Monsanto Protein Technologies/ Agracetus/Calgene	Monoclonal antibodies, human growth hormone	171–176, 276
			Novoplant	Oral-delivered MAbs	277
			Orf Genetics	GM-CSF, Interleukin-3, Stem cell factor, Erythropoietin, Beta-interferon	278
			ProdiGene	Beta-glucuronidase, avidin, trypsin, vaccines, aprotinin, laccase	23, 24, 279–291
			Saint George London John Innes Centre	Single chain antibodies, secretory antibodies	267, 292, 293
			Sembiosys (oil)	Insulin, ApoA1, hirudin, somatotropin	294–297
			Sungene	Not available	12, 298
			Syngenta	Not available	299
			Universidade de Sao Paulo, Brazil	Human Growth Hormone	300
			University of Ottawa	Human insulin-like growth factor, Human granulocyte-macrophage colony stimulating factor, glycoprotein B from human cytomegalovirus	27, 301–306

Table 5.1 (continued)

Production strategy	Material used for extraction	Host organism	Institutions	Product examples	Reference
			Ventria	Alpha antitrypsin, Lactoferrin, lysozyme	307–314
			Washington State University	Human lactoferrin, lysozyme, gelatin	315, 316
Transgenic aquatic plants, cells in liquid suspension, or tissues express-ing recombinant protein in a bioreactor	Growth media or harvested cells/tissue	Moss	Greenovation	Monoclonal antibody	317
		Duckweed (*Lamna*)	Biolex	Plasmin, Human growth hormone, Monoclonal antibodies, alpha-interferon	318
		Algae *Chlamydomo-nas reinhardti*	Phycotransgenics	Viral antigens	319
			Scripps Research Institute/Rincon Pharmaceuticals	Monoclonal antibodies	232, 320, 321
		Roots	Phytomedics	Human placental alkaline phosphatase	322–324
		Mechanically injured-induced promoter for foliage secretion	CropTech	Glucocerebrosidase, alpha-iduronidase, serum proteins and monoclonal antibodies, vaccines	325–328
		Germinated oil seed	UniCrop	Monoclonal antibodies	329, 330
		Plant cell culture	Flanders University	Antibodies	331
			Fraunhofer	Antibody fragments	310
			John Innes Centre, Norwich, UK	Antibodies	310
			National Institute of Public Health, Tokyo, Japan	Human monoclonal antibody anti-hepatitis B virus surface antigen	332
			Phytoprotein	Vaccines	333
			Protalix (Metabogal)	Glucocerebrosidase, Monoclonal antibodies	334
			ROOTec	Not available	335
			University New South Wales, Sydney, Australia	Antibodies	336–338

used as biopharmaceuticals, including monoclonal antibodies, hormones, enzymes, inhibitors, fusion proteins, vaccines, and structural proteins that have been successfully expressed in crop plants and farm animals such as corn, tobacco, alfalfa, tomato, potato, barley, rice, cows, goats, pigs, and chicken.

In this review we will first discuss why transgenic technology is being considered for the production of biopharmaceuticals, and then focus on commercialized products derived from transgenic systems and on those products with clinical experience to illustrate the potential of the technology to impact the healthcare industry. There are many publications and conference presentations describing the production technologies available for the manufacture of recombinant human proteins in transgenic systems and criteria for selection of these production systems [7–31, 233]. Some excellent reviews by Knäblein also compare transgenic plant systems on the basis of recent data [32–35]. The main point to remember is that for each particular biopharmaceutical the production technologies must be carefully analyzed in order to match specific quality, amount, marketing, regulatory, level of containment, and selling price requirements related with each product, and its medical use. Transgenic systems meeting provide new opportunities to commercialize some biopharmaceuticals, and to make others more accessible to healthcare providers.

5.2
Advantages and Disadvantages of Transgenic Systems for the Production of Biopharmaceuticals

Today, transgenic systems are being considered for the production of many diverse biopharmaceuticals, including monoclonal antibodies, hormones, therapeutic enzymes, structural proteins, and vaccines. Table 5.1 illustrates that there are about 120 academic institutions and commercial enterprises considering transgenic systems as attractive production technologies for the commercialization of more than 130 biopharmaceuticals.

There are many reasons why transgenics should be considered for biopharmaceutical production. First, crop plants and farm animals are capable of producing human proteins with similar complex post-translational modifications, folding and assembly as native human proteins. Biochemical and structural equivalency with human proteins facilitates the development, regulatory approval, and commercialization of recombinant-derived biopharmaceuticals, thereby increasing the probability of obtaining satisfactory human-like safety and efficacy profiles in clinical trials. Second, transgenic systems provide improved material traceability and source reproducibility compared with what is available for biologics obtained from natural sources such as human blood or from animals by-products. The use of transgenic plants provides improved safety by avoiding animal-derived pathogens and immunogenic contaminants. Third, transgenics offer a reliable and cost-advantageous alternative to mammalian cell culture and eukaryotic microbial systems, particularly for biopharmaceuticals required in large quantities (more than tens of metric tons) and at low cost (<$5 g^{-1}). Significant operational and capital cost savings can result from replacing the bioreactor-based biosynthesis step with farming [284]. The downstream recovery/purification facility cost remains unchanged or may be lower with transgenics as feed streams – especially in the case of milk, eggs, and seeds – are more reprodu-

Table 5.2 Comparison of capacity of production systems based on highest reported crude protein expression levels (acreage/animals/bioreactor capacity adjusted to 50% purification yield)

Production system	Optimal productivity	Requirement for 1 metric ton per yr
Potato foliage	250 kg protein acre yr^{-1} [29]	8 acres
Alfalfa/tobacco foliage	40 kg protein acre yr^{-1} [29]	50 acres
Rice/barley seed	40 kg protein acre yr^{-1} [314]	50 acres
Chicken egg	40 g protein hen yr^{-1} [31]	50 000 hens, 14 million eggs
Goat milk	12 kg protein goat yr^{-1} [340]	160 goats, 87 000 L milk (@20 g L^{-1})
Mammalian cell culture	900 kg/15 000 L yr^{-1}	2 bioreactors (20 runs/year @ 3 g L^{-1})
Microbial fermentation	2400 kg/40 000 L yr^{-1}	1 bioreactor (20 runs/year @ 3 g L^{-1})

cible and may contain fewer contaminating proteins than broth obtained from cell culture or microbial fermentation processes. It has been reported that the production cost for a recombinant protein at 1 metric ton scale is in the range of $20 to $40 per gram for transgenic animals, and $10 to $20 per gram for transgenic plants [339]. The same report estimates that, at the same scale, bioreactor-based systems using yeast results in production costs in the range of $50 to $100 per gram, while using mammalian systems costs are in the $300–5000 per gram range. Bacterial-based production systems can deliver recombinant proteins in the $1–5 per gram cost range at 1 metric ton scale because of the high volumetric productivity of these systems and the use of large bioreactors. Plants and animals provide a readily scalable production system based on accepted agronomic and farming practices that require minimal capital investment. Manufacturing capital avoidance enabled by the use of transgenic systems could enable companies with limited funding to retain their independence from large companies to commercialize biopharmaceuticals allowing them to direct their resources to support clinical trials rather than building manufacturing facilities. The impact on capacity of different transgenic systems

based on optimal productivity reported in the literature is shown in Table 5.2. As these requirements are based on optimal productivity levels, capacity requirements can be significantly higher at early development stages or with particular recombinant proteins having lower accumulation levels than the optimal productivity reported for these systems.

Transgenic systems provide enabling technology for the commercialization of some biopharmaceuticals that cannot be produced efficiently by mammalian cell culture or by microbial fermentation due to the complex multimeric assembly, or proteolysis sensitivity of these products. Transgenic systems also provide enabling technology for biopharmaceuticals to be delivered in edible form in the expression matrix, usually of plant origin, allowing their commercialization without any additional purification. Edible vaccines expressed in plants are examples of the application of this cost-effective drug delivery option. If desired, the protein in the accumulation matrix can be stabilized by freeze-drying the plant or the animal tissue containing the desired product. This results in biopharmaceuticals that do not require cold-chain storage, and is a significant advantage for the successful introduction of biopharmaceuticals into developing regions of the world. In addition, the raw materials used to produce biopharmaceuticals – namely milk, eggs, and seeds – have well-characterized and reproducible compositions that facilitate downstream process design and performance. Usually, plant extracts contain few proteins that are similar to human-derived proteins. For example, eggs contain only 12 proteins, compared with over 20 000 in traditional fermentation systems.

Manufacturing flexibility is another important advantage of using transgenic systems. The long term storage stability (e.g., seeds) can uncouples biological production from downstream processing. Costly bioreactor-based facilities need to be operated at maximum capacity to be profitable. On occasion, there has been a demand to build new facilities simply to meet projected peak demands – which sometimes do not materialize. The multi-ton scale production capability enabled by transgenics can be quickly adjusted to meet lower market demands, as producing the material in milk or seeds is very inexpensive. Grains and oil seeds are readily stored for long periods, without refrigeration, and without loss of protein activity as a result of degradation or enzyme hydrolysis. Obtaining more animals or land to plant crops in order to meet peak-projected demand is relatively simple and inexpensive. For milk, eggs, cereal grains (e.g., corn, rice, barley), for oilseeds (e.g., soybeans, canola, safflower), for some foliage (e.g., tobacco, alfalfa), and for tubers there is an efficient commercial infrastructure for their growth, harvest, transportation and storage, and this can easily be adapted to the production of biopharmaceuticals almost to any volume.

However, there is no "ideal" production technology suitable for all biopharmaceuticals. The author's qualitative ranking of

transgenic systems relative to conventional production systems for biopharmaceuticals – microbial fermentation, baculovirus-induced insect cell culture, and mammalian cell culture – is listed in Table 5.3.

Table 5.3 illustrates three categories of parameters comparing transgenic systems with other recombinant protein production systems. The first category (scalability, cost, product storage and stability) illustrates the significant advantages of transgenic systems already discussed above. The second category (endogenous pathogenicity, accumulation level, biosynthesis) illustrates that animals and plants offer different advantages and benefits that must be carefully considered based on the product to be made, its use, required purity, demand, targeted cost, and other commercial considerations. The third category (speed for system selection and development, regulatory risk, containment) shows the potential disadvantages of transgenic systems when compared with bioreactor-based production technologies.

5.2.1
Cost and Capacity Advantages

During the past five years, the cost and capacity advantage of transgenic systems compared with mammalian cell culture has been reduced by significant recent improvements in the productivity of cell culture systems and improved culture development technologies [341]. Increased productivity, coupled with increased capacity derived from the construction of several large facilities, will increase the annual mammalian cell culture the estimated capacity to 14 000 kg by 2006 at a targeted production cost below $200 g^{-1}. Currently, most commercially available human therapeutic proteins sell at prices higher than $10 000 g^{-1}, whilst the annual requirement

Table 5.3 Qualitative ranking to illustrate strengths and weakness of recombinant expression systems

Parameter	Worst	→	→	→	Best
Scalability >100 kg yr^{-1} products					
Mammalian	Insect cells	Bacteria	Yeast	Animals	Plants
Cost for <$50 g^{-1} products					
Mammalian	Insect cells	Yeast	Bacteria	Animals	Plants
Product storage stability					
Mammalian	Insect cells	Yeast	Animals	Bacteria	Plants (seeds)
Endogenous pathogenicity					
Animals	Mammalian	Insect cells	Yeast	Bacteria	Plants
Accumulation level					
Plants	Mammalian	Insect cells	Yeast	Bacteria	Animals (milk)
Folding and processing					
Bacteria	Yeast	Plants	Insect cells	Animals	Mammalian
Speed to production organism selection and development					
Animals	Plants	Yeast	Mammalian	Insect cells	Bacteria
Regulatory risk					
Plants	Animals	Insect cells	Yeast	Mammalian	Bacteria
Containment					
Plants	Animals	Yeast	Bacteria	Insect cells	Mammalian

is <10 kg. Very few biopharmaceuticals have a market demand exceeding 100 kg yr^{-1} or require a selling price <$50 g^{-1} when transgenic systems become attractive.

5.2.2
Development and Implementation Times

Most transgenic systems require considerably longer development and implementation times, and this results in a need for higher levels of development resources compared to other production systems. The selection and optimization activities for high productivity, stable, and fertile founder animals or master seeds require years rather than months, as is the case for other production systems. The low rate of live birth for transgenic animals transformed using microinjection technology, combined with a low rate of trait transfer to the offspring, limits the attractiveness of using transgenic animals for biopharmaceutical production. The possibility that the target protein might leak into the blood circulation can result in physiological perturbations and poor growth performance. For plants, the low rate of healthy generation after transformation results in a long development and productivity optimization timeline. As mentioned earlier, microbial and mammalian cell culture systems are improving productivity, cost competitiveness, and the time advantage over transgenics. The long development time and relatively high cost required for transgenic animal development is mitigated by the use of tissue-directed synthesis based on high expression promoters (e.g., casein for milk expression); this allows the highest protein accumulation ever seen in a recombinant protein production system, without affecting the overall physiological state of the productive organism. This is usually not possible in cell culture, and microbial systems that can be physiologically perturbed by high heterologous protein accumulation levels. The average accumulation of biopharmaceuticals seen in milk is 2–10 g of product per liter, compared with 0.2–2 g L^{-1} accumulation that is typically seen in microbial and mammalian systems. The implementation of nuclear transplantation leads to a more rapid development of founder animals. Some animals mature quickly. For example, chickens mature in 27 weeks, compared to 48 weeks for first-generation corn seed, 64 weeks for goats, and 140 weeks for cows [86]. Rabbits have short gestation (1 month) and sexual maturity times, and typically a rabbit can have eight to ten litters each year, thereby allowing for a quick production scale-up [38]. There is no known prion disease in rabbit, and no known serious disease transmission to humans, which makes the rabbit safer than other milk-producing animals [70]. Reports have been made that the expression of some human recombinant proteins in milk can induce a lactational phenotype resulting in the abnormal morphological, biochemical, and functional differentiation of mammary gland cells [342]. This suggests that, especially in animals, it is important to monitor the effect of the expressed protein on the host organism. The rapid reproduction cycle of hogs allows production scale-up for biopharmaceuticals from seminal fluid in 23 months. The use of seminal fluid offers a closed secretory system, so that the host is unaffected by the proteins it produces [70]. For plants, the use of fast-growing aquatic plants and viral induction can deliver products as fast as – or faster than – mammalian cell culture or microbial systems. The *Lemna* aquatic plant system allows the generation of transformed plants in 6 weeks, and has the fastest growth rate of all higher plants,

doubling its biomass every 1.5 days [318]. *Lemna* grows in aqueous solutions of simple inexpensive mineral salts and requires minimal energy inputs. Biopharmaceutical productivity is enhanced by the high protein content (35% based on a dry weight basis) and their ability to secrete biologically active, recombinant protein to the inorganic media surrounding the aquatic plants.

5.2.3
Post-translational Modification

The lack of identical human-like post-translational modification (glycosylation, sulfation, phosphorylation) constitutes an obstacle to implement the use of transgenic systems for the production of biopharmaceuticals. Transgenic systems, as with all other recombinant production systems, are not capable of identical human-like post-translational modification, specifically glycosylation [216, 343]. For example, human antithrombin III, as isolated from blood plasma, contains mainly biantennary disialylated glycosylation. The same recombinant human protein produced in goat milk contains a mixture of biantennary mono- and disialylated glycosylation with low levels of altered fucosylation, *N*-glycolylneuraminic acid, *N*-acetylgalactosamine for galactose substitution, terminal galactose, and high-mannose structures. Human proteins expressed in plants can contain a sugar not found in mammalian systems (xylose), a different fucose linkage, terminal *N*-acetylglucosamine, and plants are not capable of adding terminal galactose or sialic acid [153]. A comparison of the structures of the N-linked glycans attached to the heavy chains of a tobacco-derived monoclonal antibody with the corresponding antibody of murine origin indicated that all glycosylation sites are N-glycosylated as in mouse, but the number of glycoforms was higher in the plant than in the mammalian expression system [343]. In addition to high-mannose-type N-glycans, 60% of the oligosaccharides that are N-linked to the plant-derived antibody have $\beta(1,2)$-xylose and $a(1,3)$-fucose residues linked to the core $Man_3GlcNAc_2$. Differences in glycosylation can alter the function, stability, pharmacokinetics, immunogenicity, protease sensitivity, and consistency of biopharmaceuticals, and this results in the need for extensive bioequivalency studies if human-derived products were used originally for therapeutic indications [344, 345]. The presence of plant-specific oligosaccharide structures may not be a limitation to the use of plant-derived antibodies for topical immunotherapy, as plant-derived antibodies were shown to be biologically active. However, their immunogenic potential may raise concerns for systemic applications of plant-glycosylated antibodies in humans. Variations in immunogenicity are very difficult to detect, as animal models are not predictive of immunogenicity in humans. The induction of antibodies against the biopharmaceutical can limit its effectiveness, alter its pharmacokinetics performance, and may also lead to the generation of antibodies against the native human protein, resulting in the loss of a critical human physiological function. One study indicated that having plant-like glycosylation in a tobacco-derived murine monoclonal antibody did not elicit an immunogenic response in mice [346]. Methionine oxidation, tryptophan photo-oxidation, aspartate isomerization, asparagine and glutamine deamidation, proteolysis, and protein aggregation are other modifications that are influenced by the production system and downstream processing. Tissue-directed genetic manipula-

tion of plants and animals to clone human genes responsible for post-translational modifications can result in the development of host organisms designed to conduct human-like post-translational modifications. Research has demonstrated the induction of human-like glycosylation in plants (addition of terminal galactose) and human-like proline hydroxylation for collagen in plants and animals by expression of the corresponding human modification enzyme [212]. These results illustrate the possibility of having mammalian-like post-translational modifications in transgenic systems in a near future (see Part IV, Chapter 7).

5.2.4
Regulatory Experience

The lack of regulatory experience – since no product derived from a transgenic system has been approved for human use – is another significant limitation for transgenic systems to gain acceptance as a production technology for biopharmaceuticals, although – as discussed above – this will change very soon if antithrombin from goat milk is approved. The eventual approval by regulatory agencies (one has already been submitted for approval) of products from transgenic systems will be facilitated by the familiarity of the regulatory agencies and of the public with the safe use of crop plants and farm animals and their derived products (milk, eggs, seeds) used for biopharmaceutical accumulation. Crop plants and farm animals have genetic and breeding properties that are well-understood, and their derived products have well-characterized inherent toxins, anti-nutrients, and exogenous contaminant profiles. Technologies to be employed for producing biopharmaceuticals from transgenic systems will not differ from current agricultural, farming, and biopharmaceutical production practices. Dedicated, custom-designed equipment and facilities will be developed with procedures equivalent to the "Good Manufacturing Practices" currently in place for the production of biopharmaceuticals. Once the protein is extracted from the transgenic tissue, similar purification technologies as those used for cell culture and microbial recombinant systems can be applied. In many cases, the host-contaminating proteins and toxic metabolites such as endotoxins are present at lower concentration and diversity in transgenic systems compared to cell culture, non-transgenic tissues, or microbial production systems. Transgenic animal production will be conducted minimizing virus- and prion-transfer risks, providing a safer alternative to the extraction of these products from human or non-transgenic animal tissues. The use of plants will practically eliminate this risk. The FDA, in collaboration with the USDA/APHIS (US Department of Agriculture/Animal and Plants Health Inspection Service), published the "GUIDANCE FOR INDUSTRY": Drugs, Biologics, and Medical Devices Derived From Bioengineering Plants For Use in Humans and Animals" containing suggested norms for field growth of plants and their proper handling and transportation [347]. A similar document is available for transgenic animals [348].

5.2.5
Containment

Containment is another critical issue, especially when using crop plants such as corn that can transfer its genes to other related plants via pollen or insect-pollinating crops (e.g., alfalfa). The cost of procedures associated with containment – together with subsequent testing and monitoring and the need for dedicated equipment, waste disposal, training, security, regulatory documentation and liability insurance for the potential of mingling the transgenic-derived product with food products – can significantly reduce the cost advantage of transgenics compared with other systems. Increased production costs combine with the perception of risk could hinder the implementation of transgenic systems. The possibility of contamination of the food supply with transgenic-derived biopharmaceuticals results in the opposition to the use of transgenic systems for biopharmaceutical production from the food industry, public advocates, and environmentalists concerned about potential health and safety risks of transgenic-derived products in the environment. Animal activists oppose the use of animals for manufacturing as many animals are destroyed as part of the selection process. There have been recent incidents of corn contamination of a cloned insecticidal protein not approved for human consumption, and of soy seeds with corn residues possibly containing a biopharmaceutical. The food production industry, the public, the media, and government agencies reacted strongly and negatively to these incidents, even when the contamination could not be detected or correlated with any health threat. Industry groups such as the Grocery Manufacturers of America and grain processors are asking that the use of food crops for pharmaceutical production should be banned, while others are suggesting that crops grown to produce biopharmaceuticals should only be grown in regions where that crop is not grown for food or feed [349]. This latter practice could eliminate the attraction of transgenics, namely that existing infrastructure can be used to achieve low cost and farming flexibility. In response, several companies are developing technology for production in closed systems such as greenhouses and underground mines, or in bioreactors using transgenic-derived tissues. Gene transfer via pollen can also be avoided by using chloroplast expression systems, by growing male sterile plants, by mechanical pollen removal, or by using self-pollinating crops such as barley and rice.

The initial relatively small number of animals and acreage (see Table 5.2) required for biopharmaceutical production facilitates the implementation of pollen control, containment practices, quality assurance procedures, and monitoring systems for all aspects of animal growth and crop production, harvesting, post-harvest handling, storage, transportation, and bioprocessing [350].

5.3

Commercial Biopharmaceuticals with Human Clinical Experience for Therapeutic, Immunoprophylactic, and Medical Device Use derived from Transgenic Systems

As shown in Table 5.1, there are over 130 of products for medical applications that have been – or are being – considered as candidates for production using transgenic systems. In this section, we will discuss those biopharmaceuticals that are currently commercially available for non-human use (plant-derived avidin, trypsin, and aprotinin), and one product that has been submitted for regulatory approval, namely milk-derived antithrombin. Ten therapeutic proteins derived from transgenic systems that were (or are) included in human clinical testing will be also discussed. These products are being considered for the treatment of inherited diseases (milk-de-

rived alpha-1-antitrypsin, milk-derived C1 inhibitor, corn-derived gastric lipase, milk-derived human bile salt stimulated lipase, milk-derived acid alpha-glucosidase), auto-immune conditions (milk-derived alpha-fetoprotein), cancer (corn-derived monoclonal antibody, tobacco-derived single chain antibodies), and as anti-infective agents (milk-derived lactoferrin, tobacco-derived secretory antibodies). We will also discuss the development of oral vaccines and recombinant structural proteins used for medical devices. An understanding of these products will provide a good perspective as to how agricultural biotechnology companies and the pharmaceutical industry are considering the implementation of transgenic systems, and highlight the future impact of this technology on healthcare systems.

5.3.1

Commercial Products Derived from Transgenic Systems

The three commercial products available for medical applications derived from transgenic systems are all available from the Sigma Chemical Company, and are derived from plants.

5.3.1.1 Corn Seed-derived Recombinant Chicken Egg White Avidin to Replace Egg-derived Avidin

Avidin was the first commercially available recombinant protein (1997) to be derived from a transgenic system (corn seed) for use in medical applications [285, 286, 291]. As it forms strong, non-covalent bonds with biotin, avidin is used in medical and biochemical diagnostic kits for the detection of biotin-containing proteins and nucleic acids. Transgenic corn which accumulated avidin was developed jointly by

ProdiGene, Pioneer Hi-Bred International and the US Department of Agriculture (USDA) with its first field trials reported in 1993 [25, 285]. The product was developed to improve the quality and the consistency of this reagent at a lower production cost compared with egg-derived avidin. It was estimated that a single $2.50 bushel of corn yields the same amount of avidin as 1 ton of eggs, which costs about $1000 to produce. Avidin is available from the Sigma Chemical Company (#A8706) and is sold as a homogeneous protein of consistent quality, providing better performance and reproducibility compared to avidin produced by extraction from egg white. Avidin is purified from corn by affinity chromatography using a 2-iminobiotin agarose resin. The native source of avidin is avian, reptilian or amphibian egg white. Avidin is naturally induced by progesterone, tissue trauma, toxic agents, and infections. It is a glycoprotein with four identical subunits each of 16 kDa and 128 amino acids. There are many avidin-based analytical methods available to quantify DNA and proteins in biological samples [351].

Recombinant avidin expressed in food or in feed grain can be used as a non-selective biopesticide against a spectrum of storage-generated insect pests [352]. Expressing avidin with other traits such as herbicide resistance or co-production with another industrial protein illustrates the concept of "gene stacking" for developing agricultural crops with higher value and improved performance. Insect control effected by avidin results from the created biotin deficiency that is toxic to pests. The recombinant avidin in corn seed is not toxic to mice when administered as the sole component of their diet for 21 days. Avidin's insecticidal activity is different from other transgenic alternatives such as

Bt and from conventional insecticides. As avidin is a natural food protein found in eggs, it may be easier to register than other insecticide proteins that are not present in the food supply [352]. Beta-glucuronidase was also produced in corn seed by ProdiGene for use as a diagnostic enzyme, but it is not commercially available [285, 286].

5.3.1.2 Corn Seed-derived Recombinant Human Trypsin to Replace Bovine Pancreas-derived Trypsin

Trypsin was the first commercially recombinant enzyme for medical applications that was produced from a transgenic system (corn seed) and available in kilogram quantities [280]. Trypsin is a pancreatic enzyme which catalyzes the breakdown of proteins at specific sites as part of the digestion process. The enzyme has a molecular weight of 23–28 kDa, and in protein molecules is active only against peptide bonds adjacent to arginine and lysine. Trypsin is one of the most site-selective of all commercially available proteolytic enzymes, which makes it an ideal regent for the analysis and specific processing of proteins during manufacture. The most important industrial and medical use of trypsin is in the manufacture of insulin, as it cleaves the precursor protein into an active form by removing peptide C from proinsulin. Another bioprocessing application is for the cleavage of undesired antigens from cells in both human and veterinary vaccines manufacture. Trypsin is used in wound care as debridement treatment to reduce inflammatory edema, hematoma and pain associated with internal and external wounds [353, 354]. Proteases, including trypsin, are used in therapy for autoimmune diseases such as Type I diabetes [355]. Trypsin can also be used as a food additive to baby formula, assisting the digestion of difficult proteins and leading to improved absorption in the digestive tract. These therapeutic and nutraceutical applications could be expanded by the availability of a low-cost, abundant, and safe source of recombinant trypsin.

Corn seed-derived recombinant human trypsin was developed as an animal component-free alternative to the currently available bovine pancreas-derived trypsin for use in bioprocessing and as a cell culture media component for growth enhancement. Corn-derived recombinant trypsin is not approved for human therapeutic use. Recombinant trypsin has been produced by cell culture, bacteria and yeast, but not in quantities sufficient to meet the needs of biopharmaceutical manufacturers requiring kilogram quantities at relatively low cost [280]. The corn seed-derived recombinant protease is functionally equivalent to the native bovine pancreatic trypsin used in most current applications [286]. TrypZean®, developed by Prodi-Gene, is available from Sigma Chemical Company (#T3568) for use in the dissociation of cultured mammalian cells from plastic-ware, thereby eliminating the introduction of potential animal-derived contaminants found in bovine and porcine trypsin (see also Part IV, Chapter 1). Tryp-Zean allows the use of a consistent quality product which delivers the same kinetics for cell detachment as the current product; this results in a need for minimal protocol changes during cell preparation. Soybean trypsin inhibitors and other inhibitors have a similar inhibitory performance to TrypZean as they do with native trypsins. Recombinant trypsin is manufactured by Sigma by utilizing ProdiGene's transgenic corn protein expression system.

5.3.1.3 Cornseed and Viral-induced Tobacco Foliage-derived Recombinant Bovine Aprotinin to Replace Bovine Lung-derived Aprotinin

Tobacco-produced recombinant research-grade bovine aprotinin (Apronexin NP) [238] is also available from Sigma Aldrich. Aprotinin is a protease inhibitor which is used as a research reagent in biomanufacturing for several therapeutic applications. It has been traditionally extracted from bovine lung tissue. Aprotinin is a single, 58 amino-acid polypeptide with three disulfide bonds, and inhibits several serine proteases such as trypsin, chymotrypsin, plasmin, and kallikrein.

Bovine lung-derived aprotinin (Trasylol®, Bayer) is used in therapeutic applications as a clotting agent; its mode of action is to inhibit the degradation of fibrinogen by plasmin, thereby reducing blood loss during heart surgery and the need for blood transfusions. Aprotinin also protect platelets from damage in circulation machines during surgery [356, 357]. It can be used to treat acute pancreatitis [358]. Use of bovine-derived aprotinin can cause severe anaphylaxis in 0.5–1% of patients. In patients who receive aprotinin a second time this percentage is greater [359]. The recombinant product should have an improved safety profile when it becomes available for human use.

Apronexin NP (Sigma, #A6103) is sold as a research reagent, for cell culture, and for other biomanufacturing applications. Large Scale Biology uses its biomanufacturing facility and its Geneware gene expression technology for Apronexin NP production based on the use of recombinant tobacco mosaic viruses containing the bovine aprotinin gene to express the recombinant protein in non-transgenic tobacco plants. This process consists of growing non-transgenic tobacco in open fields or greenhouses, infecting the leaves with the recombinant plant virus to induce expression of a aprotinin in the leaves, and transporting the harvested leaves (fresh or frozen) to a closed facility for protein extraction.

Recombinant aprotinin has been prepared by yeast fermentation [360] and from corn seed, developed by ProdiGene [283] under the tradename AproliZean®. Corn seed-derived aprotinin has been grown in field trials since 1998. In addition to research and bioprocessing uses, corn-grown aprotinin can be used as avidin as a built-in insecticide based on recombinant aprotinin's trypsin-inhibiting activity. For its recovery, recombinant aprotinin – together with a native corn trypsin inhibitor – is purified using a trypsin-agarose column. The corn inhibitor binds to the column, but the recombinant human aprotinin does not. As the corn inhibitor can also be purified, co-production of valuable recombinant and non-recombinant materials from the same crop will clearly increase the value of a crop.

5.3.2
Products Derived from Transgenic Systems Submitted for Regulatory Approval

Until now, only one product derived from transgenic technology, namely goat milk-derived recombinant human antithrombin III, has been submitted for approval in Europe.

5.3.2.1 Goat Milk-derived Recombinant Human Antithrombin III to Replace Human Blood Plasma-derived THROMBATE III®

Human antithrombin III (ATryn®) was the first therapeutic recombinant protein to be produced using a transgenic system

(goat milk), tested in clinical trials (1994), and submitted for approval to a regulatory agency (Marketing Authorization Application to European regulatory agencies) in 2004. GTC Biotherapeutics, the developer of ATryn®, plans a market launch of the product in Europe by mid-2005 (see Part IV, Chapter 11).

Antithrombin III (AT-III), a single-chain glycoprotein of 58 kDa and 480 amino acids, is synthesized in the liver. It is a serine protease inhibitor, and acts as the most important inhibitor in the coagulation cascade to avoid blood clot formation. AT-III inhibits a wide spectrum of serine proteases including thrombin, factors IXa, Xa and XIa, kallikrein, plasmin, urokinase, C1-esterase, and trypsin. AT-III interacts with heparin by binding to specific sulfated and non-sulfated monosaccharide units on heparin. The binding of AT-III to heparin enhances the inhibition of factors IXa, Xa, and thrombin.

THROMBATE III® (Bayer) is a plasma-derived product that is used to treat AT-III deficiency, the first inherited trait discovered which was identified in 1965. The condition is associated with thrombophilia caused by low levels of AT-III or the presence of altered AT-III activity. Both conditions can result in excessive blood clotting. Acquired AT-III deficiency is another condition that occurs in situations with high risk of thrombosis such as trauma, burns, and sepsis. GTC Biotherapeutics conducted clinical trials with ATryn® in high-risk situations such as surgery or child delivery to prevent deep-vein thrombosis in hereditary AT-III-deficient patients and heparin-resistant patients with acquired AT-III deficiency undergoing coronary bypass. A pharmacokinetic study in patients with hereditary AT-III deficiency indicated that the administration of the recombinant product resulted in an increase in blood

AT-III activity which was similar to that seen with the plasma-derived product. Clinical use of ATryn® in patients who had surgical procedures indicated that it was well tolerated in all cases. Shortages in THROMBATE III® supplier resulted in the use of ATryn® on compassionate grounds when the plasma product was not available. No clinical evidence of thrombosis has been reported in any of these compassionate-use patients. The routine use of THROMBATE III® as replacement therapy is not generally recommended due to the high cost, the risk of infection, and the need for frequent intravenous administration. The recombinant product derived from milk should have improved safety and reduced cost, resulting in a product that is suitable for chronic use.

ATryn® production in goat milk is the only recombinant production system shown able to deliver AT-III at costs commercially acceptable. GTC has several goat production lines accumulating human AT-III in milk at 2–6 g L^{-1} concentration. Purification from milk includes tangential flow filtration to remove casein, micelles, fat globules, bacteria, somatic cells, and viruses – followed by a heparin-based affinity chromatography to remove DNA, lactose, mineral salts, vitamin, hormones, and most of the milk proteins. Two additional chromatographic steps – anion-exchange to remove additional milk proteins including lactoferrin and hydrophobic interaction to remove goat AT-III and milk proteins – complete the process. From 300 L of milk, 300 g of human AT-III are purified to 99.9% purity with an overall 55% recovery yield. The effectiveness of the purification process to remove viruses and prions was demonstrated, with the high degree of purity of ATryn® being further illustrated by the observation that no immunogenic events were detected in

clinical trials after examining 180 subjects, including healthy volunteers and patients with hereditary or acquired AT-III deficiency, at 1 day after administration and 28 days later.

In addition to ATryn®, GTC Biopharmaceuticals has several other products under development in collaboration with Centocor, Abbott, Elan, Bristol-Myers Squibbs, Alexion, Progenics Pharmaceuticals, ImmunoGen, and Merrimak Pharmaceuticals. These collaborations and internal programs involve the production of several monoclonal antibodies, recombinant human serum albumin, a malaria vaccine, fusion proteins, and a protein for the treatment of autoimmune conditions. Some of the products that are currently undergoing clinical trials and will be discussed later [361].

The commercial success of ATryn® could stimulate the use of transgenic systems for the commercialization of other plasma-derived biotherapeutics [27, 52]. Milk-derived antitrypsin and human C1 inhibitor currently in clinical trials will be discussed later. The development of tobacco foliage-derived factor VIII and factor IX for the treatment of hemophilia, and factor XIII and thrombin for wound healing therapies as clot promoters is currently in progress at Battelle [94–97]. For factor VIII it was estimated that, based on the current non-optimized process, the needs of an average 80-kg patient could be satisfied with 2–3.5 kg of tobacco foliage, or two plants. It was estimated that fewer than 4000 plants, suitable for greenhouse growth, would be needed to supply the worldwide demand for factor VIII. Thrombin, currently derived from bovine blood, is used to control bleeding associated with surgical procedures. Each year thrombic is used in over 500 000 operations in the United States. Bovine thrombin is frequently applied topically during surgery to achieve hemostasis. It also is used as a component of commercial fibrinogen preparations. The use of Bovine-derived thrombin has been associated with the development of antibodies that may cross-react with human blood proteins [362]. These antibodies could lead to serious bleeding complications. A transgenic product will have both improved safety and reduced cost.

Another important plasma-derived product considered for production using transgenics systems by several organizations (see Table 5.1) is human serum albumin (hSA). This $1.3-billion, 400-tonne (in 2000) product is currently produced from donor human blood. Its role is to maintain fluid balance in the blood and to transport amino acids, fatty acids, and hormones [169, 254, 363, 364]. Current clinical uses of hSA include blood volume replacement during shock, the treatment of serious burns, emergency surgeries, and as an excipient to maintain structural stability and activity in biological drug formulations. As with AT-III, hSA is difficult to express in conventional recombinant production systems at amounts and cost-competitively with the current blood-derived product. GTC Biotherapeutics produces hSA derived from a transgenic system based in milk expression. GTC is producing qualification batches for evaluation by potential clients for use as an excipient and to prepare a Drug Master File (DMF) allowing data related to rhSA to be used in the manufacture of therapeutic products. GTC considers that the commercialization of rhSA is possible by 2007, and estimates that the annual total production volume required to meet the needs of the excipient market will be 1–2 tonnes [40].

The success of ATryn® may also stimulate the use of transgenic technology for similar replacement therapeutic enzymes and protein-based inhibitors to treat inherited or acquired conditions caused by reduced amounts or variant activity of a critical enzyme or inhibitor. In the next section, we will discuss in detail five such products (milk-derived alpha-1-antitrypsin, milk-derived C1 inhibitor, corn-derived gastric lipase, milk-derived human bile salt stimulated lipase, and milk-derived acid alpha-glucosidase) with human clinical trial experience. Transgenic technology is attractive for the category of therapeutic replacement products that are used for lifelong enzyme/inhibitor replacement therapy in small numbers of patients who are also in need of other costly supplemental drugs and expensive medical services related to their condition. These replacement enzymes and inhibitors are expensive because they tend to be complex proteins which are difficult to prepare by recombinant technology and available only in natural form from human tissue or mammalian cell culture. The preparation of a complex product in small amounts and at a relatively low cost is difficult using conventional recombinant systems, and this creates a good opportunity for transgenic systems. Another advantage for transgenic producers is that these products can be commercialized as "orphan drugs" because the limited number of patients allows significant market protection to the developer.

5.3.3
Products Derived from Transgenic Systems with Human Clinical Experience for Therapeutic Use

At least 10 proteins developed for therapeutic use have been derived from transgenic systems, and have been used in human clinical trials according to published data. Five of these proteins (milk-derived antitrypsin, milk-derived C1 inhibitor, corn-derived gastric lipase, milk-derived bile salt-stimulated lipase, milk-derived acid alpha-glucosidase) were developed to treat inherited diseases, while the others were developed to treat autoimmune conditions (milk-derived alpha-fetoprotein) and cancer (corn-derived monoclonal antibody, tobacco-derived cancer cell antigens), or to serve as anti-infective agents (milk-derived lactoferrin, tobacco-derived secretory antibodies).

5.3.3.1 Transgenic-derived Biopharmaceuticals to Treat Inherited Conditions

As noted earlier, this is a therapeutic category with a significant need for products that need to be accessible to a small group of patients for life-long treatment at low cost. Moreover, they must be free of human/animal-derived infectious agents and of immunogenic animal components because of chronic use, ATryn® belongs to this category.

Sheep milk-derived recombinant human alpha-1-antitrypsin to replace human blood-derived Prolastin® PPL Therapeutics, in conjunction with Bayer [who manufacture human plasma-derived alpha-1-antitrypsin (AAT); Prolastin®], sought to develop sheep milk-derived recombinant human alpha-1-antitrypsin (rhAAT), also known as alpha-1-protease inhibitor. AAT is a human blood glycoprotein which is produced in the liver and inhibits several proteases, including neutrophil elastase, a lung enzyme that digests phagocyte cells and bacteria, thereby promoting tissue healing [365]. Without ATT, lung elastase is not properly regulated and can destroy the lung tissue. Many respiratory diseases, including cystic fibrosis and chronic obstructive pulmonary disease, occur as a result of the imbalance of AAT and elastase in the lungs. AAT deficiency is an inherited condition which occurs in approximately 1 in 5000 live births. Severe AAT deficiency (hereditary emphysema) affects around 150000–200000 people in the US and Europe. Prolastin® can be used to alleviate this condition, but its availability fluctuates as it is produced by only one company. Indeed, Prolastin® is the only available product for chronic replacement of AAT in congenital AAT-deficient patients with emphysema.

The efficient production of hrATT in milk was achieved by PPL Therapeutics, who reported the highest ever accumulation levels for a recombinant product, namely 35 g L^{-1} in mouse milk and 20 g L^{-1} in goat milk. PPL Therapeutics had a flock of 4000 AAT-producing transgenic sheep in New Zealand, and subsequently conducted successful clinical trials using sheep milk-derived rhATT in patients with cystic fibrosis. Data were also available from Phase I/II clinical studies in patients with congenital AAT deficiency. Studies of the treatment of emphysema and acute respiratory distress syndrome were also planned. In 2000, PPL and Bayer announced a placebo-controlled Phase III efficacy study for the treatment of AAT deficiency designed to commercialize an aerosol formulation of rhAAT. However, in 2003, it was announced by PPL that the rhAAT development program had been put on hold because of financial risk related to construction of the manufacturing facility. There were also questions regarding the purity of the AAT produced in sheep milk, as well as some minor (but unexplained) side effects that included mild coughing. Overall, the suggestion was that this product would be too expensive to develop [60].

The production of rhAAT in rice cell culture and rice plants was reported by Ventria [311, 366] (see Part IV, Chapter 8). As part of the program to produce human proteins as nutritional supplements to infant formula. Human milk contains relatively high concentrations of AAT. The stability of rice-derived rhAAT was examined by biochemical and functional assays such as elastase and trypsin inhibition, following exposure to heat, low pH, and *in vitro* digestion. Native AAT can withstand *in vitro* digestion modeled after conditions in the infant gut. Studies show that rhAAT may survive the conditions of the infant stomach and duodenum and effect protein digestion in the infant small intestine. An rhAAT expression equivalent to 20% of the total secreted proteins was achieved, and purification yielded active rhAAT with purity greater than 95% [311].

In addition to the development of rhAAT products based on transgenic technology, clinical studies to investigate the use of gene therapy based on a recombinant adeno-associated virus with the hAAT gene had been proposed to treat AAT-deficient human subjects [367]. Yeast-derived rhAAT developed for aerosol administration to treat AAT-deficient individuals was shown to be safe and resulted in increased lung anti-neutrophil elastase defenses [368].

Milk-derived recombinant human C1 inhibitor Pharming is conducting human clinical trials for recombinant human C1 inhibitor (rhC1I) in patients suffering hereditary angioedema (HAE), a condition caused by human C1 inhibitor deficiency [58]. Human C1 inhibitor is a 105 kDa protein which has a critical role in the inhibition of proteases involved in fibrinolytic and complement pathways. Deficiency of this inhibitor is manifested in acute swelling of soft tissues in the hands, feet, limbs, face, intestinal tract, mouth, or airway (larynx or trachea). If swelling closes the airway, it can be fatal. Attacks of swelling can become more severe in late childhood and adolescence. In the US and Europe, there are some 22 000 patients suffering from HAE, acute attacks of which can be treated by intravenous injection of C1 inhibitor purified from human blood plasma (C1I-Immuno; ImmunoAG, Austria). The condition may be prevented in some instances by treatment with attenuated androgenic male hormones. As with many other blood plasma products, the supply of human C1 inhibitor is limited and is prone to viral contamination, while the use of male hormones can cause severe adverse side effects. The results from a Phase II clinical trial established the efficacy, safety and tolerability of rhC1I in the treatment of acute attacks of HAE [58]. All patients treated with rhC1I showed rapid onset of relief and time to complete resolution of the attack, accompanied with acceptable pharmacokinetic and pharmacodynamic performance. After treatment, patients were monitored for 90 days to evaluate the safety of rhC1I. Treatment with milk-derived rhC1I did not elicit any allergic or clinically relevant immune responses. None of the patients treated with milk-derived rhC1I in the clinical trial experienced relapse of the initial HAE attack.

Corn seed-derived dog gastric lipase as replacement of porcine-derived Pancrelipase Corn seed-derived dog gastric lipase is under development by Meristem Therapeutics for the treatment of gastric lipase deficiency (steatorrhea), neonatal deficient pancreatic function, or when pancreatic function is compromised by diseases such as cystic fibrosis or chronic alcoholism [159]. The absence of gastric lipase in steatorrhea patients prevents the digestion of food lipids. Pre-term infants have difficulty digesting lipids and, unless they are breast-fed, have no access to this enzyme.

Cystic fibrosis – a genetic disease which affects approximately 30 000 children and adults in the US – can lead to reduced levels of gastric lipase. For all of these patients, oral delivery of active gastric lipase can significantly improve the quality of life and nutrition.

Gastric lipase could also be used as an industrial enzyme for the digestion of fats to make soaps, the tanning of leather, and the preparation of animal feedstuffs.

Currently, gastric lipase is available for oral delivery as porcine or bovine pancreatic extract capsules or tablets (Pancrelipase; Ortho/McNeil Pharmaceuticals, Axcan Pharma, Organon, and Digestive Care). The pancreatic enzyme concentrate is predominantly steapsin (pancreatic lipase), but it also contains amylases and proteases. These products are unpalatable and have low activity; hence patients are required to take large numbers of tablets daily (perhaps 30–50), resulting in poor compliance. These products are also unsuitable for pre-term infants.

Meristem selected dog lipase (a 379-amino acid enzyme with three cysteines) for expression in tobacco foliage and corn seed as it was the most active of the animal gastric lipases when tested using long-chain triacylglycerols which represent the major components of human dietary fats [159, 166]. Dog lipase is a glycoprotein with 13% carbohydrate content by weight. The carbohydrate moiety is not required for activity, but it does increase the solubility and availability of the enzyme. Meristem produced dog gastric lipase using vacuolar retention and secretion to obtain active glycosylated enzyme. The secreted enzyme had improved proteolytic maturation compared with the vacuolar retained enzyme, and this resulted in a product with higher specific activity [166]. Corn seed-derived gastric lipase has similar activity, acid resistance, and acidic optimum pH, as the native animal-derived enzyme. The transgenic-produced lipase is more active and more palatable than current lipase products.

Corn seed-derived dog gastric lipase also has improved quality and consistency, and will be significantly cheaper than other purified gastric lipases. Meristem claims a 75% reduction in production costs, and has already successfully manufactured kilogram quantities of pharmaceutical-grade gastric lipase for clinical trials. Meristem can purify 350 g of gastric lipase per week from 700 kg of corn seed in their manufacturing facility, which is capable of producing up to 20 kg of purified lipase each year.

Other recombinant versions of gastric lipases are available from yeast, *Pseudomonas* and filamentous fungi [369]. Sheep milk-derived human bile salt-stimulated lipase (BSSL) – an enzyme produced in the pancreas and in human milk – was developed by PPL Therapeutics for similar indications, and it is discussed below [63].

Sheep milk-derived human bile salt stimulated lipase to replace porcine-derived lipase
PPL Therapeutics reported positive results of its Phase II clinical trial using transgenic BSSL to treat patients with pancreatic insufficiency [63]. The milk-derived product was equally as effective as Creon®, the current market leader used to improve digestion and restore fat absorption to normal levels. Creon® capsules are administered orally and contain delayed-release microencapsulated porcine pancreatic pancrelipase. Six patients with chronic pancreatic insufficiency were successfully treated, four patients having chronic pancreatitis and one suffering from cystic fibrosis. The larger commercial opportunity for this product is for premature infants

who do not receive their mother's milk, which naturally contains BSSL. PPL developed BSSL in collaboration with AstraZeneca.

Rabbit milk-derived human acid alpha-glucosidase for the treatment of Pompe disease
Pharming conducted clinical trials to test the long-term safety and efficacy of rabbit milk-derived recombinant human alpha-glucosidase (rhAGLU) used to treat a lysosomal storage disorder: Pompe disease [370]. This disease, which has a frequency of 1 in 40000, usually results in infant death at a median age of 6–8 months. Pompe disease is caused by the absence of alpha-glucosidase activity and/or presence of deleterious mutations in the alpha-glucosidase gene. This enzyme is used to breakdown glycogen within cellular lysosomes. The absence of glycogen catabolism results in loss of muscle strength, including the heart, preventing infants from developing. Milder forms of the disease can lead to life-long metabolic and physical disorders. For a clinical trial, four patients with infantile Pompe disease showing a lack of the processed forms of alpha-glucosidase were treated. The rhAGLU was well tolerated by the patients during more than 3 years of treatment. Anti-rhAGLU immunoglobulin G titers initially increased during the first 20–48 weeks of therapy, but declined thereafter. All four patients had mature forms of alpha-glucosidase as shown by Western blot analysis, indicating that the enzyme was properly targeted to lysosomes. Pre-clinical testing using a knockout mouse model (see Part III, Chapter 4) indicated full correction of acid alpha-glucosidase deficiency in all tissues except brain after a single dose of milk-derived enzyme administration [371, 372]. Weekly enzyme infusions over a period of 6 months resulted in the degradation of lysosomal glycogen in heart, and skeletal and smooth muscle.

Products for the treatment of life-threatening lysosomal storage enzyme deficiencies provide good targets for production using transgenic systems. Large Scale Biology Corporation (LSBC) developed viral-induced tobacco foliage-derived alpha-galactosidase [239], while CropTech produced two human lysosomal enzymes, beta-glucocerebrosidase [328] and iduronidase, expressed in tobacco foliage using a mechanically induced promoter to induce secretion of the recombinant enzyme in a bioreactor. The LSBC product, alpha-galactosidase, catalyzes the hydrolysis of globotriaosylceramides and related glycosphingolipids in lysosomes [239]. This enzyme is missing or insufficient in patients suffering Fabry disease, a genetic disorder resulting in the accumulation of these glycolipids in cells within the kidneys, heart, skin, and cells lining the blood vessels, the result being early death. There are an estimated 5000 Fabry patients in Europe, and 7000 in the US. The recombinant product Fabrazyme was recently approved for use in these patients to reduce cellular lipid build-up and related neuropathic pain. Pre-clinical data with the tobacco-derived enzyme showed positive results using an animal model of Fabry disease [239]. The study showed that reductions of excess lipids characteristic of Fabry disease can be achieved in target organs with proper targeting of the enzyme to affected organs and no apparent toxicity or tissue damage. No neutralizing antibodies were observed to the glycosylated plant-derived enzyme in extended infusion studies, indicating that plant glycosylation in this protein may be non-immunogenic.

CropTech developed beta-glucocerebrosidase which catalyzes hydrolysis of the glucocerebrosides in lysosomes. This enzyme is either missing or insufficient in Gaucher's disease, the result being pain, fatigue, jaundice, bone damage, nerve damage, or even death. This condition is present in 10000–20000 patients in the US. Beta-glucocerebrosidase is available as Ceredase® derived from human placenta, or as Cerezyme® that is a mammalian cell culture recombinant product. The protein is a monomeric glycoprotein of 497 amino acids with modified carbohydrates (6% content by weight). The recombinant product requires *in vitro* altering of the sugar residues at the non-reducing ends of the oligosaccharide to obtain efficient enzymatic activity and to be recognized by carbohydrate receptors on macrophage cells. The recombinant version is delivered to patients every 2 weeks for life at an annual cost of $160000, making it one of the most expensive drugs available. This high price is mostly due to manufacturing costs. Using the CropTech process, a single tobacco plant could make enough glucocerebrosidase for one dose, resulting in significant cost reduction. Studies are in progress on carbohydrate modification of recombinant glycoproteins produced in plants to reduce or eliminate the need and cost of carbohydrate remodeling required for this enzyme [212]. CropTech also produced the lysosomal enzyme alpha-L-iduronidase responsible for mucopolysaccharidosis (MPS-I), a condition that severely affects normal growth and development.

5.3.3.2 Transgenic-derived Biopharmaceuticals in Clinical Trials to Treat Autoimmune Conditions

Goat-milk derived recombinant human alpha-fetoprotein In 2003, Merrimack Pharmaceuticals began a clinical trial to study MM-093, a recombinant version of human alpha-fetoprotein, to develop a novel therapeutic option for treatment of autoimmune diseases such as rheumatoid arthritis and multiple sclerosis. GTC Biotherapeutics provided purified MM-093 using its transgenic goat milk production technology. A transgenic system was selected, since recombinant human alpha-fetoprotein has been difficult to express and purify using traditional recombinant production systems. Merrimack recently completed dosing in its Phase I trial to determine the safety, tolerability, and pharmacokinetics of MM-093 in healthy volunteers [49]. Merrimack is planning a clinical study to assess the safety and tolerability of MM-093 in patients with rheumatoid arthritis, evaluating the relationship between pharmacokinetics and pharmacodynamic markers. This study will be eventually expanded to other autoimmune patient populations. MM-093 is the second protein using GTC Biopharmaceuticals production technology to enter human clinical studies.

5.3.3.3 Transgenic-derived Biopharmaceuticals in Clinical Trials for Cancer Treatment

Corn seed-derived huNR-LU-10 monoclonal antibody The NeoRx Corporation, in collaboration with Monsanto's Integrated Protein Technologies, conducted the first clinical trial of a transgenic (corn seed)-derived monoclonal antibody (MAb), huNR-LU-10, in 1998 [171, 373].

This full-size immunoglobulin G recognizes an epithelial cellular-adhesion molecule (EpCAM) present in colorectal cancer and other solid-tumor cancers. Serum antibodies such as immunoglobulin G are tetrameric glycoproteins composed of two identical heavy chains and two identical light chains. Antibodies are designed to recognize and bind target antigens with great specificity, making them great tools for the diagnosis, prevention, and treatment of disease. Antigen recognition depends on having the correct folding and assembly that needs to take place in the transgenic system during protein synthesis in order for antibodies to correctly recognize their antigens.

Corn seed-derived huNR-LU-10 was prepared from seeds harvested from multiple grooving sites and seasons followed by grinding in buffer, extraction, and purification based on two or three chromatographic steps. The initial chromatographic step consisted of Protein A affinity chromatography that is usually used for antibody purification. The purified antibody purity was >99% using a process with >50% recovery yield from corn seed to purified product. The process was shown to remove contaminants such as endotoxins, corn proteins and corn DNA to undetectable levels. The purified MAb was then conjugated with a radioisotope-binding complex (chelator), followed by *in vivo* conjugation with radioisotopes to allow its use for tumor imaging and for delivery of high doses of radiation directly to tumors with less exposure to normal tissue. The use of tumor-directed radiotherapy requires lower radiation exposure, potentially avoiding the need for marrow transplantation. Corn seed-derived huNR-LU-10 was genetically altered to remove the glycosylation site in the constant region of the heavy chain, thus avoiding plant-like glycosylation. Plant-like glycosylation is a potential source of structural variability and of human immunogenicity. Removing carbohydrates in MAbs such as huNR-LU-10 was shown to be possible because antigen binding is not affected by glycosylation, and the glycosylation-dependent effector functions are not required for the intended radiotherapy use of this MAb. Using binding studies, blood analysis, whole-body imaging, and cell-based assays it was determined that removal of glycosylation had no effect on antigen binding, pharmacokinetics, and *in vivo* targeting properties in both mice and humans, but resulted in reduced cellular-mediated complement activation and complete elimination of the antibody-dependent complement activation [373]. It was shown that accumulation of MAbs in corn seed allows long-term stable storage (at least 2 years) at ambient temperatures, without notable degradation or loss of activity [171]. The high stability of the product in corn seed allows the initial processing to be conducted at room temperature, which simplifies facility design and further reduces costs. Viral clearance steps are not needed, and this results in further cost savings.

The purity, consistency, potency, antigen binding, serum and urine clearance, target-tissue binding and stability for the purified antibody were reproducible and acceptable for conducting FDA-approved clinical trials. Pre-clinical studies and human clinical comparability studies with the analogous glycosylated mammalian cell culture-derived huNR-LU-10 indicated that the corn seed-derived product was as non-immunogenic and as effective as a anti-cancer agent as the mammalian-derived MAb. However, both molecules were withdrawn from development because of diarrhea and other side effects in Phase II trial patients, probably as a result of a cross-reaction of the MAb with related epitopes on the digestive system. These effects were not specific to the corn-derived antibody.

In addition to immunoglobulins, plants have been shown to be capable of producing many other antibody-related molecules. Two of those will be discussed in the next sections: antibody fragments used as antigens for cancer vaccines; and CaroRx, a secretory antibody for topical use. Secretory antibodies are dimers of serum antibodies linked by a joining peptide. In order properly to assemble a secretory antibody, plants need to express four different cloned genes. Plants have also successfully accumulated single chain and dimeric antibody fragments, bispecific fragments, diabodies, and antibody fusion proteins [121, 122]. Antibodies have been expressed successfully in many different plant systems in yields in excess of 1% of the total soluble protein. Antibodies have also been expressed in transgenic animals, and using aquatic plants, plant cell culture, and virus-infected plants (see Table 5.1).

MAbs were initially thought to be the ideal product category to implement the use of transgenic systems for the production of biopharmaceuticals. Several reasons made the use of transgenic systems attractive for MAb production. First, a eukaryotic production is needed for MAb production because of the MAb's structural complexity. Second, MAbs have demonstrated good accumulation levels in many transgenic systems with acceptable clinical performance, as discussed above. Third, MAbs are designated as well-characterized biologics facilitating the implementation of manufacturing changes expected to occur when a new technology is implemented. Finally, there are a large number of MAb products in the development pipeline, and those approved had rapid commercial acceptance in important therapeutic areas (cancer, arthritis, organ rejection, infectious disease treatment) leading to the potential requirement for large quantities of some of these MAbs. There are more than 200 MAbs under development, with nearly all of them being produced via mammalian cell culture. The need to develop transgenic technology for MAb production

was further justified by information in the late 1990s that their commercial success depended in having a low-cost, high-capacity production technology as enabled by transgenic systems. At that time, it was projected that there would be a significant shortage of biopharmaceutical production capacity. Table 5.1 illustrates that 43 companies and institutes had (or have) programs for the production of MAbs using transgenic technology. The leaders in the industry – GTC Biotherapeutics, Dow, and Monsanto – have (or had) several MAb production programs for milk or corn expression of several important commercial MAbs. GTC Biotherapeutics has programs to evaluate the production of important commercial MAbs such as Remicade® and Humira® that could be required in large quantities, and an additional six MAbs (IgG1, IgG2, IgG4) and three MAb-fusion biotherapeutics. Monsanto produced the glycosylated form of huNR-LU-10 in corn and several other immunoglobulins G in non-glycosylated (e.g., BR96, developed with Bristol-Myers Squibb Co. and Seattle Genetics) and glycosylated forms. However, recent increases in MAb accumulation in mammalian cell culture, coupled with the use of protein-free media and the expansion of the biopharmaceutical production capacity, had resulted in the realization that conventional mammalian technology can produce most of the antibody-related molecules considered for therapeutic use (see Part IV, Chapter 1).

Tobacco foliage-derived tumor-specific antigens used as individualized cancer vaccines
Large Scale Biology Corporation (LSBC) is using recombinant tobacco mosaic virus (TMV) containing genes for antigenic cancer markers to infect non-transgenic tobacco plants to deliver large amounts of individualized patient-specific antigenic pro-

tein in few weeks after cloning (see Part I, Chapter 2). These antigenic proteins are produced to manufacture personalized vaccines for use in cancer treatment [238]. These antigens are tumor-specific proteins containing the characteristics of each individual patient's cancer. LSBC has been able to manufacture these vaccines for 90% of patients qualified for clinical treatment in as little as 6 weeks from the receipt of biopsy materials.

The first vaccines currently in clinical trial are used for the treatment of non-Hodgkin's lymphoma, the most prevalent form of lymphoma and the sixth leading cause of cancer-related deaths in the US. This treatment works for lymphoma because all the cancer cells have the same surface protein, the unique antibody made by the original B cell that became malignant. The recombinant viral-infected tobacco plants are used to make a fragment from this unique antibody.

A mouse model of lymphoma was used in pre-clinical studies to validate the vaccination procedure using a tobacco-derived idiotype-specific single-chain variable region fragment of the immunoglobulin from the cancerous mouse B-cell lymphoma [236]. Non-vaccinated mice died within 3 weeks of tumor injection, while 80% of the vaccinated mice were protected from the cancer and survived.

So far, 16 individual vaccines from 16 patients have been produced from tobacco, 15 are glycosylated, and one is a non-glycosylated vaccine [374]. These vaccines were applied as 6-monthly subcutaneous injections in studies conducted at Stanford University. Excellent safety profile in all patients at all immunization times was observed, with significant cellular and humoral responses observed in 8/16 and 7/16 patients, respectively, equivalent to the response seen in previous cancer vaccine trials.

5.3.3.4 Transgenic-derived Biopharmaceuticals in Clinical Trials as Anti-infective Agents

Tobacco foliage-derived recombinant secretory antibody CaroRx® Planet Biotechnology's CaroRx® is a tobacco-derived secretory antibody in clinical trials since 1998 to prevent the adhesion of tooth decay-causing bacteria to the tooth surface [122, 191, 194]. The antibody recognizes the main adhesion protein of *Streptococcus mutans*, the oral pathogen responsible for tooth decay in humans. Tooth decay caused by bacterial infection results in about 70% of US dental expenditures, or $50 billion annually. The treatable population in the US and Europe is estimated at approximately 115 million people. CaroRx® has completed Phase I clinical trials under an approved US FDA Investigational New Drug (IND) application. A Phase II clinical trial indicated that CaroRx® on a topical application after the bacteria have been removed from the mouth helps to prevent recolonization by *S. mutans* for several months. CaroRx® can be applied either by dental hygienists or by the patients themselves after tooth cleaning. CaroRx® is expected to eliminate the decay-causing bacteria in 2 years.

CaroRx® is purified from tobacco foliage and then applied topically to the teeth. CaroRx® is a chimeric secretory immunoglobulin A/G that is produced in transgenic tobacco plants through the expression of four separate cloned genes. These genes were stacked by the sequential crossing of independent transgenic plants, each expressing a different component. Secretory antibodies such as CaroRx® consist of secreted immunoglobulin A dimers – the most abundant form of immunoglobulin in mucosal secretions, present in saliva, sweat, colostrum, and the mucosal epithelia of the human body. Secretory antibodies protect a vast surface of permanently exposed areas of the body against being attacked by exogenous pathogens, and they play a major role in host defense at mucosal surfaces by inhibiting colonization of pathogenic microorganisms. The immunoglobulin A dimers are associated with the joining J-chain that is added during secretion. Engineering plants to generate functional secretory antibodies allows the development of mucosal passive immunization. All other recombinant systems tested for the production of secretory antibodies had resulted in poor productivity and unacceptable yields of fully assembled antibodies.

Planet Biotechnology selected tobacco as the production platform for CaroRx® as the accumulation of heterologous proteins in tobacco is a well-established technology with high productivity due to tobacco's high biomass yield. In 2004, Large Scale Biology and Planet Biotechnology entered into a biomanufacturing agreement to extract and purify CaroRx® [375]. Tobacco plants expressing CaroRx® will be extracted at the Owensboro, Kentucky, LSBC manufacturing facility.

The production of secretory antibodies in plants represents an important opportunity for the commercialization of plant-derived biopharmaceuticals. Planet Biotechnology is developing two additional secretory antibodies. RhinoRx is under development for the treatment of colds due to rhinovirus, which represents about half of all common colds and over 20 million doctors' office visits a year. For the prevention of doxorubicin-induced hair loss (alopecia) – a disturbing side effect for cancer patients undergoing chemotherapy – Planet Biotechnology is developing DoxoRx. Each year in the US, over 250 000 patients receive chemotherapy that results in hair loss.

EpicCyte was developing secretory antibodies to provide products for unmet needs for sexual health (genital herpes, 45 million US patients), contraception (spermicidal, 42 million US potential users), HIV/AIDS (500 000 US potential users), respiratory conditions such as pneumonia (5 million US patients), and gastrointestinal conditions such as intestinal infections [191, 265].

Cows' milk-derived recombinant human lactoferrin Pharming completed Phase I human clinical studies using cows' milk-derived human lactoferrin as an anti-infective agent and for ophthalmic indications [376]. Pharming was the first to breed a transgenic bovine – the bull *Herman* (1990) – and *Herman*'s daughters produced milk containing lactoferrin. Lactoferrin is an iron-binding glycoprotein which is secreted into colostrum, milk and tears, and protects the new-born baby from potential infections [377]. Lactoferrin is also present in secondary granules of neutrophils deposited by these circulating cells in septic sites to attack infection and inflammation [378]. Its principal function is to act as a scavenging agent for non-protein-bound iron in body fluids and inflamed areas in order to suppress free radical-mediated damage and decrease accessibility of the metal to invading bacterial, fungal, and neoplastic cells. Potential therapeutic indications include the treatment of iron-deficiency anemia, gastrointestinal infections, dry-eye syndrome, and for use as an anti-inflammatory agent, anti-oxidant, and neutralizing agent for heparin. Iron-deficiency anemia is the most common nutritional deficiency in the world, affecting almost 25% of the world population, mostly young children and women of childbearing age [379]. Studies have shown a reduced frequency of diarrhea in breast-fed children, this being attributed to the antimicrobial action of the human milk lacto-ferrin and lysozyme by inhibiting growth of diarrhea-associated organisms such as rotavirus, *Cholera*, *Salmonella*, and *Shigella*. Lactoferrin could also be a nutritional sup-plement aimed at the prevention and treat-ment of gastrointestinal tract infections, especially for patients under immunosup-pressed conditions after chemotherapy or radiotherapy. Lactoferrin might also be ef-fective in ophthalmic and pulmonary ap-plications, as the protein is naturally pres-ent in tears and lung secretions. Lactofer-rin has been demonstrated to inhibit ma-lignant tumor growth, presumably through immunomodulation [380]. In ad-dition to these anti-infective and anti-in-flammatory uses, lactoferrin was found to increase osteoblast differentiation, reduce osteoblast apoptosis, and increase prolif-eration of primary chondrocytes, thereby indicating a role in new bone formation and a potential therapeutic use for the treatment of bone disorders [381]. Lactofer-rin could also be used in food preserva-tion, fish farming, and oral hygiene [378].

Phase I studies conducted by Pharming indicated that cows' milk-derived recombi-nant human lactoferrin (rhLF) is well tol-erated at high doses. Volunteers were in-jected intravenously with rhLF without negative side effects. An oral biodistribu-tion study in human volunteers has shown that natural and recombinant hLF behave in a similar fashion in the digestive tract.

Ventria is producing rhLF and recombi-nant human lysozyme (rhLZ) accumulat-ing in rice seed at a level of 5 g kg^{-1} flour weight [382]. Rice-derived rhLF and rhLZ were shown to be identical to the human proteins, and to have stability similar to native human lactoferrin and lysozyme when exposed to heat, pH changes, and *in vitro* digestion [308, 314, 383]. Both lacto-ferrin and lysozyme are multifunctional

proteins and play key roles in many as-pects of human health. The initial applica-tion of rhLF and rhLZ will be for the de-velopment of an oral rehydration solution to prevent and treat diarrhea in infants, babies and travelers [382]. Further applica-tions of rhLF and rhLZ include use in functional foods for health maintenance of individuals with a compromised immune system due to medical treatment of cancer, aging or HIV/AIDS, as both rhLF and rhLZ have reported immunostimulatory activity [384, 385]. Recent findings on lac-toferrin prevention of biofilm formation by microbial isolates from lung [386] and the critical role of lysozyme in pulmonary health [387] indicate that rhLF and rhLZ might be developed to treat patients with lung disease, for example those with cystic fibrosis.

Meristem is producing lactoferrin in corn seed for the treatment of dry-eye syndrome and gastrointestinal infections [164]. Lacto-ferrin is a complementary product to gastric lipase, Meristem's lead development prod-uct. Correct N-glycosylation has been deter-mined to be important to maintain lactofer-rin's stability. An analysis of corn-derived lactoferrin indicated that both N-glycosyla-tion sites are mainly substituted by typical plant-type glycans, with beta-1,2-xylose and alpha-1,3-linked fucose at the proximal *N*-acetylglucosamine. As expected, the com-plex-type glycans typical of human proteins are not present in maize recombinant lacto-ferrin [164]. Lactoferrin has also been accu-mulated in transgenic potatoes [243].

5.3.5
Transgenic-derived Oral Vaccines

Vaccines available today are injectable anti-gens made from killed or weakened ver-sions of a pathogen or of some material (usually proteins or protein fragments) derived from a pathogen. Injection of these antigens stimulates the immune system to behave as if the pathogen had infected the body; this results in a response to eliminate the pathogen and the creation of memory cells that later will repel the pathogen. Oral vaccines would be the preferred way to provide certain antigens, particularly for those pathogens that infect by oral routes of entry. Oral vaccines work by stimulating the digestive, respiratory, and/or reproductive mucosal immune system by delivering antigens in a stable matrix (usually the edible part of a plant) as described in many publications [10, 26, 112, 113, 115, 146, 192, 209, 226, 259, 260, 271, 273, 388–398]. Antigens expressed within the matrix of transgenic plants can assemble into complex structures and be stored in plant tissues; this allows them to act as effective antigens when released from the plant cell in the lower intestine. These antigenic materials are resistant to digestion and capable of reaching lymphoid tissues [146].

There are several challenges to develop effective oral vaccines that can be met by producing oral antigens in transgenic plants. Oral immunization requires larger amounts of antigen compared to injectable antigens – typically milligram rather than microgram quantities. Oral antigen cannot be produced economically in such large quantities using current vaccine production technologies (microbial fermentation, cell culture, eggs). Transgenic plants provide advantages in addition to cost and convenience, because these plants can produce significant quantities of protein and deliver oral antigens in an acceptable matrix for administration without purification. The use of plant derivatives with a long shelf-life or processed by freeze-drying can achieve antigen preservation at room temperature, avoiding cold chain transportation and storage. This represents

a significant advantage of plant-derived vaccines over current products, especially when they are to be delivered in developing regions, and due to lower costs compared to current vaccines and the reduction of hazards associated with injection.

For the successful implementation of transgenic systems producing oral vaccines, the system must accumulate stable, fully assembled, orally available antigens at high, consistent accumulation levels to enable controlled dosing. Plants that produce edible leaves, roots and fruits are the best choices for oral vaccines. Potato was selected as the first oral vaccine host because transformation and cultivation technologies were available in the late 1980s [399, 400], but today bananas, lettuce, lupine, spinach, sweet potato, corn, and tomato are all being developed for human vaccines, and alfalfa, corn and beans for animal vaccines. Emphasis is given to foods that are well-liked, consumed raw, and have a long shelf-life so that the acceptability and effectiveness of the technology are enhanced.

In 1990, plants were shown capable of expressing biologically active antigens, as demonstrated by the expression of the surface protein antigen of the dental bacterium *Streptococcus mutans* in tobacco [401–403]. When fed to mice, biologically active antibodies were induced to inhibit growth of these bacteria.

The heat-labile toxin B subunit of *E. coli* (LTB) [398], hepatitis B surface antigen [404], respiratory syncytial virus F protein [258], measles virus hemagglutinin [180], and Norwalk virus capsid protein [240, 405, 406] have each been successfully expressed in plants and delivered orally in animals or humans to determine their immunoprophylactic activity. The first account of a human clinical trial of oral vaccine based on an *E. coli* enterotoxin as antigen delivered by consuming raw potatoes was published in 1998 [407]. Ten of the 11 test subjects produced specific antibodies to the toxin used, whilst no specific antibodies were produced in the control subjects. Immunity level was comparable to that measured in volunteers exposed to live organisms. The study demonstrated that oral vaccines could survive digestion delivered in a plant matrix and effectively stimulate an immune response. Several other animal and human studies based on *E. coli* enterotoxin for the prevention of traveler's diarrhea have been reported, these having used potato, tobacco, and corn as the delivery matrix [102, 115, 207, 271, 408, 409]. In addition to these two anti-bacterial plant-derived vaccines, other immunoprophylactics delivered in transgenic plants have been tested in clinical trials, including the hepatitis B surface antigen, analogous yeast-derived Recombivax® by Merck, expressed in potato or lettuce. This antigen has been used as a prototype for vaccines and also expressed in banana, tobacco, and lupin [100, 104, 200, 226, 332, 404, 410–414]. Other examples of antigens expressed in transgenic plants for the prevention of human diseases include tomato-derived rabies glycoprotein [222], *Helicobacter pylori* antigen for preventing peptic ulcer/cancer [415], rotavirus antigen for preventing severe diarrhea [415], tobacco-derived human cytomegalovirus glycoprotein [301], tobacco and potato-derived cholera antigen [244, 246, 247], and Norwalk virus capsid protein expressed in potato, tobacco, or tomato [240, 257, 400, 405, 406, 416]. Research is also under way to create oral vaccine candidates in transgenic corn for AIDS prevention and for treatments using antigens derived from the simian immunodeficiency virus [288]. Genetically engineered spinach has been used to express HIV-suppressing proteins

in an attempt to develop a safe and inexpensive AIDS vaccine [417]. The Institut für Pflanzengenetik und Kulturpflanzenforschung has genetically engineered plants to produce human papillomavirus Type 16 virus-like particles to develop plant vaccines against cervical cancer, the third most common cancer among women worldwide [101, 146, 185, 210, 418, 419]. Dow Pharma (Midlands, IL, USA) is working with the Fraunhofer Institute and the National Institutes of Health to develop technology based on viral particle production in foliage for the rapid development of vaccines against infectious diseases, including biowarfare agents, to be delivered by capsule or nasal spray [420].

Many animal vaccines are also under development such as those based on potato-derived rabbit hemorrhagic disease virus antigen, mink enteric virus antigen, *Arabidopsis* and alfalfa-derived foot-and-mouth disease antigens for agricultural domestic animals [421, 422], and *Arabidopsis*, tobacco, corn-derived transmissible gastroenteritis coronavirus antigens for pigs [287].

Transgenic animals have also been considered for the production of vaccines. GTC Biotherapeutics is currently working with the National Health Institute to develop a malaria vaccine based on the production of the viral surface protein antigen MSP-1 in goat milk. This antigen is difficult to express in conventional recombinant production systems.

One challenge related to the use of oral vaccines is the potential risk of inducing oral tolerance to antigens, and thus the use of oral vaccines requires control and monitoring of the administration of these products as any other biopharmaceutical. However, oral tolerance may also be used for the development of orally delivered treatments for autoimmune diseases. Plant-derived "autoantigens" may be able to suppress immune activity related to autoimmune diseases. Autoimmune response is involved in several diseases, including insulin-dependent diabetes, psoriasis, systemic lupus erythematosus, Graves' disease and rheumatoid arthritis. The oral administration of disease-specific autoantigens may either prevent or delay the onset of autoimmune disease symptoms. Studies are in progress to determine if plant-derived oral autoantigens can suppress autoimmunity related to type I diabetes [244–247]. Insulin-producing beta-cell proteins can elicit autoimmunity in people predisposed to type I diabetes. A potato-based diabetes vaccine candidate was developed based on using insulin or glutamic acid decarboxylase linked to the innocuous B subunit of the cholera toxin to enhance uptake of the antigenic fusion protein by the gut-associated lymphoid tissues [244–247]. Studies showed that repeated feeding of these antigens to mice with a tendency to become diabetic was effective in suppressing autoimmune responses, and delaying the onset of high blood sugar levels, an indication of diabetes [244].

Plant-derived vaccines and toxin deactivation enzymes can also play a critical role in biodefense by enabling the availability of unlimited, low-cost proteins and other biological agents to generate stable stockpiles of preventive and therapeutic agents to protect military and civilian populations. Research is currently in progress to address the danger of smallpox, plague, Ebola virus, and anthrax virus when used as bioterrorist weapons, with those nations at risk developing programs rapidly to vaccinate part or all of their population, though this requires stockpiles to be prepared, stored, and renewed. Plant-derived vaccines can become ideal substitutes for traditional vaccines for biodefense applications. A plant-derived vaccine for anthrax, based on the anthrax protective antigen, is under development using tobacco transformed for expression of the antigen in the chloroplast [423]. The current anthrax vaccine, which was designed in 1950, causes edema and has other lethal facets that lead to harmful side effects. The use of chloroplast expression minimizes the risk of gene spread, and can lead to higher accumulation levels than with nuclear expression. Spinach infected with a recombinant virus containing the protective antigen is also being considered for the production of a vaccine against anthrax infection [424].

In a program related to biodefense, Nexia is developing jointly with US and Canadian defense agencies a recombinant version in goat milk of human butyrylcholinesterase (hBChE) enzyme (Protexia®) [15]. This agent can act as an enzymatic bioscavenger for nerve agents, such as soman, sarin, VX and tabun, to absorb and degrade organophosphate poisons before they cause neurological damage. Studies using plasma-derived hBChE have shown that increasing hBChE concentrations in the blood protects laboratory animals from the toxic effects of nerve agents. Protexia® is being developed for post-exposure (rescue) therapy and military prophylaxis to prevent against the toxic effects of nerve agents. Pharmacokinetics studies in animals have shown that a single injection of Protexia® resulted in a sustained elevation of blood BChE levels for many hours.

5.3.6
Transgenic-derived Proteins for Use in Medical Devices and Drug Delivery

Many animal- and human-derived protein including silk, elastin, fibrinogen, collagen, and gelatin, are used in medical devices and for drug delivery. Transgenic systems can provide cost effective technology in the production of structural proteins for medical use that are required to be biocompatible and stable in biological tissues.

The most advanced structural protein production program based on using a transgenic system is the production of goat milk-derived spider silk (BioSteel®) by Nexia [55]. Spider silk is the strongest fiber known, holding up to 400 000 lb per square inch $(281 \times 10^6 \text{ kg m}^{-2})$ without breaking. Dragline spider silk proteins contain iterated alanine-rich crystal-forming blocks with mechanical strength and glycine-rich amorphous blocks that provide elasticity. Recombinant production is the only alternative for producing this fiber commercially as it cannot be harvested from spiders [425]. Transgenic technology is required for its production because conventional recombinant microbial or cell culture production systems have been not been successful in expressing silk genes. These genes are large and contain many repetitive units which stress the protein synthetic machinery when these organisms are grown in bioreactors. Nexia is producing soluble recombinant spider silk using traditional goat dairy techniques for milk collection, after which the protein is extracted from milk and then spun into fibers [426]. In 2000, two transgenic goats – *Peter* and *Webster* – were born with the spider silk gene incorporated into their genetic composition, and used to generate the milking herd. BioSteel® is being developed for use in medical device products used in surgery, ophthalmic applications and as prostheses. BioSteel® can also be used in industrial applications such as in lightweight, flexible bullet-resistant body armor for the military and law enforcement agencies, as well as high-performance sporting equipment such as biodegradable fishing lines and nets. Nexia have worked with the US Army Soldier and Biological Chemical Command to develop techniques for making fibers from soluble recombinant spider silk proteins, and have demonstrated that wet spinning of fibers is possible from a concentrated aqueous solution of mammalian cell culture and goat milk-derived spider silk monofilaments. Nexia also collaborated with Acordis Specialty Fibers to develop spider silk fibers and specialty materials for industrial, textile, medical, and hygiene applications.

Synthetic spider silk has also been produced in transgenic tobacco and potato expressing the endogenous silk protein genes of the spider *Nephila clavipes* [147]. Proteins of up to 100 kDa in size, and with 90% identity to silk protein, were produced in tobacco leaves, potato leaves and potato tubers at up to 2% of the total soluble protein accumulation level [151].

Pharming and Infigen are each developing milk-derived recombinant human fibrinogen (rhFIB) to be used as tissue sealant to stop internal or external bleeding during surgery or after traumatic injury [50, 51, 427]. Human fibrinogen is a soluble blood protein that can form insoluble fibrin polymers after activation by thrombin. In 2000, the fibrinogen market was estimated at US$284 million. Commercially available fibrin sealants use fibrinogen purified from human donor plasma. As with other plasma products, there are safety, availability, quality, and reproducibility concerns related to the use of plasma-derived fibrinogen. Furthermore, a substantial shortage of human fibrinogen is anticipated as market demand is expected to increase to 500 kg per year and require $>10^6$ L of donor blood. Pharming has established a production line with high expression of rhFIB that is virtually identical to plasma fibrinogen, whilst Infigen claims an accumulation level of 2.4 g L^{-1} rhFIB in cows' milk.

Several groups have demonstrated an accumulation of recombinant human collagen-related proteins in tobacco, milk, and silkworms [50, 51, 56, 177, 376, 428–431]. Collagen available today for medical uses is mostly a by-product of the meat industry, produced mainly from bovine hooves and porcine/bovine bones. There is a mammalian cell culture-derived collagen which is available only for selected, high-value applications due to its limited availability and high cost. Current uses of collagen as a component of medical devices include hemostats, vascular/tissue sealants, implant coatings, artificial skin, bone graft substitutes, dental implants, and wound dressings. Injectable collagen solutions are used for dermal augmentation and the treatment of incontinence. Many additional applications are under development, such as a component for the tissue engineering of cartilage, bone, skin, artificial tendons, blood vessels, nerve regeneration, and for drug delivery [177]. Gelatin, made from hydrolyzed collagen, also has many medical applications such as a vaccine/biologic stabilizer and as a plasma expander. Gelatin is also used to manufacture hemostat sponges, hard capsules, soft capsules and gel tablets. Most of the collagen and gelatin available is a variable mixture of several collagens and is not highly purified; this leads to the possibility of causing inflammatory reactions in those individuals who are sensitive to animal-derived components. In addition, over the past few years there has been growing concern about the potential for contamination of bovine products with "mad cow disease" and its human variant causing Creutzfeldt–Jakob disease.

Recombinant collagen and gelatin products provide a consistent and reliable human, animal or engineered amino acid composition material which is compatible with current pharmaceutical manufacturing processes and potentially free of animal-derived components and pathogens [177]. FibroGen developed a production technology that demonstrated the use of recombinant insect cells, yeast, transgenic plants, and transgenic animals, accumulating stable recombinant human collagen and gelatin. This technology is based on expressing the genes for collagen or collagen fragments simultaneously with prolyl hydroxylase, resulting in recombinant collagen which is stable at biologically relevant temperatures. Yeast was selected for the production of recombinant collagen and gelatin for most medical applications as it accumulates fully assembled, stable collagens and gelatin fragments at high levels.

Transgenic systems could be used for the cost-effective, large-scale production of recombinant human collagens and gelatins for selected medical applications requiring large quantities at low cost. Human types I and III collagen homotrimers have been expressed in transgenic tobacco plants [177], while transgenic mice have been engineered to produce full-length type I procollagen homotrimer in milk [428, 429]. Most recently, a transgenic silkworm system was used to produce a fusion protein containing a collagenous sequence [430]. As seen in other recombinant expression systems, these transgenic systems lack sufficient endogenous prolyl hydroxylase activity to produce fully hydroxylated collagen. In mice and tobacco, this deficiency was overcome by over-expression of human prolyl hydroxylase, analogous to the procedures conducted in yeast and insect cell culture [177]. Pharming and Infigen have each demonstrated an accumulation of recombinant collagen related proteins in milk, with cows at Infigen accumulating recombinant human collagen-related molecules at a concentration of $8 \, g \, L^{-1}$ in milk [50]. Meristem has shown that human collagen can be produced in transgenic tobacco plants, and that the protein is spontaneously processed and assembled into its typical triple-helix conformation [160, 167]. The plant-derived collagen had a low thermal stability owing to the lack of hydroxyproline residues, but this was remedied by co-expressing with animal proline-4-hydroxylase [163].

Recombinant elastin has not been produced in transgenic systems as it is not currently a commercial product with high-volume demand. Elastin is used for the production of vascular grafts [432–436] which today are frequently used to replace a damaged artery or to create a new artery for improved blood flow. Whilst use of autologous vessels is preferable, synthetic grafts made from expanded polytetrafluoroethylene are used on many occasions when autologous vessels are not available. However, this synthetic material may be thrombogenic and result in smooth muscle cell hyperplasia. It has been shown that recombinant elastin fibers can form fibrillar structures and can also be cross-linked to form stable, insoluble structures similar to natural elastin [432]. These self-assembly and cross-linking abilities – combined with the biological characteristics of low platelet activation and inhibitory effects on smooth muscle cell growth – make recombinant elastin-based polypeptides ideal candidates for coating vascular grafts.

5.4
Conclusions

Transgenic systems offer production alternatives for biopharmaceuticals, and have significant advantages when compared with bioreactor-based microbial and cell culture-based recombinant production systems. However, transgenic systems remain largely untested for production biopharmaceuticals, as products derived from transgenic systems are not yet commercially available for human use. By the time this book has been published however, this situation may have changed, as antithrombin III derived from goat milk may be marketed. As discussed earlier, there are three transgenic plant-derived commercial products available for non-human use, and 10 plant- and milk-derived biotherapeutics with human clinical experience. Many vaccines are also under development from transgenic systems, some with limited human clinical data, and several structural proteins under development for potential medical uses. These diverse products represent what hopefully will be by 2010 the first wave of products from transgenic systems that will facilitate the implementation and acceptance of transgenic technology by the pharmaceutical industry, regulatory agencies, the medical community, and patients.

The benefits on healthcare of biopharmaceuticals derived from transgenic technology will be reflected by the commercial availability of new products and therapies, while facilitating the delivery, availability, and accessibility of existing biopharmaceuticals that cannot be produced using current production approaches. Likewise, the implementation of transgenic systems will enable the commercialization of complex multimeric proteins (e.g., viral particles for oral vaccines and secretory antibodies) that are difficult – if not impossible – to create using current methods. Transgenic systems will also allow the production of fusion proteins, metabolic toxic proteins, and unstable peptides that cannot be produced efficiently in bioreactor-based production systems.

The implementation of human-like post-translational processing in specific tissues of transgenic systems generating human-like processed biopharmaceuticals may result in products with improved homogeneity and engineered therapeutic performance.

By making biopharmaceuticals available practically in unlimited quantities and at low cost, transgenic systems will improve the accessibility of biopharmaceuticals to patients with life-long needs for such products. Likewise, transgenics may enable biogenerics to allow pharmaceutical companies to maintain the profit margins required to sustain product development. Unlimited, high-quality, low-cost biopharmaceuticals will also facilitate the implementation of products with multiple indications and the administration of biopharmaceuticals using non-injectable delivery routes (oral, transdermal, pulmonary). Non-injectable biopharmaceutical delivery often requires high doses due to poor bioavailability and degradation associated with these routes of administration. The therapeutic use of blood proteins derived from transgenic systems can expand the use of blood-derived factors as the availability, economics and safety parameters associated with many such current biopharmaceuticals are unattractive.

Transgenic production will facilitate the use of biopharmaceuticals and nutraceuticals in oral formats for the prevention of infectious and autoimmune diseases, and also for biodefense. The availability of biopharmaceuticals and nutraceuticals in stable matrixes that do not require refrigeration during transportation and storage will undoubtedly impact on the accessibility of these products to healthcare and nutrition crises in developing countries.

Finally, transgenic systems may permit the commercialization of functionalized, consistent quality, low-cost, protein-based biomaterials resulting in improved biocompatibility, accessibility, quality and performance for medical devices and for advanced drug delivery systems.

We can expect the implementation of transgenic technology for the production of biopharmaceuticals to take place in the future as this technology is suited to meet several critical needs confronting the healthcare industry during the next few decades. First, there is a worldwide increase in the health-conscious and physically active aging population that requires innovative high-performance health maintenance products, together with nutrition products in larger quantities and at lower cost that potentially only transgenic systems will be able to deliver. The economic improvements in many currently developing regions of the world such as Eastern Europe, China, and India will significantly increase the demand for biopharmaceuticals, as well as the growth resulting from an increasingly aging population in the US, Japan, and Western Europe. Second, the worldwide growth of nationalized healthcare systems and of managed care organizations will result in significant pressure to reduce the costs of biopharmaceuticals. In the US, major efforts are currently being made to obtain drugs from outside the country to reduce costs, and this is being encouraged by some government officials. It is clear that keeping drug prices at high levels in particular regions will be difficult to achieve in the future. The increasing use of combination therapies for diseases such as arthritis, diabetes and cancer further aggravates the need to control overall therapy costs, as only limited resources are available for each patient. Third, the pharmaceutical market is fractionated into many companies, with no single company controlling more than 15% of the market. Patent expirations leading to the eventual introduction of biogenerics (see Part VIII, Chapter 3), reduced R&D productivity, increased development costs ($800 million for a new therapeutic agent), and outsourcing activities to regions with lower costs will place increasing pressure on the pharmaceutical industry to improve its efficiency (see Part IV, Chapter 16). Transgenic technology can reduce the production cost of biopharmaceuticals, thereby allowing the industry to maintain its traditionally high profit margins that are required in a difficult business and innovative drug development environment.

Fourth, the regulatory agencies are becoming harmonized worldwide and subject to economic and political pressures to improve the diversity, cost benefits, and accessibility of drugs. Once transgenic systems become accepted by these organizations, there will be significant pressure to accelerate their implementation.

Finally, new technologies related to the rapid advancements in genomics, proteomics, bioanalytics, high-throughput screening, and protein engineering will improve the therapeutic ratio of current biotherapeutics. This in turn will create new therapies that should generate many new biopharmaceuticals requiring transgenic systems for their successful commercialization.

The success of any innovative technology resides in providing benefits to all of those involved in its implementation and commercialization. Transgenic technology will be implemented if its products meet the needs of the many entities involved in their commercialization: agricultural biotechnology companies, seed producers, farmers, processors, food producers, pharmaceutical industry, medical personnel, insurance companies, regulatory agencies, and patients. The acceptance of transgenic systems for the production of biologics could be negatively impacted by another accident such as the Starlink incident [437], which involved the inadvertent mixing of plant or animal materials containing a biopharmaceutical into the food supply, or into the environment. The government, activists, and food producers will demand that farmers and others involved in the processing of agricultural materials each test for the presence of biopharmaceuticals to very low detection levels, though no scientific evidence of harm to humans or animals may be demonstrated. The producers would have to pay for the testing, and also could lose sales abroad and domestically, even forcing food producers to reformulate their products to remove any use of potentially contaminated material. At present, producers are working on containment strategies to insure that no agricultural material from a transgenic-derived biopharmaceutical will enter the food chain or the environment (see Part IV, Chapter 7).

Procedures required for transgenic technology implementation and currently undefined regulatory requirements will affect the cost of biopharmaceuticals from transgenic systems. There are many "unknowns" in the regulatory environment, and these pose risks for those entities whose participation downstream of the transgenic production technology – specifically the pharmaceutical industry – is necessary to implement these production systems. As with any new technology, risks will have to be in balance with the benefits to patients, farmers, processors, food producers, governments, insurers, seed companies, and to the healthcare industry of the commercialization of transgenic-derived biopharmaceuticals. The success of the technology developers, processors, and farmers in managing these hazards – combined with public/industry/regulatory acceptance of the first wave of transgenic-derived biopharmaceuticals – will allow transgenic technologies to deliver significant benefits to the healthcare and agricultural industries, thus illustrating the significant value of the agricultural biotechnology as applied to human health.

References

1 Guide to Biotechnology. 2004. Biotechnology Industry Association www.bio.org.

2 DePalma A. Is a Green Plant in Your Manufacturing Future? BioPharm Int., 2003; 16(11): 24–33.

3 USDA. US Biotech Crop Plantings Fact Sheet – A Precis From The USDA Report. http://usda.mannlib.cornell.edu/reports/nassr/field/pcp-bba/acrg0604.txt 2004.

4 Peterson R, Arntzen CJ. On risk and plant-based biopharmaceuticals. Trends Biotechnol., 2004; 22(2).

5 Denman J, Hayes M, O'Day C, et al. Transgenic expression of a variant of human tissue-type plasminogen activator in goat milk: purification and characterization of the recombinant enzyme. Biotechnology (NY), 1991; 9(9): 839–843.

6 Barta A. The expression of a nopaline synthase human growth hormone chimaeric gene in transformed tobacco and sunflower callus tissue. Plant Mol. Biol., 1986; 6: 347–357.

7 Steiner U. Business Case for Plant Factories. 2003; Plant-Made Pharmaceuticals Conference, Quebec City, Quebec, Canada (on line copy available at http://www.cpmp2003.org/).

8 Baez J. State of the Science: Role of Transgenic Technology in the Biosynthesis of Bio-Pharmaceuticals and Industrial Protein. 2004 (Corn-Produced Pharmaceuticals and Industrials Risk Assessment Symposium, sponsored by the Biosafety Institute for Genetically Modified Agricultural Products Symposium, Ames, Iowa).

9 Behboodi E, Chen M, Destrempes L, et al. Principles of Cloning. Elsevier Science (USA), 2002.

10 Larrick JW, Thomas DW. Producing proteins in transgenic plants and animals. Curr. Opin. Biotechnol., 2001; 12(4): 411–418.

11 Giddings G, Allison G, Brooks D, et al. Transgenic plants as factories for biopharmaceuticals. Nature Biotechnol., 2000; 18(11): 1151–1155.

12 Herbers K, Sonnewald U. Production of new/modified proteins in transgenic plants. Curr. Opin. Biotechnol., 1999; 10(2): 163–168.

13 Okada Y. [Transgenic plants as medicine production systems]. Nippon Yakurigaku Zasshi, 1997; 110 Suppl 1: 1P–6P.

14 Schillberg S, Fischer R, Emans N. Molecular farming of recombinant antibodies in plants. Cell. Mol. Life Sci., 2003; 60(3): 433–445.

15 Twyman RM, Stoger E, Schillberg S, et al. Molecular farming in plants: host systems and expression technology. Trends Biotechnol., 2003; 21(12): 570–578.

16 Erickson L, Yu W, Brandle J, et al. (Eds), Molecular farming of plants and animals for human and veterinary medicine. Dordrecht, The Netherlands: Kluwer Academic Publishers, 2002.

17 Cunningham C, Porter A (Eds), Methods in molecular Biology – the production of clinically important recombinant proteins in plants. Totowa, New Jersey, U.S.: Humana Press, 1998.

18 Rohricht P, Ayares D. Transgenic Protein Production: Part 3, Nuclear Transfer Technology. BioPharm, 1999; 56–60.

19 Fischer R, Stoger E, Schillberg S, et al. Plant-based production of biopharmaceuticals. Curr. Opin. Plant Biol., 2004; 7(2): 152–158.

20 Fischer R, Emans N. Molecular farming of pharmaceutical proteins. Transgenic Res., 2000; 9(4/5): 279–299.

21 Houdebine L. Transgenic animal bioreactors. Transgenic Res., 2000; 9(4/5): 305–320.

22 Giddings G. Transgenic plants as protein factories. Curr. Opin. Biotechnol., 2001; 12(5): 450–454.

23 Kusnadi A, Nikolov Z, Howard JA. Production of recombinant proteins in transgenic plants: practical considerations. Biotechnol. Bioeng., 1997; 56: 473–484.

24 Horn ME, Woodard SL, Howard JA. Plant molecular farming: systems and products. Plant Cell. Rep., 2004; 22(10): 711–720.

25 Hood EE, Kusnadi A, Nikolov Z, et al. Molecular farming of industrial proteins from transgenic maize. Adv. Exp. Med. Biol., 1999; 464: 127–147.

26 Langridge WH. Edible vaccines. Sci. Am., 2000; 283(3): 66–71.

27 Ganz P, Sardana R, Dudani A, et al. In: Owen M, Pen J (Eds), Transgenic Plants: A production system for pharmaceutical proteins. New

York: John Wiley and Sons Ltd.; 1996: 281–297.

28 Molecular farming may be next wave for biotech drug manufacturers. January 31, 2003. The Food and Drug Letter Issue, No. 668.

29 Van Brunt J. Molecular Farming Factories. Signals Magazine, Online Magazine of Biotechnology Industry Analysis 2002; http://www.signalsmag.com.

30 Raskin I, Ribnicky DM, Komarnytsky S, et al. Plants and human health in the twenty-first century. Trends Biotechnol., 2002; 20(12): 522–531.

31 Meni Y, Kovacs J. In: Erickson L, Yu W, Brandle J, et al. (Eds), Molecular Farming of Plants and Animals for Human and Veterinary Medicine. Kluwer Academic; 2002: 287–317.

32 Knäblein J, McCaman M. Modern Biopharmaceuticals – Recombinant Protein Expression in Transgenic Plants. Screening. Trends Drug Discov., 2003; 6: 33–35.

33 Knäblein J. Biotech: A New Era In The New Millennium? Fermentation and Expression of Biopharmaceuticals in Plants. Screening. Trends Drug Discov., 2003; 4: 14–16.

34 Knäblein J. Biopharmaceuticals expressed in plants – a new era in the new Millennium. In: Müller R, Kayser O (Eds), Applications in Pharmaceutical Biotechnology. Wiley-VCH 2004.

35 Knäblein J. Plant-based Expression of Biopharmaceuticals. In: Meyers RE (Ed), Encyclopedia of Molecular Cell Biology and Molecular Medicine. Wiley & Sons, 2005; 2nd edition, Vol. 10: p. 385–410.

36 Stromqvist M, Houdebine L, Andersson J, et al. Recombinant human extracellular superoxide dismutase produced in milk of transgenic rabbits. Transgenic Res., 1997; 6(4): 271–278.

37 Galet C, Le Bourhis CM, Chopineau M, et al. Expression of a single beta-alpha chain protein of equine LH/CG in milk of transgenic rabbits and its biological activity. Mol. Cell. Endocrinol., 2001; 174(1/2): 31–40.

38 BioProtein Technologies Scientific Publications. http://www.bioprotein.com/gb/news_events.htm#Séquence_1 2004.

39 Transgametic® Process Gala Transgenics. http://www.gala.com/transgametic.html 2004.

40 GTC Biotherapeutics Scientific Publications. http://www.transgenics.com/publications.html 2004.

41 Echelard Y, Meade H. Toward a new cash cow. Nature Biotechnol., 2002; 20(9): 881–882.

42 Stowers AW, Chen LH, Zhang Y, et al. A recombinant vaccine expressed in the milk of transgenic mice protects Aotus monkeys from a lethal challenge with *Plasmodium falciparum*. Proc Natl Acad Sci USA, 2002; 99(1): 339–344.

43 Pollock DP, Kutzko JP, Birck-Wilson E, et al. Transgenic milk as a method for the production of recombinant antibodies. J. Immunol. Methods, 1999; 231(1/2): 147–157.

44 Young MW, Meade H, Curling JM, et al. Production of recombinant antibodies in the milk of transgenic animals. Res. Immunol., 1998; 149(6): 609–610.

45 Edmunds T, Van Patten SM, Pollock J, et al. Transgenically produced human antithrombin: structural and functional comparison to human plasma-derived antithrombin. Blood, 1998; 91(12): 4561–4571.

46 Ebert KM, Ditullio P, Barry CA, et al. Induction of human tissue plasminogen activator in the mammary gland of transgenic goats. Biotechnology (NY), 1994; 12(7): 699–702.

47 Ditullio P, Cheng SH, Marshall J, et al. Production of cystic fibrosis transmembrane conductance regulator in the milk of transgenic mice. Biotechnology (NY), 1992; 10(1): 74–77.

48 Konkle BA, Bauer KA, Weinstein R, et al. Use of recombinant human antithrombin in patients with congenital antithrombin deficiency undergoing surgical procedures. Transfusion, 2003; 43(3): 390–394.

49 Merrimack Pharmaceuticals Completes Dosing in Phase I Study of Immunomodulator, MM-093. http://www.merrimackpharma.com./news.archive.html#11_19_03 2003.

50 Infigen Transgenic Production Technology. http://www.infigen.com/sci_tech_prot.html 2004.

51 Infigen Terminates Agreement With Pharming Holding, N.V. http://www.biospace.com/ccis/news_story.cfm?StoryID=6578115&full=1 2001.

52 Niemann H, Halter R, Carnwath J, et al. Expression of human blood clotting factor VIII in the mammary gland of transgenic sheep. Transgenic Res., 1999; 8(3): 237–347.

53 Ko J, Lee C, Kim K, et al. Production of biologically active human granulocyte colony stimulating factor in the milk of transgenic goat. Transgenic Res., 2000; 9(3): 215–222.

54 Nexia Biotechnology Fundamental Biology. http://www.nexiabiotech.com/en/03_bio/index.php 2004.

55 Lazaris A, Arcidiacono S, Huang Y, et al. Spider silk fibers spun from soluble recombinant silk produced in mammalian cells. Science, 2002; 295(5554): 472–476.

56 Pharming Products. http://www.pharming.com/index.php?act=prod 2004.

57 van Berkel PH, Welling MM, Geerts M, et al. Large scale production of recombinant human lactoferrin in the milk of transgenic cows. Nature Biotechnol., 2002; 20(5): 484–487.

58 Pharming Features Clinical Results Of C1 Inhibitor At Investigator Meeting. http://www.pharming.com/index.php?act=show&pg=5 2004.

59 PPL Therapeutics Products. http://www.ppl-therapeutics.com/products/products.html 2004.

60 Vogel G. Biotechnology. Sheep fail to produce golden fleece. Science, 2003; 300(5628): 2015–2016.

61 Harris DP, Andrews AT, Wright G, et al. The application of aqueous two-phase systems to the purification of pharmaceutical proteins from transgenic sheep milk. Bioseparation, 1997; 7(1): 31–37.

62 Chen SH, Vaught TD, Monahan JA, et al. Efficient production of transgenic cloned calves using preimplantation screening. Biol. Reprod., 2002; 67(5): 1488–1492.

63 Positive clinical trial result on BSSL. http://www.ppl-therapeutics.com/news/news_2_content_15.asp 2001.

64 Van Cott K, Lubon H, Russell C, et al. Phenotypic and genotypic stability of multiple lines of transgenic pigs expressing recombinant human protein C. Transgenic Res., 1997; 6(3): 203–212.

65 Van Cott K, Lubon H, Gwazdauskas F, et al. Recombinant human protein C expression in the milk of transgenic pigs and the effect on endogenous milk immunoglobulin and transferrin levels. Transgenic Res., 2001; 10(1): 43–51.

66 Uusi M, Hyttinen J, Korhonen V, et al. Bovine as1-casein gene sequences direct high level expression of human granulocyte-macrophage colony-stimulating factor in the milk of transgenic mice. Transgenic Res., 1997; 6(1): 75–84.

67 Hematech Scientific Publications. http://www.hematech.com/hematech/publications/default.asp 2004.

68 Cloned Cows Generate Human Antibodies. Nature Biotechnol., 2002; 10: 1038.

69 Therapeutic Human Polyclonals Products. http://www.polyclonals.com/index.html 2004.

70 Birse D, Lacroix D, Dyck M, et al. Biotherapeutics from Transgenic Porcine Sources: Bioprocessing Approaches and Challenges. Bio-Processing J., 2004; March/April: 37–44.

71 Ryoo Z, Kim M, Kim K, et al. Expression of recombinant human granulocyte macrophage-colony stimulating factor (hGM-CSF) in mouse urine. Transgenic Res., 2001; 10(3): 193–200.

72 Utilizing the Bladder as a Bioreactor in Transgenic Animals. http://www.med.nyu.edu/OIL/Biotech/BSun.html 2004.

73 Kerr DE, Liang F, Bondioli KR, et al. The bladder as a bioreactor: urothelium production and secretion of growth hormone into urine. Nat Biotechnol. 1998; 16(1): 75–79.

74 Advanced Bionutrition Crustacean Expression System. http://www.advancedbionutrition.com/html/prod_platforms.html#crustacean 2004.

75 Chesapeake PERL Technology. http://www.c-perl.com/2004.

76 Advanced Cell Technology Scientific Publications. http://www.advancedcell.com/Scientific-papers.htm 2004.

77 Nextran Information. http://www.baxter.ca/htdocs/en/doctors/renal_therapies/nextran.html 2004.

78 AviGenics Technology Overview. http://www.avigenics.com/services.htm 2004.

79 Rapp JC, Harvey AJ, Speksnijder GL, et al. Biologically active human interferon alpha-2b produced in the egg white of transgenic hens. Transgenic Res., 2003; 12(5): 569–575.

80 BioAgri Products. http://www.bioagricorp.com/strategic.aspx 2004.

81 Alper J. Biotechnology. Hatching the golden egg: a new way to make drugs. Science, 2003; 300(5620): 729–730.

82 Chang K, Qian J, Jiang M, et al. Effective generation of transgenic pigs and mice by linker based sperm-mediated gene transfer. BMC Biotechnol., 2002; 2(1): 5.

83 Coghlan A. Big breakfast – Crack open an egg and cure a disease. New Scientist Magazine, 1999; 11.

84 GenWay Biotech Products. http://
www.genwaybio.com/
DesktopPage.aspx?TabID=3329&Lang=en-US
2004.

85 Etches RJ, Verrinder Gibbins AM. Strategies
for the production of transgenic chicken.
Methods Mol. Biol., 1997; 62: 433–450.

86 Linda Foster Benedict. Transforming Chickens
to Lay 'Golden' Eggs. Louisiana Agriculture
Magazine OnLine 2003; 46(4).

87 TranXenoGen Technology. http://www.tranxe-
nogen.com/NewFiles/background.html 2004.

88 Viragen Avian Technology. http://
www.viragen.com/avian_intro.htm 2004.

89 Vivalis Transgenics Information. http://
www.vivalis.com/ 2004.

90 Pain B, Chenevier P, Samarut J. Chicken embryonic stem cells and transgenic strategies.
Cells Tissues Organs, 1999; 165(3/4): 212–219.

91 Menassa R, Kennette W, Nguyen V, et al. Subcellular targeting of human interleukin-10 in
plants. J. Biotechnol., 2004; 108(2): 179–183.

92 Pacific NorthWest Laboratories Transgenic
Plants Patents. http://availabletechnolo-
gies.pnl.gov/biomedical/trans.stm 2004.

93 Dai Z, Hooker BS, Anderson DB, et al. Expression of *Acidothermus cellulolyticus* endoglucanase E1 in transgenic tobacco: biochemical
characteristics and physiological effects. Transgenic Res., 2000; 9(1): 43–54.

94 Hooker B, Dai Z, Gao J, et al., inventors;
Transgenic Plant-derived Human Blood
Coagulation Factors. 1999. PCT-World Intellec-
tual Property Organization WO 99/58699.

95 Plants make blood coagulants. Chemical &
Engineering News 1999; 44.

96 Hooker B, Dai Z, Kingsley M, et al., inventors;
Method for Producing Human Growth Factors
from Whole Plants or Plant Cell Culture.
1998. PCT-World Intellectual Property Organization WO98/21348.

97 Gao J, Hooker BS, Anderson DB. Expression
of functional human coagulation factor XIII
A-domain in plant cell suspensions and whole
plants. Protein Expr. Purif., 2004; 37(1): 89–
96.

98 Hooker B. Two Crops, One Plant. BioPharm.,
1999; 29–30.

99 Warzecha H, Mason HS. Benefits and risks
of antibody and vaccine production in transgenic plants. J. Plant Physiol., 2003; 160(7):
755–764.

100 Smith ML, Richter L, Arntzen CJ, et al.
Structural characterization of plant-derived
hepatitis B surface antigen employed in oral
immunization studies. Vaccine, 2003;
21(25/26): 4011–4021.

101 Warzecha H, Mason HS, Lane C, et al. Oral
immunogenicity of human papillomavirus-
like particles expressed in potato. J. Virol.,
2003; 77(16): 8702–8711.

102 Walmsley AM, Alvarez ML, Jin Y, et al. Expression of the B subunit of *Escherichia coli*
heat-labile enterotoxin as a fusion protein in
transgenic tomato. Plant Cell Rep., 2003;
21(10): 1020–1026.

103 Axis Genetics and Biomira sign research collaboration for the development of a cancer
vaccine. http://www.biomira.com/news/
detailNewsRelease/89/:November 9, 1998.

104 Ramirez N, Ayala M, Lorenzo D, et al. Expression of a single-chain Fv antibody fragment specific for the hepatitis B surface
antigen in transgenic tobacco plants. Transgenic Res., 2002; 11(1): 61–64.

105 Lopez L, Muller R, Balmori E, et al. Molecular cloning and nucleotide sequence of the
coat protein gene of a Cuban isolate of potato leafroll virus and its expression in *Escherichia coli*. Virus Genes, 1994; 9(1): 77–83.

106 Carrillo C, Wigdorovitz A, Trono K, et al. Induction of a virus-specific antibody response
to foot and mouth disease virus using the
structural protein VP1 expressed in transgenic potato plants. Viral Immunol., 2001;
14(1): 49–57.

107 Gil F, Brun A, Wigdorovitz A, et al. High-
yield expression of a viral peptide vaccine in
transgenic plants. FEBS Lett., 2001; 488(1/2):
13–17.

108 Gomez N, Wigdorovitz A, Castanon S, et al.
Oral immunogenicity of the plant derived
spike protein from swine-transmissible gastroenteritis coronavirus. Arch. Virol., 2000;
145(8): 1725–1732.

109 Wigdorovitz A, Carrillo C, Dus Santos MJ, et
al. Induction of a protective antibody response to foot and mouth disease virus in
mice following oral or parenteral immunization with alfalfa transgenic plants expressing
the viral structural protein VP1 Virology,
1999; 255(2): 347–353.

110 Carrillo C, Wigdorovitz A, Oliveros JC, et al.
Protective immune response to foot-and-
mouth disease virus with VP1 expressed in

transgenic plants. J. Virol., 1998; 72(2): 1688–1690.

111 Chlorogen Technology. http://www.chlorogen.com/technology.htm 2004.

112 Daniell H, Khan MS, Allison L. Milestones in chloroplast genetic engineering: an environmentally friendly era in biotechnology. Trends Plant Sci., 2002; 7(2): 84–91.

113 Daniell H, Lee SB, Panchal T, et al. Expression of the native cholera toxin B subunit gene and assembly as functional oligomers in transgenic tobacco chloroplasts. J. Mol. Biol. 2001; 311(5): 1001–1009.

114 Rachel Melcer. Creve Coeur Startup Develops Human Plasma from Tobacco. St. Louis Post-Dispatch. June 20, 2003.

115 Kang TJ, Loc NH, Jang MO, et al. Expression of the B subunit of E. coli heat-labile enterotoxin in the chloroplasts of plants and its characterization. Transgenic Res., 2003; 12(6): 683–691.

116 Cobento Biotech products. http://www.cobento.com/2004.

117 Bouquin T, Thomsen M, Nielsen L, et al. Human Anti-Rhesus D IgG1 Antibody Produced in Transgenic Plants. Transgenic Res., 2002; 11(2): 115–122.

118 Tavladoraki P, Benvenuto E, Trinca S, et al. Transgenic plants expressing a functional single-chain Fv antibody are specifically protected from virus attack. Nature, 1993; 366(6454): 469–472.

119 Benvenuto E, Ordas RJ, Tavazza R, et al. 'Phytoantibodies': a general vector for the expression of immunoglobulin domains in transgenic plants. Plant Mol. Biol., 1991; 17(4): 865–874.

120 Vine ND, Drake P, Hiatt A, et al. Assembly and plasma membrane targeting of recombinant immunoglobulin chains in plants with a murine immunoglobulin transmembrane sequence. Plant Mol. Biol., 2001; 45(2): 159–167.

121 Ma JK, Hiatt A, Hein M, et al. Generation and assembly of secretory antibodies in plants. Science, 1995; 268(5211): 716–719.

122 Ma JK, Lehner T, Stabila P, et al. Assembly of monoclonal antibodies with IgG1 and IgA heavy chain domains in transgenic tobacco plants. Eur. J. Immunol., 1994; 24(1): 131–138.

123 Hiatt A, Ma JK. Monoclonal antibody engineering in plants. FEBS Lett., 1992; 307(1): 71–75.

124 Hiatt A, Cafferkey R, Bowdish K. Production of antibodies in transgenic plants. Nature, 1989; 342(6245): 76–78.

125 ERA Biotech Technology. http://www.eraplantech.com/ang/zera.html 2004.

126 Yang IC, Iommarini JP, Becker DK, et al. A promoter derived from taro bacilliform badnavirus drives strong expression in transgenic banana and tobacco plants. Plant Cell Rep., 2003; 21(12): 1199–1206.

127 Farmacule Technology. http://www.farmacule.com/technology.htm#T2 2004.

128 Fraunhofer Center for Molecular Biology Technology. http://www.fraunhofer-cmb.org/index.cfm?act=technology 2004.

129 Stoger E, Schillberg S, Twyman RM, et al. Antibody production in transgenic plants. Methods Mol. Biol., 2004; 248: 301–318.

130 Fischer R, Schumann D, Zimmermann S, et al. Expression and characterization of bispecific single-chain Fv fragments produced in transgenic plants. Eur. J. Biochem., 1999; 262(3): 810–816.

131 de Zoeten GA, Penswick JR, Horisberger MA, et al. The expression, localization, and effect of a human interferon in plants. Virology, 1989; 172(1): 213–222.

132 Peeters K, De Wilde C, De Jaeger G, et al. Production of antibodies and antibody fragments in plants. Vaccine, 2001; 19(17–19): 2756–2761.

133 Peeters K, De Wilde C, Depicker A. Highly efficient targeting and accumulation of a F(ab) fragment within the secretory pathway and apoplast of Arabidopsis thaliana. Eur. J. Biochem., 2001; 268(15): 4251–4260.

134 De Jaeger G, De Wilde C, Eeckhout D, et al. The plantibody approach: expression of antibody genes in plants to modulate plant metabolism or to obtain pathogen resistance. Plant Mol. Biol., 2000; 43(4): 419–428.

135 De Jaeger G, Buys E, Eeckhout D, et al. High level accumulation of single-chain variable fragments in the cytosol of transgenic Petunia hybrida. Eur. J. Biochem., 1999; 259(1/2): 426–434.

136 Bruyns AM, De Jaeger G, De Neve M, et al. Bacterial and plant-produced scFv proteins

have similar antigen-binding properties. FEBS Lett., 1996; 386(1): 5–10.

137 De Neve M, De Loose M, Jacobs A, et al. Assembly of an antibody and its derived antibody fragment in *Nicotiana* and *Arabidopsis*. Transgenic Res., 1993; 2(4): 227–237.

138 Edelbaum O, Stein D, Holland N, et al. Expression of active human interferon-beta in transgenic plants. J. Interferon Res., 1992; 12(6): 449–453.

139 Rosenberg N, Reichman M, Gera A, et al. Antiviral activity of natural and recombinant human leukocyte interferons in tobacco protoplasts. Virology, 1985; 140(1): 173–178.

140 Ohya K, Itchoda N, Ohashi K, et al. Expression of biologically active human tumor necrosis factor-alpha in transgenic potato plant. J. Interferon Cytokine Res., 2002; 22(3): 371–378.

141 Ohya K, Matsumura T, Ohashi K, et al. Expression of two subtypes of human IFN-alpha in transgenic potato plants. J. Interferon Cytokine Res., 2001; 21(8): 595–602.

142 Icon Genetics Products. http://www.icongenetics.com/html/tech3_6.htm 2004.

143 Marillonnet S, Giritch A, Gils M, et al. In planta engineering of viral RNA replicons: efficient assembly by recombination of DNA modules delivered by *Agrobacterium*. Proc. Natl. Acad. Sci. USA, 2004; 101(18): 6852–6857.

144 Gleba Y, Marillonnet S, Klimyuk V. Engineering viral expression vectors for plants: the 'full virus' and the 'deconstructed virus' strategies. Curr. Opin. Plant Biol., 2004; 7(2): 182–188.

145 Borisjuk NV, Borisjuk LG, Logendra S, et al. Production of recombinant proteins in plant root exudates. Nature Biotechnol., 1999; 17(5): 466–469.

146 Biemelt S, Sonnewald U, Galmbacher P, et al. Production of human papillomavirus type 16 virus-like particles in transgenic plants. J. Virol., 2003; 77(17): 9211–9220.

147 Scheller J, Guhrs KH, Grosse F, et al. Production of spider silk proteins in tobacco and potato. Nature Biotechnol., 2001; 19(6): 573–577.

148 Fiedler U, Conrad U. High-level production and long-term storage of engineered antibodies in transgenic tobacco seeds. Biotechnology (NY), 1995; 13(10): 1090–1093.

149 Artsaenko O, Peisker M, Zur NU, et al. Expression of a single-chain Fv antibody against abscisic acid creates a wilty phenotype in transgenic tobacco. Plant J., 1995; 8(5): 745–750.

150 Conrad U, Fiedler U. Expression of engineered antibodies in plant cells. Plant Mol. Biol., 1994; 26(4): 1023–1030.

151 Scheller J, Henggeler D, Viviani A, et al. Purification of spider silk-elastin from transgenic plants and application for human chondrocyte proliferation. Transgenic Res., 2004; 13(1): 51–57.

152 Byun K, Hong J, Lee C, et al., inventors; KIST SK, assignee. Human interleukin-6 producing tobacco. 1996.

153 Matsumoto S, Ikura K, Ueda M, et al. Characterization of a human glycoprotein (erythropoietin) produced in cultured tobacco cells. Plant Mol. Biol., 1995; 27(6): 1163–1172.

154 Matsumoto S, Ishii A, Ikura K, et al. Expression of human erythropoietin in cultured tobacco cells. Biosci. Biotechnol. Biochem., 1993; 57(8): 1249–1252.

155 D'Aoust M, Busse U, et al. In: Fischer R, Schillberg S (Eds), Molecular Farming: Plant-made Pharmaceuticals and Technical Proteins. Wiley-VCH; 2004.

156 Medicago Technology. http://www2.medicago.com/en/tech_platform/proficia/ 2004.

157 Busse ULVTSeVLP. In: L. Erickson W-JYJB, Rymerson R (Eds), Molecular Farming of Plants and Animals for Human and Veterinary Medicine. Kluwer Academic Publishers; 2002.

158 D'Aoust M, Busse U, et al. In: Christou P, Klee H (Eds), Handbook of Plant Biotechnology. Wiley & Sons; 2004.

159 Meristem Products. http://www.meristem-therapeutics.com/ 2004.

160 Perret S, Merle C, Bernocco S, et al. Unhydroxylated triple helical collagen I produced in transgenic plants provides new clues on the role of hydroxyproline in collagen folding and fibril formation. J. Biol. Chem., 2001; 276(47): 43693–43698.

161 Samyn-Petit B, Wajda Dubos JP, Chirat F, et al. Comparative analysis of the site-specific N-glycosylation of human lactoferrin produced in maize and tobacco plants. Eur. J. Biochem., 2003; 270(15): 3235–3242.

162 Mokrzycki-Issartel N, Bouchon B, Farrer S, et al. A transient tobacco expression system coupled to MALDI-TOF-MS allows validation of the impact of differential targeting on structure and activity of a recombinant therapeutic glycoprotein produced in plants. FEBS Lett., 2003; 552(2/3): 170–176.

163 Merle C, Perret S, Lacour T, et al. Hydroxylated human homotrimeric collagen I in *Agrobacterium tumefaciens*-mediated transient expression and in transgenic tobacco plant. FEBS Lett., 2002; 515(1–3): 114–118.

164 Samyn-Petit B, Gruber V, Flahaut C, et al. N-glycosylation potential of maize: the human lactoferrin used as a model. Glycoconj J., 2001; 18(7): 519–527.

165 Theisen M. Production of recombinant blood factors in transgenic plants. Adv. Exp. Med. Biol., 1999; 464: 211–220.

166 Gruber V, Berna P, Arnaud T, et al. Large-scale production of a therapeutic protein in transgenic tobacco plants: effect of subcellular targeting on quality of a recombinant dog gastric lipase. Molecular Breeding, 2001; 7(4): 329–340.

167 Ruggiero F, Exposito JY, Bournat P, et al. Triple helix assembly and processing of human collagen produced in transgenic tobacco plants. FEBS Lett., 2000; 469(1): 132–136.

168 Dieryck W, Pagnier J, Poyart C, et al. Human haemoglobin from transgenic tobacco. Nature, 1997; 386(6620): 29–30.

169 Sijmons PC, Dekker BM, Schrammeijer B, et al. Production of correctly processed human serum albumin in transgenic plants. Biotechnology (NY), 1990; 8(3): 217–221.

170 Schlittler MR, Taylor DW, Russell DA, et al. Production of human somatotropin from corn seed. Am. Chem. Soc., 221st BIOT-019, 2001.

171 Baez J, Russell D, Craig J. Corn Seed Production of Therapeutic Proteins Moves Forward. BioPharm., 2000; 3.

172 Baez J, Russell D. Characterization of human monoclonal antibodies produced in plants. 219th ACS National Meeting, San Francisco, CA, March 26–30, 2000, BIOT-018, 2000.

173 Staub JM, Garcia B, Graves J, et al. High-yield production of a human therapeutic protein in tobacco chloroplasts. Nature Biotechnol., 2000; 18(3): 333–338.

174 Russell DA. Feasibility of antibody production in plants for human therapeutic use. Curr. Top. Microbiol. Immunol., 1999; 240: 119–138.

175 Zeitlin L, Olmsted SS, Moench TR, et al. A humanized monoclonal antibody produced in transgenic plants for immunoprotection of the vagina against genital herpes. Nature Biotechnol., 1998; 16(13): 1361–1364.

176 Francisco JA, Gawlak SL, Miller M, et al. Expression and characterization of bryodin 1 and a bryodin 1-based single-chain immunotoxin from tobacco cell culture. Bioconjug. Chem., 1997; 8(5): 708–713.

177 Olsen D, Yang C, Bodo M, et al. Recombinant collagen and gelatin for drug delivery. Adv. Drug Deliv. Rev., 2003; 55(12): 1547–1567.

178 Webster DE, Thomas MC, Strugnell RA, et al. Appetising solutions: an edible vaccine for measles. Med. J. Aust., 2002; 176(9): 434–437.

179 Webster DE, Cooney ML, Huang Z, et al. Successful boosting of a DNA measles immunization with an oral plant-derived measles virus vaccine. J. Virol., 2002; 76(15): 7910–7912.

180 Huang Z, Dry I, Webster D, et al. Plant-derived measles virus hemagglutinin protein induces neutralizing antibodies in mice. Vaccine, 2001; 19(15/16): 2163–2171.

181 Tawaiwa F, Katsumata K, Anzai H. Production of human lactoferrin in transgenic plants. 2nd International Molecular Farming Conference; London, Ontario, Canada, 1999.

182 Takase K, Hagiwara K. Expression of human alpha-lactalbumin in transgenic tobacco. J. Biochem. (Tokyo), 1998; 123(3): 440–444.

183 Higo K, Saito Y, Higo H. Expression of a chemically synthesized gene for human epidermal growth factor under the control of cauliflower mosaic virus 35S promoter in transgenic tobacco. Biosci. Biotechnol. Biochem., 1993; 57(9): 1477–1481.

184 NexGen Molecular Pharming Program. http://www.campnexgen.com/english/mf/mf.htm 2004.

185 Allina S, Heng Lui Y, Him S, et al. Production of a Canine oral papillomavirus vaccine in Tobacco. 2nd International Molecular Farming Conference; London, Ontario, Canada, 1999.

186 Phylogix Information. http://www.phylogix.com/about_us.htm 2004.

187 Moore JG, Fuchs CA, Hata YS, et al. A new lectin in red kidney beans called PvFRIL stimulates proliferation of NIH 3T3 cells expressing the Flt3 receptor. Biochim. Biophys. Acta, 2000; 1475(3): 216–224.

188 Planet Biotechnology Publications. http://www.planetbiotechnology.com/publications.html 2004.

189 Larrick JW, Thomas DW. Producing proteins in transgenic plants and animals. Curr. Opin. Biotechnol., 2001; 12: 411–418.

190 Daniell H, Streatfield SJ, Wycoff K. Medical molecular farming: production of antibodies, biopharmaceuticals and edible vaccines in plants. Trends Plant Sci., 2001; 6(5): 219–226.

191 Larrick JW, Yu L, Naftzger C, et al. Production of secretory IgA antibodies in plants. Biomol. Eng., 2001; 18(3): 87–94.

192 Daniell H, Streatfield SJ, Wycoff K. Medical molecular farming: production of antibodies, biopharmaceuticals and edible vaccines in plants. Trends Plant Sci., 2001; 6(5): 219–226.

193 Larrick JW, Yu L, Chen J, et al. Production of antibodies in transgenic plants. Res. Immunol., 1998; 149(6): 603–608.

194 Ma JK, Hikmat BY, Wycoff K, et al. Characterization of a recombinant plant monoclonal secretory antibody and preventive immunotherapy in humans. Nature Med., 1998; 4(5): 601–606.

195 Plantigen technology. http://www.lhsc.on.ca/plantigen/science_tech.html 2004.

196 Ma S, Huang Y, Yin Z, et al. Induction of oral tolerance to prevent diabetes with transgenic plants requires glutamic acid decarboxylase (GAD) and IL-4. Proc. Natl. Acad. Sci. USA, 2004; 101(15): 5680–5685.

197 Menassa R, Kennette W, Nguyen V, et al. Subcellular targeting of human interleukin-10 in plants. J. Biotechnol., 2004; 108(2): 179–183.

198 Ma S, Jevnikar AM. Autoantigens produced in plants for oral tolerance therapy of autoimmune diseases. Adv. Exp. Med. Biol., 1999; 464: 179–194.

199 Ma SW, Zhao DL, Yin ZQ, et al. Transgenic plants expressing autoantigens fed to mice to induce oral immune tolerance. Nature Med., 1997; 3(7): 793–796.

200 Kong Q, Richter L, Yang YF, et al. Oral immunization with hepatitis B surface antigen expressed in transgenic plants. Proc. Natl. Acad. Sci. USA, 2001; 98(20): 11539–11544.

201 Ma JK, Drake PM, Christou P. The production of recombinant pharmaceutical proteins in plants. Nature Rev. Genet., 2003; 4(10): 794–805.

202 Avesani L, Falorni A, Battista G, et al. Improved in planta expression of the human islet autoantigen glutamic acid decarboxylase (GAD65). Transgenic Res., 2003; 12(2): 203–212.

203 Tuboly T, Yu W, Bailey A, et al. Immunogenicity of porcine transmissible gastroenteritis virus spike protein expressed in plants. Vaccine, 2000; 18(19): 2023–2028.

204 Du S, Yu W, DeLange C, et al. Expression of Porcine Epidermal Growth Factor in Plants. 2nd International Molecular Farming Conference; London, Ontario, Canada, 1999.

205 Desai UA, Sur G, Daunert S, et al. Expression and affinity purification of recombinant proteins from plants. Protein Expr. Purif., 2002; 25(1): 195–202.

206 Li Q, Lawrence CB, Xing HY, et al. Enhanced disease resistance conferred by expression of an antimicrobial magainin analog in transgenic tobacco. Planta, 2001; 212(4): 635–639.

207 Rigano MM, Alvarez ML, Pinkhasov J, et al. Production of a fusion protein consisting of the enterotoxigenic *Escherichia coli* heat-labile toxin B subunit and a tuberculosis antigen in *Arabidopsis thaliana*. Plant Cell Rep., 2004; 22(7): 502–508.

208 Rigano MM, Sala F, Arntzen CJ, et al. Targeting of plant-derived vaccine antigens to immunoresponsive mucosal sites. Vaccine, 2003; 21(7/8): 809–811.

209 Sala F, Manuela RM, Barbante A, et al. Vaccine antigen production in transgenic plants: strategies, gene constructs and perspectives. Vaccine, 2003; 21(7/8): 803–808.

210 Franconi R, Di Bonito P, Dibello F, et al. Plant-derived human papillomavirus 16 E7 oncoprotein induces immune response and specific tumor protection. Cancer Res., 2002; 62(13): 3654–3658.

211 Franconi R, Roggero P, Pirazzi P, et al. Functional expression in bacteria and plants of an scFv antibody fragment against topo-

viruses. Immunotechnology, 1999; 4(3/4): 189–201.

212 Bakker H, Bardor M, Molthoff JW, et al. Galactose-extended glycans of antibodies produced by transgenic plants. Proc. Natl. Acad. Sci. USA, 2001; 98(5): 2899–2904.

213 Schouten A, Roosien J, de Boer JM, et al. Improving scFv antibody expression levels in the plant cytosol. FEBS Lett., 1997; 415(2): 235–241.

214 Schouten A, Roosien J, van Engelen FA, et al. The C-terminal KDEL sequence increases the expression level of a single-chain antibody designed to be targeted to both the cytosol and the secretory pathway in transgenic tobacco. Plant Mol. Biol., 1996; 30(4): 781–793.

215 van Engelen FA, Schouten A, Molthoff JW, et al. Coordinate expression of antibody subunit genes yields high levels of functional antibodies in roots of transgenic tobacco. Plant Mol. Biol., 1994; 26(6): 1701–1710.

216 Elbers I, Stoopen G, Bakker H, et al. Influence of growth conditions and developmental stage on N-glycan heterogeneity of transgenic immunoglobulin G and endogenous proteins in tobacco leaves. Plant Physiol., 2001; 126: 1314–1322.

217 Zhang G, Leung C, Murdin L, et al. In planta expression of HIV-1 p24 protein using an RNA plant virus-based expression vector. Mol. Biotechnol., 2000; 14(2): 99–107.

218 Agrenvec Information. http://agrenvec.com/eng/ 2004.

219 Sanchez F, Wang X, Jenner CE, et al. Strains of Turnip mosaic potyvirus as defined by the molecular analysis of the coat protein gene of the virus. Virus Res., 2003; 94(1): 33–43.

220 Jenner CE, Sanchez F, Nettleship SB, et al. The cylindrical inclusion gene of Turnip mosaic virus encodes a pathogenic determinant to the *Brassica* resistance gene TuRB01. Mol. Plant Microbe Interact., 2000; 13(10): 1102–1108.

221 Roggero P, Ciuffo M, Benvenuto E, et al. The expression of a single-chain Fv antibody fragment in different plant hosts and tissues by using Potato virus X as a vector. Protein Expr. Purif., 2001; 22(1): 70–74.

222 Yusibov V, Hooper DC, Spitsin SV, et al. Expression in plants and immunogenicity of plant virus-based experimental rabies vaccine. Vaccine, 2002; 20(25/26): 3155–3164.

223 Koprowski H, Yusibov V. The green revolution: plants as heterologous expression vectors. Vaccine, 2001; 19(17–19): 2735–2741.

224 Belanger H, Fleysh N, Cox S, et al. Human respiratory syncytial virus vaccine antigen produced in plants. FASEB J., 2000; 14(14): 2323–2328.

225 Yusibov V, Shivprasad S, Turpen TH, et al. Plant viral vectors based on tobamoviruses. Curr. Top. Microbiol. Immunol., 1999; 240: 81–94.

226 Kapusta J, Modelska A, Figlerowicz M, et al. A plant-derived edible vaccine against hepatitis B virus. FASEB J., 1999; 13(13): 1796–1799.

227 Yusibov V, Shivprasad S, Turpen TH, et al. Plant viral vectors based on tobamoviruses. Curr. Top. Microbiol. Immunol., 1999; 240: 81–94.

228 Verch T, Yusibov V, Koprowski H. Expression and assembly of a full-length monoclonal antibody in plants using a plant virus vector. J. Immunol. Methods, 1998; 220(1/2): 69–75.

229 Modelska A, Dietzschold B, Sleysh N, et al. Immunization against rabies with plant-derived antigen. Proc. Natl. Acad. Sci. USA, 1998; 95(5): 2481–2485.

230 Fraunhofer Vaccine Development Technology. http://www.fraunhofer-cmb.org/index.cfm?act=tech_VaccineDevelopment 2004.

231 Marillonnet S, Giritch A, Gils M, et al. In planta engineering of viral RNA replicons: efficient assembly by recombination of DNA modules delivered by *Agrobacterium*. Proc. Natl. Acad. Sci. USA, 2004; 101(18): 6852–6857.

232 Crabtree P. Rincon Pharmaceuticals. San Diego, Union-Tribune, May 20, 2005.

233 Fischer R, Schillberg S. Molecular Farming – Plant made pharamaceuticals and technical proteins. Weinheim, Wiley-VCH, 2004.

234 Gelderman M, Oliver K, Yazdani A, et al. Preclinical Studies with Plant-Produced alpha-Galactosidase A in Fabry Mice Show Potential for Replacement Therapy. NA. Preclinica, 2004; 2(1): 67–74.

235 Kumagai MH, Donson J, Della-Cioppa G, et al. Rapid, high-level expression of glycosylated rice alpha-amylase in transfected plants by an RNA viral vector. Gene, 2000; 245(1): 169–174.

236 McCormick AA, Kumagai MH, Hanley K, et al. Rapid production of specific vaccines for lymphoma by expression of the tumor-derived single-chain Fv epitopes in tobacco plants. Proc. Natl. Acad. Sci. USA, 1999; 96(2): 703–708.

237 Turpen TH, Reinl SJ, Charoenvit Y, et al. Malarial epitopes expressed on the surface of recombinant tobacco mosaic virus. Biotechnology (NY), 1995; 13(1): 53–57.

238 Large Scale Biology Corporation Announces Biomanufacturing and Commercial Distribution Agreement with Sigma-Aldrich Corporation. March 6, 04 Wall Street Reporter 2004.

239 Joint NIH And Large Scale Biology Corporation (LSBC) Studies With Plant-Produced Alpha-Galactosidase Show Its Potential For Enzyme Replacement Therapy In Fabry Disease. http://www.biospace.com/news_story.cfm? StoryID=14899320&full=1 2004.

240 Tacket CO, Mason HS, Losonsky G, et al. Human immune responses to a novel Norwalk virus vaccine delivered in transgenic potatoes. J. Infect. Dis., 2000; 182(1): 302–305.

241 Dai Z, Hooker BS, Anderson DB, et al. Expression of *Acidothermus cellulolyticus* endoglucanase E1 in transgenic tobacco: biochemical characteristics and physiological effects. Transgenic Res., 2000; 9(1): 43–54.

242 Farran I, Sanchez-Serrano JJ, Medina JF, et al. Targeted expression of human serum albumin to potato tubers. Transgenic Res., 2002; 11(4): 337–346.

243 Chong DK, Langridge WH. Expression of full-length bioactive antimicrobial human lactoferrin in potato plants. Transgenic Res., 2000; 9(1): 71–78.

244 Arakawa T, Yu J, Chong DK, et al. A plant-based cholera toxin B subunit-insulin fusion protein protects against the development of autoimmune diabetes. Nature Biotechnol., 1998; 16(10): 934–938.

245 Arakawa T, Langridge WH. Plants are not just passive creatures! Nature Med., 1998; 4(5): 550–551.

246 Arakawa T, Chong DK, Langridge WH. Efficacy of a food plant-based oral cholera toxin B subunit vaccine. Nature Biotechnol., 1998; 16(3): 292–297.

247 Arakawa T, Chong DK, Merritt JL, et al. Expression of cholera toxin B subunit oligomers in transgenic potato plants. Transgenic Res., 1997; 6(6): 403–413.

248 During K, Hippe S, Kreuzaler F, et al. Synthesis and self-assembly of a functional monoclonal antibody in transgenic *Nicotiana tabacum*. Plant Mol. Biol., 1990; 15(2): 281–293.

249 Klaus During. Engineered storage organs as bioreactors for protein production. 2nd International Molecular Farming Conference; London, Ontario, Canada, 1999.

250 Collins S. Potato the medical factory of tomorrow. New Zealand Herald. April 23, 2003.

251 Novoplant Technology. http://www.novoplant.de/ 2004.

252 Planton technology. http://www.planton.de/en/index.html 2004.

253 Malaysia Rubber Institute Biotechnology and Strategic Research Unit. http://www.lgm.gov.my/r&d/bsru/transgenic.html 2004.

254 Arokiaraj P, Rueker F, Carter D, et al. Expression of Human Serum Albumin in the laticifers of Transgenic *Hevea brasiliensis*. 2nd International Molecular Farming Conference; London, Ontario, Canada, 1999.

255 Yeang, H. Engineering Crop Plants for Industrial End Uses. London, UK: Portland Press; 1998: 55–64.

256 Rod Santa Ana III. Medical proteins from sugarcane. Farm Press online, USA 2003.

257 Michael Smith. Norwalk virus vaccine grown in tomatoes. United Press International. February 16, 2003.

258 Sandhu JS, Krasnyanski SF, Domier LL, et al. Oral immunization of mice with transgenic tomato fruit expressing respiratory syncytial virus-F protein induces a systemic immune response. Transgenic Res., 2000; 9(2): 127–135.

259 Jani D, Singh NK, Bhattacharya S, et al. Studies on the immunogenic potential of plant-expressed cholera toxin B subunit. Plant Cell Rep., 2004; 22(7): 471–477.

260 Jani D, Meena LS, Rizwan-ul-Haq QM, et al. Expression of cholera toxin B subunit in transgenic tomato plants. Transgenic Res., 2002; 11(5): 447–454.

261 Arazi T, Lee HP, Huang PL, et al. Production of antiviral and antitumor proteins MAP30 and GAP31 in cucurbits using the plant virus vector ZYMV-AGII. Biochem. Biophys. Res. Commun., 2002; 292(2): 441–448.

262 Cropdesign Technology. http:// www.cropdesign.com 2004.

263 Dow Plant Technology. http:// www.dow.com/plantbio 2004.

264 Frigerio L, Vine ND, Pedrazzini E, et al. Assembly, secretion, and vacuolar delivery of a hybrid immunoglobulin in plants. Plant Physiol., 2000; 123(4): 1483–1494.

265 Company of the Month: Epicyte. http:// www.biotechjournal.com/Journal/Jun2001/ juneartA2001.pdf 2001; San Diego Biotech Journal.

266 Stoger E, Vaquero C, Torres E, et al. Cereal crops as viable production and storage systems for pharmaceutical scFv antibodies. Plant Mol. Biol., 2000; 42(4): 583–590.

267 Perrin Y, Vasquero C, Gerrard I, et al. Transgenic pea seeds as bioreactors for the production of a single-chain Fv fragment antibodies used in cancer diagnosis and therapy. Molecular Breeding, 2000; 6: 345–352.

268 Vandekerckhove J, Van Damme J, Van Lijsebetten M, et al. Enkephalins produced in transgenic plants using modified 2S seed storage proteins. Bio/Technology, 1989; 7: 929–932.

269 De Jaeger G, Scheffer S, Jacobs A, et al. Evaluation of *Phaseolus vulgaris* arcelin-5I gene for single-chain antibody production in transgenic plants. European Society Plant Physiology Annual Meeting, Helsinki, Finland, 2001.

270 Mäkinen K. Production of recombinant gelatin in barley. http://www.honeybee.helsinki. fi/mmsbl/Biokemlab/makinen/research.html 2004.

271 Chikwamba R, Cunnick J, Hathaway D, et al. A functional antigen in a practical crop: LT-B producing maize protects mice against *Escherichia coli* heat labile enterotoxin (LT) and cholera toxin (CT). Transgenic Res., 2002; 11(5): 479–493.

272 Yang SH, Moran DL, Jia HW, et al. Expression of a synthetic porcine alpha-lactalbumin gene in the kernels of transgenic maize. Transgenic Res., 2002; 11(1): 11–20.

273 Eudes F, Gilbert S, Acharya S, et al. Expression of a Bovine Virus Protein in Barley: a Potential Edible Vaccine. 2nd International Molecular Farming Conference; London, Ontario, Canada, 1999.

274 Maltagene Technology. http://www.malta-gen.de/index-english.htm 2004.

275 Mokrzycki-Issartel N, Bouchon B, Farrer S, et al. A transient tobacco expression system coupled to MALDI-TOF-MS allows validation of the impact of differential targeting on structure and activity of a recombinant therapeutic glycoprotein produced in plants. FEBS Lett., 2003; 552(2/3): 170–176.

276 Schlittler MR, Taylor DW, Russell DA, et al. Production of human somatotropin from corn seed. Am. Chem. Soc., 221st BIOT-019 2001.

277 Novoplant Technology. http://www.novo-plant.de/ 2004.

278 Orf Genetics Technology. http:// www.orfgenetics.com/orf/wgorf.nsf/key2/ Technology 2004.

279 Bailey MR, Woodard SL, Callaway E, et al. Improved recovery of active recombinant laccase from maize seed. Appl. Microbiol. Biotechnol., 2004; 63(4): 390–397.

280 Woodard SL, Mayor JM, Bailey MR, et al. Maize (Zea mays)-derived bovine trypsin: characterization of the first large-scale, commercial protein product from transgenic plants. Biotechnol. Appl. Biochem., 2003; 38(Pt 2): 123–130.

281 Lamphear BJ, Streatfield SJ, Jilka JM, et al. Delivery of subunit vaccines in maize seed. J. Control. Release, 2002; 85(1–3): 169–180.

282 Streatfield SJ, Jilka JM, Hood EE, et al. Plant-based vaccines: unique advantages. Vaccine, 2001; 19(17–19): 2742–2748.

283 Azzoni AR, Kusnadi AR, Miranda EA, et al. Recombinant aprotinin produced in transgenic corn seed: extraction and purification studies. Biotechnol. Bioeng., 2002; 80(3): 268–276.

284 Evangelista RL, Kusnadi AR, Howard JA, et al. Process and economic evaluation of the extraction and purification of recombinant beta-glucuronidase from transgenic corn. Biotechnol. Prog., 1998; 14(4): 607–614.

285 Kusnadi AR, Hood EE, Witcher DR, et al. Production and purification of two recombinant proteins from transgenic corn. Biotechnol. Prog., 1998; 14(1): 149–155.

286 Hood E. From green plants to industrial enzymes. Enzyme Microbial Technol., 2002; 30(3): 279–283.

287 Lamphear BJ, Jilka JM, Kesl L, et al. A corn-based delivery system for animal vaccines: an oral transmissible gastroenteritis virus

vaccine boosts lactogenic immunity in swine. Vaccine, 2004; 22(19): 2420–2424.

288 Horn ME, Pappu KM, Bailey MR, et al. Advantageous features of plant-based systems for the development of HIV vaccines. J Drug Target., 2003; 11(8–10): 539–545.

289 Streatfield SJ, Lane JR, Brooks CA, et al. Corn as a production system for human and animal vaccines. Vaccine, 2003; 21(7/8): 812–815.

290 Kusnadi AR, Evangelista RL, Hood EE, et al. Processing of transgenic corn seed and its effect on the recovery of recombinant beta-glucuronidase. Biotechnol. Bioeng., 1998; 60(1): 44–52.

291 Hood E, Witcher DR, Maddock S, et al. Commercial production of avidin from transgenic maize: characterization of transformant, production, processing, extraction, and purification. Mol. Breed., 1997; 3: 291–306.

292 Nicholson L, Christou P, Ma J, et al. Expression of recombinant SIgA in cereals. 2nd International Molecular Farming Conference; London, Ontario, Canada, 1999.

293 Stoger E, Vaquero C, Torres E, et al. Cereal crops as viable production and storage systems for pharmaceutical scFv antibodies. Plant Mol. Biol., 2000; 42(4): 583–590.

294 Moloney MM, Holbrook LA. Subcellular targeting and purification of recombinant proteins in plant production systems. Biotechnol. Genet. Eng. Rev., 1997; 14: 321–336.

295 van Rooijen GJ, Moloney MM. Plant seed oil-bodies as carriers for foreign proteins. Biotechnology (NY), 1995; 13(1): 72–77.

296 Parmenter DL, Boothe JG, van Rooijen GJ, et al. Production of biologically active hirudin in plant seeds using oleosin partitioning. Plant Mol. Biol., 1995; 29(6): 1167–1180.

297 Hogge L, Boothe J, Maloney M, et al. Structure Confirmation of Recombinant Hirudin Using MALDI with C-Terminal Sequencing. 2nd International Molecular Farming Conference; London, Ontario, Canada, 1999.

298 SunGene Technology. http://www.sungene.de/ 2004.

299 Syngenta BioPharma. http://www.syngenta.com/en/biopharma/2004.

300 Leite A, Kemper E, Bocaccorsi E, et al. Correctly Processed Human Growth Hormone is Expressed in Seeds of Transgenic Tobacco Plants. 2nd International Molecular Farming Conference; London, Ontario, Canada, 1999.

301 Tackaberry ES, Prior F, Bell M, et al. Increased yield of heterologous viral glycoprotein in the seeds of homozygous transgenic tobacco plants cultivated underground. Genome, 2003; 46(3): 521–526.

302 Sardana RK, Alli Z, Dudani A, et al. Biological activity of human granulocyte-macrophage colony stimulating factor is maintained in a fusion with seed glutelin peptide. Transgenic Res., 2002; 11(5): 521–531.

303 Wright KE, Prior F, Sardana R, et al. Sorting of glycoprotein B from human cytomegalovirus to protein storage vesicles in seeds of transgenic tobacco. Transgenic Res., 2001; 10(2): 177–181.

304 Tackaberry ES, Dudani AK, Prior F, et al. Development of biopharmaceuticals in plant expression systems: cloning, expression and immunological reactivity of human cytomegalovirus glycoprotein B (UL55) in seeds of transgenic tobacco. Vaccine, 1999; 17(23/24): 3020–3029.

305 Panahi M, Cheng X, Callagham M, et al. Expression of human insulin-like growth factor (IGF-I) in plants. 2nd International Molecular Farming Conference; London, Ontario, Canada, 1999.

306 Panahi M, Alli Z, Cheng X, et al. Recombinant protein expression plasmids optimized for industrial E. coli fermentation and plant systems produce biologically active human insulin-like growth factor-1 in transgenic rice and tobacco plants. Transgenic Res., 2004; 13(3): 245–259.

307 Dalton R. California edges towards farming drug-producing rice. Nature, 2004; 428(6983): 591.

308 Suzuki YA, Kelleher SL, Yalda D, et al. Expression, characterization, and biologic activity of recombinant human lactoferrin in rice. J. Pediatr. Gastroenterol. Nutr., 2003; 36(2): 190–199.

309 Humphrey BD, Huang N, Klasing KC. Rice expressing lactoferrin and lysozyme has antibiotic-like properties when fed to chicks. J. Nutr., 2002; 132(6): 1214–1218.

310 Torres E, Vaquero C, Nicholson L, et al. Rice cell culture as an alternative production system for functional diagnostic and therapeutic antibodies. Transgenic Res., 1999; 8(6): 441–449.

311 Huang J, Sutliff TD, Wu L, et al. Expression and purification of functional human alpha-

1-Antitrypsin from cultured plant cells. Biotechnol. Prog., 2001; 17(1): 126–133.

312 Nandi S, Suzukib Y, Huang J, et al. Expression of human lactoferrin in transgenic rice grains for the application in infant formula. Plant Sci. 163(4): 713–722.

313 Huang J, Nandi S, Wu L, et al. Expression of natural antimicrobial human lysozyme in rice grains. Mol. Breeding, 2004; 10(1/2): 83–94.

314 Huang N. High-level Protein Expression System Uses Self-Pollinating Crops as Hosts. BioProcess Int., 2004; 54–59.

315 WSU Dieter Von Wettstein Research Interest. http://molecular.biosciences.wsu.edu/faculty/vonwettstein.html 2004.

316 Horvath H, Huang J, Wong O, et al. The production of recombinant proteins in transgenic barley grains. Proc. Natl. Acad. Sci. USA, 2000; 97(4): 1914–1919.

317 Greenovation Technology. http://www.greenovation.com/Projects_Production.htm 2004.

318 Gasdaska J, Spencer D, Dickey L. Advantages of Therapeutic Protein Production in the Aquatic Plant Lemna. BioProcessing J., 2003; 49–53.

319 Phycotransgenic Technology. http://www.phycotransgenics.com/ 2004.

320 Franklin SE, Mayfield SP. Prospects for molecular farming in the green alga *Chlamydomonas*. Curr Opin. Plant Biol., 2004; 7(2): 159–165.

321 Mayfield SP, Franklin SE, Lerner RA. Expression and assembly of a fully active antibody in algae. Proc. Natl. Acad. Sci. USA, 2003; 100(2): 438–442.

322 Borisjuk NV, Borisjuk LG, Logendra S, et al. Production of recombinant proteins in plant root exudates. Nature Biotechnol., 1999; 17(5): 466–469.

323 Gleba D, Borisjuk NV, Borisjuk LG, et al. Use of plant roots for phytoremediation and molecular farming. Proc. Natl. Acad. Sci. USA, 1999; 96(11): 5973–5977.

324 Gaume A, Komarnytsky S, Borisjuk N, et al. Rhizosecretion of recombinant proteins from plant hairy roots. Plant Cell Rep., 2003; 21(12): 1188–1193.

325 Medina-Bolivar F, Cramer C. Production of recombinant proteins by hairy roots cultured in plastic sleeve bioreactors. Methods Mol. Biol., 2004; 267: 351–364.

326 Medina-Bolivar F, Wright R, Funk V, et al. A non-toxic lectin for antigen delivery of plant-based mucosal vaccines. Vaccine, 2003; 21(9/10): 997–1005.

327 Cramer CL, Boothe JG, Oishi KK. Transgenic plants for therapeutic proteins: linking upstream and downstream strategies. Curr. Top. Microbiol. Immunol., 1999; 240: 95–118.

328 Cramer CL, Weissenborn DL, Oishi KK, et al. Bioproduction of human enzymes in transgenic tobacco. Ann. N.Y. Acad. Sci., 1996; 792: 62–71.

329 Koivu K. Contained Plant Bioreactor. 4th International Conference, Production & Economics of Biopharmaceuticals. San Diego, California, 2001.

330 Unicrop Publications. http://www.unicrop.fi/index.jsp?p=74, 79, 88 2004.

331 Geert De Jaeger Research Team Plant Cell Culture Program. http://www.vib.be/Research/EN/Research+Departments/Department+of+Plant+Systems+Biology/Geert+De+Jaeger/2004.

332 Yano A, Maeda F, Takekoshi M. Transgenic tobacco cells producing the human monoclonal antibody to hepatitis B virus surface antigen. J. Med. Virol., 2004; 73(2): 208–215.

333 Information update on Phytoprotein Biotech Pte Ltd. http://www.agenix.com/news/biotechnews11apr1.htm 2001.

334 Protalix Technology. http://www.protalix.com/html/technology_technology.htm 2004.

335 ROOTec Technology. http://www.pharmaceutical-map.com/non-member/storefront_view.php?id=26 2004.

336 Tsoi BM, Doran PM. Effect of medium properties and additives on antibody stability and accumulation in suspended plant cell cultures. Biotechnol. Appl. Biochem., 2002; 35(Pt 3): 171–180.

337 Sharp JM, Doran PM. Characterization of monoclonal antibody fragments produced by plant cells. Biotechnol. Bioeng., 2001; 73(5): 338–346.

338 Doran PM. Foreign protein production in plant tissue cultures. Curr. Opin. Biotechnol., 2000; 11: 199–204.

339 Mison D, Curling J. The industrial production costs of recombinant therapeutic proteins expressed in transgenic corn. BioPharm, 2000; 48–54.

340 Young M. Production of biopharmaceutical proteins in the milk of transgenic animals. BioPharm, 1997; 10(6): 34–38.

341 Price B. The Capacity and Capability 'Crunch': Myth or Reality. Plant-Made Pharmaceuticals conference, Quebec, Canada, 2003.

342 Palmer C, Lubon H, McManaman J. Transgenic mice expressing recombinant human protein C exhibit defects in lactation and impaired mammary gland development. Transgenic Res., 2003; 12(3): 283–292.

343 Cabanes-Macheteau M, Fitchette-Laine AC, Loutelier-Bourhis C, et al. N-Glycosylation of a mouse IgG expressed in transgenic tobacco plants. Glycobiology, 1999; 9(4): 365–372.

344 Gala FA, Morrison SL. V region carbohydrate and antibody expression. J. Immunol., 2004; 172(9): 5489–5494.

345 Rifai A, Fadden K, Morrison SL, et al. The N-glycans determine the differential blood clearance and hepatic uptake of human immunoglobulin (Ig)A1 and IgA2 isotypes. J. Exp. Med., 2000; 191(12): 2171–2182.

346 Chargelegue D, Vine N, van Dolleweerd C, et al. A murine monoclonal antibody produced in transgenic plants with plant-specific glycans is not immunogenic in mice. Transgenic Res., 2000; 9(3): 187–194.

347 United States Department of Agriculture. Field testing of plants engineered to produce pharmaceutical and industrial compounds. 2003, pp. 11337–11340. United States Federal Register 68.

348 FDA-CBER. Points to Consider in the Manufacture and Testing of Therapeutic Proteins for Human Use Derived from Transgenic Animals. http://mbcr.bcm.tmc.edu/BEP/ERMB/ptc_tga.html 1995.

349 Bair J. Food Value Chain Perspective. Corn-Produced Pharmaceuticals and Industrials Risk Assessment Symposium, sponsored by the Biosafety Institute for Genetically Modified Agricultural Products Symposium, Ames, Iowa, 2004.

350 Wall R, Kerr D, Bondioli K. Transgenic dairy cattle: genetic engineering on a large scale. J. Dairy Sci., 1997; 80: 2213–2224.

351 Masarik M, Kizek R, Kramer KJ, et al. Application of avidin-biotin technology and adsorptive transfer stripping square-wave voltammetry for detection of DNA hybridization and avidin in transgenic avidin maize. Anal. Chem., 2003; 75(11): 2663–2669.

352 Kramer KJ, Morgan TD, Throne JE, et al. Transgenic avidin maize is resistant to storage insect pests. Nature Biotechnol., 2000; 18(6): 670–674.

353 Grigorian AV, Shekhter AB, Tolstykh PI, et al. [Proteolytic enzymes in the overall treatment of wounds]. Khirurgiia (Mosk), 1979; (8): 19–23.

354 Yucel VE, Basmajian JV. Decubitus ulcers: healing effect of an enzymatic spray. Arch. Phys. Med. Rehabil., 1974; 55(11): 517–519.

355 Roep BO, van den Engel NK, van Halteren AG, et al. Modulation of autoimmunity to beta-cell antigens by proteases. Diabetologia, 2002; 45(5): 686–692.

356 Landis RC, Asimakopoulos G, Poullis M, et al. The antithrombotic and anti-inflammatory mechanisms of action of aprotinin. Ann. Thorac. Surg., 2001; 72(6): 2169–2175.

357 Landis RC, Haskard DO, Taylor KM. New anti-inflammatory and platelet-preserving effects of aprotinin. Ann. Thorac. Surg., 2001; 72(5): S1808–S1813.

358 Belorgey D, Dirrig S, Amouric M, et al. Inhibition of human pancreatic proteinases by mucus proteinase inhibitor, eglin c and aprotinin. Biochem. J., 1996; 313 (Pt 2): 555–560.

359 Laxenaire MC, Dewachter P, Pecquet C. [Allergic risk of aprotinin]. Ann. Fr. Anesth. Reanim., 2000; 19(2): 96–104.

360 Barthel T, Kula MR. Studies on the extraction of DesPro(2)-Val15-Leu17-aprotinin from the culture broth of a recombinant Saccharomyces cerevisiae. Bioseparation, 1992; 3(6): 365–372.

361 GTC Biotherapeutics Reports Second Quarter 2004 Financial Results. http://www.transgenics.com/pressreleases/pr080504.html 2004.

362 Winterbottom N, Kuo J, Nguyen K, et al. Antigenic Responses to Bovine Thrombin Exposure During Surgery: A Prospective Study of 309 Patients. J. Appl. Res. Clin. Exp. Ther., 2002; 2(1).

363 GTC Biotherapeutics Reports Fourth Quarter and Year End 2003 Financial Results. http://www.transgenics.com/pressreleases/pr030304.html 2004.

364 Melcer R. Creve Coeur Startup Develops Human Plasma from Tobacco. St. Louis Post-Dispatch, June 20, 2003.

365 Lomas DA, Parfrey H. Alpha1-antitrypsin deficiency. 4: Molecular pathophysiology. Thorax, 2004; 59(6): 529–535.

366 Chowanadisai W, Huang J, Huang N, et al. Stability of recombinant human alpha-1-antitrypsin produced in rice in infant formula. J. Nutr. Biochem., 2003; 14(7): 386–393.

367 Flotte TR, Brantly ML, Spencer LT, et al. Phase I trial of intramuscular injection of a recombinant adeno-associated virus alpha 1-antitrypsin (rAAV2-CB-hAAT) gene vector to AAT-deficient adults. Hum. Gene Ther., 2004; 15(1): 93–128.

368 Hubbard RC, McElvaney NG, Sellers SE, et al. Recombinant DNA-produced alpha 1-antitrypsin administered by aerosol augments lower respiratory tract antineutrophil elastase defenses in individuals with alpha 1-antitrypsin deficiency. J. Clin. Invest., 1989; 84(4): 1349–1354.

369 Smerdon GR, Aves SJ, Walton EF. Production of human gastric lipase in the fission yeast Schizosaccharomyces pombe. Gene, 1995; 165(2): 313–318.

370 Van den Hout JM, Kamphoven JH, Winkel LP, et al. Long-term intravenous treatment of Pompe disease with recombinant human alpha-glucosidase from milk. Pediatrics, 2004; 113(5): e448–e457.

371 Bijvoet AG, Van Hirtum H, Kroos MA, et al. Human acid alpha-glucosidase from rabbit milk has therapeutic effect in mice with glycogen storage disease type II. Hum. Mol. Genet., 1999; 8(12): 2145–2153.

372 Bijvoet AG, Kroos MA, Pieper FR, et al. Recombinant human acid alpha-glucosidase: high level production in mouse milk, biochemical characteristics, correction of enzyme deficiency in GSDII KO mice. Hum. Mol. Genet., 1998; 7(11): 1815–1824.

373 Reno J. Performance of corn-seed derived huNR-LU-10. 1998; IBC conference on Transgenic Production, San Francisco, September 1998.

374 Pogue G. Clinical and Pre-Clinical Application of Plant-Derived Glycoproteins. BIO2005 International Conference, San Francisco, 2004.

375 Large Scale Biology and Planet Biotechnology Announce Biomanufacturing Agreement. http://www.lsbc.com/pdfs/Planet-Bio.pdf 2004.

376 Pharming Group 2002 annual report. http://www.pharming.com/downloads/Annual_Report_2002.pdf 2004.

377 van Berkel PH, Welling MM, Geerts M, et al. Large scale production of recombinant human lactoferrin in the milk of transgenic cows. Nature Biotechnol., 2002; 20(5): 484–487.

378 Weinberg ED. The therapeutic potential of lactoferrin. Expert Opin. Investig. Drugs, 2003; 12(5): 841–851.

379 Bethell DR, Huang J. Recombinant human lactoferrin treatment for global health issues: iron deficiency and acute diarrhea. Biometals, 2004; 17(3): 337–342.

380 Wolf JS, Li D, Taylor RJ, et al. Lactoferrin inhibits growth of malignant tumors of the head and neck. ORL J. Otorhinolaryngol. Relat. Spec., 2003; 65(5): 245–249.

381 Cornish J, Callon KE, Naot D, et al. Lactoferrin is a potent regulator of bone cell activity and increases bone formation in vivo. Endocrinology, 2004; 145(9): 4366–4374.

382 Huang J, Nandi S, Wu L, et al. Expression of natural antimicrobial human lysozyme in rice grains. Mol. Breeding, 2002; 10(1/2): 83–94.

383 Nandi S, Suzukib Y, Huang J, et al. Expression of human lactoferrin in transgenic rice grains for the application in infant formula. Plant Sci., 2002; 163(4): 713–722.

384 Sava G. Pharmacological aspects and therapeutic applications of lysozymes. EXS, 1996; 75: 433–449.

385 Brock JH. The physiology of lactoferrin. Biochem. Cell. Biol., 2002; 80(1): 1–6.

386 Singh PK, Parsek MR, Greenberg EP, et al. A component of innate immunity prevents bacterial biofilm development. Nature, 2002; 417(6888): 552–555.

387 Markart P, Korfhagen TR, Weaver TE, et al. Mouse lysozyme M is important in pulmonary host defense against *Klebsiella pneumoniae* infection. Am. J. Respir. Crit. Care Med., 2004; 169(4): 454–458.

388 A plethora of hi-tech vaccines – genetic, edible, sugar glass, and more. CVI Forum, 1999; (18): 5–23.

389 Arntzen CJ. Edible vaccines. Public Health Rep., 1997; 112(3): 190–197.

390 Azhar AM, Singh S, Anand KP, et al. Expression of protective antigen in transgenic plants: a step towards edible vaccine against

anthrax. Biochem. Biophys. Res. Commun., 2002; 299(3): 345–351.

391 Bonetta L. Edible vaccines: not quite ready for prime time. Nature Med., 2002; 8(2): 94.

392 Bonn D. Edible vaccines tackle mucosal infections head on. Lancet Infect Dis., 2002; 2(5): 263.

393 Clough J. Edible vaccines against human papilloma virus. Drug Discov. Today, 2002; 7(17): 886–887.

394 Daniell H, Streatfield SJ, Wycoff K. Medical molecular farming: production of antibodies, biopharmaceuticals and edible vaccines in plants. Trends Plant Sci., 2001; 6(5): 219–226.

395 Fooks AR. Development of oral vaccines for human use. Curr. Opin. Mol. Ther., 2000; 2(1): 80–86.

396 Keusch GT. The National Institutes of Health agenda for international research in micronutrient nutrition and infection interactions. J. Infect. Dis., 2000; 182 Suppl 1: S139–S142.

397 Lauterslager TG, Hilgers LA. Efficacy of oral administration and oral intake of edible vaccines. Immunol. Lett., 2002; 84(3): 185–190.

398 Lauterslager TG, Florack DE, van der Wal TJ, et al. Oral immunisation of naive and primed animals with transgenic potato tubers expressing LT-B. Vaccine, 2001; 19(17–19): 2749–2755.

399 Wu YZ, Li JT, Mou ZR, et al. Oral immunization with rotavirus VP7 expressed in transgenic potatoes induced high titers of mucosal neutralizing IgA. Virology, 2003; 313(2): 337–342.

400 Mason HS, Ball JM, Shi JJ, et al. Expression of Norwalk virus capsid protein in transgenic tobacco and potato and its oral immunogenicity in mice. Proc. Natl. Acad. Sci. USA, 1996; 93(11): 5335–5340.

401 Curtiss R, Cardineau G, inventors; Washington University, Mycogen, assignees. Oral immunization by transgenic plants. US Patent US 5,654,184; US 5 679 880; US 5 686 079. 1997.

402 Koga T, Oho T, Shimazaki Y, et al. Immunization against dental caries. Vaccine, 2002; 20(16): 2027–2044.

403 Ma JK. The caries vaccine: a growing prospect. Dent. Update, 1999; 26(9): 374–380.

404 Richter LJ, Thanavala Y, Arntzen CJ, et al. Production of hepatitis B surface antigen in transgenic plants for oral immunization. Nature Biotechnol., 2000; 18(11): 1167–1171.

405 Kirk DD, Rempel R, Pinkhasov J, et al. Application of *Quillaja saponaria* extracts as adjuvants for plant-made vaccines. Expert Opin. Biol. Ther., 2004; 4(6): 947–958.

406 Kirk DD, Vonhof W, Eibner J, et al. Model Production of a Potent Plant-Made Vaccine. NFID Sixth Annual Conference on Vaccine Research, 5–8 May, Arlington, VA, 2003.

407 Tacket CO, Mason HS, Losonsky G, et al. Immunogenicity in humans of a recombinant bacterial antigen delivered in a transgenic potato. Nature Med., 1998; 4(5): 607–609.

408 Mason HS, Haq TA, Clements JD, et al. Edible vaccine protects mice against *Escherichia coli* heat-labile enterotoxin (LT): potatoes expressing a synthetic LT-B gene. Vaccine, 1998; 16(13): 1336–1343.

409 Lam D, Arntzen C, Mason H, inventors; ProdiGene, assignee. Vaccines expressed in plants. US Patent US 6034298. 2000.

410 Stiefelhagen P. [Oral vaccines – a future prospect! Hepatitis vaccination in the vegetable market?]. MMW Fortschr. Med., 2003; 145(47): 10.

411 Smith ML, Mason HS, Shuler ML. Hepatitis B surface antigen (HBsAg) expression in plant cell culture: Kinetics of antigen accumulation in batch culture and its intracellular form. Biotechnol. Bioeng., 2002; 80(7): 812–822.

412 Smith ML, Keegan ME, Mason HS, et al. Factors important in the extraction, stability and in vitro assembly of the hepatitis B surface antigen derived from recombinant plant systems. Biotechnol. Prog., 2002; 18(3): 538–550.

413 Thanavala Y, Yang YF, Lyons P, et al. Immunogenicity of transgenic plant-derived hepatitis B surface antigen. Proc. Natl. Acad. Sci. USA, 1995; 92(8): 3358–3361.

414 Mason HS, Lam DM, Arntzen CJ. Expression of hepatitis B surface antigen in transgenic plants. Proc. Natl. Acad. Sci. USA, 1992; 89(24): 11745–11749.

415 Edible vaccines for *Helicobacter pylori* and rotaviruses: Deprohealth project. http://www.flair-flow.com/consumer-docs/ffe63503.html 2004.

416 Nakata S. [Vaccine development for Norwalk virus]. Nippon Rinsho, 2002; 60(6): 1222–1227.

417 Koprowski H. [Old and new prescriptions for infectious diseases and the newest recipes for biomedical products in plants]. Arch. Immunol. Ther. Exp. (Warsz), 2002; 50(6): 365–369.

418 Bosch X. Tobacco fights back. Lancet Oncol., 2004; 5(5): 264.

419 Varsani A, Williamson AL, Rose RC, et al. Expression of Human papillomavirus type 16 major capsid protein in transgenic *Nicotiana tabacum* cv. *Xanthi*. Arch. Virol., 2003; 148(9): 1771–1786.

420 Dow enters NIH research agreement to develop rapid vaccine production system. http://www.fraunhofer-cmb.org/index.cfm? act=news 2004.

421 Hood EE, Jilka JM. Plant-based production of xenogenic proteins. Curr. Opin. Biotechnol., 1999; 10(4): 382–386.

422 Walmsley AM, Arntzen CJ. Plants for delivery of edible vaccines. Curr. Opin. Biotechnol., 2000; 11(2): 126–129.

423 Touchette N. New Company to Make Drugs from Plants. GNN News Alerts 2003.

424 Beck E. Researchers at Thomas Jefferson University developed a way to make a safer anthrax vaccine. United Press International. March 12, 2003.

425 Huang JK, Li M. [Silk protein fiber biomaterials and tissue engineering]. Zhongguo Xiu Fu Chong Jian Wai Ke Za Zhi, 2004; 18(2): 127–130.

426 Williams D. Sows' ears, silk purses and goats' milk: new production methods and medical applications for silk. Med. Device Technol., 2003; 14(5): 9–11.

427 Prunkard D, Cottingham I, Garner I, et al. High-level expression of recombinant human fibrinogen in the milk of transgenic mice. Nature Biotechnol., 1996; 14(7): 867–871.

428 Toman PD, Pieper F, Sakai N, et al. Production of recombinant human type I procollagen homotrimer in the mammary gland of transgenic mice. Transgenic Res., 1999; 8(6): 415–427.

429 John DC, Watson R, Kind AJ, et al. Expression of an engineered form of recombinant procollagen in mouse milk. Nature Biotechnol., 1999; 17(4): 385–389.

430 Tomita M, Munetsuna H, Sato T, et al. Transgenic silkworms produce recombinant human type III procollagen in cocoons. Nature Biotechnol., 2003; 21(1): 52–56.

431 Ruggiero F, Chanut H, Fichard A. Production of Recombinant Collagen for Biomedical Devices. BioPharm, 2000; 32–37.

432 Bellingham CM, Lillie MA, Gosline JM, et al. Recombinant human elastin polypeptides self-assemble into biomaterials with elastin-like properties. Biopolymers, 2003; 70(4): 445–455.

433 Bellingham CM, Woodhouse KA, Robson P, et al. Self-aggregation characteristics of recombinantly expressed human elastin polypeptides. Biochim. Biophys. Acta, 2001; 1550(1): 6–19.

434 Keeley FW, Bellingham CM, Woodhouse KA. Elastin as a self-organizing biomaterial: use of recombinantly expressed human elastin polypeptides as a model for investigations of structure and self-assembly of elastin. Philos. Trans. R. Soc. Lond. B Biol. Sci., 2002; 357(1418): 185–189.

435 Woodhouse KA, Klement P, Chen V, et al. Investigation of recombinant human elastin polypeptides as non-thrombogenic coatings. Biomaterials, 2004; 25(19): 4543–4553.

436 Yang G, Woodhouse KA, Yip CM. Substrate-facilitated assembly of elastin-like peptides: studies by variable-temperature in situ atomic force microscopy. J. Am. Chem. Soc., 2002; 124(36): 10648–10649.

437 Bucchini L, Goldman LR. Starlink corn: a risk analysis. Environ. Health Perspect., 2002; 110(1): 5–13.

438 Benowitz S. Jefferson Scientists Create Plant Factories churning out Antibodies against Tumor Cells (Press Release). Jefferson University Hospital, Philadelphia, May 3, 2005.

6
Production of Recombinant Proteins in Plants

Victor Klimyuk, Sylvestre Marillonnet, Jörg Knäblein, Michael McCaman, and Yuri Gleba

Abstract

This chapter reviews progress and challenges in the area of production of recombinant proteins, in particular biopharmaceuticals, in plants. Different expression platforms are summarized, including those based on the use of transgenic, transplastomic or transfected plants as production hosts. The quality and yield of recombinant proteins produced in and purified from plants, as well as progress in clinical trials with plant-made pharmaceutical proteins are described. The advantages, limitations and biological safety aspects of plant-based production of biopharmaceuticals are discussed.

Abbreviations

BSE	bovine spongiform encephalopathy
Bt	*Bacillus thuringiensis*
CJD	Creutzfeld–Jakob disease
EPSPS	5-enolpyruvylshikimate-3-phosphate synthase
ER	endoplasmic reticulum
GAD	glutamic acid decarboxylase
GD	Gaucher disease
GM	genetically modified
GUS	β-glucuronidase
hVEGF	human vascular endothelial growth factor
IL	interleukin
Mabs	monoclonal antibodies
MHC	major histocompatible complex
NPT	neomycin phosphotransferase
PAT	phosphinothricin acetyltransferase
PTGS	post-transcriptional gene silencing
TMV	tobacco mosaic virus
TSP	total soluble protein
vCJD	variant Creutzfeld-Jakob disease

6.1
Introduction

Numerous reviews concerning plant "molecular farming" have been published in recent years [1–6]. Analysis of these reviews and of recent research publications shows a change of priorities in the perceived advantages of plants as production hosts for recombinant proteins. Initially, the emphasis was on unlimited scalability and low cost of plant-based production, whereas yield and biosafety issues were not properly addressed. However, the last two parameters are crucial for determining the economics and, consequently the chances for commercial success of each specific plant-based system.

Modern Biopharmaceuticals. Edited by J. Knäblein
Copyright © 2005 WILEY-VCH Verlag GmbH & Co. KGaA, Weinheim
ISBN: 3-527-31184-X

A fresh boost was given to plant-based molecular farming in recent years, as the biopharmaceutical industry is trying to eliminate manufacturing processes that rely on production in animal cells due to the possible contamination of these products by human pathogens such as bovine spongiform encephalopathy (BSE) or Creutzfeld–Jakob disease (CJD, vCJD) (see-Part IV, Chapters 1, 2, 3, and 4).

In this chapter, we discuss different expression systems that are being developed. We consider the potential of each system by taking into account the impact of several parameters on economics and regulatory acceptability of the system: productivity (absolute and relative yield), biological safety (in particular transgene containment), scalability, versatility (ability to accommodate diverse proteins and to express a recombinant protein identical to the natural one), speed of research, development and commercial scalability provided by each of these systems.

6.2
Plant-based Expression Systems

Numerous plant-based expression systems that differ by the type of expression cassette and the location of such expression cassettes within a plant host cell have been developed. In the simplest way, the systems can be classified depending on how the foreign genetic material is incorporated into a plant cell. For example, systems can be based either on expression of heterologous sequences stably incorporated into nuclear (Section 6.2.1) or plastid (Section 6.2.2) genome, or on sequences transiently expressed within plant cells (Section 6.2.3).

6.2.1
Vectors Introduced via Stable Nuclear Transformation

The majority of genetically modified (GM) plants in use today have been produced by nuclear transformation. The vectors designed for such stable transformation usually include an expression cassette under control of a strong constitutive, inducible or tissue-specific promoter. The use of such promoters for expression of the protein of interest has been described in many publications (Table 6.1). The most commonly used promoters to drive transcription in dicotyledonous plants are the constitutive cauliflower mosaic virus (CaMV) 35S promoter [7] and the tuber-specific patatin promoter from potato [8–10]. Recently, a highly efficient system based on the use of the promoter-terminator of *Chrysanthemum morifolium* rbcS1 was reported, providing up to 10% of total soluble protein (TSP) in tobacco leaf for gusA expression [11]. The stress-inducible peroxidase (SWPA2) promoter was successfully used for expressing human lactoferrin in cultured ginseng cells [12]. Another group of frequently used promoters are those with seed-specific expression patterns. Such specificity is usually conserved in a heterologous host, as seed-specific promoters from monocotyledonous plants can also drive seed-specific expression in dicotyledonous plants. For example, the rice glutelin 3 promoter was used to express the recombinant glycoprotein b gene of the human cytomegalovirus in tobacco seeds [13]. Such promoters, like the hordein gene promoter, maize ubiquitin promoter and patatin promoter, were used for expression of genes of interest in barley grains [14], maize seeds [15], and potato tubers [16], respectively. However, the promoters of maize ubiquitin 1 [17] and rice

Table 6.1 Expression levels of selected recombinant proteins from nuclear transgenes

Recombinant protein	Expression level	Plant tissue	Reference(s)
β-Glucuronidase	0.4–0.7% TSP	seeds (corn)	34–36
β-Glucuronidase	10% TSP	leaves (tobacco)	11
Avidin	5.7% TSP	seeds (corn)	36
Lysozyme	5–45% TSP	seeds (rice)	32, 33
Spider silk proteins	2% TSP	tuber (potato), leaves (tobacco)	10
Spider silk proteins	2% TSP	seeds (tobacco)	10, 45
Human somatotropin	0.16% TSP	tuber (potato), leaves,	46
Human serum albumin	0.2% TSP	cell culture	47, 48
Human serum albumin	0.2% TSP	(tobacco)	47, 48
Human collagen	0.1 mg g^{-1}	tobacco (leaves)	49
Human α-lactalbumin	5 μg g^{-1}	tobacco (leaves)	50
B subunit of *E. coli* enterotoxin	13 μg g^{-1}	tuber (potato)	51
Thermostable β-glucanase	0.1–5.4% TSP	seeds (barley)	14
Human interleukin-2	115 U g^{-1}	microtuber (potato)	16
Aprotinin	0.17% TSP	seeds (corn)	52
Bovine trypsin	3.3% TSP	seeds (corn)	53
β-Casein	0.01% TSP	leaves (potato)	54
Cholera toxin B subunit	0.3% TSP	leaf, tuber (potato)	55
HIV p-24 casid protein	0.35% TSP	leaves (tobacco)	56
Hepatitis B surface antigen	0.33–16 μg g^{-1}	tuber (potato)	57, 58
Human lactoferrin	0.1% TSP	tuber(potato)	59
Human lactoferrin	0.5% $^{a)}$	seeds (rice)	60
Human lactoferrin	Unknown	seeds (corn)	61
Human lactoferrin	4.3% TSP	cell culture (tobacco)	62
Human lactoferrin	3% TSP	cell culture (ginseng)	12
Human cytomegalovirus glycoprotein B	1% TSP	seeds (tobacco)	13
Monoclonal antibodies	0.01–0.25% TSP	leaves (tobacco)	63–65
Monoclonal antibodies	0.5 mg g^{-1}	leaves (tobacco)	66, 67
Monoclonal antibodies	1% TSP	alfalfa	68
Monoclonal antibodies	11.7 μg g$^{-1\,b)}$	rhizosecretion (tobacco)	69
scFv	0.01–6.8% TSP	leaves, seeds (tobacco)	70–72
scFv	30 μg g^{-1}	leaves, seeds (wheat, rice)	37
scFv	1% TSP	*Petunia hybrida*	73
scFv	36.5% TSP	seeds (*Arabidopsis*)	38
Diabody	0.5 mg kg^{-1}	leaves (tobacco)	74
Human interferon-alpha	560 IU g^{-1}	potato	75
HPV major capsid protein L1	0.2–0.5% TSP	tuber (potato), leaves (tobacco)	76
Human interleukin-18	0.05% TSP	leaves (tobacco)	75
Human glucocerebrosidase	1 mg g^{-1}	leaves (tobacco)	77
Human α and β hemoglobin	0.05% TSP	seeds (tobacco)	78
Human a1 antitrypsin	18.2–24 mg g$^{-1\,c)}$	cell culture (rice)	79
Human placental alkaline phosphatase	20–28 μg g$^{-1\,b)}$	rhizosecretion (tobacco)	80, 81
Human vascular endothelial growth factor	30 mg L$^{-1\,d)}$	cell culture (moss *Ph. patens*)	82

a) Relative yield per gram of dry seed weight.
b) Rhizosecretion of protein per gram dry roots in 24 h.
c) Accumulation of protein in the medium per gram dry cells biomass in 50–70 h.
d) Secreted amount of protein per 1 L of culture in 24 h.

actin 1 [18] genes are the most frequently used for monocotyledonous plants.

In general, yields of recombinant proteins under control of constitutive promoters are low (ca 0.1% of TSP). Even though some high-yield expression systems have also been reported (e.g., [11]), such systems do not appear to represent broadly applicable solutions for the production of pharmaceutical proteins in plants, because constitutive expression of many biologically active proteins at high levels often compromises plant growth and development. It probably also can trigger transgene silencing.

Interesting alternatives to constitutive expression are expression systems based on inducible promoters. Such systems provide for separation of the growth and production phases, thus theoretically allowing for improved yield with proteins that are toxic or interfere with plant physiology/development. Several inducible systems based on application of small molecules-inducers have been described: the tetracycline-inducible system [19–21], the copper-inducible [22], steroid-inducible systems [23, 24], ethanol- [25, 26] or acetaldehyde-inducible [27] systems, and insecticide methoxyfenozide-inducible system [28]. A hybrid system representing a combination of two different inducers was also described. It consists of a chimeric promoter that can be switched on by the glucocorticoid hormone dexamethasone and switched off by tetracycline [29]. For the latest review on chemically inducible systems, see Ref. [30]. At present, it is still not known whether inducible systems will provide yields higher than those generated by transgenes expressed under control of constitutive or tissue-specific promoters. Also, the leakiness of such systems might be a serious problem for expression of cytotoxic proteins. Additionally, the use of

some chemical inducers such as steroids and antibiotics is not desirable for large-scale application. Potentially, the ethanol switch proposed by Caddick and colleagues could perhaps satisfy requirements for the commercial use of a chemical gene switch [25]. The use of vectors that can potentially produce high yields, such as plant viral replicons stably incorporated into plant nuclear DNA, does not provide a solution either, because such vectors are subject to transgene silencing. However, a promising approach was found by using post-transcriptional gene silencing (PTGS) suppressors to boost expression of genes of interest in plants. By crossing a transgenic tobacco plant carrying the potato virus X-based replicon with a transgenic plant providing for the viral PTGS suppressor HC-Pro, progeny expressing high level of a gene of interest (GUS) could be obtained [31]. It is clear however, that such an approach is not suitable for the expression of cytotoxic proteins, and that further improvements of the system are required. One possible improvement is the development of a tightly regulated inducible system that does not allow "leaky" expression of the transgene and the PTGS suppressor.

The use of seed-specific promoters appears to be the most promising (see Table 6.1). Seeds are also an attractive choice for molecular farming because they can be transported and stored for downstream processing, without any significant loss of yield or quality of the recombinant protein. Several seed-specific promoters from dicotyledonous and monocotyledonous plants have been isolated and used for expression of recombinant proteins in rice [32, 33], corn [34–36], barley [14], wheat [37], tobacco [13], and even *Arabidopsis* [38]. The yield of recombinant proteins in seeds can reach up to 45% of TSP, as was shown in the case of seed-specific expression of the human lysozyme gene in rice [33].

The use of major food/feed crops (rice, corn, wheat, barley, etc.) for molecular farming – and especially for the production of pharmaceutical proteins – carries the risk of uncontrolled release into the environment and of entering the food chain by contamination of non-transgenic crops. In such cases, biological safety issues should be carefully considered and given the highest priority for each specific recombinant protein. Potential problems for companies ignoring or miscalculating the problem can range from costly delays in field trials (such as one that Ventria Bioscience faced this year with transgenic rice expressing human lactoferrin and lysozyme [39–41]), to high financial liabilities once a permit for commercial production has been obtained. ProdiGene Inc. was ordered to pay US$ 3 million penalty by the United States Department of Agriculture (USDA) for contaminating soybean fields with transgenic "volunteer" corn plants [42]. The unpaid portion of this penalty was paid by Stine Seed Company as a part of ProdiGene's rescue purchase [43]. The choice of corn as a production host for monoclonal antibodies (Mabs) is one of the major reasons behind the recent failure of Epicyte Pharmaceuticals to raise new capital [44]. In contrast, several small plant biotechnology companies have chosen to use plants grown in a closed environment as the basis of their production platform. Among them, Biolex uses a small aquatic plant *Lemna* (duckweed) grown in closed environment, and Medicago Inc. uses alfalfa grown in contained glasshouses. Other companies, such as Plantigen, Planet Biotechnology, and Large Scale Biology rely on open-field cultivation, but using tobacco, a production host that cannot contaminate feed and food stocks. SemBioSys, a Canadian biotechnology company, has chosen

safflower (*Carthamus tinctrius* L.), a minor crop plant with medicinal, industrial, and food applications, as an alternative production host. The choice of a minor crop plant reduces the risks of cross-contaminating non-transgenic crops, while benefiting the developer who can rely on well-established agro-cultivation technologies.

In brief, it appears very unlikely that high yields of recombinant proteins in transgenic plants can be achieved with the use of constitutive promoters, especially in the case of recombinant proteins that have deleterious effects on plant growth and development. The use of inducible systems is a more promising approach, though this requires the development of "leakage-proof" systems for the production of cytotoxic proteins. Today, production in the seeds of transgenic plants using seed-specific promoters is the most obvious choice for those interested, and a viable solution for many recombinant proteins.

6.2.2
Plastid-based Vectors

Since the first successful transformation of tobacco chloroplasts [83], expression systems based on the transformation of plant plastids has attracted the attention of plant biotechnologists. The features that make this technology so attractive are its potential for high protein yield, along with inherent biosafety features such as limited plastid transfer via pollen (due to maternal inheritance of plastid-encoded genes) [84] and the relatively low probability of transgene movement from the chloroplast to the nucleus [85, 86].

Many other heterologous proteins have been expressed in tobacco chloroplast, and in many cases yields exceeding 5% of TSP were reported (Table 6.2). Examples of such proteins are neomycin phosphotransferase

(NPTII) (23% of TSP) [87], 5-enolpyruvyl-shikimate-3-phosphate synthase (EPSPS) (10% of TSP) [87], phosphinothricin acetyl-transferase (PAT) (7% of TSP) [88], *Bacillus thuringiensis* (Bt) toxin (41% of TSP). As expected, another large group of proteins successfully expressed in plastids are different antigens without any obvious cytotoxicity that are derived from human and animal pathogens of prokaryotic nature (Table 6.2). These data suggest that expression in plastids can be a good choice for an expression platform for these proteins.

There are numerous reviews describing the state of the art in plastid-based expression systems [89–94], so there is no need to go into their detailed description. The main point worth mentioning is that, at present, only tobacco plastids and plastids of the unicellular alga *Chlamydomonas reinhardtii* [95] can be transformed routinely. However, this situation is changing. In addition to tobacco, successful plastid transformation was reported for Brassicaceae [96, 97], potato [98], tomato [99], and cotton (H. Daniell, personal communication). Transplastomic potato plants, tomato cell lines as well as other *Nicotiana* species (*N. benthamiana, N. excelsiana* and *N. excelsior*), were also produced (H.-U. Koop, personal communication).

There are limitations on the choice of the recombinant protein to be expressed in transplastomic hosts due to the prokaryotic nature of the translational and post-translational machinery of plastids [100] (see Part

Table 6.2 Expression levels of selected recombinant proteins in plastids

Recombinant protein	Expression level [% TSP]	Host	Reference(s)
β-Glucuronidase	1.3–8.8	Tobacco	111, Icon [a]
Neomycin phosphotransferase (NPTII)	0.16–23	Tobacco	112–114
Green fluorescent protein	5	Tobacco	98
EPSP synthase	0.001–10	Tobacco	87
Phosphinothricin acetyltransferase	7	Tobacco	88
Bacillus thuringiensis toxin	2–46.1	Tobacco	115–118
Human somatotropin	0.2–7	Tobacco	102
Human serum albumin	0.02–11.1	Tobacco	119
Interferon-alpha	1	Tobacco	Icon [a]
Interferon-gamma	0.1–6	Tobacco	104
Tetanus toxin fragment C	10–25	Tobacco	120
B subunit of *E. coli* enterotoxin	2.5	Tobacco	121
Native cholera toxin B subunit	4.1	Tobacco	122
Rotavirus VP6 protein	0.6–3	Tobacco	123
2L21 peptide (virulent canine parvovirus)	23–31	Tobacco	124
Xylanase	6	Tobacco	125
Phenylalanine ammonia lyase	1.0–1.5	Tobacco	Icon [a]
Bacillus anthracis protective antigen	1.7–18.1	*C. reinhardtii*	126
lsc antibody	1	*C. reinhardtii*	127
Green fluorescent protein	0.5	*C. reinhardtii*	128
β-Glucuronidase	0.01	*C. reinhardtii*	129

a) Icon Genetics, unpublished data.

IV, Chapters 5 and 7; and Part V, Chapter 1). Whereas some constraints of translation are more easily addressed through vector design, lack of some essential post-translational capabilities (in particular N-glycosylation) cannot be easily corrected. In some cases, the latter can be an advantage if glycosylation is not essential for recombinant protein function, since N-linked oligosaccharides of plant origin might result in new immunogenic structures [101]. Also, achievement of the correct amino acid sequence at the N-terminal end of proteins that do not start by a methionine might be a problem for the expression of such proteins in plastids. For example, human somatotropin is a secreted protein with a N-terminal signal peptide cleaved off after secretion into the endoplasmic reticulum (ER), resulting in a processed protein with a N-terminal amino acid that is not a methionine. As a solution for expression of this protein in the chloroplast, an N-terminal fusion with the ubiquitin gene was made, and the fusion protein processed by cleavage of ubiquitin during the extraction from plant tissue [102]. The protein was expressed at a high level (more than 7% of TSP), but removal of N-terminal ubiquitin from the ubiquitin-human somatotropin fusion resulted in only 30–80% (depending on extraction conditions) of all molecules being correctly processed. In contrast, human somatotropin expression in tobacco seeds based on stable nuclear transformation vectors yielded the correctly processed protein with the correct amino acid sequence, but at a very low level (0.16% of seed TSP) [103].

In general, expression levels of recombinant proteins in transplastomic plants can be significantly higher than those in nuclear transformants, subject to optimal design of the expression cassette and depending on the nature of the recombinant protein of interest. Clear evidence of an impact of the expression cassette is the differences in expression levels obtained using various vectors for Bt toxin (25-fold difference), human serum albumin (5×10^2-fold difference), EPSPS (10^4-fold difference) (see Table 6.2). A significant increase in yield can also be achieved by using translational fusion of a gene of interest to another one (usually a reporter gene) which is easy to express. An example of such an increase in yield is the expression level of the interferon-a gene as translational fusion with the GUS reporter gene [104].

Another remarkable advantage of plastid transformation technology is the availability of efficient methods, similar to those used in bacterial genetics – for example, the use of homologous or site-specific recombination, for complete removal of transformation markers and other unnecessary sequences from plastid transformants. There are several approaches to achieve this result. Some approaches, also used with nuclear transformants [105], are based on site-specific recombination using the CRE/lox system [106, 107] (see Part III, Chapter 2 and Part IV, Chapters 2 and 3). Two other approaches are based on homologous recombination between two short direct repeats flanking the selectable marker [108, 109], and do not require crosses with plants that provide for the site-specific recombinase. An approach recently developed by Icon Genetics [109] has the advantage that it allows the easy selection of homoplastomic lines after the antibiotic resistance gene has been lost. Removal of the antibiotic resistance gene improves the biosafety parameters of the system, since the presence of the antibiotic resistance gene in plants grown in an open field is not desirable.

An expression system based on the plastids of the green algae *Chlamydomonas reinhardtii* represents a special case. Despite exhibiting relative yields of recombinant protein generally lower than those achieved with transplastomic tobacco, the system allows for very fast scale-up, producing gram quantities of recombinant protein within 2 months after transformation, with the possibility of further increase in geometric progression, reaching yields of up to 150 kg of recombinant protein per acre of fully contained growth area annually (S. Mayfield, personal communication). For comparison, the generation of a homoplastomic tobacco plant with a transgene stably incorporated into plastid DNA requires at least 6 months. Therefore in terms of speed – but not yield – the *Chlamydomonas* system is comparable to many known microbial systems. For the most recent review on molecular farming in plastids of *C. reinhardtii*, see Ref. [110].

In conclusion, plastid-based expression systems provide for potentially higher expression levels than the majority of nuclear expression systems, and with a higher level of transgene containment, but the platform provides limited post-translational processing choices. Also, despite reports of successful plastid transformation of new species beyond *Nicotiana*, the choice of production host is still predominantly restricted to the *Nicotiana* family. However, this limitation does not present a serious drawback for production of pharmaceutical proteins (except perhaps for "edible vaccines") since tobacco is a non-feed/non-food crop with a well-established agriculture.

6.2.3
Vectors for Transient Transformation

Transient expression [130] is a fast and convenient alternative to stable transformation because of the speed it provides for research and development and, to some extent, because of less complex regulatory hurdles, as no stably transformed plants are involved in the production process. Transient expression can be achieved by transfecting plants with viral vectors, or, on a small scale, by agroinfiltrating plant tissue with a standard expression cassette under control of a constitutive promoter, for example the 35S promoter [131]. Usually, agroinfiltration itself does not provide for high yield, but in combination with PTGS suppressors such as p19 or HcPro, protein expression levels can be increased up to 50-fold [132]. However, even such significant improvement is still well below the levels that can be achieved with some viral vectors. The two approaches (agroinfiltration and the use of viral vectors) can be combined, and we recently demonstrated that tobacco mosaic virus (TMV)-based vectors can be delivered via agroinfiltration as separate structural blocks and assembled *in planta* with the help of a site-specific recombinase [133].

Viruses first began to attract attention as a potential basis for developing expression systems almost 20 years ago. Since then, tremendous progress has been achieved in the development of viral vectors. There are numerous reviews describing different viral vectors and strategies for their use [134–139]. Comparison of the expression levels provided by different vectors (Table 6.3) makes it clear that TMV-based expression systems achieve the highest yield. In a recently published report [133], it was shown that TMV-based expression of a reporter gene (GFP or DsRed) could reach biological limits of the plant leaf system, producing up to 5 mg of recombinant protein per gram fresh leaf biomass. The relative yield with such system can approach 80% of total soluble protein. Such high absolute and relative yield translates into much more efficient and cost-effective upstream and downstream processes. Also, high relative yield is possible because of

Table 6.3 Transient expression levels of different recombinant proteins

Recombinant protein	Expression level	Vector	Production host	Reference(s)
Dihydrofolate reductase	8 µg g^{-1}	CaMV	Turnip	149
Metallothionein II	0.5% TSP	CaMV	Turnip	150
Interferon αD	2 µg g^{-1}	CaMV	Turnip	151
Interferon γ	0.5% TSP	BMV	Protoplasts (tobacco)	152
ScFv	5 µg g^{-1}	p35S[a]	Leaves (tobacco)	131
ScFv	12–30 µg g^{-1}	TMV	*N. benthamiana*	153
ScFv	0.02–0.8 mg mL^{-1}IF[b]	TMV	*N. benthamiana*	154
ScFv	0.25–1.2 mg g^{-1}	TMV	*N. benthamiana*	Icon[c]
Monoclonal antibody	Unknown	TMV	*N. benthamiana*	155
Glycoprotein D of BHV-1	20 µg g^{-1}	TMV	*N. benthamiana*	156
α-amylase	5% TSP	TMV	*N. benthamiana*	157
Pollen allergen Bet v1	0.2 mg g^{-1}	TMV	*N. benthamiana*	158
GFP, DsRed	5 mg g^{-1}	TMV	*N. benthamiana*	133
Human somatotropin	1.2 mg g^{-1}	TMV	*N. benthamiana*	Icon[c]
Diabody	1.5 mg kg^{-1}	Unknown	Leaves (tobacco)	74
Human lactoferrin	0.6% TSP	PVX	*N. benthamiana*	159

a) CaMV 35S promoter.
b) Interstitial fluid.
c) Icon Genetics, unpublished data.

virus-controlled gene amplification and because of the relatively less understood mechanism of virus-induced shut-off of host protein biosynthesis. Other features of viral vectors are their ability for cell-to-cell and systemic movement, which allows the vector to spread from the infection zone to most of the plant tissue. Fully functional viral vectors can tolerate only relatively modest heterologous inserts with an upper size limit of around 1 Kb. Significantly shorter inserts practically do not affect viral functions, and there are numerous examples of using such inserts as fusions with the viral coat protein for antigenic epitopes production [140–147]. In contrast, larger inserts affect vector stability, resulting in wild-type revertants that successfully outcompete the original recombinant vector for all viral functions – amplification rate, cell-to-cell and systemic movement. These problems have been ad-

dressed by applying the so-called "deconstructed virus" strategy – for example, by removing the missing or undesired functions of a viral vector and complementing the necessary functions *in trans* (for a review, see Ref. [139]). Such approach allows not only for the payload capacity of the viral vector to be increased, but also provides for significantly better containment of the recombinant viral vector. Using viral vectors devoid of the coat protein gene, we have achieved very high yield (up to 40% of TSP) with inserts encoding different proteins and protein fusions that are as large as 2.2 Kb (see Fig. 6.3).

Systemic viral vectors are restricted to certain parts of infected plants (usually the newly developing leaves, stem and roots), and are excluded from a significant part of plant biomass, including the mature leaves. Clearly, the creation of transgenic plants with a viral replicon precursor stably inserted on a chromosome could be a solution, but it was found that such plants produce very low amount of recombinant protein because of silencing [148]. As already mentioned in Section 6.2.1, the use of PTGS suppressors allows this problem to be addressed [31]. The problem of separating the plant growth phase from the production phase can also be resolved by developing either a tightly regulated inducible system or a virus-based transient expression system, where tight control over transgene expression is not an issue. However, viral vector delivery to each plant cell in a transient expression system is not a trivial task. *Agrobacterium*-mediated delivery of T-DNAs encoding RNA viral replicons provides only a small fraction of plant cells with active viral replicons due to low ability of the primary transcripts to leave a nucleus. The RNAs of plant plus-sense RNA viruses replicate in the cytoplasm and normally never enter the cell nucleus, and contain multiple sequence features that are likely to be impro-

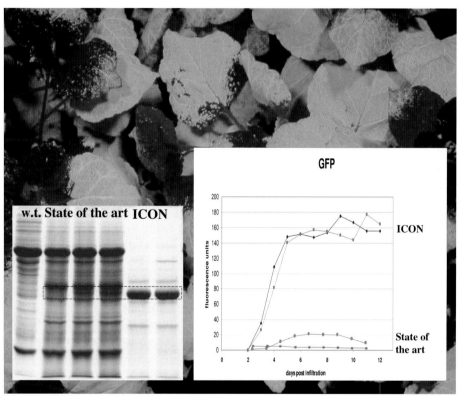

Fig. 6.1 Comparative analysis of recombinant protein expression systems: state-of-the-art versus magnICON®. Background: magnICON®-based expression of green fluorescent protein (GFP) in *Nicotiana benthamiana* (Plants are exposed to UV light). Left side insert: Coomassie blue-stained polyacrylamide gel after SDS-electrophoretic separation of total soluble protein (TSP) extracted from *N. benthamiana* leaves. The framed area contains GFP bands. w.t.: TSP extracted from the leaves of wild-type *N. benthamiana*. Right side insert: time-course of GFP expression levels for magnICON®-based (ICON) and other state-of-the-art systems.

perly recognized by the nuclear RNA processing machinery if delivered to the nucleus.

We recently developed highly active synthetic templates for delivery of RNA viral vectors as DNA precursors using *Agrobacterium*, and found that *Agrobacterium* delivery of such templates can be used to start gene amplification and obtain high-level expression in all mature leaves of a plant, simultaneously. Such a transfection route can be performed on an industrial scale by vacuum-infiltration of batches of multiple plants (Marillonnet et al., unpublished results). In this process, the bacteria assume the (formerly viral) functions of primary infection and systemic movement, whereas the viral vector itself provides for cell-to-cell (short distance) spread, amplification and high-level expression. A comparative analysis of state-of-the-art systems with our approach is illustrated in Fig. 6.1.

Assuming protein yields as mentioned above, and based on realistic yields of 100 tons of plant leaf biomass per hectare (ha) of greenhouse per year, a 1-ha facility should be capable of producing 280–400 kg of recombinant protein each year. This means that for the vast majority of pharmaceutical proteins, industrial-scale production can be carried out entirely in a partially or fully contained greenhouse facility. The whole process is a straightforward protocol, similar to existing industrial microbial technologies; it requires, in addition to well-established industrial upstream (greenhouse plant cultivation) and downstream (protein extraction and purification) components, a contained technology block that includes an apparatus for vacuum-infiltration of batches of plants and a chamber/greenhouse for short-term subsequent incubation, as well as a small bacterial fermentation apparatus. Such a block would of course require certain safety "locks", so as to eliminate the release of agrobacteria into the open environment and to protect the operating personnel.

6.3
Plant-made Recombinant Proteins available Commercially, and under Development

A relatively small number of biotechnology companies operate in the field of recombinant proteins production in plants (Table 6.4). This small number represents roughly one-third of approximately 60 dedicated plant biotechnology companies – an unfavorable comparison to the several hundred small and medium-sized businesses that use bacterial, yeast or animal-based expression platforms and operate in the sector of "red" biotechnology. The number of plant-made recombinant proteins that have reached the market is also very limited, but has a healthy tendency to grow.

Although not a biopharmaceutical, one of the first plant-derived recombinant proteins of potential importance to reach the market, trypsin, has been produced in transgenic maize by ProdiGene, and is marketed by Sigma-Aldrich Fine Chemicals as TrypZean™. ProdiGene also established production in maize of recombinant protease inhibitor aprotinin (called AproliZean™). In 2002, ProdiGene planted 400 acres of transgenic corn expressing trypsinogen, an amount sufficient (according to ProdiGene) to meet 5% of the market demand. The company planned to scale-up the cultivation and to meet full market demand in 2003, but this did not materialize. The company faced problems caused by field contamination with transgenic "volunteer" corn plants expressing recombinant proteins, and this forced the company to down-size its operations. Subsequently, ProdiGene was sold to H. Stine Seeds [43]. Based on ProdiGene's

Table 6.4 Companies using plant-based recombinant proteins production platforms

Company	Internet link	Production host
Biolex	http://www.biolex.com	Duckweed
Chlorogen	http://www.chlorogen.com	Tobacco
CropTech[a]		Tobacco
Dow Chemical	http://www.dow.com/plantbio/index.htm	Maize
Epicyte Pharmaceutical[a]	http://www.biolex.com	Maize
Greenovation	http://www.greenovation.com	Moss *Physcomitrella*
Icon Genetics	http://www.icongenetics.com	Tobacco
Large Scale Biology	http://www.lsbc.com	Tobacco
LemnaGene	http://www.lemnagene.com	Duckweed
Meristem Therapeutics	http://www.meristem-therapeutics.com	Maize, Tobacco
Medicago	http://www2.medicago.com	Alfalfa
Monsanto Protein Tech.[a]	http://www.mpt.monsanto.com	Maize
MPB Cologne[a]		Potato
Novoplant	http://www.novoplant.com/	Various
SemBioSys	http://www.sembiosys.ca	Safflower
Syngenta Biopharma	http://www.syngenta.com/en/biopharma/	Various
Phytomedics	http://www.phytomedics.com	Tobacco
Planet Biotechnology	http://www.planetbiotechnology.com	Tobacco
Plantigen	http://www.plantigen.com/	Tobacco
ProdiGene	http://www.prodigene.com	Maize
Protalix	http://www.protalix.com	Cell culture
Ventria	http://www.ventriabio.com	Rice

a) Operations terminated.

data, planting 400 acres of transgenic corn yielded ca. 1000 tons of seed expressing trypsin at a level of 58 mg kg^{-1} seed, which corresponds to a total harvest (assuming 50% recovery of 90% pure enzyme) of 11–13 kg of enzyme. Although the price of plant-made trypsin (TrypZean) in the Sigma catalogue (2004–2005) is approximately 20-fold higher than that for trypsin of animal origin, plant-made trypsin is bona fide human pathogen-free, and is safer for use in many applications. However important, this advantage does not justify such a high price, but likely reflects the very high downstream processing cost of low-expressing seed; certainly, a more efficient plant expression system is needed in order to lower the cost of plant-made trypsin to make it a commercial success.

At the start of 2004, Large Scale Biology Corporation (LSBC) began to ship test quantities of its plant-produced recombinant bovine-sequence aprotinin (rAprotinin) to customers for R&D and manufacturing applications. LSBC also has several other products in development; for example, plant-produced human therapeutic enzyme alpha-galactosidase A for enzyme replacement therapy showed positive results in preclinical studies using an animal model of Fabry disease. This enzyme is currently undergoing clinical trials. LSBC are also developing vaccines for animal health in collaboration with Schering-Plough Animal Health Corporation (SPAH), and a human papillomavirus vaccine is currently under development in collaboration with the University of Louisville

(Kentucky). The company uses a TMV-based transient expression platform that is fast and easy to apply. The platform is built on viral vectors capable of systemic movement.

California-based Ventria Bioscience (Sacramento) is at the stage of field trials of transgenic rice plants expressing recombinant lactoferrin and lysozyme in rice seed (see Part IV, Chapter 8). It is worth mentioning, that these products might reach market later than planned considering difficulties with obtaining the necessary field trial permits [39–41].

Meristem Therapeutics (Clermont-Ferrand, France) has a gastric lipase, produced in grains of corn, for the treatment of cystic fibrosis in Phase II clinical trials. The company also has a recombinant lactoferrin at the stage of preclinical trials, in addition to human serum albumin, collagen and monoclonal antibodies at R&D stage.

Several other companies have products at different stages of development (see Part IV, Chapter 5). We provide here a brief update of their performance using publicly available information, obtained predominantly from the web pages of the companies mentioned below.

SemBioSys Genetics Inc. (Calgary, Canada), in addition to insulin and apolipoprotein AI, demonstrated a proof-of-concept for proteins addressing osteoporosis, pulmonary and liver fibrosis, psoriasis and gastrointestinal disorders. Plantigen Inc. (London, Canada) has proof of principle for several products under development. Those include glutamic acid decarboxylase (GAD) for the treatment of type 1 diabetes; interleukin-10 (IL-10) for treatment of inflammatory bowel disease; interleukin-4 (IL-4) for use as adjuvant to enhance immune response; and major histocompatible complex (MHC) and cytokines for use in organ transplants.

Greenovation (Freiburg, Germany) uses the advantage of highly efficient homologous recombination in the moss *Physcomitrella* to develop a new production system for humanized antibodies with required glycosylation pattern. For a more detailed description of production of biopharmaceuticals with "humanized" glycosylation (see Part IV, Chapter 7).

The productivity of moss bioreactor for human vascular endothelial growth factor (hVEGF) reached 30 mg L^{-1} per day [82]. The company claims that further increases in yield can be achieved by optimization of production strains for specific proteins.

Protalix (formerly Metabogal, Israel) began preclinical tests on enzyme therapy for Gaucher disease (GD), a genetically linked condition that affects the metabolism of people deficient in the enzyme β-glucocerebrosidase. Without this enzyme, fatty deposits build up in the spleen, brain, liver, and bone marrow, and this leads to extreme pain and may even prove fatal. In 2000, the annual world-wide market for GD was US$ 620 million.

Among pharmaceutical proteins, the antibody market is potentially the most interesting (see Part IV, Chapter 16 and Part V, Chapter 1). Antibody production is traditionally based on animal cell culture, commanding high manufacturing costs ($500–1000 g^{-1} purified protein). To date, no plant-made antibodies are available commercially, and some analysts foresee potential problems with production capacities within next few years. Most of the antibodies available or under development are immunoglobulins that are complex glycosylated proteins, and cannot be expressed in bacterial cells. Expression in eukaryotic organisms (animal cells, insect cells, yeast, or plants) (see Part IV, Chapters 1, 2, 12, 13, and 14) results in glycoproteins that differ considerably from human antibodies because of the differences in protein glycosylation among different organisms. In many cases, glycosylation does not affect the pharmacological properties of the resultant protein, and it is not required for proper protein folding. Thus, some developers mutate glycosylation sites in antibody genes in order to avoid glycosylation altogether. Alternatively, the plant host can be engineered to provide for a required "humanized" glycosylation pattern, and this approach is currently being used by Greenovation (Freiburg, Germany), as mentioned above. A smaller proportion of immunopharmaceuticals are proteins that are less complex. Alternatively – and probably more easily – these proteins can be manufactured in non-animal cells.

Starting from the pioneering work of scientists at the Scripps Research Institute [160, 161] and the Max-Planck-Institute (Cologne, Germany) [162], numerous antibodies have been expressed in plants and shown to be properly processed and assembled into fully functional molecules with full immunological activity. Today, almost every plant biotechnology company involved in the molecular pharming of pharmaceutical proteins has antibodies included in their product portfolio. The level of expression in transgenic plants expressing heavy- and light-chain Mab polypeptides under a strong constitutive promoter are rather low, amounting usually to 10–50 mg of antibody per kg leaf biomass (tobacco) or up to 1 g kg^{-1} seed (maize), and may require up to 3 years before gram quantities of the protein of interest can be obtained for research or clinical studies.

The level of expression in plants infected using viral vectors can be higher (up to 200 mg kg^{-1} leaf biomass), and research quantities can be obtained within 4–6 weeks, although the current versions of viral vectors allow expression only of single-chain antibodies. The technology can be used immediately for applications such as the production of individualized antibody vaccines, for example vaccines for non-Hodgkin's lymphoma made by Large Scale Biology. Phase I clinical trials were successfully completed with these individualized vaccines, and subsequently Large Scale Biology Corporation built the world's first commercial-scale biopharmaceutical production facility, based at Owensboro, Kentucky.

Icon Genetics has conducted a feasibility study aimed at the expression of single-chain antibodies in plants using its viral expression technology built on a TMV-based transient expression system. Using different antibodies currently at a preclinical stage, it was possible to express them at a level of $0.25–1.2 \text{ g kg}^{-1}$ fresh leaf biomass, with relative recombinant protein amounts reaching up to 35% of TSP. Among other recombinant proteins that Icon is developing are included aprotinin, trypsin, thrombin, several viral antigens, and thaumatin.

Planet Biotechnology Inc. of Hayward, California and LSBC announced recently a biomanufacturing agreement to extract and purify a plant-made antibody to control dental caries; Phase I clinical trials were completed by Planet Biotechnology Inc. The same company is also developing a new approach for blocking infection by rhinovirus, a major cause of the common cold. In preclinical testing, the fusion protein proved highly protective against cellular damage caused by human rhinovirus infection. Planet Biotechnology also successfully completed pilot human clinical trials with neutralizing toxicity of chemotherapeutic drug doxorubicin by topically applied antibodies. The oral application of antibodies can prevent chemotherapy-induced gastrointestinal toxicity and, when applied topically in liposomes, prevents doxorubicin-induced hair loss.

Novoplant is developing, for the first time, a comprehensive portfolio of antibodies for use in veterinary medicines. These monoclonal immunoglobulin preparations are based on single-chain-antibodies produced in plant seeds, and are designed to combat typical diseases that occur in swine, poultry, and calves in animal husbandry. They are designed to offer protection against coccidiosis in chickens and scour caused by ETEC *Escherichia coli*, rotavirus and coronavirus in piglets and calves.

Medicago currently has 10 Mabs and plasma-proteins in early-stage development, with preclinical studies planned to start in 2005 for two products. Greenovation has one antibody (ABC-48) in preclinical development for the prevention of deep-vein thrombosis. The secreted IgG antibody was shown to be correctly assembled, and displayed normal binding activity to its natural ligand.

This brief account of currently available products and under development (see Table 6.4) provides evidence that, despite several drawbacks, the production of recombinant proteins in plants is steadily progressing. However, it is also evident that all expression platforms in use have certain limitations, and it appears that as yet there is no universal platform to satisfy the requirements for production of all biopharmaceuticals of choice.

6.4
Comparative Analysis of the Expression Systems and Production Platforms

In principle, all three expression platforms (see Section 6.2) provide for expression levels within essentially the same range. Comparison of expression levels (see Tables 6.1–6.3) reveals that high yields (over 5% of TSP) are feasible using either nuclear, plastid, or transient expression. However, some systems "are more equal than others" when additional requirements are introduced, such as the speed of R&D process and biosafety parameters. The transfection platform is by far the fastest: starting with the DNA construct of the protein of interest, milligram and gram quantities of recombinant protein are available in 3–4 weeks. Thus, the platform supports the highest possible speed of R&D in the industry, including the microbial and animal cells systems (Fig. 6.2). The biosafety parameter shall not be underestimated, as regulators, food producers, and almost everybody else – alerted by the mistakes of

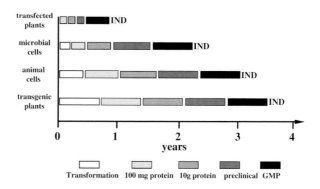

Fig. 6.2 Research & Development time course using different recombinant protein production platforms.

Prodigene – are all calling for stronger regulations to ensure that plants expressing pharmaceuticals are entirely and effectively separated from the food supply. Attempts to derive certain assumptions by monitoring the progress of plant biotechnology companies involved in pharmaceutical proteins production can be fruitful, despite the limited number of such companies (see Table 6.4).

It is already clear that the early adoption of food or feed crops – especially cereals with large production areas such as corn and rice – as production hosts for biopharmaceuticals has been a fatal choice for some players. However, we believe that there are sufficient technological achievements in the fields of plant molecular biology, genetics and biotechnology to make any technology very safe, and a more detailed discussion of these topics can be found in a recent review [163]. Nevertheless, the main issue in biosafety is cross-contamination of non-transgenic stock – an issue that can be simply and efficiently resolved by the physical containment of transgenic plants and/or by employing a production host that is a non-food, non-feed crop. Several biotech companies have chosen these approaches. Medicago uses alfalfa plants grown in high-tech glasshouses, whilst other companies (e.g., Large Scale Biology) use both approaches (glasshouse and open field) depending on the volume of product required, with the production host being tobacco, a non-food/non-feed plant. It is worthy of note that in the case of Large Scale Biology it is not stably transformed but rather transiently transformed tobacco plants that are being used. Also, the choice of a plant host such as duckweed (Biolex, Lemnagen), moss *Physcomitrella* (Greenovation), cell suspension cultures (Protalix), or rhizosecretion (Phytomedics) [80] leads to no other alternative but contained production. One Canadian company, Prairie Plant Systems Inc., offers a new concept and facilities for growing pharmaceutically active plants in subterranean growth chambers. Undoubtedly, contained production shall be the solution to many problems faced by plant biotech companies using open-field and food crops for molecular farming. However, growth in a closed environment is substantially more expensive than in open field and, may not be the solution for products required in large (tons per year) quantities at a competitive price, such as human serum albumin.

Fortunately, the majority of recombinant proteins for pharmaceutical use are required in quantities of, at most, hundreds of kilograms per year, whilst in some cases (e.g., glucocerebrosidase) even sub-kilogram or gram quantities can satisfy market needs. Although these recombinant proteins can be produced in a closed environment, in order to compensate for the space limit and improve the economics of the production process, high-yield expression systems are required. It is very unlikely that the use of technology based on seed-specific promoters and monocotyledonous plants can be the solution for indoor growth, as only a relatively small proportion of total plant biomass (seed) can be used for production. There is still no consensus which plant host shall be used for pharmaceutical protein production, but it is evident that if we favor the safer choice of glasshouse production to drastically reduce the danger of transgene escape, then preference shall be given to crops allowing for the highest possible yield of productive plant biomass per year per hectare. These crops shall be easy to transform, and shall have well-established expression systems. It is very natural that the most likely choice of crop for such production is tobacco, as this can support several harvests per year with yields of leaf biomass reaching over 100 tons per hectare – that is, four to 30 times more than other crop candidates such as alfalfa, wheat, rice, and corn [164]. Additionally, a transgenic tobacco line serving as a host for "molecular farming" can be engineered to incorporate other biosafety features, including male sterility or competence for a specific expression vector (e.g., an ability to complement viral vector function such as cell-to-cell or systemic movement), thereby further increasing biosafety parameters and drastically reducing the chance of transgene escape.

6.5
Summary and Conclusion

It is clear that transient expression is significantly faster for generating results, and requires a significantly shorter time for optimization compared to systems requiring stable transformation. For example, vectors used in transient expression systems can be evaluated within 4–7 days, whilst systems built on stable transformation require several months to obtain primary transformants, which makes the process of their optimization slower by at least an order of magnitude. This parameter shall not be underestimated in the highly competitive environment of modern biotechnology. Surprisingly enough, no efforts

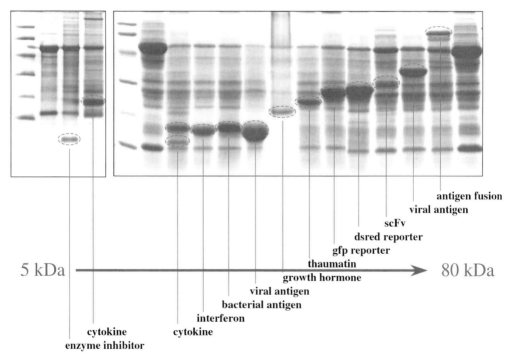

5 kDa ⟶ 80 kDa

antigen fusion
viral antigen
scFv
dsred reporter
gfp reporter
thaumatin
growth hormone
viral antigen
bacterial antigen
interferon
cytokine
cytokine
enzyme inhibitor

Fig. 6.3 magnICON® expression platform versatility: Coomassie blue-stained SDS–PAGE of total soluble protein extracted from transfected *Nicotiana* leaves. The recombinant protein bands are circled.

have been made within the biotech industry to use transient expression systems at an industrial scale, with the notable exception of Large Scale Biology Inc., that is using TMV-based vectors and transfection in tobacco as its main expression platform. Scientists at Icon Genetics have further improved the TMV-based system which, in its present form, uses highly efficient agrobacterial delivery of optimized viral vectors to practically every plant cell, thus removing the need for systemic and, to some extent, even cell-to-cell movement (unpublished data). Moreover, such vectors provide for the highest possible yield, practically reaching the biological limits of the system for many non-toxic proteins (Fig. 6.3). The speed of the system is supported by the well-established protocol for rapid optimization of recombinant protein expression level [133].

Once the expression cassette has been optimized, highly diluted agrobacterial suspension carrying the cDNA of the TMV-based vector is used for large-scale agroinfiltration of whole tobacco plants. The biomass can be harvested 7–10 days later and used for isolation of the recombinant protein of interest. In comparison with other systems, this protocol is extremely fast, provides high yields, and has the potential for unlimited scalability. In addition, our

Strength ＼ System	transient	plastid	nuclear
Yield	○ ○ ●	○ ○ ○	○
Relative yield	○ ● ●	○ ○	○
Containment	○ ● ●	○ ○ ●	○ ● ●
Insert size	○ ●	○ ○ ○	○ ○ ○
Multiple crops	○	○ ●	○ ○ ○
R&D speed	○ ○ ●	○	○

● - improvements by Icon Genetics

Fig. 6.4 Comparison of expression systems.

transfection system has almost no restriction on the size of biopharmaceutical (no limitation in gene size) to be expressed (see Section 6.2.3). However, improvement is still required for the expression of hetero-oligomeric proteins, such as full-length antibodies.

A comparison of the different expression systems is summarized in Fig. 6.4.

Acknowledgments

The authors thank Dr. Leonid Shlumukov for critical reading of this manuscript, and Dr. Stefan Herz for providing the data included in Table 6.2, and Dr. Romy Kandrzia for providing the picture for Fig. 6.3.

References

1 Horn, M. E., S. L. Woodard, and J. A. Howard, Plant molecular farming: systems and products. Plant Cell Rep, 2004; 22(10):711–720.

2 Fischer, R., et al., Plant-based production of biopharmaceuticals. Curr Opin Plant Biol, 2004; 7(2):152–158.

3 Stoger, E., et al., Antibody production in transgenic plants. Methods Mol Biol, 2004; 248:301–318.

4 Ma, J. K., P. M. Drake, and P. Christou, The production of recombinant pharmaceutical proteins in plants. Nat Rev Genet, 2003; 4(10):794–805.

5 Hellwig, S., et al., Plant cell cultures for the production of recombinant proteins. Nat Biotechnol, 2004; 22(11):1415–1422.

6 Knäblein, J., Biopharmaceuticals expressed in plants: a new era in the new Millennium. In: *Applications in Pharmaceutical Biotechnology*. R. Müller & O. Kayser (Eds.). Wiley-VCH.

7 Franck, A., et al., Nucleotide sequence of cauliflower mosaic virus DNA. Cell, 1980; 21(1):285–294.

8 Liu, X. J., et al., Cis regulatory elements directing tuber-specific and sucrose-inducible expression of a chimeric class I patatin promo-

ter/GUS-gene fusion. Mol Gen Genet, 1990; 223(3):401–406.

9 Jefferson, R., A. Goldsbrough, and M. Bevan, Transcriptional regulation of a patatin-1 gene in potato. Plant Mol Biol, 1990; 14(6):995–1006.

10 Scheller, J., et al., Production of spider silk proteins in tobacco and potato. Nat Biotechnol, 2001; 19(6):573–577.

11 Outchkourov, N. S., et al., The promoter-terminator of chrysanthemum rbcS1 directs very high expression levels in plants. Planta, 2003; 216(6):1003–1012.

12 Kwon, S. Y., et al., Transgenic ginseng cell lines that produce high levels of a human lactoferrin. Planta Med, 2003; 69(11):1005–1008.

13 Tackaberry, E. S., et al., Increased yield of heterologous viral glycoprotein in the seeds of homozygous transgenic tobacco plants cultivated underground. Genome, 2003; 46(3):521–526.

14 Horvath, H., et al., The production of recombinant proteins in transgenic barley grains. Proc Natl Acad Sci USA, 2000; 97(4):1914–1919.

15 Hood, E. E., et al., Molecular farming of industrial proteins from transgenic maize. Adv Exp Med Biol, 1999; 464:127–147.

16 Park, Y. and H. Cheong, Expression and production of recombinant human interleukin-2 in potato plants. Protein Expr Purif, 2002; 25(1):160–165.

17 Christensen, A. H. and P. H. Quail, Ubiquitin promoter-based vectors for high-level expression of selectable and/or screenable marker genes in monocotyledonous plants. Transgenic Res, 1996; 5(3):213–218.

18 McElroy, D., et al., Isolation of an efficient actin promoter for use in rice transformation. Plant Cell, 1990; 2(2):163–171.

19 Gatz, C. and P. H. Quail, Tn10-encoded tet repressor can regulate an operator-containing plant promoter. Proc Natl Acad Sci USA, 1988; 85(5):1394–1397.

20 Gatz, C., C. Frohberg, and R. Wendenburg, Stringent repression and homogeneous de-repression by tetracycline of a modified CaMV 35S promoter in intact transgenic tobacco plants. Plant J, 1992; 2(3):397–404.

21 Weinmann, P., et al., A chimeric transactivator allows tetracycline-responsive gene expression in whole plants. Plant J, 1994; 5(4):559–569.

22 Mett, V. L., L. P. Lochhead, and P. H. Reynolds, Copper-controllable gene expression system for whole plants. Proc Natl Acad Sci USA, 1993; 90(10):4567–4571.

23 Aoyama, T. and N. H. Chua, A glucocorticoid-mediated transcriptional induction system in transgenic plants. Plant J, 1997; 11(3):605–612.

24 McNellis, T. W., et al., Glucocorticoid-inducible expression of a bacterial avirulence gene in transgenic Arabidopsis induces hypersensitive cell death. Plant J, 1998; 14(2):247–257.

25 Caddick, M. X., et al., An ethanol inducible gene switch for plants used to manipulate carbon metabolism. Nat Biotechnol, 1998; 16(2):177–180.

26 Roslan, H. A., et al., Characterization of the ethanol-inducible alc gene-expression system in *Arabidopsis thaliana*. Plant J, 2001; 28(2):225–235.

27 Junker, B. H., et al., In plants the alc gene expression system responds more rapidly following induction with acetaldehyde than with ethanol. FEBS Lett, 2003; 535(1–3):136–140.

28 Padidam, M., et al., Chemical-inducible, ecdysone receptor-based gene expression system for plants. Transgenic Res, 2003; 12(1):101–109.

29 Bohner, S., et al., Technical advance: transcriptional activator TGV mediates dexamethasone-inducible and tetracycline-inactivatable gene expression. Plant J, 1999; 19(1):87–95.

30 Padidam, M., Chemically regulated gene expression in plants. Curr Opin Plant Biol, 2003; 6(2):169–177.

31 Mallory, A. C., et al., The amplicon-plus system for high-level expression of transgenes in plants. Nat Biotechnol, 2002; 20(6):622–625.

32 Yang, D., et al., Expression and localization of human lysozyme in the endosperm of transgenic rice. Planta, 2003; 216(4):597–603.

33 Huang, J., et al., Expression of natural antimicrobial human lysozyme in rice grains. Molecular Breeding, 2002; 10(1–2):83–94.

34 Kusnadi, A. R., et al., Processing of transgenic corn seed and its effect on the recovery of recombinant beta-glucuronidase. Biotechnol Bioeng, 1998; 60(1):44–52.

35 Evangelista, R. L., et al., Process and economic evaluation of the extraction and purification of recombinant beta-glucuronidase from transgenic corn. Biotechnol Prog, 1998; 14(4):607–614.

36 Kusnadi, A. R., et al., Production and purification of two recombinant proteins from transgenic corn. Biotechnol Prog, 1998; 14(1):149–155.

37 Stoger, E., et al., Cereal crops as viable production and storage systems for pharmaceutical scFv antibodies. Plant Mol Biol, 2000; 42(4):583–590.

38 De Jaeger, G., et al., Boosting heterologous protein production in transgenic dicotyledonous seeds using *Phaseolus vulgaris* regulatory sequences. Nat Biotechnol, 2002; 20(12):1265–1268.

39 Dalton, R., California edges towards farming drug-producing rice. Nature, 2004; 428(6983):591.

40 California rejects plan for drug-producing rice. Nature, 2004; 428:790–791.

41 Jacobs, P., Ventria Biosciences. http://www.checkbiotech.org/root/index.cfm?fuseaction=search&search=ventria&doc_id=7582&start=1&fullsearch=0, 2004(16 April).

42 Fox, J. L., Puzzling industry response to Prodi-Gene fiasco. Nat Biotechnol, 2003; 21(1):3–4.

43 Johnson, D., Stine Seed purchases biotech company ProdiGene. http://www.checkbiotech.org/root/index.cfm?fuseaction=search&search=prodigene&doc_id=5994&start=1&fullsearch=0, 2003.

44 Crabtree, P., Epicyte Pharmaceutical joins failing-biotech row. http://www.checkbiotech.org/root/index.cfm?fuseaction=search&search=%20epicyte&doc_id =7731&start=1&fullsearch=1, 2004.

45 Scheller, J., et al., Purification of spider silk-elastin from transgenic plants and application for human chondrocyte proliferation. Transgenic Res, 2004; 13(1):51–57.

46 Leite, A., et al., Expression of correctly processed human growth hormone in seeds of transgenic tobacco plants. Molecular Breeding, 2000; 6(1):47–53.

47 Farran, I., et al., Targeted expression of human serum albumin to potato tubers. Transgenic Res, 2002; 11(4):337–346.

48 Sijmons, P. C., et al., Production of correctly processed human serum albumin in transgenic plants. Biotechnology (N Y), 1990 8(3):217–221.

49 Ruggiero, F., et al., Triple helix assembly and processing of human collagen produced in transgenic tobacco plants. FEBS Lett, 2000; 469(1):132–136.

50 Takase, K. and K. Hagiwara, Expression of human alpha-lactalbumin in transgenic tobacco. J Biochem (Tokyo), 1998; 123(3):440–444.

51 Lauterslager, T. G., et al., Oral immunisation of naive and primed animals with transgenic potato tubers expressing LT-B. Vaccine, 2001; 19(17–19):2749–2755.

52 Azzoni, A. R., et al., Recombinant aprotinin produced in transgenic corn seed: extraction and purification studies. Biotechnol Bioeng, 2002; 80(3):268–276.

53 Woodard, S. L., et al., Maize (*Zea mays*)-derived bovine trypsin: characterization of the first large-scale, commercial protein product from transgenic plants. Biotechnol Appl Biochem, 2003; 38(Pt 2):123–130.

54 Chong, D. K., et al., Expression of the human milk protein beta-casein in transgenic potato plants. Transgenic Res, 1997; 6(4):289–296.

55 Arakawa, T., et al., Expression of cholera toxin B subunit oligomers in transgenic potato plants. Transgenic Res, 1997; 6(6):403–413.

56 Zhang, G. G., et al., Production of HIV-1 p24 protein in transgenic tobacco plants. Mol Biotechnol, 2002; 20(2):131–136.

57 Richter, L. J., et al., Production of hepatitis B surface antigen in transgenic plants for oral immunization. Nat Biotechnol, 2000; 18(11):1167–1171.

58 Joung, Y. H., et al., Expression of the hepatitis B surface S and preS2 antigens in tubers of *Solanum tuberosum*. Plant Cell Rep, 2004; 22(12):925–930.

59 Chong, D. K. and W. H. Langridge, Expression of full-length bioactive antimicrobial human lactoferrin in potato plants. Transgenic Res, 2000; 9(1):71–78.

60 Nandi, S., et al., Expression of human lactoferrin in transgenic rice grains for the application in infant formula. 2002; 163:713–722.

61 Samyn-Petit, V., et al., N-glycosylation potential of maize: the human lactoferrin used as a model. Glycoconj J, 2001; 18:519–527.

62 Choi, S. M., et al., High expression of a human lactoferrin in transgenic tobacco cell cultures. Biotechnol Lett, 2003; 25(3):213–218.

63 Stevens, L. H., et al., Effect of climate conditions and plant developmental stage on the stability of antibodies expressed in transgenic tobacco. Plant Physiol, 2000; 124(1):173–182.

64 Ko, K., et al., Function and glycosylation of plant-derived antiviral monoclonal antibody. Proc Natl Acad Sci USA, 2003; 100(13):8013–8018.

65 Ko, K., et al., Elimination of alkaloids from plant-derived human monoclonal antibody. J Immunol Methods, 2004; 286(1–2):79–85.

66 Ma, J. K., et al., Characterization of a recombinant plant monoclonal secretory antibody and preventive immunotherapy in humans. Nat Med, 1998; 4(5):601–606.

67 Ma, J. K., et al., Generation and assembly of secretory antibodies in plants. Science, 1995; 268(5211):716–719.

68 Khoudi, H., et al., Production of a diagnostic monoclonal antibody in perennial alfalfa plants. Biotechnol Bioeng, 1999; 64(2):135–143.

69 Drake, P. M., et al., Rhizosecretion of a monoclonal antibody protein complex from transgenic tobacco roots. Plant Mol Biol, 2003; 52(1):233–241.

70 Fischer, R., et al., Expression and characterization of bispecific single-chain Fv fragments produced in transgenic plants. Eur J Biochem, 1999; 262(3):810–816.

71 Fiedler, U., et al., Optimization of scFv antibody production in transgenic plants. Immunotechnology, 1997; 3(3):205–216.

72 Schouten, A., et al., The C-terminal KDEL sequence increases the expression level of a single-chain antibody designed to be targeted to both the cytosol and the secretory pathway in transgenic tobacco. Plant Mol Biol, 1996; 30(4):781–793.

73 De Jaeger, G., et al., High level accumulation of single-chain variable fragments in the cytosol of transgenic *Petunia hybrida*. Eur J Biochem, 1999; 259(1–2):426–434.

74 Vaquero, C., et al., A carcinoembryonic antigen-specific diabody produced in tobacco. FASEB J, 2002; 16(3):408–410.

75 Ohya, K., et al., Expression of two subtypes of human IFN-alpha in transgenic potato plants. J Interferon Cytokine Res, 2001; 21(8):595–602.

76 Biemelt, S., et al., Production of human papillomavirus type 16 virus-like particles in transgenic plants. J Virol, 2003; 77(17):9211–9220.

77 Cramer, C. L., et al., Bioproduction of human enzymes in transgenic tobacco. Ann NY Acad Sci, 1996; 792:62–71.

78 Dieryck, W., et al., Human haemoglobin from transgenic tobacco. Nature, 1997; 386(6620):29–30.

79 Terashima, M., et al., Utilization of an alternative carbon source for efficient production of human alpha(1)-antitrypsin by genetically engineered rice cell culture. Biotechnol Prog, 2001; 17(3):403–406.

80 Borisjuk, N. V., et al., Production of recombinant proteins in plant root exudates. Nat Biotechnol, 1999; 17(5):466–469.

81 Komarnytsky, S., et al., A quick and efficient system for antibiotic-free expression of heterologous genes in tobacco roots. Plant Cell Rep, 2004; 22(10):765–773.

82 Decker, E. L. and R. Reski, The moss bioreactor. Curr Opin Plant Biol, 2004; 7(2):166–170.

83 Svab, Z., P. Hajdukiewicz, and P. Maliga, Stable transformation of plastids in higher plants. Proc Natl Acad Sci USA, 1990; 87(21):8526–8530.

84 Scott, S. E. and M. J. Wilkinson, Low probability of chloroplast movement from oilseed rape (*Brassica napus*) into wild *Brassica rapa*. Nat Biotechnol, 1999; 17(4):390–392.

85 Huang, C. Y., M. A. Ayliffe, and J. N. Timmis, Direct measurement of the transfer rate of chloroplast DNA into the nucleus. Nature, 2003; 422(6927):72–76.

86 Stegemann, S., et al., High-frequency gene transfer from the chloroplast genome to the nucleus. Proc Natl Acad Sci USA, 2003; 100(15):8828–8833.

87 Ye, G. N., et al., Plastid-expressed 5-enolpyruvylshikimate-3-phosphate synthase genes provide high level glyphosate tolerance in tobacco. Plant J, 2001; 25(3):261–270.

88 Lutz, K. A., J. E. Knapp, and P. Maliga, Expression of bar in the plastid genome confers herbicide resistance. Plant Physiol, 2001; 125(4):1585–1590.

89 Maliga, P., Plastid Transformation in Higher Plants. Annu Rev Plant Physiol Plant Mol Biol, 2004; 55:289–313.

90 Maliga, P., Engineering the plastid genome of higher plants. Curr Opin Plant Biol, 2002; 5(2):164–172.

91 Maliga, P., Progress towards commercialization of plastid transformation technology. Trends Biotechnol, 2003; 21(1):20–28.

92 Heifetz, P. B. and A. M. Tuttle, Protein expression in plastids. Curr Opin Plant Biol, 2001; 4(2):157–161.

93 Heifetz, P. B., Genetic engineering of the chloroplast. Biochimie, 2000; 82(6–7):655–666.

94 Bogorad, L., Engineering chloroplasts: an alternative site for foreign genes, proteins, reactions and products. Trends Biotechnol, 2000; 18(6):257–263.

95 Boynton, J. E., et al., Chloroplast transformation in *Chlamydomonas* with high-velocity microprojectiles. Science, 1988; 240(4858):1534–1538.

96 Skarjinskaia, M., Z. Svab, and P. Maliga, Plastid transformation in *Lesquerella fendleri*, an oilseed Brassicacea. Transgenic Res, 2003; 12(1):115–122.

97 Hou, B. K., et al., Chloroplast transformation in oilseed rape. Transgenic Res, 2003; 12(1):111–114.

98 Sidorov, V. A., et al., Technical Advance: Stable chloroplast transformation in potato: use of green fluorescent protein as a plastid marker. Plant J, 1999; 19(2):209–216.

99 Ruf, S., et al., Stable genetic transformation of tomato plastids and expression of a foreign protein in fruit. Nat Biotechnol, 2001; 19(9):870–875.

100 Harris, E. H., J. E. Boynton, and N. W. Gillham, Chloroplast ribosomes and protein synthesis. Microbiol Rev, 1994; 58(4):700–754.

101 Wilson, I. B., Glycosylation of proteins in plants and invertebrates. Curr Opin Struct Biol, 2002; 12(5):569–577.

102 Staub, J. M., et al., High-yield production of a human therapeutic protein in tobacco chloroplasts. Nat Biotechnol, 2000; 18(3):333–338.

103 Leite, A., et al., Expression of correctly processed human growth hormone in seeds of transgenic tobacco plants. Molecular Breeding, 2000; 6:47–53.

104 Leelavathi, S. and V. S. Reddy, Chloroplast expression of His-tagged GUS-fusions: a general strategy to overproduce and purify foreign proteins using transplastomic plants as bioreactors. Mol Breed, 2003; 11:49–58.

105 Dale, E. C. and D. W. Ow, Gene transfer with subsequent removal of the selection gene from the host genome. Proc Natl Acad Sci USA, 1991; 88(23):10558–10562.

106 Corneille, S., et al., Efficient elimination of selectable marker genes from the plastid genome by the CRE-lox site-specific recombination system. Plant J, 2001; 27(2):171–178.

107 Hajdukiewicz, P. T., L. Gilbertson, and J. M. Staub, Multiple pathways for Cre/lox-mediated recombination in plastids. Plant J, 2001; 27(2):161–170.

108 Iamtham, S. and A. Day, Removal of antibiotic resistance genes from transgenic tobacco plastids. Nat Biotechnol, 2000; 18(11):1172–1176.

109 Klaus, S. M., et al., Generation of marker-free plastid transformants using a transiently cointegrated selection gene. Nat Biotechnol, 2004; 22(2):225–229.

110 Franklin, S. E. and S. P. Mayfield, Prospects for molecular farming in the green alga *Chlamydomonas*. Curr Opin Plant Biol, 2004; 7(2):159–165.

111 Staub, J. M. and P. Maliga, Accumulation of D1 polypeptide in tobacco plastids is regulated via the untranslated region of the psbA mRNA. EMBO J, 1993; 12(2):601–606.

112 Carrer, H., et al., Kanamycin resistance as a selectable marker for plastid transformation in tobacco. Mol Gen Genet, 1993; 241(1–2): 49–56.

113 Kuroda, H. and P. Maliga, Complementarity of the 16S rRNA penultimate stem with sequences downstream of the AUG destabilizes the plastid mRNAs. Nucleic Acids Res, 2001; 29(4):970–975.

114 Kuroda, H. and P. Maliga, Sequences downstream of the translation initiation codon are important determinants of translation efficiency in chloroplasts. Plant Physiol, 2001; 125(1):430–436.

115 Reddy, V. S., et al., Analysis of chloroplast transformed tobacco plants with cry1Ia5 under rice psbA transcriptional elements reveal high level expression of Bt toxin without imposing yield penalty and stable inheritance of transplastome. Mol Breed, 2002; 9:259–269.

116 McBride, K. E., et al., Amplification of a chimeric Bacillus gene in chloroplasts leads to an extraordinary level of an insecticidal protein in tobacco. Biotechnology (NY), 1995; 13(4):362–365.

117 De Cosa, B., et al., Overexpression of the Bt cry2Aa2 operon in chloroplasts leads to formation of insecticidal crystals. Nat Biotechnol, 2001; 19(1):71–74.

118 Kota, M., et al., Overexpression of the *Bacillus thuringiensis* (Bt) Cry2Aa2 protein in chloroplasts confers resistance to plants against susceptible and Bt-resistant insects. Proc Natl Acad Sci USA, 1999; 96(5):1840–1845.

119 Fernández-San Millán, A., et al., A chloroplast transgenic approach to hyper-express and purify Human Serum Albumin, a protein highly susceptible to proteolytic degradation. Plant Biotechnol J, 2003; 1:71–79.

120 Tregoning, J. S., et al., Expression of tetanus toxin Fragment C in tobacco chloroplasts. Nucleic Acids Res, 2003; 31(4):1174–1179.

121 Kang, T. J., et al., Expression of the B subunit of *E. coli* heat-labile enterotoxin in the chloroplasts of plants and its characterization. Transgenic Res, 2003; 12(6):683–691.

122 Daniell, H., et al., Expression of the native cholera toxin B subunit gene and assembly as functional oligomers in transgenic tobacco chloroplasts. J Mol Biol, 2001; 311(5):1001–1009.

123 Birch-Machin, I., et al., Accumulation of rotavirus VP6 protein in chloroplasts of transplastomic tobacco is limited by protein stability. Plant Biotechnol J, 2004; 2(3): 261–270.

124 Molina, A., et al., High-yield expression of viral peptide animal vaccine in transgenic tobacco chloroplasts. Plant Biotechnol J, 2004; 2:141–153.

125 Leelavathi, S., et al., Overproduction of an alkali- and thermo-stable xylanase in tobacco chloroplasts and efficient recovery of the enzyme. Mol. Breed., 2003; 11:59–67.

126 Watson, J., et al., Expression of *Bacillus anthracis* protective antigen in transgenic chloroplasts of tobacco, a non-food/feed crop. Vaccine, 2004; 22(31–32):4374–4384.

127 Mayfield, S. P., S. E. Franklin, and R. A. Lerner, Expression and assembly of a fully active antibody in algae. Proc Natl Acad Sci USA, 2003; 100(2):438–442.

128 Franklin, S., et al., Development of a GFP reporter gene for *Chlamydomonas reinhardtii* chloroplast. Plant J, 2002; 30(6):733–744.

129 Ishikura, K., et al., Expression of a foreign gene in *Chlamydomonas reinhardtii* chloroplast. J Biosci Bioeng, 1999; 87:307–314.

130 Fischer, R., et al., Towards molecular farming in the future: transient protein expression in plants. Biotechnol Appl Biochem, 1999; 30 (Pt 2):113–116.

131 Vaquero, C., et al., Transient expression of a tumor-specific single-chain fragment and a chimeric antibody in tobacco leaves. Proc Natl Acad Sci USA, 1999; 96(20):11128–11133.

132 Voinnet, O., et al., An enhanced transient expression system in plants based on suppression of gene silencing by the p19 protein of tomato bushy stunt virus. Plant J, 2003; 33(5):949–956.

133 Marillonnet, S., et al., In planta engineering of viral RNA replicons: efficient assembly by recombination of DNA modules delivered by *Agrobacterium*. Proc Natl Acad Sci USA, 2004; 101(18):6852–6857.

134 Awram, P., et al., The potential of plant viral vectors and transgenic plants for subunit vaccine production. Adv Virus Res, 2002; 58:81–124.

135 Pogue, G.P., et al., Making an ally from an enemy: plant virology and the new agriculture. Annu Rev Phytopathol, 2002; 40:45–74.

136 Porta, C. and G.P. Lomonossoff, Viruses as vectors for the expression of foreign sequences in plants. Biotechnol Genet Eng Rev, 2002; 19:245–291.

137 Scholthof, K.B., T.E. Mirkov, and H.B. Scholthof, Plant virus gene vectors: biotechnology applications in agriculture and medicine. Genet Eng (NY), 2002; 24:67–85.

138 Mor, T.S., et al., Geminivirus vectors for high-level expression of foreign proteins in plant cells. Biotechnol Bioeng, 2003; 81(4):430–437.

139 Gleba, Y., S. Marillonnet, and V. Klimyuk, Engineering viral expression vectors for plants: the 'full virus' and the 'deconstructed virus' strategies. Curr Opin Plant Biol, 2004; 7(2):182–188.

140 Turpen, T.H., et al., Malarial epitopes expressed on the surface of recombinant tobacco mosaic virus. Biotechnology (NY), 1995; 13(1):53–57.

141 Lomonossoff, G.P. and W.D. Hamilton, Cowpea mosaic virus-based vaccines. Curr Top Microbiol Immunol, 1999; 240:177–189.

142 Brennan, F.R., T.D. Jones, and W.D. Hamilton, Cowpea mosaic virus as a vaccine carrier of heterologous antigens. Mol Biotechnol, 2001; 17(1):15–26.

143 Chatterji, A., et al., Cowpea mosaic virus: from the presentation of antigenic peptides to the display of active biomaterials. Intervirology, 2002; 45(4–6):362–370.

144 Nemchinov, L.G., et al., Development of a plant-derived subunit vaccine candidate against hepatitis C virus. Arch Virol, 2000; 145(12):2557–2573.

145 Bendahmane, M., et al., Display of epitopes on the surface of tobacco mosaic virus: impact of charge and isoelectric point of the epitope on virus–host interactions. J Mol Biol, 1999; 290(1):9–20.

146 Wu, L., et al., Expression of foot-and-mouth disease virus epitopes in tobacco by a tobacco mosaic virus-based vector. Vaccine, 2003; 21(27–30):4390–4398.

147 Belanger, H., et al., Human respiratory syncytial virus vaccine antigen produced in plants. FASEB J, 2000; 14(14):2323–2328.

148 Angell, S.M. and D.C. Baulcombe, Consistent gene silencing in transgenic plants expressing a replicating potato virus X RNA. EMBO J, 1997; 16(12):3675–3684.

149 Brisson, N., et al., Expression of a bacterial gene in plants by using a viral vector. Nature, 1984; 310:511–514.

150 Lefebvre, D., B. Miki, and J. Laliberte, Mammalian metallothionein functions in plants. Bio-technology, 1987; 5:1053–1056.

151 De Zoeten, G.A., et al., The expression, localization, and effect of a human interferon in plants. Virology, 1989; 172(1):213–222.

152 Mori, M., et al., Efficient production of human gamma interferon in tobacco protoplasts by genetically engineered brome mosaic virus RNAs. J Gen Virol, 1993; 74 (Pt 7): 1255–1260.

153 McCormick, A.A., et al., Rapid production of specific vaccines for lymphoma by expression of the tumor-derived single-chain Fv epitopes in tobacco plants. Proc Natl Acad Sci USA, 1999; 96(2):703–708.

154 McCormick, A.A., et al., Individualized human scFv vaccines produced in plants: humoral anti-idiotype responses in vaccinated mice confirm relevance to the tumor Ig. J Immunol Methods, 2003; 278(1–2):95–104.

155 Verch, T., V. Yusibov, and H. Koprowski, Expression and assembly of a full-length monoclonal antibody in plants using a plant virus vector. J Immunol Methods, 1998; 220(1–2):69–75.

156 Perez Filgueira, D.M., et al., Bovine herpes virus gD protein produced in plants using a recombinant tobacco mosaic virus (TMV) vector possesses authentic antigenicity. Vaccine, 2003; 21(27–30):4201–4209.

157 Kumagai, M.H., et al., Rapid, high-level expression of glycosylated rice alpha-amylase

in transfected plants by an RNA viral vector. Gene, 2000; 245(1):169–174.

158 Krebitz, M., et al., Rapid production of the major birch pollen allergen Bet v 1 in *Nicotiana benthamiana* plants and its immunological in vitro and in vivo characterization. FASEB J, 2000; 14(10):1279–1288.

159 Li, Y., et al., Expression of a human lactoferrin N-lobe in *Nicotiana benthmiana* with potato virus X-based agroinfection. Biotechnol Lett, 2004; 26(12):953–957.

160 Hiatt, A., Antibodies produced in plants. Nature, 1990; 344(6265):469–470.

161 Hiatt, A., R. Cafferkey, and K. Bowdish, Production of antibodies in transgenic plants. Nature, 1989; 342(6245):76–78.

162 During, K., et al., Synthesis and self-assembly of a functional monoclonal antibody in transgenic *Nicotiana tabacum*. Plant Mol Biol, 1990; 15(2):281–293.

163 Gleba, Y., S. Marillonnet, and V. Klimyuk, Design of Safe and Biologically Contained Transgenic Plants: Tools and Technologies for Controlled Transgene Flow and Expression. Biotechnol Genet Eng Rev, 2004; 21:325–367.

164 Daniell, H., S.J. Streatfield, and K. Wycoff, Medical molecular farming: production of antibodies, biopharmaceuticals and edible vaccines in plants. Trends Plant Sci, 2001; 6(5):219–226.

7

Humanized Glycosylation: Production of Biopharmaceuticals in a Moss Bioreactor

Gilbert Gorr and Sabrina Wagner

Abstract

Genetically engineered plants are promising systems for the production of biopharmaceutical proteins. Among the different plant-based systems, the moss bioreactor shows unique properties. Mosses are cultivated as haploid, photo-autotrophically active and fully differentiated gametophytic tissue performed as suspension cultures in bioreactors. In addition, moss is the only known plant system which shows a high frequency of homologous recombination which allows for gene knock-outs, thereby opening the possibility of genetic engineering of the glycosylation pathway. Here, we present an overview of the biotechnologically relevant aspects of mosses, with a special emphasis on glycoengineering performed in *Physcomitrella patens*.

Abbreviations

ADCC	antibody-dependent cell-mediated cytotoxicity
CHO	chinese hamster ovary
ER	endoplasmic reticulum
EST	expressed sequence tag
FucT	alpha 1,3-fucosyltransferase
GlcNAC	N-acetylglucosaminyl residue
GNT I	N-acetylglucosaminyltransferase I
GNT II	N-acetylglucosaminyltransferase II
GNT	glucosaminyltransferase
hVEGF	human vascular endothelial growth factor
MALDI-TOF	matrix assisted laser desorption/ionization time-of-flight
Mbp	mega base pairs
PEG	polyethylene glycol
rHSA	human serum albumin
rhVEGF	recombinant human vascular endothelial growth factor
XylT	beta 1,2-xylosyltransferase

7.1
Introduction

Until now, biopharmaceuticals have been produced either in microorganisms such as *E. coli* or in animal cell cultures (e.g., Chinese hamster ovary (CHO) cells), if the therapeutic protein requires complex post-translational modification (see Part IV, Chapters 1, 2, 3, 5, 12 and 13).

An expected shortage of manufacturing capacities, as well as safety issues in terms of virus load and contamination with TSE,

Modern Biopharmaceuticals. Edited by J. Knäblein
Copyright © 2005 WILEY-VCH Verlag GmbH & Co. KGaA, Weinheim
ISBN: 3-527-31184-X

resulted in the search for alternative production systems [1]. Plants are the most promising production organisms for biopharmaceuticals because of their cost-efficient upstream processes [2] and excellent safety aspects. In addition to molecular pharming performed in greenhouses or on the field, secretion-based plant systems such as rhizosecretion from tobacco roots have been developed [3–5; for a review, see [6]]. Secretion of the target protein into the medium is a major improvement for the downstream process, because extraction and purification of proteins from plant tissues is a complex and costly process, as recently reviewed by Knäblein [7].

However, the major limitation for the use of biopharmaceuticals produced by plants is the glycosylation pattern. Animal and plant N-linked glycosylation patterns are identical in the core structure, but there are differences in the additional sugar residues. Plant N-glycans contain beta 1,2-linked xylose and alpha 1,3-linked fucose residues, whereas the beta 1,4-linked terminal galactose residue, which is typical for animal-derived glycoproteins (e.g., on antibodies) is not present in plants. There is also some evidence that plant-specific residues have immunogenic potential [8, 9].

Here, we present the moss bioreactor, which is based on secretion of the target protein into a simple medium [10] and, by humanization of the N-glycans, avoids plant-specific immunogenicity.

7.2
Mosses: Some General Aspects

In contrast to higher plants, the main phase of the life cycle of mosses is not the diploid sporophyte but the haploid, photosynthetically active gametophyte. The gametophyte consists of the filamentous protonema, which shows apical growth, and of the morphologically more distinct gametophore (Fig. 7.1). The differentiation steps in the development of the moss gametophyte are clearly defined. Growth of the first cell type, the chloronema, begins after germination of spores or protoplasts. The second cell-type in the branched protonema is the caulonema, on which buds and later the complete gametophore are developing. Chloronema and caulonema cells can be distinguished not only by their morphology but also by their predominant cell-cycle phases [11]. Sporophyte development occurs only under specific conditions [12], and therefore mosses are propagated in general vegetatively without sporulation (for a review, see Refs. [13, 14]). The long-term storage of mosses is performed on solid medium, as well as by cryopreservation, with regrowth rates of 100% [15, 16].

Most investigations carried out with *Physcomitrella* were based on a strain which was collected during the 1960s by H. L. K. Whitehouse. Engel [17] established the in-vitro culture by subcultivation of plant material grown from one spore. The haploid nature of the gametophytic tissue, the clearly defined differentiation pattern, and the simple cultivation parameters have provided an excellent basis for scientific studies, and consequently many genetic and physiological studies of *Physcomitrella* have been performed during the past four decades (e.g., [18]). The establishment and optimization of protoplast isolation [19, 20] opened the possibility for further development of transformation methods. Interestingly, agrobacteria-mediated transformation – one of the main transformation methods used in higher-plant technology – is not applicable to mosses due to the lack of a useful agro strain. Biolistic transformation [21] and electroporation [22] can be used, but neither method is particularly

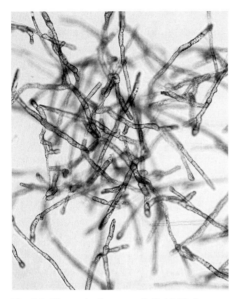 

Fig. 7.1 *Physcomitrella patens*. Left: highly homogeneous protonema tissue. Right: a leafy gametophore.

sufficient. Polyethyleneglycol (PEG)-mediated DNA transfer is the method of choice for moss transformation (for a review, see Ref. [23]), the protocol being quite different to that used for higher plants – for example, there is no requirement for a cooling step [24, 25]. For recombinant expression in mosses, in most cases constitutive active heterologous regulatory sequences such as the 35S promoter or the rice actin 1 5′ region were used [24, 26]. Inducible expression has also been described for moss; for example, Zeidler et al. [27] used the tetracycline-based Top 10 system successfully, whilst Knight et al. [28] described expression driven by the Em-promoter in *Physcomitrella*. Although many expressed sequence tag (EST) data were available for *Physcomitrella* (see below), only recently the first endogenous regulatory sequences from moss were characterized [29–31].

During the 1990s, *Physcomitrella* was discovered as a new tool for functional geno-mics and, for the first time in plants, highly efficient targeted homologous recombination into the genome was described for *Physcomitrella* [25, 32, 33]. Subsequently, numerous molecular data have been generated in this respect. The genome size was determined as 511 Mbp, which is three- to four-fold that of *Arabidopsis thaliana* [11]. Some 95% of the *Physcomitrella* transcriptome is known, based on EST data from about 25 000 protein-coding genes [34, 35]. On the basis of these data, a calculation of codon usage was possible, and this resulted in there being no significant preferences for *Physcomitrella* [36]. Thus, no codon optimization is necessary for the recombinant expression of human proteins in moss. Taking these results together, it is clear that *Physcomitrella* is indeed a well-characterized organism.

7.3
Cell Culture

Mosses can be grown on solid medium as well as in liquid suspension cultures. In general, the moss tissue is cultivated under photoautotrophic conditions. Light, air, and a simple medium based on a mineral salt composition without any sugars or plant hormones are sufficient to cultivate the fully differentiated tissue. Mosses can be grown with ammonium or with nitrate as nitrogen sources [37]. Although the optimal growth conditions are known for many moss species, the composition of medium components can be varied broadly. Mosses can also be grown under heterotrophic conditions, though the addition of sugar to the medium results in a marked production of secondary metabolites [38].

By utilizing medium supplementation with plant hormones or additives such as ammonium tartrate, the differentiation of moss tissue can be influenced in a well-defined manner. Whereas the addition of auxin promotes the development of caulonema cells, the addition of ammonium tartrate results in an arrest on the chloronema stage of the moss protonema [11]. The latter effect is of interest for biotechnological applications, because of the high homogeneity of the cells. Although filaments consisting of chloronema cells are fully differentiated, only one type of cells is present in such cultures. Moreover, these cells are arrested in the G_2/M phase of the cell cycle [11]. In conclusion, mosses can be grown as an extremely homogeneous and well-defined suspension culture, whilst the tissue material is both fully differentiated and photosynthetically active.

Liquid cultivation can be performed as suspension culture not only in Erlenmeyer flasks but also in bioreactors. Mosses were cultivated in photo-bioreactors in stirred glass tanks [39], as well as under airlift conditions [26, 40]. Supplementation with CO_2 and light are the major parameters that influence the growth rates. Since light is a limiting factor for the large-scale photoautotrophic cultivation of mosses, a glass tube reactor was developed. Based on technology already established for the large-scale photoautotrophic cultivation of algae [41], the cultivation of the filamentous protonema under sterile conditions required significant adaptations [42]. The prototype of the photobioreactor has a working volume of 30 L (Fig. 7.2).

Down-scaling of the suspension cultures to microtiter plates was successfully established for testing culture supplementations [43]. This allows efficient and rapid medium optimization, and also opens the possibility of automating the upstream development. In combination, the overall direct transformation, clonal growth without any crossing steps, and the technical ability to speed the process development will minimize the time-to-market for biopharmaceuticals produced in moss.

Fig. 7.2 A photobioreactor. The tubular photobioreactor design allows scaling of moss suspension culture up to large volumes. (Illustration courtesy of Prof. C. Posten, Karlsruhe.)

7.4
Recombinant Expression

Antibiotic resistance markers and reporter genes were the first heterologous proteins to be expressed in mosses [24, 26, 44]. The production of human proteins in moss was first shown by expression of human vascular endothelial growth factor 121 (rhVEGF). rhVEGF was successfully targeted to the secretory pathway, and this resulted in an efficient secretion of the recombinant protein into the medium. Moreover, the moss-derived rhVEGF was shown to be biologically active [45, 46]. An important criterion for the successful expression of a therapeutic protein from a recombinant cell is to obtain a transgenic plant that maintains stability, both of production and at the molecular level. Several transgenic moss strains which were aged between 2 and 7 years were examined with regard to expression of the target protein rhVEGF and neomycin phosphotransferase as an antibiotic resistance marker. Protein levels of rhVEGF were measured using an enzyme-linked immunosorbent assay (ELISA), and found to be unchanged. Furthermore, 100% of the transgenic plant material showed resistance to the antibiotic G418, even after several years of cultivation without selection pressure. In addition, analysis at the molecular level confirmed the protein data [47]. Thus, the moss appears to be an ideal production system for biopharmaceuticals under strict regulatory requirements. rhVEGF was further used to optimize expression in *Physcomitrella*, with molecular tools such as endogenous promoters, 5′ regulatory regions and signal peptides being isolated and characterized [30, 31, 48]. In the meantime, as in other plant expression systems, different human proteins were successfully expressed in mosses, including fully assembled antibodies [49] and human serum albumin (rHSA). Indeed, the co-expression of rHSA simultaneously with rhVEGF led to the development of a new approach for enhanced protein recovery [43].

Transient expression systems established for higher plants rely on virus infection or agroinfiltration [50–52] (see Part IV, Chapters 5 and 6; Part IV, Chapter 9). Due to the lack of moss viruses and suitable agrobacteria strains, none of these methods is applicable for mosses. Nevertheless, transient expression is a useful tool not only for the rapid analysis of molecular tools but also for the expression of recombinant protein for first characterization. Transient transformation was developed and optimized, and resulted in the expression of up to $10 \, \mu g \, mL^{-1}$ rhVEGF [46]. In addition, this transient expression system was used for promoter analysis [30, 31] and for the analysis of secretion capacity [46]. Analysis of different expression vectors in the transient system allows the selection of the optimal combinations (e.g., promoter, signal peptide) for each target protein in a short time. Therefore, the time-consuming generation of stably transformed plants can be focused on the optimal expression vectors at a very early stage of the process. Overall, the transient system can be used for further screening of molecular tools to improve recombinant protein expression in mosses.

On the protein level, transient expression allows feasibility studies to be conducted rapidly, as it enables production of limited amounts of protein for first analysis within weeks.

7.5
N-Glycosylation

The production of complex biopharmaceuticals is closely associated with glycosylation issues. N-glycosylation can be responsible for protein folding [53] and prevention of protein degradation in the cells, as well as metabolism in the liver of mammals [54] (see Part IV, Chapters 1 and 3). In addition, effector functions such as antibody-dependent cell-mediated cytotoxicity (ADCC) have been discussed as being linked to N-glycan structures [55–57] (see Part I, Chapter 15 and Part V, Chapter 1). The process of N-glycosylation seems to be highly conserved in most eukaryotes, and in particular a minimal core structure consisting of Man_3. $GlcNAc_2$ is common for N-glycans (see Part IV, Chapter 2 and Part VI, Chapter 2). In glycoproteins, N-glycans are covalently linked to the asparagine (Asn) residues of the tripeptide Asn-X-Ser/Thr, where X can be any amino acid except aspartic acid and proline. N-glycosylation starts in the endoplasmic reticulum (ER). An oligosaccharide precursor is transferred to the Asn residue and further processed to a high-mannose structure consisting of $Man_9GlcNAc_2$. The next step in this process occurs in the Golgi apparatus. Mannose residues are removed by alpha-mannosidase I, followed by the addition of a GlcNAc residue to the terminal mannose of one branch by the enzyme N-acetyl glucosaminyltransferase I (GNTI). The remaining mannose residues of the high-mannose structure are processed by a second, Golgi-located mannosidase (alpha-ManII). The resulting structure, Man_3.$GlcNAc_3$, is used as a substrate by GNTII, which transfers a second terminal GlcNAc residue to the N-glycan resulting in complex-type N-glycans.

Further modifications are different in plants and mammals. In plants, an alpha 1,3-linked fucosylation to the first core GlcNAc, and a beta 1,2-linked xylosylation to the first core mannose residue, are mediated by specific glycosyltransferases (for a review, see Ref. [58]). These structures are common for plants, including mosses [59, 60].

In contrast, mammalian N-glycans contain alpha 1,6 fucosyl residues linked to the first core GlcNAc. The terminal structures of mammalian N-glycans can be processed in a much more complex manner compared to plant structures. Galactose residues are attached to the terminal GlcNAcs in 1,4-linkage. In mammals, sialyltransferases use galactose-containing N-glycans as substrates for further processing. Depending on the glycoproteins, the N-glycan structures can be completely different. Whereas coagulation factors such as factor IX are highly sialylated [61], antibodies (if at all) contain only minor fractions of sialylated N-glycans [55].

The major drawback for plant-derived complex biopharmaceuticals is that the plant-specific xylosyl and fucosyl residues are attached to the core structure of N-glycans. Although Chargelegue et al. [62] observed no immunogenic effects of a plant-derived murine monoclonal antibody in an animal study based on a mouse model, both residues are described in the literature as structures with high immunogenic potential [8, 9].

Consequently, major efforts were made to overcome this limitation. One approach is based on an observation by von Schaewen et al. [63]. These authors isolated a mutant strain of *Arabidopsis thaliana* (*cgl*) which showed a loss of GNTI activity. All N-glycan structures isolated from this mutant strain were related to high mannose-type, and no complex-type structures were detected. The loss of complex-type N-glycans was accompanied by the loss of the plant-specific sugar residues on the core structure. From these data it was known that high-mannose structures, which are processed in the ER, are not substrates for the plant-specific alpha 1,3-fucosyltransferase (FucT) and the beta 1,2-xylosyltransferase (XylT). It was also shown that targeting of the recombinant protein to the ER by attaching the ER retention signal KDEL results in N-glycans of only the high-mannose type on these proteins, which showed poor stability after injection into mice [64].

A second approach was based on antisense technology (see Part I, Chapter 10; Part III, Chapter 3; and the Introduction). In this study, antisense constructs were designed for the GNTI of *Nicotiana benthamiana*. Whereas reduction of GNTI activity to 2% was very successful, only minor changes in N-glycan composition were observed. The remaining very low GNTI activity seemed to be sufficient for close to normal N-glycan processing [65]. A completely new approach was developed by Koprivova et al. [66], who isolated the gene coding for GNTI from *Physcomitrella patens*. Based on the highly efficient homologous recombination in mosses (see above), these authors performed targeted disruption of *gntI*. Although the *gntI* knock-out (ko) was successful, the remaining N-glycan structures in the ko plants were similar to that of the wild-type – including the complex-type structures. Again, the effect was compensated, which illustrates the overall complexity of the glycosylation pathway. Nevertheless, the putative genes coding for alpha 1,3-FucT and beta 1,2-XylT were isolated from *Physcomitrella* [59], and knock-out constructs were designed. To remove the immunogenic potential of both plant-specific N-glycans completely, plants containing double knock-outs were generated. MALDI-TOF analysis of the remaining N-glycans clearly demonstrated the absence of plant-specific sugar residues in the double knock-out, respectively [66]. Double knock-outs were performed not only through gene disruption but also through complete gene replacement. All approaches resulted in complete loss of the putative immunogenic sugar residues, thus confirming the dysfunctional character of the genes and the loss of corresponding glycosyltransferase activity in all cases (Fig. 7.3). Although the N-glycan structures with the xylosyl and fucosyl residues are highly conserved over the whole plant kingdom, surprisingly the double knock-out plants showed no differences in growth and differentiation. Regeneration of the plants was also similar to that of the wild-type, whilst the secretion capacity of rhVEGF was as high as in the wild-type [46].

Human-like terminal beta 1,4-galactosylation was recently described in tobacco plants [67] and in tobacco-derived BY2 suspension cell cultures [68] recombinantly expressing the mammalian enzyme. Moreover, the extracellular proteins of the transgenic BY2 cell line GT6 recombinantly expressing the human beta 1,4-galactosyltransferase also contain galactose-extended N-glycans [69]. Surprisingly, the same results were obtained by expression of the human galactosyltransferase in the gametophytic tissue of mosses which already lacks the plant-specific sugar residues due to its double knock-out character, and therefore shows extensive manipulation of the glycosylation pathway [70].

Taken together, humanized glycosylation at least sufficient for antibody production was achieved in genetically engineered moss strains, without any negative influence on growth rates in bioreactor cultures or on their secretion capacity for recombinant proteins.

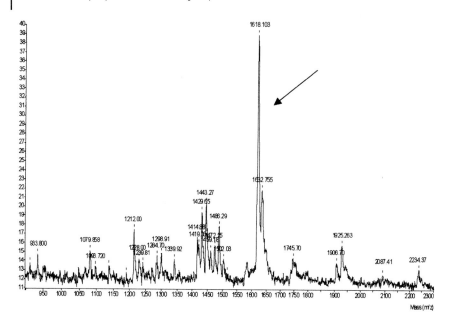

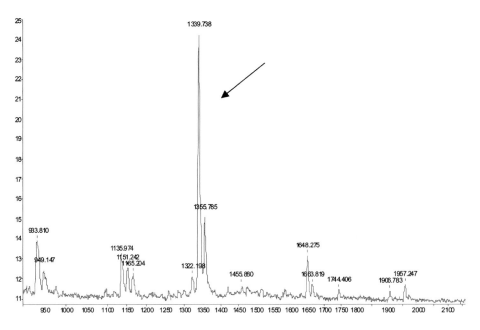

Fig. 7.3 Glycoengineering in mosses. Mass spectroscopic analysis of N-glycans in mosses (Prof. F. Altmann, Vienna) performed by matrix-assisted laser desorption ionization time of flight (MALDI-TOF) spectroscopy. Top: N-glycan analysis of wild-type. Bottom: N-glycan analysis of a transgenic moss strain lacking plant-specific sugar residues. The mass shift of the major N-glycan structure in wild-type to the complex-type GnGnXF to the major structure GnGn [lacking the plant-specific 1,3-linked fucosyl (F) and 1,2-linked xylosyl (X) residues] in the glycoengineered transgenic plant is marked by an arrow.

◀──

7.6
Conclusions and Outlook

Physcomitrella is a well-established plant system that, since the 1960s, has been cultivated under axenic conditions in many laboratories. It can be cultivated in liquid cultures, and such photobioreactors are scaleable. Transgenic moss strains show high genetic stability. Different biopharmaceuticals, including human growth factor, antibodies and non-glycosylated proteins such as HSA were successfully expressed and secreted into the medium.

Molecular tools such as promoters, other regulatory sequences and signal peptides were isolated and optimized for expression in mosses. Together with its high secretion capacity, the moss therefore allows the development of high-production strains. By using directed knock-out of plant-specific glycosyltransferases via homologous recombination, and through the simultaneous introduction of human galactosyltransferase, the humanization of glycosylation was successfully achieved in *Physcomitrella*. This not only removes the immunogenic potential of plant-produced proteins for injection, but also opens perspectives for glycosylation design. With further understanding of the function of protein glycosylation, this will include on the one hand improved effector function of monoclonal antibodies such as ADCC. On the other hand, glycosylation seems to play a role in pharmacokinetics which is only partly understood. The establishment of sialylation in plants will be a major step towards the production of "authentic" biopharmaceuticals.

References

1 J. Knäblein, Screening – Trends in Drug Discov., **2003**, *4*, 14–16.
2 J. Knäblein, M. McCaman, Screening – Trends in Drug Discov., **2003**, *6*, 33–35.
3 R. Wongsamuth, P. M. Doran, Biotechnol. Bioeng., **1997**, *54*, 401–415.
4 N. V. Borisjuk, L. G Borisjuk, S. Logendra, F. Petersen, Y. Gleba, I. Raskin, Nat. Biotechnol., **1999**, *17*, 466–469.
5 P. M. Drake, D. M. Chargelegue, N. D. Vine, C. J. van Dolleweerd, P. Obregon, J. K. Ma, Plant Mol. Biol., **2003**, *52*, 233–241.
6 J. K. Ma, P. M. Drake, P. Christou, Nat. Rev. Genet. **2003**, *4*, 794–805.
7 J. Knäblein, Biopharmaceuticals expressed in plants – a new era in the new Millennium. In: *Applications in Pharmaceutical Biotechnology*. R. Müller & O. Kayser (Eds.). Wiley-VCH.
8 G. Garcia-Casado, R. Sanchez-Monge, M. J. Chrispeels, A. Armentia, G. Salcedo, L. Gomez, Glycobiology, **1996**, *6*, 471–477.
9 M. Bardor, C. Faveeuw, A. C. Fitchette, D. Gilbert, L. Galas, F. Trottein, L. Faye, P. Lerouge, Glycobiology, **2003**, *13*, 427–434.
10 E. L. Decker, G. Gorr, R. Reski, BIOforum Europe, **2003**, *2*, 96–97.
11 G. Schween, G. Gorr, A. Hohe, R. Reski, Plant Biol., **2003**, *5*, 50–58.
12 A. Hohe, S. A. Rensing, M. Mildner, D. Lang, R. Reski, Plant Biol., **2002**, *4*, 595–602.
13 R. Reski, Bot. Acta, **1998**, *111*, 1–15.
14 R. Reski, D. J. Cove, Current Biol., **2004**, *14*, R261–R262.

15 N. H. Grimsley, L. A.Withers, Cryo-Letters, **1983**, *4*, 251–258.

16 J. Schulte, R. Reski, Plant Biol., **2004**, 6, 119–127.

17 P. P. Engel, Am. J. Bot. **1964**, *55*, 438–446.

18 D. J. Cove, A. Schild, N. W. Ashton, E. Hartmann, Photochem. Photobiol., **1978**, *27*, 249–254.

19 S. Rother, B. Hadeler, J. M. Orsini, W. O. Abel, R. Reski, J. Plant Physiol. **1994**, *143*, 72–77.

20 A. Hohe, R. Reski, Plant Sci., **2002**, *163*, 69–74.

21 W. Sawahel, S. Onde, C. Knight, D. J. Cove, Plant Mol. Biol. Rep., **1992**, *10*, 314–315.

22 D. G. Schaefer, G. Bisztray, J.-P. Zrÿd, in: *Biotechnology in Agriculture and Forestry*, Vol. 29, Plant Protoplasts and Genetic Engineering V (Y. P. S. Bajaj, Ed.). Springer-Verlag, Berlin Heidelberg, **1994**, Chapter II.11, 349–364.

23 D. G. Schaefer, Annu. Rev. Plant Physiol. Plant Mol. Biol., **2002**, 477–501.

24 M. Zeidler, E. Hartmann, J. Hughes, J. Plant Physiol., **1999**, *154*, 641–650.

25 R. Strepp, S. Scholz, S. Kruse, V. Speth, R. Reski, Proc. Natl. Acad. Sci. USA, **1998**, *95*, 4368–4373.

26 K. Reutter, R. Reski, Pl. Tissue Cult. Biotech., **1996**, *2*, 142–147.

27 M. Zeidler, C. Gatz, E. Hartmann, J. Hughes, Plant Mol. Biol., **1996**, *30*, 199–205.

28 C. D. Knight, A. Sehgal, K. Atwal, J. C. Wallace, D. J. Cove, D. Coates, R. S. Quatrano, S. Bahadur, P. G. Stockley, A. C. Cuming, Plant Cell, **1995**, *7*, 499–506.

29 V. Horstmann, C. M. Huether, W. Jost, R. Reski, E. L. Decker, BMC Biotechnol, **2004**, *4*, 13.

30 W. Jost, S. Link, V. Horstmann, E. L. Decker, R. Reski, G. Gorr, Curr. Genet., **2005**, *47*, 111–120.

31 A. Weise, M. Rodriguez-Franco, B. Timm, M. Hermann, S. Link, W. Jost, G. Gorr, Isolation of four members of a plant actin gene family with remarkable gene structures and the use of their 5' regions for high transgene expression. (submitted).

32 D. G. Schaefer, J. P. Zryd, Plant J., **1997**, *11*, 1195–1206.

33 T. Girke, H. Schmidt, U. Zahringer, R. Reski, E. Heinz, Plant J., **1998**, *15*, 39–48.

34 S. A. Rensing, S. Rombauts, Y. van de Peer, R. Reski, Trends Plant Sci., **2002**, *7*, 535–538.

35 T. Nishiyama, T. Fujita, T. Shin-I, M. Seki, H. Nishide, I. Uchiyama, A. Kamiya, P. Carninci, Y. Hayashizaki, K. Shinozaki, Y. Kohara, M. Hasebe, Proc. Natl. Acad. Sci. USA, **2003**, *100*, 8007–8012.

36 E. L. Decker, R. Reski, Curr. Opin. Plant Biol., **2004**, *7*, 166–170.

37 M. Bopp, B. Knoop, Culture methods for bryophytes, in: *Cell culture and somatic cell genetics of plants* (J. K. Vasil, Ed.), **1984**, *1*, chapter 12, 96–105.

38 P. E. Simon, J. B. Naef, Physiol. Plant., **1981**, *53*, 13–18.

39 A. Hohe, E. L. Decker, G. Gorr, G. Schween, R. Reski, Plant Cell Rep., **2002**, *20*, 1135–1140.

40 P. J. Boyd, J. Hall, D. J. Cove, in: *Methods in bryology*, Proc. Bryol. Meth. Workshop Mainz (J. M. Glime, Ed.), Hattori Bot. Lab., Nichinan, **1988**, 41–45.

41 O. Pulz, Appl. Microbiol. Biotechnol., **2001**, *57*, 287–293.

42 A. Lucumi, C. Posten, Improved *in vitro* culture of the moss *Physcomitrella patens* in a tubular pilot photobioreactor. (submitted).

43 A. Baur, R. Reski, G. Gorr, Enhanced recovery of a secreted recombinant human growth factor by stabilising additives and by coexpression of human serum albumin in the moss *Physcomitrella patens*. Plant Biotechnol. J., in press.

44 D. Schaefer, J.-P. Zryd, C. D. Knight, D. J. Cove, Mol. Gen. Genet., **1991** *226*, 418–424

45 G. Gorr, E. L. Decker, M. Kietzmann, R. Reski, Naunyn-Schmiedeberg's Arch. Pharmacol., **2001**, *363*, Suppl: R 85.

46 A. Baur, F. Kaufmann, H. Rolli, A. Weise, R. Luethje, B. Berg, M. Braun, W. Baeumer, M. Kietzmann, R. Reski, G. Gorr A fast and flexible PEG-mediated transient expression system in plants for high level expression of secreted recombinant proteins. J. Biotechnol., in press.

47 A. Baur, B. Timm, A. Weise, M. Rodriguez-Franco, G. Gorr, Genetic stability of transgenic *Physcomitrella patens* plants, Moss 2004, The 7th Annual Moss International Conference, Freiburg, Germany.

48 A. Schaaf, R. Reski, E. L. Decker, Eur. J. Cell Biol., **2004**, *83*, 145–152.

49 W. Jost, E. Schulze, M. Rodriguez-Franco, A. Weise, G. Gorr, Naunyn-Schmiedeberg's Arch. Pharmacol., **2004**, *369*, Suppl: R 80.

50 R. Fischer, C. Vaquero-Martin, M. Sack, J. Drossard, N. Emans, U. Commandeur, Biotechnol. Appl. Biochem., **1999**, *30*, 113–116.

51 C. Vaquero, M. Sack, J. Chandler, J. Drossard, F. Schuster, M. Monecke, S. Schillberg, R. Fischer, Proc. Natl. Acad. Sci. USA, **1999**, *96*, 11128–11133.

52 V. Klimyuk, S. Marillonnet, J. Knäblein, M. McCaman, Y. Gleba, Chapter 6.

53 A. Helenius, Mol. Biol. Cell, **1994**, *5*, 253–265.

54 A. G. Morell, G. Gregoriadis, I. H. Scheinberg, J. Hickman, G. Ashwell, J. Biol. Chem., **1971**, *246*, 1461–1467.

55 P. Umana, J. Jean-Mairet, R. Moudry, H. Amstutz, J. E. Bailey, Nat. Biotechnol., **1999**, *17*, 176–180.

56 R. L. Shields, J. Rai, R. Keck, L. Y. O'Connell, K. Hong, Y. G. Meng, S. H. Weikert, L. G. Presta, J. Biol. Chem. **2002**, *277*, 26733–26740.

57 T. Shinkawa, K. Nakamura, N. Yamane, E. Shoji-Hosaka, Y. Kanda, M. Sakurada, K. Uchida, H. Anazawa, M. Satoh, M. Yamasaki, N. Hanai, K. Shitara, J. Biol. Chem., **2003**, *278*, 3466–3473.

58 P. Lerouge, M. Cabanes-Macheteau, C. Rayon, A. C. Fischette-Laine, V. Gomord, L. Faye, Plant Mol. Biol., **1998**, *38*, 31–48.

59 A. Koprivova, F. Altmann, G. Gorr, S. Kopriva, R. Reski, E. L. Decker, Plant Biol., **2003**, *5*, 582–591.

60 R. Vietor, C. Loutelier-Bourhis, A. C. Fitchette, P. Margerie, M. Gonneau, L. Faye, P. Lerouge, Planta, **2003**, *218*, 269–275.

61 Y. Makino, K. Omichi, N. Kuraya, H. Ogawa, H. Nishimura, S. Iwanaga, S. Hase, J. Biochem., **2000**, *128*, 175–180.

62 D. Chargelegue, N. D. Vine, C. J. van Dolleweerd, P. M. Drake, J. K. Ma, Transgenic Res., **2000**, *9*, 187–194.

63 A. von Schaewen, A. Sturm, J. O'Neill, M. J. Chrispeels, Plant Physiol., **1993**, *102*, 1109–1118.

64 K. Ko, P. M. Rudd, D. J. Harvey, R. A. Dwek, S. Spitsin, C. A. Hanlon, C. Rupprecht, B. Dietzschold, M. Golovkin, H. Koprowski, **2003**, *100*, 8013–8018.

65 R. Strasser, F. Altmann, J. Glossl, H. Steinkellner, Glycoconj. J., **2004**, *21*, 275–282.

66 A. Koprivova, C. Stemmer, F. Altmann, A. Hoffmann, S. Kopriva, G. Gorr, R. Reski, E. L. Decker, Plant Biotechnol. J., **2004**, *2*, 517–523.

67 H. Bakker, M. Bardor, J. W. Molthoff, V. Gomord, I. Elbers, L. H. Stevens, W. Jordi, A. Lommen, L. Faye, P. Lerouge, D. Bosch, Proc. Natl. Acad. Sci. USA, **2000**, *98*, 2899–2904.

68 N. Q. Palacpac, S. Yoshida, H. Sakai, Y. Kimura, K. Fujiyama, T. Yoshida, T. Seki, Proc. Natl. Acad. Sci. USA, **1999**, *96*, 4692–4697.

69 R. Misaki, Y. Kimura, N. Q. Palacpac, S. Yoshida, K. Fujiyama, T. Seki, Glycobiology, **2003**, *13*, 199–205.

70 C. M. Huether, O. Lienhart, A. Baur, C. Stemmer, G. Gorr, R. Reski, E. L. Decker, Glyco-engineering of moss lacking plant-specific sugar residues. Plant Biol., in press.

8

ExpressTec: High-level Expression of Biopharmaceuticals in Cereal Grains

Ning Huang and Daichang Yang

Abstract

ExpressTec has been developed to produce biopharmaceuticals both cost-effectively and in large quantities. ExpressTec is successful because it utilizes the latest developments in plant molecular biology with the use of strong, endosperm-specific promoters; signal peptides targeting the subcellular compartments to prevent proteolytic degradation of the recombinant protein; optimized codons to maximize translational efficiency; and transcriptional activators that increase target gene transcription and control of the expression of competitive molecules. Several recombinant proteins have been expressed using the ExpressTec system. The expression level of these proteins is between 0.1 to 1% of brown rice weight, or 25–60% of soluble protein. Data shows that both the transgenes and their expression are stable over 5 years and 10 generations. The physical and biochemical properties of the recombinant proteins are the same as for native proteins. Scale-up processing has shown that recombinant proteins are easily extracted from cereal grains, and economical analysis has placed the cost of biopharmaceuticals produced by ExpressTec at about US$ 6 per gram.

Abbreviations

CFU	colony-forming units
ER	endoplasmic reticulum
GC/MS	gas chromatography/mass spectrometry
hITF	human intestine trefoil factor
hLF	human lactoferrin
hLZ	human lysozyme
LPS	lipopolysaccharides
nhLF	native human lactoferrin
PBF	prolamin-box binding factor
PMP	plant made pharmaceuticals
PSV	protein storage vacuole
rhLF	recombinant human lactoferrin
rhLZ	recombinant human lysozyme
TSP	total soluble protein

8.1
Introduction

The 1990s and the beginning of the twenty-first century mark a new era of transgenic biopharmaceutical production in plant and animal cells. Researchers all over the world explore various ways to produce biopharmaceuticals in large volumes at a low cost. Scientists at Ventria Bioscience have developed a protein expression system, ExpressTec, which expresses recombinant proteins, enzymes and sec-

Modern Biopharmaceuticals. Edited by J. Knäblein
Copyright © 2005 WILEY-VCH Verlag GmbH & Co. KGaA, Weinheim
ISBN: 3-527-31184-X

ondary metabolites in cereal grains. In ExpressTec, target genes are codon-optimized conforming to the codon preference of the host genes. The codon-optimized target genes are then linked to the strong endosperm-specific promoters and the signal peptides derived from the storage proteins. The signal peptide leads the target protein through the endoplasmic reticulum (ER) where the signal peptides are cleaved and the high level recombinant protein produced. The mature proteins are accumulated in cell compartments such as protein bodies in the endosperm. The target proteins are isolated from these cereal grains for various applications.

While ExpressTec can be used in all cereal grains, Ventria has focused on the use of rice and barley grains. There are several advantages in using rice and barley grains as the host to produce biopharmaceuticals:

1. Rice and barley grains are generally regarded as safe for consumption. In many countries, rice flour is the first solid food for infants. Rice-based infant formulas are commercially available and rice is considered hypoallergenic. Thus, cereal grains such as rice are particularly suitable for the production of recombinant protein for oral applications.
2. The storage proteins in rice and barley grains are synthesized during grain maturation and stored in protein bodies for use in the germination and seedling growth of the next generation. Thus, protein accumulation in rice and barley grain is a natural process and suitable for recombinant protein production.
3. Cereal grains can be produced in large quantity at very low cost. There is essentially no scale limitation.
4. Cereal grains can be stored for years without loss of functionality, and therefore downstream processing can be conducted independent from the growing season.
5. Production of recombinant proteins in cereal grain will be devoid of any animal pathogen – a risk present in transgenic animal systems.
6. Rice and barley are both self-pollinating crops. The pollen viability and out-crossing rates are very low, reducing the segregation requirement and the chance of gene flow via pollen.

8.2
Development of ExpressTec for High-level Expression of Recombinant Proteins in Cereal Grains

Plant expression systems can be generally categorized into three groups: 1) whole-plant systems producing proteins in the leaves or the entire plant; 2) cell culture systems where the protein can be produced in the culture cells or secrete into culture media; and 3) the seed/fruit or tuber systems with proteins expressed in storage organs. Regardless of which expression system is used to produce biopharmaceuticals, the expression level of the active protein is the foundation and one of the most critical factors impacting the commercialization of plant made pharmaceuticals (PMP). In general, gene expression is regulated at four different levels: transcription; post-transcription; translation; and post-translation. In order to achieve high-level expression of recombinant proteins, our strategy is to focus on increasing transcription, enhancing translational efficiency, and improving the protein targeting and trafficking (Fig. 8.1).

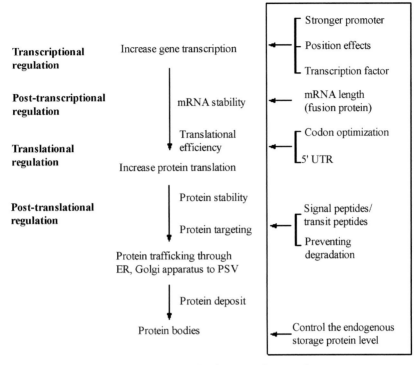

Fig. 8.1 Schematic diagram of ExpressTec development. The control points on gene expression regulation are illustrated; the technologies listed in the boxed panel represent areas where major efforts have been made to increase recombinant protein expression.

8.2.1
Increasing Protein Expression by Boosting Gene Transcription

Transcription is controlled by promoter activity and regulated by the *cis* elements on the promoters. The promoter activity is enhanced by transcriptional factors that interact with *cis* elements. To search for the strong promoters, we have examined and screened various promoters from storage protein genes via both transient expression and transgenic analysis [1, 2]. GUS and human lysozyme genes are used as reporter genes in transient and transgenic analysis, respectively. Fig. 8.2 shows one such comparison study. Glutelin 1 (*Gt1*) promoter from the rice glutelin gene and globulin

(*Glb*) promoter from the rice globulin gene show the strongest promoter activities. The expression level of lysozyme in R_1 seeds reaches, on average, 10.5% (*Glb*) and 13% (*Gt1*) total soluble protein (TSP). The best line shows that the expression level is up to 60% TSP which is derived from a *Gt1*-based construct. On a Coomassie blue-stained gel, the lysozyme band is the most abundant band among all the rice protein bands (Fig. 8.3B, lanes 1 to 3) indicating the strength of the *Gt1* promoter activity.

Transcription can be enhanced by transcriptional factors that bind to *cis* elements on the promoter. We tested the effects of various transcriptional factors on recombinant protein expression. The factors include REB binding to the rice globulin

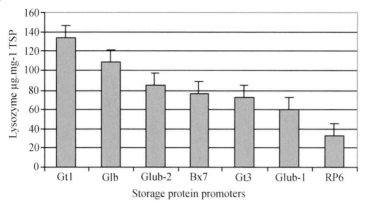

Fig. 8.2 Comparison of the promoter activities from the promoters of different storage protein genes. Seven promoters were tested using a transgenic approach, and human lysozyme was used as reporter gene. *Gt1, Gt3, Glub-1* and *Glub-2* are from the rice glutelin gene family; *Glb* is from the rice globulin gene; *Rp6* is from the rice prolamin gene; and *Bx-7* is from the wheat high molecular-weight glutenin gene. Human lysozyme was quantified by turbidimetric assay.

promoter, PBF (Prolamin-box Binding Factor) from maize and Opaque 2 from maize. Rice plants containing the human lysozyme gene were generated both with and without the transcription factor. The results show a 3.7-fold increase of human lysozyme expression when co-expressing a Glb promoter-specific REB transcriptional factor with the *Glb-lys* construct [3]. A significant increase of human lysozyme was observed when co-expressing PBF with the *Gt1-lys* construct (Fig. 8.3 A and B lane 4). Furthermore, transient analysis shows that PBF and O2 can act additively to enhance the expression of the GUS reporter gene in immature rice endosperm [4].

8.2.2
Increasing Protein Expression by Enhancing Translational Efficiency

Abundant transcripts would produce abundant mRNA which sets the foundation of effective protein translation. The untranslational leader sequence is one of the key elements to translation initiation, a deter-

mining factor on the number of peptides produced from each mRNA. We assume that native 5′ untranslational sequence of a strongly expressed gene would be the best for recombinant protein expression; thus, the native 5′ untranslational sequence was used in our expression cassettes. These cassettes were used in both transient expression analysis [2] and transgenic analysis [5, 6]. High-level protein expression is achieved with the use of native 5′ untranslational sequence.

The translational efficiency is one of the other important elements that affect protein synthesis and accumulation. The translational efficiency is highly impacted by the use of genetic codons of the genes. Due to genetic codon degeneracy, codon usage has high diversity among different organisms. In triple-letter genetic codons, whilst the first and second positions are largely conserved among organisms, the third position is quite diverse. The preferred codons in rice genes at the third position are 100% G or C. This is, however, not the case for other organisms. For ex-

A   **B**

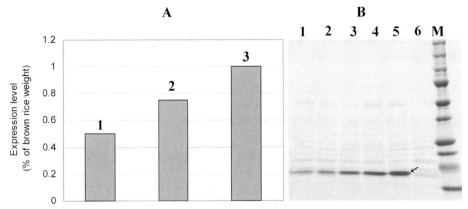

Fig. 8.3 Human lysozyme expression level was improved through using different strategies during ExpressTec development. (A) Milestones of increasing human lysozyme level in rice endosperm. The lysozyme expression level was increased to 1% of dry weight by selection stronger promoters (1), use of transcription factor (2), and space-emptying strategy (3). (B) Human lysozyme protein in a Coomassie blue-stained gel from differ- ent approaches. Lanes 1–3 shows the effects of selecting the stronger promoters, *Glb* promoter/ *Glb* signal peptide (lane 1), *Glb* promoter/*Gt1* signal peptide (lane 2) and *Gt1* promoter/*Gt1* signal peptide (lane 3). Lane 4 shows the effects of using the transcription factor. Lane 5 represents the effects of using the space-emptying strategy. Lane 6 is for non-transgenic TP309. Lane M is molecular weight marker.

ample, the preferred codons in *Arabidopsis* genes are 15% G or C at the third position of the codons. Therefore, when expressing foreign genes, using the preferred codons of the host can maximize the translational efficiency. This has been confirmed in our laboratory and in other laboratories [7–9]. In producing the human blood protein, alpha-1-antitrypsin, in rice culture cells, expression of the codon-optimized gene is several folds higher than that of the native gene [10, 11]. In expressing another protein, subtilisin, expression of the codon-optimized gene is over 100-fold higher than that of native gene. All data point to a very important conclusion that foreign genes must be codon-optimized to match the codon preference of the host for high-level expression.

8.2.3
Increasing Protein Expression by "Managing the Space" for Recombinant Protein Deposition

8.2.3.1 **Protein Bodies in Endosperm Cells Provide a Safe Environment for Foreign Protein Accumulation**

Post-translational regulation mainly includes the signal peptide cleavage, glycosylation, phosphorylation, and proper protein folding while the protein is translocated into the ER and transported to the Golgi apparatus. Then, the protein is targeted and trafficked to the protein storage vacuole (PSV), ER-derived protein bodies, and other organelles, or secreted to cytosol [12–15]. In those steps, the proteins could either be accumulated in high amounts or rapidly turned over due to protease activity, depending on the destination of the protein targeting. It has been shown that PMPs expressed at very low levels in plant cells without a targeting destination [16–18] because of the protein(s) exposure to protease(s) degradation. Therefore, a promising gene expression system for PMP proteins should use a particular targeting signal to deposit the recombinant proteins to certain organelles or cell compartments to prevent their degradation.

In rice, grain endosperm, two protein bodies – PBI and PBII – are considered to be the "sink" for protein storage during endosperm development. We hypothesize that protein bodies provide a "safe" environment for the deposition and accumulation of PMPs, because there is limited protease activity within the protein bodies. Targeting PMPs to protein bodies in cereal endosperm cells can be achieved by attaching a signal sequence to a mature peptide of the PMP, which can guide PMPs through the inner membrane system instead of to the cytosol. As soon as the gene is transcribed and processed, mRNA is bound to the subdomains of the ER, which determines where the protein targets [19–21]. Then, the synthesized recombinant protein targets to the protein bodies through the protein trafficking pathway during endosperm development [22]. Comparison studies with and without signal peptide confirms this hypothesis. In the expression of heat stable beta-glucanase in barley grain, signal peptide from hordien D, a barley storage protein, was used. The expression level of beta-glucanase with hordein signal peptide is several fold higher than the same construct without the signal peptide [23]. In expressing recombinant protein in rice grain, we use strong promoters and signal peptides from two rice storage proteins, glutelin and globulin, to achieve high-level expression of the PMP [2]. The use of *Gt1* signal peptide and its 5′ untranslated sequence promote higher expression than that from Glb (Fig. 8.3 B, lanes 1, 2 and 3).

To confirm that recombinant human lysozyme (rhLZ) is present in protein body, immature rice endosperm from LZ264 grain was harvested and sectioned. The sections were incubated with anti-lysozyme and anti-glutelin antibody. The anti-lysozyme antibody derived from sheep was specifically recognized by IgG conjugated with green fluorescence; hence, the presence of rhLZ would appear green. The anti-glutelin antibody derived from rabbit was specifically recognized by IgG conjugated with red fluorescence; hence, the presence of glutelin would appear red. A yellow color would appear if rhLZ and glutelin were to be co-localized. As seen in Fig. 8.4, panel A indicates the presence of rhLZ shown in green, while panel B shows the presence of glutelin shown in red. Panel C shows yellow spots, indicating that rhLZ and glutelin are co-localized. Panel D is a linear scan to show that the green and red peak at the same position rather than showing shadows from a nearby color. Since glutelin is a storage protein known to be stored in the protein body of endosperm cells, Fig. 8.4 confirms that rhLZ is targeted to the protein body in LZ264 grain [24]. The deposit of rhLZ in protein bodies is further confirmed by electronic microscopic analysis [24].

8.2.3.2 Increasing Protein Expression using Different Translational Machineries

To improve human lysozyme expression, we attempted to improve heterologous recombinant protein expression levels by: 1) co-transformation of *Glb* and *Gt1* expression cassettes; and 2) crossing of two independent transgenic lines expressing lysozyme protein from *Gt1* and *Glb* cassettes. These experiments failed to improve lysozyme expression. This implies that the gene transcription and copy number are

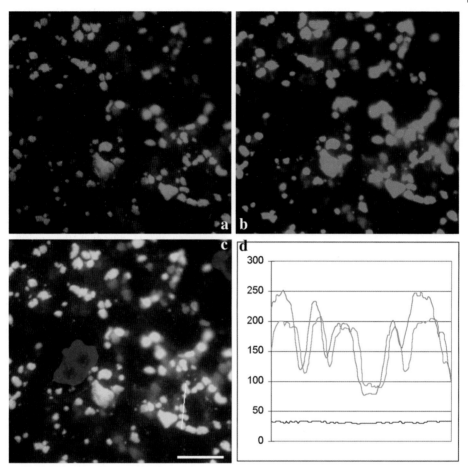

Fig. 8.4 Evidence of rhLZ targeting to protein bodies in rice endosperm. The immature endosperm at 14 days after pollination was used for fluorescence microscopic studies. The endosperm section was incubated with antibodies against human lysozyme (a) and rice glutelin (b). A merger of the two images from (a) and (b) is shown in (c). (d) Diagram of the fluorescence scans along the white line shown in (c). Red line=glutelin; green line=human lysozyme, showing both lysozyme and glutelin co-localized in the same protein bodies.

not the bottle-neck for improvement of human lysozyme expression. Other possible limiting factors include the translation efficiency, protein trafficking, and targeting. Therefore, we attempted to improve human lysozyme expression by using different promoters/signal peptides which can bind to different ER subdomains than that of *Gt1* and *Glb* promoters/signals would.

We hypothesized that we could achieve higher expression by using different targeting signals to use different ER subdomains. A wheat puroindoline b promoter and signal peptide have been tested. Co-expression of both constructs (*Gt1* promoter, its signal peptide plus the human lysozyme gene and the puraindoline b promoter, its signal peptide and the human lysozyme gene), resulted in an increase in the expression of human lysozyme by 79% to 8 mg kg^{-1} rice grain flour. Electron microscopy studies show that the puroindoline-based construct directed rhLZ to both protein body I and II.

8.2.3.3 Reduce Endogenous Protein Expression to Reserve the Space to Recombinant Protein Storage

When we examined the protein body structure of the lines expressing high levels of recombinant human lysozyme, we observed that rice endosperm generated novel storage vesicles or protein body variants for recombination protein deposition [24]. It also indicated that rice endosperm cells are capable of generating novel storage vesicles for recombinant protein deposition when large amounts of recombinant protein are expressed in the endosperm cells. Furthermore, we also observed a negative correlation between native storage proteins and the recombinant protein expression. We hypothesized that the protein bodies are the "sink". When more "source" proteins are available, they compete for the "sink" causing an imbalance between the "sink" and the "source". Reduced native storage protein in high lysozyme-expressing lines indicates that the recombinant protein can partially compete for ER sub-domains with native storage proteins and chaperones during the trafficking. This implies that human lysozyme expression could be further increased by shutting down native storage protein expression, making more "sink" space available to recombinant protein deposition. Thus, we call this strategy "space-emptying". This concept was tested by reducing the endogenous protein expression via antisense technology. The antisense constructs of glutelin and globulin were introduced into the transgenic line that expressed high

levels of human lysozyme using gene stacking. The expression of recombinant human lysozyme in the best lines expressing the antisense gene is increased from 5 to 10 mg g^{-1} rice flour (Fig. 8.3).

In summary, we conclude that ExpressTec is a promising biomanufactory system for expressing PMPs in rice grain as well as other cereal grain endosperms. In addition to all the advantages of other plant expression systems, it has a higher capacity for obtaining higher expression levels of PMPs. Its core technology is to target the recombinant proteins to protein bodies so that the biopharmaceutical can be protected from protease degradation and accumulated at a very high level. Moreover, the technology is improved by boosting transcription, enhancing translational efficiency, improving protein trafficking and deposition to maximize PMP expression in cereal grains (Fig. 8.3). The recombinant protein expression can be as high as 1% of flour weight.

8.3 High-level Expression of Biopharmaceuticals in Cereal Grain using ExpressTec

The success of developing ExpressTec laid the foundation for production of various recombinant polypeptides, multipeptide proteins and secondary metabolites. To express small peptide, a fusion strategy is used to achieve high level expression.

8.3.1 Expression of Human Lysozyme in Rice Grain

Human lysozyme (hLZ) hydrolyzes 1,4-beta-linkage between *N*-acetylmuramic acid and *N*-acetyl-D-glucosamine residue in peptidoglycan. Human LZ exhibits antibacterial, antiviral, antifungal and antiparasitic activities, and has also been implicated as an anti-inflammatory/anti-oxidant agent or direct binding to lipopolysaccharides (LPS) for immunomodulation [25]. Human LZ is found in human secretions, such as milk, tears and saliva, and consists of an ungly-cosylated polypeptide chain with 130 amino acid residues, giving it a molecular weight of 14.5 kDa.

To express hLZ in rice grain, the hLZ gene was codon-optimized. A total of 92 codons out of 130 codons was modified, resulting in the G + C content being raised from 46 to 68%. The synthetic hLZ gene was cloned to produce pAPI159, which contains the *Gt-1* promoter, *Gt-1* signal sequence and *nos* terminator. After transformation, over 500 transgenic rice R_0 plants were generated and seeds from fertile rice plants were analyzed via LZ activity assay, Coomassie blue-stained gel and Western blot analysis to determine the amount of recombinant human lysozyme (rhLZ) in

the endosperm. As shown in Fig. 8.3 B, a dominating band corresponding to the position of a protein with a molecular mass of rhLZ was detected in the salt-soluble fraction of crude extracts from the transgenic rice grains, while it is absent in untransformed rice. The identity of the protein was confirmed by Western blotting analysis and verified further by N-terminal analysis [5]. One of the lines, named LZ159 was selected and advanced for 10 generations from 1999 to 2003. Both the transgenes and expression levels are stable. The expression level of LZ159 remains at 5 g kg^{-1} rice flour, amounting to 60% total soluble protein. To determine the bactericidal activity of rhLZ, an *E. coli* strain, K12, was used. Bacterial culture with the addition of rhLZ at 20 μg mL^{-1} resulted in significantly fewer colony forming units (CFU) than those where rhLZ was not added (Fig. 8.5), thus proving that rhLZ is biologically active. Further studies

A   **B**

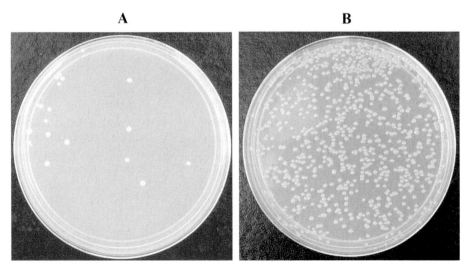

Fig. 8.5 Bactericidal effect of purified rhLZ. *E. coli* (10^5 CFU) was incubated for 120 min with buffer plus 20 μg mL^{-1} purified rhLZ (A) or buffer only (B). At the end of the incubation period, numbers of CFUs were determined by plating a sample of the incubation mixture.

show that rhLZ are thermal stable and active in a wide range of pH [5].

8.3.2
Expression of Human Lactoferrin in Rice Grain

Human lactoferrin (hLF) is an 80-kDa iron-binding glycoprotein, and another major component found in human milk (average 1–2 mg mL^{-1}); lower concentrations are present in the exocrine fluids of glandular epithelium cells such as bile, tears and saliva (0.1–0.3 mg mL^{-1}). LF has been suggested to have several biological activities, including antimicrobial, regulation of iron absorption, immunomodulation, protection from pathogen infection, and cellular growth-promoting activity [26].

To express hLF, the gene was synthesized based on the codon-preference of rice genes. Of the 692 codons for the mature peptide of the hLF gene, 413 codons were changed. The codon-optimized hLF gene was expressed using ExpressTec. Total soluble protein extracted from rice grains is analyzed by SDS-PAGE (Fig. 8.6 A, lane 4). Among the protein bands, the recombinant human lactoferrin (rhLF) band was the strongest, indicating that rhLF is the most abundant soluble protein extracted from rice grain. Quantitative analysis by ELISA indicated that up to 25% of soluble protein or 0.5% flour weight was rhLF. The expression level of rhLF in the best line reaches 5.0±0.5 g kg^{-1} and is stable through 10 generations (Fig. 8.6 B).

In order to characterize the biochemical properties of rhLF expressed in rice grain, rhLF from transgenic rice grain was purified to homogeneity. The N-terminal sequence of rhLF was identical to the corresponding region of hLF, indicating that the rice signal peptidase recognized and cleaved at the junction between the *Gt1* signal peptide sequence and the mature peptide of rhLF. The isoelectric point (pI) of hLF and rhLF was similar, indicating that both have similar surface charges. Both native human lactoferrin (nhLF) and rhLF can reach iron-saturation by picking up iron from a solution to form holo-LF. The stability of iron-binding by rhLF toward low pH was analyzed and compared to that of nhLF (Fig. 8.6 C). Iron release began at about pH 4, was completed around pH 2, and was similar for both proteins.

The antimicrobial effect of rhLF was tested against a Gram-negative strain of *E. coli*, DH5α (Fig. 8.6 D). *E. coli* at a concentration of 10^5 CFU was mixed with and without rhLF. After incubation at 37 °C for 120 min, the CFU after treatment with rhLF were reduced by 90%, while CFU without rhLF remained unchanged.

8.3.3
Expression of Fibrinogen in Rice Grain

Fibrinogen is a multi-chain protein involving the assembly of three different polypeptides with a molecular mass of 340 kDa. The molecule is arranged as a dimer with each half-molecule containing a set of each of the three different chains. The subunits and the chains are linked together by three disulfide bonds at the N-terminal portions of the polypeptides and form a symmetrical trinodular structure. There are two symmetrical bonds that are located between adjacent γ chains and another bond between α chains. In addition, there are 29 inter- and intra-chain disulfide bonds interspersed throughout the molecule that are responsible for maintaining proper structure. Fibrinogen is a blood plasma protein that serves as one of the main components in blood clotting.

Expression of multipolypeptide protein in rice grain posed a new challenge to ExpressTec. After gene codon optimization,

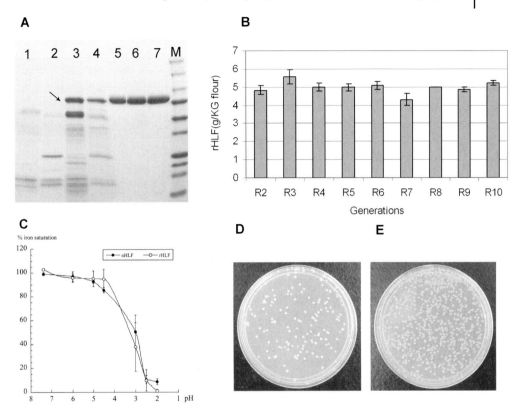

Fig. 8.6 Biochemical properties and biological activity of rhLF. (A) rhLF in Coomassie blue-stained gel. Lanes 1 and 2 represent non-transgenics of Golden Promise (barley) and TP309 (rice), respectively; lanes 3 and 4 show rhLF from the endosperm extracts of transgenic barley and rice, respectively; lanes 5–7 indicate native human LF standard, titrated to 6, 8 and 10 μg per lane, respectively. M indicates molecular mass marker.

(B) Stable expression of rhLF over 10 generations as determined by ELISA. (C) pH-dependent iron release of recombinant and native human LF (see Ref. [6]). (D, E) Bactericidal effect of purified rhLF. *E. coli* (1×10^5 CFU) was incubated for 120 min with buffer plus 1 mg mL^{-1} purified rhLF (D) or buffer only (E). At the end of the incubation period, numbers of CFUs were determined by plating a sample of the incubation mixture.

genes for individual chain of fibrinogen (α, β, and γ,) were delivered into the rice cell by co-transformation. Transgenic plants were obtained and rice grains analyzed for the expression of the fibrinogen polypeptide via Western blot analysis. All three polypeptides were expressed in the same cell. Compared to the positive control, it is estimated that the expression level of the three polypeptide reaches about 0.4% brown rice flour weight.

8.3.4
Expression of Intestine Trefoil Factor in Rice Grains

Human intestine trefoil factor (hITF) consists of 75 amino acids. After cleavage of a signal peptide, the resulting mature hITF contains 60 amino acids [27, 28]. Human ITF is present in both monomer and dimer forms in gastrointestinal tissue [29]. Several biological functions of ITF have been identified, including the promotion of wound healing, stimulation of epithelial cell migration and protection of the intestinal epithelial barrier. It is thus believed that the preparation of ITF can be used in the prevention and treatment of these disease conditions.

In general, the expression of a peptide of less than 100 amino acids proves to be difficult. When hITF is directly expressed in rice grain using the *Gt1* promoter/sig-

nal peptide, the expression level is about 1 µg per grain, or 0.005% grain weight. To obtain a higher expression level, a modification to the expression system was made. The relative low expression level of ITF was not due to lower transcription based on Northern blot analysis. This suggested that it could be post-translation modification and protein trafficking. Using a fusion partner will generally increase the expression of a peptide. To apply this specifi-

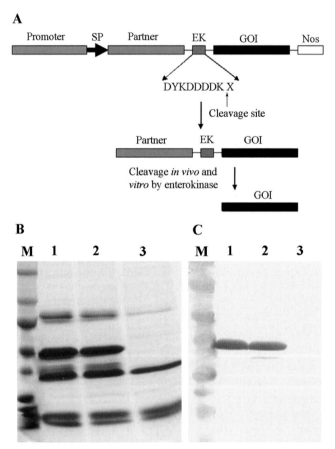

Fig. 8.7 A fusion strategy to express small peptide in rice endosperm. (A) Schematic diagram of the fusion strategy to express small peptide. GOI = gene of interest. The enterokinase recognition site was used as a linker between the fusion partner and GOI. (B, C) ITF fusion protein in Coomassie blue-stained gel and in Western blot using anti-ITF antibody. Lanes 1 and 2 represent individual transgenic lines; lane 3 indicates non-transgenic TP309. M indicates molecular mass marker.

cally to our rice seed expression system, proteins such as globulin are selected as fusion partners. These proteins are selected because they are relatively small, they have a high expression level in rice grain, they are targeted to the protein body, and they are water- and/or salt-soluble. These characteristics are important for increasing the expression of a peptide as well as the extraction and purification of the peptide (Fig. 8.7 A).

The human ITF DNA sequence based on the GenBank accession number L08044 [28] was codon-optimized with rice genetic codon preference. The codon-optimized gene was then linked to a rice globulin gene. Between the codon-optimized gene and the rice globulin gene was a segment of DNA encoding for a five amino-acid peptide, which is an enterokinase recognition site (Fig. 8.7 A).

Transgenic rice grains carrying the ITF gene were analyzed by Coomassie blue-stained gel and Western blot analysis (Fig. 8.7 B and C). A strong band was observed which is absent in the non-transgenic plant TP309 (Fig. 8.7 B); this band was confirmed to be ITF fusion by Western analysis using anti-ITF antibody (Fig. 8.7 C) and anti-GLB antibody. This band is the strongest and stronger than the native globulin band, indicating high-level expression of the fusion protein. Using a reference marker, it is estimated that the expression level of the fusion protein was about 60 μg per grain. Since ITF is about one-quarter of the fusion protein, about 15 μg ITF per grain (or 0.075% flour weight) was achieved.

8.3.5
Expression of Lignans in Rice Grain via Metabolic Engineering

Plant lignans are secondary metabolites which are most commonly found in woody stems, roots, seeds, oils, and leaves, and exist in low levels in cereal endosperm [30]. Matairesinol and secoisolariciresinol are two typical plant lignans which are essentially not detectable in rice endosperm. Plant lignans, once consumed, are then converted to mammalian lignans by fermentation in the large intestine, where matairesinol is converted to enterodiol and secoisolariciresinol to enterolactone. Enterolactone and enterodiol are the major lignans found in humans, and are present in the serum, urine, bile, and seminal fluid [31]. Studies have shown that lignans can prevent the development of cancer, and human populations that consume high quantities of lignans have a lower incidence of hormonally dependent cancers than do other populations consuming high-fat diets [30, 32, 33].

Plant lignans are derived from a process called the shikamate-chorismate pathway in phenylpropanoid metabolism, which leads to the production of coniferyl alcohol and other metabolites [34]. Specific oxidative coupling of coniferyl alcohol under the action of laccase and dirigent protein generates pinoresinol (Fig. 8.8 A). Pinoresinol, through the intermediate lariciresinol, is then converted to secoisolariciresinol, which is finally modified to form matairesinol [35]. Rice endosperm contains essentially no matairesinol (Fig. 8.8 B); this may be due to a lack of one or more of the four genes involved in the lignan biosynthetic pathway (Fig. 8.8 A), or that the expression of the genes in rice endosperm is deregulated.

In order to elevate lignan concentrations in rice endosperm, Ventria Bioscience and

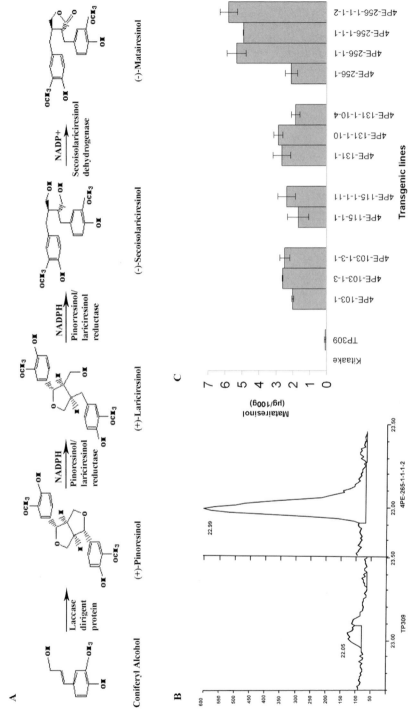

Fig. 8.8 Expression of lignan in transgenic rice endosperm. (A) Lignan biosynthesis pathway derived from *Forsythia intermedia* [35]. (B) Matairesinol level in transgenic grain (4PE-256) and non-transgenic wild-type (TP309). GC/MS scan shows accumulation of lignan in transgenic rice endosperm. (C) Matairesinol level in transgenic rice grains over three generations as measured by GC/MS scan.

Washington State University at Pullman collaborated to engineer a lignan synthesis pathway using ExpressTec. The four genes for laccase, dirigent protein, pinoresinol/lariciresinol reductase and secoisolariciresinol dehydrogenase involved in lignan synthesis pathway were fused to the *Gt1* promoter. Native signal peptides from the genes of the lignan synthetic pathway were used. The four constructs, along with a plasmid containing a plant-selectable marker, were delivered into rice cells via particle bombardment. Over 400 transgenic plants were generated. Transgenic seeds harvested from fertile plants were analyzed using gas chromatography/mass spectrometry (GC/MS) for elevated lignan, matairesinol, in the endosperm. The highest of these (4PE-256-1) had a matairesinol level approximately 15-fold that in the corresponding wild-type rice. Other families tested (4PE-103-1, 4PE-115-1, 4PE-131-1) had matairesinol levels which were up to 3- to 5-fold those in the wild-type. A typical GC/MS profile is shown in Fig. 8.8 B. In order to determine the generational stability of lignan expression in rice endosperm, a test was carried out using four lines with R_1, R_2, and R_3 seeds tested simultaneously. This test showed levels in the R_1 seeds to be similar to the lower levels seen in the R_2 and R_3 seeds, and the levels of matairesinol to remain fairly consistent across all three generations. The 4PE-256-1 line was the best of all the transgenic lines tested, and consistently gave results well above 3 ng per 100 mg for matairesinol. This line also showed consistency when the R_1, R_2, and R_3 seeds were tested simultaneously, as all three generations showed elevated levels of matairesinol. An elevated lignan level in transgenic rice endosperm not only proved that ExpressTec could be used to engineer metabolic pathways to produce secondary metabolites, but also

provided a line expressing a high level of lignan which could be used to provide a source of lignans in food for the benefit of human health.

8.3.6
Protein Expression using ExpressTec in Barley and Wheat Grains

In addition to expressing polypeptides and metabolites in rice grains, ExpressTec has been used to obtain high expression levels of proteins in different hosts such as barley or wheat grain. Recombinant hLZ is expressed at about 0.5% flour weight in barley and 0.5% flour weight in wheat endosperm. Similarly, rhLF is expressed at 0.7% flour weight in barley endosperm (see Fig. 8.6 A, lane 3).

8.4
Impact of Expression Level on the Cost of Goods

Expression level has a profound impact on cost of goods in biopharmaceutical production. One way to study this issue is to perform computer simulations based on data from plot production in order to obtain an early projection of production costs as a function of the expression level (Nandi et al., unpublished results). By utilizing the crop production and process data from the bench-scale (2 kg per batch), which was subsequently verified on a pilot scale at 180 kg per batch, an rhLF and rhLZ recovery and purification process is simulated. With an annual production of 600 kg and expression of 0.005% flour weight, rhLF can be produced at $ 382 g^{-1} (Fig. 8.9); however, when the expression level increases to 0.5% of flour weight (100×), as was achieved using ExpressTec, the cost was only $ 5.90 g^{-1}. Hence, an approxi-

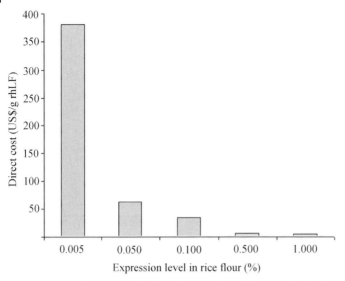

Fig. 8.9 Impact of expression level on the cost of goods. Based on the procedures of purifying rhLZ and rhLF from rice grain, a computer simulation was conducted assuming an expression level from 0.005 to 1% with the assumption of annual production capacity of 600 kg of pharmaceutical grade lactoferrin and lysozyme.

mately 7-fold reduction in direct costs could be achieved for each 10-fold incremental increase in the expression level. It is generally believed that the cost of pharmaceutical production in mammalian cell culture is about $ 200 g^{-1}. PMPs would lose competitiveness if this cost were close to $ 200 g^{-1}, particularly if the market size were to be small. The present analysis indicates that this would require a protein expression level of at least 0.05% cell mass.

8.5
Perspectives of Expressing Biopharmaceuticals in High Plants

Historically, the cost of recombinant biopharmaceutical production has not been a limiting factor, as drug manufacturers can easily pass the cost on to the consumers. Whilst this is still true in some cases, an in-

creasing demand for large volume and necessary low cost for some applications has forced the drug manufacturers to reduce their production costs and to make pharmaceuticals more affordable. It is believed that PMPs will be able to meet this demand. Although efforts made during the past decade have not produced satisfactory results, one reason for this is that expression levels in most plant systems have been very low. However, with the development of ExpressTec, a solution has been found to the problem of low expression level.

Despite obtaining high expression levels for certain proteins using ExpressTec, not all proteins are expressed to the same extent, and this suggests that expression level is in fact protein-dependent. Both, the biochemical and biophysical properties of the biopharmaceutical impact upon individual protein expression in cereal endosperm, these factors including protein folding, conformation, trafficking, deposition, and accumulation, whilst in some cases the recombinant proteins showed changes in solubility when expressed in rice grain. While the challenge remains that all soluble recombinant proteins and multiple polypeptides should be correctly assembled, these factors will undoubtedly be the main topics of research in this area in the future.

References

1 Hwang, Y.-S., Yalda, D., McCullar, C., Wu, L., Chen, L., Pham, P., Nandi, S. and Huang, N. Plant Cell Rep. **2002**, 20, 842–847.
2 Hwang, Y.-S., McCullar, C. and Huang, N. Plant Sci. **2001**,161, 1107–1116.
3 Yang, D., Wu, L., Hwang, Y.S., Chen, L. and Huang, N. Proc Natl Acad Sci USA **2001**, 98, 11438–11443.
4 Hwang, Y.-S., Ciceri, P., Parsons, R., Moose, S.P., Schmidt, R.J. and Huang, N. Plant Cell Physiol. **2004**, 45, 1509–1518.
5 Huang, J., Nandi, S., Wu, L., Yalda, D., Bartley, G., Rodriguez, R.L., Lonnerdal, B. and Huang, N. Molec. Breed. **2002**, 10, 83–94.
6 Nandi, S. et al. Plant Sci. **2002**, 163, 713–722.
7 Akashi, H. Curr. Opin. Genet. Dev. **2001**, 11, 660–666.
8 Davis, B.K. Prog. Biophys. Molec. Biol. **1999**, 72, 157–243.
9 Rouwendal, G.J.A., Mendes, O., Wolbert, E.J.H. and Boer, A.D.d. Plant Molec. Biol. **1997**, 33, 989–999.
10 Huang, J., Sutliff, T.D., Wu, L., Nandi, S., Benge, K., Terashima, M., Ralston, A.H., Drohan, W., Huang, N., Rodriguez, R.L. Biotechnol Prog. **2001**, 17, 126–133.
11 Terashima, M., Murai, Y., Kawamura, M., Nakanishi, S., Stoltz, T., Chen, L., Drohan, W., Rodriguez, R.L., Katoh, S. Appl. Microbiol. Biotechnol. **1999**, 52, 516–523.
12 Neuhaus, J.M. and Rogers, J.C. Plant Molec. Biol. **1998**, 38, 127–144.
13 Marty, F. Plant Cell **1999**, 11, 587–600.
14 Muntz, K. Plant Molec. Biol. **1998**, 38, 77–99.
15 Vitale, A. and Raikhel, N.V. Trends Plant Sci. **1999**, 4, 149–155.

16 Giddings, G., Allison, G., Brooks, D. and Carter, A. Nature Biotechnol. **2000**, 18, 1151–1155.
17 Larrick, J.W., Yu, L., Naftzger, C., Jaiswal, S. and Wycoff, K. Biomol. Eng. **2001**, 18, 87–94.
18 Schillberg, S., Fischer, R. and Emans, N. Cell. Mol. Life Sci. **2003**, 60, 433–445.
19 Choi, S.B., Wang, C., Muench, D.G., Ozawa, K., Franceschi, V.R., Wu, Y. and Okita, T.W. Nature **2000**, 407, 765–767.
20 Li, X.X., Franceschi, V.R. and Okita, T.W. Cell **1993**, 72, 869–879.
21 Okita, T.W. and Choi, S.B. Curr. Opin. Plant Biol. **2002**, 5, 553–559.
22 Vitale, A. and Galili, G. Plant Physiol. **2001**, 125, 115–118.
23 Horvath, H., Huang, J., Wong, O., Kohl, E., Okita, T., Kannangara, C.G. and von Wettstein, D. Proc. Natl. Acad. Sci. USA **2000**, 97, 1914–1919.
24 Yang, D., Guo, F., Liu, B., Huang, N. and Watkins, S.C. Planta **2003**, 216, 597–603.
25 Sava, G. in: Lysozyme: model enzyme in biochemistry and biology, Jolles, P. (Ed.) **1996**.
26 Brock, J.H. Biochem. Cell. Biol. **2002**, 80, 1–6.
27 Hauser, F., Poulsom, R., Chinery, R., Rogers, L.A., Hanby, A.M., Wright, N.A. and Hoffmann, W. Proc. Natl. Acad. Sci. USA **1993**, 90, 6961–6965.
28 Podolsky, D.K., Lynch-Devaney, K., Stow, J.L., Oates, P., Murgue, B., DeBeaumont, M., Sands, B.E. and Mahida, Y.R. J. Biol. Chem. **1993**, 268, 6694–6702.
29 Chinery, R., Bates, P.A., De, A. and Freemont, P.S. FEBS Lett. **1995**, 357, 50–54.
30 Nilsson, M., Harkonen, H., Hallmans, G., Knudsen, K.E.B., Mazur, W. and Adlercreutz, H.J. Sci. Food Agric. **1997**, 73, 143–148.
31 Borriello, S.P., Setchell, K.D., Axelson, M. and Lawson, A.M. J. Appl. Bacteriol. **1985**, 58, 37–43.
32 Thompson, L.U., Seidl, M.M., Rickard, S.E., Orcheson, L.J. and Fong, H.H. Nutr. Cancer **1996**, 26, 159–165.
33 Adlercreutz, H. and Mazur, W. Ann. Med. **1997**, 29, 95–120.
34 Lewis, N.G. and Sarkanen, S. Lignin and lignan biosynthesis. Oxford University Press, Washington, DC, USA, **1998**.
35 Dinkova-Kostova, A.T., Gang, D.R., Davin, L.B., Bedgar, D.L., Chu, A. and Lewis, N.G. J. Biol. Chem. **1996**, 271, 29473–29482.

9

Biopharmaceutical Production in Cultured Plant Cells

Stefan Schillberg, Richard M. Twyman, and Rainer Fischer

Abstract

The use of plants for the production of recombinant proteins has received a great deal of recent attention, but production systems that utilize whole plants lack several of the intrinsic benefits of cultured cells, including the precise control over growth conditions, batch-to-batch product consistency, the high level of containment and the ability to produce recombinant proteins in compliance with current good manufacturing practice (cGMP). Plant cell cultures combine the merits of plant-based systems with those of microbial and animal cell cultures, particularly in terms of downstream processing, a section of the production pipeline which is rarely given appropriate prominence when different production systems are compared. In this chapter, we discuss the benefits of plant cell cultures compared to other systems, the technological requirements for producing biopharmaceutical proteins in plant cells, and the unique aspects of downstream processing which are applied to this expression platform.

Abbreviations

Asn	asparagine
ATPS	aqueous two-phase systems
BSA	bovine serum albumin
BY-2	*Nicotiana tabacum* cv. Bright Yellow 2
CaMV	cauliflower mosaic virus
cGMP	current good manufacturing practice
DMSO	dimethylsulfoxide
EBA	expanded bed adsorption
ER	endoplasmic reticulum
FDA	Food and Drug Administration
FW	fresh weight
GM-CSF	granulocyte-macrophage colony-stimulating factor
HBsAg	hepatitis B surface antigen
hGM-CSF	human granulocyte-macrophage colony-stimulating factor
HSA	human serum albumin
Ni-NTA	nickel nitriltriacetic acid
NT-1	*Nicotiana tabacum* 1
PEG	polyethylene glycol
PVP	polyvinylpyrrolidone
scFv	single chain fragment variable
TSP	total soluble protein
Ubi1	ubiquitin-1

Modern Biopharmaceuticals. Edited by J. Knäblein
Copyright © 2005 WILEY-VCH Verlag GmbH & Co. KGaA, Weinheim
ISBN: 3-527-31184-X

9.1
Introduction

Plant cells combine some of the more beneficial features of other production systems and therefore occupy a unique position among the emerging platforms for commercial recombinant protein production [1, 2]. Like microbial cells, plant cells are inexpensive to maintain. They have simple nutrient requirements, and can be grown under controlled and defined conditions in accordance with current good manufacturing practice (cGMP). However, unlike microbes, plant cells are higher eukaryotic systems and have the ability to produce complex, multimeric proteins and glycoproteins (see Part IV, Chapter 7). The N-glycans synthesized in plants are not exactly the same as those synthesized in mammals, so human glycoproteins produced in plants do not contain native glycan profiles [3]. However, plant-derived recombinant proteins are more similar to their mammalian counterparts than proteins synthesized in bacteria (which do not glycosylate proteins at all) or in yeast and filamentous fungi (which produce very different glycans). Other advantages of plant cells include the lack of endotoxins that are often present in bacteria, and the absence of human pathogens such as viruses or prions (which may be present in mammalian cell lines) (see Part IV, Chapter 1). This high level of safety makes plant cells suitable for the production of biopharmaceuticals [4].

Many different plant-based expression systems are now available for the production of recombinant proteins. Those utilizing whole plants have been extensively reviewed, and will not be discussed in detail here (see Refs. [5–8]). Other systems, based on cultured plant cells or organs, include hairy roots [9], shooty teratomas [10],

immobilized cells [11] and suspension cell cultures [12]. With the exception of suspension cells, these systems are heterogeneous especially when scaled up, and hence are difficult to maintain under cGMP conditions. Therefore, suspension cells have attracted the most attention since they are amenable to cGMP and they can be cultivated in large-scale bioreactors [13, 14] (see Part IV, Chapter 7).

Suspension cell cultures have been derived from a number of different plant species, including the widely-used laboratory model *Arabidopsis thaliana* [15], plants such as *Catharanthus roseus* and *Taxus cuspidata* which are used to produce valuable secondary metabolites [16, 17], and important domestic crops such as tobacco, rice, alfalfa, tomato, and soybean [18–22]. Because cell lines from domestic crop species are well-characterized, they have been the most frequently used for recombinant protein production. The most popular cell lines include those derived from the tobacco cultivars Bright Yellow 2 (BY-2) (Fig. 9.1) and *Nicotiana tabacum* 1 (NT-1) [2].

Plant cell suspensions are typically derived from undifferentiated callus tissue which has been induced from tissue explants growing on solid medium. Friable callus pieces are transferred into liquid medium and then agitated on rotary shakers or in fermenters to break the callus into small aggregates and single cells. The correct balance of plant hormones is present in the medium to maintain the undifferentiated state and promote rapid growth. Transgenic cell suspensions can be generated by agitating callus derived from transgenic plant tissue, or transformation can take place after the cell suspension has been prepared. In the latter case, transformation is usually achieved either by cocultivation with *Agrobacterium tumefaciens* [23, 24] or particle bombard-

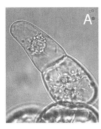

Fig. 9.1 Tobacco BY-2 suspension cells (A; original magnification ×400) are cultivated in shake flasks (B) and bioreactors (C) (2-L stirred reactor) under sterile and controlled conditions, allowing the production of recombinant proteins according to current pharmaceutical production standards.

ment [25]. Plant cell suspensions can be cultivated using conventional fermenter equipment and the same running modes as applied to microbial cultures, for example, batch, fed-batch, perfusion, and continuous fermentation [26, 27].

9.2
Recombinant Proteins Produced in Plant Cell Suspension Cultures

The first recombinant protein expressed in cultured plant cells was human serum albumin, produced in tobacco suspension cells derived from transgenic plants [28]. Since then, many different proteins have been produced in suspension cells from a variety of plant species, with a focus on pharmaceutical proteins such as antibodies, cytokines, growth factors, hormones, and enzymes. A selection of these proteins is listed in Table 9.1, which also provides details of the expression construct used in each case and the final product yield. An overview of plant-based biopharmaceuticals, their indication and status in clinical trials is provided by Knäblein [52]. Tobacco cell lines have been used the most widely because tobacco is the most popular

Table 9.1 Recombinant proteins expressed in cultured plant cells

Expression host	Expressed protein	Promoter	Localization, yield	Reference(s)
Tobacco suspension cells initiated from transgenic plants	Human serum albumin (HSA)	Modified CaMV 35S	Secretion/apoplast targeting, 0.25 µg mg^{-1} protein in supernatant	28
Tobacco suspension cells initiated from transgenic plants	ScFv antibody fragment	CaMV 35S	Secretion, up to 0.5 µg L^{-1}, up to 0.5% of TSP	29
Tobacco cv BY-2 suspension cells	Human erythropoietin	CaMV 35S	Secreted, 1 pg g^{-1} FW	30, 31
Tobacco cv NT-1 suspension cells	Mouse monoclonal Heavy chain	CaMV 35S	Native heavy-chain secretion signal, ca. 10 µg L^{-1} extracellular without PVP, 350 µg L^{-1} with PVP	32
Tobacco cv NT-1 suspension cells	Heavy chain mAb	CaMV 35S	Secreted up to 10 µg L^{-1}, with stabilization up to 350 µg L^{-1}	33
Tobacco cv NT-1 suspension cells	Bryodin 1	CaMV 35S	Secreted up to 30 mg L^{-1}	34
Tobacco cv NT-1 suspension cells	Human interleukin-2 and interleukin-4 (hIL-2 and hIL-4)	CaMV 35S	Secreted (native signal peptides), 8–180 µg L^{-1} of culture broth	35
Tobacco suspension cells	Recombinant ricin	CaMV 35S	25–37.5 µg L^{-1}	36
Rice cv Bengal callus cells	ScFv antibody fragment	Maize ubiquitin-1	Apoplast targeting (optimized Ig leader peptides) and ER-retention, up to 3.8 µg g^{-1} callus FW	37
Tobacco cv Petite Havana SR1 suspension cells initiated from transgenic plants	Mouse IgG-2b	Enhanced CaMV 35S	0.3% of TSP or 15 µg g^{-1} wet weight	38
Rice cv Taipei 309 suspension cells	Human α_1-antitrypsin (hAAT)	RAmy3D	Secreted, 85 mg L^{-1} in shake flask, 25 mg L^{-1} in bioreactor	39
Tobacco cv BY-2 suspension cells	BiscFv antibody fragment	Enhanced CaMV 35S	Cytosolic (at detection limit), apoplast-targeted (up to 0.0064% of TSP), ER-retained (up to 0.064% of TSP)	40
Tobacco cv NT-1 suspension cells	Human granulocyte-macrophage colony-stimulating factor (hGM-CSF)	CaMV 35S	Secreted/targeted to the apoplast, ca. 250 µg L^{-1} extracellular, ca. 150 µg L^{-1} intracellular (based on culture volume)	41

Table 9.1 (continued)

Expression host	Expressed protein	Promoter	Localization, yield	Reference(s)
Tobacco suspension cells initiated from transgenic plants	ScFv antibody fragment	CaMV 35S	Apoplast targeting (sporamin secretion signal) 1 mg L^{-1} extracellular, 5 mg L^{-1} intracellular	42
Rice suspension cells	Human a_1-antitrypsin (hAAT)	*RAmy3D*	Up to 200 mg L^{-1} (calli suspended to 40% (v/v) cell density in induction medium)	43
Soybean cv Williams 82 and tobacco cv NT-1 suspension cells	Hepatitis B surface antigen (HBsAg)	*(ocs)₃mas*	Intracellular up to 22 mg L^{-1} in soybean ca. 2 mg L^{-1} in tobacco	44
Tobacco suspension cells	Human granulocyte-macrophage colony-stimulating factor (hGM-CSF)	CaMV 35S	1.6–6.6 µg mL^{-1} upon homogenizing the entire culture broth	45
Rice cv Taipei 309 suspension cells	Human lysozyme	*RAmy3D*	Intracellular (although *RAmy3D* signal peptide was used), up to 3–4% of TSP	46
Tobacco cv Petite Havana SR1 suspension cells	IL-12	Enhanced CaMV 35S	Secreted, up to 800 µg L^{-1} of supernatant	47
Tomato cv Seokwang suspension cells	Human granulocyte-macrophage colony-stimulating factor (hGM-CSF)	Enhanced CaMV 35S	Secreted, up to 45 µg L^{-1} of supernatant	21
Tobacco cv NT-1 suspension cells	Hepatitis B surface antigen (HBsAg)	*Arabidopsis* ubq3	Secreted, up to 10 µg L^{-1} of particulate HBsAg	48
Tobacco cv BY-2 suspension cells	MAb against HBsAg	CaMV 35S	Secreted, ca. 50/50 between supernatant and cells, total max ca. 15 mg L^{-1}	49
Tobacco cv BY-2 suspension cells	*Desmodus rotundus* Salivary plasminogen activator $a1$ (DSPA$a1$)	Enhanced CaMV 35S	Intracellular, up to 1.5 µg g^{-1} FW and degraded when secreted to the supernatant (3 different signal peptides were used)	50
Tobacco cv BY-2 suspension cells	Thrombomodulin derivate Solulin™	Enhanced CaMV 35S	Intracellular, up to 27 µg g^{-1} FW and secreted, up to 2.1 mg L^{-1} of supernatant (3 different signal peptides were used)	51

TSP: total soluble protein, FW: fresh weight.

whole-plant system for recombinant protein production, and robust expression cassettes are available. However, rice cell lines are now becoming popular due to the availability of the inducible a-amylase promoter, which in direct comparisons performs better than any tobacco promoter [53]. The a-amylase system has been used for the production of GM-CSF [53], alpha-1-antitrypsin [39, 43, 54, 55] and lysozyme [46]. Other proteins have been produced in soybean and tomato suspension cells [21, 44].

9.3
Challenges and Solutions for the Production of Recombinant Proteins

Despite their great promise, several challenges remain to be addressed before plant cell cultures can become commercially viable as a production system. As shown in Table 9.1, many of the proteins that have been produced in cultured plant cells have shown relatively poor yields (<10 mg L^{-1}). In many cases, proteins have been produced efficiently for extended periods (e.g., [32]), but sometimes the yields decrease over time, probably as a result of somaclonal variation and other forms of genetic instability (e.g., [44]). It may be possible to circumvent such limitations by the careful choice of stable, high-producing callus lines (e.g., [46]). Technical obstacles include poor growth rates (at least compared to microbial cultures), problems with culture morphology (e.g., aggregation, tendency for cells to grow on the vessel walls) and shear sensitivity [56–58]. Some of these issues can be minimized through improved fermenter design and culture conditions [59], or though modification of the nutrient supply [60–66]. However, it may also be necessary to consider expression strategies in a broader sense, including the design of the expression construct, the medium composition, and the extraction and purification strategy. We consider these factors in more detail below.

9.3.1
Design of the Expression Construct

In any platform for the expression of recombinant proteins, it is necessary to optimize gene expression to promote accumulation of the desired product. In plant cells, this means fine-tuning all stages of gene and protein expression, starting with transcription and finishing with protein targeting and post-translational modification (see Part IV, Chapter 5). Typically, a strong and constitutive promoter is preferred to maximize the rate of transcription. The cauliflower mosaic virus (CaMV) 35S promoter has been used in most cell lines from dicotyledonous plants [2], although a hybrid *ocs-mas* promoter from *A. tumefaciens* has been used to express hepatitis B virus surface antigen in tobacco and soybean cells [44]. For monocotyledonous plants, in which the CaMV promoter has a lower activity, the constitutive maize ubiquitin-1 (Ubi1) promoter can be used [37]. However, the most popular monocot promoter by far is the rice inducible a-amylase promoter. The native a-amylase protein is secreted abundantly from cultured rice cells in response to sucrose starvation, so the promoter represents an ideal inducible system for use with rice suspension cells (inducible promoters are discussed in a recent review; see Ref. [67]). Proteins that have been expressed using this promoter system are listed in Table 9.1.

While transcriptional activity is important for high product yields, the destination of the recombinant protein probably has a more significant and direct impact (see Part IV, Chapter 6). The protein's destination is governed by targeting information in the expression construct, specifically the presence or absence of a signal peptide (which directs the protein into the secretory pathway), plus other signals that determine whether the protein is actually secreted or instead accumulates within the cell. Various different signal peptides have been used in plant cell culture systems, and those from plants and animals appear to work with equal efficiency. Therefore, when mammalian proteins that are normally secreted from their native cells are expressed in plant cells, the native signal peptide is often retained and used to target the recombinant protein for secretion in its new environment [10, 31, 37, 45]. Where there is no native signal peptide, perhaps because the protein is not secreted from its source tissue, a heterologous plant leader may be included upstream of the transgene [29]. Where the rice α-amylase promoter is used, the rice α-amylase signal sequence is generally also included [39, 46, 53, 54].

Proteins targeted to the secretory pathway in cultured plant cells are generally secreted into the culture medium, a strategy which simplifies protein recovery and purification. In the absence of other sorting information, proteins directed to the secretory pathway in cultured cells will eventually reach the apoplast, from where they will either diffuse through the cell wall and into the culture medium or lodge in the cell wall matrix. The fate of the protein depends mostly on its size, since the pores in the plant cell wall allow the passage of globular proteins less than 20 kDa in size. However, some larger pores are present, and larger proteins can therefore diffuse through the cell wall, albeit at a slow rate [68]. Recombinant proteins with molecular weights significantly in excess of 20 kDa have been recovered efficiently from the culture medium [10, 47, 51, 54], while in other cases even fairly small proteins have remained trapped in the cell wall and have needed to be released by disruption or mild enzymatic digestion [31, 69]. Enzymatic digestion is preferable to cell disruption because the latter releases phenolic substances and proteases that can reduce protein yield, and hence introduces additional processing steps.

9.3.2
Post-translational Modification of Recombinant Proteins in Plants

The destination of a recombinant protein within the cell not only affects the yield of product but also how it is modified. In particular, it is possible to influence how a protein is glycosylated by targeting it for retrieval to the endoplasmic reticulum (ER) (an early part of the secretory pathway), since universal-type high-mannose glycosylation takes place here, whereas all plant-specific modifications take place downstream in the Golgi apparatus [70]. Glycans serve a number of structural and functional roles in a glycoprotein, and often affect protein stability, solubility, folding, biological activity, longevity, interactions with cellular receptors, and immunogenicity (see Part IV, Chapters 2 and 3 and Part VI, Chapter 2). Therefore, where the precise glycan structures are important for a particular glycoprotein, it may be necessary to sacrifice yield and recover the recombinant protein from the intracellular environment by cell disruption rather than secrete an inappropriately modified protein into the medium.

Mammalian glycoproteins produced in plants are glycosylated on the same asparagine (Asn) residues as they are in mammalian cells. The first stage of N-glycan synthesis begins in the ER, involving the co-translational addition of an oligosaccharide precursor ($Glc_3Man_9GlcNAc_2$) to specific Asn residues within the consensus sequence Asn-X-Ser/Thr. The N-glycans then undergo several maturation reactions, including the removal of certain residues in the ER and the addition of further residues in the ER and Golgi apparatus. The steps occurring in the ER are conserved in mammals and plants, but they diverge in the late Golgi apparatus so that core $a(1,6)$-linked fucose and terminal sialic acid residues are added in mammals, whereas bisecting $\beta(1,2)$-xylose and core $a(1,3)$-fucose residues are added in plants. Most of the evidence for differential glycosylation has been obtained through the analysis of antibodies produced in cultured mammalian cells and transgenic plants, but there remains some controversy over the exact capabilities of plants in terms of glycan synthesis (see Part IV, Chapter 5). For example, initial reports suggested that antibodies produced in tobacco were glycosylated in an heterogeneous manner, comprising a mixture of high-mannose-type and complex-type N-glycans, whereas those produced in alfalfa were more homogeneous and contained mostly complex-type glycans [71]. However, more recent studies suggest that this observation may reflect properties of the antibody itself rather than the plant (E. Stoger, unpublished data). Recently, it has been reported that sialylated glycoconjugates can be produced in *Arabidopsis thaliana* suspension cells [72], but this has been disputed and the exact capability of plant cells regarding the synthesis of sialic acid remains unclear [73, 74].

9.3.3
Factors that Affect Protein Stability

Maximizing protein expression and accumulation is only half of the story when considering how to improve the yield of recombinant proteins in plant cell cultures. It is also necessary to minimize protein turnover, which can occur due to protein degradation within the cell and in the extracellular environment [10]. Intracellular degradation probably results from the release of proteases during homogenization, although incomplete folding or assembly may also occur where multimeric or particularly complex proteins are expressed [10]. Extracellular degradation, particularly in the culture medium when the product is secreted, is likely to reflect a collection of underlying factors. Significant causes may include the presence of proteases in the medium [10, 39, 45, 47, 50, 53], the tendency for some proteins to precipitate or aggregate under non-physiological conditions [10], and the adsorption of certain charged proteins to the walls of the reactor vessel [32]. The activity of proteases in the culture medium can be reduced using various strategies such as adding protease-inhibitors or gelatin to act as an alternative substrate [47]. A very successful strategy is to use rice cell suspensions with the transgene under the control of the inducible a-amylase promoter. Rice cells are thought to secrete less protease into the culture medium than tobacco cells, and the promoter can be used to restrict protein expression to a defined production phase which is separate from the growth phase, when most proteases are produced [53].

The most commonly used media for plant cell cultures are MS medium [75], Gamborg's B5 medium [76], and White's medium [77]. The composition of these media has been modified in order to optimize the growth of plant cell cultures for the production of secondary metabolites [78, 79] or recombinant proteins [41]. The recovery of recombinant proteins from plant cell cultures can be improved by adding agents to the medium that interfere with protein degradation and other undesirable processes, while enhancing protein synthesis and secretion (Table 9.2). Because the exact circumstances and requirements differ according to the cell line, expression strategy and product, it is difficult to predict the effect each reagent may have under any particular set of conditions. Medium additives must therefore be tested empirically. Substances that have been used include simple inorganic compounds [81], amino acids [38], dimethylsulfoxide (DMSO) [82], polyethylene glycol [45, 47], heamin [69], polyvinylpyrrolidone (PVP) [10, 31, 32, 45, 47, 81], plant hormones such as gibberellic acid [69], and proteins such as gelatin [45, 47, 81] and bovine serum albumin (BSA) [41]. Some of these products enhance cell growth or survival,

Table 9.2 The effect of medium additives on product yields in cultured plant cells

Expression host	Expressed protein	Medium additive	Effect	Reference(s)
Tobacco cv NT-1 suspension cells	Heavy-chain monoclonal antibody	PVP	Addition of 0.75% PVP 360 000 increased secreted product accumulation 35-fold	32, 33
Tobacco cv BY-2 suspension cells	HGM-CSF	BSA, NaCl	Enhanced secreted product accumulation 100% (BSA), or 50% (NaCl)	41
Tobacco suspension cells initiated from transgenic plants	Mouse IgG$_1$	Brefeldin A	Inhibited secretory pathway and prevented degradation of secreted protein. Increased mAb accumulation 2.7-fold	10
Tobacco suspension cells initiated from transgenic plants	Mouse IgG$_1$	Reducing manganese	In manganese-depleted medium the stability and accumulation of secreted IgG$_1$ was increased ca. 1.7 fold	69
Tobacco cv Petite Havana SR1 suspension cells	Human granulocyte-macrophage colony-stimulating factor (hGM-CSF)	Pluronic antifoam, PEG	Pluronic antifoam increased the growth rate almost 2-fold, PEG-8000 increased hGM CSF accumulation 4-fold	80
Tobacco cv Petite Havana SR1 suspension cells	HGM-CSF	Gelatin, PVP, PEG	2% gelatin resulted in 4.6-fold improvement in yield, PVP and PEG showed no effect	45
Tobacco cv Petite Havana SR1 suspension cells	IL-12 heterodimer	Gelatin, PVP, PEG	2% gelatin increased IL-2 accumulation 7-fold	47

some enhance protein synthesis, some interfere either positively or negatively with secretion, some act to stabilize extracellular proteins, and some either inhibit protease activity or provide alternative substrates for proteases. Care must be exercised because the presence of certain reagents may be counterproductive. For example, BSA and gelatin inhibit proteolysis but can interfere with downstream processing. Additionally, these are derived from animal sources and thus possess the risk of contamination with human pathogens (see Part I, Chapter 6 and Part VII, Chapter 1).

9.4
Process Engineering

Improvements in product yields can be gained not only by construct design and modification of the media, but also by variations in the growth and harvest conditions. Culture process development was initially directed towards dedicated bioreactor designs suitable for very large working volumes or providing adequate agitation at high cell densities with low shear stress [83, 84]. As the focus shifted towards the production of recombinant proteins, different fermentation strategies (batch, fed-batch, repeated batch or draw/fill, continuous culture and perfusion culture) were applied to see if a standardized approach would be useful, particularly vis-à-vis regulatory approval (see Part VII, Chapter 4) [27, 85, 86]. It was soon discovered that genetic stability was difficult to maintain in large-scale cultures [2], so continuous and perfusion fermentation strategies, which are long-term processes, were generally unsuitable. Therefore, batch and fed-batch processes are the most likely to be favored when plant suspension cultures are used

for the production of recombinant proteins on a commercial scale. Another issue arising from this consideration is the absolute necessity to establish cryopreservation techniques and master and working cell banks for commercial processes. Therefore, process development for plant cell cultures has focused on the optimization of media, medium supplements and physical parameters.

Another way to improve yields is continuous product removal by the addition of an inert solid absorber to the fermentation broth, or by using an absorber in a bypass. In the case of recombinant antibodies, this can be achieved simply by cycling the medium through an in-line Protein G column during the fermentation, and eluting the bound antibody at the end of the production cycle (see Part IV, Chapter 16 and Part V, Chapter 1). Using this procedure, the yields of recombinant antibody can be increased up to eight-fold, depending on variables such as the flow rate, culture period and column pH [87]. Similar principles can be applied to any protein expressed with a suitable affinity tag – for example, an epitope tag, a specific affinity partner such as glutathione-S-transferase, or a His_6 tag. The latter has been used to purify granulocyte-macrophage colony-stimulating factor (GM-CSF) produced in tobacco suspension cells by continuous cycling of the culture medium through a column packed with metal affinity resin, resulting in a two-fold increase in yields compared to cultures that were harvested by conventional methods [41, 87].

One aspect of plant cell culture that plays a critical role in determining product yields is the rate of oxygen transfer. Inadequate aeration has a limiting effect on cell growth, and has been shown to limit product yields in, for example, an antibody-producing tobacco cell line grown in shake flasks [88]. However, while increased aeration in a stirred 5-L bioreactor helped to increase the protein yields, excessive aeration caused foaming and had a counterproductive effect on both cell growth and product yields [88].

9.5
Downstream Processing

Downstream processing refers to those post-culture steps which result in the isolation and purification of the product. Regardless of the production system, downstream processing is expensive, accounting for up to 80% of overall production costs (although this depends on the level of purity required and is highest for clinical-grade materials). Downstream processing involves multiple steps, beginning with stages that are host-specific but product-generic and ending with the opposite situation – processes that are host-generic but product-specific (see Part IV, Chapter 16). That is, the early stages of downstream processing are generally similar regardless of the product being purified, but must take into account differences between alternative expression platforms, whereas the last stages are defined very much by the product itself and how it can be separated from unwanted contaminants (see Part I, Chapter 6 and Part VII, Chapter 1). In many cases, it is necessary to develop specific processing steps for each product, although certain classes of product can be isolated using a standardized approach (e.g., the use of affinity chromatography to isolate recombinant antibodies or recombinant proteins expressed with integral epitope tags). Several aspects of early downstream processing must be customized specifically for plant systems, including steps for the removal of fibers, oils and

other by-products, and process optimization for the treatment of different plant species and tissues. However, these issues are much less important in plant cell cultures, which tend to lack the phenolics and other metabolites produced by whole plant tissues, and do not accumulate oils and fibers since these tend to be the products of specialized storage tissues.

In plant cell cultures, the main process decision is whether to extract the protein directly from the cell biomass, or to recover it from the medium. As stated above, secreted proteins are advantageous from a processing perspective because they circumvent several unit operations during the purification process (e.g., cell disruption, which results in the release of proteases, and the removal of debris, which can clog filters and chromatography media therefore interfering with downstream operations) (see Part IV, Chapters 6 and 7). Furthermore, the culture supernatant is much simpler than the intracellular milieu: it has fewer contaminating (competing) proteins and metabolites. On the other hand, the target protein is highly diluted in this environment, so larger volumes of raw material must be processed to extract the product.

Where the product is secreted, it should form the major proteinaceous component of the harvest broth. The best time to harvest will depend on the dynamics of protein expression, secretion and stability, but in general it is beneficial to seek an appropriate harvesting window in the production cycle where product expression and accumulation are maximized but degradation is limited. For example, in batch or fed-batch processes, peak production of the recombinant protein often occurs in the exponential phase, and begins to decline at the point where the cell mass and total protein content of the culture reach their maximum levels [10]. The next operations include clarification of the medium, concentration of the product, and its isolation. Large-scale clarification is generally carried out by a combination of dead-end and/or cross-flow filtration, sometimes preceded by bulk cell mass removal using a decanter, plate separator or (semi)continuous centrifuge. Of these methods, cross-flow filtration generally provides the best clarified feed for packed-bed chromatography, but it is also requires the greatest amount of fine-tuning. Hollow-fiber systems may be more suitable for clarification if the cell line produces substantial amounts of extracellular polysaccharides, as is the case for some transformed BY-2 cell lines. Polysaccharides in the medium can coat membrane surfaces and form a gel, which blocks the filters and traps the desired protein within the gel matrix. Another alternative is expanded bed adsorption (EBA), which may help to capture the target protein from particulate feed material. This technique has been used to capture recombinant proteins from several industrial-scale microbial and animal cell cultures (see Part IV, Chapters 4, 12, and 13), and has recently been adapted for use with plant cells [89]. With further design improvements, EBA could facilitate the simultaneous clarification, concentration, and initial purification of proteins from plant cell fermentation broth or cell extracts [90]. Another alternative is the use of aqueous two-phase systems (ATPS) for cultivation and, eventually, in-situ extraction from plant cell cultures [91].

Where the product must be captured from the intracellular environment, the first downstream processing steps are cell disruption and the removal of debris. Several methods have been used for disruption, including sonication, pressure homogenization, enzymatic treatment, and wet milling. These are all efficient techniques, but the resulting feedstream is much denser and more particulate than decanted medium, and the cell debris and fines generated during homogenization may be more difficult to remove than intact cells. The advantages of intracellular expression for initial downstream processing lie in the smaller volume of the starting material and the generally higher concentration of the target protein, while the major disadvantage is the more complex composition of the feedstream and the liberation of proteolytic and oxidizing substances.

The final operations inevitably involve liquid chromatography steps that selectively elute the desired protein. For some products, such as antibodies, highly selective capture can be achieved immediately using Protein A or Protein G affinity columns. Similarly, proteins with affinity tags can be captured using columns charged their respective ligands (e.g., His_6 and Ni-NTA resin). For other proteins, which may need to be purified using non-affinity-based methods, several rounds of ion-exchange and/or reversed-phase chromatography may be required. As stated above, the first chromatographic steps need the most development with regard to the specific production system, whereas the final steps need tuning to the specific product. For large-scale production, robust and inexpensive chromatography media are used in the initial steps, and some loss of selectivity and resolution is anticipated [92].

9.6
Regulatory Considerations

For the production of clinical-grade proteins, plant cell cultures and all downstream processing operations need to meet the standards that have been set for other biopharmaceutical production systems [93] (see Part VII, Chapter 4). Regulatory guidance for biopharmaceutical production in plants currently exists only as draft legislation, and this does not explicitly discuss plant cell cultures since the legislation was drafted to deal with the growing use of terrestrial plants [94–96]. Furthermore, the use of plant cell cultures also falls under the general GMP guidelines for biotechnological production (e.g., Annex 18 of the EU Guide to Good Manufacturing Practice) (see Part I, Chapter 6 and Part VII, Chapter 1). One of the most important requirements is a thorough description of the genetic background and genetic stability of the host cell, as well as the precise documentation of all events associated with the introduction of the transgene into the plant cell (see Part IV, Chapter 6). However, most of the plant cell lines currently used for recombinant protein expression – including BY-2 and NT-1 – have a long history in the public domain and have not been characterized sufficiently to fulfill GMP requirements, and are not deposited in dedicated cell banking facilities. Since cell banking is a prerequisite for the reliable supply of well-defined starting material, it will be necessary to identify suitable wild-type and transformed cell lines for banking. Also, routine procedures for the cryopreservation of plant cells will have to be developed further [97].

9.7
Conclusions

The use of plant suspension cells for the production of biopharmaceutical proteins has many advantages, including safety, defined growth conditions, containment, and the ability to use continuous-culture strategies. However, some challenges remain to

be met before the commercial success of plant cells is assured. Such challenges include improving the current low yields, optimization of downstream processing, and demonstration of biological equivalence. If these issues can be addressed, then plant cells offer a real potential to compete not only with transgenic plants, but also with other fermenter systems for the commercial production of biopharmaceuticals.

References

1 Doran PM (2000) Foreign protein production in plant tissue cultures. Curr Opin Biotechnol 11: 199–204.
2 Hellwig S, Drossard J, Twyman RM, Fischer R (2004) Plant cell cultures for the production of recombinant diagnostic and therapeutic proteins. Nature Biotechnol 22: 1415–1422.
3 Gomord V, Faye L (2004) Posttranslational modification of therapeutic proteins in plants. Curr Opin Plant Biol 7: 171–181.
4 Knäblein J (2004) Biopharmaceuticals expressed in plants? A new era in the new millennium. In: Müller R, Kayser O (Eds.), Applications in Pharmaceutical Biotechnology, Wiley-VCH, pp. 35–56.
5 Twyman RM, Stoger E, Schillberg S, Christou P, Fischer R (2003) Molecular farming in plants: host systems and expression technology. Trends Biotechnol 21: 570–578.
6 Ma JKC, Drake PMW, Christou P (2003) The production of recombinant pharmaceutical proteins in plants. Nature Rev Genet 4: 794–805.
7 Fischer R, Stoger E, Schillberg S, Christou P, Twyman RM (2004) Plant-based production of biopharmaceuticals. Curr Opin Plant Biol 7: 152–158.
8 Twyman RM, Schillberg S, Fischer R (2005) The transgenic plant market in the pharmaceutical industry. Expert Opin Emerg Drugs 10: 185–218.
9 Hilton MG, Rhodes MJC (1990) Growth and hyoscyamine production of hairy root cultures of *Datura stramonium* in a modified stirred tank reactor. Appl Microbiol Biotechnol 33: 132–138.

10 Sharp JM, Doran PM (2001) Strategies for enhancing monoclonal antibody accumulation in plant cell and organ cultures. Biotechnol Prog 17: 979–992.

11 Archambault J (1991) Large-scale (20-L) culture of surface-immobilized *Catharanthus roseus* cells. Enzyme Microbial Technol 13: 882–892.

12 Kieran PM, MacLoughlin PF, Malone DM (1997) Plant cell suspension cultures: some engineering considerations. J Biotechnol 59: 39–52.

13 Schlatmann JE, ten Hoopen HJG, Heijnen JJ (1996) Large-scale production of secondary metabolites by plant cell cultures. In: DiCosmo F, Misawa M (Eds.), *Plant cell culture secondary metabolism: Toward industrial application*. CRC Press, Boca Raton, FL, pp. 11–52. Appl Biochem Biotechnol 50: 189–216.

14 Wen WS (1995) Bioprocessing technology for plant cell suspension cultures.

15 Desikan R, Hancock JT, Neill SJ, Coffey MJ, Jones OT (1996) Elicitor-induced generation of active oxygen in suspension cultures of *Arabidopsis thaliana*. Biochem Soc Trans 24: 199S.

16 Seki M, Ohzora C, Takeda M, Furusaki S (1997) Taxol (Paclitaxel) production using free and immobilized cells of *Taxus cuspidata*. Biotechnol Bioeng 53: 214–219.

17 Van Der Heijden R, Verpoorte R, ten Hoopen HJG (1989) Cell and tissue cultures of *Catharanthus roseus* (L) Don G. – A literature survey. Plant Cell Tiss Org Cult 18: 231–280.

18 Chen MH, Liu LF, Chen YR, Wu HK, Yu SM (1994) Expression of α-amylases, carbohydrate-metabolism, and autophagy in cultured rice cells is coordinately regulated by sugar nutrient. Plant J 6: 625–636.

19 Daniell T, Edwards R (1995) Changes in protein methylation associated with the elicitation response in cell cultures of alfalfa (*Medicago sativa* L.). FEBS Lett 360: 57–61.

20 Hoehl U, Upmeier B, Barz W (1988) Growth and nicotinate biotransformation in batch cultured and airlift fermenter grown soybean cell suspension cultures. Appl Microbiol Biotechnol 28: 319–323.

21 Kwon TH, Kim YS, Lee JH, Yang MS (2003) Production and secretion of biologically active human granulocyte-macrophage colony stimulating factor in transgenic tomato suspension cultures. Biotechnol Lett 25: 1571–1574.

22 Nagata T, Nemoto Y, Hasezawa S (1992) Tobacco BY-2 cell line as the HeLa cell in the cell biology of higher plants. Int Rev Cytol 132: 1–30.

23 Horsch RB, Fry JE, Hoffmann NL, Eichholtz D, Rogers SG, Fraley RT (1985) A simple and general method for transferring genes into plants. Science 227: 1229–1231.

24 Koncz C, Schell J (1986) The promoter of TL-DNA gene 5 controls the tissue-specific expression of chimeric genes carried by a novel type of *Agrobacterium* binary vector. Mol Gen Genet 204: 383–396.

25 Christou P (1993) Particle gun-mediated transformation. Curr Opin Biotechnol 4: 135–141.

26 Hooker BS, Lee JM, An GH (1990) Cultivation of plant-cells in a stirred vessel – effect of impeller design. Biotechnol Bioeng 35: 296–304.

27 Ten Hoopen HJG, van Gulik WM, Heijnen JJ (1992) Continuous culture of suspended plant cells. In Vitro Cell Dev Biol Plant 28: 115–120.

28 Sijmons PC, Dekker BMM, Schrammeijer B, Verwoerd TC, van den Elzen PJM, Hoekema A (1990) Production of correctly processed human serum albumin in transgenic plants. Bio/Technology 8: 217–221.

29 Firek S, Draper J, Owen MRL, Gandecha A, Cockburn B, Whitelam GC (1993) Secretion of a functional single-chain Fv protein in transgenic tobacco plants and cell suspension cultures. Plant Mol Biol 23: 861–870.

30 Matsumoto S, Ishii A, Ikura K, Ueda M, Sasaki R (1993) Expression of human erythropoietin in cultured tobacco cells. Biosci Biotechnol Biochem 57: 1249–1252.

31 Matsumoto S, Ikura K, Ueda M, Sasaki R (1995) Characterization of a human glycoprotein (erythropoietin) produced in cultured tobacco cells. Plant Mol Biol 27: 1163–1172.

32 Magnuson NS, Linzmaier PM, Gao JW, Reeves R, An GH, Lee JM (1996) Enhanced recovery of a secreted mammalian protein from suspension culture of genetically modified tobacco cells. Protein Express Purif 7: 220–228.

33 LaCount W, An GH, Lee JM (1997) The effect of polyvinylpyrrolidone (PVP) on the heavy chain monoclonal antibody production from plant suspension cultures. Biotechnol Lett 19: 93–96.

34 Francisco JA, Gawlak SL, Miller M, Bathe J, Russell D, Chace D, Mixan B, Zhao L, Fell

HP, Siegall CB (1997) Expression and characterization of bryodin 1 and a bryodin 1-based single-chain immunotoxin from tobacco cell culture. Bioconjug Chem 8: 708–713.

35 Magnuson NS, Linzmaier PM, Reeves R, An GH, HayGlass K, Lee JM (1998) Secretion of biologically active human interleukin-2 and interleukin-4 from genetically modified tobacco cells in suspension culture. Protein Expr Purif 13: 45–52.

36 Sehnke PC, Ferl RJ (1999) Processing of preproricin in transgenic tobacco. Protein Expr Purif 15: 188–195.

37 Torres E, Vaquero C, Nicholson L, Sack M, Stoger E, Drossard J, Christou P, Fischer R, Perrin Y (1999) Rice cell culture as an alternative production system for functional diagnostic and therapeutic antibodies. Transgenic Res 8: 441–449.

38 Fischer R, Liao YC, Drossard J (1999) Affinity-purification of a TMV-specific recombinant full-size antibody from a transgenic tobacco suspension culture. J Immunol Methods 226: 1–10.

39 Terashima M, Ejiri Y, Hashikawa N, Yoshida H (1999) Effect of osmotic pressure on human alpha(1)-antitrypsin production by plant cell culture. Biochem Eng J 4: 31–36.

40 Fischer R, Schumann D, Zimmermann S, Drossard J, Sack M, Schillberg S (1999) Expression and characterization of bispecific single-chain Fv fragments produced in transgenic plants. Eur J Biochem 262: 810–816.

41 James EA, Wang CL, Wang ZP, Reeves R, Shin JH, Magnuson NS, Lee JM (2000) Production and characterization of biologically active human GM-CSF secreted by genetically modified plant cells. Protein Express Purif 19: 131–138.

42 Ramirez N, Lorenzo D, Palenzuela D, Herrera L, Ayala M, Fuentes A, Perez M, Gavilondo J, Oramas P (2000) Single-chain antibody fragments specific to the hepatitis B surface antigen, produced in recombinant tobacco cell cultures. Biotechnol Lett 22: 1233–1236.

43 Huang JM, Sutliff TD, Wu LY, Nandi S, Benge K, Terashima M, Ralston AH, Drohan W, Huang N, Rodriguez RL (2001) Expression and purification of functional human alpha-1-antitrypsin from cultured plant cells. Biotechnol Prog 17: 126–133.

44 Smith ML, Mason HS, Shuler ML (2002) Hepatitis B surface antigen (HBsAg) expression

in plant cell culture: Kinetics of antigen accumulation in batch culture and its intracellular form. Biotechnol Bioeng 80: 812–822.

45 Lee JH, Kim NS, Kwon TH, Jang YS, Yang MS (2002) Increased production of human granulocyte-macrophage colony stimulating factor (hGM-CSF) by the addition of stabilizing polymer in plant suspension cultures. J Biotechnol 96: 205–211.

46 Huang JM, Wu LY, Yalda D, Adkins Y, Kelleher SL, Crane M, Lonnerdal B, Rodriguez RL, Huang N (2002) Expression of functional recombinant human lysozyme in transgenic rice cell culture. Transgenic Res 11: 229–239.

47 Kwon TH, Seo JE, Kim J, Lee JH, Jang YS, Yang MS (2003) Expression and secretion of the heterodimeric protein interleukin-12 in plant cell suspension culture. Biotechnol Bioeng 81: 870–875.

48 Sunil Kuma GB, Ganapathi TR, Revathi CJ, Prasad KSN, Bapat VA (2003) Expression of hepatitis B surface antigen in tobacco cell suspension cultures. Protein Expr Purif 32: 10–17.

49 Yano A, Maeda F, Takekoshi M (2004) Transgenic tobacco cells producing the human monoclonal antibody to hepatitis B virus surface antigen. J Med Virol 73: 208–215.

50 Schiermeyer A, Schinkel H, Apel S, Fischer R, Schillberg S (2005) Production of *Desmodus rotundus* salivary plasminogen activator α1 (DSPAα1) in tobacco is hampered by proteolysis. Biotechnol Bioeng 89: 848–858.

51 Schinkel H, Schiermeyer A, Soeur R, Fischer R, Schillberg S (2005) Production of an active recombinant thrombomodulin derivative in transgenic tobacco plants and suspension cells. Transgenic Res (in press).

52 Knäblein J (2003). Biotech: A New Era In The New Millennium? Fermentation and Expression of Biopharmaceuticals in Plants. SCREENING – Trends in Drug Discovery 4: 14–16.

53 Shin YJ, Hong SY, Kwon TH, Jang YS, Yang MS (2003) High level of expression of recombinant human granulocyte-macrophage colony stimulating factor in transgenic rice cell suspension culture. Biotechnol Bioeng 82: 778–783.

54 Terashima M, Murai Y, Kawamura M, Nakanishi S, Stoltz T, Chen L, Drohan W, Rodriguez RL, Katoh S (1999) Production of functional human alpha(1)-antitrypsin by plant cell

culture. Appl Microbiol Biotechnol 52: 516–
523.

55 Trexler MM, McDonald KA, Jackman AP
(2002) Bioreactor production of human al-
pha(1)-antitrypsin using metabolically regulated plant cell cultures. Biotechnol Prog 18:
501–508.

56 Meijer JJ, ten Hoopen HJG, Vangameren YM,
Luyben KCAM, Libbenga KR (1994) Effects of
hydrodynamic stress on the growth of plant-
cells in batch and continuous-culture. Enzyme
Microbial Technol 16: 467–477.

57 Offringa R, Degroot MJA, Haagsman HJ,
Does MP, Vandenelzen PJM, Hooykaas PJJ
(1990) Extrachromosomal homologous recom-
bination and gene targeting in plant cells after
Agrobacterium-mediated transformation.
EMBO J 9: 3077–3084.

58 Yu SX, Kwok KH, Doran PM (1996) Effect of
sucrose, exogenous product concentration,
and other culture conditions on growth and
steroidal alkaloid production by *Solanum avi-
culare* hairy roots. Enzyme Microbial Technol
18: 238–243.

59 Bohme C, Schroder MB, Jung-Heiliger H,
Lehmann J (1997) Plant cell suspension culture in a bench-scale fermenter with a newly
designed membrane stirrer for bubble-free
aeration. Appl Microbial Biotechnol 48: 149–
154.

60 Sakamoto K, Iida K, Sawamura K, Hajiro K,
Asada Y, Yoshikawa T, Furuya T (1993) Effects
of nutrients on anthocyanin production in cultured cells of *Aralia cordata*. Phytochemistry
33: 357–360.

61 Sato K, Nakayama M, Shigeta J (1996) Cultur-
ing conditions affecting the production of anthocyanin in suspended cell cultures of straw-
berry. Plant Sci 113: 91–98.

62 Ishikawa A, Yoshihara T, Nakamura K (1994)
Jasmonate-inducible expression of a potato ca-
thepsin-D inhibitor-GUS gene fusion in tobacco cells. Plant Mol Biol 26: 403–414.

63 Ketchum REB, Gibson DM, Croteau RB, Shuler L (1999) The kinetics of taxoid accumulation in cell suspension cultures of *Taxus* following elicitation with methyl jasmonate. Biotechnol Bioeng 62: 91–105.

64 McCormack BA, Gregory ACE, Kerry ME,
Smith C, Bolwell GP (1997) Purification of an
elicitor-induced glucan synthase (callose
synthase) from suspension cultures of French
bean (*Phaseolus vulgaris* L.): purification and

immunolocation of a probable M^r-65000 subunit of the enzyme. Planta 203: 196–203.

65 Mirjalili N, Linden JC (1996) Methyl jasmonate induced production of taxol in suspension cultures of *Taxus cuspidata*: Ethylene interaction and induction models. Biotechnol
Prog 12: 110–118.

66 Yukimune Y, Tabata H, Higashi Y, Hara Y
(1996) Methyl jasmonate-induced overproduction of paclitaxel and baccatin III in *Taxus* cell
suspension cultures. Nature Biotechnol 14:
1129–1132.

67 Padidam M (2003) Chemically regulated gene
expression in plants. Curr Opin Plant Biol 6:
169–177.

68 Carpita N, Sabularse D, Montezinos D, Del-
mer DP (1979) Determination of the pore size
of cell walls of living plant cells. Science 205:
1144–1147.

69 Tsoi BMY, Doran PM (2002) Effect of medium
properties and additives on antibody stability
and accumulation in suspended plant cell cultures. Biotechnol Appl Biochem 35: 171–180.

70 Gomord V, Fitchette AC, Lerouge P, Faye L
(2004) Glycosylation of plant-made pharmaceuticals. In: Fischer R, Schillberg S (Eds.),
*Molecular Farming: Plant-made pharmaceuticals
and technical proteins*. Wiley-VCH Verlag
GmbH & Co. KGaA, pp 233–250.

71 Bardor M, Loutelier-Bourhis C, Paccalet T, Co-
sette P, Fitchette AC, Vezina LP, Trepanier S,
Dargis M, Lemieux R, Lange C, Faye L, Lerouge P (2003) Monoclonal C5-1 antibody produced in transgenic alfalfa plants exhibits a
N–glycosylation that is homogenous and suitable for glyco-engineering into human-compa-
tible structures. Plant Biotechnol J 1: 451–462.

72 Shah MM, Fujiyama K, Flynn CR, Joshi L
(2003) Sialylated endogenous glycoconjugates
in plant cells. Nature Biotechnol 21: 1470–1471.

73 Seveno M, Bardor M, Paccalet T, Gomord V,
Lerouge P, Faye L (2004) Glycoprotein sialylation in plants? Nature Biotechnol 22: 1351–
1352.

74 Shah MM, Fujiyama K, Flynn CR, Joshi L
(2004) Glycoprotein sialylation in plants? Re-
ply. Nature Biotechnol 22: 1352–1353.

75 Murashige T, Skoog F (1962) A revised medium for rapid growth and bio assay for tobacco
tissue cultures. Physiol. Plant. 15: 473–497.

76 Gamborg OL, Miller RA, Ojima K (1968) Nutrient requirements of suspension culture of
soyabean root cells. Exp Cell Res 50: 151–158.

77 White PR (1963) *The cultivation of animal and plant cells.* Ronald Press, NY

78 Gao WY, Fan L, Paek KY (2000) Yellow and red pigment production by cell cultures of *Carthamus tinctorius* in a bioreactor. Plant Cell Tiss Org Cult 60: 95–100.

79 Raval KN, Hellwig S, Prakash G, Ramos-Plasencia A, Srivastava A, Buchs J (2003) Necessity of a two-stage process for the production of azadirachtin-related limonoids in suspension cultures of *Azadirachta indica.* J Biosci Bioeng 96: 16–22.

80 Lee SY, Kim DI (2002) Stimulation of murine granulocyte macrophage-colony stimulating factor production by pluronic F-68 and polyethylene glycol in transgenic *Nicotiana tabacum* cell culture. Biotechnol Lett 24: 1779–1783.

81 Wongsamuth R, Doran PM (1997) Production of monoclonal antibodies by tobacco hairy roots. Biotechnol Bioeng 54: 401–415.

82 Wahl MF, An GH, Lee JM (1995) Effects of dimethylsulfoxide on heavy-chain monoclonal antibody production from plant cell culture. Biotechnol Lett 17: 463–468.

83 Singh G, Curtis WR (1995) Reactor design for plant cell suspension cultures. In: Shargool PD, Ngo TT (Eds.), *Biotechnological applications of plant cultures.* CRC Press, Boca Raton, FL, pp 151–183.

84 Tanaka H, Nishijima F, Suwa M, Iwamoto T (1983) Rotating drum fermenter for plant-cell suspension-cultures. Biotechnol Bioeng 25: 2359–2370.

85 Chang, HN, Sim SJ (1995) Extractive plant cell culture. Curr Opin Biotechnol 6: 209–212.

86 van Gulik WM, ten Hoopen HJG, Heijnen JJ (2001) The application of continuous culture for plant cell suspensions. Enzyme Microbial Technol 28: 796–805.

87 James E, Mills DR, Lee JM (2002) Increased production and recovery of secreted foreign proteins from plant cell cultures using an affinity chromatography bioreactor. Biochem Eng J 12: 205–213.

88 Liu F, Lee JM (1999) Effect of culture conditions on monoclonal antibody production from genetically modified tobacco suspension cultures. Biotechnol Bioprocess Eng 4: 259–263.

89 Valdes R, Gomez L, Padilla S, Brito J, Reyes B, Alvarez T, Mendoza O, Herrera O, Ferro W, Pujol M, Leal V, Linares M, Hevia Y, Garcia C, Mila L, Garcia O, Sanchez R, Acosta A, Geada D, Paez R, Vega JL, Borroto C (2003) Large-scale purification of an antibody directed against hepatitis B surface antigen from transgenic tobacco plants. Biochem Biophys Res Commun 308: 94–100.

90 Bai Y, Glatz CE (2003) Capture of a recombinant protein from unclarified canola extract using streamline expanded bed anion exchange. Biotechnol Bioeng 81: 855–864.

91 Choi JW, Cho GH, Byun SY, Kim DI (2001) Integrated bioprocessing for plant cell cultures. Adv Biochem Eng Biotechnol 72: 63–102.

92 Bai Y, Glatz CE (2003) Bioprocess considerations for expanded-bed chromatography of crude canola extract: sample preparation and adsorbent reuse. Biotechnol Bioeng 81: 775–782.

93 Fahrner RL, Knudsen HL, Basey CD, Galan W, Feuerhelm D, Vanderlaan M, Blank GS (2001) Industrial purification of pharmaceutical antibodies: development, operation, and validation of chromatography processes. Biotechnol Genet Eng Rev 18: 301–327.

94 FDA (2002) Guidance for industry. Drugs, biologics, and medical devices derived from bioengineered plants for use in humans and animals. Food and Drug Administration.

95 CPMP (2002) Points to consider on quality aspects of medicinal products containing active substances produced by stable transgene expression in higher plants (CPMP/BWP/764/02), The European Agency for the Evaluation of Medicinal Products (EMEA).

96 Knäblein J, McCaman M (2003) Modern Biopharmaceuticals – Recombinant Protein Expression in Transgenic Plants. SCREENING – Trends in Drug Discovery 6: 33–35.

97 Knäblein J (2005) Plant-based Expression of Biopharmaceuticals. In: Meyers RA (Ed.), *Encyclopedia of Molecular Cell Biology and Molecular Medicine,* 2nd ed., Vol. 10, Wiley & Sons, pp 489–410.

10

Producing Biopharmaceuticals in the Desert: Building an Abiotic Stress Tolerance in Plants for Salt, Heat, and Drought

Shimon Gepstein, Anil Grover, and Eduardo Blumwald

Abstract

Transgenic plant production technology, though relatively a new arena in modern plant science, has an immense potential to change the shape of agriculture and offer new opportunities in the field of modern agrobiotechnology. The use of transgenic plants in expressing recombinant foreign proteins in crop plants demonstrates the feasibility of using this approach for the production of biopharmaceutical products. However, several environmental conditions such as salinity, drought, and extreme temperatures are major limiting factors for plant growth and crop productivity. High-temperature stress is highly unfavorable in optimal growth of plants, but nearly 25% of total arable land is affected by heat and drought stress. It is estimated that more than one-third of all of the irrigated land in the world is presently affected by salinity, and this is *exclusive* of those regions already classified as arid and desert lands (which comprise 25% of the total land of our planet). The problem of drought stress is even more severe and economically damaging. Drought and salinity is predicted to cause serious salinization of more than 50% of all arable land by the year 2050. Moreover, the problem is that conventional plant-breeding tools have been of only partial help so far in alleviating these abiotic stress problems. Recent microarray studies have been employed to examine expression profiles of the whole genomes of some plants in response to saline and drought stress. The use of microarray techniques has significantly accelerated efforts in assigning the functional role of genes involved in plant responses to saline and drought stresses and shed light on possible involvement of regulatory pathways in stress tolerance. This information is used for planning new strategies to produce abiotic, stress-tolerant, transgenic plants and eventually to produce biopharmaceuticals in areas which are today not farmable at all.

Abbreviations

ABA	abscisic acid
ABRE	ABA responsive element
APX	ascorbate peroxidase
BADH	betaine aldehyde dehydrogenase
CaMV	cauliflower mosaic virus
CAT	catalases
CDH	choline dehydrogenase
CE	coupling element
CMO	choline monooxygenase
COD	choline oxidase
COX	cytochrome oxidase

Modern Biopharmaceuticals. Edited by J. Knäblein
Copyright © 2005 WILEY-VCH Verlag GmbH & Co. KGaA, Weinheim
ISBN: 3-527-31184-X

DRE	dehydration response element
GPX	glutathione peroxidase
GSH	glutathione sulfhydryl
GST	glutathione S-transferase
HMW	high-molecular weight
HSE	heat shock element
HSF	heat shock factor
Hsp	heat shock protein
HSR	heat shock response
HVA	LEA family protein
KIRC	potassium inward rectifying channel
KORC	potassium outward rectifying channel
LEA	late embryogenesis abundant
LMW	low-molecular weight
MAS	marker-assisted selection
NAD	nicotinamide adenine dinucleotide
P5CR	pyrroline-5-carboxylate reductase
P5CS	pyrroline-5-carboxylate synthase
PEAMT	phosphoethanolamine N-methyltransferase
ProDH	proline dehydrogenase
PS II	photosystem II
QTL	quantitative trait loci
ROS	reactive oxygen species
SOD	superoxide dismutase
TCA	transgene combining ability
UTR	untranslated region
VIC	Voltage-independent cation channels

10.1
General Comments on Abiotic Stresses

Plants have an inherent ability to adjust to circadian and seasonal fluctuations in environmental variables. These variables are, in fact, advantageous to plants, and plants have selected these as decisive factors in controlling their various physiological attributes such as timing of seed germination, length of the period of vegetative growth, onset of reproductive cycle, flowering intensity, fruit set or whole plant senescence. However, apart from the regular circadian and seasonal variations, there may be certain irregular, rapid and unpredicted disturbances in the environment, and these may result in stressful conditions. For example, a paucity of water for long periods due to lack of irrigation, infrequent rains or lowering of the water table causes drought stress. On the other hand, excess water through rain, cyclones or frequent irrigation results in flooding (also called water-logging, submergence or anaerobic stress). The cultivation of plants on saline soils or frequent irrigation with ground water leads to a build-up in excess salt levels, causing salinity stress. Sudden atmospheric heating or cooling due to transient changes in wind patterns, cloud formation or excessive sunlight results in temperature stress.

Most crop plants have been selected by breeders for yield and yield-related parameters, and not for meeting exigencies caused by different abiotic stress factors. Therefore, crops have a limited ability to adapt to extreme abiotic stresses.

However, abiotic stress such as drought, salinity and extreme temperatures pose serious threats to agriculture and to natural vegetation. The loss of farmable land due to abiotic stresses is directly in conflict with the needs of the world's population, which is projected to increase by 1.5 billion during the next 20 years, and the challenge of maintaining the world food supplies.

Although famine in the world nowadays is originated by complex problems and not only by an insufficient production of food, there is no doubt that the gains in food production provided by the Green Revolution have reached their ceiling, while the world population continues to rise. Therefore, increasing the yield of crop plants in normal soils and in less-productive lands – including salinized and semi-arid lands and in regions of extreme climates – is an absolute requirement for feeding the world.

10.2
Drought and Salt Tolerance

Agricultural productivity is severely affected by water availability, and the damaging effects of water stress have influenced ancient and modern civilizations.

Water stress in its broadest context refers to both, drought and salt stress, which affect virtually almost every aspect of plant physiology and metabolism. Drought due to lack of rainfall or irrigation reduces the amounts of water available for plant growth, whereas the presence of high salt concentrations in saline habitats makes it more difficult for the plant roots to absorb water from the environment. Because plant responses to salt and drought stress are closely related, and the mechanisms for drought and salt tolerance are partially overlapped, both aspects will be reviewed together.

It is estimated that more than one-third of all of the irrigated land in the world is presently affected by salinity. This is exclusive of the regions classified as arid and desert lands (which comprise 25% of the total land of our planet). The problem of drought stress is even more severe and economically damaging. Drought and salinity is predicted to cause serious salinization of more than 50% of all arable land by the year 2050. Stress also plays an important role in determining how soil and climate limit the distribution of plant species. Thus, an understanding of the physiological and biochemical processes that underlie stress injury, adaptation, and acclimation mechanisms of plants to environmental stresses is of immense importance for both the agriculture and the environment.

The need to produce stress-tolerant crops was evident even in ancient periods [1]. However, efforts to improve crop performance under environmental stresses have not yet been very fruitful, mainly because the fundamental mechanisms of stress tolerance in plants remain to be completely understood. A genetic approach to the development of specific stress-tolerant crop varieties requires as a pre-requisite the identification of key genetic determinants of stress tolerance-related genes or quantitative trait loci (QTL). The existence of salt-tolerant plants (halophytes) and differences in salt tolerance between genotypes within salt-sensitive plant (glycophytes) species clearly indicates that there is a genetic basis to salt response.

Two basic genetic approaches are currently being utilized to improve stress tolerance:

- The exploitation of natural genetic variations, either through direct selection in stressful environments or through the mapping of QTLs and subsequent marker-assisted selection.
- The generation of transgenic plants to introduce novel genes or alter expression levels of the existing genes to affect the degree of stress tolerance.

Here, we discuss these approaches and focus on the recent experimentation with transgenic plants that has led to increased salinity and drought tolerance.

10.2.1
Genetics of Salt and Drought Tolerance

Tomato has been a valuable species to analyze the genetic basis of salinity and drought tolerance, because making successful crosses between wild and cultivated tomato plants are relatively simple. Lyon [2] made one of the first attempts to evaluate the inheritance of salt tolerance. An interspecific cross of *Lycopersicon esculentum* and *L. pimpinellifolium* showed that the fruit yield of the hybrid was more sensitive to increasing salt (Na_2SO_4) than either parent. Other crosses of wild and cultivated tomato also suggested a complex genetics. Stem elongation was a dominant trait in hybrids with *L. pennellii*, but not with *L. cheesmanii* as the parent. Total dry matter production of another F_1 hybrid between *L. esculentum* and *L. pennellii* showed hybrid vigor under saline conditions [3].

Analysis of other plant species has also suggested that the genetics of salt tolerance is complex. In rice, sterility – an important factor in yield under saline conditions – is determined by at least three genes [4, 5]. In a genetic analysis, the effects of salinity on the seedling stage and on sterility suggested both additive and dominance effects, some with high heritability [6, 7]. There is also evidence of dominance in the salt tolerance of sorghum. Diallele analysis, based on assessing root tolerance to NaCl in salt-treated as compared with control plants, showed that there were both additive and dominance effects of NaCl [8]. The above examples clearly indicate that salinity tolerance, as in case of growth and yield, appear to be a complex trait.

Early attempts to evaluate the genetic basis of stress tolerance in plants were restricted to simple genetic models. However, with the development of molecular markers, evaluating the inheritance of salinity tolerance became a more tractable problem since specific QTLs could be identified. Through the development of molecular markers, it has become possible not only to determine the genetic basis of the trait, but also to map specific chromosomal segments or QTL and to determine the relative contribution of each QTL to the variance observed for the trait. Genomic maps have been constructed in various crops to exploit genetic diversity [9], tag qualitative and quantitative traits [10], and analyze the stability of detected QTL across different environments [11]. Stable and consistent QTLs provide an excellent opportunity to improve the efficiency of selection, especially for traits controlled by multiple genes and highly influenced by the environment, as is the case for salinity [12]. The efficiency of marker-assisted selection (MAS) is dependent on a number of factors such as the distance of observed QTL from marker loci [12] and the proportion of the total additive variance explained by the QTL [13]. There is considerable evidence to support the view that salt tolerance and its sub-traits are determined by multiple QTLs. In an inter-generic cross of tomato, several QTLs were found associated with fruit yield in plants growing under saline conditions [14], although some of the QTL identified were later shown to be dependent on the parentage of the cross [15]. Some QTL associated with specific aspects of fruit yield were found regardless of whether the plants were grown with or without salt; others were detected only under saline or under non-saline conditions [16]. Other crosses have also identified both stress- (salt and cold) specific and stress-non-specific QTLs. The stress-non-specific QTL generally exhibited larger individual effects and accounted for a greater portion of the total phenotypic variation under each condition than the stress-specific QTL [17]. Recently, Gur and Zamir evaluated the progress in breeding for increased tomato (*Solanum lycopersicum*) yield under drought conditions using genotypes carrying a pyramid of three independent yield-promoting genomic regions introduced from the drought-tolerant, green-fruited wild species *Sola-*

num pennellii. Yield of hybrids was more than 50% higher than that of a control market leader variety under both wet and dry field conditions that received 10% of the irrigation water [18].

10.2.2
Physiological and Biochemical Strategies used by Plants to Maintain Water Balance

Adaptation and acclimation to environmental stresses result from integrated events occurring at all levels of organization, from the molecular and cellular level to the physiological anatomical and morphological level. At the biochemical level, plants alter metabolism in various ways to accommodate environmental stresses, including producing osmoregulatory compounds and altering ion concentrations in the different cellular compartments.

10.2.2.1 Osmotic Adjustment
Osmotic adjustment is based on solute accumulation. Under drought conditions, when the soil dries, plants can extract water only as long as their water potential is lower than that of the soil water. Osmotic adjustment is one of the strategies that plants use to maintain water balance under drought or salt stress. Osmotic adjustment is the process by which accumulation of solutes by cells causes increase in osmotic pressure and, as a result, this prevents water loss and the accompanying reduction in cellular turgor. Most of the adjustments can be attributed to increase in concentrations of various regular solutes, including inorganic ions (especially K^+) and organic molecules such as, sugars, organic acids, and amino acids. Since cytosolic enzymes can be severely inhibited by high concentrations of ions, most of the ions accumulated in the vacuoles during

osmotic regulation and are kept away from cytosolic and other subcellular organelles. However, in order to maintain water potential equilibrium within the cell, the synthesis and accumulation of specific organic solutes occurs. These solutes – designated as compatible solutes – are organic compounds that do not interfere with enzyme functions.

Similarly to drought stress, salinity imposes a water-deficit that results from the relatively high solute concentrations in the soil, but in addition it may cause ion-specific stresses resulting from altered K^+/Na^+ ratios, and also may lead to a build-up in Na^+ and Cl^- concentrations that are detrimental to plants. Plants respond to salinity using two different types of responses. Salt-sensitive plants restrict the uptake of salt and adjust their osmotic potential by the synthesis of compatible solutes (proline, glycinebetaine, sugars, etc.) [19].

Salt-tolerant plants sequester and accumulate salt into the cell vacuoles, controlling the salt concentrations in the cytosol and maintaining a high cytosolic K^+/Na^+ ratio in their cells [20]. Clearly, ion-exclusion mechanisms could provide a degree of tolerance to relatively low NaCl concentrations, but would not function at high salt concentrations; this would result in the inhibition of key metabolic processes, with concomitant growth inhibition. Here, we discuss three key processes that contribute to salt tolerance at the cellular level: 1) the establishment of cellular ion homeostasis; 2) the synthesis of compatible solutes for osmotic adjustment; and 3) the increased ability of the cells to neutralize reactive oxygen species generated during the stress response (Fig. 10.1).

Ion homeostasis This is a strategy for adaptation to salinity stress and for the creation of transgenic plants. Although

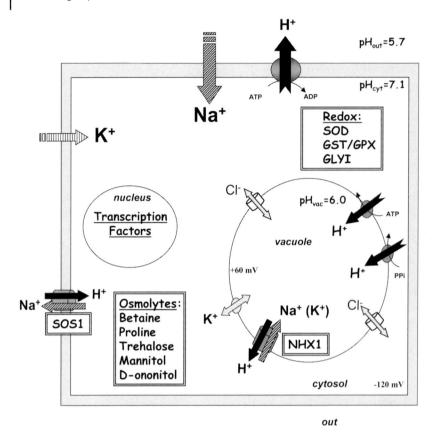

Fig. 10.1 Overview of plant responses to salinity and drought stresses. Stress signals activate several molecular responses to re-establish cellular homeostasis. The stress-responsive mechanisms include, selective ion transport across membranes of different cellular compartments, gene activation by specific transcription factors, synthesis of compatible compounds (betaine, proline, etc.) and activation of detoxification mechanisms against reactive oxygen species (ROS). An important aspect of salinity stress in plants is the stress-induced production of ROS, including superoxide radicals (O_2^-), hydrogen peroxide (H_2O_2) and hydroxyl radicals (OH·).

Na^+ is required in some plants (particularly halophytes [20]), a high NaCl concentration is a toxic factor for plant growth. The alteration of ion ratios in plants is due to the influx of Na^+ through pathways that function in the acquisition of K^+ [21]. The sensitivity to salt of cytosolic enzymes is similar in both glycophytes and halophytes, indicating that the maintenance of a high cytosolic K^+/Na^+ concentration ratio is a key requirement for plant growth in high salt [20]. Strategies that plants could use in order to maintain a high K^+/Na^+ ratio in the cytosol include: 1) extrusion of Na^+ ions out of the cell; and 2) vacuolar compartmentation of Na^+ ions. Under typical physiological conditions, plants maintain a high cytosolic K^+/Na^+ ratio. Given the negative membrane potential difference at the plasma membrane (–140 mV) [22], a rise in extracellular Na^+ concentration will establish a large electrochemical gradient, favoring the passive transport of Na^+ into the cells.

Three classes of low-affinity K^+ channels have been identified. Inward rectifying channels (KIRC), such as AKT1 [23], activate K^+ influx upon plasma-membrane hyperpolarization and they display a high K^+/Na^+ selectivity ratio. A knockout mutant of *AKT1* in *Arabidopsis* (akt1-1) displayed similar sensitivity to salt as the wild-type, suggesting that this channel does not play a role in Na^+ uptake [24]. K^+ outward rectifying channels (KORCs) could play a role in mediating the influx of Na^+ into plant cells. KORC channels showed a high selectivity for K^+ over Na^+ in barley roots [25], and a somewhat lower K^+/Na^+ selectivity ratio in *Arabidopsis* root cells [26]. These channels, which open during the depolarization of the plasma membrane (i.e., upon a shift in the electrical potential difference to more positive values), could mediate the efflux of K^+ and the influx of Na^+ ions [27]. Voltage-independent cation channels (VIC) in plant plasma membranes have been reported [28, 29]. These channels have a relatively high Na^+/K^+ selectivity, are not gated by voltage, and provide a pathway for the entry of Na^+ into plant cells [27].

Sodium ions can enter the cell through a number of low- and high-affinity K^+ carriers, amongst which is AtHKT1 from *Arabidopsis*. This was shown to function as a selective Na^+ transporter and, to a lesser extent, to mediate K^+ transport [30]. AtHKT1 was identified as a regulator of Na^+ influx in plant roots. This conclusion was based on the capacity of *hkt1* mutants to suppress Na^+ accumulation and sodium hypersensitivity in a *sos3* (salt-overly-sensitive) mutant background [31], suggesting that AtHKT1 is a salt-tolerance determinant that controls the entry of Na^+ into the roots.

Na^+ extrusion from plant cells is powered by the operation of the plasma membrane H^+-ATPase generating an electrochemical H^+ gradient that allows plasma membrane Na^+/H^+ antiporters to couple the passive movement of H^+ inside the cells, along its electrochemical potential, to the active extrusion of Na^+ [21]. Recently, *AtSOS1* from *Arabidopsis thaliana* has been shown to encode a plasma membrane Na^+/H^+ antiport with significant sequence similarity to plasma membrane Na^+/H^+ antiporters from bacteria and fungi [32]. The overexpression of *SOS1* improved the salt tolerance of *Arabidopsis*, demonstrating that improved salt tolerance can be attained by limiting Na^+ accumulation in plant cells [33] (Table 10.1). The compartmentation of Na^+ ions into vacuoles also provides an efficient mechanism to avert the toxic effects of Na^+ in the cytosol. The transport of Na^+ into the vacuoles is mediated by a Na^+/H^+ antiporter that is driven by the electrochemical gradient of protons generated by the vacuolar H^+-translocating enzymes, the H^+-ATPase and the H^+-PPiase [44]. The overexpression of a *AtNHX1*, a vacuolar Na^+/H^+ antiporter from *Arabidopsis*, in *Arabidopsis* resulted in transgenic plants that were able to grow in high salt concentrations [34]. The paramount role of Na^+ compartmentation in plant salt tolerance has been further demonstrated in transgenic tomato plants overexpressing AtNHX1 [35]. The transgenic tomato plants grown in the presence of 200 mM NaCl were able to grow, flower, and set fruit. Although the leaves accumulated high sodium concentrations, the tomato fruits displayed very low amounts of sodium [35]. Similar results were obtained with transgenic *Brassica napus* (Canola) overexpressing *AtNHX1* [36]. Leaves of transgenic plants grown in the presence of 200 mM NaCl accumulated sodium to up to 6% of their dry weight, but the seed yields and

Table 10.1 Salt tolerance in transgenic plants expressing genes involved in ion transporters

Gene	Gene product	Source	Cellular role(s)	Target plant	Parameter studied	Reference
AtNHX1	Vacuolar Na^+/ H^+ antiporter	*A. thaliana*	Na^+ vacuolar sequestration	*Arabidopsis*	Biomass	34
AtNHX1	Vacuolar Na^+/ H^+ antiporter	*A. thaliana*	Na^+ vacuolar sequestration	Tomato	Biomass, and fruit production	35
AtNHX1	Vacuolar Na^+/ H^+ antiporter	*A. thaliana*	Na^+ vacuolar sequestration	*B. napus*	Biomass, and oil production	36
AtNHX1	Vacuolar Na^+/ H^+ antiporter	*A. thaliana*	Na^+ vacuolar sequestration	Maize	Biomass	37
GhNHX1	Vacuolar Na^+/ H^+ antiporter	*Gossypium hirsutum*	Na^+ vacuolar sequestration	Tobacco	Biomass	38
AgNHX1	Vacuolar Na^+/ H^+ antiporter	*Atriplex gmelini*	Na^+ vacuolar sequestration	Rice	Biomass	39
AtSOS1	Plasma membrane Na^+/H^+ antiporter	*A. thaliana*	Na^+ extrusion	*Arabidopsis*	Biomass	33
AVP1	Vacuolar H^+-pyrophosphatase	*A. thaliana*	Vacuolar acidification	*Arabidopsis*	Biomass	40
HAL1	K^+/Na^+ transport regulation	*S. cerevisiae*	K^+/Na^+ homeostasis	Tomato melon *Arabidopsis*	Ion content, plant growth	41–43

oil quality were not affected, demonstrating the potential use of this technology for agricultural use in saline soils. Similar results have been reported in other species. The introduction of a vacuolar Na^+/H^+ antiporter from the halophyte *Atriplex gmelini* conferred salt tolerance in rice [39]. Most recently, the overexpression of *GhNHX1* from cotton into tobacco plants [38] and the overexpression of *AtNHX1* in maize [37] resulted in enhanced salt tolerance.

Additional evidence supporting the role of vacuolar transport in salt tolerance has been provided by *A. thaliana* plants overexpressing a vacuolar H^+-PPiase [40]. Transgenic plants overexpressing *AVP1*, coding for the vacuolar H^+-pyrophosphatase, displayed enhanced salt tolerance that was correlated with the increased ion content of the plants. These results suggest that the enhanced vacuolar H^+-pumping in the transgenic plants provided additional driving force for vacuolar sodium accumulation via the vacuolar Na^+/H^+ antiporter.

Synthesis of compatible solutes and production of stress-tolerant plants The cellular response of salt- and drought-tolerant organisms to both long- and short-term salinity stresses includes the synthesis and accumulation of a class of osmoprotective compounds known as compatible solutes. These relatively small, organic osmolytes include amino acids and derivatives, polyols and sugars, and methylamines. The osmolytes stabilize proteins and cellular structures, and can also increase the osmotic pressure of the cell [45]. This response is homeostatic for cell water status and protein integrity, which is perturbed in the face of soil solutions containing higher amounts of NaCl and the consequent loss of water from the cell. The accumulation of osmotically active compounds in the cytosol increases the osmotic pressure to provide a balance between the apoplastic (extracellular) solution, which itself becomes more concentrated with Na^+ and Cl^- ions, and the vacuolar lumen, which in halophytes can accumulate up to 1 M Na^+ (and Cl^-). For a short-term stress, this may provide the cells with the ability to prevent water loss. However, for continued growth under salinity stress, an osmotic gradient (towards the cytosol) must be kept in order to maintain turgor, water uptake and facilitate cell expansion.

The enhancement of proline and glycinebetaine synthesis in target plants has received much attention [46]. Two themes have emerged from the results of these efforts:

- there are metabolic limitations on the absolute levels of the target osmolyte that can be accumulated; and
- the degree to which the transformed plants are able to tolerate salinity stress is not necessarily correlative with the amounts of the osmoprotectants attained.

The metabolic limitations on increasing the concentration of a given osmoprotectant are well illustrated with both proline and glycinebetaine. Initial strategies aimed at engineering higher concentrations of proline began with the overexpression of genes encoding the biosynthetic enzymes pyrroline-5-carboxylate (P5C) synthase (P5CS) and P5C reductase (P5CR) that catalyze the two steps between the substrate (glutamic acid) and the product (proline). P5CS overexpression in tobacco dramatically elevated free proline in transgenic to-

bacco [47] (Table 10.2). However, the regulation of free proline does not appear to be straightforward. Proline catabolism, via proline dehydrogenase (ProDH), is up-regulated by free proline, and there exists a strong evidence that free proline inhibits P5CS [64]. Further, a two-fold increase in free proline was achieved in tobacco plants transformed with a P5CS modified by site-directed mutagenesis [65]. This modification alleviated the feedback inhibition by proline on the P5CS activity and resulted in an improved germination and growth of seedlings under salt stress. Free cellular proline levels are also transcriptionally and translationally controlled. P5CR promoter analysis revealed that P5CR transcripts have reduced translational initiation. A 92-bp segment of the 5′ untranslated region (UTR) of P5CR was sufficient to provide increased mRNA stability and translational inhibition under salt stress to the GUS reporter gene that was ligated at the 3′ end to this small region [66]. These results highlighted the complex regulation of P5CR during stress, and emphasized the importance of stability and translation of P5CR mRNA during salt stress. An alternative approach to attain significant free proline levels, where antisense cDNA transformation was used to decrease ProDH expression, was utilized [55]. Levels of proline in the transgenic Arabidopsis were twice (100 µg g^{-1} fresh weight) that of control plants grown in the absence of stress, and three times higher (600 µg g^{-1} fresh weight) than in control plants grown under stress. The high levels of proline were correlated with an improvement in tolerance to salinity, albeit for a short duration exposure to 600 mM NaCl.

Considerably more experimentation has been directed at the engineering of glycinebetaine synthesis than for any other compatible solute. Unlike proline, glycine-

Table 10.2 Salt and drought tolerance in transgenic plants expressing genes involved in osmolyte biosynthesis

Gene	Gene product	Source	Cellular role(s)	Target plant	Parameter studied	Reference
betA	Choline dehydrogenase	E. coli	Glycine-betaine	Tobacco	Dry weight	48, 49
BADH	Betaine dehydrogenase	Atriplex hortensis	Glycine-betaine	Tomato	Root growth	50
EctA, EctB, EctC	L-2,4-diamino-butyric acid acetyltransferase, L-2,4-diamino-butyric acid trans-aminase, L-ectoine synthase	Halomonas elongata	Ectoyne	Tobacco	Salinity tolerance	51
OtsA OtsB	Trehalose-6-P synthase Trehalose-6-P phosphatase	E. coli	Trehalose	Tobacco Rice	Increased biomass, morphogenesis, growth	52, 53
TPS1	Trehalose-6-P synthase	S. cerevisiae	Trehalose	Tobacco	Improved drought tolerance	54
P5CS	$\Delta\mu^1$-Pyrroline-5-carbo xylate synthase	V. aconitifolia	Proline	Tobacco	Increased proline; plant growth	47
ProDH	Proline dehydro-genase	Arabidopsis thaliana	Proline	Arabidopsis	Inflorescence lodging in response to NaCl stress	55
IMT1	Myo-inositol-O-methyl transferase	M. chrystalli-num	D-Ononitol	Tobacco	Seed germi-nation	56, 57
COD1; COX	Choline oxidase	A. globiformis; A. panescens	Glyc-inebetaine	Arabidopsis Rice Brassica	Seed germi-nation; plant growth	58–62
HAL3	FMN-binding protein	S. cerevisiae	K^+/Na^+ homeostasis	Arabidopsis	Seedlings	63

betaine degradation is not significant in plants [67], but the problems of metabolic fluxes, compounded with the compartmentation of the substrate and product pools, has made the engineering of appreciable levels of glycinebetaine problematic. In plants that are naturally glycinebetaine accumulators (e.g., spinach and sugarbeet), synthesis of this compound occurs in the chloroplast, with two oxidation reactions from choline to glycinebetaine. The first oxidation to betaine aldehyde is catalyzed by choline monooxygenase (CMO), an iron-sulfur enzyme. Betaine aldehyde oxidation to glycinebetaine is catalyzed by betaine aldehyde dehydrogenase (BADH), a non-specific soluble aldehyde dehydrogenase [68]. In *E. coli*, these reactions are cytosolic; in this species, the first reaction is catalyzed by the protein encoded by the *betA* locus choline dehydrogenase (CDH), which is an NAD^+-dependent enzyme, and BADH in *E. coli*, is encoded by the *betB* locus. In *Arthrobacter globiformis*, the two oxidation steps are catalyzed by one enzyme choline oxidase (COD), which is encoded by the *codA* locus [69]. The codA gene of *A. globiformis* offers an attractive alternative to the engineering of glycinebetaine synthesis as it necessitates only a single gene transformation event. This strategy was employed for engineering glycinebetaine synthesis in *Arabidopsis* [70]. The 35S promoter driven construct for transformation included the transit peptide for the small subunit of Rubisco so that the COD protein would be targeted to the chloroplast. Improved salinity tolerance was obtained in transgenic *Arabidopsis* that accumulated, as a result of the transformation, $1 \, \mu mol \, g^{-1}$ fresh weight glycinebetaine. The same construct was used by for transformation of *Brassica juncea* [62] and tolerance to salinity during germination and seedling establishment was improved markedly in the transgenic lines. COX from *Arthrobacter panescens*, which is homologous to the *A. globiformis* COD, was used to transform *Arabidopsis*, *Brassica napus* and tobacco [61]. This set of experiments differs from those above in that the COX protein was directed to the cytoplasm and not to the chloroplast. Improvements in tolerance to salinity and drought and freezing were observed in some transgenics from all three species, but the tolerance was variable. The levels of glycinebetaine in the transgenic plants were not significantly higher than those of wild-type plants, but increased significantly with the exogenous supply of choline to plants, suggesting that the supply of choline is a significant constraint on the synthesis of glycinebetaine [61].

Two important issues emerge from the results of the above discussion. The first is that the concentrations of glycinebetaine in the transgenic plants were much lower than the concentrations noted in natural accumulators. Despite the fact that these levels are not high enough to be osmotically significant, a moderate (and significant) increase in tolerance to salinity and other stresses was conferred. This raises the possibility that the protection offered by glycinebetaine is not only osmotic, which is a point raised by several of the above groups. [71]. Compatible solutes (including mannitol) may also function as scavengers of oxygen radicals, which may be supported by the results of Alia et al. [72], where the protection to photosystem II in plants expressing codA was observed. An alternative possibility, not necessarily exclusive of the first, is that the increased level of peroxide generated by the COD/COX oxidation of choline causes an up-regulation of ascorbate peroxidase and catalase [49] which may also improve the tolerance to salinity stress [46]. The second issue is that the level of glycinebetaine production in the transgenics is limited by choline. Because betaine synthesis takes place in the chloroplast, the free choline pool may not reflect its availability to the chloroplast, which may be limited in this compartment by the activity and/or abundance of choline transporters. However, a dramatic increase in glycinebetaine levels (to $580 \, \mu mol \, g^{-1}$ dry weight in *Arabidopsis*) was shown in the transgenic plants when they were supplemented with choline in the growth medium [61]. This limitation was not explored in the transgenic tobacco expressing *E. coli* enzymes CDH and BADH in the cytoplasm [49]. Although these transgenic plants demonstrated an improved tolerance to salinity, glycinebetaine levels were on the order of those mentioned above. Sakamoto and Murata [69] also asserted that despite the similarities in tolerance exhibited by transgenic plants engineered to synthesize betaine in either the chloroplast or cytoplasm, the site of synthesis of betaine may play a role in the degree of tolerance shown. Indeed, if the betaine present in these plants is localized primarily in the chloroplast, it may be present at significant concentrations (50 mM) [58]. However, Sakamoto and Murata [69] downplayed the limitation of the metabolic pool of choline on the levels of glycinebetaine obtained in the engineered plants, by suggesting that the choline-oxidizing activity may be the limiting factor. This argument seems to be supported by Huang et al. [61], who found that the levels of glycinebetaine correlated with the levels of COX activity measured in each plant. The increase in glycinebetaine with exogenous choline argues against this notion. Stronger evidence for the limitations of choline metabolism have been presented by McNeil et al. [73]. By over-expressing spinach phosphoethanolamine N-methyltransferase (PEAMT), which catalyzes the three methylation reactions required for the conversion of phosphoethanolamine to phosphocholine, up to a 50-fold increase in free choline was obtained. This led to an increase in glycinebetaine levels (+60%) in plants that were expressing spinach CMO and BADH in the chloroplast. Further, the addition of ethanolamine to the plant growth medium caused an increased choline and glycinebetaine levels, showing that the metabolic flux through this pathway is also limited by the supply of ethanolamine. As PEAMT is itself inhibited by phosphocholine, further engineering efforts need to include:

- the modification of PEAMT to remove this inhibition [73];

- increasing the supply of ethanolamine by overexpression of serine decarboxylase; and
- resolving the compartmentation problem of choline supply and choline oxidation, either by use of choline oxidation in the cytoplasm or by finding the appropriate transporters to improve choline supply to the chloroplast [46].

Finally, as the compatible solutes are nontoxic, the interchangeability of these compounds between species has held much interest (see Table 10.1). The recent examples include the engineering of: 1) ectoine synthesis with enzymes from the halophilic bacterium *Halomonas elongata* in plants [51,74]; and 2) trehalose synthesis which occurs in bacteria, yeast, and in extremely desiccation-tolerant plants [75] into potato [76] and rice [53]. An intriguing report on the improved tolerance to salinity in tobacco expressing yeast invertase in the apoplast highlights the potential of manipulating sucrose metabolism [77]. The overexpression of polyols, such as mannitol [78] and D-ononitol [57], have been shown to contribute to enhanced drought and salt tolerance in transgenic tobacco plants.

10.2.2.2 Antioxidant Protection Improves Salt Tolerance

An important aspect of salinity stress in plants is the stress-induced production of reactive oxygen species (ROS), including superoxide radicals (O_2^-), hydrogen peroxide (H_2O_2), and hydroxyl radicals ($OH\cdot$). ROS are a product of altered chloroplast and mitochondrial metabolism during stress. These ROS cause oxidative damage to different cellular components including membrane lipids, proteins, and nucleic acids [79]. The alleviation of this oxidative damage could provide enhanced plant resistance to salt stress. Plants use low molecular mass antioxidants such as ascorbic acid and reduced glutathione (GSH) and employ a diverse array of enzymes such as superoxide dismutases (SOD), catalases (CAT), ascorbate peroxidases (APX), GSH-S-transferases (GST), and GSH peroxidases (GPX) to scavenge ROS. Transgenic rice overexpressing yeast mitochondrial Mn-dependent SOD displayed enhanced salt tolerance [80] (Table 10.3). The overexpression of a cell wall peroxidase in tobacco plants improved germination under osmotic stress [82]. Transgenic tobacco plants overexpressing both GST and GPX displayed improved seed germination and seedling growth under stress [84]. Subsequent studies [83] demonstrated that, in addition to increased GST/GPX activities, the transgenic seedlings contained higher levels of GSH and ascorbate than wild-type seedlings, showed higher levels of monodehydroascorbate reductase activity, and the GSH pools were more oxidized. These results would indicate that the increased

GSH-dependent peroxidase scavenging activity and the associated changes in GSH and ascorbate metabolism led to reduced oxidative damage in the transgenic plants and contributed to their increased salt tolerance. During salt stress, plants display an increase in the generation of H_2O_2 and other ROS [83,85]. The major substrate for the reductive detoxification of H_2O_2 is ascorbate, which must be continuously regenerated from its oxidized form. A major function of GSH in protection against oxidative stress is the reduction of ascorbate via the ascorbate-GSH cycle, where GSH acts as a recycled intermediate in the reduction of H_2O_2 [86]. Ruiz and Blumwald [87] investigated the enzymatic pathways leading to GSH synthesis during the response to salt stress of wild-type and salt-tolerant *Brassica napus* L. (Canola) plants overexpressing a vacuolar Na^+/H^+ antiporter [36]. Wild-type plants showed a marked increased in the activity of enzymes associated with cysteine synthesis (the crucial step for assimilation of reduced sulfur into

Table 10.3 Salt tolerance in transgenic plants expressing genes involved in redox reactions

Gene	Gene product	Source	Cellular role(s)	Target plant	Parameter studied	Reference
MnSOD	Superoxide dismutase	S. cerevisiae	Reduction of O_2 content	Rice	Photosynthetic electron transport	80
GlyI	Glyoxylase	B. juncea	S-D-Lactoyl-glutathione	Tobacco	Chlorophyll content of detached leaves	81
TPX2	Peroxidase	N. tabaccum	Change cell wall properties	Tobacco	Germination; water retention in seed walls	82
GST GPX	Glutathione S-transferase Glutathione peroxidase	N. tabaccum N. tabaccum	ROS scavenging	Tobacco	Germination and growth	83

organic compounds such as GSH) resulting in a significant increase in GSH content. On the other hand, these activities were unchanged in the transgenic salt-tolerant plants, and their GSH content did not change with salt stress. These results clearly showed that salt stress induced an increase in the assimilation of sulfur, and the biosynthesis of cysteine and GSH aimed to mitigate the salt-induced oxidative stress. The small changes seen in the transgenic plants overexpressing the vacuolar Na^+/H^+ [35] suggested that the accumulation of excess Na^+ in the vacuoles (and the maintenance of a high cytosolic K^+/Na^+ ratio) greatly diminished the salt-induced oxidative stress, highlighting the important role of Na^+ homeostasis during salt stress.

10.2.2.3 Expression of Stress-responsive Genes may Confer Improved Stress Tolerance

The developments of stress-resistant transgenic plants are largely predicted on the capability of the products of overexpressed stress-responsive genes to protect otherwise sensitive plants. However, it must be noted that not all the identified stress-responsive genes have proven successful in conferring stress resistance.

A large group of genes, the expression of which is regulated by osmotic stress, was first identified in seeds during maturation and desiccation, and are known to encode for so-called LEA proteins (named for late embryogenesis abundant). Some of these proteins are known also to increase in vegetative tissues of plants exposed to stresses related to water-deficit phenomena. Although the exact defense mechanism and their function is still unknown, the overexpression of HVA1 – an LEA III family protein in rice and wheat – has been shown to increase tolerance to water-deficit stresses [88, 89]. The LEA proteins are hydrophilic, and their protective role might be associated with their abilities to retain water and to prevent the crystallization of major proteins during desiccation processes.

Recent investigations have focused on the large-scale identification of genes associated with root growth in corn, and results have demonstrated the spatial expression pattern of specific set of genes under water-deficit conditions. These genes are predicted to be involved in processes related to stress tolerance, and also in basic cellular processes associated with cell expansion [90]. The products of the candidate genes related to stress can be classified into two groups: 1) those that can directly protect against stress; and 2) those that are associated with signal transduction in the stress response.

Several of the stress-responsive genes reported in the literature are regulated by abscisic acid (ABA), a plant hormone known to play a crucial role in plant responses to water stress, and its levels increase under these conditions. Studies of the promoters of several stress-induced genes have led to the identification of specific regulatory sequences for genes associated with water-deficient stresses. Some of the promoters contain a six-nucleotide sequence element referred as the ABA responsive element (ABRE) which binds transcriptional factors involved in ABA-dependent gene activation [91]. Additional promoter sequences, called coupling elements (CEs), function with the ABRE to control the expression of ABA-regulated genes. CEs may provide additional regulatory element involved in tissue-specific and temporally regulated patterns.

Not all of these water-stressed responsive genes are induced by ABA. The cis-acting promoter element, the dehydration response element (DRE), plays an important role in regulating gene expression in response to drought, salt and freezing stresses, but not by ABA. It has been shown that the transcription factor DREB1A specifically interacts with the DRE and induces expression of stress-tolerance genes. Moreover, overexpression of the cDNA encoding DREB1A in transgenic *Arabidopsis* and tobacco plants activated the expression of many of these stress-tolerance genes and resulted in improved tolerance to drought, salt loading, and freezing [92, 93]. However, use of the strong constitutive 35S cauliflower mosaic virus (CaMV) promoter for driving the expression of DREB1A resulted also in a problem of severe growth retardation under normal growing conditions. In contrast, expression of DREB1A under the control of the stress-inducible rd29A promoter caused minimal inhibitory effects on plant growth while providing an even greater tolerance to stress conditions than did expression of the gene from the CaMV promoter. These results suggest an optimal combination of a gene and promoter for the production of stress-tolerant transgenic plants. Use of the stress-inducible rd29A promoter rather than the constitutive promoter for overexpressing the DREB1A is an excellent example for improving abiotic stress tolerance of agriculturally important crops by gene transfer [93].

Recent microarray studies have been used to examine the expression profiles of whole genomes of some plants in response to saline and drought stress [94–96]. The use of microarray techniques has significantly accelerated efforts in assigning the functional role of genes involved in plant responses to saline and drought stresses, and will shed light on the possible involvement of regulatory pathways in stress tolerance. This information will be used for planning new strategies to produce abiotic stress-tolerant transgenic plants.

10.3
High-temperature Stress

High-temperature stress is very unfavorable for the optimal growth of plants. Almost 25% of the total arable land worldwide is affected by heat and drought stress. The annual mean air temperature of 23% of the Earth's land surface is above 40 °C [97], and with the increasing concentration of greenhouse gases the Earth's surface temperature is expected to increase further by 1.2–5.8 °C by 2100 AD (*http://www.epa.gov*). This rise in ambient temperature would clearly create a warmer climate in most parts of the world.

In the main, all stages of crop growth are vulnerable to high-temperature stress. Processes such as seed germination, seedling emergence and vigor and survival of the seedling/plants are adversely affected by high-temperature stress during the vegetative phase. High-temperature stress counteracts the developmental process by causing damage to leaf photosynthesis, protein metabolism (including enzyme activities), membrane stability, respiration, and transpiration. The reproductive phase in crops is highly susceptible to supra-optimal temperatures, and results in a diminished crop yield. Pollen viability is one of the principal factors affected negatively by high-temperature stress.

In general, cool-season crops are more sensitive to heat stress than are warm-season annuals. In barley, a high temperature leads to a reduction in tiller number, a shortened duration of the tillering stage, abortion of spikelets, and a decline in sucrose synthase activity and starch deposition. The abortion of flowers is a conspicuous effect of heat stress in brassicas. In common bean, high-temperature stress causes an increase in flower abscission, a reduction in pollen viability, a shriveled appearance of pollen grains, decrease in number of pollen tubes penetrating stigma, and a decline in fertility, poor pod-setting and seed-setting and a high number of parthenocarpic pods. In rice, the major processes affected by heat stress include reduction in pollen viability and floret fertility, a shortened duration of grain filling, a reduction in starch content of grains, and increase packing of starch within the endosperm causing a chalky appearance of kernels.

10.3.1
Heat Shock Response and Heat Shock Proteins

High-temperature stress is, in general, simulated in laboratory experiments by subjecting biological systems to heat shock (HS) treatment. Plants mount resistance to HS stress by eliciting specific metabolic adjustments which, as a whole, are referred to as a heat shock response (HSR) (Fig. 10.2). The temperature required to induce the HSR varies amongst different plant species, but an increase of 5–10 °C over and above the ambient temperature is sufficient to elicit HSR. The molecular basis of HS response was revealed for the first time when Ritossa [98] reported that temperature elevation brings about altered puffing pattern of polytene chromosomes in *Drosophila*. Tissieres et al. [99] for the first time showed that the HS condition results in an altered protein profile in *Drosophila* cells. Further research established that almost all organisms – ranging from bacteria to human – respond to HS by synthesizing a new set of proteins called heat shock proteins (Hsps). The first re-

ports on HS-induced alterations in the protein profile of plants appeared during the 1980s [100].

Hsps are broadly classified on the basis of their molecular weights as high- molecular weight (HMW, 70–100 kDa) and low-molecular weight (LMW, 15–20 kDa) Hsp. The level of accumulation of different Hsps is often variable; for example, under stress conditions the LMW-Hsps may comprise up to 1% of the cellular proteins. During the past 25 years of research, Hsp have been extensively analyzed for their physiological, biochemical, cellular, and molecular properties. Consequently, it has been shown that Hsp are highly conserved proteins – that is, plant Hsps are similar to Hsps reported from animal and microbial cells. Detailed characterization of Hsp in terms of their molecular weights, different inducers that trigger Hsp synthesis, cellular locations, expression characteristics, synthesis under field conditions, cellular levels, and conservation of amino acid sequence have been discussed at length elsewhere [101, 102].

Besides HS, selective plant Hsps are induced in response to different abiotic stresses such as heavy-metal stress, water stress, wounding stress, salt stress, cold shock, and anoxia stress. Plants, in general, survive lethal temperature stress more efficiently after prior exposure to a mild stress as against a direct response to lethal stress. This phenomenon is termed "acquired thermotolerance" [103]. Hsp are believed to be important for the protection of cells against heat injury both in basal thermotolerance (i.e., thermotolerance shown without prior heat shock) as well as in acquired thermotolerance responses.

Over the years, a large number of genes that encode Hsp (referred to as heat shock genes or *hsp*) have been isolated, sequenced, and cloned. This has been

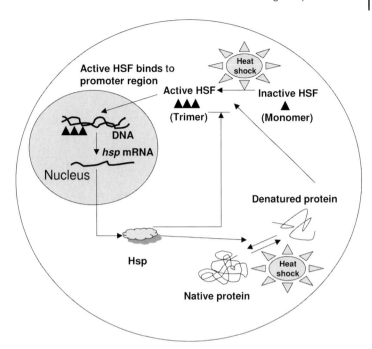

Fig. 10.2 Synthesis and regulation of cytoplasmic heat shock proteins. Plants mount resistance to heat shock (HS) stress by eliciting specific metabolic adjustments, together referred to as heat shock response (HSR). The temperature for the induction of the HSR varies amongst different plant species, but an increase of 5–10 °C over and above the ambient temperature is sufficient to elicit HSR. Heat shock proteins (Hsp) are synthesized in the cytoplasm using hsp mRNA transcribed in the nucleus. Most Hsp act as chaperones and protect cellular proteins from the heat shock (HS)-based denaturation process. Synthesis of hsp mRNA from hsp genes is governed by heat shock factor (HSF). HSF is present in the cytoplasm as inactive molecule (monomer state) that is unable to bind to DNA. Under high-temperature stress, HSF is activated (trimer state). Accumulation of denatured proteins also leads to activation of HSF. Active HSF binds to the promoter regions of *hsp* genes and mediate *hsp* gene transcription. The activity of HSF is down-regulated by Hsp.

achieved from a large number of plant species representing diverse taxonomic classes. The availability of the complete genomic sequence data of *Arabidopsis* has provided vast information on different families of the *hsp* [104,105]. As the genomic sequence of rice is almost complete, and programs to sequence the genomes of several other plant species are in progress, enormous amounts of information on plant *hsp* can be expected in the future. The induction of *hsp* genes is noted to be mediated by specific cis-acting sequences present in the promoters of heat shock genes. These cis-acting sequences are specifically termed as heat shock elements (HSEs). There is clear evidence to show that HSEs interact with positively-acting regulatory proteins called heat shock factors (HSFs) to bring about increased transcription of heat shock genes.

10.3.2
Production of High-temperature-tolerant Transgenic Plants

A detailed understanding of the molecular mechanisms that underlie HSRs in plants (including heat shock genes/proteins, promoters, trans-acting factors and signaling components) has played a vital role in the production of high-temperature-tolerant transgenic plants, as discussed and summarized (Table 10.4) in the following sections.

10.3.2.1 Transgenics Made for Altered Levels of Hsps

Cellular proteins reportedly lose their biological activity upon HS due to aggregation and/or misfolding. There is evidence that the stress-induced accumulation of aggregated and misfolded proteins proves deleterious to the cells, and that the abnormal state of proteins triggers the HSR in living organisms [106]. HS is also known to enhance the synthesis of certain proteases involved in the degradation of abnormal proteins. Hsp are reported to function as molecular chaperones that cooperate as a functional-network in protecting cells against heat damage. Hsp16.9, Hsp17.1, Hsp17.3, and Hsp18.1 are shown to prevent the aggregation or denaturation of proteins during heat shock [107–110]. Hsp100 is shown to rescue the heat-induced protein aggregates by their resolubilization during the recovery phase [111]. Certain other Hsps such as Hsp40, Hsp60, Hsp70, and Hsp90 (alone or in cooperation) are known to stabilize the heat-denatured proteins [111–113].

The level of expression of different Hsps has been genetically manipulated, the aim being to achieve an enhanced thermotolerance in plants. Malik et al. [114] produced transgenic carrot lines and plants in which carrot small Hsp17.7 was overexpressed under the control of CaMV35S promoter (high-strength, constitutive promoter mainly used for expressing alien genes mainly in dicotyledonous plants). Thermotolerance was assessed in terms of cell viability and growth, as well as electrolyte leakage in plants after severe stress. The modified expression of Hsp17.7 in this experiment resulted in an enhanced survival of transgenic cell lines and plants at high temperature. Park and Hong [115] raised transgenic tobacco plants overexpressing tobacco small *hsp*. In these studies, seedlings transformed with sense construct (that leads to synthesis of sense transcript) showed a higher cotyledon opening rate compared to seedlings transformed with antisense construct (that leads to synthesis of antisense transcript). Transgenic rice plants overexpressing *Oshsp17.7* gene have shown an increased thermotolerance as well as increased resistance to UV-B irradiation [116]. Tomato *Lehsp* (M) gene overexpressed in tobacco showed that transgenics transformed with sense construct were more thermotolerant at 48 °C than those transformed with antisense construct [117]. Sun et al. [118] overexpressed *Arabidopsis hsp17.6A* gene in *Arabidopsis*. As a matter of contrast, this study showed that transformants are more tolerant to osmotic stress, but not heat stress.

Apart from LMW-Hsp, there are selective instances where levels of HMW-Hsp have also been manipulated. Queitsch et al. [119] reported the production of transgenic *Arabidopsis* plants by modifying level of Hsp100 protein. In this study, 14-day-old transgenic plants were tested for their performance at high temperature. Transgenic plants survived up to 45 °C temperature stress for 1 h as they showed vigorous growth after the removal of stress. The

Table 10.4 Selective reports on production of high-temperature, stress-tolerant transgenic crops

Gene	Protein	Source	Cellular role(s)	A. Regulatory genes		Comments	Reference
				Trans-host	Promoter used		
Transcription factor genes							
hsf1	Heat shock transcriptional factor 1	A. thaliana	Transcription factor	A. thaliana	CaMV 35S	Transformants exhibited thermotolerance and there was constitutive expression of the *hsp* genes at normal temperature	107
hsf3	-do-	A. thaliana	Transcription factor	A. thaliana	CaMV 35S	Increased thermotolerance	121
hsf1	-do-	L. esculentum	Transcription factor	L. esculentum	CaMV 35S	Overexpressing line showed increased thermotolerance while post-transcriptionally silenced line plants and fruits were thermosensitive	123
hsf3	-do-	A. thaliana	Transcription factor	A. thaliana	CaMV 35S	Increased thermotolerance	122
hsf1b	-do-	A. thaliana	Transcription factor	Lycopersicum (Tomato inbred line L4783)	CaMV 35S	Transgenics showed enhanced thermotolerance and chilling tolerance	124
Signal transduction component genes							
dbf 2	Cell cycle regulated phospho-protein	A. thaliana	Protein kinase	A. thaliana	CaMV 35S	Transformants showed striking tolerance to heat, salt, cold and osmotic stress upon overexpression	125
Structural genes						*Detoxification component genes*	
apx1	Ascorbate peroxidase	H. vulgaris	Peroxisomal ascorbate peroxidase	A. thaliana	CaMV35S	Transformants were significantly more tolerant to heat stress as compared to wild-type	131

Table 10.4 (continued)

Gene	Protein	Source	Cellular role(s)	A. Regulatory genes		Comments	Reference
				Trans-host	Promoter used		
PHGPx	Glutathione peroxidase	L. esculentum	Phospholipid hydroperoxide glutathione peroxidase	N. tabacum (transient expression)	Duplicated PVX coat protein promoter	Transient expression of LePHGPx protected tobacco leaves from salt and heat stress Moreover, stable expression of LePHGPx in tobacco conferred protection against the fungal phytopathogen Botrytis cinerea	132
Fatty acid metabolism genes							
fad 7	Omega-3 fatty acid desaturase	A. thaliana	Causes reduction of trienoic fatty acids and hexadecatrienoic acid	N. tabacum	CaMV 35S	Transformants showing silencing of the gene were bette rable to acclimate to higher temperature	130
Heat shock genes							
hsp17.6A	Heat shock Protein 17.6A	A. thaliana	Molecular chaperone (in vitro)	A. thaliana	CaMV35S	Transformants were tolerant to osmotic stress but not heat stress	118
hsp 17.7	Heat shock protein 17.7	D. carota	Heat shock protein	D. carota	CaMV 35S	Transformants expressed the hsp17.7 gene in the absence of heat shock and showed increased thermotolerance	114
hsp101	Heat shock protein 101	A. thaliana	-do-	A. thaliana	CaMV 35S	Transformants constitutively expressing hsp101 tolerated sudden shifts to extreme temperature better than the controls	119
-do-	-do-	-do-	-do-	O. sativa	Ubi1	Transformants expressing hsp101 showed enhanced tolerance to high temperature	120

Table 10.4 (continued)

Gene	Protein	Source	Cellular role(s)	A. Regulatory genes		Comments	Reference
				Trans-host	Promoter used		
Cytosolic small hsp Sense and antisense	Heat shock protein	N. tabacum	Heat shock protein	N. tabacum	CaMV 35S	Sense seedlings showed higher cotyledon opening rate compared to the antisense seedlings	115
Oshsp17.7	-do-	O. sativa	Heat shock protein	O. sativa	CaMV 35S	Increased thermotolerance, increased resistance to UV-B	116
Lehsp (M) (both sense and antisense)	-do-	L. esculentum	Heat shock protein	N. tabacum	CaMV 35S	Sense transgenics were thermotolerant at 48°C while Antisense plants were severely damaged	117

vector-transformed control plants could not regain growth during the post-stress recovery period. The critical role of Hsp100 in providing thermotolerance was thus shown in this experiment. Katiyar-Agarwal et al. [120] successfully raised transgenic basmati rice overexpressing *hsp100* cDNA. After exposure to different levels of high-temperature stress, the survival rate of transgenic rice plants was compared with that of untransformed control plants. Hsp100 overexpression was seen to provide a distinct growth advantage to transgenic rice during the post-stress recovery period.

10.3.2.2 Transgenics Made for Altered Levels of Regulatory Proteins

As mentioned above, the transcriptional regulation of *hsp* is mediated by HSEs located in the promoter region. HSFs comprise a complex and highly-conserved family of proteins that bind to HSEs and coordinate binding of the RNA polymerase and other related factors so as to actively transcribe *hsp* transcripts. The *hsp* gene induction system has emerged as a powerful target for manipulating levels of Hsps in transgenic experiments. Lee et al. [107] altered the expression of *hsf1* gene in *Arabidopsis thaliana*. Transgenic *Arabidopsis* plants produced in this study were shown to express Hsp even at normal temperatures. It was further noted that transgenic plants could tolerate temperatures up to 50 °C, while wild-type plants were killed above 46 °C stress. Prandl et al. [121] overexpressed *Arabidopsis hsf3* in *Arabidopsis* using a CaMV35 promoter. Transgenic plants in this experiment showed a clearly enhanced thermotolerance. In an independent study, *hsf3* gene overexpressed in *Arabidopsis* also resulted in increased thermotolerance [122]. Mishra et al. [123] overexpressed tomato *hsfA1* gene

in tomato plants. In these studies, overexpressing lines showed an increased thermotolerance, while transgenic lines in which trans-gene was silenced were thermosensitive. By contrast, Li et al. [124] overexpressed *Arabidopsis hsf1B* in tomato and demonstrated an enhanced thermotolerance and chilling tolerance in transgenic plants.

Apart from the transcription factor genes, attempts have also been made to alter signal transduction component genes to manipulate levels of the Hsp. In one such study, Lee et al. [125] raised transgenic *Arabidopsis* plants that overexpressed *Atdbf2* gene encoding for cell cycle-regulated phosphoprotein (that has a cellular role as protein kinase). The transformants in these studies demonstrated a striking tolerance to heat, salt, cold, and osmotic stress.

10.3.2.3 Transgenics Made for Altered Levels of Osmolytes

Manipulating levels of Hsps through *hsp* genes or through regulatory genes that change the expression levels of Hsps (as mentioned in the above two categories) are not the only means by which thermotolerance appears to be governed. Indeed, several other target sites have been identified which, upon manipulation in transgenic experiments, are important in governing the thermotolerance property. As mentioned for salinity and for drought tolerance, there are certain low molecular-weight compounds such as amino acids (e.g., proline), polyamines (e.g., putrescine), quaternary ammonium compounds (e.g., glycinebetaines), sugars (e.g., mannitol) and sugar alcohols (e.g., polyols) that help plants to acclimatize against osmotic stresses. Glycinebetaine protects the photosynthetic machinery by stabilizing the oxygen-evolving photosystem II (PS II) complex [126]. As mentioned earlier, metabolic

engineering for the biosynthesis of glycinebetaine in *Arabidopsis* has been achieved
by introducing the bacterial *codA* gene
(that encodes choline oxidase protein).
When the effect of high-temperature stress
on these transgenics was examined at the
imbibition and germination stage of seeds,
the seeds of transgenic plants were seen to
be more tolerant to heat stress than the
wild-type [127]. The overproduction of glycinebetaine also provided a significant advantage to the growth of young transgenic
seedlings at supra-optimal temperatures.

10.3.2.4 Transgenics Made for Altered Levels of Membrane Lipids

Living cells adapt to changes in extracellular temperature by altering the composition of their membrane lipids. Many years
ago, it was established that membrane
fluidity increases due to increased unsa-
turation of membrane lipids in response
to low temperature [128]. Conversely, the
saturation of membrane lipids is noted to
increase when cells are subjected to supra-
optimal temperatures, thereby increasing
membrane rigidity [129]. Murakami et al.
[130] produced transgenic tobacco plants
in which the gene encoding the chloro-
plast-localized fatty acid desaturase was silenced. Significant reduction in the
amount of trienoic fatty acids in homozygous transgenic lines in comparison to
wild-type plants was observed. Thermotolerance assays revealed that transgenic tobacco plants were resistant to high-temperature stress (41 °C for 2 h), whereas
wild-type plants could not survive such extreme temperature treatment. This report
showed that lowering the content of unsa-
turated fatty acid reduces the sensitivity of
plants to heat stress.

10.3.2.5 Transgenics Made for Altered Levels of Cell Detoxification Proteins

Detoxification pathways play a major role
when plants are subjected to different
abiotic stresses. The components of cell
detoxification mechanisms have been used
in specific experiments to alter thermotolerance responses in transgenic plants.
Overexpression of barley *hvapx1* gene (encoding for peroxisomal ascorbate peroxidase) in *Arabidopsis* caused an increased
thermotolerance in transgenic plants as
compared to wild-type plants [131]. In a recent study, tomato gene encoding for glutathione peroxidase was overexpressed in
tobacco; subsequent transient expression
of the transgene protected transgenic
leaves from salt and heat stress [132].

10.4 Conclusions and Perspectives

The use of a variety of transgenic plants
for the production of biopharmaceuticals
is feasible. However, several environmental
conditions such as salinity, drought, and
extreme temperatures are major limiting
factors for plant growth and crop productivity. High-temperature stress is highly
unfavorable for the optimal growth of
plants, yet almost 25% of the total arable
land worldwide is affected by heat and
drought stress. Moreover, it is estimated
that more than one-third of all irrigated
land worldwide is presently affected by
salinity. The problem of drought stress is
even more severe and economically damaging, the main difficulty being that
drought and salinity together is predicted
to cause serious salinization of more than
50% of all arable land by the year 2050.

Conventional breeding programs for
raising abiotic stress-tolerant genotypes

have met with limited success. In evaluating the possibility of improving stress tolerance in plants, a number of considerations should be considered. First, whilst it has been recognized by many researchers that dramatic changes in gene expression are associated with all types of stresses, the promoters that are most commonly used for transgene introductions are primarily constitutively expressed, including the CaMV35S promoter, ubiquitin, and actin promoters [133]. Recent studies have noted that the overexpression of specific stress-induced genes under the control of stress-induced or tissue-specific promoters often display a better phenotype [134, 135]. Second, whilst there have been a number of successes in the production of abiotic stress-tolerant plants using tobacco or *Arabidopsis*, there is a clear need to begin to introduce these tolerance genes into crop plants. Third, it is likely that the effectiveness of a specific transgene will be based on the specific genetic background into which it is transformed. One component of this is the well-known phenomenon of "position effect", although in addition the ability of a transgene to function may well be determined by the overall genetic background, independently of the chromosomal location of the transgene, referred to as "Transgene Combining Ability" (TCA).

Finally, we also need to establish better comparative systems. At the same time, we need to look at rational concepts for combining genes, just as the researchers of disease resistance are currently doing with gene stacking. For example, the overexpression of AtSOS1 in meristems (non-vacuolated cells) and AtNHX1 (for vacuolar Na^+ accumulation), together with the overproduction of compatible solutes, would not only provide the ability of using NaCl as an osmoticum during vegetative growth but also would provide the seed-lings with the ability to reduce Na^+ toxicity during early growth and seedling establishment. Wherever applicable, genes for protection against oxidative stress must be combined, particularly in actively photosynthesizing cells that are prone to more chloroplast damage due to ROS.

While progress in improving stress tolerance has been slow in the past, there are a number of opportunities and reasons for optimism. Over the past ten years, there has been the development of a number of the functional tools that can allow us to dissect many of the fundamental questions associated with stress tolerance. These include: 1) the development of molecular markers for gene mapping and the construction of associated maps; 2) the development of EST libraries; 3) the complete sequencing of plant genomes including *Arabidopsis*, rice, and maize; 4) the production of T-DNA or transposon-tagged mutagenic populations of *Arabidopsis*; and 5) the development of a number of forward genetics tools that can be used in gene function analysis such as TILLING [136]. And indeed, of late, transgenic plants have been raised that in fact show increased resistance to abiotic stresses such as heat, drought, and salinity. For example, transgenic *Arabidopsis* plants expressing Hsp could tolerate temperatures up to $50\,^{\circ}C$, while wild-type plants were killed above $46\,^{\circ}C$ stress, and transgenic, drought-tolerant, tomatoes showed 50% higher yields under dry field conditions with only 10% of the irrigation water. Also, transgenic plants overexpressing a Na^+/H^+ antiporter could be obtained with remarkable salt tolerance. Transgenic tomato plants for example grown in the presence of 200 mM NaCl were able to grow, flower, and set fruit. Similar results could be obtained with transgenic canola leaves which grew in the presence of 200 mM NaCl and accumulated sodium up to 6% (!) of their dry weight. However, the seed yields and oil quality were not affected – an impressive demonstration of the application of such technology for agricultural use in saline soils.

Clearly, we need to counteract against a changing environment and climate: increases in ambient temperature, drought and salinity. Therefore, we must focus on examining the comparative effects and interaction of specific transgenes within a defined genetic background, combine the improved genetic elements, and determine the efficacy of these approaches in the field.

In the future, we will be able to produce even high-value traits – for example, biopharmaceuticals – in areas of the world which, today, are not farmable at all. Moreover, as recommended by Knäblein, this is what we should focus on because then, at the dawn of this new millennium, we would for the first time be capable of producing sufficient amounts of biopharmaceuticals to treat everybody on our planet [137].

References

1 T. Jacobsen, R. M. Adams, *Science* **1958**, 128, 1251–1258.
2 C. Lyon, *Bot. Gazette* **1941**, 103, 107–122.
3 Y. Saranga, D. Zamir, A. Marani, J. Rudich, *J. Am. Soc. Hort. Sci.* **1991**, 116, 1067–1071.
4 M. Akbar, T. Yabuno, S. Nakao, *Jap. J. Breeding* **1972**, 22, 277–284.
5 M. Akbar, T. Yabuno, *Jap. J. Breeding* **1977**, 27, 237–240.
6 S. Moeljopawiro, H. Ikehashi, *Euphytica* **1981**, 30, 291–300.
7 M. Akbar, G. S. Khush, R. Hille, D. Lambers, *IRRI* **1986**, 399–409.
8 F. M. Azhar, T. McNeilly, *Plant Breeding* **1988**, 101, 114–121.
9 C. E. Thormann, M. E. Ferreira, L. E. A. Camargo, J. G. Tivang, T. C. Osborn, *Theor. Appl. Genet.* **1994**, 88:973–980.
10 D. V. Butruille, R. P. Guries, T. C. Osborn, *Genetics* **1999**, 153, 949–964.
11 S. Hittalmani, H. E. Shahidhar, P. G. Bagali, N. Huang, J. S. Sidhu, V. P. Singh, G. S. Khush, *Euphytica* **2002**, 125, 207–214.
12 J. W. Dudley, *Crop Sci.* **1993**, 33, 660–668.
13 R. Lande, R. Thompson, *Genetics* **1990**, 124, 743–756.
14 M. P. Breto, M. J. Asins, E. A. Carbonell, *Theor. Appl. Genet.* **1994**, 88, 395–401.
15 J. Monforte, M. J. Asins, E. A. Carbonell, *Theor. Appl. Genet.* **1997**, 95, 284–293.
16 J. Monforte, M. J. Asins, E. A. Carbonell, *Theor. Appl. Genet.* **1997**, 95, 706–713.
17 M. R. Foolad, G. Y. Lin, F. Q. Chen, *Plant Breeding* **1999**, 118, 167–173.
18 A. Gur, D. Zamir, *PLoS Biol.* **2004**, 2(10): e245, 0001–0006.
19 H. Greenway, R. Munns, *Annu. Rev. Plant Physiol.* **1980**, 31, 149–190.
20 E. P. Glenn, J. J. Brown, E. Blumwald, *Crit. Rev. Plant Sci.* **1999**, 18, 227–255.
21 E. Blumwald, G. S. Aharon, M. P. Apse, *Biochim. Biophys. Acta* **2000**, 1465, 140–151.
22 N. Higinbotham, *Annu. Rev. Plant Physiol.* **1973**, 24, 25–46.
23 H. Sentenac, N. Bonneaud, M. Minet, F. Lacroute, J. M. Salmon, F. Gaymard, C. Grignon, *Science* **1992**, 256, 663–665.
24 E. P. Spalding, R. E. Hirsch, D. R. Lewis, Q. Zhi, M. R. Sussman, B. D. Lewis, *J. Gen. Physiol.* **1999**, 113, 909–918.
25 L. H. Wegner, K. Raschke, *Plant Physiol.* **1994**, 105, 799–813.
26 F. J. M. Maathuis, D. Sanders, *Planta* **1995**, 197, 456–464.
27 F. J. M. Maathuis, A. Amtmann, *Ann. Bot.* **1999**, 84, 123–133.
28 S. K. Roberts, M. Tester, *J. Exp. Bot.* **1997**, 48, 839–846.
29 H. de Boer, L. H. Wegner, *J. Exp. Bot.* **1997**, 48, 441–449.
30 N. Uozumi, E. J. Kim, F. Rubio, T. Yamaguchi, D. Muto, A. Tsuboi, E. P. Bakker, T. Nakamura, J. I. Schroeder, *Plant Physiol.* **2000**, 122, 1249–1259.
31 A. Rus, S. Yokoi, A. Sharkhuu, M. Reddy, B. H. Lee, T. K. Matsumoto, H. Koiwa, J. K. Zhu, R. A. Bressan, P. M. Hasegawa, *Proc. Natl. Acad. Sci. USA* **2001**, 98, 14150–14155.
32 H. Shi, M. Ishitani, C. Kim, J. K. Zhu, *Proc. Natl. Acad. Sci. USA* **2000**, 97, 6896–6901.

33 H. Shi, B. H. Lee, S. J. Wu, J. K. Zhu, *Nature Biotechnol.* **2003**, 21, 81–85.

34 M. P. Apse, G. S. Aharon, W. A. Snedden, E. Blumwald, *Science* **1999**, 285, 1256–1258.

35 H. X. Zhang, E. Blumwald, *Nature Biotechnol.* **2001**, 19, 765–768.

36 H. X. Zhang, J. N. Hodson, J. P. Williams, E. Blumwald, *Proc. Natl. Acad. Sci. USA* **2001**, 98, 6896–6901.

37 Z. Y. Yin, A. F. Yang, K. W. Zhang, J. R. Zhang, *Acta Bot. Sin.* **2004**, 46, 854–861.

38 A. Wu, G. D. Yang, Q. W. Meng, C. C. Zheng, *Plant Cell Physiol.* **2004**, 45, 600–607.

39 M. Ohta, Y. Hayashi, A. Nakashima, A. Hamada, A. Tanaka, T. Nakamura, T. Hayakawa, *FEBS Lett.* **2002**, 532, 279–282.

40 R. A. Gaxiola, J. Li, S. Unurraga, L. M. Dang, G. J. Allen, S. L. Alper, G. R. Fink, *Proc. Natl. Acad. Sci. USA* **2001**, 98, 11444–11449.

41 M. Bordas, C. Montesinos, M. Dabauza, A. Salvador, L. A. Roig, R. Serrano, V. Moreno, *Trans. Res.* **1997**, 6, 41–50.

42 C. Gisbert, A. M. Rus, M. C. Bolarin, J. M. Lopez-Coronado, I. Arrillaga, C. Montesinos, M. Caro, R. Serrano, V. Moreno, *Plant Physiol.* **2000**, 123, 393–402.

43 S. X. Yang, Y. X. Zhao, Q. Zhang, Y. K. He, H. Zhang, H. Luo, *Cell Res.* **2001**, 11, 142–148.

44 E. Blumwald, *Physiol. Plant.* **1987**, 69, 731–734.

45 P. H. Yancey, M. E. Clark, S. C. Hand, R. D. Bowlus, G. N. Somero, *Science* **1982**, 217, 1214–1222.

46 D. Rontein, G. Basset, A. D. Hanson, *Metabolic Eng.* **2002**, 4, 49–56.

47 P. B. K. Kishor, Z. Hong, G. H. Miao, C. A. A. Hu, D. P. S. Verma, *Plant Physiol.* **1995**, 108, 1387–1394.

48 G. Lilius, N. Holmberg, L. Bulow, *Biotechnology* **1996**, 14, 177–180.

49 K. O. Holmstrom, S. Somersalo, A. Mandal, T. E. Palva, B. Welin, *J. Exp. Bot.* **2000**, 51(343), 177–185.

50 X. Jia, Z. Q. Zhu, F. Q. Chang, Y. X. Li, *Plant Cell Rep.* **2002**, 21, 141–146.

51 H. Nakayama, K. Yoshida, H. Ono, Y. Murooka, A. Shinmyo, *Plant Physiol.* **2000**, 122, 1239–1247.

52 E. A. H. Pilon-Smits, N. Terry, T. Sears, H. Kim, A. Zayed, S. B. Hwang, K. Van Dun, E. Voogd, T. C. Verwoerd, R. W. H. Krutwagen, O. J. M. Goddijn, *J. Plant Physiol.* **1998**, 152, 525–532.

53 K. Garg, J. K. Kim, T. G. Owens, A. P. Ranwala, Y. Do Choi, L. V. Kochian, R. J. Wu, *Proc. Natl. Acad. Sci. USA* **2002**, 99, 15898–15903.

54 C. Romero, J. M. Belles, J. L. Vaya, R. Serrano, F. A. Culianez-Macia, *Planta* **1997**, 201, 293–297.

55 T. Nanjo, M. Kobayashi, Y. Yoshiba, Y. Kakubari. K. Yamaguchi-Shinozaki, K. Shinozaki, *FEBS Lett.* **1999**, 461, 205–210.

56 D. M. Vernon, M. C. Tarczynski, R. G. Jensen, H. J. Bohnert, *Plant J.* **1993**, 4, 199–205.

57 E. Sheveleva, W. Chmara, H. J. Bohnert, R. G. Jensen, *Plant Physiol.* **1997**, 115(3), 1211–1219.

58 H. Hayashi, L. Alia Mustardy, P. Deshnium, M. Ida, N. Murata, *Plant J.* **1997**, 12(1), 133–142.

59 Alia, H. Hayashi, A. Sakamoto, N. Murata, *Plant J.* **1998**, 16(2), 155–161.

60 A. Sakamoto, Alia, N. Murata, *Plant Mol. Biol.* **1998**, 38(6), 1011–1019.

61 J. Huang, R. Hirji, L. Adam, K. L. Rozwadowski, J. K. Hammerlindl, W. A. Keller, G. Selvaraj, *Plant Physiol.* **2000**, 122(3), 747–756.

62 V. S. K. Prasad, P. Sharmila, P. A. Kumar, P. P. Saradhi, *Mol. Breeding* **2000**, 6, 489–499.

63 A. Espinosa-Ruiz, J. M. Belles, R. Serrano, F. A. Culiaez-Macia, *Plant J.* **1999**, 20, 529–539.

64 N. H. Roosens, R. Willem, Y. Li, I. I. Verbruggen, M. Biesemans, M. Jacobs, *Plant Physiol.* **1999**, 121, 1281–1290.

65 Z. L. Hong, K. Lakkineni, Z. M. Zhang, D. P. S. Verma, *Plant Physiol.* **2000**, 122, 1129–1136.

66 X. J. Hua, B. Van De Cotte, M. Van Montagu, N. Verbruggen, *Plant J.* **2001**, 26, 157–169.

67 L. Nuccio, S. D. McNeil, M. J. Ziemak, A. D. Hanson, R. K. Jain, G. Selvaraj, *Metabol. Eng.* **2000**, 2, 300–311.

68 B. Rathinasabapathi, *Ann. Bot.* **2000**, 86, 709–716.

69 A. Sakamoto, N. Murata, *J. Exp. Bot.* **2000**, 51, 81–88.

70 H. Hayashi, Alia, L. Mustardy, P. Deshnium, M. Ida, N. Murata, *Plant J.* **1997**, 12(1), 133–142.

71 H. J. Bohnert, B. Shen, *Sci. Hort.* **1999**, 78, 237–260.

72 Alia, Y. Kondo, A. Sakamoto, H. Nonaka, H. Hayashi, P. P. Saradhi, T. H. H. Chen, N. Murata, *Plant Mol. Biol.* **1999**, 40, 279–288.

73 S. D. McNeil, M. L. Nuccio, M. J. Ziemak, A. D. Hanson. *Proc. Natl. Acad. Sci. USA* **2001**, 98, 10001–10005.

74 H. Ono, K. Sawads, N. Khunajakr, T. Tao, M. Yamamoto, M. Hiramoto, A. Shinmyo, M. Takano, Y. Murooka, *J. Bacteriol.* **1999**, 181, 91–99.

75 O. J. M. Goddijn, K. Van Dun, *Trends Plant Sci.* **1999**, 4, 315–319.

76 E. T. Yeo, H. B. Kwon, S. E. Han, J. T. Lee, J. C. Ryu, M. O. Byu, *Molecules and Cells* **2000**, 10, 263–268.

77 E. Fukushima, Y. Arata, T. Endo, U. Sonnewald, F. Sato, *Plant Cell Physiol.* **2001**, 42, 245–249.

78 M. Tarczynski, R. Jensen, H. Bohnert, *Science* **1993**, 259, 508–510.

79 B. Halliwell, J. M. C. Gutteridge, *Arch. Biochem. Biophys.* **1986**, 246, 501–514.

80 Y. Tanaka, T. Hibin, Y. Hayashi, A. Tanaka, S. Kishitani, T. Takabe, S. Yokota, T. Takabe, *Plant Sci.* **1999**, 148, 131–138.

81 J. Veena, V. S. Reddy, S. K. Sopory, *Plant J.* **1999**, 17, 385–395.

82 I. Amaya, M. A. Botella, M. De La Calle, M. I. Medina, A. Heredia, R. A. Bressan, P. M. Hasegawa, M. A. Quesada, V. Valpuesta, *FEBS Lett.* **1999**, 457, 80–84.

83 V. P. Roxas, S. A. Lodhi, D. K. Garrett, J. R. Mahan, R. D. Allen, *Plant Cell Rep.* **2000**, 41, 1229–1234.

84 V. P. Roxas, R. K. Smith, E. R. Allen, R. D. Allen, *Nature Biotechnol.* **1997**, 15, 988–991.

85 Y. Gueta-Dahan, Z. Yaniv, B. A. Zilinskas, G. Ben-Hayyim, *Citrus. Planta* **1997**, 203, 460–469.

86 C. Foyer, B. Halliwell, *Planta* 1976, 133, 21–25.

87 M. Ruiz, E. Blumwald, *Planta* **2002**, 214, 965–969.

88 D. Xu, X. Duan, B. Wang, B. Hong, T. Ho, R. Wu, *Plant Physiol.* **1996**, 110(1), 249–257.

89 E. Sivamani, A. Bahieldin, J. M. Wraithc, T. Al-Niemia, W. E. Dyera, T. H. D. Hod, R. Qu, *Plant Sci.* **2000**, 155, 1–9.

90 M. Bassani, P. M. Neumann, S. Gepstein, *Plant Mol. Biol.* **2004**, 56(3), 367–380.

91 M. J. Guiltinan, W. R. Marcotte, R. S. Quatrano, *Science* **1990**, 250, 267–271.

92 M. Kasuga, Q. Liu, S. Miura, K. Yamaguchi-Shinozaki, K. Shinozaki, *Nature Biotechnol.* **1999**, 17, 287–291.

93 M. Kasuga, S. Miura, K. Shinozaki, K. Yamaguchi-Shinozaki, *Plant Cell Physiol.* **2004**, 45(3), 346–350.

94 S. Kawasaki, C. Borchert, M. Deyholos, H. Wang, S. Brazille, K. Kawai, D. Galbraith, H. J. Bohnert, *Plant Cell* **2001**, 13, 889–905.

95 M. Seki, J. Ishida, M. Narusaka, M. Fujita, T. Nanjo, T. Umezawa, A. Kamiya, M. Nakajima, A. Enju, T. Sakurai, M. Satou, K. Akiyama, K. Yamaguchi-Shinozaki, P. Carninci, J. Kawai, Y. Hayashizaki, K. Shinozaki, *Funct. Integr. Genomics* **2002**, 2(6), 282–291.

96 B. Sottosanto, A. Gelli, E. Blumwald, *Plant J.* **2004**, 40(5), 752–71.

97 Y. Klueva, E. Maestri, N. Marmiroli, H. T. Nguyen, In: A. S. Basra (Ed.), *Crop Responses and Adaptations to Temperature Stress.* Food Products Press, Binghamton, NY, **2001**, pp. 177–217.

98 M. Ritossa, *Experientia* **1962**, 18, 571–573.

99 A. Tissieres, H. K. Mitchell, U. M. Tracey, *J. Mol. Biol.* **1974**, 84, 389–398.

100 T. M. Barnett, C. Altohuler, N. McDaniel, J. P. Mascarenhas, *Dev. Genet.* **1980**, 1, 331–340.

101 M. Agarwal, N. Sarkar, A. Grover. *J. Plant Biol.* **2003**, 30(2), 141–149.

102 W. Wang, B. Vinocur, O. Shoseyov, A. Altman, *Trends Plant Sci.* **2004**, 9(5), 244–252.

103 S. L. Singla, A. Pareek, A. Grover, In: M. N. V. Prasad (Ed.), *Plant Ecophysiology.* John Wiley and Sons, Inc., New York, **1997**, pp 101–127.

104 M. Agarwal, S. Katiyar-Agarwal, C. Sahi, D. R. Gallie, A. Grover, *Cell Stress Chaperone* **2001**, 6(3), 219–224.

105 A. Grover, *Cell Stress Chaperone* **2002**, 7(1), 1–5.

106 J. Ananthan, A. L. Goldberg, R. Voellmy, *Science* **1986**, 232, 522–524.

107 J. H. Lee, A. Hubel, F. Schoffl. *Plant J.* **1995**, 8(4), 603–612.

108 C. H. Yeh, P. F. L. Chang, K. W. Yeh, W. C. Lin, Y. M. Chen, C. Y. Lin, *Proc. Natl. Acad. Sci. USA* **1997**, 94, 10967–10972.

109 L. S. Young, C. H. Yeh, Y. M. Chen, C. Y. Lin, *Biochem. J.* **1999**, 344, 31–38.

110 D. Low, K. Brandle, L. Nover, C. Forreiter, *Planta* **2000**, 211, 575–582.

111 J. R. Glover, S. Lindquist, *Cell* **1998**, 94, 73–82.

112 F. U. Hartl, R. Hlodan, T. Langer, *Trends Biochem. Sci.* **1994**, 19, 20–25.

113 J. Buchner, *Trends Biochem. Sci.* **1999**, 24, 136–141.

114 K. Malik, J. P. Slovin, C. H. Hwang, J. L. Zimmerman. *Plant J.* **1999**, 20, 89–99.

115 S. M. Park, C. B. Hong, *J. Plant Physiol.* **2002**, 159, 25–30.

116 T. Murakami, S. Matsuba, H. Funatsuki, K. Kawaguchi, H. Saruyama, M. Tanida, Y. Sato, *Molecular Breeding* **2004**, 13 (2), 165–175.

117 K. Sanmiya, K. Suzuki, Y. Egawa , M. Shono, *FEBS Lett.* **2004**, 557, 265–268.

118 W. Sun, C. Bernard, B. van de Cotte, M. Van Montagu, N. Verbruggen, *Plant J.* **2001**, 27 (5), 407–415.

119 C. Queitsch, S. W. Hong, E. Vierling, S. Lindquist, *Plant Cell* **2000**, 12(4), 479–492.

120 S. Katiyar-Agarwal, M. Agarwal, A. Grover, *Plant Mol. Biol.* **2003**, 51(5), 677–686.

121 R. Prändl, K. Hinderhofer, G. Eggers-Schumacher, F. Schöffl, *Mol. Gen. Genet.* **1998**, 258, 269–278.

122 I. Panchuk, R. A. Volkov, F. Schoffl, *Plant Physiol.* **2002**, 129(2), 838–853.

123 S. K. Mishra, J. Tripp, S. Winkelhaus, B. Tschiersch, K. Theres, L. Nover, K. D. Scharf, *Genes Dev.* **2002**, 16, 1555–1567.

124 Y. Li, C. S. Chang, L. S. Lu, C. A. Liu, M. T. Chan, Y. Y. Chang, *Bot. Bull. Acad. Sin.* **2003**, 44, 129–140.

125 H. Lee, M. V. Montagu, N. Verbruggen, *Proc. Natl. Acad. Sci. USA*, **1999**, 96, 5873–5877.

126 G. C. Papageorgiou, N. Murata, *Photosynth Res.* **1995**, 44, 243–252.

127 Alia, H. Hayashi, A. Sakamoto, N. Murata, *Plant J.* **1998**, 16, 155–161.

128 N. Murata, *Plant Cell Physiol.* **1983**, 24, 81–86.

129 G. Thomas, P. J. Dominy, L. Vigh, A. R. Mansourian, P. J. Quinn, W. P. Williams, *Biochim. Biophys. Acta* **1986**, 849, 131–140.

130 Y. Murakami, M. Tsuyama, Y. Kobayashi, H. Kodama, K. Iba, *Science* **2000**, 287(5452), 476–9.

131 W. M. Shi, Y. Muramoto, A. Ueda, T. Takabe, *Gene*, **2001**, 273, 23–27.

132 S. Chen, Z. Vaghchhipawala, W. Li, H. Asard, M. B. Dickman, *Plant Physiol.* **2004**, 135(3), 1630–1641.

133 A. Grover, P. K. Aggarwal, A. Kapoor, S. Katiyar-Agarwal, M. Agarwal, A. Chandramouli, *Curr. Sci.* **2003**, 84, 355–367.

134 B. Zhu, J. Su, M. Chang, D. P. S. Verma, Y.-L. Fan, R. Wu, *Plant Sci.* **1998**, 139, 41–48.

135 M. Kasuga, Q. Liu, S. Miura, K. Yamaguchi-Shinozaki, K. Shinozaki, *Nature Biotechnol.* **1999**, 17(3), 287–291.

136 T. Colbert, B. J. Till, R. Tompa, S. Reynolds, M. N. Steine, A. T. Yeung, C. M. McCallum, L. Comai, S. Henikoff, *Plant Physiol.* **2001**, 126, 480–484.

137 J. Knäblein, Biopharmaceuticals expressed in plants – a new era in the new Millennium. In: R. Müller, O. Kayser (Eds.), *Applications in Pharmaceutical Biotechnology.* Wiley-VCH, **2004**.

11

The First Biopharmaceutical from Transgenic Animals: ATryn®

Yann Echelard, Harry M. Meade, and Carol A. Ziomek

Abstract

Antithrombin (AT) concentrates derived from pooled human plasma have been used for the management of hereditary and acquired AT deficiencies since the early 1980s. The development of a recombinant version of AT would alleviate supply and safety concerns associated with the use of the plasma-derived biotherapeutic. However, the complex structure of the AT molecule, and the large doses usually required in supplementation treatments, have precluded the use of traditional bacterial and cell culture bioreactors for commercial production. GTC Biotherapeutics has applied the transgenic animal expression system to the production of recombinant AT (tradename ATryn®). A herd of transgenic dairy goats expressing high levels of human AT in their milk was characterized and expanded, providing a homogeneous, well-defined and abundant supply of AT. This chapter describes the clinical development of ATryn® (including eight clinical studies), and the production of this modern biopharmaceutical in transgenic goats.

Abbreviations

ACT	activated clotting time
ARDS	adult respiratory distress syndrome
AT	antithrombin
BSE	bovine spongiform encephalopathy
CABG	coronary artery bypass grafting
CD	circular dichroism
CHO	Chinese hamster ovary
CJD	Creutzfeld–Jakob disease
CPB	cardiopulmonary bypass
DIC	disseminated intravascular coagulation
DVT	deep-vein thrombosis
ELISA	enzyme linked immunosorbent assay
FFP	fresh-frozen plasma
FISH	fluorescence in-situ hybridization
HD	hereditary-deficient
HSPG	heparan sulfate proteoglycans
HUVEC	human umbilical vein endothelial cells
IL	interleukin
LC/MS	liquid chromatography/mass spectrophotometry
LPS	lipopolysaccharide
nvCJD	new-variant Creutzfeld-Jakob disease
PCR	polymerase chain reaction

Modern Biopharmaceuticals. Edited by J. Knäblein
Copyright © 2005 WILEY-VCH Verlag GmbH & Co. KGaA, Weinheim
ISBN: 3-527-31184-X

rhAT   recombinant human anti-
        thrombin
TAT   thrombin–antihrombin (com-
        plex)
TSE   transmissible spongiform ence-
        phalopathy

11.1
Introduction

In the fall of 1992, a goat kid was born at Tufts University School of Veterinary Medicine: a normal healthy male goat (Fig. 11.1) with added genetic material in his chromosomes. It was a DNA construct that would imbue his progeny with a unique ability to produce a recombinant human protein, antithrombin, in their milk. The offspring of this genetically engineered goat are living bioreactors that produce a recombinant protein much more efficiently than traditional cell culture bioreactors. This is the story of how the small biotech company that engineered that animal proceeded down the road to testing the product in the clinic. It is the story of this transgenic product's development path towards market.

The serpin antithrombin (AT), also known as antithrombin III or AT-III, is the main regulator of many serine proteases generated during activation of the clotting cascade (see Part II, Chapters 1 and 3). AT is a 58-kDa single-chain glycoprotein composed of 432 amino acids with four oligosaccharide chains, and with a plasma concentration of approximately $150 \, \mu g \, mL^{-1}$ (2–3 μM). It is a key factor in the prevention of clotting. In the presence of heparin, it is not only a strong inhibitor

Fig. 11.1 (A) Aerial view of the GTC Biotherapeutics farm in Charlton (MA. USA) with rhAT-producing transgenic goats. (B) 155-92, the transgenic founder. (C) rhAT transgenic does, descendants of 155-92.

of thrombin and Factor Xa, but it is an equally effective inhibitor of Factor IX, as well as, to a lesser extent, Factors XIa, XIIa (and its fragment), trypsin, plasmin, and kallikrein [1–6] (see Part III, Chapter 6). AT also weakly neutralizes Factor VIIa [7, 8]. AT inhibits thrombin by forming a 1:1 stoichiometric complex between the two components via a reactive site (arginine)-active center (serine) interaction. Heparin, the first-line anticoagulant in cardiovascular medicine, works as an anticoagulant because its binding to AT induces an AT configuration change, making it more than 1000-fold more active towards thrombin [9]. Once the equimolar thrombin–antithrombin (TAT) complex is formed, both molecules are incapacitated. The TAT complex is then removed by hepatic receptor identified as the LDL receptor-related protein [10]. The half-life of native AT is much longer (55 h) than that of the TAT complex (<5 min) [11].

Hereditary deficiency of AT is associated with an increased risk of venous thromboembolism, often beginning in adolescence and frequently accompanied by pulmonary emboli [12–15]. It is a rare disease, with a prevalence of 1:2000 to 1:5000 [13] and is characterized by a decreased AT activity of 25–60% of normal. The pattern of inheritance is autosomal dominant. The null mutation, in mice, is a fetal lethal [16] (see Part III, Chapter 4). The majority of patients with a hereditary deficiency disorder experience at least one thrombotic episode between the ages of 10 and 35 years, usually triggered in high-risk situations such as surgery, birth delivery, pregnancy, trauma, or severe infection [17]. Currently, plasma-derived human AT concentrates are being used prophylactically for the prevention of thromboembolic complications in these high-risk situations in patients with hereditary AT deficiency. The development

of rhAT will provide a stable supply of antithrombin and reduce the perceived viral risks associated with the current plasma-derived products.

Acquired deficiencies of AT are much more frequently encountered, either due to increased consumption [sepsis, disseminated intravascular coagulation (DIC), surgery, extensive thrombosis], increased loss (burns, trauma, nephrotic syndrome), drug treatment (heparin treatment, estrogen use, L-asparaginase treatment), or decreased production (prematurity, liver diseases) [18–20]. In several of these acquired conditions, the decrease in AT levels is associated with a poor outcome, especially for septic patients [18, 20–25]. The value of AT supplementation with concentrates in the case of acquired deficiencies is controversial. Numerous small trials have shown a favorable impact of AT replacement therapy on the severity of sepsis-induced DIC [26–30]. However, the small size of these trials did not permit a determination of whether AT treatment had a significant impact on mortality. The only large clinical trial that examined the effect of AT in the severe sepsis patient was the KyberSept trial that used plasma-derived product. This was an international, multi-center, placebo-controlled, double-blind, randomized clinical trial involving over 2300 patients [31, 32]. The KyberSept trial did not meet its primary end-point of a significant reduction in 28-day, all-cause mortality, the mortality rate being equivalent in the treated and placebo groups. These results were particularly disappointing, in that treatment with antithrombin had been convincingly shown to be of survival advantage in several Phase II trials [28, 29], and small non-randomized clinical studies [33, 34]. Further review of the KyberSept trial data revealed a strong treatment interaction between antithrombin and heparin.

When the subgroup of patients treated with antithrombin without heparin was compared with placebo patients that did not receive heparin, the 90-day mortality rate was significantly reduced ($p < 0.05$) in the antithrombin group [31, 35]. Further large trials will be required to demonstrate the therapeutic value of AT supplementation in sepsis as well as for other acquired deficiency indications.

11.2
Recombinant Production of AT

11.2.1
The Milk Expression Technology

This protein expression technology is based on producing recombinant proteins in the milk of transgenic animals. It begins with linkage of the promoter region of a milk-specific gene to a gene of inter-est. This DNA construct is then introduced into the chromosomes of a mammalian embryo to be transferred to surrogate mothers. Animals carrying this transgenic are then capable of producing the gene product in their milk (Fig. 11.2). During lactation, the recombinant protein is synthesized in the mammary gland and secreted into the milk. The product can then be purified from the milk to pharmaceutical-grade purity [36–45].

One advantage of the milk-expression system is that the constructs generated are expressed both in the mouse model system and in the large dairy animal production system at similar levels. This allows optimization of the expression transgene in the mouse system, before embarking on the large-animal portion of the project.

The predominant protein in goat milk is beta casein ($10–20 \text{ mg mL}^{-1}$); this is thought to comprise 25–50% of the total protein ($\sim 30 \text{ mg mL}^{-1}$). The beta casein

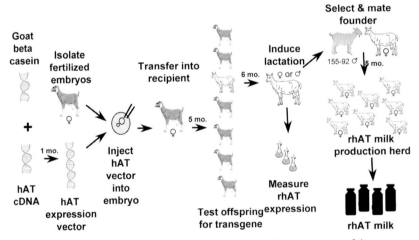

Fig. 11.2 Representation of the production process for rhAT. The hAT cDNA was linked to caprine mammary gland-specific regulatory elements. This transgene was microinjected into goat embryos that were then transferred to the oviducts of recipients and carried to term. Offspring were tested for the presence of the transgene. Female and male transgenic founders were induced to lactate to evaluate hAT expression in milk. The selected founder, 155-92, was mated to non-transgenic females to generate the rhAT production herd.

gene is regulated in both a tissue-specific and temporal fashion, with maximal mammary gland expression occurring post parturition. The goat beta casein gene was cloned as an 18.5 kilobase pairs (kb) fragment in a lambda EMBL3 vector from a Saanen goat genomic library and characterized in transgenic mice [46]. The coding sequence of the goat beta casein was replaced with an *Xho*I cloning site immediately upstream of the initiating ATG codon. The 3′ downstream region begins at the end of exon 7 and continues through the 3′ untranslated region through 6.2 kb of downstream genomic sequence. This type of construct has now been used to express nearly 100 different proteins in the milk of mice. It also has been now used successfully to secrete high levels (g L^{-1}) of 14 of these proteins in the milk of transgenic goats. It has a unique ability to express cDNA constructs at gram levels, which has been a challenge for other milk promoters.

11.2.2
Generation of the AT Founder

11.2.2.1 The AT Transgene

The human AT cDNA was obtained from Dr. G. Zettlemeissl (Behringwerke A.G., Marburg, Germany) in plasmid pβAT6. The sequence of the cDNA is the same as that published by Bock et al., 1982 [47], with the exception of the silent nucleotide changes at bp 1050 (T→C), 1317 (C→T) and bp 1371 (A→G). The cDNA was engineered with an *Xho*I site at the 5′ end by site-directed mutagenesis to allow excision of the cDNA as a 1.45 kb *Xho*I to *Sa*1I fragment. The transgene (pBC6; Fig. 11.3) was then assembled by linking the AT cDNA to the *Xho*I site of the goat beta casein vector. The resulting 14.8 kb transgene was excised from bacterial sequences by digesting pBC6 with *Not*I and *Sal*I, restriction sites which flank the eukaryotic sequences. The DNA fragment was purified by agarose gel electrophoresis. The resulting fragment was then used to generate transgenic animals.

Transgenic mice were used to test expression of the AT transgene. The generation of transgenic mice has become rou-

AT Vector Construction

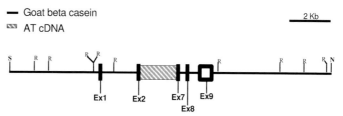

Fig. 11.3 Schematic representation of the BC6 transgene for recombinant expression of rhAT in the milk of transgenic animals. The cDNA encoding hAT (striped box) replaces the coding region of the caprine beta-casein gene. This cDNA is flanked by the promoter and the by 3′ untranslated caprine beta-casein sequences. Black boxes indicate the non-coding exons of the casein gene. R, restriction enzyme *Eco*RI; N, restriction enzyme *Not*I; S, restriction enzyme *Sal*I.

tine in academic and commercial laboratories (see Part III, Chapter 4). However, in the early 1990s it could still be a challenge to generate transgenic mice. Microinjection of pronuclear preimplantation embryos was used to introduce the heterologous DNA (BC6 transgene) into the genome. Of more than 200 founder animals born from microinjected embryos, only one transgenic mouse was identified. Females belonging to the line derived from this founder consistently produced nearly $1 \, \text{mg mL}^{-1}$ of recombinant human AT (rhAT) in their milk, as measured by Western analysis. With this success, it was decided to generate transgenic goats with the same construct.

11.2.2.2 Pronuclear Microinjection

Genzyme Corporation (Genzyme Transgenics Corporation, the progenitor of GTC Biotherapeutics was spun off by Genzyme in 1993) had established a research collaboration with Dr. Karl Ebert at Tufts University School of Veterinary Medicine. Dr. Ebert had carried out the microinjections of the mice to test many of the early milk vector constructs. Dr. Ebert also led the team that adapted the pronuclear microinjection technique to development of transgenic goats [48]. The AT microinjections were started early in 1991. In a process that is very similar to that used to generate transgenic mice, Dr. Ebert microinjected goat embryos. Female goats were superovulated and then mated. One-cell, pronuclear stage embryos were collected surgically from these does and submitted to the microinjection procedure (see Part I, Chapter 11). (This technique is impressively shown on the supplement CD-ROM in a microscopic video kindly provided by Professor Hwang.) A few picoliters of a DNA solution containing the purified

transgene fragment was injected into one or both pronuclei of the embryo using micromanipulators and a micropipette. The microinjected embryos were then transferred into the oviduct of suitably prepared surrogate female recipients that carried the progeny to term.

As the microinjected DNA randomly integrates into the embryonic genome at a low rate, through an unknown mechanism, it is believed that the integration event can take place at anytime as the embryo divides. This may give rise to "mosaicism" in which not all of the cells of the resulting transgenic animal will carry the transgene. In addition, the transgene may integrate in several locations in the genome. The transgenic male that was identified as the founder for the AT program had these characteristics; he was mosaic with multiple chromosomal integration sites.

The microinjection process was carried out on 139 caprine embryos, which were transferred into 57 surrogate mothers. Of the 70 progeny born, five were identified as being transgenic for the beta casein hAT construct (Table 11.1), BC6. Two founder females (223-92, 225-92) and three males (155-92, 222-92, and 224-92) were identified. A later series of injections with the genomic version of hAT gave rise to one founder female (227-92). The procedure to identify transgenic animals required both ear tissue and blood cells to be analyzed. DNA was isolated from each source and a polymerase chain reaction (PCR) specific for the hAT gene was used to determine which animals carried the BC6 construct. In addition to PCR, Southern blot analyses were also performed to confirm the integrity of the coding sequence and the region of the beta casein vector flanking the hAT gene.

Hormones were used to induce peri- or pre-pubertal goats to lactate. At the age of

Table 11.1 Summary of BC6 founder goat information

Animal ID	Date of birth	Sex	Germline transmission	Induced lactation rhAT level [g L^{-1}]	Natural lactation rhAT level [g L^{-1}]
155-92	02/05/92	M	4/23	1.3	ND
222-92	09/30/92	M	ND	ND	ND
223-92	10/07/92	F	1/5	0.03	<0.15
224-92	10/07/92	M	ND	ND	ND
225-92	10/07/92	F	0/2	Undetectable	Undetectable

ND, Not determined.

5 months, does could produce up to a few liters of milk following induction. Analysis of the milk for expression of rhAT allowed prediction of whether the transgene would be expressed during natural lactation, which would not occur for another 5 months following successful breeding. The 223-92 doe was induced to lactate and produced 0.5 mg mL^{-1} of rhAT in its milk, whereas the other female, 225-92, produced nothing. It is also possible to induce approximately 50% of males to lactate by using the same hormonal regime. Only young male 155-92 produced a few milliliters of milk, and this contained 1–2 mg mL^{-1} of rhAT, as determined by Western blot analysis. This male served as the genetic founder for the hAT transgenic production herd. Milk obtained from female transgenic goats, the offspring derived from 155-92, is the source material for purification of rhAT.

11.2.3
Establishment of the hAT Transgenic Herd

One of the challenges to breeding a transgenic production herd is that the expression level of the founder needs to be assured before committing significant effort to herd expansion. It is assumed that if the founder can produce the recombinant protein, then the progeny that inherit the transgene will also express the protein. During the 14-year period that we have carried out these types of program, this has been the case.

11.2.3.1 Germline Transmission
In order to establish the hAT production herd, it was necessary to expand the number of animals by breeding. The founder male, 155-92, was bred to 14 does during the first year. Of the progeny generated, only five kids carried the transgene – less than 50% inheritance of the transgene marker. This indicated that the founder 155-92 was mosaic. Mosaicism can be an issue in transgenic herd expansion since it increases the number of animals that must be mated to yield a sufficient number of transgenic females. Even with 50% transmission (non-mosaic), only 25% of the offspring will be the desired transgenic females. Several of the transgenic male offspring are kept for herd expansion, since they are not mosaic and would pass the transgene onto 50% of their progeny.

In the case of the hAT transgenic herd, it was necessary to identify a male carrying a single expressively active integration site. Southern blot analysis of the 155-92 founder showed four to six copies of the transgene. However, when the progeny of 155-92 were analyzed, four of the five carried four to six copies, but one had only one copy. This was initially disturbing, since it suggested that there was instability of the transgene, similar to what had been shown with the alpha-1 antitrypsin transgenic founder sheep [49]. To establish the genetic pattern of these animals, fluorescence in-situ hybridization (FISH) was used to visualize and follow the chromosomal integra-

tion sites throughout the breeding program [50]. Semi-quantitative FISH showed that the founder male, 155-92, carried at least four integration sites, each with a different extent of mosaicism (Fig. 11.4). The most highly represented site, located on chromosome 5 (C5), carried four copies of the transgene. The other sites on chromosomes 1, 12, and 23 (C1, C12, C23) each appeared

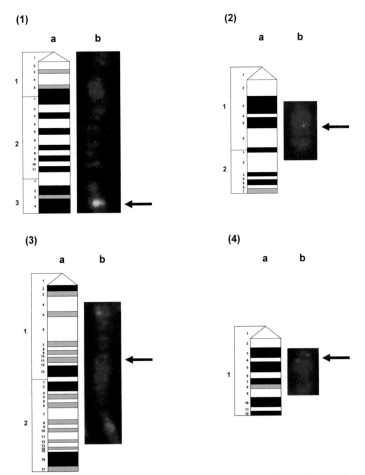

Fig. 11.4 FISH analysis of BC6 transgene integration sites carried by offspring of the 155-92 founder. Integrations located on chromosome 5 (1), chromosome 12 (2), chromosome 1 (3), and chromosome 23 (4) are respectively represented.

For each panel, (a) is the respective chromosome ideogram, while (b) shows a representative picture of each of the founder's corresponding chromosome, with the arrow indicating the transgene-specific signal.

to carry only a single copy of the transgene. During breeding, these chromosomes with their transgenes can segregate independently. The progeny of 155-92 have different combinations of these sites. Analysis of the milk from those progeny in which these sites showed full or partial segregation confirmed that it was only the four- to six-copy C5 transgene integration site that was expressed in the 155 line. Transgenic offspring that carried any combination of the C1, C12, or C23 sites in the absence of the C5 integration never expressed rhAT, indicating that these integration sites are non-functional.

11.3
Characterization of rhAT

11.3.1
Purification of rhAT

A process (summarized in Fig. 11.5) was designed for purification of rhAT from goat milk with a cumulative yield greater than 50% [51]. The source material for purification is milk produced by transgenic goats expressing the rhAT protein at approximately $2 \, g \, L^{-1}$. Goats typically lactate for 300 days each year, producing 2–4 L of milk daily. The process normally produces 300 g of purified rhAT per batch from no more than 375 L of milk containing approximately 600 g of rhAT. Milk containing rhAT is diluted with EDTA buffer and is then clarified by tangential flow filtration with a nominal 500-kDa pore-size Hollow Fiber membrane filter (Step 1). The filter permeate is cycled through a closed loop linking the filtration system to the Heparin-Hyper D column (Step 2) until >90% of the rhAT is captured (about eight volume cycles). Heparin affinity chromatography is also used routinely in the production of all commercially available hpAT in the European Union [52]. The Heparin-Hyper D column is washed and then eluted with a 2.5 M sodium chloride buffer. Once the Heparin-Hyper D el-

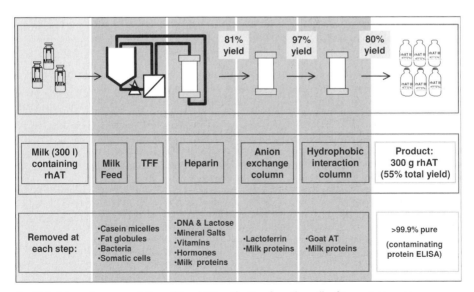

Fig. 11.5 Schematic representation of the rhAT purification from the milk of transgenic goats.

uate is obtained, it is transferred into a downstream processing area.

The eluate is then filtered through a Pall DV-20 viral removal filter (Step 3), concentrated, and diafiltered by membrane filtration to adjust the ionic strength for the application of the rhAT onto the ANX-Sepharose column. After loading, the ANX-Sepharose column is washed and the rhAT is eluted with 0.32 M buffer (Step 4). The ANX-Sepharose eluate is conditioned with sodium citrate and applied to the Methyl HyperD column (Step 5). The Methyl HyperD column is washed and the rhAT eluted from the resin with a lower concentration sodium citrate buffer.

The final formulation is achieved by concentration and diafiltration into a citrate, glycine, sodium chloride buffer with the proper ionic strength and dilution to the final protein concentration of approximately 25 mg mL^{-1}. The product is filled into vials (10 mL, containing ~250 mg protein), lyophilized, and then treated in a validated terminal viral inactivation step (Step 6).

11.3.2
Structural Characterization

The recombinant human AT purified from transgenic goat milk is indistinguishable from plasma-derived AT (hpAT) with the exception of the carbohydrates (see Part IV, Chapters 2 and 7). Recombinant human AT made in the milk of transgenic goats contains the same 432 amino acids as hpAT preparations as determined by amino-terminal sequence analysis, peptide mapping, and liquid chromatography/ mass spectrophotometry analysis (LC/MS) [51]. N-terminal sequence analysis confirmed that the rhAT had the correct N-terminal sequence. The reduced and pyridyl-ethylated peptide map of rhAT was essen-

tially identical to that of Thrombate® III (a plasma-derived preparation of AT commercialized in the United States by Bayer Corporation). The only differences noted were in the regions of the glycopeptides due to the glycosylation heterogeneity in the rhAT. The primary sequence of rhAT was confirmed by on-line LC/MS analysis of an endoproteinase Lys-C digest. Both rhAT and hpAT contain the same three disulfide bonds (Cys 8–128, Cys 21–95, Cys 247–430) as determined by peptide mapping under non-reducing conditions [51].

AT contains four methionine residues, which may be prone to oxidation under forced conditions *in vitro*. Normally, there is low oxidation of AT. In a comparative study of rhAT, Thrombate® III and Kybernin® (a commercial plasma-derived AT preparation), all three AT preparations were found to have similar low levels of methionine oxidation [53]. It was also shown that methionine oxidation had little impact on the inhibitory activity of rh or hpAT.

The conformation of rhAT was analyzed further by circular dichroism (CD) spectroscopy [51]. The far-UV CD spectrum was similar for both rhAT and hpAT proteins, and was characterized by two negative bands and a positive maximum indicative of the presence of both *a*-helix and *β*-sheet, which was consistent with the crystal structure data of hpAT [54]. In the far-UV CD spectrum, the addition of heparin produced little change, which suggested that secondary structures were not altered. The near-UV spectra of both proteins were also similar [51], and in excellent agreement with previously published spectra of AT derived from human and bovine plasma [55]. The near-UV spectrum showed a dramatic increase in band intensity across the whole region when heparin was added to both proteins. This increase was attributable to the conformational change of buried and exposed tryptophan residues upon heparin binding.

11.3.3
Glycosylation of rhAT

Using on-line LC/MS analysis of an endo-proteinase Lys-C digest, the only post-translational modifications detected were at the known N-glycosylation sites on either rhAT or hpAT [51]. Both rhAT and hpAT contain the same four N-linked glycosylation sites (Asn96, 135, 155, 192), as determined by peptide mapping and by LC/MS. No evidence of O-linked glycosylation was observed during LC/MS analysis of both proteins. Human plasma AT lacks glycosylation at the Asn 135 (the β-form) in 5–15% of the total AT found in plasma [56, 57]. LC/MS data indicated that the rhAT had glycosylation at Asn135 greater than 80% of the time.

As inferred previously from peptide maps, the monosaccharide composition of rhAT was different from that of Thrombate® III [51]. hpAT has predominantly identical oligosaccharides on the four N-linked glycosylation sites [58, 59], although 15–30% of the chains may lack terminal sialic acid [60, 61]. The main glycosylation differences observed for the rhAT were the presence of fucose and GalNAc, a higher level of mannose, and a lower level of galactose and sialic acid. There is also substitution of 40–50% of the N-acetylneuraminic acid with N-glycolylneuraminic acid. As expected from the monosaccharide compositional analysis, the LC/MS analysis was more complex for all the rhAT glycopeptides than for hpAT. The terminal sialic acid in the rhAT contained the same 2,6 linkage found in hpAT [61, 62]. Several laboratories have determined that differences in glycosylation of AT do not affect the intrinsic rate

constant of the uncatalyzed or heparin-catalyzed inhibition of thrombin, indicating that the carbohydrate chains solely affect heparin binding and not heparin activation or proteinase binding functions [63–65]. Thus, glycosylation differences do not impact the major biological activity of AT which is thrombin inhibition, but did explain the differences in affinity for heparin and in pharmacokinetics.

11.3.4
Inhibitor Activity and Heparin Binding Affinity of rhAT

Antithrombin is a serine protease inhibitor that inhibits thrombin and Factors IXa and Xa and, to a lesser extent, Factors VIIa, XIa, XIIa, as well as proteases such as trypsin, plasmin, and kallikrein [1–6]. The addition of heparin increases its inhibitory activity 300- (Factor Xa) to 1000- (thrombin) fold by inducing an allosterically transmitted conformational change in the reactive center loop of the AT molecule [66]. Thrombin and AT interact to produce a tightly bound TAT complex that is essentially irreversible and is cleared quickly from the circulation [10, 67]. Heparin binding to the AT molecule plays a catalytic role in increasing the inhibitory activity of AT toward thrombin and Factor Xa. There are two forms of AT in human plasma; these have different heparin affinities, but the same inhibitory activity toward thrombin [56, 57]. Some 85–90% of circulating hpAT has glycosylation on four Asn residues; this fully glycosylated form is referred to as the α-form. Between 5 and 15% of circulating hpAT (referred to as the β-form) lacks glycosylation at Asn135, and has a 3- to 10-fold higher heparin affinity than the α-form [56].

The specific activity of rhAT was identical to that of hpAT (Thrombate III) in an *in vitro* thrombin inhibition assay in the presence of excess heparin (~6 IU mg^{-1}) [51], and very similar to that reported for Atenativ® (Pharmacia) [52, 68]. Equivalent inhibition for rhAT and Thrombate III was also seen in an *in vitro* Factor Xa inhibition assay in the presence of excess heparin [51]. Heparin cofactor activation of rhAT versus hpAT was determined by varying the amount of heparin used in either inhibition assay. A lower concentration of heparin was required for rhAT than hpAT for inhibition of both enzymes, similar to the β-form of hpAT. Thus, rhAT closely resembles hpAT with respect to its activity for both thrombin and Factor Xa in the presence of saturating levels of heparin [51]. By using a tryptophan fluorescence assay, a four-fold higher affinity for heparin was observed with rhAT when compared with Thrombate III, but similar to that reported for the β-form of hpAT. The fluorescence values at saturating heparin were indistinguishable for rhAT and Thrombate III [51]. RhAT has a three- to four-fold higher overall heparin affinity than the α-form of hpAT, due to the glycosylation differences between these molecules. However, in the presence of excess heparin moieties, as found on the surface of vascular endothelial cells or with exogenous heparin supplementation, the alpha, beta, and recombinant forms of AT have identical inhibitory activities against thrombin [56] because the heparin is not itself involved in the inhibition.

in vitro studies were also conducted to assess the behavior of rhAT in assays used routinely to monitor hpAT in patient therapy. Cooper et al. [69] compared rhAT to dilutions of normal pooled plasma in three assays: 1) a thrombin-based assay using 60, 180 and 300 s incubation of thrombin (Dade Behring kit); 2) a Factor Xa-based assay (Chromogenix kit); and 3) an hAT

ELISA (Dako antibodies). Antithrombin level in the rhAT concentrate was assayed against the 8th British Blood Coagulation Factors Standard to compare the stated dose with the assayed dose per vial. Heparin binding was also assessed by two-dimensional electrophoresis with or without heparin in the first dimension, and by heparin-Sepharose gel filtration. The concentration of rhAT in the vial was 89% of the stated value by this thrombin-based assay, 85% by Xa-based assay, and 119% by the antigen assay.

11.3.5
Viral and Prion Safety

Since it is obtained from the milk of closed herd of transgenic goats, rhAT is inherently unlikely to transmit human blood-borne viruses and other human infectious agents. Moreover, no human-derived protein is added during the production, isolation, or formulation of rhAT. The production goat herd is closed and highly controlled. A high level of donor control and testing is a key parameter in the viral safety strategy for rhAT. In addition, the rhAT-containing milk is assessed *in vitro* on three cell cultures (human MRC-5, monkey Vero, and goat turbinate) to screen for adventitious viruses, including any emerging unknown virus, that may be present in the milk pool used as the starting material for production of rhAT. To date, all milk pools from the GTC Farm have tested negative in the *in vitro* cell line screening. In addition, a nanofiltration step has been incorporated into the current rhAT process for viral removal between the heparin affinity column and the ion-exchange column. The validated viral removal capacity of the rhAT purification process is summarized in Table 11.2. The viral validation studies demonstrated that a significant virus reduction of ≥8.5 to ≥25.3 log$_{10}$ was accom-

Table 11.2 Validation of $\log_{10}$ viral removal or inactivation by the RhAT process

Process step	Pseudorabies virus	Xenotropic murine retrovirus	Human adenovirus	Porcine parvovirus	Poliovirus[a]	Mouse adenovirus[a]
VirA/Gard 500-kDa	≥5.1	≥3.7	≥5.3	2.4	4.1	3.5
Heparin Hyper-D	1.8	1.2	<1.0	1.4	4.0	2.3
Pall DV-20 Filter	≥4.8	≥3.8	≥6.3	≥3.7	ND	ND
ANX-Sepharose	3.6	1.0	≥7.1	1.1	2.4	<1.0
Methyl HyperD	≥5.6	≥3.5	≥4.8	≥5.7	≥5.2	≥2.7
Heat treatment	2.8	≥5.0	≥1.8	2.4	≥1.9	ND
Total reduction	≥23.7	≥18.2	≥25.3	≥16.7	≥17.6	≥8.5

a) Poliovirus and mouse adenovirus were only used once in preliminary validation runs. All the others were run in duplicate and, in some cases, three times.
ND, Not determined.

plished across the distinctly different modes of the rhAT process.

Transmissible spongiform encephalopathies (TSE), such as new-variant Creutzfeld-Jakob disease (nvCJD) in humans, bovine spongiform encephalopathy (BSE) in cattle and scrapie in sheep and goats, also must be considered in assuring the safety of products made from human or ruminant sources. Human donors are monitored for CJD and nvCJD, and potentially contaminated blood, plasma pools and products made from them have been recalled or traced when a contributing donor has been diagnosed with CJD. All GTC goats are certified free of scrapie in the 5-year United States Department of Agriculture (USDA) Voluntary Scrapie Flock Certification Program, and various risk-minimization measures have been instituted to reduce any potential risk from this TSE in this highly controlled, closed donor goat population. In addition, the rhAT purification process has been validated for its ability to remove ≥11.3 $\log_{10}$ scrapie.

In aggregate, these data strongly support the conclusion that the rhAT manufacturing and production process is safe for human use with respect to potential viral and prion adventitious agent contamination.

11.4
Preclinical Studies

Antithrombin is a complex protein with multiple biologically important activities. It is the most critical modulator of coagulation, and has potent anti-inflammatory properties independent of its effects on coagulation [35, 70–73]. Several in vitro and in vivo animal studies, some of which used hpAT as a direct comparator, were undertaken to evaluate the potency of rhAT.

11.4.1
In vitro Efficacy Studies

Antithrombin has been shown to promote the release of prostacyclin from endothelial cells [74] and to attenuate ischemia-induced leukocyte extravasation [75]. In addition, chemotactic effects of AT on neutrophils and signaling via interaction with cell-surface heparan sulfate proteoglycans

(HSPG), possibly via syndecan-4, have been reported [76].

A study was undertaken to characterize the mechanism by which AT regulates the migration of neutrophils, which are involved in a variety of conditions including inflammatory diseases. (A dramatic video animation of neutrophils is available on the supplement CD-ROM.) Human neutrophils were obtained from healthy volunteers, and migration was measured in modified Boyden chambers. Either Kybernin P (a commercial hpAT preparation) or rhAT was used as an attractant. rhAT was at least as effective in deactivating neutrophil chemotaxis as Kybernin P [77]. To investigate the role of intact HSPG on the neutrophil surface for AT-induced cell migration, neutrophils were pretreated with heparinase or chondroitinase. Chemotactic effects of hpAT (1 U mL^{-1}) or rhAT (1 U mL^{-1}) were completely abolished by pretreatment with both agents. Antibodies to syndecan-4 also inhibited rhAT-induced migration of neutrophils. Collectively, these data suggest that AT regulates neutrophil migration via effects on its heparin-binding site on cell surface syndecan-4. Aspects of these investigations with rhAT confirm the findings of previous studies of the binding of hpAT to syndecans on the cell surface and AT inhibition of neutrophil chemotaxis. Similar effects were observed with rhAT and hpAT on human eosinophils [78].

In another study [79], the same group of investigators tested whether rhAT might influence lipopolysaccharide (LPS)-induced enhancement of adhesion of neutrophils to human endothelial cells and the involvement of glycosaminoglycans. Using human umbilical vein endothelial cells (HUVEC), experiments were performed aiming to quantitate fluorometrically the adhesion of neutrophils. Treatment with LPS and interleukin-1 (IL-1) increased neutrophil adhesion to HUVEC, which was inhibited by rhAT. Concomitant incubation of rhAT and an inhibitory pentasaccharide reversed rhAT's effects on adhesion. Treatment of endothelial cells with heparinase and chondroitinase to release HSPG which normally bind AT led to higher neutrophil adhesion. Treatment of endothelial cells with antibodies to syndecan-4, which is a receptor for AT, enhanced the adhesion of neutrophils. Western blotting studies showed that LPS-induced signaling was diminished by rhAT, and that the effect was reversible by chondroitinase or heparinase. From these results, it was concluded that LPS-induced adhesion of leukocytes to endothelium is reversed by ligation of rhAT with syndecan-4. Complementary experiments showed that rhAT attenuated the expression the beta-2 integrin CD11/CD18 on activated neutrophils and monocytes [80], providing a potential direct molecular mechanism for the effect of AT on adhesion.

11.4.2
In vivo Efficacy Studies

Animal disease models for acquired AT deficiency, such as *Escherichia coli*-induced sepsis, were used to assess the biological consistency of the rhAT compared to hpAT. Although in these models the exact contributions of the anti-coagulant and anti-inflammatory properties of AT are still somewhat controversial [81], they do allow for some comparison of the properties of rhAT with hpAT. Human plasma-derived AT has been shown to prevent the lethal effects of experimentally induced sepsis in several animal models [82–89], and to block cytokine production *in vitro*. To date, several dose– response studies have been performed with rhAT in rats (GTC Biotherapeutics, unpublished results [90]) and baboons [91] that have been lethally challenged with *E. coli*, *K. pneumoniae* or LPS. Both rhAT and hpAT preparations were equally effective in preventing sepsis and septic shock in these models. Additional studies have been performed demonstrating protective effects of rhAT in a smoke inhalation sepsis model in sheep, and in a xenotransplantation model in primates.

Recombinant human AT was initially tested by Taylor et al. in a lethal *E. coli* baboon sepsis model. Infusion of rhAT was found significantly to improve the survival of treated baboons [91]. This was also observed previously, by the same group, with hpAT [82]. Five adult animals were used, and the dosing regimen involved administration of 500 U kg^{-1} rhAT as 0.5-h infusions at t=–1 h and at t=+3 h and 250 U kg^{-1} rhAT as a bolus at the time of *E. coli* challenge (t=0 h). Administration of rhAT protected three of the five baboons from the challenge. One of the two baboons that died was not administered the third dose of rhAT; this animal died due to sepsis. The cause of death of the other baboon was attributed to capillary leakage in the lungs consistent with adult respiratory distress syndrome (ARDS); there was no evidence of DIC. These results indicated that, when given in the appropriate doses, rhAT protects against DIC and that, in three of four cases, rhAT protects against death from a lethal dose of *E. coli*. The protective effect of rhAT was due to a combination of anticoagulation and anti-inflammatory effects. As previously noted for the hpAT study, rhAT attenuated the coagulation response and fibrinolysis. Significantly, treatment with rhAT also markedly attenuated the release of the pro-inflammatory cytokines IL-6 and IL-8.

Further evidence of the anti-inflammatory activity of rhAT was found in an endotoxemic rat model [90]. In this model, rhAT treatment reduced endotoxin-mediated mesenteric venule leukocyte adhesion and small intestine mucosal barrier injury. Rolling and firm adhesion of leukocytes in mesenteric venules of endotoxemic rats was measured using intravital microscopy. Endotoxemia was induced by intravenous administration of 10 mg kg^{-1} of endotoxin, after which rats were treated either with saline or rhAT (250 and 500 U kg^{-1}). Following anesthesia, the distal ileum was exteriorized and the mesentery inserted in an intravital microscopy chamber fitted with a video camera system; the mesenteric circulation was then examined. Flux of rolling leukocytes was measured as the number of white blood cells that could be seen rolling past a fixed perpendicular line in the venule during a 1-minute interval. Quantification of venular endothelium leukocyte adherence was performed off-line by playing back videotaped images and counting the number of leukocytes that stuck and remained stationary for a period of >30 seconds. rhAT was shown to attenuate both endotoxin-induced venular leukocyte rolling and adhesion in a dose-dependent manner. Pretreatment with indomethacin, a prostaglandin synthesis inhibitor, completely abolished the effect on leukocytes rolling and adhesion, suggesting that the effect of AT could be mediated by an effect on prostacyclin production.

This effect on leukocyte adhesion obtained with rhAT is similar to the activity of hpAT observed in related models. For example, in the skinfold of endotoxemic Syrian hamster, multiple injections of 250 U kg^{-1} of hpAT attenuated LPS-induced arteriolar and venular leukocyte adhesion [92, 93]. Here again, this effect was completely abolished by pretreatment with indomethacin. Previously, in a feline mesentery ischemia/reperfusion model using intravital microscopy to monitor leukocyte rolling and adhesion, pretreatment with hpAT (250 U kg^{-1}) reduced neutrophil rolling and adhesion to pre-ischemic levels during reperfusion [75].

The previously described baboon study showed survival after *E. coli* challenge when rhAT was used in a pretreatment regime. Murakami et al. [94] investigated the effect of post-treatment with rhAT on sepsis after smoke inhalation in sheep. Acute lung injury frequently arises after smoke inhalation complicated by pneumonia induced by *Pseudomonas aeruginosa*. Previously, these investigators had shown that high-dose hpAT administration attenuated endotoxin-induced acute lung injury in rats. In addition, they found that hpAT administration promoted prostacyclin production, which inhibited leukocyte activation.

In the present study, *Pseudomonas aeruginosa* was instilled into the lungs of anesthetized sheep after insufflations of cool cotton smoke. One group of sheep also received rhAT by continuous infusion, starting 1 hour after injury and continuing for the next 24 hours at 1000 U kg^{-1} per 44-hour period. Plasma AT levels fell significantly in the control animals, but were maintained at baseline levels in rhAT-treated animals, as was previously demonstrated in other sepsis models. In addition, rhAT attenuated septic shock in these animals and acute lung injury was improved histologically, with a reduction of cast formation in the airways. All animals receiving rhAT were negative for fibrin degradation products, in contrast to untreated animals. Platelet levels fell in control animals; however, platelet counts in the rhAT-treated group at 24 hours were not different from baseline values, which suggested that rhAT administration attenuated the coagulation abnormalities observed with

sepsis. The investigators concluded that post-treatment with rhAT was effective in sepsis after smoke inhalation in sheep.

In a subsequent study [95], nebulized rhAT was used in the same model and was even more effective than intravenously administered rhAT at half the dose. In addition, pulmonary gas exchange, shunt fraction and lung wet:dry weight ratio were significantly attenuated by AT nebulization, thereby underscoring the protective effect of rhAT in this sepsis model.

The hypothesis that treatment with rhAT would prevent or at least delay the onset of rejection and coagulopathy was tested using a life-supporting pig-to-baboon renal xenotransplantation model [96]. Non-immunosuppressed baboons were transplanted with transgenic pig kidneys expressing the human complement regulators CD55 and CD59. The baboons were treated with rhAT by intravenous infusion every 8 hours, with or without heparin. No bleeding complications were observed. RhAT-treated baboons had preservation of normal renal function for 4–5 days, which was twice as long as untreated animals, and developed neither thrombocytopenia nor significant coagulopathy during this period. Thrombin clotting times were relatively normal in the rhAT-treated baboons for 4–5 days, and platelets and clotting factors were not consumed faster than they could be replaced. The relative importance of the anticoagulant and anti-inflammatory properties of AT in the xenografts setting remains to be determined.

11.4.3
Toxicology, Pharmacokinetic, and Mutagenicity Studies

A series of single-dose and repeat-dose toxicological studies were conducted to examine the safety of rhAT administered intravenously to rats, dogs and non-human primates at doses up to 10 times the highest anticipated dose in man. In addition, the safety of rhAT and heparin administered intravenously to rats was evaluated in a single-dose toxicological study. In all of these studies, rhAT was well tolerated. Most clinical observations and/or adverse reactions were related to the pharmacological anticoagulant properties of the test article, and were seen only at the highest doses tested.

Pharmacokinetic studies were performed to determine if gender, dose and/or repeated administration affect the kinetic disposition of rhAT. The single-dose pharmacokinetic studies were performed in Sprague-Dawley rats, beagle dogs, cynomolgus monkeys, and baboons. The repeat-dose pharmacokinetic studies were performed in Sprague-Dawley rats and cynomolgus monkeys. The results indicate that the kinetic disposition of rhAT was non-linear in all species examined. Clearance decreased as a function of increasing dose concurrent with an increase in half-life. Single or repeated administration of rhAT did not alter the kinetics of rhAT, nor

were there any gender-related effects associated with this compound.

Three studies were performed to evaluate the potential mutagenicity or genotoxicity of rhAT. The studies performed included an Ames assay that evaluated mutagenicity in five different *Salmonella typhimurium* strains and two *E. coli* strains, a mouse *in vivo* micronucleus assay that evaluated genotoxicity, and a CHO cell *in vitro* assay that evaluated chromosomal aberration. The results obtained did not show any potential mutagenicity or genotoxicity associated with rhAT.

11.5
Clinical Trials with rhAT

At this point, eight clinical studies have been undertaken with ATryn® (Table 11.3). Two clinical indications were pursued:
- Heparin resistance in patients undergoing cardiac surgery involving cardiopulmonary bypass (CPB).
- Prevention of deep-vein thrombosis (DVT) in patients who have a hereditary

Table 11.3 Summary of clinical studies

Phase	Indication	Type of study	Total patients [n]
I	Not applicable	Open-label, PK in normal healthy volunteers	17
I	Not applicable	Open-label, cross-over PK trial in normal healthy volunteers	26
I/II	Hereditary deficiency	Open-label PK in patients with hereditary deficiency	15
II	Heparin resistance	Dose-ranging trial, nine dose groups, CABG	36
III	Heparin resistance	Double-blind, placebo-controlled in heparin-resistant patients/CPB	54
III	Heparin resistance	Double-blind, placebo-controlled in heparin-resistant patients/CPB	52
III	Heparin resistance	Active-controlled trial in heparin-resistant patients	47
III	Hereditary deficiency	Open-label in hereditary deficient patients	14

CABG, Coronary artery bypass grafting; CPB, Cardiopulmonary bypass; PK, Pharmacokinetics.

deficiency of AT and who are in high-risk situations such as birth delivery or surgery.

In addition, rhAT was used in a compassionate-use program for United States AT hereditary deficient patients undergoing a high-risk event and that could not access hpAT due to lack of availability. In all the human studies completed to date, rhAT has proved safe and met the primary endpoints of that study. All studies showed that rhAT was well-tolerated and safe in these patient populations.

11.5.1
Single-dose Pharmacokinetic Study in Healthy Volunteers

A single-center, randomized, parallel group Phase I study was conducted in 17 male volunteers aged between 20 and 28 years. The subjects were divided into four groups each of three volunteers, with individual subjects receiving either 50, 100, 150, or 200 U AT per kg body weight. An additional five volunteers were enrolled to donate plasma for the determination of endogenous baseline AT concentrations. There were no serious adverse events. Review of all hematology and biochemistry profiles, urinalysis, vital signs, electrocardiogram, activated clotting time (ACT) and coagulation parameters revealed no clinically significant changes in any group, or any differences between those subjects receiving active drug and those that did not receive AT. There were four mild adverse events. One subject in the 150 U kg^{-1} dose group experienced a non-serious headache which was judged as possibly being related to the drug. Patients were tested before and after treatment for the presence of antibodies to rhAT. No antibodies could be detected either before or after treatment.

Lu et al. [97] investigated the pharmacokinetics of rhAT in these study subjects. The concentrations of AT, after initial doses given over 30 minutes, were best described by a weight-normalized, two-compartment model. The fast compartment volume was 41.1 ml kg^{-1} and the volume of distribution was 115.4 ml kg^{-1}. Intercompartmental clearance was 0.0763 ml kg^{-1} min^{-1}, and elimination clearance was 0.0383 ml kg^{-1} min^{-1}. These variables are equivalent to a distribution half-life of 196 minutes and an elimination half-life of 2568 minutes. Approximately 75% of the supplemental dose was removed from plasma by the initial distribution process.

In conclusion, rhAT – when given as a 30-minute infusion at doses up to 171 U kg^{-1} – was shown to be safe and well-tolerated in healthy male volunteers.

11.5.2
Heparin Resistance

11.5.2.1 Heparin Resistance in Coronary Artery Bypass Graft Patients: Dose-finding Study

Acquired AT deficiency may render heparin less effective during cardiac surgery and CPB. This Phase II study was designed to examine the pharmacodynamics and optimal dose of rhAT needed to maintain normal AT activity during CPB, to optimize the anticoagulant response to heparin, and to attenuate excessive activation of the hemostatic system in patients undergoing coronary artery bypass grafting (CABG). During CPB, AT activity frequently decreases to as little as 30–50% of normal. Low AT concentrations during cardiac surgery are likely to develop because of the preoperative use of heparin, the effect of hemodilution on the pump, and CPB- associated excessive hemostatic system activation [98]. Anticoagulation is used during cardiac surgery to prevent thrombosis of the extracorporeal circuit and to minimize CPB-related activation of the hemostatic system. In some cases, when heparin alone is not effective, either fresh-frozen plasma (FFP) [99] or hpAT concentrates [100–102] have been used in patients that show an appreciable heparin resistance prior to initiation of CPB. However, AT concentrate has not been approved for this indication in the US.

A single-center, open-label, single-dose, dose-escalation study was conducted in 36 patients, aged between 18 and 80 years, admitted for primary cardiac surgery requiring CPB [103]. All patients underwent elective primary CABG and had been receiving heparin therapy for at least 12 hours prior to surgery. Thirty patients received rhAT, and six received placebo. Patients receiving active drug were divided into groups of three, and assigned to one of nine dosing cohorts. The individual treatment dosing cohorts were 10, 25, 50, 75, 100, 125, 150, 175, and 200 U kg^{-1} rhAT. A tenth, placebo, cohort was added which included an additional three patients.

None of the patients that had post-drug samples taken developed circulating antibodies to AT following treatment. Supplementation of rhAT significantly (p<0.0001) improved heparin responsiveness, as measured by an increase in the ACT (844+191 s) as compared to heparin administration alone (531+180 s). Furthermore, AT supplementation resulted in significantly (p=0.001) better inhibition of thrombin (as measured by a decrease in fibrin monomer) and fibrinolysis (as measured by a decrease in D-dimer) at doses up to 125 U kg^{-1}. There was also a reduced impairment of platelet function after CPB, which is thought to be the most important hemostatic defect after CPB. Results suggest that single rhAT doses of 75 U kg^{-1} and higher will maintain the AT activity level at greater than 100% throughout the course of CPB.

Based on the results of this Phase II trial, two Phase III studies were conducted to test the potential benefit of supplementing rhAT levels in patients with an acquired deficiency state who demonstrate heparin resistance.

11.5.2.2 Heparin Resistance in CABG Patients

Two identical Phase III placebo-controlled, double-blind, multicenter studies were conducted in heparin-resistant patients scheduled for cardiac surgery requiring CPB. The study objective was to establish whether heparin-resistant patients who receive rhAT are less likely to require FFP as a source of AT to achieve an ACT >480 seconds, as compared to patients receiving placebo [104, 105]. The secondary objectives were to compare the effect of rhAT and FFP on laboratory measures of plasma levels of AT, thrombin activity, and fibrinolysis. Specifically, the trials addressed whether rhAT, at a dose of 75 U kg^{-1}, increases plasma AT levels and inhibits thrombin activity and fibrinolysis more effectively, than 2 units of FFP alone. One study (AT97-0502) enrolled 54 patients (rhAT:placebo=27:27). The second study (AT97-0504) enrolled 52 patients (rhAT:placebo=28:24).

One dosing cohort received 75 U kg^{-1} rhAT, and the other cohort received a normal saline placebo (both as single bolus intravenous injection). Heparin resistance was defined as failure to achieve an ACT of >480 seconds after receiving a total dose of 400 U kg^{-1} heparin intravenously after anesthesia induction and surgical incision, but just prior to CPB. The proportion of rhAT patients who required administration

of 2 U FFP to achieve an ACT of >480 seconds was significantly less (p <0.001) than
that of placebo patients (19% versus 81%
in the AT97-0502 study; 21% versus 91%
in the AT97-0504 study).

In summary:

- Administration of rhAT precluded the
  need for FFP for 44/55 patients versus
  only 7/50 patients treated with placebo.
- The mean AT activity level and change
  in AT activity level was significantly
  greater after rhAT administration than
  placebo throughout the study period.
- RhAT replacement therapy maintained
  AT activity within or above normal range
  throughout the study period.
- AT activity continued to decline in the
  placebo group due to continued consumption and hemodilution, despite 2
  units of FFP not providing adequate AT
  replacement therapy.
- ACTs, the standard measure of coagulation status, were significantly increased
  in the rhAT-treated group compared to
  the placebo group.

- Compared to placebo patients, patients
  who received rhAT showed significant
  inhibition of the generation of two
  markers of thrombin activation – pro-
  thrombin fragment 1.2 and thrombin–
  antithrombin complex.
- Some trends were observed for a decrease in the production of D-dimer
  after rhAT treatment compared to placebo treatment.
- The safety profiles of the rhAT and placebo groups were comparable.
- No evidence for an immune response
  (measured by patient immune response
  assays) was observed.

11.5.3
Hereditary Deficiency

11.5.3.1 Compassionate Use of rhAT for Hereditary AT-deficient Patients in High-risk Situations

Hereditary AT deficiency is associated with
a significant risk of venous thrombosis in
high-risk situations such as birth delivery

Table 11.4 Hereditary AT-deficient patients treated with rhAT in a compassionate-use basis

Subject no.	Age [years]	Sex	Baseline AT level [%]	Surgery type	Total rhAT received [U]	Vascular duplex ultra-sound	Clinical evidence of thrombosis	Anti rhAT antibody
1a	22	M	40	Laparoscopic splenectomy	74 480	Negative (d5)	Negative	Negative
1b	22	M	40	Bilateral hip replacement	294 000	Not tested	Negative	Negative
2	47	F	49	Hysterectomy	101 921	Negative (d7)	Negative	Negative
3	36	F	58	C-section	65 000	Negative (w6)	Negative	Negative
4	72	M	52	Coronary artery bypass	39 200	Not tested	Negative	Not tested
5	71	M	53	Knee replacement	73 480	Negative	Negative	Negative

and surgery. Plasma-derived AT concentrate (Bayer Corp.; Thrombate® III) has been approved in the United States for replacement when anticoagulation is interrupted in these patients. However, Thrombate III supplies have been limited and there have been periods when it was not available.

Five patients with hereditary AT deficiency and a prior history of thromboembolism were treated with rhAT on a compassionate use basis for six surgical procedures [106]. One patient had two surgical procedures 6 weeks apart, and received rhAT on each occasion. Patients were treated perioperatively, receiving multiple doses of rhAT for 2–16 days. Dosing was determined individually by the investigators, with the goal of maintaining an AT activity of 80–150% of normal. Patients were followed for clinical evidence of thrombosis, bleeding, adverse events, and development of antibodies to the rhAT.

All six surgical events were successfully treated with rhAT, as shown by the absence of clinical evidence of thrombosis for these patients that all had an history of thromboembolism. In two patients, where initial pre-and post-treatment levels were available, there was a 1.69 and 1.66% per U per kg increase, which is similar to the 1.39 and 2.05% per U per kg reported for hpAT. There was no clinical evidence of thrombosis or bleeding, and no adverse events related to the drug were reported. Four of the six surgical events were followed-up by vascular duplex ultrasound of the lower extremities, with no clinical evidence of acute thrombosis (Table 11.4). Four of the five patients, who receive multiple doses of rhAT, were also screened for antibody formation against rhAT several weeks postoperatively. None of the patients developed detectable antibodies to the rhAT.

11.5.3.2 rhAT Use for Hereditary AT-deficient Patients (009 and 02001)

Two trials studying the use of rhAT in hereditary-deficient (HD) patients were recently completed [107, 108; see also von Depka et al., in preparation].

Pharmacokinetic study The Phase I/II (009) trial was a pharmacokinetic study to obtain clearance data from asymptomatic HD patients after single-dose administration of rhAT, and to use these data to simulate steady-state concentration profiles and develop dosing regimens. Fifteen HD patients were administered a 50 or 100 IU kg^{-1} dose of rhAT as a bolus dose. Plasma samples were collected over a 72-hour period and analyzed for AT activity. In this study, the non-compartmental pharmacokinetics (mean ± SD) were: C_{max} (100 U kg^{-1}) 212.67 ± 25.51%; C_{max} (50 U kg^{-1}) 147.33 ± 25.83%; T_{max} 0.077 ± 0.083 h; K_{el} 0.184 ± 0.266 L h^{-1}; $T_{1/2}$ 10.49 ± 7.19 h; Vd area 167.65 ± 122.25 mL kg^{-1}; MRT 15.66 ± 10.02 h;, Cl 15.78 ± 12.44 mL h^{-1} kg^{-1}, and incremental recovery 1.53 ± 0.34 (IU mL^{-1})/(IU kg^{-1} administered).

The data were also modeled using a two-compartment model. Median values for the modeled baseline, terminal half-life, Vd area, K_{10}, K_{12} and K_{21} were 66.87%, 2.4 h, 66.33 mL kg^{-1}, 0.30 h^{-1}, 0.31 h^{-1}, and 7.87 h^{-1}, respectively. These median model parameters were used to simulate plasma concentrations when AT is given once, twice, three, and four times per day and as a continuous infusion (with a loading dose) at various doses and patient baseline values. The results indicated that a combination of a short infusion loading dose and a continuous infusion maintenance dose appeared to be the optimal dosing regimen to maintain AT levels between 80 and 120%.

Efficacy study The 02001 Phase III efficacy study was open-label, blinded evaluator trial of rhAT in at least 12 hereditary AT-deficient patients being treated prior to, during, and following high-risk events. The trial assessed the incidence of DVT following prophylactic intravenous rhAT administration to hereditary AT-deficient patients during situations associated with a high-risk event. Dosing of rhAT was achieved by a loading infusion of rhAT followed by continuous infusion for at least 3 days, with the objective of maintaining AT plasma activity between 80% and 120% of normal.

The primary study end-point was incidence of DVT and other thromboembolic events assessed clinically and by both locally and centrally assessed ultrasonography. The duplex-ultrasounds of the lower extremities were used to confirm or exclude the occurrence of DVT. These procedures were performed and interpreted by qualified specialists within the same hospital/institution on a real-time basis for the timely and appropriate clinical care of the patient. Furthermore, duplex ultrasound studies were videotaped for a subsequent standardized blinded interpretation by a qualified, independent laboratory that provided an unbiased evaluation of the incidence of DVT. Thromboembolic events other than DVT that occurred during the study period were assessed, and the investigator established the clinical relationship of the event to treatment with rhAT. Secondary end-points were safety, adverse events and immunogenicity.

The study was initiated in December 2003, and completed in the fourth quarter of 2003, enrolling 14 patients comprising nine birth deliveries and five surgeries. None of the patients showed clinical signs of DVT or thromboembolism, nor developed antibodies against rhAT. Upon local review of duplex-ultrasounds, one patient was evaluated to have an acute DVT which was resolved by day 7 after treatment. Upon centralized review, an additional patient was evaluated to have a DVT which was resolved by day 30 after treatment. Neither patient exhibited clinical signs of thrombosis. The patient in whom DVT was detected only with central review was a birth delivery patient who was evaluated locally to be exhibiting chronic changes, and no special treatment was initiated. For the other patient in whom DVT was detected both centrally and locally, the findings developed following a hip replacement (a highly thrombogenic procedure, even in non-AT-deficient patients), treatment with rhAT in combination with therapeutic low molecular-weight heparin and vitamin K antagonists was continued. The patient remained without symptoms, and the DVT resolved.

11.6
Conclusions

Following completion of the 02001 efficacy trial, a European regulatory filing was submitted in January 2004, for the use of rhAT in the prophylaxis of DVT in hereditary AT-deficient patients in a high-risk situation. If this application is approved, this will constitute the first approval of a transgenically produced biopharmaceutical. Indeed, this will constitute the first approval of a biologic manufactured in a new recombinant production system since approval of the first product manufactured in cell culture in the early 1990s. The development of a recombinant option for antithrombin will provide a safe and reliable supply of this important factor, and will facilitate the resumption of clinical trials aimed at acquired deficiencies of antithrombin such as cardiovascular surgery, severe burns, and severe sepsis. Other transgenically expressed recombinant products which are currently in development include human albumin, C1-esterase inhibitor and alpha-1 antitrypsin, as well as several monoclonal antibodies. In summary, the emergence of this transgenic manufacturing platform will provide an attractive option for the recombinant production of complex biopharmaceuticals that are needed in large amounts.

Acknowledgments

The authors are deeply indebted to their colleagues at GTC Biotherapeutics and Genzyme. The development of a new drug is truly an enormous endeavor, and entire scientific teams have dedicated years of their professional lives to the filing of ATryn. In addition, the authors extend special thanks to Suzanne Groet, Tom Newberry, and Dick Scotland for their critical reading and constructive suggestions, as well as to Jennifer Williams and Merry Harvey for their generous help with the illustrations.

References

1 U. Abildgaard, *Scand. J. Clin. Lab. Invest.* **1969**, *24*, 23–27.
2 E. T. Yin, S. Wessler, P. J. Stoll, *J. Biol. Chem.* **1971**, *246*, 3712–3719.
3 R. D. Rosenberg, P. S. Damus, *J. Biol. Chem.* **1973**, *248*, 6490–6505.
4 J. S. Rosenberg, P. McKenna, R. D. Rosenberg, *J. Biol. Chem.* **1975**, *250*, 8883–8888.
5 K. Kurachi, K. Fujikawa, G. Schmer, E. W. Davie, *Biochemistry* **1976**, *15*, 373–377.
6 N. Stead, A. P. Kaplan, R. D. Rosenberg, *J. Biol. Chem.* **1976**, *251*, 6481–6488.
7 S. Kondo, W. Kisiel, *Thromb. Res.* **1987**, *46*, 325–335.
8 M. A. Blajchman, R. Austin, F. Fernandez-Rachubinski, W. P. Sheffield, *Blood* **1992**, *80*, 2159–2171.
9 L.-O. Andersson, L. Engman, E. Henningson, *J. Immunol. Methods* **1977**, *14*, 271–281.
10 M. Z. Kounnas, F. C. Church, W. S. Argraves, D. K. Strickland, **1996**, *271*, 6523–6529.
11 S. V. Pizzo, *Am. J. Med.* **1989**, *87 (Suppl. 3B)*, 10S.
12 O. Egeberg, *Thromb. Diath. Hemorrh.* **1965**, *14*, 473–489.
13 E. Thaler, K. Lechner, *Clin. Haematol.* **1981**, *10*, 369–390.
14 K. Lechner, P. A. Kyrle, *Thromb. Haemost.* **1995**, *73*, 340–348.
15 H. H. van Boven, D. A. Lane, *Semin. Hematol.* **1997**, *34*, 188–204.
16 K. Ishiguro, T. Kojima, K. Kadomatsu, Y. Nakayama, A. Takagi, M. Suzuki, N. Takeda, M. Ito, K. Yamamoto, T. Matsushita, K. Kusugami, T. Muramatsu, H. Saito, *J. Clin. Invest.* **2000**, *106*, 873–878.
17 D. Menache, J. P. O'Malley, J. B. Schorr, B. Wagner, C. Williams & The Cooperative Study Group, *Blood* **1990**, *75*, 33–39.
18 H. R. Buller, J. W. ten Cate, *Am. J. Med.* **1989**, *87(3B)*, 44S–48S.
19 E. F. Mammen, *Semin. Thromb. Haemost.* **1995**, *32*, 2–6.
20 W. F. Jr. Baker, R. L. Bick, *Semin. Thromb. Haemost.* **1999**, *25*, 387–406.
21 N. Smith-Erichsen, A. O. Aasen, M. J. Gallimore, E. Amundsen, *Circ. Shock.* **1982**, *9*, 491–497.
22 R. F. Wilson, A. Farag, E. F. Mammen, Y. Fujii, *Am. Surg.* **1989**, *55*, 450–456.
23 F. Fourrier, C. Chopin, J. Goudemand, S. Hendrycx, C. Caron, A. Rime, A. Marey, P. Lestavel, *Chest* **1992**, *101*, 816–823.
24 R. F. Wilson, E. F. Mammen, J. G. Tyburski, K. M. Warsow, S. M. Kubinec, *J. Trauma* **1996**, *40*, 384–387.
25 J. T. Owings, M. Bagley, R. Gosselin, D. Romac, E. Disbrow, *J. Trauma* **1996**, *41*, 396–405.
26 B. Blauhut, S. Necek, H. Vinazzer, H. Bergmann, *Thromb. Res.* **1982**, *27*, 271–278.
27 F. Fourrier, C. Chopin, J. J. Huart, I. Runge, C. Caron, J. Goudemand, *Chest* **1993**, *104*, 882–888.
28 B. Eisele, M. Lamy, L. G. Thijs, H. O. Keinecke, H. P. Schuster, F. R. Matthias, F. Fourrier, H. Heinrichs, U. Delvos, *Intensive Care Med.* **1998**, *24*, 663–672.

29 F. Baudo, T.M. Caimi, F. de Cataldo, A. Ravizza, S. Arlati, G. Casella, D. Carugo, G. Palareti, C. Legnani, L. Ridolfi, R. Rossi, A. D'Angelo, L. Crippa, D. Giudici, G. Gallioli, A. Wolfler, G. Calori, *Intensive Care Med.* **1998**, *24*, 336–342.

30 H. Vinazzer, *Semin. Thromb. Haemost.* **1999**, *25*, 257–263.

31 B.L. Warren, A. Eid, P. Singer, S.S. Pillay, P. Carl, I. Novak, P. Chalupa, A. Atherstone, I. Penzes, A. Kubler, S. Knaub, H.O. Keinecke, H. Heinrichs, F. Schindel, M. Juers, R.C. Bone, S.M. Opal, *JAMA* **2001**, *286*, 1869–1878.

32 D. Rublee, S.M. Opal, W. Schramm, H.O. Keinecke, S. Knaub, *Crit. Care* **2002**, *6*, 349–356.

33 S.M. Opal, *Crit. Care Med.* **2000**, *28 (9 Suppl.)*, S34–S37.

34 F. Fourrier, M. Jourdain, A. Tournoys, *Crit. Care Med.* **2000**, *28 (9 Suppl.)*, S38–S43.

35 S.M. Opal, C.M. Kessler, J. Römisch, S. Knaub. *Crit. Care Med.* **2002**, 30 *(Suppl.)*, S325–S331.

36 J.P. Simons, M. McClenaghan, A.J. Clark, *Nature* **1987**, *328*, 530–532.

37 K. Gordon, E. Lee, J.A. Vitale, A.E. Smith, H. Westphal, L. Hennighausen, *Biotechnology (NY)* **1987**, *5*, 1183–1187.

38 H. Meade, L. Gates, E. Lacy, N. Lonberg, *Biotechnology (NY)* **1990**, *8*, 443–446.

39 A.J. Clark, *J. Mammary Gland Biol. Neopl.* **1998**, *3*, 337–350.

40 H.M. Meade, Y. Echelard, C.A. Ziomek, M.W. Young, M. Harvey, E.S. Cole, S. Groet, T.E. Smith, J.M. Curling, *Gene Expression Systems: Using Nature For The Art of Expression*, Academic Press, USA, **1998**.

41 D.P. Pollock, J.P. Kutzko, E. Birck-Wilson, J.L. Williams, Y. Echelard, H.M. Meade, *J. Immunol. Methods* **1999**, *137*, 147–157.

42 L.M. Houdebine, *Transgenic Res.* **2000**, *9*, 305–320.

43 E.A. Maga, J.D. Murray, *Biotechnology (NY)* **1995**, *13*, 1452–1457.

44 A.L. Archibald, M. McClenaghan, V. Hornsey, J.P. Simons, A.J. Clark, *Proc. Natl. Acad. Sci. USA* **1990**, *87*, 5178–5182.

45 J. Denman, M. Hayes, C. O'Day, T. Edmunds, C. Bartlett, S. Hirani, K.M. Ebert, K. Gordon, J.M. McPherson, *Biotechnology (NY)* **1991**, *9*, 839–843.

46 B. Roberts, P. DiTullio, J. Vitale, K. Hehir, K. Gordon, *Gene* **1992**, *121*, 255–262.

47 S.C. Bock, K.L. Wion, G.A. Vehar, R.M. Lawn, *Nucleic Acids Res.* **1982**, *10*, 8113–8125.

48 K.M. Ebert, J.P. Selgrath, P. DiTullio, J. Denman, T.E. Smith, M.A. Memon, J.E. Schindler, G.M. Monastersky, J.A. Vitale, K. Gordon, *Biotechnology (NY)* **1991**, *9*, 835–838.

49 A.S. Carver, M.A. Dalrymple, G. Wright, D.S. Cottom, D.B. Reeves, Y.H. Gibson, J.L. Keenan, J.D. Barrass, A.R. Scott, A. Colman, et al., *Biotechnology (NY)* **1993**, *11*, 1263–1270.

50 J. Williams, F.A. Ponce de Leon, P. Midura, M. Harrington, H. Meade, Y. Echelard, *Theriogenology* **1998**, *49*, 398.

51 T. Edmunds, S.M. Van Patten, J. Pollock, E. Hanson, R. Bernasconi, E. Higgins, P. Manavalan, C. Ziomek, H. Meade, J.M. McPherson, E.S. Cole *Blood* **1998**, *91*, 4561–4571.

52 P. Hellstern, U. Mober, M. Ekblad, C.U. Anders, B. Faller, S. Muller, *Haemostasis* **1995**, *25*, 193–201.

53 S.M. Van Patten, E. Hanson, R. Bernasconi, K. Zhang, P. Manavalan, E.S. Cole, J.M. McPherson, T. Edmunds, *J. Biol. Chem.* **1999**, *274*, 10268–10276.

54 H.A. Schreuder, B. de-Boer, R. Dijkema, J. Mulders, H.J. Theunissen, P.D. Grootenhuis, W.G. Hol, *Nat. Struct. Biol.* **1994**, *1*, 48–54.

55 B. Nordenman, C. Nystrom, I. Bjork, *Eur. J. Biochem.* **1977**, *78*, 195–203.

56 B. Turk, I. Brieditis, S.C. Bock, S.T. Olson, I. Bjork, *Biochemistry* **1997**, *36*, 6682–6691.

57 J. Swedenborg, *Blood Coagul. Fibrinolysis* **1998**, *9 (Suppl. 3)*, S7–S10.

58 L.-E. Franzen, S. Svensson, *J. Biol. Chem.* **1980**, *255*, 5090–5093.

59 T. Mizouchi, J. Fujii, K. Kurachi, A. Kobata, *Arch. Biochem. Biophys.* **1980**, *203*, 458–465.

60 G. Zettlmeissl, H.S. Conrad, M. Nimtz, H.E. Karges, *J. Biol. Chem.* **1989**, *264*, 21153–21159.

61 B. Fan, B.C. Crews, I.V. Turko, J. Choay, G. Zettlmeissl, P. Gettins, *J. Biol. Chem.* **1993**, *268*, 17588–17596.

62 E. Munzert, J. Muthing, H. Buntemeyer, J. Lehmann, *Biotechnol. Prog.* **1996**, *12*, 559–563.

63 E. Ersdal-Badju, A. Lu, X. Peng, V. Picard, P. Zendehrouh, B. Turk, I. Bjork, S.T. Olson, S.C. Bock, *Biochem. J.* **1995**, *310*, 323–330.

64 S.T. Olson, A.M. Frances-Chmura, R. Swanson, I. Bjork, G. Zettlmeissl, *Arch. Biochem. Biophys.* **1997**, *341*, 212–221.

65 H. Biescas, M. Gensana, J. Fernandez, P. Ristol, M. Massot, E. Watson, F. Vericat, *Haematologica* **1998**, *83*, 305–311.

66 J. L. Meagher, J. M. Beechem, S. T. Olson, P. G. W. Gettins, *J. Biol. Chem.* **1998**, *273*, 23283–23289.

67 D. H. Perlmutter, G. I. Glover, M. Rivetna, C. S. Schasteen, R. J. Fallon, *Proc. Natl. Acad. Sci. USA* **1990**, *87*, 3753–3757.

68 T. W. Barrowcliffe, C. A. Eggleton, M. Mahmoud, *Br. J. Haematol.* **1983** *55*, 37–46.

69 P. C. Cooper, S. Cooper, S. Kitchen, M. Makris, *J. Thromb. Haemost.* **2003**, *1 (Suppl. 1)*, 1016.

70 K. Okajima, *Immunol. Rev.* **2001**, *184*, 258–274.

71 V. De Palo, C. Kessler, S. M. Opal, *Adv. Sepsis* **2001**, *1*, 114–124.

72 J. Römisch, E. Gray, J. N. Hoffmann, C. J. Wiedermann, *Blood Coagul. Fibrinolysis* **2002**, *13*, 657–670.

73 C. J. Wiedermann, J. Römisch, *Acta Med. Austriaca* **2002**, *29*, 89–92.

74 M. Uchiba, K. Okajima, K. Murakami, H. Okabe, K. Takatsuki, *Thromb. Res.* **1995**, *80*, 201–208.

75 L. Ostrovsky, R. C. Woodman, D. Payne, D. Teoh, P. Kubes, *Circulation* **1997**, *96*, 2302–2310.

76 S. Dunzendorfer, N. C. Kaneider, A. Rabensteiner, C. Meierhofer, C. Reinisch, J. Römisch, C. J. Wiedermann, *Blood* **2001**, *97*, 1079–1085.

77 N. C. Kaneider, P. Egger, S. Dunzendorfer, C. J. Wiedermann, *Biochem. Biophys. Res. Commun.* **2001**, *287*, 42–46.

78 C. Feistritzer, N. C. Kaneider, D. H. Sturn, C. J. Wiedermann, *Clin. Exp. Allergy* **2004**, *34*, 696–703.

79 N. C. Kaneider, E. Förster, B. Mosheimer, D. H. Sturn, C. J. Wiedermann, *Thromb. Haemost.* **2003**, *90*, 1150–1157.

80 D. Gritti, A. Malinverno, C. Gasparetto, C. J. Wiedermann, G. Ricevuti, *Int. J. Immunopathol. Pharmacol.* **2004**, *17*, 27–32.

81 B. Risberg, *Blood Coagul. Fibrinolysis* **1998**, *9 (Suppl. 3)*, S3–S6.

82 F. B. Taylor, T. E. Emerson, R. Jordan, A. K. Chang, K. E. Blick. *Circ. Shock.* **1988**, *26*, 227–235.

83 G. Dickneite, E. P. Paques, *Thromb. Haemost.* **1993**, *69*, 98–102.

84 T. E. Emerson, *Blood Coagul. Fibrinolysis* **1994**, *5 (Suppl. 1)*, S37–S45.

85 C. M, Kessler, Z. Tang, H. M. Jacobs, L. M. Szymanski, *Blood* **1997**, *89*, 4393–4401.

86 G. Dickneite, *Semin. Thromb. Hemost.* **1998**, *24*, 61–69.

87 M. Uchiba, K. Okajima, K. Murakami, *Thromb. Res.* **1998**, *89*, 233–241.

88 G. Dickneite, B. Leithaüser, *Arterioscler. Thromb. Vasc. Biol.* **1999**, *19*, 1566–1572.

89 G. Dickneite, M. Kroez, *Blood Coagul. Fibrinolysis* **2001**, *12*, 1–9.

90 R. Nevière, A. Tournoys, S. Mordon, X. Maréchal, F. Song, M. Jourdain, F. Fourrier, *Shock* **2001**, *15*, 220–225.

91 M. C. Minnema, A. C. Chang, P. M. Jansen, Y. T. Lubbers, B. M. Pratt, B. G. Whittaker, F. B. Taylor, C. E. Hack, B. Friedman, *Blood* **2000**, *95*, 1117–1123.

92 J. N. Hoffmann, B. Vollmar, D. Inthorn, F. W. Schildberg, M. D. Menger, *Am. J. Physiol. Cell Physiol.* **2000**, *279*, C98–C107.

93 J. N. Hoffmann, B. Vollmar, J. Römisch, D. Inthorn, F. W. Schildberg, M. D. Menger, *Crit. Care Med.* **2002**, *30*, 218–225.

94 K. Murakami, R. McGuire, R. A. Cox, J. M. Jodoin, F. C. Schmalstieg, L. D. Traber, H. K. Hawkins, D. N. Herndon, D. L. Traber, *Crit. Care Med.* **2003**, *31*, 577–583.

95 K. Murakami, P. Enkhbaatar, R. A. Cox, H. K. Hawkins, L. D. Traber, D. L. Traber, *Am. J. Physiol.* (in press)

96 P. J. Cowan, A. Aminian, H. Barlow, A. A. Brown, K. Dwyer, R. J. A. Filshie, N. Fisicaro, D. M. A. Francis, H. Gock, D. J. Goodman, J. Katsoulis, S. C. Robson, E. Salvaris, T. A. Shinkel, A. B. Stewart, A. J. F. d'Apice, *Am. J. Transplant.* **2002**, *2*, 520–525.

97 W. Lu, T. G. K. Mant, J. H. Levy, J. M. Bailey, *Anesth. Analg.* **2000**, *90*, 531–534.

98 W. Dietrich, M. Spannagl, W. Schramm, W. Vogt, A. Barankay, J. A. Richter, *J. Thorac. Cardiovasc. Surg.* **1991**, *102*, 505–514.

99 A. H. Sabbagh, G. K. T. Chung, P. Shuttleworth, B. J. Applegate, W. Gabrhel. *Ann. Thorac. Surg.* **1984**, *37*, 466–468.

100 W. Dietrich, A. Schroll, A. Barankay, J. A. Richter, *Anesthesiology* **1985**, *63*, 3A.

101 K. Hashimoto, M. Yamagishi, T. Sasaki, M. Nakano, H. Kurosawa, *Ann. Thorac. Surg.* **1994**, *58*, 799–804.

102 M. R. Williams, A. B. D'Ambra, J. R. Beck, T. B. Spanier, D. L. Morales, D. N. Helman, M. C. Oz, *Ann. Thorac. Surg.* **2000**, *70*, 873–877.

103 J. H. Levy, G. J. Despotis, F. Szlam, P. Olson, D. Meeker, A. Weisinger, *Anesthesiology* **2002**, *96*, 1095–1102.

104 H. van Aken, G. Hartlage, E. Martin, J. Motsch, D. E. Birnbaum, C. Schmidt, E. Ott, R. O. Feneck, R. D. Latimer, A. Vuylsteke, J. B. Streisand, J. H. Levy, G. J. Despotis, *European Association of Cardiothoracic Anesthesiologists Annual Meeting*, Aarhus, Denmark June 2000. Meeting abstract.

105 M. S. Avidan, J. H. Levy, H. van Aken, R. O. Feneck, R. D. Latimer, E. Ott, E. Martin, D. E. Birnbaum, J. Bonfiglio, G. J. Despotis. *Anesthesiology* **2005**, *102*, 276–284.

106 B. A. Konkle, K. A. Bauer, R. Weinstein, A. Greist, H. E. Holmes, J. Bonfiglio, *Transfusion* **2003**, *43*, 390–394.

107 R. C. Tait, M. Morfini, I. D. Walker, M. Makris, G. Dolan, R. Menon, P. K. Noonan, J. Frieling, J. Bonfiglio, *Blood* **2002**, *100*, Abstract 3970.

108 R. C. Tait, B. A. Konkle, K. A. Bauer, J. Bonfiglio, G. Dolan, J. Frieling, A. Greist, H. E. Holmes, M. Makris, M. Morfini, B. Noonan, J. B. Streisand, R. Weinstein, I. D. Walker, Y. Echelard, *Ann. Hematol.* **2003**, *82 (Suppl. 1)*, S84.

Alea Non Iacta Est – Improving Established Expression Systems

12
Producing Modern Biopharmaceuticals:
The Bayer HealthCare Pharma Experience with a Range
of Expression Systems

Heiner Apeler

Abstract

Today, recombinant protein production involves many options. In addition to *E. coli*, several yeast systems (see Part IV, Chapter 13), insect cells (see Part IV, Chapter 14), different mammalian expression systems (CHO, BHK, NS0, HKB11, PER.C6) (see Part II, Chapter 3 and Part IV, Chapters 1 and 3) other alternative expression systems are currently under development for the production of biopharmaceuticals. These include transgenic animals or plants, and will be discussed in Part IV, Sub-Part 2 of this book. This chapter will focus on *E. coli*, a still-modern secretory *Saccharomyces cerevisiae* system, and the recently developed mammalian HKB11 expression system. An *E. coli* host/vector system is described that was originally developed for the efficient production of an interleukin-4 variant. Later, it transpired that this system is ideally suited to the expression of other proteins and Fab fragments. The secretory

S. cerevisiae system has long been known in biotechnology, but remains highly attractive; the expression of a protease inhibitor will be presented as an example. Recombinant human glycoprotein therapeutics should at best have structural identity with the natural product. A hybrid clone, designated HKB11 (hybrid of human kidney and B cells) is a favorable cell host for the production of human biopharmaceutical proteins. The host/vector system supports the production of gram quantities of proteins in a large-scale transient transfection format, as well as the development of stable cell lines. These systems, together with the baculovirus insect system, are used routinely within BayerHealthCare Pharma Biotechnology to produce biopharmaceutical proteins for research purposes [ultra high-throughput screening (uHTS)], for POC (proof-of-concept) studies, and for therapeutic applications.

Modern Biopharmaceuticals. Edited by J. Knäblein
Copyright © 2005 WILEY-VCH Verlag GmbH & Co. KGaA, Weinheim
ISBN: 3-527-31184-X

Abbreviations

BPTI   bovine pancreatic trypsin inhibitor
CAI   codon adaptation index
cDNA   complementary DNA
Dhfr   dihydrofolate reductase
EBV   Epstein-Barr virus
Fab   fragment antigen binding
GMP   good manufacturing practice
HAT   hypoxanthine-aminopterin-thymidine
HKB11   hybrid of human kidney and B cells
IL   interleukin
IL-2 SA interleukin 2-selective agonist
IPTG   isopropyl-β-D-thiogalactopyranoside
MBP   maltose-binding protein
MFα mating factor α
MTX   methotrexate
PEG   polyethyleneglycol
POC   proof-of-concept
Rop   repressor of primer
RSCU   relative synonymous codon usage
uHTS   ultra high-throughput screening
UV   ultra violet

12.1
The *Escherichia coli* Expression Platform

Mature human interleukin-4 (IL-4) is composed of 129 amino acids. IL-4 variants that are able to block both IL-4 and IL-13 activities have been described [1]. These antagonistic properties are regarded as useful for the treatment of diseases which involve T_{H2} development and/or IgE production [2]. For the set-up of a viable commercial production system for an IL-4 variant (IL-4 v), a broad screening of host organisms and expression systems on a small scale was conducted (data not shown).

After this screening, it became clear that intracellular expression of IL-4 v forming inclusion bodies in *E. coli* was by far the most suitable and productive system. With many of the evaluated systems it was possible to express IL-4 v, but the yield was generally rather low as compared to the *E. coli* inclusion body expression system. Therefore, it was clear that this system should be optimized for the intracellular expression of IL-4 v inclusion bodies in *E. coli*.

The main criteria for an efficient and safe expression system are:
• a high product yield,
• regulatable stable expression, and
• stability of the expression vector.

The initial production process for Bay IL-4 v relied upon a thermoinducible system involving the λP_R-promoter and the heat-sensitive cI857 repressor [1]. In comparison to a heat-induced expression system, an isopropyl-β-D-thiogalactopyranoside (IPTG) or lactose-induced system has several advantages (no heat shock proteins are induced, modulation of induction is possible, and *E. coli* can be grown at its temperature optimum). Promoters that are inducible through IPTG or lactose include the trc, T5, and T7 promoters. All of these promoters in different vector backgrounds were tested for the expression of IL-4 v. In addition to the promoter, several other features of an expression plasmid are important for the criteria listed above. These are:
• ribosomal binding site (rbs),
• codon usage of the corresponding gene,
• transcriptional terminator,
• resistance gene,
• regulation of expression, and
• origin of replication (ori).

Expression plasmids for IL-4 v with modifications in all of these elements were gen-

Table 12.1 Criteria for evaluation of an *E. coli* inclusion body expression system

Criterion	Experimental evaluation
Yield of IL-4 v	SDS-PAGE and capillary gel electrophoresis
Plasmid stability	Replica plating over 78 generations without antibiotic selection
Maintenance of induction capability	Small-scale expression studies (SDS-PAGE analysis) over 78 generations without antibiotic selection
Performance of the fermentation process	Fermentation at the 10-L scale
Performance of the renaturation and purification process	Routine renaturation and purification of material derived from 10 L fermentation

erated during the course of the development process. The quality and suitability of the corresponding expression system was ranked mainly according to five criteria shown in Table 12.1.

The finally selected expression vector pRO2.1.O, in combination with the host cell *E. coli* W3110, ideally fulfills all criteria. The expression vector pRO2.1.O contains the following elements.

12.1.1
T5 Promoter

The *E. coli* phage T5 promoter, together with two lac operator, sequences is derived from the pQE30 plasmid belonging to the pDS family of plasmids [3, 4].

12.1.2
T7 g10 Ribosomal Binding Site

The ribosomal binding site (rbs) is derived from the region upstream from gene 10 of the phage T7 (T7 g10 leader). Gene 10 of phage T7 codes for the coat protein, which is the major protein expressed after T7 infection. The T7 g10 rbs was obtained from the vector pET-9a [5]. It is one of the strongest rbs known. The T7 g10 leader spans a region of about 100 bp. In the final construct pRO2.1.O the region up-

stream of the XbaI site is deleted [6]. Olins et al. described that the section of mRNA between the *Xba*I site within the T7 g10 leader and the initiator methionine codon is sufficient for its function. The T7 g10 leader sequence in pRO2.1.O now spans 42 bp and harbors one base exchange from G to A in position 16 (starting from the *Xba*I site).

12.1.3
Codon Usage of the IL-4 v Gene

As an effective measure of synonymous codon usage bias, the codon adaptation index (CAI), can be useful for predicting the level of expression of a given gene [7, 8]. The CAI is calculated as the geometric mean of relative synonymous codon usage (RSCU) values corresponding to each of the codons used in a gene, divided by the maximum possible CAI for a gene of the same amino acid composition. RSCU values for each codon are calculated from very highly expressed genes of a particular organism (e.g., *E. coli*), and represent the observed frequency of a codon divided by the frequency expected under the assumption of equal usage of the synonymous codons for an amino acid. Highly expressed genes (e.g., genes encoding ribosomal proteins) have generally high CAI values

≥0.46. Poorly expressed genes (e.g., *lacI* and *trpR* in *E. coli*) have low CAI values ≤0.3 [7].

The calculated *E. coli* CAI value for the natural IL-4 v gene is 0.733. This means that the natural gene should be – and in fact is – well-suited for high-level expression in *E. coli*. Nevertheless it was felt that a synthetic gene with optimal *E. coli* codon usage (CAI value = 1) has the potential to further increase the expression level.

Often, restriction sites are introduced into the synthetic gene sequence to allow for the assembly and cloning of small sections of the gene. In this case, however, internal restriction sites were omitted in order to make no compromises regarding the optimal codon usage. An *Nde*I and a *Bam*HI site were added to the 5′ and 3′ ends of the synthetic gene, respectively. The *Nde*I site includes the ATG start codon, and the *Bam*HI site immediately follows the TGA stop codon.

12.1.4
Transcriptional Terminator

A T7 DNA fragment containing the transcription terminator Tφ is derived from the vector pET-9a [5]. Transcriptional terminators determine the points where the mRNA-RNA polymerase–DNA complex dissociates, thereby ending transcription. The presence of a transcriptional terminator at the end of a highly expressed gene has several advantages: they minimize sequestering of RNA polymerase that might be engaged in unnecessary transcription; they restrict the mRNA length to the minimal, thus limiting energy expense; as strong transcription may interfere with the origin of replication, a transcriptional terminator increases plasmid stability due to copy number maintenance [9].

12.1.5
Resistance Gene

The kan resistance gene is derived from the vector pET-9a [5]. Originally, this is the kan gene of Tn903 from the vector pUC4 KISS [10]. In the final vector pRO2.1.O the kan gene and the IL-4 v gene have opposite orientations, so there should be no increase in the kan gene product after induction due to read-through transcription from the T5 promoter. Kanamycin was chosen as selective marker because it is the preferred antibiotic for GMP-purposes. In addition, kan gene-based vectors are more stable than ampicillin-resistant (bla) plasmids. Ampicillin selection tends to be lost in cultures as the drug is degraded by the secreted β-lactamase enzyme. In contrast to that, the mode of bacterial resistance to kanamycin relies upon an aminoglycoside phosphotransferase that inactivates the antibiotic.

12.1.6
Regulation of Expression

Controlled gene expression is absolutely necessary for the set-up of a stable plasmid system, particularly if the protein of interest is deleterious to the host cell. The expression vector pRO2.1.O uses a lac-based inducible system consisting of a lac repressor gene (lacI) and two synthetic lac operator sequences fused downstream to the *E. coli* phage T5 promoter. The lacI^q promoter and the lacI structural gene were isolated from the vector pTrc99A [11]. I^q is a promoter mutation which leads to overproduction of the lacI repressor. The wild-type lac repressor is a tetrameric molecule comprising four identical subunits of 360 amino acids each. The lac repressor tetramer is a dimer of two functional dimers. The four subunits are held together by a four-helix bundle formed from residues 340 to 360. Due to the isolation of the lacI gene from the vector pTrc99A by a *Nar*I restriction enzyme cut, the residues beyond amino acid 331 are deleted and 10 amino acids not normally encoded in the lacI gene are added. It is known that mutations or deletions that occur in the C-terminal part of lacI, beyond amino acid 329, result in functional dimers that appear phenotypically similar to the wild-type repressor [12].

12.1.7
Origin of Replication (ori)

The origin of replication (ori) of the expression plasmid pRO2.1.O is derived from the vector pET-9a, the ori of which originates from pBR322. pRO2.1.O therefore carries the pMB1 (ColE1) replicon. Plasmids with this replicon are multicopy plasmids that replicate in a "relaxed" fashion. A minimum of 15–20 copies of plasmid are maintained in each bacterial cell under normal growth conditions. The actual number for pRO2.1.O has been determined to be 14–25 copies per cell. Replication of pRO2.1O from its ColE1-type ori is initiated by a 555-nucleotide RNA transcript, RNA II, which forms a persistent hybrid with its template DNA near the ori. The RNA II–DNA hybrid is then cleaved by RNase H at the ori to yield a free 3' OH that serves as a primer for DNA polymerase I. This priming of DNA synthesis is negatively regulated by RNA I, a 108-nucleotide RNA molecule complementary to the 5' end of RNA II. Interaction of the antisense RNA I with RNA II causes a conformational change in RNA II that inhibits binding of RNA II to the template DNA and consequently prevents the initiation of plasmid DNA synthesis. The binding between RNAs I and II is enhanced by

a small protein of 63 amino acids (the Rop protein, *R*epressor *o*f *p*rimer), which is encoded by a gene located 400 nucleotides downstream from the origin of replication [13, 14]. Deletion of the *rop* gene leads to an increase in copy number and, due to a gene dosage effect, to enhanced expression levels of the plasmid-encoded heterologous gene. This observation was also made for the IL-4 v expression vectors tested. However, it transpired that *rop⁻*-plasmids are instable and lost very rapidly during fermentation under non-selective conditions. Therefore, the replicon of pRO2.1.O contains the *rop* gene to ensure high plasmid stability.

The vector pRO2.1.O lacks the *mob* gene that is required for mobilization and is therefore incapable of directing its own conjugal transfer from one bacterium to another. The nic/bom site is present in pRO2.1.O.

In the process of cloning of pRO2.1.O, all elements not necessary for plasmid replication, resistance and regulatable expression were deleted. Therefore, pRO2.1.O can be termed a "cleaned-up vector". This is important for regulatory purposes.

With this expression system (W3110/pRO2.1.O) very high IL-4 v yields of 10.2 g L^{-1} and a specific IL-4 v content of 248 mg g^{-1} cell dry weight are obtained under high cell density conditions. With other IL-4 v expression vectors these yields were never achieved (see Table 12.2). Table 12.2 also summarizes the results from the plasmid stability tests. The vector pRO2.1.O remains fully stable over 78 generations without antibiotic selection (Fig. 12.1). In contrast, the expression vector pRO21.1.O, which is based on the pET-30a plasmid, is lost very rapidly. Only 16% of the colonies contain the plasmid after 78 generations without kanamycin selection. In addition, it is very important that the capability of

Table 12.2 Important features of selected IL-4 v expression plasmids

Vector	Remarks[a)]	Strain	Plasmid stability[a)]	Expression level [mg IL-4 g^{-1} dry weight]	Fermentation yield [g L^{-1}][b)]	Expression before induction
pWü4A2	λP_R-promoter	JM103 recA$^-$	++++	86	0.2 low cell density complex medium	No
pAD49	T7-promoter rop$^+$ pET-9a	W3110 (DE3)	++++	85	4.4	No
pRO17.4.M	T7-promoter rop$^-$ cleaned up[c)]	W3110 (DE3)	+++	230	6.3	Yes
pRO17.12.M	T7-promoter lacIq rop$^+$ cleaned up	W3110 (DE3)	+++	140	3.7	Yes
pRO21.1.O	T7-promoter lacIq rop$^+$	W3110 (DE3)	++	209	7.2	No
pRO2.1.O	T5-promoter lacIq rop$^+$ cleaned up	W3110	++++	248	10.2	No

a) ++++ 100% stability after 80 generations; +++ 80–100% stability after 80 generations; ++ 60–80% stability after 80 generations.
b) Fed-batch fermentation, mineral medium, 10-L scale.
c) All elements not necessary for plasmid replication and IL-4 v expression removed.

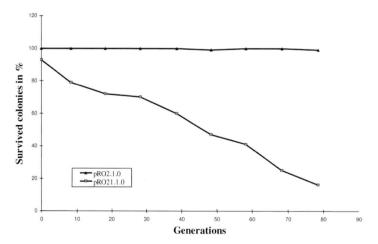

Fig. 12.1 Plasmid stability of the IL-4 v expression vectors pRO21.1.O (T7-promoter based) and pRO2.1.O (T5 promoter-based). For details, see Table 12.2.

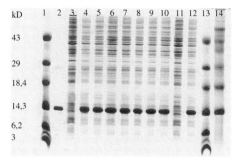

Fig. 12.2 Maintenance of induction and expression capability of the IL-4 v expression vector. SDS-PAGE (15%) analysis of total extracts from the *E. coli*/pRO2.1.O strain grown for different numbers of generations without antibiotic selection after 5 h of induction with 0.5 mM IPTG. The gel was run under reducing conditions and stained with Coomassie brilliant blue. Lane 1: Molecular weight marker (Low Range, Life Technologies), lane 2: IL-4 v (5 μg); lane 3: *E. coli*/ pRO2.1.O (eight generations) before induction; lane 4: eight generations after induction; lane 5: 18 generations after induction; lane 6: 28 generations after induction; lane 7: 38 generations after induction; lane 8: 48 generations after induction; lane 9: 58 generations after induction; lane 10: 68 generations after induction; lane 11: 78 generations before induction; lane 12: 78 generations after induction; lane 13: molecular weight marker (Low Range, Life Technologies); lane 14: molecular weight marker (High Range, Life Technologies).

selective conditions. After 78 generations without kanamycin selection the expression system is as effective as a fresh culture (Fig. 12.2).

In conclusion, with the development of the W3110/pRO2.1.O expression system, a viable production system for an IL-4 variant has been established. In addition, the newly developed host/vector system is ideally suited for the expression of other cytokines and cytokine muteins. The vector has been adapted to harbor N- and C-terminal affinity tags as well as solubility tags such as maltose-binding protein (MBP), NusA and thioredoxin for the high-throughput production of proteins derived from genomics and proteomics approaches. Data have also been acquired which show that the system can be used for the periplasmic expression of Fab-fragments. In our hands, this system is the one of choice for the expression of many different proteins.

12.2
The *Saccharomyces cerevisiae* Expression Platform

Aprotinin – which is also known as bovine pancreatic trypsin inhibitor (BPTI) – belongs to the family of Kunitz-type inhibitors, and inhibits serine proteases such as trypsin, chymotrypsin, plasmin, and plasma kallikrein [15]. Aprotinin consists of 58 amino acids. The aprotinin variant (DesPro(2)-Ser(10)-Arg(15)-Ala(17)-Asp(24)-Thr(26)-Glu(31)-Asn(41)-Glu(53)-aprotinin) was designed by means of rational mutagenesis, and differs from aprotinin by two amino acids in the active site and by seven amino acids in the backbone. The changes in the active site of the aprotinin variant increase the potency towards inhibition of plasma kallikrein, whereas the inhibition of plasmin is only marginally reduced.

regulatable induction and high yield expression is maintained under non-selective conditions. During development of the IL-4 v expression vector, it became clear that some of the T7-promoter based plasmids are very stable (see for example the vector pAD49 in Table 12.2), but that the ability of expression of the desired protein is lost very rapidly. This is due to mutations in the T7 RNA polymerase gene within the DE3 prophage (H. Wehlmann, unpublished data). With the vector pRO2.1.O, expression of IL-4 v as well as regulatable induction with the inductor IPTG, are very stable under non-

Expression of recombinant aprotinins has been achieved in *E. coli* K12 as a fusion with parts of the MS-2 polymerase gene [16]. In this case the fusion protein is deposited as inclusion bodies. Functionally active aprotinin can be obtained after solubilization and purification of the fusion protein, CNBr (cyanobromide) cleavage and renaturation of the aprotinin moiety. Low-level periplasmic expression of native, properly folded aprotinin has been shown in *E. coli* employing the *E. coli* alkaline phosphatase signal sequence [17]. With respect to expression level and ease of purification, it transpired that secretory expression in the yeast *Saccharomyces cerevisiae* is by far the most attractive system for the production of this aprotinin variant [18]. In addition, due to the absence of an N-glycosylation site there are no problems with non-human glycosylation of the protein in yeast.

The backbone of the yeast expression vector is derived from the commercially available vector pYES2 (Invitrogen). The GAL1 promoter and f1 ori region were re-moved from this vector and replaced by the *S. cerevisiae* mating factor *α*1 (MF*α*1) promoter and part of the MF*α*1 precursor sequence (the MF*α*1 pre-pro-sequence).

The DesPro2-Ser10-Arg15-Ala17-Asp24-Thr26-Glu31-Asn41-Glu53 aprotinin variant cDNA was fused to the *S. cerevisiae* *α*-mating factor pre-pro-signal sequence. The yeast expression vector for the aprotinin variant and a detailed description of the MF*α*1 leader processing is shown in Fig. 12.3. Processing of the pre-pro-*a*-factor aprotinin variant fusion construct requires two different proteolytic activities. A signal peptidase, localized in the endoplasmic reticulum, first cleaves between the pre and the pro-sequence. The glycosylated pro-*a*-factor aprotinin variant is subsequently cleaved by an endoproteinase at the carboxyl side of the dibasic sequence Lys-Arg, thereby releasing the aprotinin variant into the supernatant. This protease is the product of the KexII gene [19].

Yeast cells transformed with the corresponding vector are capable of secreting large amounts of correctly processed apro-

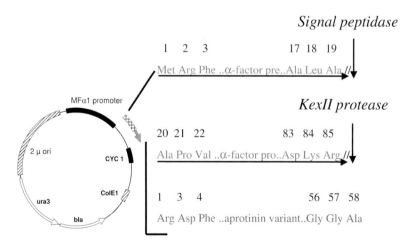

Fig. 12.3 Yeast expression vector for the DesPro2-Ser10-Arg15-Ala17-Asp24-Thr26-Glu31-Asn41-Glu53 aprotinin variant and processing of the MF*α*1 leader peptide.

tinin variant into the supernatant. In shaking-flask cultures, yields between 20 and 40 mg L^{-1} can be obtained by employing the MFα1 promoter. Fed-batch fermentations at the 10-L scale resulted in product concentrations in the supernatant of 150–230 mg L^{-1}.

Aprotinin variants with a deletion of the amino acid proline in position 2 (DesPro2) can be effectively produced using the described system.

Expression of recombinant aprotinin variants with the natural N-terminal sequence "Arg-Pro-Asp" is not possible because the KexII protease is unable to cleave off the MFα1 pre-pro signal peptide in the corresponding context.

Aprotinin variants with the natural N-terminal sequence "Arg-Pro-Asp" can be secreted from yeast by completely removing the α-factor-pro sequence and fusing the aprotinin variant gene directly to the α-factor pre sequence. In this construct, the processing of the α-factor aprotinin fusion only depends on the signal peptidase and avoids the "classical" KexII protease activity. Large amounts of homogeneously processed aprotinin variants can be obtained using this approach.

In conclusion, the described yeast system is well suited for the secretory expression of small disulfide-bonded, non-glycosylated proteins. It should be emphasized that the performance of yeast cells for expression of a particular protein can be further optimized by conventional strain development programs using UV-irradiation or treatment with mutagenic compounds.

12.3
The HKB11 Expression Platform

HKB11 is a human/human hybrid cell line that was generated by PEG-fusion of human embryonic kidney cells (293S) and Burkitt's lymphoma cells (2B8). 2B8 is a G418-resistant, HAT (hypoxanthine-aminopterin-thymidine)-sensitive clone derived from HH514-16. HH514-16 harbors a non-transforming and heterogeneous Chet/DNA-free EBV (Epstein-Barr virus) genome that does not produce EBV particles. Generation of the cell line has been described in detail [20]. The initial purpose of establishing a hybrid cell line of 293S and a human suspension B-cell line (2B8) was to resolve the aggregation problem of 293S cells, which tend to clump when grown in suspension culture. The HKB11 cell line was selected for non-aggregating properties. Adaptation of HKB11 cells to serum-free suspension conditions is readily possible. Compared to 293S cells, HKB11 cells form smaller and looser aggregates under serum-free suspension conditions. Although the transformation or hybridization events, required in most cases to produce stable cell lines, may result in altered glycosylation profiles, it was found that HKB11 cells possess typical human glycosylation enzymes such as alpha(2,3) and alpha(2,6) sialyl transferases. Moreover, proteins produced from transfected HKB11 cells were found to be capped with sialic acid of alpha(2,3) and alpha(2,6) linkages. The characteristics of the HKB11 cell line are summarized in Table 12.3.

Due to the expression of EBNA1, HKB11 cells are an excellent cell host for oriP expression vectors. Expression of an IL-2 selective agonist (IL-2 SA) using an oriP episomal vector (pCEP4 from Invitrogen) has been described [21]. The pCEP4 vector harbors (besides oriP and EBNA1) a

Table 12.3 Comparison of HKB11 with parental cell lines

Characteristics	293S	2B8	HKB11
Chromosome number	64	47	90–110
EBV genome	Negative	Positive	Positive
EBNA	–	Positive	Positive
VCA	–	ND	Negative
VCA after transfection with BZLF1	–	ND	Positive (0.2% of cell population)
EBV transformation	–	Negative	ND
het DNA	–	Negative	ND
Surface Ig-μ	Negative	Positive	Negative
Aggregates	Large (tight)	Small	Small (loose)
Sialyltransferase	$a(2,3)$, $a(2,6)$	ND	$a(2,3)$, $a(2,6)$
MSX sensitivity (μM)	ND	ND	150
MTX sensitivity (nM)	ND	ND	50

ND = not determined; EBNA = Epstein-Barr virus nuclear antigen; VCA = EBV capsid antigen; BZLF1 = EBV latency-interrupting gene involved in the viral lytic cycle; MSX = methionine sulfoximine; MTX = methotrexate.

functional hygromycin resistance gene and a CMV enhancer/promoter. Expression was further optimized by incorporating the HIV transactivator Tat and its response element TAR into this vector [21]. A synergistic effect was observed on the expression of IL-2 SA when Tat/TAR and oriP/EBNA1 elements were present in a single vector employing a transient transfection approach.

Tat and TAR sequences were subcloned into different TAR-oriP and Tat-oriP vectors due to the large plasmid size of the Tat/TAR-oriP vector (12.1 kbp) and to further optimize the cloning procedure by using smaller-sized plasmids [22]. Methods have been developed to use these vectors for the production of milligram and gram quantities of proteins and recombinant antibody candidates in a large-scale transient transfection scheme. Usually, transfection is performed with 500 µg of plasmid DNA containing the gene of interest and DMRIE-C transfection reagent (Invitrogen) in a 500-mL suspension culture of HKB11

cells. At 3 days post transfection, the culture from the shake-flask is transferred to a Cellbag (Wave Biotech LLC) and the culture volume sequentially increased to 10 L. At 10 days post transfection, the cells or supernatant are harvested for purification of proteins [22]. Under hygromycin B drug selection this process can be extended, and gram quantities of proteins and recombinant antibodies can be obtained in a matter of weeks.

Although HKB11 is not a dhfr (dihydrofolate reductase)-negative cell line, stable clones can be generated in a selection medium containing 100 nM MTX (methotrexate). Amplification of the gene cassette with increasing concentrations of MTX is possible comparable to the CHO dhfr⁻ system.

In conclusion, the described HKB11 host/vector system supports the rapid production of milligram and gram quantities of proteins and recombinant antibodies in a large-scale transient transfection format to generate material for proof of concept

(POC) studies. In addition, the HKB11 expression system can be used to develop cell lines for stable production and clinical manufacturing of biopharmaceutical products.

12.4
Outlook and Conclusion

Although, during the past few years, mammalian cells have surpassed microbial systems for the production of biopharmaceuticals, it is extremely important to have established different expression platforms. In addition to the systems described in this chapter (*E. coli* inclusion body, secretory *S. cerevisiae* and HKB11), we routinely use *E. coli* for periplasmic expression of Fab fragments, insect cells (only for research purposes) and CHO cells in batch and perfusion cultures. Based on experience with different classes of proteins, a broad screening of all available expression systems is in most cases not warranted. Yeast (*S. cerevisiae*) is an option with rather small secreted and non-glycosylated proteins. The engineering of yeast systems (*Pichia*, *Hansenula*) for human glycosylation or defined glycosylation (e.g., for tissue targeting) will clearly boost the utilization of this system for therapeutic applications (see Part IV, Chapter 7). *E. coli* (and other emerging bacterial expression systems) can also in the future be the system of choice, for example in the secretory expression of Fab fragments or single chain antibodies. It may even be worthwhile considering inclusion body-based processes for this type of biopharmaceutical. Whereas small peptides are in most cases produced by chemical synthesis, *E. coli* offers the possibility for the production of large amounts of peptides, and in the future even peptides containing non-protei-

nogenic amino acids may be accessible. Higher cell densities in bioreactors, engineered host cells and the application of high-throughput screening equipment for the screening and selection of high-producer clones will be the driving force for mammalian cells. Scaled-up approaches for transiently transfected cells – such as the HKB11 system described here – are able to produce grams of recombinant protein and antibody in a relatively short time frame. These systems may offer advantages compared to the conventional procedures, and may therefore become more applicable on a broader basis, perhaps with respect to personalized medicine or the development of other modern biopharmaceuticals.

References

1 N. Kruse, H. P. Tony, W. Sebald, EMBO J. **1992**, 11, 3237–3244.
2 J. J. Ryan, J. Allergy Clin. Immunol. **1997**, 99, 1–5.
3 H. Bujard, R. Gentz, M. Lanzer, D. Stüber, M. Müller, I. Ibrahimi, M. T. Häuptle, B. Dobberstein, Methods Enzymol. **1987**, 155, 416–433.
4 D. Stüber, H. Matile, G. Garotta, in: I. Lefkovits and B. Pernis (Eds), Immunological Methods. Academic Press, Inc., **1990**, Vol. IV, pp. 121–152.
5 F. W. Studier, A. H. Rosenberg, J. J. Dunn, J. W. Dubendorff, Methods Enzymol. **1990**, 185, 60–89.
6 P. O. Olins, C. S. Devine, S. H. Rangwala, K. S. Kavka, Gene **1988**, 73, 227–235.
7 P. M. Sharp, W.-H. Li, Nucleic Acids Res. **1987**, 15, 1281–1295.
8 H. Apeler, U. Gottschalk, D. Guntermann, J. Hansen, J. Mässen, E. Schmidt, K.-H. Schneider, M. Schneidereit, H. Rübsamen-Waigmann, Eur. J. Biochem. **1997**, 247, 890– 895.
9 P. Balbas, F. Bolivar, Methods Enzymol. **1990**, 185, 14–37.
10 F. Barany, Gene **1985**, 37, 111–123.
11 E. Amann, B. Ochs, K.-J. Abel, Gene **1988**, 69, 301–315.

12 H.C. Pace, M.A. Kercher, P. Lu, P. Markie-
wicz, J.H. Miller, G. Chang, M. Lewis, Trends
Biochem. Sci. **1997**, 22, 334–339.

13 J. Sambrook, E.F. Fritsch, T. Maniatis, Molec-
ular Cloning: A Laboratory Manual, 2nd edn.,
Cold Spring Harbor Laboratory, Cold Spring
Harbor, NY **1989**, Vol. 1.

14 R. Lahijani, G. Hulley, G. Soriano, N.A. Horn,
M. Marquet, Hum. Gene Ther. **1996**, 7, 1971–
1980.

15 W. Gebhard, H. Tschesche, H. Fritz, in: A.J.
Barrett and G.S. Salvesen (Eds), Proteinase
Inhibitors. Elsevier Science Publ. BV **1986**,
pp. 375–387.

16 E.-A. Auerswald, D. Hörlein, G. Reinhardt,
W. Schröder, E. Schnabel, Biol. Chem. Hoppe
Seyler **1988**, 369, Suppl., 27–35.

17 C.B. Marks, M. Vasser, P. Ng, W. Henzel,
S. Anderson, J. Biol. Chem. **1986**, 261 (16),
7115–7118.

18 H. Apeler, J. Peters, W. Schröder, K.-H.
Schneider, G. Lemm, V. Hinz, G.J. Rossouw,
K. Dembowsky, ArzneimForsch – DrugRes
2004, 54 (8), 483–497.

19 A.J. Brake, Methods Enzymol. **1990**, 185,
408–421.

20 M.S. Cho, H. Yee, S. Chan, J. Biomed.
Science **2002**, 9, 631–638.

21 M.S. Cho, H. Yee, C. Brown, K.-T. Jeang,
S. Chan, Cytotechnology **2001**, 37, 23–30.

22 M.S. Cho, H. Yee, C. Brown, B. Mei, C. Mi-
renda, S. Chan, Biotechnol. Prog. **2003**, 19,
229–232.

13

Advanced Expression of Biopharmaceuticals in Yeast at Industrial Scale: The Insulin Success Story

Asser Sloth Andersen and Ivan Diers

Abstract

Expression of insulin precursors in the yeast *Saccharomyces cerevisiae* has in recent years formed the basis for the manufacture of the majority of the human recombinant insulin and insulin analogues sold by Novo Nordisk A/S. In this chapter, we describe the composition of the yeast expression system and review a strategy for modification of the insulin precursors in order to improve the expression system. This strategy involves adaptation of the insulin precursor for efficient secretion and processing, and has highlighted the importance of addressing every event leading to secretion of the insulin precursor from the initial translocation and folding in the endoplasmic reticulum via transport through the Golgi apparatus, to maturation by kexin and ultimately secretion via correct localization to secretory vesicles. The modifications of the insulin precursors described in this chapter have led to significant improvements in the expression system, resulting in robust scalable expression systems amenable to industrial fermentation and downstream processing of the insulin precursor.

Abbreviations

ER   endoplasmic reticulum
FLP family of lambda phage (FLIP)
POT plasmid encoded selection
SV   secretory vesicles
YPS1 yapsin 1 aspartyl endoprotease

13.1
Introduction

13.1.1
Insulin and Diabetes

Insulin is a naturally occurring peptide hormone produced by the β-cells in the Langerhans islets of the pancreas in response to hyperglycemia. Insulin facilitates the entry of glucose into target tissues – such as muscle, adipose tissue and liver – by binding to and activating specific membrane receptors on these cells. Diabetes mellitus is a group of metabolic diseases characterized by high blood sugar (glucose) levels, which result from defects in insulin secretion, or action, or both. In Type 1 diabetes this may be due to β-cell destruction, and in Type 2 diabetes to a combination of β-cell failure and resistance of target tissues to insulin action (insulin resistance). The latter disease can, in its

Modern Biopharmaceuticals. Edited by J. Knäblein
Copyright © 2005 WILEY-VCH Verlag GmbH & Co. KGaA, Weinheim
ISBN: 3-527-31184-X

early stages, be helped by low calorie non-sugar diet and/or treatment with oral anti-diabetic drugs, while the later stages and Type 1 diabetes require insulin treatment (see also Part VI, Chapter 4). During the first 60 years after the discovery of insulin by Frederick Banting and Charles Best in 1921–1922, and their successful treatment of diabetics, insulin extracted from bovine or porcine pancreases was the drug available to treat Type 1 diabetics.

13.1.2
Human Insulin

With the rapid increase in the incidence of diabetes among the population, it is no longer possible to satisfy the pharmaceutical requirement (estimated to be 15–20 tonnes per year in 2005) from animal sources. Furthermore, the animal-extracted insulins are slightly different from human insulin, which might cause formation of insulin-binding antibodies and allergic reactions. Porcine insulin, which differs from human insulin only by a single amino acid in position B30, can be converted to human insulin in a transpeptidation reaction, in which an alanine is replaced with a threonine [1].

13.1.3
Biosynthetic Human Insulin

Recent developments in molecular biology and biotechnology have opened up new possibilities, among which is the biosynthesis of human insulin. Insulin is composed of two disulfide-linked peptide chains referred to as the A chain and B chain. The first recombinant approach used *Escherichia coli* as host and expression of A- and B-chains as fusion-proteins in separate strains [2]. In a later approach in *E. coli*, proinsulin (B-chain-Connecting-peptide-A-chain) was ex-

pressed also as a fusion protein [3]. In both of these systems the fusion-proteins were isolated as inclusion bodies, and several chemical steps were needed for dissolution, cleavage, folding, and formation of disulfide bridges. Later, Thim et al. [4] reported the successful expression of single chain insulin precursors with a mini C-peptide produced and secreted in the yeast *Saccharomyces cerevisiae*, and which containing the correct disulfide bridges. Subsequently, Markussen et al. [5] reported the successful expression and conversion of other mini C-peptide insulin precursors to human insulin, with minimal post-fermentation chemistry, and purification of the product. The *S. cerevisiae* insulin expression system [5] could be scaled up for the stable large-scale production of insulin [6].

13.1.4
Novel Administration Methods and Insulin Analogues

With the introduction of novel routes for administration of insulin, such as the intrapulmonary route (see Part VI, Chapter 4), where bioavailability is markedly lower than with subcutaneous administration, the demand for insulin will increase. The demand for a more optimal treatment of the patient has called for the design and development of new fast- and slow-acting insulin analogues [7–10] (see Introduction) and has required alterations of the yeast process. The development and optimization of the *S. cerevisiae* expression system for biosynthesis of human insulin has recently been extensively reviewed [11, 12].

13.1.5
S. cerevisiae Host Expression System

The yeast *S. cerevisiae* (known as baker's yeast) has played a major role in food production for several thousand years by its use in wine, beer, and bread making. In later years, *S. cerevisiae* has also been developed as an efficient eukaryotic expression system for biotechnology (for a review, see [13]). *S. cerevisiae* secretes very few endogenous proteins, which makes the purification of secreted heterologous proteins easier. In addition, it is robust and well adapted to the physical stress in large-scale cultivation.

Here, we review the possibilities for optimization of the production process for human insulin and its analogues in a specific *S. cerevisiae* host system described below and used by Novo Nordisk A/S.

The *S. cerevisiae* host cell (reviewed in [12]) is a diploid with the genotype (*MATa/ MATα pep4-3/pep4-3 leu2-3,112/leu2-3,112 HIS4/his4 tpi1::LEU2/tpi1::LEU2* cir⁺). The *pep4-3* mutation impairs the normal proteolytic activities in the vacuole [14], and is thought to improve heterologous protein production. The deletion of *TPI1* allows autoselection of the *POT* expression *S. cerevisiae-E. coli* 2μ-pBR322 shuttle plasmid (Fig. 13.1) and the use of the strong con-

stitutive TPI1 promoter and terminator as initiating and termination signals for the transcription of DNA encoding heterologous proteins inserted between the two. The non-transformed host strain grows poorly, when glucose is the only carbon source. Thus, transformation with the *POT* expression plasmid allows selection by the ability to grow on glucose [15, 16]. The *Schizosaccharomyces pombe POT* gene is homologous to *TPI1*, but is poorly expressed in *S. cerevisiae*, and multiple copies of the *POT* plasmid are therefore required to generate sufficient gene product to allow growth on glucose as the sole carbon source [15, 16]. The *POT* plasmid copy number is probably influenced by counteracting selective forces, including the need for complementation of the host marker (*tpi1Δ*) by the plasmid-encoded selection gene (*POT*) and a selective advantage of cells with a low copy number enabling them to limit the burden of fusion protein expression and secretion [12]. The 2μ part of the plasmid has been slightly modified to inhibit recombination events. The *FLP*-gene has been truncated and the inverted

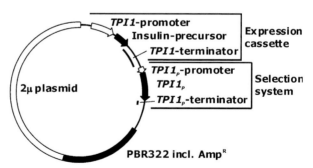

Fig. 13.1 *S. cerevisiae* expression vector used for production of recombinant insulin. The *S. cerevisiae/E. coli* shuttle vector is composed of the following genetic units: 1) The transcription promoter and terminator of the *S. cerevisiae TPI1* (triose phosphate isomerase) gene flanking the DNA encoding the leader-insulin precursor. 2) The *TPI1* gene from *Schizosaccharomyces pombe* (*TPI1ₚ*)

used as a selectable marker. 3) Part of pBR322 containing *E. coli* origin of replication to ensure replication in *E. coli* as well as the bacterial ampicillin gene (AmpR). 4) Major parts of the 2μ plasmid of *S. cerevisiae* (including the origin of replication) are included to ensure replication, amplification and inheritable stability in yeast.

repeats mutated. The pBR322 part of the shuttle vector contains the *bla*-gene adding ampicillin resistance, when grown in *E. coli*. Methods to delete this gene, which is an environmental issue with a high public concern, have been developed [17].

The host cell itself could be thought of as a target for further mutations or other genetic modifications that could improve its secretory, folding and posttranslational modification abilities. Deletion of specific genes that enhance protein secretion has been suggested [18, 19]; additionally, the overexpression of *pse1* [21] and *sso2* [22] improved the yield of other heterologous protein products. However, these observations were related to the expression of recombinant proteins unrelated to insulin. More specifically, Kozlov et al. [20] found improvement of proinsulin expression in respiratory-deficient *rho⁻ S. cerevisiae* strains.

13.2
Design and Optimization
of the Insulin Precursor Molecule

13.2.1
The Insulin Precursor Molecule: Background

The efficient secretion of heterologous proteins from eukaryotic cells often requires more than just a signal peptide N-terminally fused to the appropriate protein. Transport through the yeast secretory pathway includes translocation through the endoplasmic reticulum (ER) membrane into the ER-lumen, attachment of N-linked carbohydrate chains and folding and oxidation of disulfides, transport from the ER to the Golgi apparatus, post-translational modifications in the Golgi apparatus, transport by secretory granules to the cell membrane, and finally exocytic exit to the extracellular space. Especially small pep-

tides such as the insulin molecule require molecular assistance for efficient synthesis and secretion in *S. cerevisiae*. Proinsulin, fused to the *a*-factor leader (see below), was not readily expressed and secreted in *S. cerevisiae* [4] (Fig. 13.2). However, this *a*-leader fusion protein could be modified for successful expression and secretion in *S. cerevisiae* by substitution of insulin's long C-peptide with a dibasic amino acid sequence, creating insulin precursors that subsequently could be converted to human insulin by the action of the enzymes trypsin and carboxypeptidase B [4]. In a more subtle, optimized form this insulin fusion protein was changed by deletion of threonine^{B30} of the insulin precursor and lysine^{B29} was connected to glycine^{A1} either directly or using a short C-peptide containing a C-terminal lysine (e.g., AAK or SK) and fused to the *a*-factor leader [5]. Such single-chain insulin precursors, lacking threonine^{B30}, could be converted into human insulin by transpeptidation using porcine trypsin (see Section 13.3.2.2, Transpeptidation), but with differences in overall yields and conversion rates.

13.2.2
Optimization of the Insulin Precursor

In the following sections, optimization of the individual segments constituting the signal peptide-leader-insulin precursor will be discussed (see Fig. 13.2).

13.2.2.1 Signal Peptide
Eukaryotic signal peptides have universal properties, and many can be used as substitutes for the *a*-factor signal peptide in directing heterologous proteins to the secretory pathway in yeast [13]. The yapsin 1 aspartyl endoprotease (*YPS1*) signal peptide has been shown to provide efficient secre-

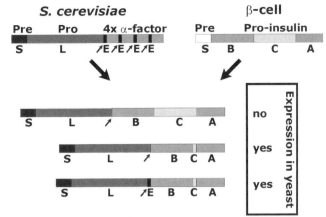

Fig. 13.2 Schematic representation of pre-pro-α-factor and pre-pro-insulin. The α-factor is expressed in multiple copies as a pre-pro-polypeptide precursor. Each α-factor is released by kexin cleavage, followed by removal of the N-terminal spacer peptide (E) by the dipeptidyl aminopeptidase Ste13p. Insulin is expressed in the β-cell as pre-pro-insulin. The C-peptide of pro-insulin is removed by the action of prohormone convertases. Combining the α-factor pre-pro with pro-insulin does not lead to secretion of pro-insulin from yeast; however, the introduction of very short C-peptides allows insulin precursor secretion. Kexin cleavage can be optimized by N-terminal extensions on the insulin precursor. S = Signal peptide; L = leader; E = spacer peptide; B = insulin B-chain; A = insulin A-chain (together denoted I in the text); C = connecting peptide. Small arrows denote the dibasic kexin cleavage sites.

tion of heterologous proteins (especially insulin) in yeast in combination with a propeptide [23]. The hydrophobicity of the signal peptide can be decisive for co- or post-translational translocation of the pre-pro-fusion protein into the ER [24], and a previously unrecognized diversity in the signal peptides was found which carries specific structural information that serves to identify the translocation route [25].

13.2.2.2 Leader Peptide

S. cerevisiae cells of mating type *a* secrete a 13-residue peptide pheromone (a-factor) [26]. The α-factor is the product of the *MFa1* gene which encodes a 165-residue polypeptide (pre-pro-α-factor) consisting of a signal peptide, a leader and four repeats of a pro-α-factor (see Fig. 13.2), which are matured during secretion [26, 27] (see Part IV, Chapter 12). It was early recognized that the *S. cerevisiae* α-factor signal-leader, was able to confer secretory competence on proteins expressed in *S. cerevisiae* [28, 29]. Although other leaders have been found, the α-factor leader has become the primary leader choice for secretory expression in *S. cerevisiae*, and also in alternative yeast species [13].

The presence of a leader sequence has been shown to be necessary for insulin precursor secretion, since a signal peptide fused directly to the insulin precursor does not result in secretion of insulin. In order to explore the possibility of developing leaders conferring greater secretory competency onto the insulin precursor, a stepwise approach was initiated starting from a minimal designed leader composed of a signal peptidase site, an N-glycosylation site, and a kexin processing site [30]. This minimal leader allowed secretion of small amounts of insulin precursor, though after extensive optimization it was possible to identify a number of synthetic leaders equivalent to or better than the α-factor leader at facilitating export of the insulin precursor to the culture supernatant [30, 31].

The three N-linked glycosylation sites of the α-leader have been shown to be important – but not essential – for the ability of the α-factor leader to secrete α-factor [32, 33]. In addition, mutation of all three-consensus sites for N-linked glycosylation decreased the quantity of secreted insulin precursor to ≈ 10% [34]. In contrast to what was found with the α-leader, elimination of the two consensus sites for N-linked glycosylation in one of the newly developed leaders, actually improved the ability to facilitate secretion of the insulin precursor [31].

13.2.2.3 Spacer Peptide

In the pro-α-factor (see Fig. 13.2), each 13-residue α-factor is preceded by a dibasic kexin processing site (KR) connected to a spacer peptide with two to three dipeptidyl repeats (EA or DA) designed for subsequent processing with the Ste13p dipeptidyl peptidase [26, 27, 35]. When the α-factor leader with spacer peptide initially was employed for recombinant insulin expression, the efficiency of the dipeptidyl aminopeptidase was shown to be a limiting step in maturation of insulin [4]; therefore, the spacer peptide was deleted and the protein fused directly to the α-leader in most of the initial studies. However, this resulted in the secretion of substantial amounts of uncleaved hyperglycosylated leader-insulin precursor from the yeast cell (see Fig. 13.3 for details of the secretion process). The introduction of a spacer peptide (EA)₃K, resembling the α-factor spacer

peptide, but with a lysine (K) introduced C-terminally to allow tryptic digestion at a later step in the production process, significantly improved expression of the insulin precursor [36]. One of the reasons for this was an improved cleavage in the Golgi of the fusion-protein by kexin, which allowed more maturated insulin precursor to be secreted. However, the previously described inefficiency of Ste13p resulted in the secretion of a mixed population of insulin precursors, complicating recovery and conversion to insulin. The addition of a glutamic acid residue to the spacer peptide N-terminus (to give E(EA)₃K) solved this problem by preventing processing by Ste13p. When this secreted insulin precursor was analyzed it was found that the extension had been cleaved off, resulting in the presence of maturated insulin precursor. This pointed to the presence of a specific protease activity [36], and characterization of this proteolytic activity indicated that the aspartyl endoprotease yapsin 1 (*YPS1*) [37] was responsible for removing the spacer peptide from the insulin precursor. A deletion of *yps1* is non-essential and solves the problem [38]; however, an alternative solution was found in that the spacer peptide could be further modified by insertion of different amino acid residues N-terminally to the lysine residue, thereby preventing proteolytic cleavage by yapsin 1. A proline residue was found to be most efficient to inhibit yapsin 1 activity and at the same time allow tryptic *in vitro* digestion C-terminally to the lysine residue [36]. Different modifications of the spacer peptide were subsequently generated (e.g., EEGEPK) which further increased the insulin precursor fermentation yield, without losing the *in vivo* stability and at the same time supporting *in vitro* conversion to insulin [12]. The *in vitro* conversion of the single-chain precursor that is required to obtain human

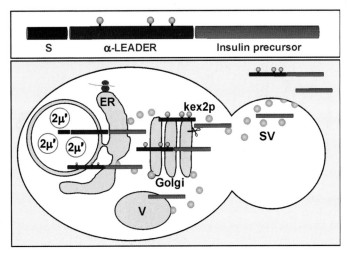

Fig. 13.3 Schematic representation of the *S. cerevisiae* insulin precursor expression system. The expression plasmids (2μ') are located in the nucleus. mRNA is transcribed and translated into Signal-leader-insulin precursor which subsequently enters the endoplasmic reticulum (ER). The signal peptide (S) is cleaved off by a signal peptidase and the leader core-glycosylated (yellow filled circles), insulin folds and disulfides are formed. Leader-insulin precursor is transported through Golgi, where glycosylations are extended and upon entering late-Golgi kexin cleaves at the dibasic site thereby separating insulin precursor from leader. The insulin precursor enters secretory vesicles (SV) and is secreted from the yeast cell. Part of the insulin precursor may misroute to the vacuole. Inefficient kexin cleavage leads to secretion of hyperglycosylated leader-insulin precursor; this can be remedied by inclusion of an N-terminal extension on the insulin precursor.

insulin can be an advantage in the recovery and purification process, because this conversion changes pI, hydrophobicity, hydrophilicity and the size of the peptide, thus providing optimal conditions for the purification of insulin – that is, the removal of impurities.

13.2.2.4 Connecting Peptide (C-peptide) and Insulin Aspart

The primary optimized connecting minipeptides, which support secretion and *in vitro* conversion, was briefly described above. Further optimization in this region, leading to improved expression, was obtained by fixing K in the N-terminal position to G^{A1}, and using a C-peptide with 2–16 residues and the most flexible amino acid glycine immediately N-terminal to K. Later, it was shown that especially glutamic or aspartic acid in C1 with G^{C2} K^{C3} results in an optimized version of a mini C-peptide [40].

A special and interesting case has recently been described [41], dealing with an insulin analogue substituted in B28 (proline to aspartic acid). This insulin analogue precursor can be converted to a fast-acting insulin aspart analogue by transpeptidation (see Section 13.3.2 and Fig. 13.4; see also the Introduction). The insulin analogue precursor is not well expressed with the AAK or the B29-A1 peptide bond, as mini-C-peptide and extractive refolding of the precursor was studied as one solution to increase yield [42]. However, precursor optimization was also successfully pursued. It

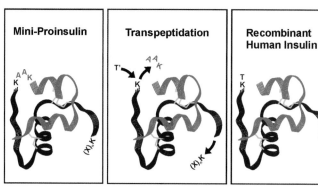

Fig. 13.4 Conversion of insulin precursor to recombinant human insulin using transpeptidation. The C-peptide is removed from the partially purified insulin precursor using a serine protease (e.g., trypsin). By inclusion of a threonine ester (T') and using appropriate reaction conditions, it is possible to couple threonine to Lys^{B29} using trypsin as a transpeptidase, thereby obtaining recombinant human insulin. The N-terminal extension (spacer peptide in text) on the B-chain of insulin (X)$_n$K is removed in the transpeptidation step. The insulin A-chain and B-chain are shown in red and blue, respectively. Disulfide bonds are depicted in yellow.

was hypothesized – based on knowledge of the 3-D structure of Asp^{B28} insulin [43] and the location of a phenol binding pocket in the insulin hexamer – that the side chain of an aromatic amino acid localized in the mini C-peptide could improve the stability of the insulin precursor [41] (see below, Stability). The interaction of this hydrophobic amino acid side chain with the hydrophobic core of the insulin precursor molecule could be energetically favorable and consequently enhance the folding stability of the precursor. Folding stability is important for secretory efficiency [12] (see Section 13.2.2.5), and therefore it was further hypothesized that enhancing the folding stability of the precursor would facilitate transport through the secretory pathway and thus increase the overall productivity. The introduction of an aromatic amino acid (X) into the mini C-peptide (e.g., EXK) did indeed increase the expression yield of the insulin aspart precursor. Tryptophan had the greatest positive influence (5-fold) on the expression yield, whereas phenylalanine and tyrosine increased the expression yield approximately 3.5-fold [41]. This is an example of a highly specific calculated optimization; Asp^{B28} is disturbing the ordinary C-peptide turn and the overall folding, which is more than compensated by introduction of an aromatic amino acid in C2 in the C-peptide.

13.2.2.5 Structural Mutations and Relationship between Expression Yields and *In vitro* Folding Stability

The insulin fold is composed of two A-chain a-helices connected by a loop, and one central B-chain a-helix flanked by an N-terminal sequence in extended conformation (in the structural T-state of insulin) and a β-sheeted C-terminal. Enhanced folding stability can for example be engineered by mutations, leading to the removal of hydrophobic side-chains at the protein surface as well as substitutions near the N- and C-terminal of the a-helices stabilizing these helices [44]. Experiments examining the expression of a series of insulin analogues in *S. cerevisiae* found a positive correlation between insulin analogue *in vitro* folding stability and the fermentation yield of the corresponding insulin analogue precursor (expressed fused to the *a*-factor leader and with the AAK mini C-peptide connecting B29 and A1) [12]. Stabilization of folding of individual parts of the precursor molecule is important for the overall stability, and this seems to have a positive impact on the fermentation yield. This principle can probably be extended to cover all parts of the signal peptide-leader-insulin precursor.

13.2.2.6 Polymerization of the Insulin Precursor Influences Secretion Yield and Retention in the Vacuole

The polymerization of insulin and proinsulin to dimers and hexamers is a very important process which takes place in the pancreas and impacts upon the pharmaceutical application of insulin, because the dissociation of the polymers is the rate-limiting processes in the absorption and action in the tissues of the biological active monomeric insulin [45]. The polymerization diminishes osmotic pressure and hydrophobicity and improves solubility. A positive correlation between expression yield and the degree of polymerization was found [46] that supports yeast *in vivo* polymerization of insulin precursors and accentuates its importance. However, polymerization can also account for a drawback by retention of product in the vacuole [47]. Intracellular retention of a substantial quantity of the synthesized insulin precursor indicated that the insulin precursor followed two different intracellular routes in the late secretory pathway [12, 47]. Constitutive secretion to the culture supernatant may reflect saturation of a sorting mechanism in the late Golgi due to overexpression. The kexin cleaves the leader-insulin precursor peptide in a late Golgi compartment to yield free insulin precursor, and

this change is provoking accumulation in the vacuole [47]. Zhang et al. [47] applied a genetic screen to identify genes influencing insulin precursor accumulation and identified several *vps*-mutants important for accumulation of insulin precursors and concomitant possible targets for regulation. The uncleaved leader-precursor was not accumulated, but was secreted to the medium. Thus, the leader seems to hide the sorting signal or prevent its formation. The sorting mechanism was suggested to be a consequence of endoproteolytic maturation and multimeric assembly of insulin [47].

13.3
Production of Insulin

A *S. cerevisiae* strain optimized for the production of insulin by some of the recombinant strategies outlined previously must comply with the cultivation conditions prevailing at industrial scale.

13.3.1 Fermentation

A prerequisite for industrial strains to be used for continuous cultivations is the ability to remain stable for extended periods of time when grown at large scale (see also the Introduction and Part IV, Chapters 1 and 12). This has previously been shown to be the case for *S. cerevisiae* strains based on a POT-selection plasmid and an insulin precursor transcribed by the constitutive TPI1-promoter [6]. During growth of *S. cerevisiae* on carbohydrates as the major carbon source, ethanol formation should be avoided since formation of ethanol leads to lower biomass yields and consequently decreasing product titers. This issue of fermentative metabolism is often solved by monitoring the oxygen consumption and carbon dioxide evolution rates, and using these variables as inputs for control algorithms adjusting the fermentation process parameters.

13.3.2
Transpeptidation

The fermentation broth leaving the fermenter is traditionally clarified for yeast cells by means of centrifugation or filtration processes, followed by purification by chromatography. Conversion of the insulin precursor lacking Thr^{B30} to human insulin is elegantly achieved through the use of enzymatic processes based on serine endopeptidases, for example porcine trypsin (EC 3.4.21.4) or lysyl endopeptidase from *Achromobacter lyticus* (EC 3.4.21.50) (Fig. 13.4). When used initially as hydrolases in water-rich solvents, the C-peptide is removed by cleavage at the peptide bond of Lys^{B29} and Lys^{A1-1}, the lysine that joins the C-peptide to the insulin A-chain. Following this hydrolysis the des(B30)-insulin molecule is coupled to a threonine ester, this time exploiting the enzyme's capability to function in mixtures of water and organic solvents [5]. Changing the equilibrium from hydrolysis to synthesis is accomplished by adjusting the solvent composition and the reaction conditions (pH, substrate concentration and water activity). Transpeptidation, where the equilibrium can be displaced towards peptide bond synthesis, can be obtained by a one-step procedure with a pre-programmed adjustment of the water activity prevailing in the reaction mixture. Spacer peptides, if applied, are elegantly removed in the same operations that remove the C-peptide. The intermediate product with an esterified carboxy-terminus of Thr^{B30} is purified by chromatography, and finally the human insulin-ester is hydrolyzed leading to human insulin.

13.4
Conclusions and Future Aspects

The expression of insulin precursors in the yeast *S. cerevisiae* has, in recent years, formed the basis for the manufacture of the majority of human recombinant insulin and insulin analogues sold by Novo Nordisk A/S (see Introduction). Studying the expression of insulin in *S. cerevisiae* has highlighted the importance of addressing every event leading to secretion of the insulin precursor from the initial translocation and folding in the ER, transport through the Golgi apparatus, to maturation by kexin and ultimately secretion via correct localization to secretory vesicles. Modification of the insulin precursor, as described in this chapter, has led to significant improvements in the expression system, resulting in robust scalable expression systems which are amenable to industrial fermentation and downstream processing of the insulin precursor. Some observations indicate that there are limitations in the secretory capacity of *S. cerevisiae* [12, 48] and that these point to alternative initiatives. Focus on the actual secretion machinery in *S. cerevisiae* and alternative hosts systems such as *Saccharomyces kluyveri* [49] or *Pichia pastoris* [50, 51] are options for further exploitation and the large-scale production of other biopharmaceuticals.

References

1 Markussen, J. US Patent 4343898 (1982).
2 Chance, R. E., Hoffmann, J. A, Kroeff, E. P., Johnson, M. G., Schirmer, E. W., Bromer, W. W., Ross, M. J., and Wetzel, R. In: *Peptides. Synthesis-Structure-Function.* Proceedings of the Seventh American Peptide Symposium (D.H. Rich and E. Gross, eds). Pierce Chemical Company, p 721 (1981).

3 Frank, B.H., Pettee, J.M., Zimmerman, R.E., and Burck, P.J. In: *Peptides. Synthesis-Structure-Function*. Proceedings of the Seventh American Peptide Symposium (D.H. Rich and E. Gross, Eds). Pierce Chemical Company, p 729 (1981).

4 Thim, L., Hansen, M.T., Norris, K., Hoegh, I., Boel, E., Forstrom, J., Ammerer, G. and Fiil, N.P. *Proc. Natl. Acad. Sci. USA* 83, 6766–6770 (1986).

5 Markussen, J., Damgaard, U., Diers, I. Fiil, N., Hansen, M.T., Larsen, P., Norris, F., Norris, K., Schou, O., Snel, L., Thim, L., and Voight, H.O. In: *Peptides 1986* (D. Theodoropoulos, Ed.), pp 189–194. Berlin: Walter de Gruyter & Co. (1987).

6 Diers, I.V., Rasmussen, E., Larsen, P.H., and Kjaersig, I.-L. In: *Drug Biotechnology Regulation, Scientific Basis and Practices* (Y.-Y.H. Chiu and J.L. Gueriguian, Eds), pp 166–176, New York, NY: Marcel Dekker, Inc. (1991).

7 Brange, J., Ribel, U., Hansen, J.F., Dodson, G., Hansen, M.T., Havelund, S., Melberg, S.G., Norris, F., Norris, K., Snel, L., Sørensen, A.R., and Voigt, H.O. *Nature* 333, 679–682 (1988).

8 Markussen, J., Hougaard, P., Ribel, U., Sørensen, A.R., and Sørensen, E. *Protein Eng.* 1, 205–213 (1987).

9 Markussen, J., Diers, I., Engesgaard, A., Hansen, M.T., Hougaard, P., Langkjaer, L., Norris, K., Ribel, U., Sørensen, A.R., and Sørensen, E. *Protein Eng.*, 1, 215–223 (1987).

10 Markussen, J., Diers, I., Hougaard, P., Langkjaer, L., Norris, K., Snel, L., Sørensen, A.R., Sørensen, E., and Voigt, H.O. *Protein Eng.* 2, 157–166 (1988).

11 Kjeldsen, T. *Appl. Microbiol. Biotechnol.* 54, 277–386 (2000).

12 Kjeldsen, T., Balschmidt, P., Diers, I., Hach, M., Kaarsholm, N.C., and Ludvigsen, S. *Biotechol. Genet. Eng. Rev.* 18, 89–121 (2001).

13 Romanos, M.A., Scorer, C.A., and Clare, J.J. *Yeast* 8, 423–488 (1992).

14 Ammerer, G., Hunter, C.P., Rothman, J.H., Saari, G.C., Valls, L.A., and Stevens, T.H., *Mol. Cell. Biol.* 6, 2490–2499 (1986).

15 MacKay, V.L., Yip, C., Welch, S., Gilbert, T., Seidel, P., Grant, F., and O'Hara, P. In: *Recombinant Systems in Protein Expression* (K.K. Alitalo, M.-L. Huhtala, J. Knowles, and A. Vaheri, Eds), pp. 25–36. Amsterdam: Elsevier Science Publishers BV (1990).

16 Kawasaki, G.H. and Bell, L. US Patent 5,871,957 (1999).

17 Kjeldsen, T. and Vad, K. US Patent 6358705 (2002).

18 Smith R.A., Duncan, M.J., and Moir, D.T. *Science* 229, 1219–1224 (1985).

19 Bartkeviciute, D. and Sasnauskas, K. *FEMS Yeast Res.* 4, 833–840 (2004).

20 Kozlov, D.G., Prahl, N., Efremov, B.D., Peters, L., Wambut, R., Karpychev, I.V., Eldarov, M.A., and Benevolensky, S.V. *Yeast* 11, 713–724 (1995).

21 Chow, T.Y., Ash, J.J., Dignard, D., and Thomas, D.Y. *J. Cell Sci.* 101, 709–719 (1992).

22 Ruohonen, L., Toikkanen, J., Tieaho, V., Outola, M., Soderlund, H., and Keranen, S. *Yeast* 13, 337–351 (1997).

23 Christiansen, L. and Petersen, J.G.L. (1994) Patent application WO95/02059.

24 Ng, D.T., Brown, J.D., and Walter, P. *J. Cell Biol.* 134, 269–278 (1996).

25 Plath, K., Mothes, W., Wilkinson, B.M., Stirling, C.J., and Rapoport, T.A. *Cell* 94, 795–807 (1998).

26 Thorner, J. In: *The Molecular Biology of the Yeast Saccharomyces cerevisiae: Life Cycle and Inheritance* (J.N. Strathern, E.W. Jones, and J.R. Broach, Eds), pp. 143–180. New York NY: Cold Spring Harbor Laboratory (1981).

27 Kurjan, J. and Herskowitz, I. *Cell* 30, 933–943 (1982).

28 Bitter, G.A., Chen, K.K., Banks, A.R., and Lai, P.-H. *Proc. Natl. Acad. Sci. USA* 81, 5330–5334 (1984).

29 Brake, A.J., Merryweather, J.P., Coit, D.G., Heberlein, U.A., Masiarz, F.R., Mullenbach, G.T., Urdea, M.S., Valenzuela, P., and Barr, P.J. *Proc. Natl. Acad. Sci. USA* 81, 4642–4646 (1984).

30 Kjeldsen, T., Pettersson, A.F., Hach, M., Diers, I., Havelund, S., Hansen, P.H., and Andersen, A.S. *Protein Expr. Purif.* 9, 331–336 (1997).

31 Kjeldsen, T., Hach, M., Balschmidt, P., Havelund, S., Pettersson A.F., and Markussen, J. *Protein Expr. Purif.* 14, 309–316 (1998).

32 Julius, D., Schekman, R., and Thorner, *Cell* 36, 309–318 (1984).

33 Caplan, S., Green, R., Rocco, J., and Kurjan, J. *J. Bacteriol.* 173, 627–635 (1991).

34 Kjeldsen, T., Andersen, A.S., Hach, M., Diers, I., Nikolajsen J., and Markussen, *Biotechnol. Appl. Biochem.* 27, 109–115 (1998).

35 Julius, D., Blair, L., Brake, A., Sprague, G., and Thorner. *Cell* 32, 839–852 (1983).

36 Kjeldsen, T., Brandt, J. Andersen, A. S., Egel-Mitani, M., Hach, M., Pettersson, A. F., and Vad, K. *Gene* 170, 107–112 (1996).

37 Egel-Mitani, M., Flygenring, H. P., and Hansen, M. T. *Yeast* 6, 127–137 (1990).

38 Egel-Mitani, M., Brandt, J., and Vad, K. US Patent 6110703 (2000).

39 Kjeldsen, T. and Ludvigsen, S. US Patent 6777207 (2004).

40 Kjeldsen, T. and Ludvigsen, S. WO200279251 (2002).

41 Kjeldsen, T., Ludvigsen, S., Diers, I., Balschmidt, P., Sørensen, A. R., and Kaarsholm, N. C. *J. Biol. Chem.* 277, 18245–18248 (2002).

42 Diers, I. Patent application WO 200146453 (2000).

43 Whittingham, J. L., Edwards, D. J., Antson, A. A., Clarkson, J. M., and Dodson, G. G. *Biochemistry* 37, 11516–11523 (1998).

44 Kaarsholm, N. C., Norris, K., Jørgensen, R. J., Mikkelsen, J., Ludvigsen, S., Olsen, H. B., Sør-ensen, A. R., and Havelund, S. *Biochemistry* 32, 10773–10778 (1993).

45 Dodson, E. J., Dodson, G. G., Hubbard, R. E., Moody, P. C. E., Turkenburg, J., Whittingham, J., Xiao, B., Brange, J., Kaarsholm, N., and Thogersen, H. *Phil. Trans. R. Soc. Lond. A* 345, 153–164 (1993).

46 Kjeldsen, T. and Pettersson, A. F. *Protein Expr. Purif.* 27, 331–337 (2003).

47 Zhang B., Chang, A. Kjeldsen, T. B., and Arvan, P. *J. Cell Biol.* 153, 1187–1198 (2001).

48 Egel-Mitani, M., Hansen, M. T., Norris, K., Snel, L., and Fiil, N. P. *Gene*, 73, 113–120 (1988).

49 Srivastava, A., Piskur, J., Nielsen, J., and Egel-Mitani, M. PCT patent application WO 200014258 (2000).

50 Kjeldsen, T., Pettersson, A. F., and Hach, M. *Biotechnol. Appl. Biochem.* 29, 79–89 (1999).

51 Wang, Y., Liang, Z.-H., Zhang, Y. S., Yao, S.-Y., Xu, Y.-G., Tang, Y. H., Zhu, S. Q., Cui, D. F., and Feng, Y.-M. *Biotechnol. Bioeng.* 73, 74–79 (2001).

14

Baculovirus-based Production of Biopharmaceuticals using Insect Cell Culture Processes

Wilfried Weber and Martin Fussenegger

Abstract

Baculovirus-based protein expression in insect cell culture represents today a ripened method for the rapid production of proteins up to pilot scale. Recent advances on all levels of the production process such as optimized expression vectors, chemically defined media, optimized nutritional and kinetic parameters as well as the design of novel cultivation systems contribute to drastically increased protein yields with concomitant reduction of process time and costs. This chapter provides a comprehensive overview on recent research and achievements on the successive steps in a protein expression process, with a focus on how to integrate these findings for achieving overall optimized performance. A case study on the production of a secreted protein illustrates the impact of the different optimization steps on the overall process yields.

Abbreviations

AcNPV	*Autographa californica* nuclear polyhedrosis virus
BEVS	baculovirus expression vector system
BTK	brutons tyrosine kinase variant
DO	dissolved oxygen
MOI	multiplicity of infection
ORF	open reading frame
OTR	oxygen transfer rate
OUR	oxygen uptake rate
PCD	peak cell density
s-ICAM-1	secreted intercellular adhesion molecule-1
TOH	time of harvest
TOI	time point of infection

14.1
Introduction

The vast variety of different drug targets and yet-to-be-characterized open reading frames (ORFs) emerging from post-genomic research initiatives has substantiated the need for rapid and generic pilot-scale production of desired proteins for biochemical/functional studies as well as for target validation. The baculovirus expression vector system (BEVS) developed for heterologous protein production in insect cell cultures almost three decades ago is one of today's preferred pilot-production technology owing to BEVS' superior protein titers and unmatched from-gene-to-protein process speed, which still surpasses currently available mammalian cell-based production processes.

Modern Biopharmaceuticals. Edited by J. Knäblein
Copyright © 2005 WILEY-VCH Verlag GmbH & Co. KGaA, Weinheim
ISBN: 3-527-31184-X

The baculovirus which prevails in today's production processes belongs to the Eubaculoviridae [1], a subfamily of double-stranded DNA viruses. Following infection of host arthropods, these viruses produce proteinaceous structures known as "occlusion bodies" representing up 50% of the total cellular protein [1]. The baculovirus's superior protein production capacity resides in the viral very late polyhedrin and p10 promoters, which are often engineered to express desired transgenes rather than structural virus genes forming occlusion bodies [2, 3]. During the past decade, biotechnologically relevant BEVS research has focused on improving process engineering parameters including nutrient/oxygen supply, infection kinetics and heterologous protein production (for a review, see Ref. [4]). In-depth understanding of BEVS-related process parameters associated with maximum production levels has resulted in generic protocols for the rapid implementation of large-scale bioreactor operation in an assembly line-like production scenario [5].

With product yield at its near maximum, the BEVS community is currently focusing on the improvement of product quality by humanizing insect cell glycosylation patterns. The unique combination of transient expression implementation, high-yield protein production capacity and the prospect of human glycoprofiles in insect cultures indicates a bright future for BEVS technology in the production of biopharmaceuticals [4, 6, 7].

14.2
Molecular Tools for the Construction of Transgenic Baculoviruses

The most widely used baculoviral expression systems are derived from the *Autographa californica* nuclear polyhedrosis virus (AcNPV) [1, 8]. Owing to its extended genome of 129 kb, baculoviruses are not amenable to standard cloning procedures and require multi-step split-genome strategies for the design of transgenic viral vectors. The classic BEVS capitalizes on a small helper vector encoding the transgene expression unit flanked by viral sequences, which recombines onto the baculoviral genome following co-transfection into insect cells. BEVS are commercially available in different variants, including β-galactosidase-encoding helper vectors, which facilitate straightforward screening for desired recombinant baculovirus genomes (e.g., BacPAK, BD Biosciences; pPBac, Stratagene; Bac-N-Blue, Invitrogen). Recent progress in transgenic baculovirus design enabled helper vector–baculovirus genome recombination in *Escherichia coli*, from which transgenic genomes are recovered and prepared for subsequent transfection and virus production in insect cells (Bac-to-Bac, Invitrogen). The latest generation of baculovirus design includes lambda integrase-mediated *in vitro* recombination of the transgene expression unit onto an engineered baculovirus genome. Such *in vitro* design technology has further been refined so that the transgene expression unit replaces a thymidine kinase expression unit on the baculovirus chromosome, thereby enabling gancyclovir-mediated selection of desired recombinants (BaculoDirect, Invitrogen).

Baculovirus stocks are typically produced by the aforementioned methods, followed by optional amplification rounds and titration. The traditional titration technology is based on plaque assays where viral stock dilutions are applied on confluent insect cell cultures and clonal plaques scored by visual inspection after several days. More rapid technologies include: 1) immunodetection of viral proteins ([9], BacPAK Rapid Titer Kit, BD Biosciences); 2) fluorescent virus-encoded marker proteins profiled by microscopy [10]; or 3) quantitative real-time PCR specific for viral sequences [11]. Despite advanced quantification technologies, only plaque assays enable direct profiling of infectious particles; all other methods have relied on empiric correction factors when considering non-infectious/defective baculoviral particles.

14.3
Insect Cell Culture

While first-generation baculovirus-based expression technology included infection of entire insect larvae [2, 12], advanced technologies take advantage of insect-derived serum-free suspension cultures for baculovirus maintenance and protein production [4, 13, 14]. The most prominent insect cell lines include the *Spodoptera frugiperda*-derived ovarian cell lines Sf-21 and its subclone Sf-9, as well as the *Trichoplusia ni* egg cell homogenate-derived cell line BI-TN-5B1-4, known as High Five™ cells. While High Five™ cells provide increased specific and volumetric productivities of secreted proteins [15–17], Sf-9 continues to be the preferred cell line for expression of intracellular or membrane proteins, while Sf-21 prevailed for isolation and propagation of viral stocks. Cell culture media requirements for insect cells are comparable to those for mammalian cells, with the exception of a lower pH (6.2–6.9) and the tolerance of higher free amino acid and glucose levels without switching to overflow metabolism [14].

14.4
Insect Cell Glycosylation and Glycoengineering

Since insect cells fail to synthesize complex N-glycans and produce potentially allergenic structures of the Fucα(1,3)GlcNAc-Asn type, BEVS-based production of protein pharmaceuticals for clinical use remains compromised [7]. Several strategies for overcoming such limitations are currently competing for industrial implementation:

- Screening for insect cell lines which produce more complex glycoforms. The *Danau plexippis*-derived DpN1 cell line showed up to 26% of complex glycoforms compared to less than 5% for High Five™ cells, yet its overall product titers were less competitive [7, 17].

- Sophisticated feeding strategies based on N-acetylmannosamine (ManNAc) [18] supplementation or addition of specific glycosyltransferase inhibitors [19]: Watanabe and co-workers reported the capacity of High Five™ for interferon-γ N-glycan sialylation following cultivation of these insect cells in the presence of an hexosaminidase inhibitor (2-AND).

- Metabolic engineering of insect cells: Ectopic expression of epimerases and kinases, such as the bifunctional enzyme SAS and UDP-GlcNAc 2-epimerase/ManNAc kinase initiating sialic acid biosynthesis in mammalian cells, resulted in N-acetylneuraminic acid (Neu5Ac) synthesis in the absence of any media supplements [18]. A particular Sf-9-derived insect cell line [6], engineered for simultaneous expression of five different glycosyltransferases, produced biantennary terminally sialylated N-glycans reminiscent of glycoprofiles typically found in mammalian cells (Mimic™ Sf-9 cells, Invitrogen [6]).

14.5
Nutrient and Kinetic Considerations for Optimized BEVS-based Protein Production

In a baculovirus-infected insect cell culture, nutrients are converted into biomass, virus progeny and desired product protein, as well as diverse by-products [20, 21]. The relative quantities of these culture products can be influenced by modification of critical kinetic process parameters, including the time point of infection (TOI, typically expressed as cell density at infection) and the multiplicity of infection (MOI, number of infective viral particles per cell) [20–22]. Different strategies to modulate TOI and MOI along with nutrient supply and timely infection are covered in the following sections.

14.5.1
Nutritional Parameters

Nutrient and oxygen supply appear to be the only limiting factors for BEVS to reach high cell densities and recombinant protein titers [23, 24]; no other limiting effects such as contact inhibition or accumulation of toxic metabolites (including alanine or ammonium) have been observed for Sf-9 cells [23]. Several feeding strategies have been devised to increase cell density and product titer. Complete medium exchange [25, 26] or perfusion systems [27, 28] initially appeared to be promising for supplying additional nutrients, but these approaches turned out to require tedious and time-consuming separation steps or special reactor configurations, which are often incompatible with streamline recombinant protein production. Fed-batch systems, which supply specific nutrient concentrates to increase cell density, seem to prevail in current insect cell-based production processes. The influence of several nutrient cocktails on insect cell densities has been quantified either by scoring specific nutrient consumption rates [26, 29] or by fractional (factorial) experimental design [5, 30, 31] consisting of individual nutrient effects being evaluated individually. Maximum insect cell densities are reached most efficiently following medium supplementation with yeastolate and lipid concentrates [32], amino acids, vitamins and trace elements [30, 31], as well as glucose and glutamine [23].

Beyond cell densities of 1×10^6 cells mL^{-1}, oxygen supply is a critical issue [23, 33–35], in particular for baculovirus-infected cultures which require 30–40% more oxygen compared to standard insect cell cultures [36, 37]. Typically, dissolved oxygen (DO) levels above 30% air saturation are considered to be sufficient to prevent oxygen limitation [33, 35, 36]. As well as oxygen supply, CO_2 removal is a critical issue for insect culture, since high concentrations of dissolved hydrogen-carbonate not only modulate the culture's pH, but also negatively impact on cell growth [38]. For example, CO_2 accumulation was suggested to be a major limiting factor for the expression of TGF-β receptor production on a 150-L scale. Although the culture medium was sparged with pure oxygen to minimize shear stress, the low volumetric gas flow was insufficient to strip CO_2 off the bioreactor, which resulted in decreased TGF-β receptor production [39].

By integrating of all the aforementioned parameters into a sophisticated feeding scheme for glucose, amino acids, yeastolate, lipids, vitamins and trace elements combined with optimized oxygen supply, Elias and co-workers [40] reached 5.2×10^7 Sf-9 cells mL^{-1}, the current record in maximum density for this cell line operated in fed-batch mode.

14.5.2
Infection-related Kinetic Parameters

Recombinant protein production using baculovirus-infected insect cell cultures has a well-defined endpoint: lysis of the host cell. Therefore, maximum product yield can only be achieved when baculovirus infection is precisely timed so that maximum cell numbers complete protein production just before the nutrients become limiting [21, 22]. Owing to Poisson-like infection distribution, infection and maximum cell density timing are best synchronized when insect cells are infected at an MOI of at least five [41]. However, infection of high cell density cultures at an MOI of five requires substantial virus amounts, resulting in undesired addition of conditioned medium to the production culture. In order to halt this vicious circle, concentrated high-titer viral stocks are typically produced prior to infection [40]. Alternatively, an *in situ* virus amplification step can be initiated by infecting a low-density cell culture at a low MOI (e.g., 0.1), so that the few infected cells produce progeny virus (approx. 500 virus per infected cell; [22]), which subsequently infects the remaining culture. Using *in situ*

virus amplification, baculovirus infections can be performed at MOIs as low as 3×10^{-5} [42, 43]. Yet, if several *in situ* amplification steps are performed, small errors in virus quantification or fluctuations in cell culture kinetics are also amplified, and may lead to premature infection of the entire culture or to a large proportion of uninfected cells should nutrients become limiting [42]. When making the choice between high-MOI infections and low-MOI infections using *in situ* virus amplification, several protein-specific parameters should be considered (see Table 14.1).

14.5.3
Protease Activities in Infected Insect Cell Cultures

Proteases released during the infection process may lead to protein inhomogeneities and product degradation [4, 5]. Several baculovirus-encoded proteases and chitinases have been identified which, upon insect larvae infection, induce liquefaction of the infected host [44, 45]. The cysteine protease *v-cath* identified in *A. californica* shows functional homologies to mammalian cathepsin L [46]. Several studies have focused on the neutralization of

Table 14.1 Decision-making parameters for determination of a high or low multiplicity of infection (MOI) strategy

Low MOI infection with *in situ* virus amplification	High MOI infection
Stable protein	Unstable protein
Intracellular protein	Secreted protein
Non-toxic protein	Toxic/growth-retarding protein
Known infection kinetics/cell growth rate	Unknown infection kinetics/cell growth rate
Only low-titer/limited virus stocks available	Concentrated virus stocks available
Bioreactor operation at high infection cell densities at or beyond $1-2 \times 10^6$ cells mL^{-1} (fed-batch processes)	Bioreactor operation at medium to low cell densities (batch processes)

baculovirus-derived proteolytic activities by: 1) genetic engineering of the viral genome (BacVector-3000, [47]); 2) using earlier viral promoters exemplified by the basic protein promoter [48]; or 3) addition of cysteine protease-specific inhibitors including leupeptin or pepstatin A [35, 49]. All of these strategies resulted in decreased protease activities, which correlated with increased protein titers and improved product homogeneities, as required for high-end protein applications as in crystallography or NMR studies. A comprehensive overview of the different strategies for improving nutrient supply and optimizing infection kinetics is provided in Table 14.2.

14.6
Scaling-up Baculovirus-based Protein Production

While laboratory-scale BEVS-based protein production can be achieved in standard cell culture flasks, manufacturing of milligram to gram quantities of a desired protein requires more sophisticated production hardware, including roller bottles, classical stirred-tank and perfusion bioreactors, or the recently developed rotating wall vessel and Wave^{TM} bioreactors. Although perfusion reactors [27, 28] or rotating-wall vessels [4] reach higher peak cell densities and show low shear forces, they remain limited by complex hardware management and the lack of generic operation protocols. While stirred-tank bioreactors remain the preferred cultivation systems for insect cells up to several hundred-liter scale [39], the recently developed Wave^{TM} bioreactor system is gathering decisive momentum in this territory [4, 5, 50]. In capitalizing on a completely disposable reactor chamber, the Wave^{TM} system is particularly efficient for short-batch processes (3–6 days) as typical

bioreactor maintenance processes including disassembly, cleaning, assembly and sterilization become obsolete [5]. Wave^{TM} bioreactor operation is characterized by low shear forces and high volumetric oxygen transfer rates mediated by the continuous renewal of the medium surface. Although Wave^{TM} and classic stirred-tank bioreactors reached identical titers of a secreted model product protein when operated at a 10-L scale, the operating costs for Wave^{TM} were 40% lower compared to the stirred-tank bioreactor [5, 51]. The latest-generation Wave^{TM} bioreactors managing culture volumes of up to 500 L are now at eye level with standard stirred-tank bioreactors as far as scale is concerned [52] (Wave Biotech, Tagelswangen, Switzerland).

14.7
Generic Protocol of Optimized Protein Production

There are three major parameters determining the overall product yield and quality of BEVS processes: 1) abundance of nutrients; 2) timing of baculovirus infection; and 3) harvest timing. In considering the state-of-the-art know-how related to these three parameters, we provide a genetic protocol for the rapid determination of optimized BEVS-based process parameters. Optimal parameters for small-scale processes are derived by the following two-step procedure:

1. Optimize nutritional factor to supply BEVS with maximum substrate quantities. Peak cell density (PCD) is considered to be the best lumped value for the assessment of nutritional parameters as it integrates all factors modulating growth and viability [29].

2. With nutritional supply at its optimum, the kinetic parameters can be adjusted

Table 14.2 Analysis and optimization of nutritional and kinetic parameters in baculovirus-based protein production

Parameter(s) studied	Result	Reference
MOI, TOI, TOH, yeastolate	Generic protocol for fast determination of optimum parameters with subsequent scale-up for production runs.	5
MOI, TOI	Modeling of the relationship between MOI and TOI and their correlation with protein titer.	21
MOI, TOI	Experimental validation and extension of the models developed by Licari and Bailey [21].	20
MOI, pH, protease activity	Proteolytic activities as a function of MOI. High MOI preferable when protease activity is critical. pH-dependent activities of proteases.	56
MOI, TOI	Experimental validation of the models developed by Licari and Bailey [21].	57
MOI, TOI	Extremely low MOIs (≤0.00003) feasible but difficult to control.	42
MOI, protease activity	Simulation and experimental validation of the relationship between MOI and proteolytic activities.	53
Low MOI infection	Virus-like particle production at low MOI.	43
Protease activity	Determination of protease activity at different time points after infection.	58
Diverse nutrients	Multiparameter study to determine optimum nutrient supply to reach 3×10^7 Sf-9 cells mL^{-1}.	24
Diverse nutrients, MOI	Factorial experimental design for determination of optimal protein production.	30
MOI, TOI	Modeling of infection and protein expression kinetics with experimental validation. Determination of critical rate constants.	22
Optimization of feeding regime	Fed-batch on demand resulting in 5.2×10^7 Sf-9 cells mL^{-1}. Production of β-galactosidase at high cell density.	40
Glucose, amino acids	Scale-up for High Five™ cells based on nutrient consumption analysis and specific nutrient supplementation.	29
Multi-parameter study	Analysis of optimum conditions using factorial experimental design and response surface experiments.	31
TOI	Effect of infection time point in suspension cultures and scale-up in bioreactor.	25
Nutrient consumption rates	Determination of specific nutrient consumption rates for sugars and amino acids prior to and after infection.	26
Yeastolate	Production and investigation of yeastolate for insect cell culture.	32
Medium development	Review of insect cell culture medium development.	14
Pluronic F-68	Protective effects of Pluronic F-68 on insect cells.	59

MOI, multiplicity of infection; TOI, time of infection; TOH, time of harvest.

to the desired product protein by selecting the best MOI, TOI, and time of harvest (TOH) [21, 53].

14.7.1
Optimization of Nutritional Parameters

Fed-batch processes are the method of choice to reach maximum cell densities and product yield because of their straightforward setup and optimal compatibility with standard bioreactor hardware. Recent research has highlighted a variety of medium supplements including yeastolate [31, 32], amino acids [24, 30] and lipid mixes [31,] which, when combined with sophisticated feeding strategies, resulted in higher cell densities of up to 5.2×10^7 Sf-9 cells mL^{-1} [40]. However, for high-throughput pilot-scale production of different biopharmaceutical proteins a well-balanced protocol, which considers maximum cell density and product titer as well as bioreactor operation parameters exemplified by the oxygen transfer rate, is required. A generic protocol for routine bioreactor-based fed-batch operation has recently been outlined for the Wave^{TM} bioreactor [5, 50] and Sf-9 cells, where the SF900II medium was supplemented once with $4\,g\,L^{-1}$ yeastolate at cell densities of 3–4×10^6 cells mL^{-1}, thereby resulting in peak cell densities of up to 8.8×10^6 cells mL^{-1}. Oxygen supply was optimally adjusted by regulating the inflow gas composition, as well as the agitation speed by means of control software integrating a predictive mass balance-based model of oxygen transfer and consumption of insect cells in a Wave^{TM} bioreactor.

14.7.2
Optimization of Infection Kinetics

In-vivo and *in silico* research has highlighted the need for the optimal adjustment of infection kinetics [5, 20–22, 26]. Infection kinetics are predominantly determined by the MOI and TOI. Whereas an optimum MOI is best chosen based on the criteria in Table 14.1, the correlating optimum TOI must be determined experimentally for each product protein. A protocol for rapid determination of optimum TOI in small-scale batch cultures and its subsequent translation into large-scale fed-batch production configurations has recently been established [5, 50]. This generic protocol is based on findings by Radford [26, 54], Bédard [24], Elias [40], Chan [31] and Power [22], whose models suggest that: 1) peak cell densities in uninfected cultures as well as maximum product titers in infected cultures are limited by the same nutrients; and 2) that the different steps of viral infection, replication and protein synthesis practically show the same kinetic behavior at different cell densities, provided that the insect cell density exceeds 1.5×10^6 cells mL^{-1} at infection. Equation (14.1) [5] describes the correlation between the peak cell density (PCD) of uninfected insect cell populations cultivated at various nutrient levels using different feeding strategies with the optimum TOI [5].

$$\frac{TOI_{low-nutrient}}{PCD_{low-nutrient}} = \frac{TOI_{high-nutrient}}{PCD_{high-nutrient}} \quad (14.1)$$

Using Eq. (14.1), the optimum TOI for small-scale cell culture operation in simple batch settings can first be determined and then translated into large-scale fed-batch bioreactor management using a high nutrient supply for the final production run, provided that the uninfected peak cell densities have been determined for both systems beforehand.

14.7.3
Determination of the Optimum TOH

Determination of the optimum TOH is critical to enable sufficient time for maximum protein production before cell leakage and proteolysis occur. When performing small-scale production experiments for the assessment of optimum TOI and MOI (as suggested in Section 14.7.2), different TOHs should also be evaluated. Translation of the optimum TOH from small-scale to large-scale production scenarios is not straightforward, as cell growth rate as well as protein production kinetics may differ between cell culture plates and bioreactors [29]. Trypan blue-based staining was suggested for monitoring the infection and determination of the TOH [5]. During the infection process the cell membrane integrity decreases [22], and this results in protein leakage [54] and trypan blue transfer into the cells. In contrast to mammalian cells, which are stained exclusively by trypan blue when dead, trypan blue staining scores the progress of baculovirus infection in insect cells [36]. Using the fraction of trypan blue-stained cells as a measure of the culture's infection state, TOH translation from small- to large-scale production processes becomes possible [5].

14.8
Case study: Rapid Optimization of Expression Conditions and Large-scale Production of a Brutons Tyrosine Kinase Variant (BTK)

In this section we will exemplify implementation of the aforementioned generic protocols for production optimization of

BTK (720 mg purified protein) within 2 weeks, starting from a virus master stock.

14.8.1
Determination of Optimum MOI, TOI, and TOH in Small-scale Roller Bottles

Sf-9 cells were cultivated in 450 cm^2 roller bottles (70 mL Sf900II medium) and infected with different MOIs at different TOIs (expressed in cell density at the infection time point). Relative protein titer and protein quality were examined by: 1) Coomassie blue-stained SDS–PAGE; 2) Western blot analysis at 48 h and 72 h after infection; and 3) trypan blue staining. The results at 48 h post infection are shown in Table 14.3; at 72 h, a lower protein quality and titers were observed post infection (data not shown). These data suggested infection of insect cells at an MOI of 1 and as the cell density reaches 31% (1.7×10^6 cells mL^{-1}) of the uninfected peak cell density under the same culture conditions (5.5×10^6 cells mL^{-1}), and that the protein should be harvested when 76% of the cells are stained by trypan blue.

14.8.2
Scale-up to 10-L Wave^{TM} Bioreactors

The BTK production run was performed in two parallel 10-L Wave^{TM} bioreactors (Fig. 14.1) operated in fed-batch mode with 4 g L^{-1} yeastolate supplementation at infection, an MOI of 1, and a TOI of 2.8×10^6 cells mL^{-1} (32% of 8.8×10^6 cells mL^{-1}, the peak cell density under fed-batch conditions). During the growth and production phase, oxygen levels were controlled by adjusting the in-gas composition and the agitation speed based on an oxygen control software simulating oxygen transfer and consumption in a Wave^{TM} bioreactor. The software's underlying model is displayed

Table 14.3 Analysis of recombinant protein titer and the fraction
of trypan blue-stained cells as a function of MOI and TOI at 48 h
post infection

Run no.	TOI $[\times 10^6$ cells mL$^{-1}]$	MOI	Protein titer [IOD]	Stained cells [%]
1	1.0	0.1	56.5	82
2	1.7	0.1	74.6	ND
3	2.4	0.1	100.6	80
4	3.5	0.1	87.7	76
5	1.0	1.0	93.8	82
6	1.7	1.0	117.0	76
7	2.4	1.0	110.8	60
8	3.5	1.0	108.6	73

IOD, integrated optical density derived from quantification of Western blot bands.
ND, not detected.

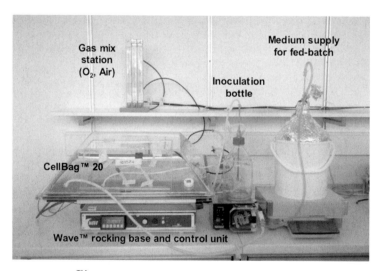

Fig. 14.1 Wave™ bioreactor system equipped with inoculation bottle as well as medium supply for fed-batch mode.

in Fig. 14.2, and the corresponding volumetric oxygen transfer rates are shown in
Table 14.4. The culture was harvested
when 77% of the cells were stained with
trypan blue, resulting in 720 mg BTK after
purification. The entire protocol was finalized within 2 weeks, with optimization
taking place in the first week and production in the second. A detailed overview on
the process as well as operation parameters for the Wave™ system are shown in
Tables 14.5 and 14.6.

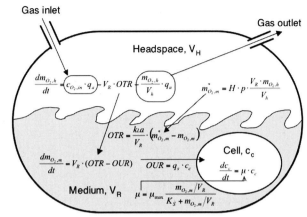

Fig. 14.2 Oxygen transfer and consumption for insect cell culture in a Wave™ bioreactor. The model can be used to adjust rocking rate, rocking angle and the in-gas composition to achieve optimum oxygen supply in the different growth phases. μ, specific growth rate [h^{-1}]; μ_{max}, maximum specific growth rate without oxygen limitations [h^{-1}]; K_S, Dissolved oxygen concentration at which $\mu = 0.5\mu_{max}$ (Monod kinetic) [g mL^{-1}]; H, Henry's constant [Pa^{-1}]; kLa, volumetric oxygen mass transfer coefficient [h^{-1}]; $c_{O2,in}$, oxygen concentration in in-gas [g mL^{-1}]; $m^*_{O2,m}$, dissolved oxygen in medium under saturation conditions [g]; $m_{O2,h}$, oxygen in reactor headspace [g]; $m_{O2,m}$, dissolved oxygen in medium [g]; OTR, volumetric oxygen transfer rate [g mL^{-1} h^{-1}]; OUR volumetric oxygen uptake rate [g mL^{-1} h^{-1}]; p, atmospheric pressure [Pa]; q_a, aeration rate [mL h^{-1}]; q_s, specific oxygen uptake rate [g cell^{-1} h^{-1}]; c_c, cell density [cell mL^{-1}]; q_{O2}, specific oxygen uptake rate [g cell^{-1} h^{-1}]; t, time [h]; VH, reactor headspace volume [mL]; VR, reactor liquid volume [mL].

Table 14.4 Volumetric oxygen mass transfer coefficients ($k_L a$ [h^{-1}]) in a Wave™ bioreactor[a] at different rocking rates and rocking angles

Rocking angle [°]	Rocking rate [min^{-1}]		
	20	24	28
5	1.15	1.89	5.77
7.5	2.87	5.51	7.33
10	5.25	6.29	8.18

a) The Wave™ reactor was equipped with a 20-L-Cellbag™ filled with 10 L of water.

14.8.3
Process Cost-based Choice of the Large-scale Cultivation System

For detailed cost analysis, a 10-L roller bottle culture, a conventional stirred-tank reactor as well as a Wave™ system were compared. Since recombinant protein titers for a secreted model glycoprotein (secreted intercellular adhesion molecule 1; s-ICAM-1) were comparable for all three systems (93 mg L^{-1}, 1241 mg L^{-1} and 113 mg L^{-1}, respectively), a detailed process cost analysis was performed [5, 50].

Table 14.5 Protocol for cultivation of Sf-9 cells in 20-L Cellbags^{TM}

Time	Operation $^{a)}$	Settings	Remark
Monday	Per 20-L CellBag^{TM} inoculate four roller bottles (850 cm^2) with 3×10^5 Sf-9 cells mL^{-1} in 250 mL Sf-900II medium. Assemble and sterilize sample adapter, inoculum bottle and medium tank. Fill medium tank with 9 L Sf-900II medium plus 0.1% Pluronic F68 and incubate for 2 days at 28 °C for sterility control.		
Thursday	Connect sample adapter, inoculum bottle and medium reservoir to CellBag^{TM} under laminar flow. Place CellBag^{TM} on rocking unit, inflate with air and fill with 9 L medium.	Rocking rate = 20 min^{-1}; angle = 7.5°; temperature = 28 °C; turn off aeration and close all inlets.	Insert DO-probe if needed
Friday	Control inoculum cultures and CellBag^{TM} for sterility. Fill inoculum into inoculum bottle under laminar flow. Inoculate the culture with final density of approx. 3×10^6 cells mL^{-1}. Start stop watch.	Rocking rate = 20 min^{-1}; angle = 7.5°; temperature = 28 °C; set air flow to 3 L h^{-1}.	Inoculum culture should be at 2.5–3.0$\times10^6$ cells mL^{-1}
Sunday	Count cells and perform Medium analysis.	Set air flow to 12 L h^{-1} to meet increased oxygen requirements.	
Monday	Count cells and perform Medium analysis.	Adjust rocking rate and aeration parameters to meet oxygen requirements.	
Tuesday	Count cells and perform Medium analysis. Add virus and nutrients at desired cell density via inoculum bottle.	Adapt oxygen concentration in inlet gas to maintain DO concentration between 50 and 100% of saturation.	

a) The aeration parameters (air and oxygen flow rates, rocking rate and rocking angle) can be determined by calculating the oxygen flux using the model described in Fig. 14.2 and Table 14.4.

Table 14.6 Protocol for BEVS-based recombinant protein production. When sufficient virus stock is available the entire procedure (determination of optimum expression conditions and expression at large scale) can be performed within 2 weeks

Time	Operation	Remark
Thursday evening	Inoculate four roller bottles (850 cm^2) with 3×10^5 cells mL^{-1} in 250 mL Sf-900II medium each	
Monday morning	Inoculate four roller bottles (850 cm^2) with 3×10^5 cells mL^{-1} in 250 mL Sf-900II medium each	Inoculum cultures for Wave^{TM} Bioreactor
	Assemble and sterilize sample port, inoculum bottle and medium tank. Fill with 9 L Sf-900II medium plus 0.1% Pluronic F-68 and incubate for 2 days at 28 °C for sterility control.	Preparation for Wave^{TM} Bioreactor
	Inoculate eight roller bottles (490 cm^2) with 0.7; 1.4; 2.1 and 2.8×10^6 cells mL^{-1}, two cultures per cell density, culture volume: 80 mL Sf-900II medium. Infect the eight roller cultures with two different MOI per cell density (e.g., 0.1 and 1 MOI).	Cultures for determination of optimum expression conditions (different MOI and TOI are tested)
	Inoculate two roller bottles (490 cm^2) with 1.8×10^6 cells mL^{-1} in 80 mL Sf-900II medium	Cultures for determination of peak cell density [a]
Wednesday	Determine fraction of trypan blue-stained cells in infected roller cultures (mix samples with trypan blue and wait 3 min before counting)	Infected cells become slowly colored, counting directly after mixing results in underestimation of stained cells.
	Take 1.5-mL samples from each infected culture, centrifuge, and store protein containing fraction appropriately.	Filtration is also possible when secreted proteins are produced
	Count the two cultures for PCD determination [a]	
Thursday	Same as on Wednesday	
	Perform electrophoresis with samples from kinetic studies on two gels, use one for Western blotting, the other for Coomassie blue staining. Block the Western blot over night, dry the Coomassie blue-stained over night.	Test on expression level and protein integrity
	Connect sample port, inoculum bottle and medium reservoir to CellBag^{TM} under laminar flow.	Preparation of CellBag^{TM}
Friday morning	Finish Western blot from Thursday. Perform quantification of bands on the gel. If available, use more exact analytical methods (ELISA, HPLC).	
	Determine optimum MOI, TOI and harvest time point with regard to expression level and protein integrity.	

Table 14.6 (continued)

Time	Operation	Remark
Friday afternoon	Install CellBag^{TM} on rocking unit, fill with 9 L medium and wait until temperature of 28 °C is stable. Inoculate then with starter cultures from Monday.	Install CellBag^{TM}
Monday morning	Calculate optimum infection cell density with Eq. (14.1) based on a PCD of 8.8×10^6 cells mL^{-1} (Sf-9 cells with yeastolate addition) Perform culture analysis in CellBag^{TM} according to above protocol.	
Following days	Perform culture analysis in CellBag^{TM} and adjust culture parameters according to above protocol. Perform infection when optimum infection cell density is reached. Supply 1× yeastolate or yeastolate ultrafiltrate at approx. 3×10^6 cells mL^{-1} [b]	Use yeastolate for intracellular proteins and yeastolate ultrafiltrate for secreted and membrane-bound ones.
	Determine fraction of trypan blue-stained cells twice a day. Harvest product when reaching the same fraction as on the optimum time point of harvest in optimization studies.	Harvest time typically on Thursday or Friday.

a) Determination of peak cell density can be omitted if this value is known from former experiments. Peak cell density is slightly dependent on inoculum density, therefore an inoculum cell density is chosen which is typically the average of the tested TOIs.
b) If infection occurs at cell densities $< 3 \times 10^6$ cells mL^{-1}, add yeastolate or yeastolate ultrafiltrate at the time point when an uninfected culture would reach this density.

Integration of fixed costs, consumables and manpower requirements on a typical industry cost basis showed that the Wave^{TM} system was far the most economical solution (€ 1297 per run), while roller bottles (€ 1789 per run) and the stirred-tank reactor (€ 2224 per run) were up to 70% more cost-intensive (Fig. 14.3, Table 14.7).

14.9
Conclusion

Today, the baculovirus expression vector system represents a well-established technology for routine pilot production of small to medium quantities of desired product proteins, mainly for research purposes. The characteristics of BEVS, including culture media design, virus production and quantification as well as large-scale cell culture, have been extensively studied during the past decade, and this has resulted in

Fig. 14.3 Process costs for three different cultivation systems based on a 10-L culture scale. Fixed costs, consumables and manpower costs were calculated on a typical industry cost basis.

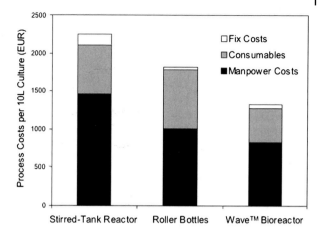

Table 14.7 Process cost calculations for the three different cell culture systems operated at 10 L scale

System	Roller bottles 20 Roller bottles with 500 mL culture volume	Stirred-tank reactor 10 L Stirred-tank reactor	Wave™ bioreactor Wave Bioreactor with 20-L CellBag™
Fix costs[a]	Incubator 8900,– € Costs per day 5,– € Process time 7 d	Bioreactor and process control unit 35 600,– € Costs per day 20,– € Process time[b] 9 d	Wave™ reactor and equipment 12 460,– € Costs per day 7,– € Process time 7 d
Total Fix costs	35,– €	180,– €	49,– €
Consumables process set-up	20 Roller bottles 150,– € 15 Pipettes 4,– € Sterile filter 6,– €	DO-Probe membrane 8,– € O-rings 13,– € Tubing, fittings 46,– € Sterile filters 25,– € Sterilization 200,– €	20-L CellBag™ 220,– € Tubing, fittings 25,– € Sterile filters 12,– € Sterilization 50,– €
	Total set-up 160,– €	Total set-up 292,– €	Total set-up 307,– €
Consumables per day	5 Pipettes 1.5,– €		
	5 Medium analysis 80,– € 3 L oxygen 0.04,– € Total material per day 82,– € Culture time/process 7 d Total daily 574,– €	Medium analysis 16,– € 370 L oxygen 5,– € Total material per day 21,– € Culture time/process 7 d Total daily 147,– €	Medium analysis 16,– € 60 L oxygen 0.8,– € Total material per day 17.80,– € Culture time/process 7 d Total daily 125,– €
Cleaning	Sterilization[c] 50,– € Total cleaning 50,– €	Sterilization[d] 100,– € Washing machine 95,– € Total cleaning 195,– €	Sterilization[e] 16,– € Total cleaning 16,– €
Total Consumables	784,– €	634,– €	448,– €

Table 14.7 (continued)

System	Roller bottles 20 Roller bottles with 500 mL culture volume	Stirred-tank reactor 10 L Stirred-tank reactor	Wave^{TM} bioreactor Wave Bioreactor with 20-L CellBag^{TM}
Manpower Process set-up		Bioreactor assembly 3.2 h	CellBag^{TM} assembly 1.0 h
		Bioreactor installation 1.0 h Probe calibration 0.8 h	CellBag^{TM} installation 0.5 h
	Inoculation 1.0 h Virus addition 0.4 h Feeding 0.4 h Total set-up 1.8 h	Inoculation 1.5 h Virus addition 0.3 h Feeding 0.3 h Total set-up 7.1 h	Inoculation 1.0 h Virus addition 0.3 h Feeding 0.3 h Total set-up 3.1 h
Manpower per day	Medium analysis and cell counting (for 5 rollers)$^{f)}$ 0.8 h Flushing rollers with oxygen 0.3 h	Medium analysis, cell counting 0.5 h	Medium analysis, cell counting 0.5 h
	Total per day 1.1 h Typical process time 7 d Total 7.7 h	Total per day 0.5 h Typical process time 7 d Total 3.5 h	Total per day 0.5 h Typical process time 7 d Total 3.5 h
Manpower cleaning	Sterilization 0.2 h	Sterilization 0.5 h	Sterilization 0.5 h
		Disassembly 1 h Cleaning 2 h	Disassembly 0.4 h Cleaning 0.5 h
	Total 0.2 h	Total 3.5 h	Total 1.4 h
Total Man-power costs	Total manpower 9.7 h	14.1 h	8.0 h
	Manpower costs per hour$^{g)}$ 100,– € 970,– €	100,– € 1410,– €	100,– € 800,– €
Total Process costs	1789,– €	2224,– €	1297,– €

a) Depreciation period of 5 years (1780 days).
b) 2 days for washing and assembling included.
c) 25% of autoclave volume required.
d) 50% of autoclave volume required.
e) 8% of autoclave volume required.
f) Per day, five roller bottles out of the 20 are sampled to obtain
   a representative value for dissolved oxygen concentration, cell
   number and viability.
g) Labor costs include costs for lab technician as well as all other
   material, administration and infrastructure costs. Costs which
   are common to all systems (consumables and manpower for
   medium preparation, harvesting) were not considered.

straightforward implementation protocols which are applicable to the vast majority of product proteins. Future research and development in the area of BEVS will likely shift from culture medium design and process engineering to metabolic engineering for more efficient nutrient utilization [55] and/or production of mammalian-like glycosylation profiles [56] (see Part IV, Chapters 2 and 7). A new era of genetically optimized insect cells and baculoviral vectors will almost certainly boost the success of previous process engineering investigations, and this may eventually result in BEVS-produced biopharmaceuticals.

Acknowledgments

The authors thank Cornelia Weber for critical comments on the manuscript. These studies were supported by Cistronics Cell Technology GmbH, Einsteinstrasse, P.O.B. 145, CH-8093 Zurich, Switzerland, the Swiss National Science Foundation (grant no. 631-065946) and the Swiss Federal Office for Education and Science (BBW) within EC Framework 6.

References

1 L. A. King, R. D. Possee: The baculovirus expression system. Chapman & Hall, London, **1992**.

2 V. A. Luckow, M. D. Summers, Bio/Technology **1988**, 6, 1406–1410.

3 R. D. Possee, Curr. Opin. Biotechnol. **1997**, 8, 569–572.

4 L. Ikonomou, Y. J. Schneider, S. N. Agathos, Appl. Microbiol. Biotechnol. **2003**, 62, 1–20.

5 W. Weber, E. Weber, S. Geisse, K. Memmert, Cytotechnology **2002**, 38, 77–85.

6 J. Hollister, E. Grabenhorst, M. Nimtz, H. Conradt, D. L. Jarvis, Biochemistry **2002**, 41, 15093–15104.

7 N. Tomiya, M. J. Betenbaugh, Y. C. Lee, Acc. Chem. Res. **2003**, 36, 613–620.

8 J. E. Vialard, M. A. Basil, C. D. Richardson: Introduction to the molecular biology of baculoviruses. In: Methods in molecular biology, C. D. Richardson (Ed.). Humana Press; **1995**: 203–224. Baculovirus expression protocols, vol. 39.

9 M. S. Kwon, T. Dojima, M. Toriyama, E. Y. Park, Biotechnol. Prog. **2002**, 18, 647–651.

10 B. Philipps, M. Forstner, L. M. Mayr, Biotechniques **2004**, 36, 80–83.

11 H. R. Lo, Y. C. Chao, Biotechnol. Prog. **2004**, 20, 354–360.

12 S. Maeda, Annu. Rev. Entomol. **1989**, 34, 351–372.

13 L. Ikonomou, G. Bastin, Y. J. Schneider, S. N. Agathos, In Vitro Cell. Dev. Biol. Anim. **2001**, 37, 549–559.

14 E. J. Schlaeger, Cytotechnology **1996**, 20, 57–70.

15 T. J. Wickham, T. Davis, R. R. Granados, M. L. Shuler, H. A. Wood, Biotechnol. Prog. **1992**, 8, 391–396.

16 T. R. Davis, T. J. Wickham, K. A. McKenna, R. R. Granados, M. L. Shuler, H. A. Wood, In Vitro Cell Dev. Biol. Anim. **1993**, 29A, 388–390.

17 L. A. Palomares, C. E. Joosten, P. R. Hughes, R. R. Granados, M. L. Shuler, Biotechnol. Prog. **2003**, 19, 185–192.

18 K. Viswanathan, S. Lawrence, S. Hinderlich, K. J. Yarema, Y. C. Lee, M. J. Betenbaugh, Biochemistry **2003**, 42, 15215–15225.

19 S. Watanabe, T. Kokuho, H. Takahashi, M. Takahashi, T. Kubota, S. Inumaru, J. Biol. Chem. **2002**, 277, 5090–5093.

20 K. T. K. Wong, C. H. Peter, P. F. Greenfield, S. Reid, L. K. Nielsen, Biotechnol. Bioeng. **1996**, 49, 659–666.

21 P. Licari, J. E. Bailey, Biotechnol. Bioeng. **1992**, 39, 432–441.

22 J. F. Power, S. Reid, K. M. Radford, P. F. Greenfield, L. K. Nielsen, Biotechnol. Bioeng. **1994**, 44, 710–719.

23 R. A. Taticek, M. L. Shuler, Biotechnol. Bioeng. **1997**, 54, 142–152.

24 E. Bedard, S. Perret, A. Kamen, Biotechnol. Lett. **1997**, 19, 629–632.

25 A. W. Caron, J. Archambault, B. Massie, Biotechnol. Bioeng. **1990**, 36, 1133–1140.

26 K. M. Radford, S. Reid, P. F. Greenfield, Biotechnol. Bioeng. **1997**, 56, 73–81.

27 V. Jäger, Cytotechnology **1996**, 20, 191–198.

28 E. Chico, V. Jäger, Biotechnol. Bioeng. **2000**, 70, 574–586.

29 J.D. Yang, P. Gecik, A. Collins, S. Czarnecki, H.H. Hsu, A. Lasdun, R. Sundaram, G. Muthukumar, M. Silberklang, Biotechnol. Bioeng. **1996**, 52, 696–706.

30 C. Bedard, A. Kamen, R. Tom, B. Massie, Cytotechnology **1994**, 15, 129–138.

31 L.C. Chan, P.F. Greenfield, S. Reid, Biotechnol. Bioeng. **1998**, 59, 178–188.

32 B. Maiorella, D. Inlow, A. Shauger, D. Harano, Bio/Technology **1988**, 6, 1406–1410.

33 P.E. Cruz, A. Cunha, C.C. Peixoto, J. Clemente, J.L. Moreira, M.J. Carrondo, Biotechnol. Bioeng. **1998**, 60, 408–418.

34 M.Y. Wang, T.R. Pulliam, M. Valle, V.N. Vakharia, W.E. Bentley, J. Biotechnol. **1996**, 46, 243–254.

35 Y.C. Hu, W.E. Bentley, Biotechnol. Prog. **1999**, 15, 1065–1071.

36 A. Kamen, C. Bedard, R. Tom, S. Perret, B. Jardin, Biotechnol. Bioeng. **1996**, 50, 36–48.

37 A. Kamen, R. Tom, A.W. Caron, C. Chavarie, B. Massie, J. Archambault, Biotechnol. Bioeng. **1991**, 38, 619–628.

38 C. Mitchell-Logean, D.W. Murhammer, Biotechnol. Prog. **1997**, 13, 875–877.

39 A. Garnier, R. Voyer, R. Tom, S. Perret, B. Jardin, A. Kamen, Cytotechnology **1996**, 22, 53–63.

40 C.B. Elias, A. Zeiser, C. Bedard, A.A. Kamen, Biotechnol. Bioeng. **2000**, 68, 381–388.

41 K.U. Dee, M.L. Shuler, Biotechnol. Prog. **1997**, 13, 14–24.

42 J.M. Liebmann, D. LaScala, W. Wang, P.M. Steed, BioTechniques **1999**, 26, 36–42.

43 L. Maranga, T.F. Brazao, M.J. Carrondo, Biotechnol. Bioeng. **2003**, 84, 245–253.

44 T. Ohkawa, K. Majima, S. Maeda, J. Virol. **1994**, 68, 6619–6625.

45 R.E. Hawtin, T. Zarkowska, K. Arnold, C.J. Thomas, G.W. Gooday, L.A. King, J.A. Kuzio, R.D. Possee, Virology **1997**, 238, 243–253.

46 J.M. Slack, J. Kuzio, P. Faulkner, J. Gen. Virol. **1995**, 76 (Pt 5), 1091–1098.

47 S.A. Monsma, M. Scott, Innovations **1997**, 16–19.

48 B.C. Bonning, P.W. Roelvink, J.M. Vlak, R.D. Possee, B.D. Hammock, J. Gen. Virol. **1994**, 75 (Pt 7), 1551–1556.

49 L.E. Pyle, P. Barton, Y. Fujiwara, A. Mitchell, N. Fidge, J. Lipid Res. **1995**, 36, 2355–2361.

50 W. Weber, E. Weber, S. Geisse, K. Memmert, Catching the Wave: the BEVS and the Biowave, E. Lindner-Olsson, N. Chatzissavidou, E. Lüllau (Eds) Dordrecht, Kluwer Academic, **2001**.

51 V. Singh, Cytotechnology **1999**, 30, 149–158.

52 L.N. Pierce, P.W. Shabram, BioProcessing J. **2004**, 3, 1–5.

53 P. Licari, J.E. Bailey, Biotechnol. Bioeng. **1991**, 37, 238–246.

54 K.M. Radford, C. Cavegn, M. Bertrand, A.R. Bernard, S. Reid, P.F. Greenfield, Cytotechnology **1997**, 24, 73–81.

55 C.B. Elias, E. Carpentier, Y. Durocher, L. Bisson, R. Wagner, A. Kamen, Biotechnol. Prog. **2003**, 19, 90–97.

56 M.J. Betenbaugh, N. Tomiya, S. Narang, J.T. Hsu, Y.C. Lee, Curr. Opin. Struct. Biol. **2004**, 14, 601–606.

15

Robust and Cost-effective Cell-free Expression of Biopharmaceuticals: *Escherichia Coli* and Wheat Embryo

Luke Anthony Miles

Abstract

The widespread use of cell-free systems in biomedical research laboratories reflects their usefulness in producing functional proteins. However, cell-free methods have typically yielded only nanogram to microgram quantities of proteins, which has limited their utility to functional studies. Cell-free systems derived from many cell types have been described in the scientific literature. For a small number of these cell types, significant advances made in recent years have seen the development of robust, cost-effective and highly efficient cell-free expression systems suitable for the preparation of proteins in milligram quantities. In this chapter, the advantages of cell-free protein synthesis methods, with particular emphasis on applications to structural biology, will be discussed. *Escherichia coli* and wheat embryo systems are the best-characterized prokaryotic and eukaryotic high-efficiency systems, respectively, and are the focus here. The bulk of the chapter will be devoted to a discussion of recent advances in cell-free synthesis methods that have facilitated the production of proteins in high yield that are soluble, intact, and functional. Recent advances in cell-free expression systems that are amenable to automation and high-throughput screening, and therefore well-suited for the development of biopharmaceuticals, will also be discussed.

Abbreviations

AMP	adenosine monophosphate
ARS	aminoacyl-tRNA synthetase
ATP	adenosine triphosphate
CAT	chloramphenicol acetyl-transferase
CDP	cytosine diphosphate
CTP	cytidine triphosphate
DHFR	dihydrofolate reductase
dNTP	desoxynucleotide triphosphate
DTT	dithiothreitol
EF	elongation factor
GSH	reduced glutathione
GSSG	oxidized glutathiones
HEPES	2-[4-(2-hydroxyethyl)-1-piperazinyl] ethanesulfonic acid
HTS	high-throughput screening
IAM	iodoacetamide
IF	initiation factor
MAD	multi-wavelength anomalous dispersion
MTF	methionyl-tRNA transformylase factor
MWCO	molecular weight cut-off
NAD	nicotinamide adenine dinucleotide
NMR	nuclear magnetic resonance

Modern Biopharmaceuticals. Edited by J. Knäblein
Copyright © 2005 WILEY-VCH Verlag GmbH & Co. KGaA, Weinheim
ISBN: 3-527-31184-X

PA	plasminogen activator
PCR	polymerase chain reaction
PDI	protein disulfide isomerase
PEG	polyethylene glycol
PEP	phosphoenol pyruvate
pIVEX	plasmid for In Vitro EXpression
PURE	protein synthesis using recombinant elements
RIL	plasmid in E. coli strain BL21
RRF	ribosome recycling factor
SDS-PAGE	sodium dodecyl sulfate-polyacrylamide gel electrophoresis
TCA	trichloroacetic acid
UTR	untranslated region
WG	wheat germ

15.1
Introduction

15.1.1
Background

Cell-free protein synthesis refers to a family of techniques in which ribosomes and translation factors are isolated from cells to synthesize polypeptides *in vitro* from a messenger RNA template. Proteins can also be expressed *in vitro* from a DNA template by exploiting the ability of bacteriophage RNA polymerases to synthesize mRNA transcripts from DNA.

Cell-free systems have long been used to test the expression of constructs and to assess the properties of the translation product such as solubility, stability, and activity. Instability and low efficiency limited their usefulness to small-scale synthesis, and *in vivo* methods were relied on for up-scaled protein production to meet the demands of consuming techniques such as those used in structural biology.

Since the early investigations of Zubay on developing *Escherichia coli* extracts for *in vitro* translation [1], much effort has been spent on increasing the yields from E. *coli* cell-free systems. Many laboratories have made modifications to extract preparation protocols and reaction conditions that have seen substantial increases in stability and efficiency of these systems, and none more so than the group of Yokoyama (RIKEN, Yokohama, Japan). This group has developed an *E. coli* cell-free system capable of producing proteins at a rate approaching $500 \ \mu g \ mL^{-1}$ reaction mixture in the first hour of reaction [2, 3].

A significant leap forward was made in 1988 by Spirin et al. [4], who developed a continuous-flow apparatus for the continuous supplementation of reaction mixtures with substrates required for protein synthesis and continuous removal of reaction by-products. In this way, it was shown that the activity of a cell-free system could be sustained for many hours, compared with batch-mode reactions which became inactive after approximately 45 minutes. Since then, many reports have been made describing the use of simple dialysis systems [3, 5] that can maintain the high productivity of a reaction over many hours, without the use of a cumbersome apparatus.

A combination of these developments has seen expression yields as high as $6 \ mg \ mL^{-1}$ achieved over 20 hours in an *E. coli* cell-free system [3], making cell-free methods a serious alternative to *in vivo* methods of large-scale protein production.

A push has been made more recently to develop productive systems that do not rely on flow apparatus or dialysis membranes, in order to make the methods compatible with robot-controlled formats. This has seen the emergence of fed-batch mode reactions where reactions are periodically supplemented with aliquots of reaction substrates [6]. Such a system has been shown to prolong synthesis activity for up to 3 hours, producing 750 μg of protein per mL reaction mixture [7].

Endo and co-workers at Ehime University, Matsuyama, Japan, have led the development of the most promising eukaryotic cell-free system to date, based on wheat embryos. A significant advance made by this group was the development of pEU expression vectors that have overcome many of the difficulties associated with mRNA synthesis for translation in a eukaryotic system [8]. In addition to extensive optimization of reaction conditions that have seen improvements in protein synthesis rates, Endo and colleagues have improved wheat extract embryo preparation protocols to enhance the stability of these systems to a remarkable extent [9]. When coupled with the dialysis mode of reaction, Endo et al. were able to maintain translational activity in a coupled transcription/translation wheat embryo reaction for 150 hours, producing 5 mg of enzymatically active protein per mL reaction mixture [10]. This again represents a serious alternative to *in vivo* methods of large-scale protein production.

15.1.2
Some Advantages to Cell-free Synthesis

Cell-free synthesis is often used to express troublesome proteins. For example, it is well-suited to the expression of toxic proteins because there is no need to consider the action of a gene product on the viability of a host cell. The open nature of cell-free methods makes them useful for preparing proteins prone to mis-folding, aggregation into inclusion bodies, and proteolytic digestion. This is because the expression reaction conditions can be readily manipulated to promote proper protein folding and inhibit proteolysis. In addition,

polymerase chain reaction (PCR)-generated DNA templates can be used in cell-free systems, obviating the need for time-consuming cloning procedures in the initial steps of DNA template and reaction condition optimization. This also makes cell-free synthesis well-suited to high-throughput and robot-controlled formats. Unpurified reaction mixtures can often be used directly in functional assays, greatly expediting the analysis of expression trials.

In the broadest terms, cell-free synthesis offers advantages over traditional *in vivo* methods of recombinant protein expression, as it enables the experimentalist strictly to control conditions under which expression reactions are performed. Benefits of such an open expression system are exemplified by the fact that cell-free reaction conditions can be sufficiently modified to support co-translational folding of active disulfide-bonded proteins.

15.1.3
Cell-free Synthesis in Structural Biology

Many techniques used in structural biology and biophysics rely on preparing proteins incorporating costly isotope and heavy-atom labels. The group of Kainosho prepared protein samples that were isotopically labeled with a chemically synthesized amino acid using *in vitro* and *in vivo* E. coli expression systems [2]. Consumption of the expensive amino acid was shown to be two orders of magnitude lower in the cell-free system than in the *in vivo* expression system. Low levels of endogenous amino acids in cell-free systems also result in high-level incorporation of labeled amino acids [2]. In addition, scrambling effects resulting from the metabolism of amino acids *in vivo* from one type to another is not observed *in vitro*. For all these reasons, cell-free synthesis methods have become indispensable to many structural biology and biophysics laboratories (see Part VII, Chapter 2).

Crystals from selenium-methionine- and/or selenium-cysteine-labeled proteins can be studied by multi-wavelength anomalous dispersion (MAD) phasing techniques that can facilitate the solution of an X-ray crystal structure from a single crystal form [11]. However, if *in vivo* expression systems are used to prepare selenium-labeled proteins, amino acid metabolism and the toxicity of Se-methionine can result in low protein yields and low incorporation rates.

All but the smallest proteins studied by nuclear magnetic resonance (NMR) need to be labeled with isotopes of hydrogen, nitrogen, or carbon (^{2}H, ^{15}N, ^{13}C), or combinations thereof. Cell-free systems can be used for uniform, selective, or site-specific labeling of proteins with isotopically enriched amino acids, and relatively inexpensive algal amino acids can be used for this purpose. Unlike the *in vivo* expression of labeled proteins in minimal media, expression conditions and synthesis yields are the same for expression of natural abundance and isotopically enriched proteins. Selective labeling with single amino acid types can be carried out in tandem to help make rapid and unambiguous resonance assignments from NMR spectra.

15.2
Transcription

Transcription and translation are coupled in the cytoplasm of prokaryotes. In eukaryotes, transcription of DNA and subsequent processsing of the messenger RNA transcript both occur in the nucleus, while translation of the mature mRNA template occurs in the cytoplasm. Cell-free systems lack mechanisms for mRNA processing, and therefore require fully matured mRNA templates for translation. DNA templates are similarly transcribed by bacteriophage RNA polymerases (T7, SP6, and T4) (see Part III, Chapter 2) for translation in both *E. coli* (prokaryotic) and wheat embryo (eukaryotic) systems. However, these systems require differently structured mRNA templates for efficient translation to occur. Details of the construction and treatment of different DNA templates for generating translatable mRNA strands are discussed here.

15.2.1
Plasmid DNA Templates for *E. coli* Systems

Any plasmid designed to express recombinant proteins in *E. coli* under the control of a bacteriophage RNA polymerase promoter can be used in an *E. coli* cell-free system. As a rule of thumb, constructs that express well *in vivo* tend to express well *in vitro*. T7 RNA polymerase is often considered intrinsically more efficient than other RNA polymerases (see Part IV, Chapter 12). However, this differential efficiency can often be attributed to differential sensitivity to salt concentration. Because of the relative robustness and efficiency of the T7 polymerase, plasmids under the control of a T7 promoter are almost exclusively used in *E. coli* cell-free systems.

pIVEX (plasmid for In Vitro EXpression) vectors (Roche) have been developed specifically for the optimal expression in *E. coli* extracts, and are available with a range of affinity tags for detection and purification. pIVEX vectors are very high copy number, which is useful for large-scale protein synthesis *in vitro* where large quantities of highly purified DNA template are required. However, one limitation of pIVEX vectors is that they are non-inducible and therefore cannot be used for induced expression trials in bacteria.

15.2.2
Plasmid DNA Templates for Wheat Embryo Systems

Efficient translation in eukaryotic systems traditionally required messenger RNA template with a capped 5′ end, optimally structured 5′ and 3′ untranslated regions (UTRs), and a polyadenylated tail in the 3′-UTR of the transcript. 5′-capped mRNA has traditionally been synthesized *in vitro* by transcribing the DNA template in the presence of cap analogues such as 7-methylguanosine in addition to the dNTPs. The ability of 5′ cap analogues to bind initiation factor eIF-4E means that free cap analogue can also occupy eIF-4E and render it unavailable for translation. In fact, researchers who work to develop improved cap analogues test the activity of a candidate molecule for its ability to inhibit *in vitro* translation in its free form. Naturally, the use of 5′ cap analogues to produce capped mRNA dictates that transcription and translation reactions must be performed separately. Their use also necessitates careful optimization of cap analogue concentration in the transcription reaction or purification of the mRNA template before use in a translation reaction.

In an effort to eliminate the requirement for 5′ cap analogue incorporation, Endo and Sawasaki explored the genomes of plant RNA viruses that have positive-sense RNA lacking 5′ caps and 3′ polyA tails [8]. In those studies, the sequence encoding the 5′-UTR in a pSP65 vector was replaced with the omega 71 (Ω71) sequence of tobacco mosaic virus. Uncapped mRNA from this construct encoding for luciferase proved to be highly active in a wheat embryo cell-free assay. Further enhancement in activity was achieved by modifying the 5′ end of the Ω71 sequence to GAA. With the 5′-UTR comprised of GAA-Ω71, Endo and Sawasaki systematically modified the length and sequence of the 3′-UTR that works synergistically with the 5′-UTR to enhance transcript stability and initiation of translation. It was found that the activity of transcripts were more sensitive to the length of the 3′-UTR than to variations in the 3′-UTR sequence. mRNA synthesized with a 1626-nucleotide 3′-UTR, and the GAA-Ω71 5′-UTR translation enhancer proved to be almost as active as the capped and polyadenylated mRNA transcribed from the unmodified pSP65 vector.

As a result of those studies, Endo et al. developed the pEU series of vectors with various affinity tags for efficient expression in wheat embryo cell-free systems without the need for expensive cap analogues. pI-VEX WG (wheat germ) vectors are now commercially available (Roche) which similarly rely on translation-enhancing untranslated regions flanking the multi-cloning site.

15.2.3
Linear DNA Templates

Some regions of a plasmid vector such as the origin of replication and antibiotic resistance genes are not required for gene expression. Linear DNA constructs are suitable templates for *in vitro* expression if the elements that regulate transcription (RNA polymerase promoter/terminator) and translation (ribosome binding site, start/stop codons) in a vector are appended to the gene to be expressed. These templates are generated and amplified by the PCR and can be used to rapidly screen protein expression and folding without the need for cloning into a plasmid vector. PCR products can generally be used directly in a cell-free reaction, without the need for purification of the template.

Kits for the generation of DNA templates by PCR can be purchased from several suppliers, including *E. coli* and wheat embryo pIVEX linear DNA template kits (Roche). Similarly, Endo and Sawasaki [8] have demonstrated efficient translation in a wheat embryo system from PCR-generated linear DNA templates based on the structure of their pEU plasmid series.

PCR-generated DNA templates are excellent for rapid batch-mode screening of domain boundaries, product solubility and expression levels, and for generating mutants. PCR conditions must be optimized to ensure that a single product is obtained, and high-fidelity enzymes should be used to minimize the introduction of random mutations. For scale-up to dialysis mode, linear templates are less useful as they are prone to degradation by nucleases and plasmids are less expensive to amplify. Successful PCR-generated constructs can be sub-cloned directly into an appropriate plasmid vector.

15.2.4
Plasmid Purification

DNA template preparations are a major source of contamination in cell-free synthesis. *In vitro* translation reactions are sensitive to contamination by salts, RNases, detergents, and alcohol, all of which are typically used in commercial plasmid purification kits. As such, plasmids prepared by resin-based purification protocols must be further purified by phenol/chloroform and chloroform extraction. DNA should be precipitated with isopropanol, washed in ethanol, dried, and taken up in RNase-free water to a concentration of about 1 mg mL^{-1}. The DNA should be stored frozen, in small aliquots.

15.2.5
Transcription Reaction

Bacteriophage RNA polymerases are available from a range of commercial suppliers, and transcription-only reactions should be performed according to the manufacturer's instructions. The concentration of mRNA added to a subsequent translation reaction must be established empirically. Conditions for coupled transcription/translation are described in Section 15.6.

15.3
Translational

15.3.1
The PURE System

An important contribution to fundamental cell-free science has come from the laboratory of Ueda et al. at the University of Tokyo, Japan [12]. This group reconstituted a cell-free system with purified recombinant translation factors used by *E. coli*, producing $160 \text{ μg mL}^{-1} \text{ h}^{-1}$ protein in a simple batch-mode reaction. That system is referred to as the PURE (*P*rotein synthesis *U*sing *R*ecombinant *E*lements) system.

Ribosomes needed for translation in the PURE system are isolated from *E. coli* using sucrose-density gradient centrifugation. The protein factors necessary for translation in *E. coli* are recombinantly expressed as His-tagged fusions, and purified to homogeneity. These include the factors for initiation (IF1, IF2, and IF3), elongation (EF-G, EF-Tu, EF-Ts), peptide chain release (RF1 and RF3), ribosome recycling (RRF), methionyl-tRNA transformylase (MTF) for formylation of the initial Met-tRNA, and the 20 aminoacyl-tRNA synthetases (ARSs) for transfer RNA (tRNA) recy-

Table 15.1 Composition of the PURE cell-free systems from *E. coli*

Component		Concentration
Ribosomes		240 pmol mL^{-1}
Initiation Factor 1 (IF1)	His-tagged	20 µg mL^{-1}
Initiation Factor 2 (IF2)	His-tagged	40 µg mL^{-1}
Initiation Factor 3 (IF3)	His-tagged	15 µg mL^{-1}
Extension Factor G (EF-G)	His-tagged	20 µg mL^{-1}
Extension Factor Ts (EF-Ts)	His-tagged	20 µg mL^{-1}
Extension Factor Tu (EF-Tu)	His-tagged	20 µg mL^{-1}
Release Factor 1 (RF1)	His-tagged	10 µg mL^{-1}
Release Factor 3 (RF2)	His-tagged	10 µg mL^{-1}
Ribosome recycling factor (RRF)	His-tagged	10 µg mL^{-1}
20 Aminoacyl-tRNA synthetases (ARSs)	His-tagged	600–6000 U mL^{-1}
Methionyl-tRNA Transformylase (MTF)	His-tagged	6000 U mL^{-1}
T7 RNA polymerase	His-tagged	10 µg mL^{-1}
Mixture of 46 tRNAs		56 A$_{260}$ U mL^{-1}
Creatine kinase		4 µg mL^{-1}
Myokinase		3 µg mL^{-1}
Nucleoside-diphosphate kinase		1.08 µg mL^{-1}
Pyrophosphatase		2 U mL^{-1}
Creatine phosphate		10 mM
Magnesium acetate		9 mM
Potassium phosphate		5 mM
Potassium glutamate		93 mM
Ammonium chloride		5 mM
Calcium chloride		0.5 mM
Spermidine		1 mM
Putrescine		8 mM
Dithiothreitol (DTT)		1 mM
ATP and GTP		2 mM
CTP and UTP		1 mM
Folinic acid		10 µg mL^{-1}
Each of 20 amino acids		0.1 mM

cling. These components are combined with RNA polymerase (His-tagged) for transcription, NTPs, 46 tRNAs, folinic acid, amino acids and factors involved in ATP regeneration. The complete composition of a PURE transcription/translation reaction is given in Table 15.1.

The PURE system is important as it definitively established the minimum requirements for high-level expression in an *E. coli* cell-free system. The benefits of this system include the elimination of pro-

teases and nucleases, and stability of the ATP concentration. All of these factors make the system well-suited to fed-batch reaction formats for robot-controlled synthesis. Since most of the proteins in the PURE system are His-tagged, they are readily removed from the reaction by chelating resins, and purification of a non-His-tagged translation product is dramatically simplified.

The main drawback of the PURE system is that preparing the reaction components requires cloning, expression, and purification of more than 30 proteins. In addition, the system lacks factors endogenous to different cells that might enhance translation and facilitate proper protein folding, such as chaperones. For these reasons, most cell-free systems are derived from crude cell extracts containing ribosomes, translation factors, other endogenous proteins, and tRNAs.

15.3.2
Cell Extracts

The best-characterized cell-free systems derived from crude extracts are the *E. coli* (prokaryotic) and wheat embryo (eukaryotic) systems. The extract preparation protocols used are central to the productivity of any cell-free system.

15.3.2.1 *E. coli* Extracts
As with *in vivo* expression of recombinant proteins in *E. coli*, different cell lines can be used to prepare *E. coli* extracts for *in vitro* expression. A19 *E. coli* is commonly used for *E. coli* preparation. BL21 lines exhibit reduced protease activity, and are preferred when expressing proteolytically labile species. Yokoyama and co-workers [13] recommend the use of BL21 codon-plus RIL (BL21 CP, Stratagene) *E. coli* that contain extra copies of the *argU*, *ileY*, and *leuW* tRNA genes. BL21 CP extracts are enriched in tRNAs that recognize minor codons for arginine (AGA and AGG), isoleucine (AUA), and leucine (CUA). Extracts prepared from BL21 CP can increase expression yields up to 50% on those obtained from BL21 or A19.

Many variations on the protocol for preparing *E. coli* extract have appeared in the literature. However, Kigawa et al. [13] recently published the most comprehensive documented protocol for *E. coli* extract preparation since that described by Pratt [14]. The success of Yokoyama's group in determining the structures of more than 100 proteins prepared by cell-free synthesis in recent years recommends their protocol as the most reliable guide for *E. coli* extract preparation.

The conditions under which *E. coli* are grown and harvested have a marked impact on the activity of the *E. coli* extract obtained. The choice of growth media, temperature, aeration conditions and point-of-harvest must be optimized to suit the choice of cell-line, and to compensate for different laboratory conditions. Once optimized, growth conditions must be strictly adhered to in order to minimize batch-to-batch variation. Kigawa et al. [13] described the growth of BL21 CP strain in 500 mL lots of 2YT medium (Table 15.2), incubated in 2-L baffled flasks at 37 °C with circular shaking at 160 r.p.m., and harvested at $OD_{600} \cong 3$ by centrifugation.

Cells are washed at least three times in S30 buffer (see Table 15.2) containing 0.5 mL L^{-1} 2-mercaptoethanol. Pelleted cells are then frozen in liquid nitrogen and stored at –80 °C for up to 3 days. Thorough washing of the cells is difficult under these conditions, as *E. coli* tend clump together in this buffer; however, a Polytron cell homogenizer (Kinematica AG, PT-MR3100) was used to overcome this problem.

Washed cells are thawed at 4 °C in S30 buffer (50–100 mL) containing 0.5 mL L^{-1} 2-mercaptoethanol. Cells are again pelleted (16000×g, 30 min, 4 °C) and resuspended in S30 buffer. The volume of S30 buffer used to resuspend cells is dependent upon the pellet mass, but should be close to 1.3 mL per gram *E. coli*. Cells are then lysed in this suspension.

Several lysis methods can be used, depending on the facilities available. Sonica-

Table 15.2 Composition of media and buffers for *E. coli* and wheat embryo extract preparations.

Reagent	Concentration
2YT Growth Media	
Tryptone	16 g L^{-1}
Yeast extract	10 g L^{-1}
NaCl	5 g L^{-1}
De-ionized water	Make up to 1 L
5 N NaOH	Adjust pH to 7.0
S30 Buffer	
Magnesium acetate	14 mM
Potassium acetate	60 mM
Tris-acetate	10 mM
DTT	1 mM
***E. coli* Lysate Pre-incubation Buffer**	
Pyruvate kinase	6.7 U mL^{-1}
Tris-acetate (pH 8.2)	293 mM
Magnesium acetate	9.2 mM
ATP	13.2 mM
Phosphoenol pyruvate	84 mM
DTT	4.4 mM
Each amino acid	40 µM
Wheat germ (WG) Buffer	
HEPES, pH 7.6	40 mM
Potassium acetate	100 mM
Magnesium acetate	5 mM
Calcium chloride	2 mM
DTT	4 mM
Each amino acid	300 µM

tion can be used, but should be regarded as a last resort because of poor reproducibility and problems with local sample heating. Multiple passes through a French press or cell crusher can be used to reproducibly prepare highly active *E. coli* extracts, as long as every effort is made to minimize sample heating and eliminate RNases from the apparatus. The method of lysis involves disruption of cells (7 g in 8.9 mL S30 buffer) with 22.7 g of glass beads (B. Braun Melsungen, AG 854-150/

7, ∅ = 0.17–0.18 mm) in a Multi-beads Shocker Type MB301 (Yatsui Kikai, Japan), switched on for three, 30-s periods separated by 30 s at rest. The authors claim that this method of lysis leads to greater reproducibility than the French press method.

The lysate is then centrifuged (30 000×g, 30 min, 4 °C) to remove glass beads and/or cell debris. The supernatant is carefully decanted off and subjected to further centrifugation (30 000×g, 30 min, 4 °C). The supernatant obtained is incubated with 0.3 volumes of pre-incubation buffer (see Table 15.2) at 37 °C for 80 min, and dialyzed four times against 50 volumes of S30 buffer (45 min each at 4 °C). The final *E. coli* extract is the supernatant obtained after centrifugation (4000×g, 10 min, 4 °C) of the dialyzed mixture. This extract is immediately frozen in liquid nitrogen in small aliquots, and stored in liquid nitrogen or at –80 °C until required (it is stable for at least one year).

15.3.2.2 Wheat Embryo Extracts
Selected strains of *E. coli* can be grown under controlled conditions in a highly reproducible manner. Wheat embryo extracts suffer from batch-to-batch variation due to differences in the conditions under which the wheat is grown, harvested, milled, and stored. This source of variation in the activity of wheat embryo extracts can be offset by the simplicity of the preparation protocol. Small amounts of extract can be quickly prepared from embryos obtained from different sources, and the activity of each assessed. Large batches of extract can then be prepared from the best embryo source and stored at –80 °C until required. As with *E. coli* extract, wheat embryo extracts are stable for more than a year if stored in liquid nitrogen, or at –80 °C.

Freshly milled raw wheat germ is readily available from mills or granaries. Intact embryos are separated from damaged embryos and debris by solvent flotation. Carbon tetrachloride or chloroform is mixed with cyclohexane in a 600:240 mL ratio until Schlieren mixing lines are no longer visible. Wheat germ (40 g) is added to the solvent mixture and stirred thoroughly to remove endosperm from embryos. The suspension is allowed to stand until good separation is achieved. Floating embryos are harvested in a Buchner funnel, air-dried, and set aside in a fume hood. The remaining solvent is filtered for re-use, and the process repeated until the desired quantity of embryos is obtained. Embryos are left to stand overnight in a fume hood.

The protocol for extract preparation given here is slightly modified from that described by Endo and colleagues [9]. At this stage, the wheat embryos are covered with endosperm which contains ribosome-inactivating proteins such as RNA N-glycosidase and tritin [9]. Most endosperm is removed by washing the embryos three times in 10 volumes of water. Residual endosperm is removed by sonicating for 3 min in a 0.5% Nonidet P-40 solution and for a further 3 min in RNase-free water. Clean embryos are then ground in liquid nitrogen in a mortar and pestle. A portion (5 g) of the ground powder is suspended in 5 mL ice-cold 2× WG Buffer (see Table 15.2), vortexed, and centrifuged at 30 000×g for at least 30 min at 4 °C. The supernatant is then desalted on a G25 (fine) column and re-centrifuged (30 000×g for 10 min, 4 °C). Endo et al. recommend dilution of the extract with WG Buffer to 200 A260 mL^{-1}, and storage in small aliquots at –80 °C.

15.4
Treatment of Extracts for Synthesis of Disulfide-bonded Proteins

The reducing environment of the *E. coli* cytoplasm prevents the formation of disulfide (S–S) bonds that are often essential to correct folding and functioning of proteins and complexes. As a consequence, S–S-bonded proteins can misfold when expressed in bacteria. Misfolded polypeptide chains are proteolytically labile, susceptible to aggregation, and are often found packed into inclusion bodies.

Recovering misfolded proteins can be time-consuming because of the many parameters that need to be optimized for solubilization and then refolding. Conditions that might need to be screened in a re-fold include temperature, protein concentration, redox potential, pH, ionic strength, and the presence of chaperones, ligands and cofactors. Alternatively, the protein of interest can be expressed conjugated to a signal peptide that will see the protein exported to the *E. coli* periplasm where S–S bonding is supported. However, periplasmic folding typically results in substantially lower protein yield.

Recent attempts have seen modifications made to *E. coli* and wheat embryo cell-free systems for synthesizing proteins in an environment suitable for co-translational formation of disulfides. In general, these studies have involved translation under conditions typically screened in refolding experiments and assessing their effect on total and functional protein yield. The focus has been on batch-mode reaction systems because, as with refolding, optimum conditions for producing correctly folded proteins vary between constructs and batch-mode reactions are compatible with high-throughput condition screening.

Only a limited number of reports describing the synthesis of a few model S–S-bonded proteins have appeared in the literature. However, this reflects the relative novelty of producing significant quantities of S–S-bonded proteins by cell-free methods, rather than the potential of these methods to serve this purpose.

15.4.1
Sulfhydryl Redox Potential

Reducing agents such as dithiothreitol (DTT) are added at the early stages of cell-free lysate preparation to stabilize lysate components during preparation and storage. A strongly reducing environment is usually maintained over the course of a translation reaction. Naturally, this environment inhibits S–S-bond formation. An oxidizing environment is created in the translation reaction by omitting reducing agents from reaction mixtures, removing reducing agents from cell lysates by gel filtration, or by dialysis just prior to their use, and by adjusting the sulfhydryl redox potential with the glutathione redox pair.

In one example [15], genes for the heavy and light chains of a catalytic antibody, 6D9, were simultaneously expressed in a batch E. coli coupled transcription/translation system that was depleted of reducing agents and buffered with oxidized and reduced glutathione (GSSG/GSH). The two proteins formed a single intermolecular disulfide bond, co-translationally, yielding an enzymatically active Fab construct. This treatment seems sufficient for S–S-bonds to form in cases where only one or two disulfides are present, or where a dialysis system is used to buffer the redox potential.

Enzymes such as glutathione reductase and thioredoxin reductase that are responsible for maintaining a reducing environment in the E. coli cytoplasm are active in extracts prepared from standard E. coli strains. This is demonstrated by the fact that GSSG is rapidly reduced when incubated in E. coli lysate. This makes it impossible to maintain a stable redox potential optimal for S–S-bond formation. Further modification of the E. coli extract is necessary for synthesis of multiple S–S-bonded proteins in their functional forms, particularly in batch mode.

A strategy has been devised [16] to inactivate disulfide-reducing enzymes in E. coli lysates by covalently blocking free sulfhydryl groups present. This is done by incubating a standard E. coli extract with 2 mM iodoacetamide (IAM) as alkylating reagent at room temperature for 30 minutes. Residual IAM is then removed by extensive dialysis against S30 buffer, or by gel filtration. The same authors showed that IAM-treated E. coli extract can maintain a 4:1 GSSG:GSH ratio over 18 hours of incubation. Only a small loss (15%) in translation activity of the IAM-treated E. coli extract was observed relative to untreated extract in batch mode. Further studies from the Swartz laboratory indicated that the use of a lower IAM concentration (1 mM) disables reduction pathways sufficient for the purpose of maintaining an oxidizing sulfhydryl redox potential, and obviates the need for extensive dialysis to remove excess IAM [17].

15.4.2
pH

The initial pH of a cell-free reaction should be around 7.6 for maximum translation activity, though this may need to be adjusted with Tris-acetate buffer. However, the initial pH can be varied between 6.5 and 8.5 to improve the activity/solubility of a particular translation product. The formation of disulfide bonds is enhanced at the higher end of this pH range. The serine protease domain of murine urokinase, containing six disulfides, was expressed in an IAM-treated *E. coli* batch reaction [16]. The yield of functional protein was almost doubled by increasing the pH from 7.5 to 8.2, despite a small drop in total protein yield.

15.4.3
Chaperones

Co-expression of chaperones such as GroEL/GroES, DnaK/DnaJ/GrpE can enhance the solubility of proteins expressed by cell-free or *in vivo* methods. Chaperones that enhance solubility and enzymes that promote disulfide bond formation can be added in a controlled way to cell-free reactions. The impact of these agents on the yield of functional protein still needs to be determined empirically. In the case of murine urokinase [16], the addition of DsbC disulfide isomerase from *E. coli* to an IAM-treated *E. coli* reaction increased the rate of functional urokinase formation and doubled the yield of functional protein. When protein disulfide isomerase (PDI) from humans was added to the same expression system, the impact on urokinase activity was negligible. The same group expressed a plasminogen activator (PA) protein containing nine disulfide bonds in a similar reaction system containing DsbC [17]. It was found that addition of the *E. coli* periplasmic chaperone Skp not only increased yield of active protein, but also extended the life of the expression reaction. Similarly, co-expression of the heavy and light chains of the catalytic antibody 6D9 showed that addition of the glutathione redox pair and PDI increased both yield of soluble 6D9 Fab fragment and total expression yield [15].

15.4.4
Wheat Embryo System

Endo and colleagues [18] expressed a single-chain antibody variable fragment (scFv) in a wheat embryo batch-mode reaction with DTT eliminated from the reaction mixture. Synthesized scFv consisting of heavy and light chain variable domains of an antibody connected by a linker was shown to be active, indicating the correct formation of both intradomain disulfide bonds. The addition of PDI had a positive impact on product solubility, showing that wheat embryo cell-free systems can be used to prepare active disulfide-bonded proteins in batch mode. Unfortunately, the productivity of the system fell by almost 50% when DTT (usually at 4 mM) was excluded from the reaction. This is a marked response compared with the 15% drop in activity for the IAM-treated, non-reducing *E. coli* system described elsewhere [16]. When it is considered that wheat embryo systems are much more stable but much less efficient than *E. coli* cell-free systems, this significant fall in activity on removing DTT suggests that scale-up to dialysis-mode protein synthesis in a non-reducing wheat embryo system is likely to be problematic.

15.5
ATP Regeneration Systems

Adenosine triphosphate (ATP) is the primary energy source for cell-free synthesis reactions. Rapid depletion of ATP leads to the cessation of translation in under an hour in batch mode. ATP regeneration systems are used to extend the life of translation reactions by maintaining a stable ATP concentration. ATP is regenerated by the enzyme-catalyzed transfer of high-energy phosphate bonds from a secondary energy store to ADP. The productivity of any cell-free system relies on the efficiency and stability of the ATP regeneration system.

Creatine phosphate (**1**) is most commonly used as the secondary energy store in both *E. coli* and wheat embryo cell-free systems. However, other secondary energy stores (**2–5**) can be used to replace or supplement the creatine phosphate:

1 creatine phosphate + ADP $\xrightarrow{\text{creatine kinase}}$ creatine + ATP

2 phosphoenol pyruvate (PEP) + ADP $\xrightarrow{\text{pyruvate kinase}}$ pyruvate + ATP

3 acetyl phosphate + ADP $\xrightarrow{\text{acetate kinase}}$ acetate + ATP

4 ADP + ADP $\xrightarrow{\text{myokinase}}$ AMP + ATP

5 CTP + ADP $\xrightarrow{\text{CDP kinase}}$ CDP + ATP

Pyruvate dehydrogenase and pyruvate-formate lyase are enzymes that are endogenous to *E. coli* and which process pyruvate into acetyl coenzyme A. Acetyl phosphate is then produced from acetyl coenzyme A by phosphotransacetylase, also endogenous to *E. coli*. In this way, pyruvate can be processed in *E. coli* to yield acetyl phosphate, a phosphate donor (**3**). The activities of these pathways in *E. coli* lysate were tested [7] by supplying a batch-mode reaction with pyruvate in the absence of an alternative secondary energy store.

Protein synthesis was observed in the pyruvate-only system. However, addition of cofactors involved in pyruvate processing (0.33 mM NAD, 0.27 mM coenzyme A) enhanced protein synthesis to 70% of that observed when a PEP/pyruvate kinase (**2**) secondary energy store was used. Furthermore, supplying these cofactors and a PEP synthetase inhibitor (oxalate) to a PEP/pyruvate kinase batch reaction substantially improved ATP regeneration and protein synthesis activity.

The improved PEP/pyruvate system developed by Kim and Swartz [7] yielded 350 µg mL^{-1} of chloramphenicol acetyltransferase (CAT) in the first hour of an *E. coli* batch reaction. The same reaction system yielded 750 µg mL^{-1} CAT in 3 hours when periodically fed with amino acids, PEP, and magnesium acetate [6]. This is remarkable efficiency for a system lacking a dialysis membrane, and is a feature that well suits robotic handling and high-throughput screening (HTS) strategies.

15.6
Reaction Conditions

15.6.1
E. coli Systems

Conditions for performing coupled transcription/translation reactions in an *E. coli* cell-free system are listed in Table 15.3. These conditions have been described by two prominent groups in the development of *E. coli* cell-free systems [13, 16, 17]. Although most of the reaction conditions are similar in the two examples given in Table 15.3, they differ in three important aspects. The conditions reported initially [2] are for standard protein synthesis, whereas those reported by others [17] have been optimized for the production of disulfide-bonded proteins. These systems also differ in their choice of stabilizing reagents (PEG8000 or spermidine/putrescine) and the ATP regeneration systems.

The conditions given in Table 15.3 refer to the cell-free reaction mixture. In dialysis mode, this reaction mixture is dialyzed against a buffer of the same composition but lacking *E. coli* extract, RNA polymerase, creatine kinase/pyruvate kinase, and mRNA/DNA template. If a dialysis membrane with a molecular weight cut-off

Table 15.3 Conditions for coupled transcription/translation systems from *E. coli* under reducing and oxidizing conditions

Reagent	Concentration Reducing [a]	Oxidizing [b]	Units
HEPES/KOH	55	57.2	mM
DTT	1.7		mM
ATP	1.2	1.2	mM
CTP	0.8	0.86	mM
GTP	0.8	0.86	mM
UTP	0.8	0.86	mM
Creatine phosphate	80		mM
Creatine kinase	250		$\mu g\,mL^{-1}$
PEG8000	4		% (w/v)
Spermidine		1.5	mM
Putrescine		1.0	mM
3′,5′-cyclic AMP	0.64	0.64	mM
Folinic acid	68	57	μM
E. coli total tRNA	175	170.6	$\mu g\,mL^{-1}$
Potassium glutamate	210	230	mM
Ammonium acetate	27.5	80	mM
Magnesium acetate	10.7	16	mM
Each amino acid	1	2	mM
T7 RNA polymerase	93	70	$\mu g\,mL^{-1}$
E. coli extract	30	24	% (v/v)
Plasmid DNA	6	6.8	$\mu g\,mL^{-1}$
Sodium azide	0.05		% (w/v)
Phosphoenol pyruvate		30	mM
NAD		0.33	mM
CoA		0.27	mM
Oxalic acid		2.69	mM
Reduced glutathione		1	mM
Oxidized glutathione		4	mM

a) From Ref. [13].
b) From Ref. [17].

(MWCO) of 50 kDa is used, tRNA must be included in the dialysis buffer or the reaction will be depleted of tRNA. Dialysis membranes with MWCOs of 15 or 10 kDa are commonly used, and under those circumstances tRNA can be excluded from the dialysis buffer.

Polyethylene glycol (PEG) is used as a crowding agent, and is thought to stabilize mRNA. Excluding PEG from the reaction mixture can lower protein synthesis yield by around 40% [17]. These authors found that spermidine and putrescine could be used in place of PEG with no loss in synthesis activity, but with some enhancement in the yield of functional disulfide-bonded PA protein (discussed in Section 15.4.3). One benefit of excluding PEG from the reaction mixture is that it interferes with analysis by sodium dodecyl sulfate-polyacrylamide gel electrophoresis (SDS-PAGE), causing streaking and background staining. If included in the reaction mixture, PEG can be removed from samples for SDS-PAGE analysis by either trichloroacetic acid (TCA) or acetone precipitation.

In general, most of the components of the cell-free system do not need to be varied during expression optimization trials. It is essential, however, to establish optimum concentrations of RNA polymerase and mRNA/DNA template for each expression construct. An optimum magnesium concentration should only need to be established once for each batch of extract.

Antibody detection methods can be used to test expression conditions in batch/dialysis mode, and Coomassie blue-stained SDS-PAGE is often sufficient for an analysis of dialysis mode reactions. However, the simplest method for the rapid screening of expression conditions is to measure radiolabeled ^{35}S-met/^{35}S-cys incorporation into the expressed protein. Using this approach, unpurified cell-free reaction mixtures are separated by SDS-PAGE, after which the polyacrylamide gel is exposed to X-ray film or a phosphor imaging screen.

Because cell-free methods are highly efficient in their incorporation of amino acids, only minute quantities of radiolabeled amino acids are required for analysis by autoradiography. Background translation of endogenous mRNA is also extremely low. This is advantageous because expression conditions can be rapidly screened for their effect on aggregation, solubility, and proteolytic susceptibility of the expressed construct. Low background translation also means that total and soluble protein yields can be analyzed by auto-

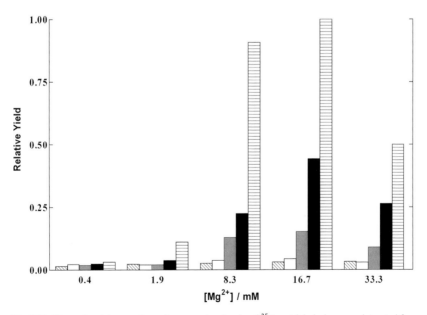

Fig. 15.1 Normalized (to maximum) expression levels of ^{35}S-met-labeled α-synuclein at different Mg^{2+} and DNA template concentrations as determined by autoradiography. DNA concentrations (μg mL^{-1}) studied were 0.6 (diagonal lines), 2.7 (white solid), 10.6 (checked), 26.5 (black solid), and 53.0 (horizontal lines).

radiography of dot-blots, or by scintillation counting of TCA-precipitable material.

The results of expression trials for ^{35}S-met labeled *a*-synuclein at different DNA template and Mg^{2+} concentrations are shown in Fig. 15.1. The height of each bar in the graph represents the relative amount of *a*-synuclein synthesized in 1 hour, normalized to the highest-yielding reaction, as measured by autoradiography. The figure demonstrates the sensitivity of expression yields on both DNA and Mg^{2+} concentrations. Similar DNA concentration dependence at each magnesium concentration suggests that these parameters are mutually exclusive.

15.6.2
Wheat Embryo System

The conditions for performing a wheat embryo cell-free translation reaction from an mRNA template are listed in Table 15.4. Points discussed for *E. coli* optimization in Section 15.6.1 are relevant to expression in wheat embryo extracts. However, it is worth noting that the wheat embryo system is better suited to the translation of added mRNA template, whereas coupled transcription/translation is better in the *E. coli* system. The principal reason for this difference is that transcription with bacteriophage RNA polymerases requires a relatively high Mg^{2+} concentration (ca. 16 mM); *E. coli* translation-only reac-

Table 15.4 Conditions for wheat embryo translation systems under reducing and oxidizing conditions

Reagent	Concentration		Units
	Reducing [a]	Oxidizing [b]	
HEPES/KOH	24 (pH 7.8)	28 (pH 7.8)	mM
DTT	2		mM
ATP	1.2	1.2	mM
GTP	0.25	0.25	mM
Creatine phosphate	16	15	mM
Creatine kinase	0.45	1.8	$\mu g\ mL^{-1}$
Spermidine	0.4	0.4	mM
Wheat embryo deacylated tRNA	50	50	$\mu g\ mL^{-1}$
Potassium acetate	100	53	mM
Magnesium acetate	2.5	1.6	mM
Calcium chloride		0.6	mM
Each amino acid	0.3	0.23	mM
Embryo lysate	24	24	% (v/v)
mRNA	144	200	$\mu g\ mL^{-1}$
Sodium azide	0.005		% (w/v)
Nonidet P-40	0.05		% (v/v)
E-64 proteinase inhibitor	1		μM
RNasin (ribonuclease inhibitor)		0.4	$U\ \mu L^{-1}$
Biotin (for biotinylation only)		19.5	μM
Biotin ligase (for biotinylation only)		19.5	$\mu g\ mL^{-1}$

a) From Ref. [9].
b) From Ref. [18].

tions occur optimally at a similar Mg^{2+} concentration. In contrast, wheat embryo translation-only reactions perform best when the Mg^{2+} concentration is an order of magnitude lower (ca. 1.6 mM). For this reason, coupled transcription/translation best suites analytical-scale expression, whereas uncoupled reactions are best used for preparative-scale synthesis.

15.6.3
Temperature

Once optimal conditions have been achieved for total expression, yield factors that influence folding and solubility must be examined; these include the addition of cofactors, ligands, and chaperones. The reaction temperature may also have a dramatic effect on protein solubility; lowering it leads to lower rates of protein synthesis, but this can be compensated for by extending the reaction time. For example, expression of dihydrofolate reductase (DHFR) at 37 °C in a dialysis-mode *E. coli* reaction generally yields 2–3 mg of insoluble DHFR in 8 hours, but the same yield of mostly functional protein is obtained after 24 hours at 30 °C.

15.6.4
Protease Inhibitors

Inhibition of proteolysis is best achieved by promoting proper folding. As discussed above, this is achieved by manipulating expression conditions such as pH, temperature, and the addition of ligands and chaperones. The cell-free systems discussed here are fairly tolerant to the addition of protease inhibitors commonly used in the purification of proteolytically labile species. Protease inhibitor cocktails can be used as long as they are free of EDTA, which may chelate magnesium ions and inhibit

mRNA/protein synthesis. EGTA can be added to approximately 10 mM in order to inactivate Ca^{2+}-dependent enzymes, though the Mg^{2+} concentration must be optimized under these conditions.

15.6.5
RNase Inhibitors

Every effort must be made to minimize RNase contamination during cell extract preparation, and in the preparation of reaction mixtures. These efforts include chloroform washing and subsequent baking of glassware prior to use, and the use of diethyl pyrocarbonate (0.1%)-treated deionized water in the preparation of extracts and reaction mixtures. Gloves should be worn at all times and, ideally, RNase-free work areas should be dedicated to cell-free expression. RNase inhibitors such as human placental RNase inhibitors can be included in the reaction mixture to minimize the degradation of mRNA and ribosomes. However, the use of these inhibitors adds substantially to the cost of a reaction, and they should be used sparingly. The concentration of RNase inhibitor used should be optimized in dialysis-mode to identify the lowest concentration required to protect a given mRNA template.

15.7
Conclusion

An SDS-PAGE analysis of optimized dialysis-mode expression (10 vols dialysis solution) in an *E. coli* system is shown in Fig. 15.2 for:
- Peptide chain elongation factor Ts (EF-Ts) from *E. coli* at 37 °C after 1, 3, 6, and 12 hours in lanes 1, 2, 3, and 4, respectively; and

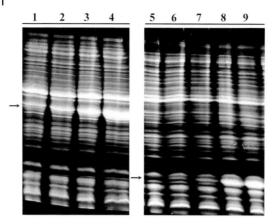

Fig. 15.2 SDS-PAGE analysis of optimized dialysis-mode expression in an *E. coli* cell-free system for: (A) peptide chain elongation factor Ts (EF-Ts) from *E. coli* at 37 °C after 1, 3, 6, and 12 hours in lanes 1, 2, 3, and 4, respectively; and (B) α-synu-clein at 30 °C after 0, 1, 3, 12, and 24 hours in lanes 5, 6, 7, 8, and 9, respectively. Expression bands are indicated by arrows to the left of each gel.

- α-synuclein at 30 °C after 0, 1, 3, 12, and 24 hours in lanes 5, 6, 7, 8, and 9, respectively. These expression profiles correlate with yields of several milligrams of protein per mL reaction mixture.

Such excellent outcomes are a result of careful and systematic optimization of DNA constructs and expression conditions. There are occasions, however, where proteins fail to express, or are expressed in an inactive form. Other proteins express well in one expression system, but not in another, and this is of course also true of all *in vivo* expression systems. A true benefit of cell-free synthesis over *in vivo* methods is that the "open" nature of the systems presents the researcher with unparalleled opportunities to affect the outcome of an expression experiment. As such, cell-free synthesis methods represent an invaluable addition to any laboratory investigating the expression of biopharmaceuticals.

References

1 Zubay, G. (1973) *Annu. Rev. Genet.* 7: 267–287.
2 Yokoyama, S., Matsuo, Y., Hirota, H, Kigawa, T., Shirouzu, M., Kuroda, Y., Kurumizaka, H., Kawaguchi, S., Ito, Y., Shibata, T., Kainosho, M., Nishimura, Y., Inou, Y., and Kuramitsu, S. (2000) *Prog. Biophys. Mol. Biol.* 73: 363–376.
3 Kigawa, T., Yabuki, T., Yoshida, Y., Tsutsui, M., Ito, Y., Shibata, T., and Yokoyama, S. (1999) *FEBS Lett.* 442: 15–19.
4 Spirin, A. S., Baranov, V. I., Ryabova, L. A., Ovodov, S. Y., and Alakhov, Y. B. (1988) *Science* 242: 1162–1164.
5 Kim, D. M. and Choi, C. Y. (1996) *Biotechnol. Prog.* 12 (5): 645–649.
6 Kim, D.-M. and Swartz, J. R. (2000) *Biotechnol. Prog.* 16: 385–390.
7 Kim, D.-M. and Swartz, J. R. (2001) *Biotechnol. Bioeng.* 74 (4): 309–316.
8 Endo, Y. and Sawasaki, T. (2004) *J. Struct. Funct. Genomics* 5: 45–47.
9 Madin, K., Sawasaki, T., Ogasawara, T., and Endo, Y. (2000) *Proc. Natl. Acad. Sci. USA* 97 (2): 559–564.
10 Sawasaki, T., Hasegawa, Y., Tsuchimochi, M., Kasahara, Y., and Endo Y. (2000) *Nucleic Acids Symp. Ser.* 44: 9–10.
11 Hendrickson, W. A., Horton, J. R., and LeMaster, D. M. (1990) *EMBO J.* 9: 1665–1672.

12 Shimizu, Y., Inoue, A., Tomari, Y., Suzuki, T., Yokogawa, T., Nishikawa, K., and Ueda, T. (2001) *Nature Biotechnol.* 19 (8): 751–755.

13 Kigawa, T., Yabuki, T., Matsuda, N., Matsuda, T., Nakajima, R., Tanaka, A., and Yokoyama, S. (2004) *J. Struct. Funct. Genomics* 5: 63–68.

14 Pratt, J. M. (1984) In: Hames, B. D. and Higgins, S. J. (Eds), *Transcription and Translation*. IRL Press, Oxford, Washington, pp. 179–209.

15 Jiang, X., Ookubo, Y., Fujii, I., Nakano, H., and Yamane, T. (2002) *FEBS Lett.* 514: 290–294.

16 Kim, D.-M. and Swartz, J. R. (2004) *Biotechnol. Bioeng.* 85 (2): 122–129.

17 Yin, G. and Swartz, J. R. (2004) *Biotechnol. Bioeng.* 86 (2): 188–195.

18 Kawasaki, T., Guoda, M. D., Sawasaki T., Takai, K., and Endo, Y. (2003) *Eur. J. Biochem.* 270 (23): 4780–4786.

When Success Raises its Ugly Head – Outsourcing to Uncork the Capacity Bottleneck

16

Contract Manufacturing of Biopharmaceuticals Including Antibodies or Antibody Fragments

J. Carsten Hempel and Philipp N. Hess

Abstract

The manufacture of monoclonal antibodies or antibody fragments requires suitable expression systems. Whilst whole antibodies are currently produced in expensive mammalian cell culture systems, for some applications the expression of antibody fragments in prokaryotes, yeast, or filamentous fungi is an alternative. Currently, a large portion of the worldwide fermentation capacity for the production of monoclonal antibodies or antibody fragments is provided by contract manufacturing organizations (CMOs). Among the different methods, continuous perfusion cell culture (in suspension) is a proven method to manufacture monoclonal antibodies and recombinant proteins in quantities of up to several hundred kilograms. However, that method appears to be less common in the industry compared to fed-batch cell culture. Based on a large number of industry interviews, this chapter highlights the key factors of outsourcing projects. Outsourcing reflects a wide range of benefits such as time-to-market, avoidance of capital expenditure, cost containment and flexibility. Key issues of CMO/client contracting are discussed. This chapter also provides insights into successful strategies for the management of technology transfer processes. The management of technology transfer requires careful preparation and stringent definition of all project steps and scopes covering the process, the analytical procedures, the technical equipment, and the qualification and validation protocols. The ability to meet demands for biopharmaceutical production capacity – whether through in-house manufacturing or outsourced contract manufacturing – carries strategic and financial implications for biopharmaceutical companies. FDA approval policy will have a major impact on future capacity requirements for the production of biopharmaceuticals.

Abbreviations

ATF	alternating tangential filtration
CHO	Chinese hamster ovary
CMC	chemistry manufacturing and control
CMO	contract manufacturing organization
CRO	contract research organisation
DSP	downstream processing
FDA	Food and Drug Administration
GMP	good manufacturing practice
IPC	in-process-control
MAB	monclonal antibody
P	packaging
QA/QC	quality assurance/quality control support
RTP	Research Triangle Park
USP	upstream processing
WCB	working cell bank

16.1
Introduction

Although this chapter will focus on antibodies and fragments, the content clearly also applies to all biopharmaceuticals in the same way. Since the famous Köhler and Milstein [1] experiment in 1975, the success story of monoclonal antibodies was always linked to the development of appropriate expression systems and production technologies.

Immortalized mouse cell lines secreting one single type of antibody with a unique binding affinity against virtually any type of antigen (proteins, carbohydrates, or nucleic acids) initiated heavy competition in the development of faster, cheaper, and safer procedures. In order to guarantee the safety and efficacy of these pharmaceutically active products, all regulatory issues need to be addressed and fulfilled.

Whilst expression rates in the 1980s achieved product concentrations of only about 100 mg L^{-1}, some ten years later several immunoglobulins were being produced at concentrations which were between two- to seven-fold higher [2]. The reasons for this ongoing improvement are linked to more advanced production procedures and better expression systems. Indeed, in 2005 the hurdle of 5 g L^{-1} heterologous protein expressed in a high-yielding CHO cell culture process is likely to be exceeded [3] (see also Part IV, Chapters 1 and 4).

The growing market demand for all kinds of monoclonal antibodies or antibody fragments is a key factor for the pharmaceutical industry. Antibody-based therapeutics cover a wide range of indications such as inflammation and oncology, in addition to different types of infectious diseases. High dosage requirements for the treatment of chronic diseases in large markets necessitate the development of high-yielding, low-cost and large-scale manufacturing processes that consistently deliver high-quality product.

At present, there are more than 370 biotech therapeutics undergoing clinical trials, and as many as 125 new drugs may reach the market during the next five to seven years. In this respect, efficient expression systems and biopharmaceutical manufacturing capacities will have to be developed.

In this chapter, novel aspects of the manufacturing procedures of monoclonal antibodies from a process engineering point of view will be highlighted. The manufacture of monoclonal antibodies in mammalian cell culture requires certain technologies and capacities, which will be described within the first section of the chapter. Technical hurdles linked to these still-expensive procedures and aimed towards the development for more economical eukaryotic or even prokaryotic manufacturing systems will be highlighted in the second section.

Since the availability of manufacturing/fermentation capacity at present remains a major challenge for the pharmaceutical industry, the third section of the chapter relates to the management of outsourcing processes.

16.2
Expression Systems and Manufacturing Procedures

The manufacture of monoclonal antibodies or antibody fragments requires suitable expression systems. Advances in molecular biology of higher eukaryotes, together with recent developments in the field of bioengineering, have made the choice of an appropriate expression system more complex. Some 20 years ago there were only a few different options, and systems incorporating recombinant *Escherichia coli*, *Saccharomyces cerevisiae* or Chinese hamster ovary (CHO) cells have each presented their own challenge, having been the pioneers of recombinant protein expression (see also Part IV, Chapter 12).

The situation has changed during the past few years, however, and a variety of products that are either undergoing clinical trials or are already on the market have been produced by alternative systems. Transgenic animals and plants will complement existing cell culture-based expression, as described in Part IV, Chapters 5–11 (for a recent review, see Knäblein J, Biopharmaceuticals expressed in plants – a new era in the new Millennium. In: R. Müller and O. Kayser (eds), *Applications in Pharmaceutical Biotechnology*. Wiley-VCH. ISBN 3-527-30554-8). Manufacturing processes using insect cell lines (see Part IV, Chapter 14) have been developed to commercial scale. Within the literature, optimistic projections can also be found for transgenic animals

(see Part IV, Chapter 11), yeasts (see Part IV, Chapter 13) and fungi.

In process development, the selection of an appropriate expression system affects important factors such as:

- product characteristics
- regulatory hurdles
- cost of goods
- intellectual property

Time to market – the most important issue when focusing on commercial success – is a direct function of the above-mentioned factors.

Nowadays, the systems of choice for the production of monoclonal antibodies or antibody fragments are either eukaryotic cell lines or, in some cases, prokaryotes. The following sections will highlight certain manufacturing aspects which are linked to these expression systems.

16.2.1
Manufacture of Monoclonal Antibodies

All monoclonal antibodies currently approved by the US regulatory authority (Food and Drug Administration, FDA) as drug substances are prepared in cell culture (see Table 16.1). Among the biopharmaceutical products made in cell culture, monoclonal antibodies (MAbs) represent the single most production capacity demanding group. Currently, a significant portion of the worldwide cell-culture capacity is provided by contract manufacturing organizations (CMOs).

Compared to microbial cultures, cell cultures still operate at low cell densities. While medium development and feed strategies during fed-batch cell culture increases cell densities and productivities, the fact that unused media components and products accumulate in the reactor represents a natural limitation.

Table 16.1 Therapeutic monoclonal antibodies approved by the
US FDA. (Products with an annual turnover of more than US$
300 million are in italics)

Trade-name (generic name)	Sponsor company	Approval date
Orthoclone OKT3® (muromomab-CD3)	Ortho Biotech, Inc. (subsidiary of Johnson & Johnson)	June 1986
ReoPro™ (abciximab) – sales >300 Mill$/year	*Centocor, Inc. (subsidiary of Johnson & Johnson) and Eli Lilly and Company*	*December 1994*
Rituxan™ (rituximab) – sales >100 Mill$/year	*IDEC Pharmaceuticals Corp. and Genentech, Inc.*	*November 1997*
Zenapax® (daclizumab)	Hoffmann-La Roche, Inc., and Protein Design Labs	December 1997
Simulect® (basiliximab)	Novartis Pharmaceutical Corp.	May 1998
SYNAGIS™ (palivizumab) – sales >100 Mill$/year	*MedImmune, Inc.*	*June 1998*
Remicade® (infliximab) – sales >100 Mill$/year	*Centocor, Inc. (subsidiary of Johnson & Johnson)*	*August 1998*
Herceptin® (trastuzumab) – sales >100 Mill$/year	*Genentech, Inc./Roche*	*September 1998*
Mylotarg™ (gemtuzumab ozogamicin)	Celltech Pharmaceuticals and Wyeth	May 2000
Campath® (alemtuzumab)	Ilex Oncology, Inc., Millennium Pharmaceuticals, Inc., and Berlex Laboratories, Inc.	May 2001
Zevalin™ (ibritumomab tiuxetan) – sales >100 Mill$/year	IDEC Pharmaceuticals Corp./Schering AG	February 2002
HUMIRA™ (adalimumab) – sales >100 Mill$/year	Cambridge Antibody Technologies and Abbott Laboratories	December 2002
Bexxar® (Tositumomab and I-131 tositumomab)	Corixa Corp. and GlaxoSmithKline	June 2003
Xolair® (Omalizumab; rec. Immunoglobin-E) – sales >100 Mill$/year	Genentech, Tanox, Inc. and Novartis Pharmaceuticals	June 2003
Raptiva™ (efalizumab; selective, reversible T-cell blocker [subcutaneous injection; self-administered])	Xoma, Ltd. and Genentech, Inc.	October 2003

Perfusion cell culture overcomes that limitation and is, therefore, particularly attractive for inhibiting or unstable products (which is typically not the case for MAbs). Furthermore, perfusion cell culture achieves higher cell densities and volumetric productivities, even when effects of the media components are not fully understood. However, that comes at a cost of a more complex bioreactor system (Figs. 16.1 and 16.2).

Perfusion cell culture (in suspension) is a proven method for the manufacture of MAbs and recombinant proteins in quantities of up to several hundred kilograms (e.g., ReoPro, Remicade). However, this method appears to be less common in the industry compared to fed-batch cell culture. Very few companies manufacture their clinical and approved MAb products by using perfusion cell-culture techniques.

Fig. 16.1 General schematic for fed-batch and perfusion cultivation. X = biomass; S = substrate; P = product; F = flow.

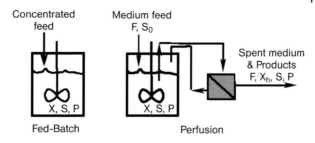

For new processes, even these companies aim to develop efficient fed-batch processes, and if MAb concentrations of $>0.7 \text{ g L}^{-1}$ could be achieved in fed-batch systems, they would no longer use the perfusion cell-culture approach.

The arguments to move to fed-batch, and to abandon perfusion cultivation – as given by various companies (e.g., Medarex and Serono) – included:

• better availability of fed-batch manufacturing capacity at contract organizations, and
• already high concentrations (e.g., 4.2 g L^{-1}, as reported by Lonza Biologics; see Part IV, Chapter 4); and
• shorter processes and therefore shorter validation cycles.

However, there are arguments not to abandon perfusion cultivation:

• Recently developed cell retention devices such as alternating tangential filtration (ATF) or an acoustic device (Biosep) can be fitted to existing bioreactors at contract organizations.
• High productivities – Crucell/DSM Biologics reported 0.9 g L^{-1} per day at 120–140×10^6 cells mL^{-1} (see also Part IV, Chapter 3).
• A proven strategy to achieve shorter validation cycles during perfusion cell culture processes is to include a short perfusion time (e.g., 15 days) in the regulatory filing, and to increase the perfusion time to 50 or 60 days as processes

Fed-batch culture

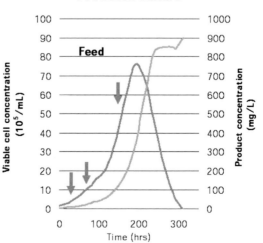

Fig. 16.2 Typical fed-batch culture, recently reported up to 4.2 g L^{-1}. (Source: Lonza Biologics; public domain)

change (that the long-term perfusion clearly should be investigated at small scale before entering that strategy).

16.2.2
Manufacture of Antibody Fragments in Microbial Systems

While whole antibodies are currently produced in expensive mammalian cell culture systems, for some applications the expression of antibody fragments in prokaryotes, yeast or filamentous fungi is an alternative. Tissue penetration is facilitated, and clearance from the blood or kidney is faster using antibody fragments. In addition, PEGylation allows the modification of the half-life, which is another advantage of antibody fragments (see also Part VI, Chapter 2).

The variety of mechanisms mediating the therapeutic effect do not always require the complete antibody glycoprotein. For some applications the optimal structure is not necessarily the whole immunoglobulin, but is rather a distinctive fragment containing the specific antigen-binding domain. Recently, it has been shown for example that their small size permits them to penetrate tissues or solid tumors more rapidly than whole antibodies [4].

However, the advantages that these fragment antigen binding (Fabs) provide have not only been limited to therapy, as reducing the requirements from the complex immunoglobulin expression towards undemanding expression of antibody fragments permits the application of simple expression systems.

For the large-scale production of antibody fragments, the expression system must be:
- economical,
- accessible for genetic modifications,
- easily scalable,
- safe.

Therefore, *E. coli* is one of the preferred hosts for the expression of antibody fragments [5]. Genetic modification of bacterial cell lines – especially *E. coli* – is rapid and trouble-free, and the media and nutrient supply is simple. The metabolic potential of microbial cell lines allows the use of serum-free, chemically defined and cheap media. Expensive viral removal steps can be avoided, whilst the removal of pyrogens does not usually cause any problems.

The volumetric productivity achievable in simple bioreactor systems when using bacterial expression systems is superior to that of mammalian expression systems. Growth rates are higher, and the ease of scale-up of the fermentation process enables manufacturing at scales up to five-fold larger compared to mammalian systems.

In 2005, the worldwide manufacturing capacity for microbial fermentation is expected to exceed $250\,000\ \mathrm{m}^3$, thus providing a potential solution to the problem of limited capacity available on the market for mammalian cultivation systems [6].

The secretion of soluble antibody fragments in the periplasm allows the use of simple down-stream technologies. Simple, but efficient, purification steps reduce the technical hurdles for large-scale production.

The use of *E. coli* as an expression system is limited to those proteins where post-translational modification such as the glycosylation or galactosylation of antibody fragments is not required. Inclusion body formation can be another disadvantage that occurs with this prokaryotic model, as these insoluble protein aggregates demand laborious and cost-intensive *in vitro* refolding (denaturation and renaturation) and purification steps.

Antibody fragments and fusion proteins can also be produced cost-effectively in large-scale production using yeasts or fungal fermentations. An extensive knowledge covering the upstream and downstream processing used for the bulk production of several other recombinant proteins is available. Filamentous fungi and also yeasts are accessible for genetic modification, and the protein of interest may be secreted into the culture broth. *Saccharomyces cerevisiae*, the genome of which was fully sequenced in 1996, has been genetically modified to produce a wide range of proteins and antibody fragments. The absence of pyrogenic compounds, toxins or viral contaminations offers an additional advantage of these expression systems.

16.3
Outsourcing and Contract Manufacturing

Expertise and knowledge along the whole value chain, from high-throughput screening, lead identification and process development to clinical development, manufacturing and marketing is available only in large companies.

One potential alternative to in-house expertise are partnerships – perhaps between smaller companies specialized in certain fields such as drug development, manufac-

turing companies, or contract research organizations. This trend is usually referred to simply as *outsourcing*.

Outsourcing reflects a wide range of benefits such as time-to-market, avoidance of capital expenditure, cost containment and flexibility. Indeed, focusing on the outsourcing of manufacturing processes involved in the need for production is the major driver for such a decision. Some aspects of outsourcing strategy, together with the management of these processes, will be discussed in the following sections.

16.3.1
Integrating Contract Manufacturing into the Manufacturing Strategy

An increasing number of companies include contract manufacturing into their plans for developing a bio-pharmaceutical drug (Table 16.2). Although, initially an appropriate quantity of material is required for toxicology studies (the pre-clinical phase), much larger quantities will be required further down the development chain as the clinical phase approaches (Fig. 16.3).

Table 16.2 Manufacturing development plan for a particular project in a medium-size biotech company required to implement its own manufacturing capabilities. As a first step the company needs enough material to characterize the protein and to further develop the process. Thereafter, the company needs material manufactured under GMP conditions for safety studies, clinical studies and, within 2–3 years, a commercial scale manufacturing facility to fulfill marketing requirements.

Function	Phase			
	Preclinical	I/II	III Launch	Commercial
Fermentation	In-house	Contract	Contract	Mixed
Purification	In-house	Contract	Contract	Mixed
Filling/Finishing	Contract	Contract	Contract	Contract
Testing	Mixed	Mixed	Mixed	Inhouse

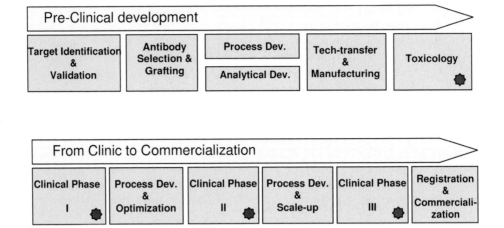

● Step requires supply of appropriate material (Quantity, Quality)

Fig. 16.3 The development of a biopharmaceutical drug.

Key reasons, to include contract manufacturing into the development strategy, are to increase cash flow/minimize fixed assets by avoiding investments into manufacturing facilities, and to accelerate the drug development by creating access to additional expertise.

During the past few years there has been a capacity shortage in particular for large-volume commercial manufacturing. Since then, all major CMOs have expanded their capacities, and it is therefore now easier to identify the correct CMO. Nevertheless, outsourcing of the majority of the manufacturing development should be well prepared in advance.

16.3.2
Typical Capacity Requirements

The availability of bioreactors of a certain size is an important factor in choosing the correct CMO, as downstream process equipment does not vary very much from product to product. The capacity require-

ments are dependent on typical dose ranges, the size of the clinical studies, and market projections. These requirements – and the productivity of the cell culture – determine the scale of operation. For example, a successful therapeutic monoclonal antibody product would require about 200–300 g for clinical Phases I/II, 500 g per indication (2 × indications studied) for clinical Phase III, and for the first year supply more than 5 kg. Depending on the dosage – which typically is milligrams per kilogram body weight – and repeated treatment, the commercial annual supply could easily reach 80–200 kg. If the antibody is targeted at a rare disease which receives an orphan drug status, the requirements might be much lower; for example, for clinical Phases I/II 100–150 g, for clinical Phase III about 250 g, and for the first year supply about 2.5 kg. Depending on dosage, the commercial annual supply might reach only 5 kg.

These capacity requirements, together with assumptions for the productivity of

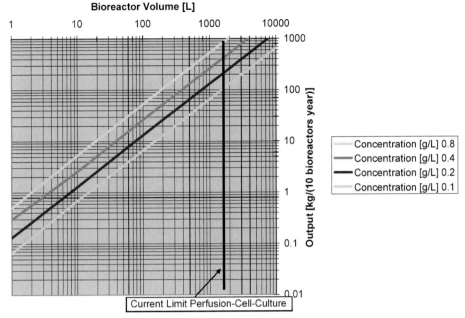

Fig. 16.4 Annual output using 10 × perfusion bioreactors (x-axis liquid volume of each single bioreactor) with the following assumptions: 50 day's perfusion rate at 0.5 L per day; overall yield 46.6%; inoculation and capture in decoupled work-centers; 90% occupation.

the cell culture, determine the bioreactor sizes (Figs. 16.4 and 16.5).

16.3.3
Selecting a CMO

Once the decision to outsource the manufacturing of a biopharmaceutical has been taken, a suitable partner must be identified, and for this numerous aspects must be taken into account. Some typical considerations in finding a CMO include:

- What are the typical capacity requirements to manufacture the products? On the basis of market development and future expectations, a first guess should be made. Knowing the expression level of an appropriate system will then allow quantification of the manufacturing capacity needed.

- Which of the manufacturing technologies is unique and should be developed in-house? In some cases, cost-effective manufacturing depends on highly sophisticated manufacturing know-how and expertise. The assessment of the technology position (weak, strong, dominant) will have an impact on the outsourcing strategy.

- For which phases shall a CMO be used? The manufacturing capacity which is needed to support clinical trials differs from the commercial-scale production. During the clinical development phases smaller amounts of the product will be needed.

- Who are the main players on the contract manufacturing market for cell culture and microbial fermentation?

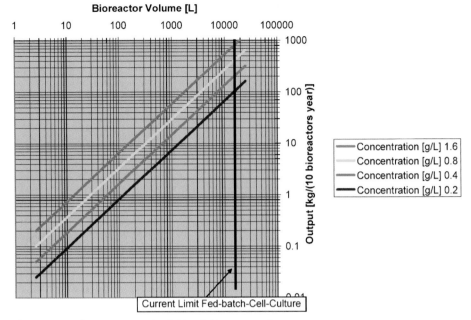

Fig. 16.5 Annual output using 10 × fed-batch bioreactor trains (x-axis liquid volume of largest single bioreactor of each train) with the following assumptions: growth rate 1 per day; overall yield 46.6%; capture in decoupled work-centers; 90% occupation.

- What technologies do they offer?
- What are their strengths and weaknesses? Management skills, availability of trained and experienced technical and production staff should be analyzed. An excellent technical facility may not necessarily implicate experienced scientific or technical staff.
- What are the prevailing price levels?

At the start of such a search a detailed market analysis is required. The assessment of the technology position is needed, and the company's own strategy must be defined. To help in this decision making it is useful to collect the knowledge of the whole organization into a decision tree on which decisions are clearly framed and the criteria for making any particular decision are made clear [11]. The next section pro-

vides an overview of the major players in the CMO market.

16.3.4
Major Players

A list of Biologic's CMOs (developmental organizations included) can easily reach about a hundred entries (see Table 16.3) Among the selection presented in the table, the *underlined* companies can be considered major players and are fully dedicated to contract manufacturing. The list is constantly changing: a number of CMOs were recently acquired by other companies (e.g., Covance RTP by Diosynth, Collaborative Bioalliances by Dow, Alpha Bioverfahrenstechnik by Siegfried, Lek by Sandoz) or have even stopped business (e.g., Accentus).

Table 16.3 Contract manufacturing organizations (CMOs) utilized by Biologic

Abio, Singapore
Abbott, USA
Siegfried Biologics (Alpha Bioverfahrenstechnik), Germany
Archport, UK
Avecia, UK
Avid Bioservices, USA
BioInvent, Sweden
BioReliance (Invitrogen), USA, Germany, United Kingdom
Boehringer Ingelheim, Germany, USA
BTC, Singapore
Cambrex, USA
Chiron, USA
Cobra Therapeutics, UK
CMC, Denmark
Diosynth (now again part of Organon), The Netherlands, USA
DOW, USA
DSM Biologics, The Netherlands, Canada, USA
Eurogentec, Belgium
Excell Biotech (Q-biogene), UK
Girindus, Germany, USA
Goodwin Biotech, USA
Henogen, Belgium
ICOS, USA
Innogenetics, Belgium
Lonza Biologics, Switzerland, United Kingdom and USA
Novartis Hunigue, France
Rentschler Biotechnologie, Germany
Sandoz Kundl, Austria and Lek, Slovenia
SynCo-Biopartners, The Netherlands

Companies in italics are major players and fully dedicated to contract manufacture.

All major CMOs offer the following benefits to their clients:
- Experience in technology transfer.
- Pilot-scale cGMP manufacturing availability to supply product for clinical Phases I and II.
- Large-scale cGMP licensed manufacturing availability to supply product for clinical Phase III and early marketing requirements.
- Compliance to worldwide cGMPs.

Some CMOs offer additional benefits such as:
- Process development capabilities.
- Cell banking capabilities.
- Proprietary expression systems.
- Formulation development capabilities.
- QC development capabilities.
- Protein analytical chemistry for product characterization.
- Clinical development services (CRO).
- Large-scale cGMP licensed commercial manufacturing capacity.
- Complete manufacturing process chain including fill/finishing.
- Assistance in worldwide registration.

As mentioned above, *all* major CMOs have recently expanded their capacity.

16.3.5
Expression Systems for Manufacturing

Among the drivers of manufacturing costs, productivity of the cell culture is the major factor, while purification, yield, media, disposables and virus testing costs represent minor factors.

The productivity of the cell culture part is a function of the expression level (specific productivity) and the cell density in the bioreactor system. Typical specific productivities in cell culture are in the range of 10 pg per cell·day for recombinant protein (e.g., MAb), whilst high-yield expression systems (high-producing cell lines generated with a range of productivity-enhancing selection methods) achieve 15–40 pg per cell·day. Typical cell densities are 5 to 20×10^6 cells mL^{-1} in fed-batch culture, and 20 to 140×10^6 cells mL^{-1} in perfusion cell culture. Thus, a key decision before selecting a CMO is to determine whether the project requires an improved expression system.

Proprietary expression systems provided by CMOs already possess the experience of a "standard" selection and manufacturing process which may shorten the project time and enhance process performance significantly. On the other hand, typical licensing terms for those expression systems include license fees and royalties, therefore making it more difficult for small biotech companies to "sell" the project at an appropriate and competitive price. The licensing fees increase as the project progresses from preclinical research through the clinical phases. The total licensing fees and milestone payments for a typical seven-year development project up to approval could easily exceed € 10^6. Negotiations related to royalties typically start in the range of 0.5–3% of net sales; thus, license and royalty-free options should be investigated thoroughly.

Examples of mammalian protein expression systems/cell lines currently available at CMOs include:

- BIHEX/CHO – *Boehringer Ingelheim High Expression System/Chinese Hamster Ovary Cell*, Boehringer Ingelheim, Germany.
- GS/CHO (or GS/NS0) – *Glutamine Synthetase System/Chinese Hamster Ovary Cell*, Lonza Biologics, Switzerland.
- Several vectors, e.g., pcDNA3002Neo/PER.C6–PER.C6® human cell line technology by Crucell and manufacturing by DSM Biologics, The Netherlands.
- Several vectors/ MARtech/CHO – *Matrix Attachment Region Technology* by Selexis and manufacturing by Diosynth, The Netherlands and USA.
- CHEF1/CHO by ICOS, USA.

16.3.6
Typical Scope

Once a general framework for the project is determined, the to-do list will be:

- Contact short-listed companies in order to determine generic capacity load and capabilities.
- Provide general conditions of project (time, quantity, etc.) to CMO and let company express their interest.
- Develop a scope document: a detailed manufacturing project plan with all the requirements (mark requirements which can be handled internally or outsourced independent of the main CMO as "OPTION").
- Define and rank selection criteria.
- Decide on final shortlist of CMOs.
- Exchange confidential information disclosure agreement.
- Visit companies for pre-qualification audits and discussions of a pre-qualification questionnaire.
- Issue scope document and request for quotation to selected CMO(s).
- Receive and evaluate quote.
- Evaluate CMO according to predefined selection criteria.
- Enter contract negotiations.
- Reevaluate CMO.

A typical scope document for a CMO might include the following items:
- Technology transfer.
- Cell line development.
- Purification development.
- Raw materials sourcing and testing.
- Manufacturing technical support.
- Transfer of assays/QC development.
- Assay validation.
- Product specific validation.
- Manufacturing.
- Quality testing.
- Managing of external testing laboratories.

- Release of drug substance.
- Fill/finishing.
- Release of drug product.
- Stability studies.
- Inventory building.
- Shipment and logistic support.

During the scope development, do not forget to include process improvements into the capacity planning. Typical selection criteria during a formal evaluation of a CMO have the following priorities:
- Highest priority (or weight factor)
  - Experience with required manufacturing technology
  - Expression system (if applicable)
  - Capacity, scale
  - cGMP compliance
  - Cooperation, client's input regarding manufacturing operations
- High priority (or weight factor)
  - Experience with cGMP and/or commercial manufacturing
  - QA/QC support
  - Regulatory support
  - Project schedule
  - Project cost
  - On a case by case basis items from the section "CMOs offer additional benefits"
- Low priority (or weight factor)
  - Location

These criteria might change depending on the scope and size of the project. However, once the technical and quality issues are found to be adequate, the most important one is finding the CMO which truly becomes a partner.

16.3.7
Key Issues of CMO/Client Contracting

Key issues during the contract negotiations are:
- *Reservation fees* required by the CMO to assure loading their GMP capacity. The terms vary from no payment (payment to secure a manufacturing slot only if another client is interested) to 30% or more of the price paid, 6 months or up to 18 months ahead of the start of manufacturing a particular batch.
- *License fees* (and future royalties) when the CMOs expression system is part of the contract.
- Process specific *laboratory and validation work* becomes either part of the price or will be reimbursable on a time and material basis.
- *Intellectual property rights* brought in or developed during the project.
- *Liability* – CMO to comply with cGMP and to be liable for operational risk; client to be liable for the suitability of the product for the intended use.
- *Performance criteria* – particularly when the CMOs expression system is part of the contract.
- *Payment schedule* – incentives tied to time and performance.
- *Quantity loss* – quantity loss during re-performance will trigger a price reduction.
- *Termination* – if CMO does not comply with cGMP, general termination clause.
- *Change of ownership* – in case the CMO is sold to another company.
- *Assignment* – to make sure the project can be licensed out in the future.
- Demarcation of *pharmaceutical responsibilities* to document the responsibilities towards the regulatory authorities.
  As mentioned above,
- *Price* is – especially during the early clinical phases – just one consideration

among others. While a typical MAb manufacturing project up to clinical Phase II may cost € 3 million, the overall cost of clinical production, consistency lots including activities related to the chemistry manufacturing and control (CMC) section of the registration dossier may exceed € 15 million. However, clinical trials are typically more expensive.

The absolute latest point in time to "marry" the CMO is after clinical Phase IIb.

As a change to another CMO will take about three years (as post-approval change), it is important to select a CMO which will not only supply the material for clinical Phase III but is willing to deliver product for the initial three years of marketing requirements. Typically, that will require scale-up of the process to larger bioreactor(s) and purification unit operations.

Therefore, key issues of CMO/Client contracting for clinical Phase III and launch requirements are:

- Capacity – the CMO must be capable of supplying quantity for at least the first year; client's commitment.
- Scale-up performance criteria.
- Regulatory support.
- Launch requirements – typically manufactured with consistency lots.
- Transferability – to transfer the process to own facility, a strategic partner (or to another CMO)

and for a typical long-term manufacturing agreement.

- $/gram pricing (e.g., based on ranges for fermentation output, DSP yield and success rate).
- Raw materials cost – reimbursable.
- Minimum volumes.
- Shared savings for yield improvements.
- Contract duration 7 years+.
- Royalties (if applicable).

16.3.8
Technology Transfer

Whenever process development leaves the laboratory scale, the know-how which is well established within the research group must be transferred to staff in the development department. Both groups are facing only minor obstacles as they know each other very well, and communication is established between the key persons. This know-how shift during an in-house project is usually accompanied with a technology-driven scale-up process. In many cases, leaving laboratory-scale to enter the pilot scale means re-engineering the whole process.

These two factors – the proximity of both groups and the simultaneous development of a pilot process – often conceal the complexity of a technology transfer once a third party is engaged.

The transfer of a biopharmaceutical process between a research-driven process owner and a CMO is usually a challenging and often time-consuming and therefore expensive project. Underestimating the process of technology transfer often causes an unexpected delay of the ambitious time-table.

Management of the technology transfer requires careful preparation and stringent definition of all project steps and scopes:

- The process itself needs to be defined in detail: All the relevant process information must be described and documented. Technology transfer of chemical processes is rather simple compared to biotech processes due to well-understood, robust and predictable chemical reactions. In contrast, technology transfers of bio-processes often cause difficulties due to the lack of process understanding/information. These areas, where process knowledge is rather limited, also need to be identified up front, in order to minimize the efforts.

- The analytical procedures which have been developed especially to control and to monitor the efficacy of each bio-process step need also to be defined and to be transferred (see Part VII, Chapter 1). Protocols must be written that describe each in-process-control (IPC) step from the analysis of the WCB cell line, the (on-line) monitoring of the cultivation to the analysis of the purification need to be established. Due to the complexity of bioprocesses, the need for good biochemical analytical methods and capabilities cannot be overemphasized.

- Technical equipment: The robustness of a bioprocess and the yield of a bioproduct are often directly linked to the process equipment used for the manufacturing. Reactor geometry, shear stress, mixing and rheology have an impact on the biology. The metabolic activity of the cell line during cultivation not only depends on standardized media but is also a function of mass and energy transfer. Temperature profiles are a key factor for induction processes, and depend on the very reaction system. Harvesting after cultivation is also challenging and requires strong expertise. Batch centrifuges give tight, dry pellets, whereas continuous disk-stack separators produce a heavy phase "slurry", either carrying over supernatant or leaving product in the heavy phase.

- Continuous-perfusion cell culture poses additional challenges to the team.

- Qualification and validation protocols. Comparability of the product and consistency of the process transferred from a client to a CMO needs to be demonstrated. A given reference standard must be met under predefined protocols to proof the ability of the CMO to manufacture the biopharmaceutical consistently.

A consensus about all features of the transfer process needs to be achieved between the receiving party and the process owner. In order to guide the technology transfer successfully through all phases, a stringent project management structure must be established. There are at least two different possibilities to structure this project organization, and Fig. 16.6 illustrates a parallel structure between the CMO and the technology provider. This structure focuses on a liaison (window) person or a (liaison) window team.

Although the additional interfaces within this structure can cause difficulties and misunderstandings, this structure can often be found in smaller hierarchical companies. To channel information via one central person prevents direct communication, inhibits information flow, and causes delay. An extended project timeline is the consequence.

An alternative organization will be presented below. Since information flow is the key for success, the formation of working teams is crucial. These teams, representing all technical aspects, must be brought together in an initial kick-off meeting. Each function needs to know his/her counterpart:

- Quality control
- Quality assurance
- Biology (cell culture, cell banking)
- Process (the complexity of the process operations often requires a larger team, with appropriate technical skills)
  - Upstream processing
  - Downstream processing
  - Polishing/Filling/Packaging

For a well-structured project management it is essential that both parties understand

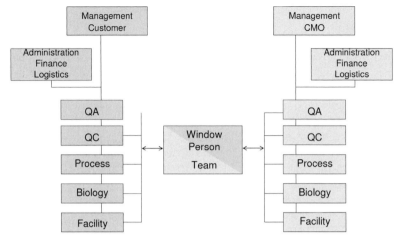

Fig. 16.6 Organization of a technology transfer using a window person or a window team.

each other as one integrated team, which can only achieve its target together. Failure on either side of this transfer project leads to a failure of the whole project. This well-established partnership is a key factor for the success.

On top of this working level, a steering committee representing the senior management of both partners should supervise the whole project. It is their responsibility to solve problems escalated from team level, to monitor progress and milestones of

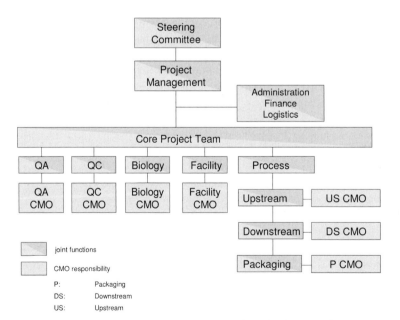

Fig. 16.7 Successful project management structure for technology transfer project.

the project, and to supervise the time-schedule and financing. While the working team should function as one single unit, it is essential for the steering committee to meet on a regular basis. With this procedure the senior management is in the position to act before reaction is needed. Fig. 16.7 provides an example of a successful technology transfer project organization.

Within this structure, each member of the team communicates directly with the respective partner on the other side. The teams should meet on a regular basis to discuss progress, constraints and timelines.

This hierarchical structure facilitates communication and ensures that the required information can flow directly from one partner to another. These communications are particularly important to ensure timely information which is required across all levels of the project organization. Delay due to long-lasting information flow can be avoided.

16.3.9
Time to Market, and the Bottom Line

It is almost impossible to comment on the average time that a biopharmaceutical product needs to progress from laboratory bench to market. For a MAb, from discovery to FDA approval and commercial launch, it costs about US$ 800 million, and takes 12–15 years [10]. However, a recent FDA whitepaper reported that the required investment for one successful drug launch (discovery through launch) is US$ 1.7 billion [12]. In addition, one must be aware of the fact that only one in 5000 candidates will make it all the way to commercial success.

Biopharmaceutical development projects "generally" take between 20 and 65 months to complete. This number corresponds to between 12 000 and 800 000 man-hours, depending on the complexity of the process [7]. In fact, the slowest project took 2.7 times longer to complete and consumed almost twice as many resources hours compared to the fastest. An analysis of different development projects revealed that the efficiency plays a major role for commercial success. More efficient companies complete two to three times more projects during the same period of time, and using the same resources, than do companies with a lower efficiency [7].

The reduction in development time leads to a significant competitive advantage, especially as getting a new biopharmaceutical to the market faster allows a market share to be established before competitors.

Fig. 16.8 illustrates an approximate model calculating the cost of the lost opportunity that a company faces for each working day a bio-therapeutic remains in process development rather than being on the market.

While 130 biopharmaceuticals accounted for sales of approximately US$42.6 billion in 2002, the average revenue potential of a single product can be calculated at US$328 million per year, or US$1.3 million per day. The calculation of the daily "burn rate" that a biopharmaceutical compound has during its development phase is based on certain estimates. On the basis of the above-mentioned data, the average number of man-hours required for a development project is 406 000, within a period of 42.5 months. At an average personnel cost of US$225 per hour, the burn rate of a biopharmaceutical product during the development phase accounts for $107 000 per day [8, 9].

These figures illustrate very well the need for efficient process development. Moreover, improving collaboration between

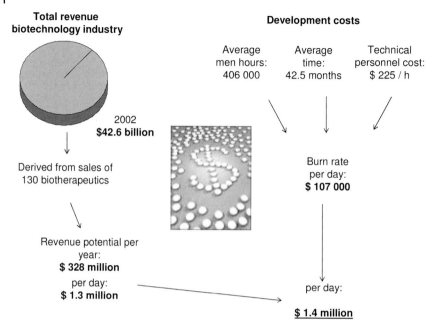

Fig. 16.8 Process development cost versus revenue potential of biopharmaceuticals.

the CMO and the process provider during process development in order to shorten the timeline could also significantly reduce costs.

16.4
Summary and Outlook

The structural complexity of a specific biopharmaceutical protein, together with the amount required, determine the choice of an appropriate expression system. While antibody fragments can easily be produced in yeasts, fungi, or *E. coli* cells, the expression of whole antibodies still requires complex mammalian cell culture processes.

Although recombinant mammalian cell culture involves expensive growth media and complex purification steps, it is the only production alternative for many products. As therapy with these potent biopharmaceutical drugs often requires high doses, the manufacturing capacity can indeed be a limiting factor for the process development.

Much of the expense in the biopharmaceutical manufacturing of antibodies or antibody fragments derives from the large capital investment required to build and operate a manufacturing facility. In this respect, the outsourcing of certain services is an alternative to building in-house capacities.

Projecting market demand for specific classes of therapeutics in development will remain a major planning issue for biopharmaceutical companies. The ability to meet the demand for biopharmaceutical production capacity, whether through in-house manufacturing or outsourced contract manufacturing, carries strategic and financial implications. Also, FDA approval policy and the development of the biogenerics market will have a major impact on future capacity requirements (see Part VII, Chapter 4).

References

1 Milstein C, Köhler G. Continuous cultures of fused cells secreting antibody of predefined specificity. *Nature* **1975**, 256: 495–497.

2 Cacciuttolo MA, et al. Large-Scale Production of a Monoclonal IgM in a Hybridoma Suspension Culture. *BioPharm* **1998**, 11(4): 20–27.

3 Wurm FM. Development of High-Yielding Cell Culture processes for medical applications: How to meet the challenge. Presented at Scale-Up of Biopharmaceutical Products, NL-Amsterdam, IBC Life Science UK, 26–27 January, 2004.

4 Yokota T, Milenic DE, Whitlow M, and Schlom J. Rapid tumor penetration of a single-chain F$_v$ and comparison with other immunoglobulin forms. *Cancer Res* **1992**, 52: 3402–3408.

5 Harrison JS and Keshavarz-Moore E. Production of antibody fragments in *Escherichia coli*. *Ann NY Acad Sci* **1996**, 782: 143–158.

6 Kara B. Challenges and solutions: Scale-up of microbial production systems. Basel Switzerland, 24 January 2003 (IBC Life Sciences, London, UK).

7 Pisano GP. The Development Factory: Unlocking the potential of process innovation. Harvard Business School Presentation, Boston, MA, 1997, pp. 1–343.

8 DiMasi JA. The value of improving the productivity of the drug development process: Faster times and better decisions. *Pharmacoeconomics* **2002**, 20 (Suppl.) 3: 1–10.

9 Neway JO. How the data ecosystem affects process development and the bottom line. *BioProcess Int* **2004**, 2(5): 82–91.

10 DiMasi JA, Hansen RW, and Grabowski HG. The price of innovation: New estimates of drug development costs. *J Health Econ* **2003**, 22: 151–185.

11 Booth R. Extending your enterprise with outsourcing. *BioProcess Int* **2003**, 1(10): 14–18.

12 US Department of Health and Human Services Food and Drug Administration: Challenge and Opportunity on the Critical Path to New Medical Products, March 2004.

Part V
Biopharmaceuticals used for Diagnositics and Imaging

From Hunter to Craftsman –
Engineering Antibodies with Nature's Universal Toolbox

1
Thirty Years of Monoclonal Antibodies:
A Long Way to Pharmaceutical and Commercial Success

Uwe Gottschalk and Kirsten Mundt

Abstract

Only about 30 years ago, Köhler and Milstein set the stage for one of the key technologies revolutionizing human life: The invention of monoclonal antibodies (MAbs) provided the basis for new tools in biochemical research, and applications spread throughout all medical areas. The 1980s were stamped by a hype that antibodies as "magic bullets" would provide a major breakthrough in oncology. Drug-targeting, immunotoxins, and radioimmunotherapy were key words that led to the foundation of research programs, expert groups, and eventually also to production companies. Since then, various drawbacks have hit the sector, and antibodies have survived in niches, as research agents, diagnostic tools and for highly specific indications. Today, with the maturation of the whole biotech industry, MAbs face a significant renaissance and appear stronger than ever. Today, MAbs represent the fastest growing pharmaceutical market segment, with a potential to reach worldwide sales of US$ 20 billion by the year 2010. Since 1982, some 20 MAb-based biopharmaceuticals have been approved for the treatment of chronic and life-threatening disease, and there are hundreds of second-generation products under clinical investigation (see Part VII, Chapter 4). This development is further accelerated by validated disease targets that are becoming accessible through the human genome sequence and many innovative research avenues. While the commercialization of MAbs is gathering momentum, the sector is facing a worldwide shortfall of available biomanufacturing capacity that is becoming a critical strategic limitation, especially for companies without established market access (see Part IV, Chapter 16 and Part VIII, Chapter 1). As a result of early clinical failures, a number of strategies were introduced to address some of the underlying limitations. Recombinant DNA technologies, humanized molecules with low immunogenic potential (see Part V, Chapter

Modern Biopharmaceuticals. Edited by J. Knäblein
Copyright © 2005 WILEY-VCH Verlag GmbH & Co. KGaA, Weinheim
ISBN: 3-527-31184-X

2), individualized treatment strategies on the basis of diagnostic tests (see Part I, Chapters 2 and 5), and genetic profiles as well as recent improvements in cell biology and fermentation are now providing the basis for pharmaceutical and commercial success (see Part IV, Chapter 1). The next generation of MAbs will be developed on the basis of tailor-made, miniaturized molecules with optimized pharmacokinetics, and given to patients most likely to benefit from the respective therapy. Although the basic principle of GMP manufacturing is identical for all pharmaceuticals, macromolecules such as MAbs are subject to some specifics with regard to their manufacturing costs and potency profile. Since the products are large and complex molecules with isoforms and microheterogeneities, the production process itself must meet the highest requirements with regard to consistency and reproducibility, and process changes are monitored with a demanding comparability policy. Improvements are due in all areas of the pharmaceutical supply chain to manage current manufacturing challenges, and are therefore vital for the long-term success of MAbs as biopharmaceuticals.

Abbreviations

ADCC	antibody-dependent cellular cytotoxicity
AML	acute myeloid leukemia
BHK	baby hamster kidney cells
BLA	biologics license application
BRMs	biological response modifiers
CAGR	compound annual growth rate
CDC	complement-dependent cytotoxicity
CDR	complementarity-determining region
CEA	carcinoembryonic antigen
cGMP	current good manufacturing practice
CHO	Chinese hamster ovary
CIPP	capturing-intermediate purification-polishing
CMC	chemistry manufacturing and controls
COG	cost of goods
DMARDs	disease-modifying anti-rheumatic drugs
DOE	design of experiments
DTPA	diethylentriaminopenta-acetate
EGF	epidermal growth factor
ELISA	enzyme-linked immunosorbent assay
ER	endoplasmic reticulum
FACS	fluorescence-activated cell sorting
HAMA	human anti mouse antibody
HTST	high-temperature short-time
IBD	inflammatory bowel disease
Ig	immunoglobulin
IGIV	immune globulin intravenous
IL	interleukin
IL-1ra	IL-1 receptor antagonist
IND	Investigational New Drug
MAbs	monoclonal antibodies
MWCB	Master Working Cell Bank
NHL	non-Hodgkin lymphoma
NSCLC	non-small cell lung cancer
PCR	polymerase chain reaction
PEG	polyethylene glycol
PI	primary immunodeficiency
PPV	porcine parvovirus
RA	rheumatoid arthritis
RIA	radioimmunoassay
RSV	respiratory syncitial virus
SIP	selectively infective phages
TAA	tumor-associated antigens
TNF	tumor necrosis factor
YAC	yeast artificial chromosome variable domain of single antibodies

1.1
Introduction

1.1.1
Immunoglobulins: Form Follows Function

Antibodies represent a class of flexible molecular adaptors that play a vital role in the adaptive immune systems of vertebrates [1]. First discovered by von Behring and Kitasato in 1890 [2], it was not before 1960 that their basic structure was determined by Porter and Edelmann [3]. With their inherent diversity and heterogeneity, antibodies mediate humoral and cellular immune responses, and directly execute biochemical mechanisms such as antigen recognition, complement-dependent cytotoxicity (CDC) and antibody-dependent cellular cytotoxicity (ADCC) [4]. In most higher mammals, antibodies exist in five distinct classes (IgG, IgA, IgM, IgD, and IgE) that differ in both form (size, charge, amino acid composition, carbohydrate content) and function [5].

Immunoglobulins are bifunctional molecules with a basic symmetrical structure consisting of pairs of identical heavy (μ, δ, γ, ε, a) and light chains (κ, λ) linked through disulfide bridges (Fig. 1.1) [6].

The class and subclass of an antibody are determined by its heavy chain type, and are linked to different physiological functions. The individual chains form globular domains that are either highly conserved between different immunoglobulins (constant region) or contain sequence variability (variable region) and are stabilized by intrachain disulfide bridges. Fc-related properties such as complement activation and lymphocyte binding reside on domains of the constant region, and are essential for the cellular immune response [7, 8]. Other structural features are

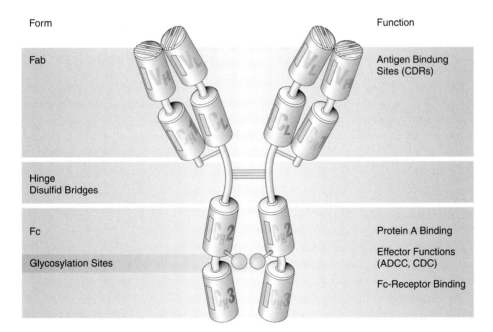

Fig. 1.1 Form follows function: the domain structure of an IgG antibody.

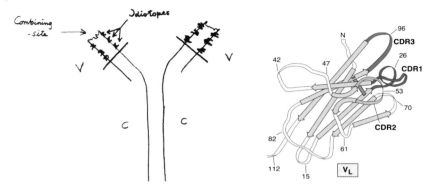

Fig. 1.2 The variable part of an antibody, as seen, 1984 and 2004. The CDR loops are highlighted. Left: Original drawing from the Nobel Prize lecture of Niels Jerne (© The Nobel Foundation). Right: Reproduced with permission of Hoffmann/Fischer RWTH Aachen.

N-linked carbohydrate moieties at Asn297 of the C_H2 domain and a protein A binding site between C_H2 and C_H3 [9, 10].

The two antibody-binding sites are formed by the interaction of six hypervariable loop regions that are exposed near the N-terminus of the polypeptide chains. These complementarity-determining regions (CDRs) are surrounded by relatively invariant framework residues (Fig. 1.2). Their diversity (idiotypic variation) represent the central aspect of the humoral immune response. Variability – and thus flexible and efficient molecular antigen recognition – arises at several levels, and is the result of somatic mutation and recombination of multiple genetic elements during B-cell differentiation [11].

The binding of antigen to antibody (epitope–paratope interaction) is a result of multiple non-covalent interactions which, in combination, lead to a considerable binding energy with thermodynamic affinity constants in the nM range. Avidity – the binding strength of a multivalent antibody – is usually greater than the sum of the monovalent affinities, and is the more relevant term under physiological conditions. Upon primary antigen stimulation, secreted

IgM is released from B cells. During maturation of the immune response, isotype switching occurs upon support through T cells, leading to the secretion of IgG, IgA, or IgE. A T-cell-dependent immune response is also accompanied by an affinity maturation which is basically the selection of high-affinity B-cell clones that are preferentially triggered to divide and differentiate, plus the result of somatic hypermutation to further increase the binding energy of the immunecomplex formation.

Antigen recognition shows a high level of specificity, and cross-reactivity can only occur if determinants are shared between different antigens. The natural immune response represents a powerful mixture of polyclonal antibodies directed to various epitopes of the antigen, which is exploited in the secondary antibody response after a vaccination.

The two functions of an antibody are therefore represented by the variable domain binding to the antigen, and the constant Fc domain that mediates recruitment of effector cells to establish a cellular immune response.

In a life-saving therapeutic application, plasma-derived antibodies are widely used for the treatment of various primary immunodeficiency (PI) diseases [12], and their full potential is just beginning to be discovered in a wide range of indications [13]. Polyclonal antibodies from human serum is a multi-tonne product isolated from approximately 30×10^6 L of plasma each year. Immune globulin intravenous (IGIV) are now also available as second-generation products with chromatographic purification and a strict safety concept considering all known human pathogens [14, 15]. For a comprehensive overview of the IGIV products available commercially, see Ref. [16].

1.2
Making Monoclonal Antibodies

For the specific recognition of a single epitope, however, polyclonal antibodies are not the right tool. It was therefore good news in 1975 when George F. Köhler and César Milstein described a method to generate monoclonal antibodies (MAbs) [17]. In Milstein's group at the MRC Laboratory of Molecular Biology in Cambridge, Köhler had cloned an immortalized cell line secreting Sp1, a mouse IgM antibody. The original process is based on the injection of an antigen into a mouse to induce a specific immune response. Differentiated B lymphocytes are isolated from the spleen and fused with immortalized myeloma cells. The resulting hybridoma cells are then selected for their potential of immortalized growth and screened for specific antibody production. The development of MAbs starts from single cell clones that – after propagation – are laid down in a cell bank and characterized for routine production in repeated manufacturing trains. Amazingly, the full significance of this invention was not immediately recog-

nized, and was not even patented (Fig. 1.3).

However, together with Niels K. Jerne, George D. Köhler and César Milstein were later awarded the 1984 Nobel Prize in Medicine "…for theories concerning the specificity in development and control of the immune system and the discovery of the principle for production of monoclonal antibodies".

The traditional method of MAb generation has a number of drawbacks and limitations. Firstly, these antibodies are murine. Despite many similarities and structural homologies, the possibilities of therapeutic use are rather limited due to the short serum half-life in humans. In addition, the "foreign" protein elicits an immune response after repeated administration, with the most discouraging being the human anti mouse antibody response (HAMA) – a possibly life-threatening immunological event that is directed against the idiotype and the isotype of murine antibodies [18].

New strategies to generate MAbs that are less immunogenic and eventually fully human have since been invented, however.

1.2.1
Generation of Human Antibodies

A variety of possibilities to reduce the risk of immunogenicity of a therapeutic antibody exist, and have been employed.

Chimeric antibodies are hybrid molecules combining the antigen-specific variable domain of the mouse antibody fused to the constant regions of a human IgG molecule (Fig. 1.4). This reduces the risk of immunogenicity somewhat, and the human Fc domain prolongs the serum half-life and is more effective in triggering the effector systems of complement and Fc receptors.

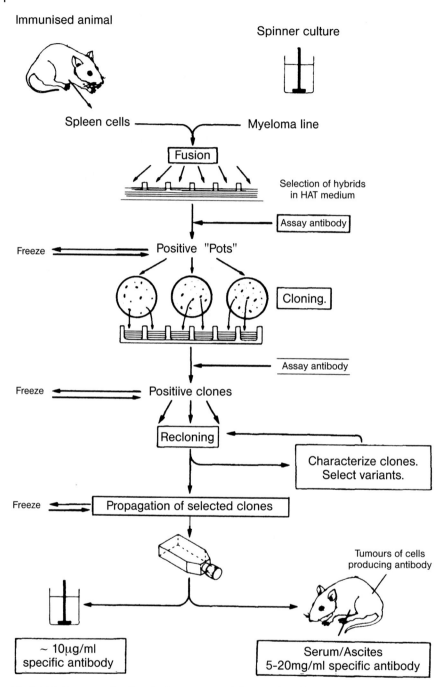

Fig. 1.3 Making monoclonal antibodies. An original drawing from the Nobel Prize lecture of César Milstein (© The Nobel Foundation).

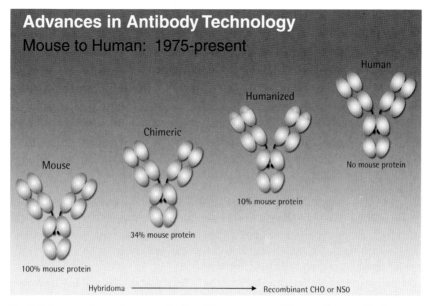

Fig. 1.4 From murine to human antibodies. (Figure courtesy of Steven Chamow, Bio Process Consultant.)

Further genetic engineering has led to the development of humanized antibodies that retain the antigen-specific CDR of the original antibody, grafted onto a fully human acceptor framework [19]. However, this grafting process is not always straightforward, and certain framework residues have been identified that are essential for correct presentation of the CDRs. Despite advanced molecular modeling efforts, grafting still requires the experimental validation, to arrive at a molecule that retains the same specificity and affinity as the parent antibody [20]. Further efforts have therefore gone into alternative technologies that allow the generation of fully human antibodies [21].

1.2.1.1 Transgenic Mice

One powerful technology is the development of transgenic mice that have functionally replaced the mouse antibody genes with the human equivalents. This was pos-

sible through technologies which allowed the insertion of large yeast artificial chromosome (YAC) fragments into the germlines of mice [22]. As these mice raise an immune response following immunization, they generate fully human antibodies, which can then be isolated with the traditional hybridoma technology [23]. Currently, Abgenix, Medarex, and Kirin exploit this technology commercially.

1.2.1.2 Recombinant Libraries

Another approach employs large combinatorial libraries based on human VH and VL genes. These libraries have the advantage that clonal selection against self antigen does not take place, which can be a problem when raising an antibody against a highly conserved epitope in an animal.

With recombinant libraries, the challenge is to link the phenotype (physical binding of the antibody protein to the antigen) to the genotype, the cDNA encoding the protein. The most commonly employed method to isolate new binders from a recombinant antibody library is that of phage display [24, 25].

Phage display technology

This method has been employed for the isolation of new scFvs from naive libraries, as well as for affinity maturation. To select for scFvs binding to a particular antigen, the scFv is fused to a minor coat protein, typically pIII (g3p) of filamentous M13 phage. During phage panning, the scFv on the phage is bound to immobilized antigen and enriched during consecutive cycles of binding, elution, and amplification after the infection of bacteria. However, a disproportional amplification of non-specifically binding phages can occur, since non-specifically and specifically bound phages can infect the bacterial cell. Therefore, it is of interest to minimize non-specific adsorption of the phages in the binding procedure and to specifically enrich those phages, which display high-affinity molecules over highly abundant, low-affinity binders. The method of selectively infective phages (SIP) addresses this problem. In the SIP procedure, the desired antigen–antibody interaction itself is essential for restoring infectivity in an otherwise non-infective phage displaying the scFv. The coat protein pIII (g3p) is lacking the N-terminal domain responsible for infectivity. A fusion between the antigen and the missing domains of the coat protein restores infectivity, thereby linking binding and infectivity. In contrast to phage display, where the interacting scFvs bind to an antigen on a solid-phase surface (e.g., an affinity column), no solid-phase interaction is necessary in SIP, thus avoiding the problem of non-specific interactions. Furthermore, SIP is a one-step procedure

(binding and infection are coupled, and elution becomes unnecessary) and therefore faster and easier to perform [26]. Phage display – as well as SIP – is a powerful tool used to isolate new antibodies and select for increasing antibody affinity. A number of companies exploit this powerful, but heavily patent-protected technology; these include Morphosys (see Part V, Chapter 2), Cambridge Antibody Technology (CAT), and Dyax. Humira® is the first phage display-derived MAb on the market. On the other hand, about 30% of all human antibodies currently in clinical development have been generated using this technology [27].

Yeast Surface Display and Ribosome Display

Alternative methods for selection of interacting proteins, including antibody–antigen interactions include yeast surface display and ribosome display (Fig. 1.5).

Despite its relatively late advent in the antibody engineering field, *yeast cell surface display* also presents an attractive an powerful method to isolate affinity-matured antibodies [28]. Unlike phages, yeast cells are large enough to be screened and separated using flow cytometry. In the yeast cell surface display the scFv is delivered to the yeast cell wall by fusion to the cell surface protein Aga2p. The binding of fluorescent or biotinylated antigen to a specific scFv on the cell surface can be measured using flow cytometry, after which the individual cells are sorted by fluorescence-activated cell sorting (FACS). This system was used successfully for the affinity maturation of antibodies and receptors, and allows very fine discrimination between mutants of slightly different affinity.

One disadvantage of both phage and yeast surface display-technologies lies in the fact that the "displayed" peptide or protein is rather small in size compared to

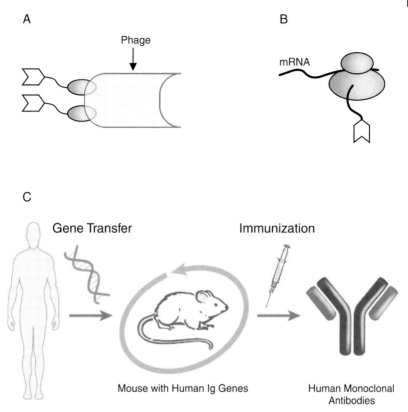

Fig. 1.5 Methods to generate fully human antibodies. A) Phage display; B) ribosomal display; C) transgenic mice. (www. Esbatech.com; www.Medarex.com.)

the carrier (phage particle, or yeast cell). This can lead to false positives due to unspecific interactions. *Ribosome display* is an *in vitro* method that avoids this "display problem", since it links the peptide directly to the genetic information (mRNA): an scFv cDNA library is expressed *in vitro* using a transcription–translation system. The translated scFvs are stalled to the translating ribosome by the addition of puromycin, and thus linked to the encoding mRNA. The scFv is then bound to the immobilized antigen and unspecific ribosome complexes are removed by extensive washes. The remaining complexes are eluted and the RNA is isolated, reverse-transcribed to cDNA, and subsequently re-amplified by polymerase chain reaction (PCR). The PCR product is then used for the next cycle of enrichment, and the mutations that are introduced through the error-prone reverse transcriptase can actually contribute to affinity evolution [29].

1.3
Other Antibody Formats: Antibody Fragments

Although full-sized IgG molecules are the naturally occurring format of antigen-binding molecules, fragments and derivatives

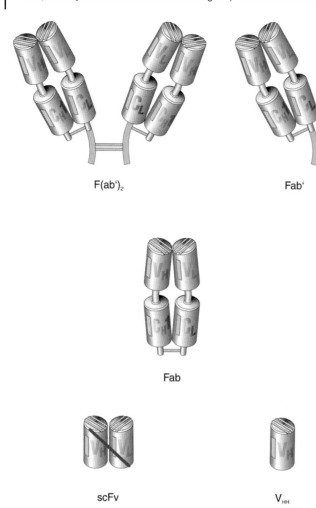

F(ab')₂

Fab'

Fab

scFv

V$_{HH}$

Fig. 1.6 Antibody-fragments under evaluation.

Table 1.1 Pharmacological characteristics of different antibody formats

Parameter	mAB	Fab	scFv
MW	150 kDa	50 kDa	27 kDa
Glomerular filtration (< 50–60 kDa)	–	(+)	+
Tumor/tissue penetration	+	++	+++
α-elimination half-life in blood (equilibrium half-life)	> 1 h	30 min	8–10 min
β-elemination half-life in blood (clearance half-life)	1–3 weeks	5–6 h	3–4 h
Biological effects of Fc part (e.g., ADCC; CDC)	+	–	–

are also developed and employed for a variety of applications. These include Fab fragments, single chain antibodies (scFv), and even single domain antibodies (Fig. 1.6).

Advances in genetic engineering methods have enabled the generation of a variety of different antibody fragments and fusion proteins, and allowed the construction of molecules, the properties of which are tailored for the specific requirements of the intended therapeutic application [30]. These properties focus on efficacy – that is, the biological activity that addresses the mechanism of action. In addition, issues such as biochemical and biophysical stability, solubility, and production yield which influence the cost of goods (COG), and the pharmacokinetic properties as well as the immunogenicity of the protein, play an important role (see Table 1.1).

As described earlier, antibody-binding to its cognate antigen can elicit an immune response through the binding of the Fc part of the antibody to the Fc receptor on various effector cells, and by activation of the complement system. This can be an undesired effect, in particular in inflammatory diseases, where generally only neutralizing antigen-binding activity is required. Antibody constructs that eliminate the constant Fc domains can therefore be an advantageous format. A number of antibody fragments have been developed specifically to address the individual applications and their requirements; for a comprehensive overview, see Ref. [31].

The Fab fragment contains the variable domain of the heavy and the light chain and the adjacent constant domains C_H1 and C_L, linked through a disulfide bond, that naturally links these two chains together [32,33]. Although the specificity and selectivity of the monovalent Fab fragment is the same as that of the parent antibody,

its avidity is lower, because it has only one antigen-binding site. Through different linker constructs at the hinge region of the antibody, dimeric and even trimeric Fab constructs have been generated [34].

The Fv fragment represents the smallest antibody domain that retains an acceptable affinity. It is usually very unstable, and aggregates as a result of only weak, non-covalent forces, but can be stabilized with a linker polypeptide yielding a single chain Fv antibody fragment (scFv) and allows expression from a single gene in a number of hosts. It also facilitates additional genetic engineering such as fusion to effector domains [35]. Through variation in the linker size, not only multimeric variants of the scFv such as diabodies and tribodies but also other linear antibody fragments can be generated that possess a higher avidity than monovalent versions [36, 37].

A variation thereof is the generation of bispecific antibodies [38]. In certain applications, these are designed to bind to a cancer-specific surface molecule with one binding domain, while the other binding domain recruits cytotoxic T cells to the pathogenic cell, inducing T-cell-dependent cytotoxicity [39].

Single-chain antibodies that consist of only a single variable domain with three CDRs occur naturally in camels and llamas – species that are lacking the light chains of antibodies with still high stability and affinities [40, 41]. Human versions have been isolated and are being developed – a process called camelization [42]. The resulting VHHs are the smallest available intact antigen-binding fragment, with a molecular weight of approx. 15 kDa, 118–136 residues that remain highly soluble and stable with affinities in the nanomolecular range [43]. Secretion in the endoplasmic reticulum (ER) is enhanced through hydrophilic amino acids.

1.3.1
Pharmacokinetics of Antibody Fragments

One problem associated with small-sized antibody fragments is that of rapid renal clearance from the bloodstream that leads to a reduction in half-life to hours, rather than weeks (as is the case for full-sized antibodies).

This may be an advantage for some acute indications such as myocardial infarction or acute infections. It has also been shown, that the much smaller protein size allows faster tissue penetration, which might be advantageous for solid tumors and local applications [44]. For systemic applications, however, the possibility of avoiding rapid elimination via glomerular filtration is to increase the size of the molecule, for example by PEGylation. The addition of polyethylene glycol (PEG) to a random or defined site of the protein increases its apparent hydrodynamic size, such that the protein becomes more stable and possibly even less immunogenic, as the large parts of the protein surface are no longer accessible by the immune system [45, 46] (see also Part VI, Chapter 2). Examples of therapeutic fragment antibodies are CDP 870 (an anti-TNF-alpha PEGylated Fab fragment for the treatment of Crohn's disease), and the complement inhibitor PexelizuMAb, which is currently under clinical development [47, 48].

The production of full-sized antibodies is restricted to mammalian cell systems that possess the correct production and glycosylation machinery (see Part IV, Chapter 1). Antibody fragments, on the other hand, can also be expressed at much lower cost in microbial expression systems, such as bacterial, yeast, or fungal fermentation [49, 50]. For the high-level expression of antibody fragments, *E. coli* fermentation provides a well-established technology basis with periplasmic expression levels at 1–2 g L^{-1} during high cell density fermentation [51].

1.4
Medical Application Areas for MAbs

The 30-year history of MAbs has been a rollercoaster ride to success. From hype to depression, and back to hype, is probably the shortest summary of what has happened. MAbs have been rapidly introduced into a number of applications within and outside the medical field (Table 1.2) [52–54]. Analytical *in vitro* methods such as enzyme-linked immunosorbent assay (ELISA), radioimmunoassay (RIA), various blotting techniques, flow-cytometry, immunofluorescence, confocal imaging, and immunohistochemistry are each dependent upon the use of polyclonal or monoclonal antibodies.

Table 1.2 Application areas for monoclonal antibodies

In-vitro use
Research reagent in drug discovery
Analytical tool in drug development and manufacturing
In-vitro diagnostics (biochemical, histological, pathological)
Immunoaffinity chromatography
Biosensors
Catalytic antibodies
In-vivo use
Immunoscintigraphy
Isotypic immunotherapy
Idiotypic immunotherapy (antagonistic and agonistic)
Drug targeting with immunotoxins and immunoconjugates
Radioimmunotherapy
Edible vaccines

Based on their physiological role, unmodified MAbs can trigger different immunological reactions that are used in medical applications [55]. These reactions may vary from passive immunization through effector functions to selective cytotoxic effects with bispecific antibody constructs that recruit cells of the immune system to destroy identified targets [56].

The vast majority of today's antibody treatment concepts rely on active immunization directed towards specific disease targets. The effect can either be antagonistic through the neutralization of a signaling molecule or its specific receptor or agonistic with rather enzyme-like support of a physiological function [57]. Antagonistic treatment concepts typically require high doses in the region of >1 mg kg^{-1}, and they include the treatment of inflammatory disorders such as rheumatoid arthritis, inflammatory bowel disease (IBD), psoriasis, allergies, and autoimmune diseases (e.g., multiple sclerosis and Crohn's disease). Soluble cytokines such as interleukin (IL)-1 and tumor necrosis factor (TNF) are targets to prevent pro-inflammatory mechanisms.

Following marketing authorization of the first therapeutic antibody OKT3 in 1986 for the treatment of acute transplant rejection, premature hopes and unrealistic expectations were raised, and MAbs had a difficult time in living up to their original promise (see Introduction and Part II, Chapter 4). During the ensuing period, antibodies survived in niches as research reagents and diagnostic enabling tools, and it was not until 1995 that Centocor's ReoPro® (Abciximab) – a chimeric MAb fragment – won approval for the prevention of thrombotic side effects in patients undergoing coronary artery angioplasty [58]. ReoPro binds to a glycoprotein receptor on human platelets and prevents their aggregation.

1.5
From Initial Failure to Success: Getting the Target Right

The initial lessons learned were derived from a number of spectacular failures during the development of MAbs antibodies, and these in turn led the way to success. This intermezzo is in fact one important reason for the first downturn in biotechnology when investors sent the sector into some of its leanest times and a good learning lesson for the companies involved.

A closer look at antibodies against endotoxins as a cause of septic shock may serve as an example [59]. Septic shock continues to be a severe problem, with mortality rates up to 70% and TNFα as a primary mediator that can be stimulated by endotoxins [60, 61]. Drug treatment standards for septic shock remain poor [62, 63]. Currently, Synergen is the leader among a long list of biotech companies that have suffered badly from attempts to achieve sepsis treatment. Their drug Antril® – a recombinant form of the naturally occurring 23 kDa IL-1 receptor antagonist (IL-1ra) – was the first recombinant protein that failed to improve significantly the survival of sepsis patients in large-scale clinical trials following positive intermediate results [64]. The company built a production facility based on the promising results of early Phase II trials, but these could not be reproduced in subsequent studies [65]. Monoclonal antibodies for sepsis treatment lost favor in 1992 when the FDA refused to approve Centoxin® (Centocor), a human monoclonal IgM antibody binding to lipid A type endotoxins from Gram-negative bacteria [66]. Centoxin® was voluntarily withdrawn from the European market in 1993 after it showed a lack of benefit in post-approval studies in septic shock treatment [67]. Besides Centoxin®, another anti-endotoxin antibody was investigated in large-scale clinical trials. This was E5®, a MAb from Xoma which also failed to demonstrate any benefit after encouraging initial results, and was not developed further [68]. The basic problem with anti-endotoxin antibodies has been recognized as being the small therapeutic window before inflammatory cytokine expression occurs [69]. Other strategies therefore focused on TNFα as the central mediator of inflammation following various stimuli, including endotoxins. Intracellular signal transduction occurs upon clustering of TNFα receptors, and this can be prevented by high-affinity neutralizing antibodies to the circulating TNFα trimers. This approach has been followed by Bayer with a murine MAb that also failed to demonstrate a significantly positive effect on patient mortality [70].

Failures such as those described above can ruin a company financially, or at least shatter its faith in biotechnology, as the development costs for a new biological entity are usually in the range of US$ 500–800 million, with late-stage clinical trials consuming the majority of this money (see Part IV, Chapter 16; Part VII, Chapter 4; and Part VIII, Chapter 1).

The fallout from these early cases persisted for a long time, and it took the entire biotech sector many years to recover from this negative impact. As a result of this consolidation, however, the sector is now stronger than ever, is generating a very positive news flow with a number of biopharmaceuticals that entered the market recently, and once again it is the antibody sector that is leading the way:

– TNF-α: this is still an important disease target, and a number of drugs are now on the market. In fact, these biological response modifiers (BRMs), directed towards specific cytokines, have changed the treatment of rheumatoid arthritis (RA) drastically and now represent the therapy standard for this condition [71]. Globally, some five million people have RA, with 1.25 million suffering from moderate-to-severe symptoms. Biological drugs represent a second-line therapy prescribed to the 50% of those patients who fail disease-modifying anti-rheumatic drugs (DMARDs). The therapies approved comprise three antibody-based drugs that target TNF-α [Enbrel® (Etanercept); Amgen Inc.; Remicade® (InflixiMAb); Centocor Inc., and Humira® (AdalimuMAb), Abbott), and one IL-1 receptor agonist that targets IL-1 (Kineret® (Anakinra), Amgen, Inc.) [72, 73]. (In the interim, Synergen was acquired by Amgen, and Antril was launched as Kineret to treat RA.) Interestingly, the efficacy and safety profiles for all four of these materials are different [74].

– Enbrel®: (etanercept) a recombinant human fusion protein of two soluble TNF-α type II receptor molecules and part of the Fc portion of an IgG1-antibody to circumvent rapid clearance and facilitate effector functionality. This represents an interesting class of fusion proteins that carry antibody domains for their specific function [75]. Inhibiting TNF-α has proven to provide a highly effective approach to the treatment of chronic inflammatory illnesses. Originally approved for RA, etanercept has now gained approval for additional indications, including psoriatic arthritis, juvenile rheumatoid arthritis, and ankylosing spondylitis. Enbrel was also a learning example as to what can happen when demand exceeds capacity and shortage of materials is limiting a drug's growth potential. Without sufficient manufacturing capacity, Immunex Corp. (the inventor of Enbrel) was forced into a takeover by Amgen.

– Another representative of this class of molecules is Amevive® (alefacept), a fusion protein consisting of the first LFA-3 extracellular domain fused to the hinge CH2 and CH3 regions of human IgG1 developed by Biogen [76, 77]. The molecule has immunomodulatory properties because it is targeting the co-stimulatory role of CD2 in T-cell activation as well as cell adhesion and the migration of lymphocytes into tissue.

A closer look at currently available MAbs inhibiting the inflammatory effects of TNF-a in diseases such as RA, Crohn's disease, psoriatic arthritis, bacterial septic shock, as well as in the prevention of alloreactivity and graft rejection, confirm their high potential [78]. Strategies include neutralization of the cytokine via either anti-TNF antibodies, soluble receptors, or receptor fusion proteins. Rheumatoid arthritis is currently addressed by no less than 10 antibody products that are either on the market or in late stage clinical development. The competitive edges for these candidates are certainly a low to non-existing antigenicity profile with no neutralizing antibodies detected in patients, and also a convenient mode of administration such as subcutaneous depot forms.

The anti-TNF biologics class is expected to continue steady growth within the RA market, and Lehman Brothers analysts predict that it will peak at sales of US$ 5–6 billion per annum in the US. From the pipeline, CDP-870 is the largest threat to the established brands.

Compared with RA, the market for biologics in psoriatic arthritis is in its infancy. According to Lehman Brothers, peak annual sales may reach US$ 2–3 billion. Many of the same biologics from the RA market are also now targeting psoriasis. Enbrel received approval for this indication in the

US in early 2002, and both Humira® and Remicade® are in pivotal Phase III trials.

Other lessons learned from the several high-profile clinical trial failures were the limitations of fully murine antibodies that may eliminate the therapeutic efficacy (neutralization, rapid clearance) or even lead to serious side effects (serum sickness, anaphylactic shock) [79]. Although a number of murine antibodies are still commercially available, the focus has shifted to chimeric, humanized and eventually human antibodies that are now becoming available. Whether maximizing the human sequence in an antibody is minimizing antigenicity is questionable, however. In fact, data on the generation of neutralizing antibodies in patients have demonstrated that some of the marketed humanized antibodies still exhibit significant immunogenicity – a phenomenon that is also known for other protein therapeutics of human origin and is probably related to other host cell-derived differences in the manufacturing process.

Other strategies to reduce the immunogenicity in therapeutic antibodies comprise the covalent attachment of PEG to mask antigenic sites through extensive hydration (see Part VI, Chapter 2), veneering (removal of exposed B-cell epitopes in the framework region), and removal of T-cell epitopes to avoid proliferation of helper T cells and stimulation of a mature immune response [80].

1.6
The Market Perspective

To date, a total of 22 MAbs and MAb-related proteins has been approved, and are being produced in different annual amounts ranging from the multi-gram to the hundreds of kilogram scale, and with

Table 1.3 Approved therapeutic monoclonal antibodies

Marketed product	Company	Antibody type	Year of first approval	Molecular target/ main indication
Anti-cytokine/anti-AB MAbs				
Remicade	Centocor/Schering Plough	Chimeric	1998	TNFα; Crohn's disease, Rheumatoid arthritis
Enbrel	Amgen/Immunex	TNF-receptor/Fc (IgG1) fusion protein	1998	TNFα; Rheumatoid arthritis
Humira	Abbott/CAT	Human	2002	TNFα; Rheumatoid arthritis
Xolair	Genentech/Novartis/ Tanox	Humanized	2003	IgE; Allergic asthma
Anti-cytokine-receptor MAbs				
Herceptin	Genentech/Roche	Humanized	1998	EGFR2; Breast cancer
Avastin	Genentech/Roche	Humanized	2004	VEGF; Various cancers
Erbitux	ImClone, BMS, E. Merck	Chimeric	2004	EGFR; Various cancers
MAbs to lymphocyte surface antigens				
OrtoClone OKT3	Johnson & Johnson	Murine	1986	CD3; Transplant rejection
Rituxan/ MabThera	BiogenIdec/Schering	Chimeric	1997	CD20; Non-Hodgkin Lymphoma (NHL)
Zevalin	BiogenIdec/Schering	Murine, linked to ^{90}Y	2002	CD 20; NHL
Bexxar	Corixa/GSK	Murine, linked to ^{131}I	2003	CD20; NHL
Zenapax	PDL/Roche	Humanized	1997	CD25 (IL-2 receptor); Transplant rejection
Simulect	Novartis	Chimeric	1998	CD25; Transplant rejection
ReoPro	Centocor/Eli Lilly	Fab Chimeric	1995	Platelet receptor; Anti-thrombotic
Antilfa	Immunotech/SangStat	Murine	1997	Anti-LFA-1a; CD11a; Transplant rejection
Mylotarg	Celltech/Wyeth	Humanized; linked to Calecheamicin	2000	CD33; AML
Campath	ILEX/Schering	Humanized	2001	CD52; B-CLL
Raptiva	Genentech/Xoma/ Serono	Humanized	2003	CD11a; Psoriasis
Amevive	BiogenIdec	LFA3-IgG1 fusion protein	2003	CD2; Psoriasis
MABs to cancer-associated antigens				
Panorex	Centocor/GSK	Murine	1995	Anti-HEA (EpCAM); Colorectal cancer
MAbs to pathogens				
Synagis	Medimmune/Abbott	Humanized	1998	RSV; RSV infection1

ton requirements coming into sight (Table 1.3).

The first humanized MAb was Synagis® (Medimmune), directed against an epitope on the surface of the respiratory syncitial virus (RSV). This was launched in 1998, and the first fully human MAb is now Humira® (Abbott/CAT), approved in 2002 for the treatment of RA.

Many MAbs are directed towards tumor-associated antigens or receptors that are up-regulated in malignancy (see Introduction and Part II, Chapter 4). In 1997, RetuxiMAb (Rituxan®, MabThera®) became the first cancer therapeutic monoclonal antibody [81]. In 1991, rituximab was developed as a chimeric anti-CD 20 antibody with non-Hodgkin lymphoma (NHL) as the original indication. Clinical trials were initiated in 1993 after Investigational New Drug (IND) filing in 1992, and lasted no longer than 3 years. The antibody binds to B cells where it induces CDC, ADCC, and apoptosis. It has been shown that Rituxan® functions via the recruitment of cytotoxic T cells to the tumor cells. Patients with a polymorphism in the Fc receptor that has higher affinity to the Fc part, have significantly better responses, demonstrating the importance of this interaction for therapeutic efficacy [82] (see Part I, Chapter 2). Today, the drug is being investigated for the treatment of various autoimmune disorders, and it is said to have high additional potential [83].

Another target is the receptor for the epidermal growth factor (EGF) that is involved in a number of solid tumor forms. Anti-EGF receptor antibodies are applied, together with traditional cytotoxic drugs.

One promising new approach to tumor treatment is the inhibition of angiogenesis as one primary disease target to slow down tumor growth and metastasis. Released pro-angiogenic factors trigger vascu-

larization of the primary tumor to provide nutrients and oxygen for the rapidly proliferating cancer cells. The first in class therapeutic is Avastin®, a humanized MAb that received market approval in 2004 for the treatment of first-line metastatic colorectal cancer, but only after some drawbacks in initial studies [84].

Herceptin® (trastuzumab) is a humanized antibody with antitumoral activity through the antagonistic inhibition of EGF-receptor activation [85] (see Part I, Chapter 5). EGF (also called HER1–4) represents a family of cytokine-regulated, tyrosine kinase-type receptors regulating an intracellular signaling cascade that can initiate cell regulation or apoptosis [86, 87]. HER1 and HER2 require dimerization for transduction signals to occur, and are therefore targets for MAbs [88, 89]. Herceptin binds to HER2 and triggers various immunological defense mechanisms. Today, Herceptin is approved as the first-line treatment for metastatic breast cancer in combination with chemotherapy for receptor-positive patients [90]. Erbitux® (cetuximab) is another representative of the EGFR-antagonist MAbs. This is a chimeric antibody that binds to HER1, where it prevents ligand-fixation, homodimerization and intracellular signal transduction, which in turn leads to cell death and tumor regression [91].

In addition, with Zevalin®, Bexxar®, and Mylotarg®, the first drug targeting devices gained market access for the specific treatment of neoplastic diseases such as NHL and acute myeloid leukemia (AML) [92].

Some of the now commercially available antibody-based products have "blockbuster" potential, and the overall development of the sector is very strong. MAbs represent the fastest growing segment within biopharmaceuticals, and are outperform-

■ Monoclonal Antibodies USD bn   ■ Estimated Demand kg

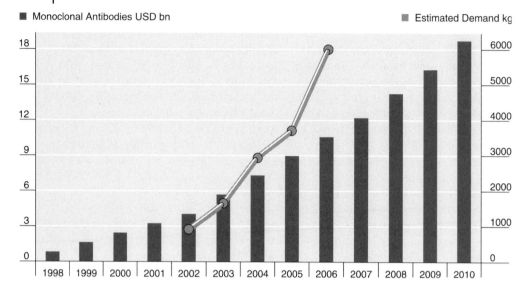

Fig. 1.7 Annual sales and manufacturing demands for marketed MAbs. (Data see Ref. [93].)

ing recombinant proteins with a compound annual growth rate (CAGR) of 20% [93] (see Part VIII, Chapter 1). In addition, there are hundreds of second- and third-generation antibody-based products in pre-clinical and clinical development, and even under the assumption of growing attrition rates, substitution pressure and margin squeeze, MAbs will probably reach a stable plateau of US$ 20 billion within 10 years from now, and the required amounts of material will rise accordingly [94].

Among the MAbs in late-stage clinical investigation, or queuing for regulatory approval, are Antegren® (Elan, Biogen MS), Tarceva® (Genentech, OSI Pharmaceuticals, and Roche) and CPD 870 (Celltech). Antegren is a monoclonal antibody for the treatment of multiple sclerosis and Crohn's disease, and received fast track review status in the US after only one year of Phase III human trials. Antegren® attaches to alpha-4-integrin and renders it unable to stick to VLA-4, thereby preventing chemo-taxis of helper T cells into the endothelium. Tarceva® blocks the EGF-receptor, and is under evaluation for the treatment of advanced non-small cell lung cancer (NSCLC). The drug has also undergone early-stage trials for a range of other tumor types, such as ovarian, colorectal, head and neck, glioma and gastrointestinal cancers. CPD 870 is an anti-TNFα antibody fragment with the indication for RA and Crohn's disease (both in Phase III trials).

1.7
Drug Targeting: The Next Generation in Cancer Treatment

Despite impressive advances in the treatment of specific cancer forms, the therapy of most solid tumor forms is still missing a major breakthrough, and is in a striking misbalance to the in-depth knowledge on carcinogenesis, tumor development and metastasis [95]. In traditional chemotherapy, physiological mechanisms (and the fact that the overall metabolism and proliferation of cancer cells are increased) are used for selective treatment. Selectivity, however, is rather poor because tumor-specific differences are quantitative rather than qualitative. This in turn leads to a therapeutic window that is very small, including the well-known side effects on all proliferating tissue such as bone marrow, lymphatic tissue, and gastrointestinal epithelium. Malignant growth is generated through transformation of the body's own cells. During malignant transformation, cells lose their differentiated status as well as their physiological role, including the basis for intracellular communication. Among various other changes, the expression of cell surface receptors is typically up- or down- regulated, and tumor-associated antigens (TAA) such as oncofetal antigens are presented [96]. These antigens differ in their copy number or structure, with glycosylation being the predominant change – a phenomenon that is also building the basis for the organotrophy of metastasis [97]. TAA may, on the other hand, provide molecular targets that are suitable docking stations for the binding and internalization of antibody–drug conjugates and immunotoxins [98]. Since the first appearance of MAbs, a large number have been developed as specific diagnostic agents for the detection as well as the serological, biochemical, and histological characterization of cancer cells and tissues [99]. Immunoscintigraphy with radiolabeled MAbs and antibody fragments is an appropriate tool for the evaluation of their biodistribution, and is a standard diagnostic method for the detection of primary tumor and metastasis with tumor markers such as the carcinoembryonic antigen (CEA) [100] (see Part V, Chapter 6). An ac-

ceptable compromise of rapid penetration and thus localization and clearance rate must be found for every application. The radioisotopes used provide a rapid decay rate (e.g., ^{99m}Tc, $t_{1/2} = 6.1$ h; ^{111}In, $t_{1/2} = 2.8$ days), and are covalently linked to the antibody with complexing agents such as diethylentriaminopenta-acetate (DTPA) [101]. Typically, less than 1% of the applied radioactivity is localized at the tumor after 6–8 hours. In that respect, immunoscintigraphy is a good tool for investigating the possibilities, as well as the limitations, of a therapeutic use [102, 103] (see Part V, Chapter 4).

Drug targeting with MAbs has always been a fascinating theoretical approach to the rational treatment of acquired diseases such as cancer, according to the concept of "Magic Bullets" – a term created by Paul Ehrlich who suggested the coupling of toxic natural compounds to specific vectors that are able to find their way to the disease and thereby increase the pharmaceutical index [104].

Specificity, dosimetric considerations as well as various pharmacological and toxicological problems, have so far limited the application of immunotoxins and radioimmunotherapy to life-threatening diseases where other treatment regimes have failed [105].

In the payload approach, different classes of cytotoxic agents have been immobilized to MAbs (Table 1.4) [106, 107]. Their toxic potential varies widely, and this must be considered in dosimetric calculations. With a stoichiometric mode of action, sufficient antibody-conjugates must recognize and enter malignant cells to facilitate cell killing, through whatever mechanism [108]. Tumor localization – and especially diffusion into solid tumors – is however relatively inefficient for large molecules, and huge doses must be administered in order to generate a therapeutic effect. Protein toxins are

Table 1.4 Cytotoxic drugs and toxins linked to monoclonal antibodies in drug-targeting studies

Substance	Origin
Small-molecule toxins and antimetabolites:	
Methotrexate (MTX)	Synthetic
Cytosin-1 β-D-arabinoside (ARA-C)	Synthetic
Daunomycin (DNM)	*Streptomcyces* sp.
Mitomycin C (MMC)	*Streptomyces* sp.
Protein toxins (plant-derived):	
Ricin A chain	*Ricinus communis*
Abrin	*Abrus precatorius*
Gelonin	*Gelonium multiflorum*
Momordin	*Momordica charantia*
Pokeweed Antiviral Protein (PAP)	Pikeweed
Protein toxin (bacterial-derived):	
Diphteria Toxin	*Cornyebacterium diphtheriac*
Pseudomonas Exotoxin	*Pseudomonas*
Radiomimetic drugs	
Bleomycin (BLM)	*Streptomyces verticillus*
Neocarcinostratin (NCS)	*Streptomyces carcinostaticus*
Calicheamicin	*Micromonospora echinospora*
Esperamicin	*Actinomadura verrucosospora*

usually very potent inhibitors of ribosomal protein translation and – due to their enzymatic mechanism of action – a single molecule can kill a cell [109]. Limitations arise from the fact that the slightest systemic cross-reaction has drastic consequences [110, 111]. In addition, protein toxins are highly immunogenic.

The ideal cytotoxic drug is a small molecule that is still very potent. A promising class of cytotoxic agents is a group of radiomimetic substances called enediyenes; these are secondary metabolites from *Streptomyces* that are minor groove DNA-binders and introduce double-strand breaks that may lead to apoptosis [112–114] (Fig. 1.8). Their anti-tumor potential has been investigated in a number of experimental studies [115–117].

The ideal antibody vector recognizes a tumor-specific antigen that facilitates receptor-mediated endocytotic uptake, and releases the cytotoxic drug inside the cell upon cleavage of the chemical linkage. Even if *in vitro* experiments demonstrate impressive selective toxicity, a number of additional drawbacks develop from a variety of pharmacological problems. Malignant tissue is not monoclonal (tumor cell heterogeneity), and cells from different parts of the tumor may escape damage through lack of antigen, or for anatomical reasons. In addition, certain antigens – shed from the tumor cell surface into the bloodstream – can neutralize immunoconjugates anywhere in the body. Serum stability is reduced due to proteolytic degradation.

The ideal linker for the covalent attachment of toxins to antibodies is a device that is stable in the bloodstream and cleavable upon uptake into the cellular compartment, where action is thought to take

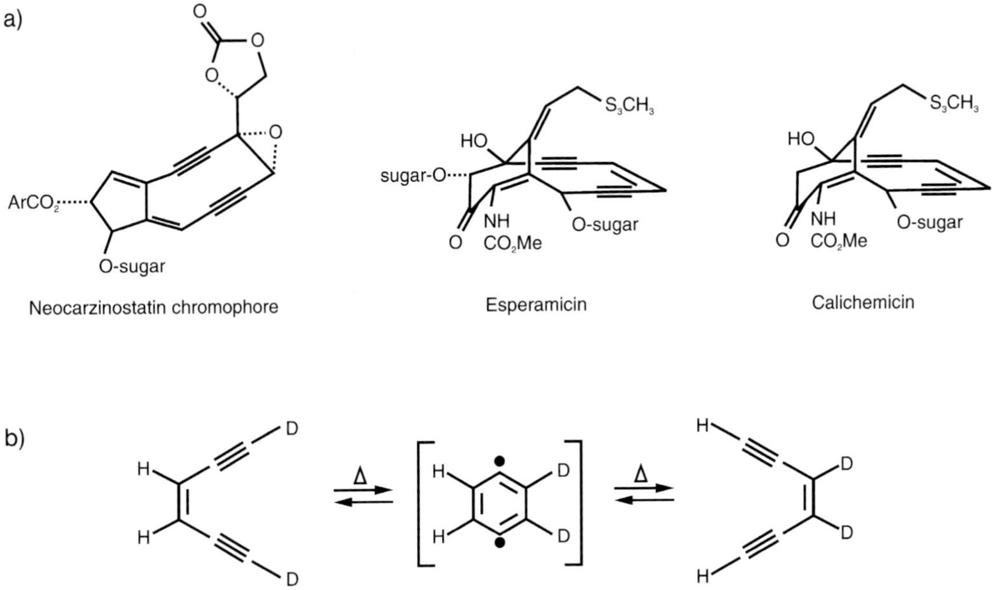

a)

Neocarzinostatin chromophore

Esperamicin

Calichemicin

b)

Fig. 1.8 (a) Structural DNA-inactivating motives in radiomimetic drugs; (b) Bergmann reaction of dehydrobenzenes [119, 120].

place. A number of approaches have been used based on heterobifunctional coupling agents; among these, an acid-labile hydrazone chemistry-based linker has shown the most promise. This principle represents the basis for Mylotarg®, a first-in-class conjugate of calicheamicin and a anti-CD33 antibody for the treatment of AML [118].

First-generation immunotoxins were hybrid molecules of bacterial toxins covalently linked to antibodies. In a number of studies, these turned out to be insufficient, with dose-limiting toxicities such as vascular leak syndrome, thrombocytopenia, and organ damage. Modern immunotoxins are recombinant fusion products of, for example, the variable domain (Fv) of a MAb, and a truncated bacterial toxin [121]. Recombinant immunotoxins are currently being developed to target a number of known hematologic and solid tumor antigens [122]. Various new strategies have been investigated, such as attacking tumor vascularization with immunotoxins [123].

Studies with radioactive isotopes have been performed to facilitate a therapeutic effect – a strategy called radioimmunotherapy [124]. Radiolabels include β-emitters such as ^{131}I, ^{47}Sc, and ^{90}Y [125] (see Part V, Chapters 4 and 5). When compared to immunoscintigraphy, the applied doses are higher in order to establish a cytotoxic effect. Even cells within the tumor that are inaccessible to the antibody can be killed without prior internalization (the "bystander effect").

The number of additional approaches using MAbs in experimental cancer therapy is very high, including bispecific antibodies, pro-drugs, and different forms of fragments and fusion proteins. In the pro-drug approach, a non-toxic precursor molecule is administered systemically and is converted into a toxic principle with the use of a pre-targeting antibody that is localized at the tumor site [126].

1.8
Developing a Manufacturing Process for MAbs

Biomanufacturing is a risky and fixed-cost-driven business, with the COG following a clear relationship with annual production demand. For a MAb with an yearly material requirement of approximately 100 kg, that results in an overall cost range in the area of 100 US$ per gram. Approximately 75% of this is related to fixed costs, and the remainder accounts for variable costs derived from process consumables [127] (see Part IV, Chapter 16).

Process development follows an overall development plan that coordinates all the various disciplines involved. It is typically performed in three cycles to allocate resources and thus manage product pipelines efficiently, and involves a scale-up of up to 1000-fold and more [128]. During the exploratory phase, a research method is used to generate small amounts of the respective MAbs to investigate target-specific effects *in vitro*. Important technical goals at this stage are feasibility and speed.

For the manufacture of clinical material, an IND-enabling method must be established under current Good Manufacturing Practice (cGMP) with validation of some critical elements. The whole process should be designed for robustness and scalability very early on.

A Biologics License Application (BLA)-enabling process is typically used for the manufacture of clinical Phase III material and the full-scale market supply after launch out of the final plant. This process

contains the full validation package for the Chemistry Manufacturing and Controls (CMC) part of the application dossier, including a virus safety concept (see Part I, Chapter 6). With transfer into the process scale, there is a shift of emphasis towards quality and process economy, and this results in an integrated, robust, cGMP-compliant manufacturing process. For macromolecules such as MAbs and recombinant proteins, reliable manufacturing including rational acceptance criteria is extremely important. Since the products are large and complex molecules with isoforms and microheterogeneities, the production process itself must be designed in a way that it meets the highest requirements with regard to consistency and reproducibility requested by regulatory authorities (see Part VII, Chapter 4).

In modern process development, process trains are compiled with generic modules that are predeveloped at scale rather than de-novo design for every new antibody. State-of-the-art tools such as design of experiments (DOE), process modeling and simulation are employed to adapt and optimize the individual steps, and to identify potential bottlenecks in the overall process which is vital for the best engineering solution. Process development is, however, a highly complex field that involves a number of different disciplines. Clear tasks and project structures are necessary to manage the various interfaces, including a rationale change control system towards the end of the development cycle. Given the complex nature of an antibody product, most process changes within an authorized manufacturing process are major ones leading to biochemical comparability and clinical bridging studies, if not to a new license application.

1.9
Routine Manufacture of MAbs

Monoclonal antibodies as biopharmaceuticals, and their manufacturing processes, are subject to the same basic requirements as any other drug, and the assessment of a licensing application focuses on the quality, safety, efficacy, and environmental risk of the product. Process-related questions such as the characterization of raw materials, in-process controls, change policies as well as process and test validation, are very demanding [129, 130]. As with other biological products, antibodies have considerably larger sizes and complexities when compared to chemicals, and an analysis of the finished product is not sufficient to control their quality and safety. A suitable process management, in-process control of

all critical parameters identified within process validation are decisive factors (see Part VII, Chapter 1).

The routine production of MAbs under cGMP became feasible with the implementation of cell culture and downstream processing in industrial biotechnology.

From the economic point of view, procedures for the manufacture of MAbs must be scalable and efficient and, at the same time, simple and robust.

Traditional methods such as the generation of mouse ascites are no longer accepted for quality reasons, and nowadays are banned in most countries. Antibody-secreting mammalian cells are cultivated in bioreactors under optimized conditions, with various parameters monitored on a continuous basis. The expression of recombinant MAbs in Chinese hamster

Mab concentration in mg/l

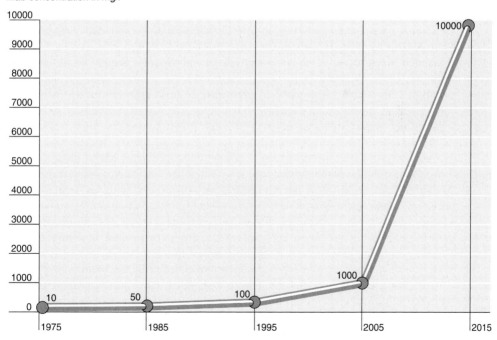

Fig. 1.9 Development of MAb expression rate in mammalian cell culture. (Data adapted from Ref. [131].)

Table 1.5 Physical properties of monoclonal antibodies compared to typical process-derived contaminants

Protein	Molecular weight	Isoelectric point
Monoclonal antibody	150 000	7.7–8.3
BSA	65 000	4.5–4.8
Transferrin	76 000	5.8
Insulin	5 700	5.3
Protein A	42 000	5.1
DNA	1 000–1 000 000	Highly negative
Pyrogens	100 000–1 000 000	Highly negative
Viruses		<7 for most strains

BSA: Bovine serum albumin.

ovary (CHO) cells is the industry standard [131] (see Part IV, Chapter 1). Other commonly used immortalized cell lines are the mouse myeloma-derived NS0, baby hamster kidney cells (BHK), and PerC6 derived from human kidney [132] (see Part IV, Chapters 1, 2, 3, 4, and 12). Fed-batch fermentation with expression rates in the gram MAb per L range is today the benchmark, and much higher levels are in sight (Fig. 1.9) [133]. Compared to the original processes, this represents a 1000-fold improvement in volumetric productivity, and in less time. Recent developments in cell culture have led to both higher cell densities all well as higher specific production rates. Despite the enormous advances that have been made in this field, there is still room for improvement, and developments towards higher biomass and more efficient gene transfer using the whole molecular genetic repertoire have only just begun [134].

In the downstream arena, process trains with the orthogonal combination of predeveloped generic modules are state of the art. Basic principles such as the Capturing-Intermediate Purification-Polishing (CIPP) strategy, as well as overall process design issues, are key in order to optimize quality, yield, and overall productivity [135]. From high dilution to high purity is the principle of a good downstream process, with the priority to concentrate and thus stabilize the product very early on and to remove critical impurities throughout the process. In this way, the focus is on selectivity and resolution towards the final steps of bulk manufacturing (drug substance). Although antibodies even from the same class and subclass are considerably different, the basic elements of the purification strategy have a generic character and allow for a rationale approach in process development (Fig. 1.10). Typically, every process step addresses a certain task in the overall strategy, and redundancy is avoided wherever possible.

Even closely related antibody molecules vary in their solubility and their stability to chemical and physical influences. Isoelectric points are typically between 5 and 9, mainly due to the different sialic acid contents. Purification techniques involved are not considerably different from the bioseparation of other proteins as they focus mainly on bioaffinity, charge, hydrophobicity and size (for an extensive review, see Ref. [136]). In modern bioseparation processes, initial recovery (capturing) and polishing are the most critical phases as they address the key features of biopharmaceuticals. A high dilution of the target molecule after biosynthesis is directly translating into higher cost of goods as compared to chemical drugs. In modern bioseparation processes, this is addressed by a rapid and efficient isolation of the product in a robust and productive capturing step to facilitate both rapid volume reduction as well as separation, and therefore stabilization of the antibody. High throughput and high dynamic capacity demands are driving the ap-

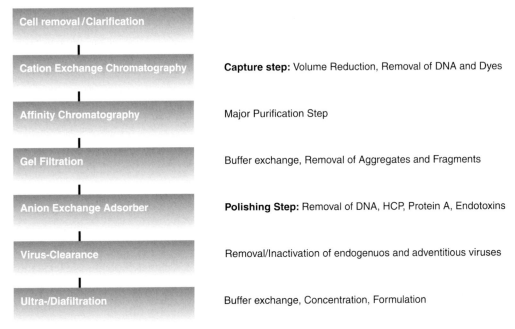

Fig. 1.10 A generic manufacturing strategy for monoclonal antibodies.

plications towards innovative "high-end" technology on the one hand, but also towards revisiting robust technology of the "low-end" type on the other hand.

In a typical procedure for the manufacture of a MAb, the first step is batch fermentation of mammalian cells from a comprehensively characterized "Master Working Cell Bank" (MWCB). The cells are harvested for further processing of the cell-free feed stream. The UF-TCF is applied to a "capture step". From a processing standpoint, the harvest is a highly diluted feed stream that causes many handling issues as long as its volume is not reduced by a productive, high-throughput technology. Modern capturing supports thus combine some selectivity for initial purification with excellent flow properties (low back-pressure at high linear flow rates), and can be used for feed streams of up to 500 column volumes (cv) and with

linear velocities far above $1000 \, \text{cm h}^{-1}$. Concepts in the initial recovery of MAbs are numerous, but none of these has yet provided a major breakthrough.

Early on or as part of the intermediate purification, an affinity chromatography step is always the workhorse that provides very high selectivity for the target molecule [137, 138].

The benchmark for the purification of most IgG species is Protein A, a 42 kDa molecular weight protein derived from a strain of *Staphylococcus aureus*. The natural molecule is located on the outer membrane surface of *Staphylococcus aureus*, and is linked to the cell surface through its C-terminal region [139]. This allows the pathogen to bind the Fc part of the IgG and re-direct the antigen-binding domain. This prevents interaction of the Fc part of the antibody with effector cells, thereby circumventing the activation of the immune system.

Table 1.6 Typical release specifications for monoclonal antibodies [28]

Host cell and media proteins (HCP)
• ppm-level

Process-related impurities (e.g., Protein A)
• ppm-level

Nucleic acids
• 10–100 pg/dose

Endotoxin (LPS)
• 5.0 EU kg^{-1} body weight h^{-1}

Microorganisms
• absence of detectable bacteria, fungi, yeast and mycoplasma

---

Protein A consists of a single polypeptide chain which is structurally and functionally composed of two parts. The N-terminal region contains four homogeneous subregions, each of which can bind human IgG molecules, with a total of two active binding sites at a time (approx. $K_a = 10^8$ M^{-1}) [140].

Today, however, a recombinant form of Protein A which is a genetically truncated version with a molecular mass of ca. 32 kDa has been developed. Non-essential regions were removed from the C-terminus, resulting in a protein containing 301 amino acids, 28 of which are lysines. It exhibits the same affinity for IgG molecules as the native protein, but has considerably lower non-specific binding properties. Originally, the main step in the isolation process of natural Protein A was an affinity purification on immobilized virus-inactivated human IgG. With recombinant Protein A, this is no longer required, and has led to an increased virus safety of the product as compared to the conventional material.

Immobilized Protein A has been used extensively as an affinity support for the purification of a wide variety of IgG molecules from many different species of mammals [141].

For chromatographic use as an affinity ligand, Protein A is chemically immobilized to a base matrix through NH$_2$- or SH-groups of the protein [142]. The protein is genetically engineered, allowing for a site-directed immobilization through introduced, N-terminal bridging groups.

Protein A as a biological reagent is still raising concerns as to the high cost of the matrix, the problem of leakage into the product, the cleanability after repeated use, and especially the sensitivity to caustic solutions that are widely used as bacteriostatic agents [143]. Most recent variants of protein A supports focus on the long-term stability in caustic cleaning agents such as 0.1 M NaOH and increased dynamic capacity [144, 145].

A number of other immunoglobulin-binding proteins have also been identified and characterized, including Protein G from *Streptococcus pyogenes* and Protein L from *Peptostreptococcus magnus*, and the genetically engineered Protein Z which offer different specificities and can be used for other antibody classes and also fragments [146–148]. However, protein-affinity sorbents form the basis of a significant part of the variable costs in manufacturing, and a number of strategies have been evaluated to identify chemical pseudoaffinity ligands from which MAbs can be bound and eluted selectively [149–151]. Various biomimetics directed to the Protein A binding domain in the C$_H$2-domain based on peptide libraries, combinatorial chemistry and rational ligand design have been synthesized and studied for MAb purification, and also to increase their visibility [152].

For the polishing step, it is vital to achieve final purification. The biomanufacturing environment provides an excellent

Conventional bead   **Membrane Adsorber**

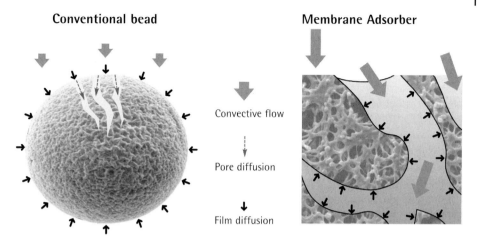

Convective flow

Pore diffusion

Film diffusion

Fig. 1.11 Structural features of membrane adsorbers versus resin-based media. (From Ref. [154].)

basis for the growth of all kinds of organisms and their metabolic products such as viruses, DNA, host cell proteins and endotoxins, as well as process-derived contaminants and impurities. In addition, the products are complex macromolecules with isoforms and microheterogeneities that require state-of-the-art production processes to meet consistency and reproducibility demands. Reliable polishing tools must provide a generic platform type of tool that works flexibly, reliably, and according to highest safety standards.

Polishing can be realized with high resolution; for example, in the removal of isoforms and other microheterogeneic variants, or through selective removal for host cell-derived or process-related contaminants such as host cell proteins (HCP), DNA, RNA, endogenous viruses, adventitious viruses, media components, endotoxins, Protein A, microorganisms, prions, and other human pathogenic agents [153]. Most contaminants have an acidic PI, and can be removed through either binding the antibody on a cation-exchange (CEX) support and further downstream in a flow through mode of an anion-exchange (AEX)

chromatography at very high linear velocities (Fig. 1.11) [154].

As chromatographic supports are diffusion-limited, and flow-through columns must be designed based on the flow rate rather then on binding capacity, membrane adsorbers with a more open structure and virtually no diffusion limitation provide a robust alternative to remove a whole set of contaminants in one step. In this respect, a number of concepts have been described (Fig. 1.11) [155]. The maximum separation power with membrane chromatography is achieved at high dilution of the target molecules and thus in capturing and polishing [156]. In addition, these membrane-based tools can be discarded after a single use; this makes them of great interest from a cleanability, as well as from a process economy standpoint [157].

For final product analysis, evidence must be provided with validated bioanalytical quality control methods that, besides correct identity and homogeneity, critical impurities have been reduced below specified limits [158, 159] (see Part I, Chapter 6 and Part VII, Chapter 1). For a validated, product-related HCP assay, mock fermentation and purification runs are performed to obtain an antigen for the production of an antiserum with a sensitivity in the 10–100 p.p.m. range [160].

Monoclonal antibodies derived from mammalian cell culture require a reliable virus safety concept that comprises three levels:

- Characterization of host cell line and raw material sources.
- Testing of product from various manufacturing stages.
- Validation of the virus clearance capability of the production process.

Since most immortalized cell lines have been shown to express endogenous retrovirus-like particles that may or may not be replication-competent and infectious, and other virus classes may also be present in mammalian cells, the risk for a virus contamination of the end product is inherent. In addition, adventitious viruses may come into contact with the product during processing. In any case, cell culture processes must demonstrate the robust and reliable ability to eliminate viruses in a risk-based approach [162] (see Part IV, Chapter 1). Virus validation studies employ model viruses that are relevant in representing known risks from the sources involved [163, 164].

For the clearance of enveloped and non-enveloped viruses, today's requirements ask for an orthogonal combination of methods that are based on the different physical principles of removal and inactivation, and are complementary to each other [165]. Virus filtration and solvent/detergent treatment are state of the art for removal and inactivation [166, 167]. Partitioning steps are considered less robust, as they are somewhat influenced by the actual process conditions. In any case, scaled-down models must be designed

carefully to represent the process conditions; moreover, aging of the support during column lifetime must be considered appropriately [168].

Among the different methods for virus inactivation, high-temperature short-time (HTST) has been used for retrovirus clearance, whereas ultra-violet irradiation is most powerful for the elimination of small, non-enveloped and otherwise very resistant viruses such as porcine parvovirus (PPV) [169–171].

1.10
Glycosylation and Other Post-translational Modifications

Immunoglobulins are glycoproteins that contain 3–12% carbohydrates [172]. In an IgG molecule, the sugar part is N-linked to a highly conserved site at Asn297 in the C_H2 domain of both heavy chains [173]. The complex carbohydrate has a biantennary structure with a pentasaccharide core and variable sugar residues (fucose, mannose, GlcNAc, sialic acid) [174] (see Part IV, Chapters 2 and 7). Although N-linked glycosylation does not interfere with antigen recognition, a number of implications are linked to this functionality such as stability, pharmacokinetics, antigenicity, Fc-related effector functions, and serum stability of antibodies [175]. For example, the removal of sialic acid variants results in a drastically reduced half-life and increased liver uptake through the asialoglycoprotein receptor which is responsible for the recycling of mature glycoconjugates. High-mannose variants lead to rapid clearance *in vivo*.

Different post-translational modification may lead to the synthesis of different glycosylation variants within a cell clone, and even within a single cell. The glycosylation

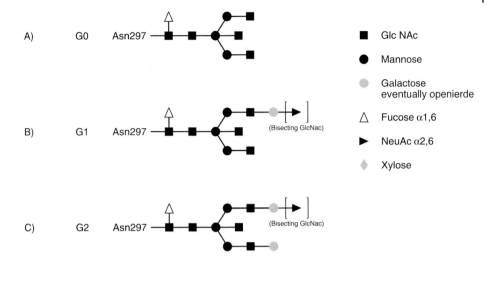

Fig. 1.12 Glycosylation pattern of an human IgG-antibody. (A–C) The Asn297 N-linked oligosaccharide is a biantennary core complex (GlcNac2-Man3GlNac) with variations such as a third bisecting GlNac, different numbers of terminal ga-lactose residues, and attachment of a terminal sialic acid that influence the effector functions and pharmacological properties of the antibody [179]. Plant core glycosylation is shown for comparison (D).

pattern of a MAb is therefore a sensitive tool to demonstrate the effective control of critical parameters within a manufacturing process, and is also a central element in biochemical comparability studies during scale-up from laboratory scale to 10 000-L fermentation runs (Fig. 1.12) [176, 177]. This is further driven by modern protein analytical tools, as well as the requirement for multi-site production of large amounts of MAbs to supply the growing markets. Glycosylation as a complex biosynthetic event is largely influenced by cell culture conditions such as the cell bank, the fermentation process with its various control parameters, and the medium composition [178]. Other intrinsic differences arise from the endogenous properties of the expression system related to the type of gly-coenzymes and sialic acids they are using.

Even within a validated manufacturing process, microheterogeneities in the glycosylation pattern with truncated forms and different overall monosaccharide compositions must be accepted, as long as predetermined acceptance criteria with specified ranges are met. It is vital that these specifications are valid throughout the clinical development and remain unchanged after the market launch of a biopharmaceutical. This requirement usually excludes major changes in the manufacturing process, such as a different expression system or cultivation method, unless new preclinical and clinical studies are performed which basically results in a new product development.

Whether a full-length antibody (including the full carbohydrate functionality) is required for the biological function of an antibody should eventually determine the decision for the appropriate expression and manufacturing system. In its aglycosylated form, an antibody can support all antagonistic functions that are required, without the need for a human-like glycosylation. Mutants can easily by expressed in a number of host cells, without the need for additional post-translational modifications. If, however, the carbohydrate functionality is required, hybridoma or myeloma expression systems (for genetically engineered antibodies) are the expression systems of choice for full-length antibodies [180]. Most of these cell lines are of rodent origin, and this accounts for various host cell-related differences in the glycan pattern, with the use of different forms of sialic acid being the predominant factor. Today, CHO cells represent the industry standard for the production of therapeutic proteins, as this cell line is able to perform most glycosylation steps comparable to the human profile. Recent progress in fermentation development has reported the possibility of routine antibody production in the $3–5 \text{ g L}^{-1}$ range, with further room for improvement. Increase of cell growth and yield optimization is interfering with the overall metabolism, including the induction of glycosyltransferases and oligosaccharide formation. Even the use of human cell lines cannot guarantee a reproducible pattern because of the high influence of the manufacturing process parameters. One aspect of modern cell line development focuses on the glycosylation pattern, as overexpressed glycoprotein tend to contain truncated oligosaccharide chains due

to incomplete post-translational modification [181]. Cell lines with overexpressed glycosyltransferases have the potential to generate a greater homogeneity through the reduction of terminal GlcNAc and increase of sialylation [182].

1.11
Emerging Issues in MAb Production

Where do we go from here in the manufacture of MAbs, and what are the major trends? Full pipelines further driven by genomic research and still unmet medical needs, limited GMP manufacturing capacity, growing competition between companies and products, economic problems of the healthcare systems, higher quality demands: these are the driving forces towards higher efficiency and productivity, and this also holds true for MAbs [183, 184]. As the biotech industry is maturing and facing a significant consolidation, efficient development and use of technology is an important factor in both upstream and downstream processing. Furthermore, the complexity of biotech products is reflected in the rapid progress and growing complexity of analytical development that allows for comparability studies between different production versions, scales and manufacturing sites, and this will eventually build the basis for biogeneric antibodies. Product development and revenue generation is a major driver for the growth of biopharmaceutical companies. Low cost of goods, higher expression rates, optimized product yields and robust, scalable manufacturing operations without compromising product quality are the key aspects.

The current situation in biomanufacturing is characterized by the fact that fermentation development is setting the pace in terms of productivity – a fact that will be further accentuated with the use of transgenic organisms. Innovative downstream processing technology that has the potential to accommodate these improvements is desperately needed, and trends are in sight to tackle the current backlog in bioseparation. Since process economics matters more and more, it can no longer be ignored that downstream processing costs account for up to 80% of the overall production costs [185].

A number of drivers force the downstream technology development towards innovative "high-end" applications on the one hand, but also towards revisiting robust technology of the "low-end" type that works well with small molecules and commodity proteins on the other hand. Examples are precipitation, crystallization, and filtration to name but a few. A robust downstream process must be simple, with quality, yield, productivity, and overall economics as the primary goals and an integrated design with compatible unit operations.

In addition, whole-process design with the integration of upstream and downstream processing is becoming the preferred scenario. The adoption of simple generic platform technology modules rather than the de-novo design of enabling processes provides the basis for economically viable biomanufacturing, and will eventually be vital for the success of MAbs and recombinant proteins.

At present, the sector is lacking the infrastructure for the manufacture of antibodies to come. Despite major investments however, the high demands and long lead times prohibit a short-term solution other than a significant increase in overall productivity to close the current gaps in biopharmaceutical supply chains.

Alternative expression systems for antibodies and fragments may provide the basis for high productivity and very large-scale applications [186]. For antibody fragments, microbial expression in *Escherichia coli*, *Pichia pastoris* and *Hansenula polymorpha* is state of the art [187–190]. Transgenic animals and plants combine the advantage of full-length expression of MAbs with very high biomass generation, and thus a cell density which is two orders of magnitude higher than that in any bioreactor [191, 192]. Some of these concepts were introduced years ago, but in particular animals suffer from certain unsolved issues such as pathogen threat and complex matrices that are difficult to process [193]. As a result, no biopharmaceutical obtained from a transgenic source is yet available commercially [194], though this will change as soon as ATryn® is approved (see Part IV, Chapter 11).

Species-specific glycosylation is another problem with transgenic production of human MAbs where Fc-related properties are required. Plant-derived glycoproteins contain beta-linked xylose sugars that may have some immunogenic potential [195]. With growing volumetric demands for antagonistic antibodies and economic concerns, plant-based expression is certainly an option, as discussed in detail in a recent overviews [196, 197]. Full-length expression of aglycosylated MAbs in non-edible plants such as alfalfa, tobacco, lemna, safflower, and moss will therefore become a vital alternative to other production systems [198, 199]. Industrial-scale culture of plant cells is another option [200]. Very large-scale applications for MAbs with annual productions of more than 1000 kg are in sight for the chronic treatment of autoimmune diseases, and also for topical and oral administration as vaccines and anti-infective agents. Uses are also likely outside the medical field, in industrial areas such as catalytic and immunopurification reagents [201]. It is unlikely that the mammalian cell culture which serves as the standard technology of today is capable of overcoming some of the inherent limitations such as the scale-up, long culture periods, high capital costs and sterility requirements of these very large-scale applications.

Another remarkable paradigm shift in biomanufacturing has been a clear trend towards disposable manufacturing [202]. Change-over procedures and cleaning validation is an area of concern for auditors, and it is not surprising that approximately 40% of all 483 citations fall into this category (see Part VII, Chapter 4). Several additional issues are also apparent, with reusable equipment such as upfront hardware investment on the basis of preliminary data, inflexibility in the use of space, handling, downtime during validation and change-over, and utility consumption with WFI being the greatest cost drivers [203]. The use of disposables such as bags (see Part IV, Chapter 14) instead of tanks, plastic tubing instead of hard piping, and membrane adsorber capsules instead of chromatography hardware, is not the only answer to today's downstream problems, but it may be a life-saving step for companies that cannot afford to lock-up their resources into long-term hardware investments.

1.12
The Future of MAbs

What the future will hold for MAbs is open to speculation. At present, MAbs are under investigation in a variety of tumor forms, allergic disorders, and in diseases characterized by chronic inflammation. In addition, a number of strategies are focusing on the prevention of infectious diseases.

Current limitations in the therapeutic use of MAbs are derived from the fact that they are pathogen-specific and that, for the time being, they require systemic, parenteral, and sometimes chronic application. In addition, antagonistic principles require rather large doses in excess of $1 \, \text{mg} \, \text{kg}^{-1}$ body weight and this results in treatment costs of up to US$ 10 000 per patient and therapy. For example, Avastin® treatment is priced at US$ 4400 per month.

Although efficacy data relating to biopharmaceuticals are typically fewer in number than for small molecules, MAbs address chronic and life-threatening diseases where the treatment standard is often poor, and even small benefits can be of high added value to the patient.

A number of areas are beginning to contribute to the overall success story however, among which are new strategies emerging from pharmacogenomics that allow for the redefinition of disease targets and a focus on subsets of patients for optimized benefit. Further progress in the treatment of multifactorial diseases will be possible through the consideration of individual genetic profiles in the design of clinical studies. Together with diagnostic tests to screen for an overexpressed target receptor, rational therapies can be directed towards patient populations, and with a high response rate (see Part I, Chapters 2 and 5).

The engineering of antibody affinity offers another approach to reduce the amount of MAb administered systemically to generate a therapeutic effect (see Part V, Chapter 2). Complement activation, Fc-receptor binding, and antigen recognition are the basic functions to be optimized. A number of affinity maturation methods have been described to generate MAbs with sub-nanomolar affinities to optimize their neutralizing activity or their biological function, although a direct correlation does not exist in any case [204, 205]. Cytotoxic antitumor MAbs such as Herceptin® (see Part I, Chapter 5) and Rituxan® rely on a complex interaction with different Fcγ receptor subtypes [206]. Selective improvement of these effector functions involves the engineering of the peptide backbone in the Fc-part, as well as the glycosylation pattern of the Fc region [207, 208]. A number of experimental studies have been conducted to influence complement binding and ADCC through specific engineering of the carbohydrate moiety in order to investigate the efficacy of therapeutic antibodies [209–211].

In future, many antibody-derived molecules will be seen in many indications, in addition to single chain Fv fragments (scFv). Dimeric and tetrameric mini-antibodies, diabodies, single antibody domains, Fv fragments fused to toxins, cytokines, antibiotics or radionuclides, protein scaffolds with hypervariable antibody domains, and related molecules such as anticalines will emerge to complement the toolbox of affinity-based agents and enhance their therapeutic efficacy.

Moreover, development towards application routes, formulation and dosing offer additional therapeutic potential, and will also improve tolerability and decrease treatment costs.

With further maturation of the antibody sector, the areas of indication will increasingly involve non life-threatening disorders. In particular, disease areas with a high level of unmet medical needs and the potential for billion dollar markets are attractive for second- and third-generation products. On the basis of robust large-scale manufacturing processes, MAbs can be produced in bulk quantities, and have the potential eventually to be used as therapeutic agents in commodities such as toothpastes and shampoos [212]. In overall terms, antibodies have a bright future and are the mainstay of modern biopharmaceuticals.

References

1 Jerne, K. J., The Immune System. *Sci. Am.* **1973**, *229*, 52–60.
2 Von Behring E, Kitasato S, *Deutsche Med. Wochenschr.* **1890**, *16*, 1113.
3 Edelmann, G. M, Antibody Structure and Molecular Immunology. *Science* **1973**, *180*, 830–834.
4 Roitt, L., Rabson, A., Essential immunology. *Blackwell Publishing*, UK, **2000**.
5 Tonegawa, S, The molecules of the immune system. *Sci. Am.* **1985**, *253*, 104–116.
6 Padlan, E. A, Anatomy of the antibody molecule. *Mol. Immunol.* **1994**, *31*, 169–217.
7 Hulett, M. D, Hogarth, P. M, Molecular basis of Fc receptor function. *Adv. Immunol.* **1994**, *57*, 1–127.
8 Radaev, S., Motyka, S., Fridmann, W. H., Sautes-Fridmann, C., Sun, P. D., The structure of a human type III Fcgamma receptor in complex with Fc. *J. Biol. Chem.* **2001**, *276*, 16469–16477.
9 Jefferis, R., Lund, J., Glycosylation of antibody molecules. Structure and functional significance. *Chem. Immunol.* **1997**, *65*, 111–128.
10 Suralia, A., Interaction of Protein A with the domains of the Fc-Fragment. *Trends Biochem. Sci.* **1982**, *7*, 318–323.
11 Tonegawa, S., Somatic generation of antibody diversity. *Nature* **1983**, *302*, 573.
12 Ballow, M., Berger, M., Bonilla, F. A., Buckley, R. H., Cunningham-Rundles, C. H., Fireman, P., Kaliner, M., Ochs, H. D., Skoda-Smith, S., Sweetser, M. T., Taki, H., Lathia, C., Pharmacokinetics and tolerability of a new intravenous immunoglobulin preparation, IGIV-C, 10%, Gamunex, 10%. *Vox Sanguinis* **2003**, *3*, 202–210.
13 Bayry, J., Misra, N., Latry, V., Prost, F., Delignat, S., Lacroix-Desmazes, S., Kazatchkine, M. D., Kaveri, S. V., Mechanisms of action of intravenous immunoglobulin in autoimmune and inflammatory diseases. *Transfusion Clinique et Biologique* **2003**, *10(3)*, 165–169.
14 Lebing, W., Remington, K. M., Schreiner, C., Paul, H.-I., Properties of a new intravenous immunoglobulin, IGIV-C, 10%, produced by

virus inactivation with caprylate and column chromatography. *Vox Sanguinis* **2003**, *84(3)*, 193–201.

15 Trejo, S. R., Hotta, J. A., Lebing, W., Stenland, C., Storms, R. E., Lee, D. C., Li, H., Petteway, S., Remington, K. M., Evaluation of virus and prion reduction in a new intravenous immunoglobulin manufacturing process. *Vox Sanguinis* **2003**, *84(3)*, 176–187.

16 http://www.gamunex.com/web_docs/ IGIV%20Comp%20Table%2Epdf

17 Köhler, G., Milstein, C., Continuous cultures of fused cells secreting antibody of predefined specificity. *Nature* **1975**, *256*, 495–497.

18 Shawler, D. L., Bartholomew, R. M., Smith, L. M., Human immune response to multiple injections of murine monoclonal IgG. *J. Immunol.* **1985**, *135*, 1–12.

19 Ewert, S., Honegger, A., Pluckthun, A., Stability improvement of antibodies for extracellular and intracellular applications, CDR grafting to stable frameworks and structure-based framework engineering. *Methods* **2004**, *34(2)*, 184–199.

20 Clark, M., Antibody humanization, A case of the 'Emperor's new clothes'? *Immunol. Today* **2000**, *21(8)*, 397–402.

21 Little, M., Kipriyanov, S. M., Le Gall, F., Moldenhauer, G., Of mice and men, hybridoma and recombinant antibodies. *Immunol. Today* **2000**, *21(8)*, 364–370.

22 Lonberg, N., Taylor, L. D., Harding, F. A., Trounstine, M., Higgins, K. M., Schramm, S. R., Kuo, C. C., Mashayekh, R., Wymore, K., McCabe, J. G., Antigen-specific human antibodies from mice comprising four distinct genetic modifications. *Nature* **1994**, *368*, 856–859.

23 Green, L. L., Antibody engineering via genetic engineering of the mouse, XenoMouse strains are a vehicle for the facile generation of therapeutic human monoclonal antibodies. *J. Immunol. Methods* **1999**, *231(1-2)*, 11–23.

24 Winter, G., Griffith, A. D., Hawkins, R. E., Hoogenboom, H. R., Making antibodies by phage display technology. *Annu. Rev. Immunol.* **1994**, *12*, 433–455.

25 Hoogenboom, H. R., Overview of antibody phage-display technology and its applications. *Methods Mol. Biol.* **2002**, *178*, 1–37.

26 Arndt, K. M., Jung, S., Krebber, C., Pluckthun, A., Selectively infective phage technology. *Methods Enzymol.* **2000**, *328*, 364–388.

27 Ostendorp, R., Frisch, C., Urban, M., Generation, engineering and production of human monoclonal antibodies using HuCAL®. In: *Antibodies, Volume 2, Novel Technologies and Therapeutic Use*, G. Subramanian (Ed.), Kluwer Academic New York, **2004**.

28 Feldhaus, M. J., Siegel, R. W., Yeast display of antibody fragments, a discovery and characterization platform. *J. Immunol. Methods* **2004**, *290(1-2)*, 69–80.

29 Schaffitzel, C., Hanes, J., Jermutus, L., Pluckthun, A., Ribosome display, an *in vitro* method for selection and evolution of antibodies from libraries. *J. Immunol. Methods* **1999**, *231(1-2)*, 119–135.

30 Stockwin, L. H., Holmes, S., The role of therapeutic antibodies in drug discovery. *Biochem. Soc. Trans.* **2003**, *31(2)*, 433–436.

31 Glover, D., Humphreys, D., Antibody Fragments. Production, purification and formatting for therapeutic applications. In: *Antibodies, Volume 1, Production and Purification*, G. Subramanian (Ed.), Kluwer Academic New York, **2004**.

32 Hudson, P. J., Souriau, C., Engineered antibodies. *Nat. Med.* **2003**, *9(1)*, 129–34.

33 Fersht, A., Winter, G., Protein engineering. *Trends Biochem. Sci.* **1992**, *17(8)*, 292–295.

34 Weir, A. N., Nesbitt, A., Chapman, A. P., Popplewell, A. G., Antoniw, P., Lawson, A. D., Formatting antibody fragments to mediate specific therapeutic functions. *Biochem. Soc. Trans.* **2002**, August 30, 512–516

35 Wörn, A., Plückthun, A., Stability engineering of antibody single-chain Fv fragments. *J. Mol. Biol.* **2001**, *305*, 989–1010.

36 Kortt, A. A., Dolezal, O., Power, B. E., Hudson, P. J., Dimeric and trimeric antibodies, high avidity scFvs for cancer targeting. *Biomol. Eng.* **2001**, *18*, 95–108.

37 Zapata, G., Ridgway, J. B. B., Mordenti, J., Osaka, G., Wong, W. L. T., Bennett, G. L., Carter, P., Engineering linear F(ab')$_2$ fragments for efficient production in *Escherichia coli* and enhanced antiproliferative activity. *Prot. Eng.* **1995**, *8*, 1057–1062.

38 Kriangkum, J., Xu, B., Nagata, L. P., Fulton, R. E., Suresh, M. R., Bispecific and bifunctional single chain recombinant antibodies. *Biomol. Eng.* **2001**, *18*, 31–40.

39 Kufer, P., Lutterbüse, R., Baeuerle, P. A., A revival of bispecific antibodies. *Trends Biotechnol.* **2004**, *22*, 238–244.

40 Hamers-Castermann, C., Atarhouch, T., Muyl-dermans, S., Robinson, G., Hamers, C., Son-ga, E.B., Bendahman, N., Hamers, R., Natu-rally occurring antibodies devoid of light chains. *Nature* **1993**, *363*, 446–448.

41 Nguyen, V.K., Desmyter, A., Muyldermanns, D., Functional heavy-chain antibodies in Ca-melidae. *Adv. Immunol.* **2001**, *79*, 261–296.

42 Muyldermanns, S., Single domain camel anti-bodies: current status. *J. Biotechnol.* **2001**, *74*, 277–302.

43 Holt, L.J., Herring, C., Jespers, L.S., Woolven, B.P., Tomlinson, I.M., Domain antibodies, proteins for therapy. *Trends Biotechnol.* **2003**, *21*, 484–490.

44 Batra, S.K., Jain, M., Wittel, U.A., Chauhan, S.C., Colcher, D., Pharmacokinetics and bio-distribution of genetically engineered antibod-ies. *Curr. Opin. Biotechnol.* **2002**, *13*, 603–608.

45 Chapman, A.P., Antoniw, P., Spitali, M., West, S., Spehens, S., King, D.J., Therapeutic antibody fragments with prolonged in vivo half-lives. *Nat. Biotechnol.* **1999**, *17*, 780–783.

46 Chapman, A.P., PEGylated antibodies and antibody fragments for improved therapy: a re-view. *Adv. Drug Deliv. Rev.* **2002**, *54*, 531–545.

47 Rose-John, S., Schooltink, H., CDP-870. Cell-tech/Pfizer. *Curr. Opin. Investig. Drugs* **2003**, *4*, 588–592.

48 Whiss, P.A., PexelizuMAb Alexion. *Curr. Opin. Investig. Drugs* **2002**, *3*, 870–877.

49 Verma, R., Boletti, E., George, A.J.T., Anti-body engineering: comparison of bacterial, yeast, insect and mammalian expression sys-tems. *J. Immunol. Methods* **1998**, *216*, 165–181.

50 Fischer, R., Drossard, J., Emans, N., Comman-deur, U., Hellwig, S., Towards molecular farm-ing in the future: *Pichia pastoris*-based produc-tion of single-chain antibody fragments. *Bio-technol. Appl. Biochem.* **1999**, *30*, 117–120.

51 Humphreys, D.P., Production of antibodies and antibody fragments in *Escherichia coli* and comparison of their functions, uses and modi-fications. *Curr. Opin. Drug Discov. Dev.* **2003**, *6*, 188–196.

52 Borrebaeck, C.A., Antibodies in diagnostics – from immunoassays to protein chips. *Immu-nol. Today* **2000**, *21*, 379–382.

53 Holt, L.J., Enever, C., de Wildt, R.M., Tomlin-son, I.M., The use of recombinant antibodies in proteomics. *Curr. Opin. Biotechnol.* **2000**, *11*, 445–449.

54 Wade, H., Scanlan, T.S., The structural and functional basis of antibody catalysis. *Annu. Rev. Biophys. Biomol. Struct.* **1997**, *26*, 461–493.

55 Groner, B., Hartmann, C., Wels, W., Thera-peutic antibodies. *Curr. Molec. Med.* **2004**, *4(5)*, 539–547.

56 Jung, C., Müller Eberhard, H.J., An *in vitro* model for tumour immunotherapy with anti-body heteroconjugates. *Immunol. Today* **1988**, *9*, 257–265.

57 Stahel, R., Pass, M., Zangemeister-Wittke, U., Cell surface antigens as therapeutic targets. *Lung Cancer* **1997**, *18*, 8–18.

58 Proimos, G., Platelet aggregation inhibition with Glycoprotein IIb–IIIa inhibitors. *J. Thrombosis Thrombolysis* **2001**, *25*, 99–110.

59 Wenzel, R., Anti-endotoxin monoclonal anti-bodies – a second look. *N. Engl. J. Med.* **1992**, *326*, 1151–1153.

60 Parillo, J.E., Management of septic shock, present and future. *Ann. Intern. Med.* **1991**, *115*, 491–493.

61 Selby, P., Hobbs, S., Viner, C., Tumour necro-sis factor in man, clinical and biological obser-vations. *Br. J. Cancer* **1987**, *56*, 803–808.

62 Oh, H.M.L., Emerging therapies for sepsis and septic shock. *Ann. Acad. Med. Singapore* **1998**, *27*, 738–743.

63 Shulmann, R., Current drug treatment of sep-sis. *Hospital Pharmacist* **2002**, *9*, 97–101.

64 Ohlsoon, K., Bjork, P., Bergenfeldt, M., Hage-man, R., Thompson, R.C., Interleukin-1 re-ceptor antagonist reduces mortality from en-dotoxin shock. *Nature* **1990**, *348*, 550–552.

65 Fischer, C.J., Chainaut, J.F., Opal, S.M., Balk, R.A., Slotman, G.J., Iberti, T.J., Recombinant human interleukin-1 receptor antagonist in the treatment of patients with sepsis syn-drome, results from a randomised, double-blind, placebo-controlled trial. *JAMA* **1994**, *271*, 1836–1843.

66 Ziegler, E.J., Fisher, C.J., Sprung, C.L., Straube, R.C., Sadoff, J.C., Foulke, G.E., Wor-tel, C.H., Fink, M.P., Dellinger, R.P., Teng, N.N., Treatment of gram-negative bacteremia and septic shock with HA-1A human mono-clonal antibody against endotoxin. A random-ized, double-blind, placebo-controlled trial. The HA-1A Sepsis Study Group. *N. Engl. J. Med.* **1991**, *324*, 429–436.

67 Medicines Control Agency, Drug Alert EL, 93, A/6. London, MCA, **1993**.

68 Bone, R. C., Balk, R. A., Fein, A. M., Perl, T. M., Wenzel, R. P., Reines, H. D., A second large controlled clinical trial of E5, a monoclonal antibody to endotoxin. Results of a prospective, multicentre, randomised, controlled trial. *Crit. Care Med.* **1995**, *23*, 994–1005.

69 Cross, A. S., Antiendotoxin Antibodies – A dead end? *Ann. Intern. Med.* **1994**, *121*, 58–60.

70 Abraham, E., Wunderink, R., Silvermann, H., Perl, T. M., Nasraway, S., Levy, H., Efficacy and safety of monoclonal antibody to human tumour necrosis factor in patients with septic syndrome. *JAMA*, **1995**, *273*, 934–941.

71 Biologic Battlefields, Changes in the US Markets for Rheumatoid Arthritis and Psoriatic Arthritis Pharmaceutical & Diagnostic Innovation, **2004**, vol. 1, no. 3, pp. 16–20.

72 Dayer, J.-M., The pivotal role of interleukin-1 in the clinical manifestations of rheumatoid arthritis. *Rheumatology* **2003**, *42* (Suppl. 2), ii3–ii10.

73 Fleischmann, R., Stern, R., Iqbal, I., Anakinra, an inhibitor of IL-1 for the treatment of rheumatoid arthritis. *Exp. Opin. Biol. Ther.* **2004**, *4(8)*, 1333–1344.

74 Fleischmann, R. M., Stern, R., Iqbal, I., Considerations with the use of biological therapy in the treatment of rheumatoid arthritis. *Exp. Opin. Drug Safety* **2004**, *3*, 391–397.

75 Nanda, S., Bathon, J. M., Etanercept, a clinical review of current and emerging indications. *Exp. Opin. Pharmacother.* **2004**, *5(5)*, 1175–1186.

76 Amevive: First to Enter Psoriasis Market – Competitor Products Closing In Quickly. *Pharmaceutical & Diagnostic Innovation* **2003**, *1(2)*, 23–26.

77 Feldman, S. R., Menter, A., Koo, J. Y., Improved health-related quality of life following a randomized controlled trial of alefacept treatment in patients with chronic plaque psoriasis. *Br. J. Dermatol.* **2004**, *150*, 317–326.

78 Asche, G. V., Rutgeerts, P., Anti-TNF agents in Crohn's disease. *Exp. Opin. Investig. Drugs* **2000**, *9(1)*, 103–111.
Adair, F., Monoclonal antibodies – magic bullets or a shot in the dark? *Drug Discovery World* **2002**, Summer, 53–59.

79 Padlan, E. A., A possible procedure for reducing the immunogenicity of antibody variable domains while preserving their ligand-binding properties. *Molec. Immunol.* **1991**, *28*, 489–498.

80 Grillo-Lôpez, A. J., Rituximab. Clinical development of the first therapeutic antibody for cancer. In: *Pharmaceutical Biotechnology – Drug Discovery and Clinical Applications.* O. Kayser, R. H. Müller (Eds). Wiley-VCH, **2003**.

81 Cartron, G., Dacheux, L., Salles, G., Solal-Celigny, P., Bardos, P., Colombat, P., Watier, H., Therapeutic activity of humanized anti-CD20 monoclonal antibody and polymorphism in IgG Fc receptor FcgammaRIIIa gene. *Blood* **2002**, *99*, 754–758.

82 Tobinai, K., RituxiMAb and other emerging monoclonal antibody therapies for lymphoma. *Exp. Opin. Emerging Drugs* **2002**, *7(2)*, 289–302.

83 Glade-Bender, J., Kandel, J. J., Yamashiro, D. J., VEGF blocking therapy in the treatment of cancer. *Exp. Opin. Biol. Ther.* **2003**, *3*, 263–276.

84 Carter, P., Presta, L., Gorman, C. M., Ridgway, J. B., Henner, D., Wong, W. I., Rowland, A. M., Kotts, C., Carver, M. E., Shepard, H. M., Humanization of an anti-p185HER2 antibody for human cancer therapy. *Proc. Natl. Acad. Sci. USA* **1992**, *89*, 4285–4289.

85 Carpenter, G., Cohen, S., Epidermal growth factor. *J. Biol. Chem.* **1990**, *265*, 7709–7712.

86 Yarden, Y., Ullrich, A., Growth factor receptor tyrosine kinases. *Annu. Rev. Biochem.* **1988**, *57*, 443–478.

87 Yarden, Y., The EGFR family and its ligands in human cancer. Signalling mechanisms and therapeutic opportunities. *Eur. J. Cancer* **2001**, *37*, 3–8.

88 Cardiello, F., Tortora, G., A novel approach in the treatment of cancer, targeting the epidermal growth factor receptor. *Clin. Cancer Res.* **2001**, *7*, 2958–2970.

89 Slamon, D. J., Leyland-Jones, B., Shak, S., Fuchs, H., Paton, V., Bajamonde, A., Fleming, T., Eiermann, W., Wolter, J., Pegram, M., Baselga, J., Norton, L., Use of chemotherapy plus a monoclonal antibody against HER2 for metastatic breast cancer that overexpresses HER2. *N. Engl. J. Med.* **2001**, *344*, 783–792.

90 Huang, S. M., Bock, J. M., Harari, P. M., Epidermal growth factor blockade with C225 modulates proliferation, apoptosis, and radiosensitivity in squamous cell carcinomas of the head and neck. *Cancer Res.* **1999**, *59*, 1925–1940.

91 White, C. A., Rituxan® immunotherapy and Zevalin® radioimmunotherapy in the treatment of non-Hodgkin's lymphoma. *Curr. Pharm. Biotechnol.* **2003**, *4*, 221–238.

92 Robinson, K., An industry comes of age. *BioPharm Int.* **2002**, *15*, 20–24.

93 Milroy, D., Auchincloss, C., Monoclonal Antibodies – on the Crest of a Wave. *Horizons*
2003, 6.

94 Cohen, M.M., Diamond, J.M., Are we losing
the war on cancer? *Nature* **1986**, *323*, 488–
492.

95 Sell, S., Cancer Marker, Past, present and fu-
ture. In: *Monoclonal Antibodies and Cancer
Therapy*. R.A. Reisfeld (Ed.), Alan R. Liss,
New York, **1985**.

96 Hakomori, S.I., Tumour associated carbohydrate antigens. *Annu. Rev. Immunol.* **1984**, *2*,
103–111.

97 Lloyed, K.O., Human tumour antigens. Tar-
gets for mouse monoclonal antibodies. In:
Immunoconjugates. C.W. Vogel (Ed.), Oxford
University Press, New York, Oxford, **1987**.

98 Stamp, G.W.H., The rational use of monoclonal antibodies in tumour diagnosis. *Arch.
Geschwulstforsch.* **1990**, *60*, 71–78.

99 Mach, J.P., Carrel, S., Forni, J., Tumor localization of radiolabeled antibodies against carcinoembryonic antigen in patients with carcinoma. *N. Engl. J. Med.* **1980**, *303*, 1384–
1390.

100 Vogel, C.W., *Immunoconjugates. Antibody
conjugates in radioimaging and therapy of cancer*. Oxford University Press, New York, Oxford, **1987**.

101 Souriau, C., Hudson, P.J., Recombinant antibodies for cancer diagnosis and therapy.
Exp. Opin. Biol. Ther. **2003**, *3(2)*, 305–318.

102 Goldenberg, D.M., Deland, F.H., History
and status of tumour imaging with radiolabeled monoclonal antibodies. *J. Nucl. Med.*
1985, *25*, 538–549.

103 Ehrlich, P., Chemotherapie. In: *The collected
papers of Paul Ehrlich*. F. Himmelweit (Ed.),
Pergamon Press, London, **1969**.

104 Hudson, P.J., Recombinant antibody constructs in cancer therapy. *Curr. Opin. Immunol.* **1999**, *11*, 548–557.

105 Vogel, C.W., *Immunoconjugates, Antibody
conjugates in radioimaging and therapy of cancer*. Oxford University Press, New York,
1987.

106 Blair, A.H. Ghose, T.I. Linkage of cytotoxic
agents to immunoglobulins. *J. Immunol.
Methods* **1983**, *59*, 129–155.

107 Garnett, M.C., Targeted drug conjugates,
principles and progress. *Adv. Drug Deliv. Rev.*
2001, *53*, 171–216.

108 Olsnes, S., Sandvig, K., How protein toxins
enter and kill cells. In: *Immunotoxins*. A.E.
Frankel (Ed.), Martinus Nijhoff Publishing,
Boston, **1984**, pp. 1–48.

109 FitzGerald, D.J., Kreitman, R., Wilson, W.,
Squires, D., Pastan, I., Recombinant immunotoxins for treating cancer. *Int. J. Med. Microbiol.* **2004**, *293*, 577–582.

110 Ng, H.C., Khoo, H.E., Cancer homing toxins. *Curr. Pharmaceut. Design* **2002**, *8*, 1973–
1985.

111 Meienhofer, J.H., Maeda, H., Gaser, C.B.,
Czomboz, J., Kuromizu, K., Primary structure of neocarzinostatin, an antitumor protein. *Science* **1972**, *178*, 875–878.

112 Poon, R., Beerman, T.A., Goldberg, I.H.,
Characterization of the DNA-strand-breakage
in vitro by the antitumor protein neocarzinostatin. *Biochemistry* **1977**, *16*, 486–492.

113 Zein, N., Sinha, A.M., McGahren, W.J., El-
lestad, G.A., Calicheamicin gammaI I. An
antitumor antibiotic that cleaves double-
stranded DNA site specifically. *Science* **1988**,
240, 1198–1192.

114 Maeda, H., Neocarzinostatin in cancer chemotherapy. *Anticancer Res.* **1981**, *1*, 175–189.

115 Gottschalk, U., Garnett, M.C., Ward, R.K.,
Maibücher, A., Köhnlein, W., Increased serum stability and prolonged biological half-
life of Neocarzinostatin covalently bound to
monoclonal antibodies. *J. Antibiotics* **1991**,
44, 1148–1154.

116 Nitin, K., Damle, E., Tumour-targeted chemotherapy with immunoconjugates of calicheamicin. *Exp. Opin. Biol. Ther.* **2004**, *4(9)*,
1445–1452.

117 Hamann, P.R., Hinman, L.M., Beyer, C.F.,
Lindh, D., Upeslacis, J., Flowers, D.A., Bern-
stein, I., An anti-CD33 antibody calicheamicin conjugate for treatment of acute myeloid
leukaemia. Choice of Linker. *Bioconj. Chem.*
2002, *13*, 40–46.

118 Nagata, R., Yamanaka, H., Okazaki, E., Saito, I., Biradical formation from acyclic conjugated eneyne-allene system related to Neocarzinostatin and Esperamicin-calicheami-
cin. *Tetrahedron Lett.* **1989**, *30*, 4995–4998.

119 Lockhart, T.P., Bergman, R.G., Evidence for
the reactive spin state of 1,4-dehydroben-
zenes. *J. Am. Chem. Soc.* **1981**, *103*, 4091–
4096.

120 Reiter, Y., Brinkmann, U., Lee, B., Pastan, I.,
Engineering antibody Fv fragments for can-

cer detection and therapy, disulfide-stabilized Fv-fragments. *Nat. Biotechnol.* **1996**, *14*, 1239–1245.

121 Choo, A. B. H., Dunn, R. D., Broady, K. W., Raison, R. L., Soluble expression of a functional recombinant cytolytic immunotoxin in insect cells. *Protein Express. Purif.* **2002**, *24(3)*, 338–334.

122 Yoshioka, Y., Tsutsumi, Y., Nakagawa, S., Mayumi, T., Recent progress on tumor missile therapy and tumor vascular targeting therapy as a new approach. *Curr. Vasc. Pharmacol.* **2004**, *12*, 259–270.

123 Goldenberg, D. M., Targeted therapy of cancer with radiolabeled antibodies. *J. Nucl. Med.* **2002**, *43*, 693–713.

124 Cobb, L. M., Humm, J. L., Radioimmunotherapy of malignancy using antibody targeted radionuclides. *Br. J. Cancer* **1986**, *54*, 863–868.

125 Xu, G., McLeod, H. L., Strategies for enzyme/prodrug cancer therapy. *Clin. Cancer Res.* **2001**, *7*, 3314–3324.

126 Steiner, U., The Business Case for Plant Factories. Plant Conference, Quebec, **2003**.

127 Rathore, A., Velayudhan, A., Guidelines for optimisation and scale up in preparative chromatography. *BioPharm* **2003**, January, 34–42.

128 Federici, M. M., The quality control of biotechnology products. *Biologicals* **1994**, *22*, 151–159.

129 Office of Biological Research and Review, US Food and Drug Administration. **1985**. Points to consider in the production and testing of new drugs and biologicals produced by recombinant DNA technology.

130 Little, M., Kipriyanov, S. M., Le Gall, F., Moldenhauer, G., Of mice and men, hybridoma and recombinant antibodies. *Immunol. Today* **2000**, *21*, 364–370.

131 Wurm, F. M., Production of recombinant protein therapeutics in cultivated mammalian cells. *Nature Biotechnol.* **2004**, *22*, 1393–1398.

132 Wurm, F. M., Griffith, J., Mammalian Cell Culture. In: *The Encyclopedia of Physical Science and Technology*, 3rd edn. R. A. Meyers (Ed.), Academic Press, **2002**, 9, pp. 31–47.

133 Wurm, F. M., Jordan, M., Gene transfer and gene amplification in mammalian cells. In: *Gene Transfer and Expression in Mammalian Cells*. S. C. Makrides (Ed.), New Comprehensive Biochemistry, **2003**, 38, pp. 309–335.

134 Grund, E., Advances in Downstream Processing. Presented at "Scaling up of Biopharmaceutical Proteins", IBC Life Sciences, Basel, 23rd January **2003**.

135 Jungbauer, A., Preparative chromatography of biomolecules. *J. Chromatogr.* **1993**, *639*, 3–16.

136 Wilchek, M., Chaiken, I., An overview of affinity chromatography. *Methods Molec. Biol.* **2000**, *147*, 1–6.

137 Lowe, C. R., Lowe, A. R., Gupta, G., New developments in affinity chromatography with potential application in the production of biopharmaceuticals. *J. Biochem. Biophys. Methods* **2001**, *49*, 561–574.

138 Sjöquist, J., Movitz, J., Johansson, I.-B., Localization of protein A in *Staphylococcus aureus*. *Eur. J. Biochem.* **1972**, *30*, 190.

139 Suralia, Interaction of Protein A with the domains of the Fc-Fragment. *Trends Biochem. Sci.* **1982**, 7, 318.

140 Huse, K., Böhme, H. J., Scholz, G. H., Purification of antibodies by affinity chromatography. *J. Biochem. Biophys. Methods* **2002**, *51*, 217–231.

141 Mechanism of activation of Sepharose and Sephadex by cyanogen bromide. *Enzyme Microb. Technol.* **1982**, 4, 161.

142 Iyer, H., Henderson, F., Cunningham, E., Webb, J., Hanson, J., Bork, C., Conley, L., Considerations during development of a protein A-based antibody purification process. *Biopharm* **2002**, *20*, 14–20.

143 Mottaqui-Tabar, A., Birath, K., Johanson, H. J., A base-tolerant resin for Monoclonal Antibody purification. 3rd International Symposium on Downstream Processing of Genetically Engineered Antibodies and Related Molecules, **2004**, Nice, France.

144 Bergander, T., Chirica, L., Ljunglöf, A., Malmquist, G., Novel high capacity Protein A affinity chromatography media. 3rd International Symposium on Downstream Processing of Genetically Engineered Antibodies and Related Molecules, **2004**, Nice, France.

145 Raeder, R., Boyle, M. D., Analysis of immunoglobulin G-binding-protein expression by invasive isolates of *Streptococcus pyogenes*. *Clin. Diagn. Lab. Immunol.* **1995**, *2*, 484–486.

146 Bjorck, L., Protein, L.. A novel bacterial cell wall protein with affinity for IgL chains. *J. Immunol.* **1988**, *140*, 1194–1197.

147 Gulich, S., Uhlen, M., Hober, S., Protein engineering of an IgG-binding domain allows milder conditions during affinity chromatography. *J. Biotechnol.* **2000**, *76*, 233–244.

148 Curling. J., Affinity Chromatography – from Textile Dyes to Synthetic Ligands by Design, Parts I and II. *BioPharm Int.* **2004**, July/August.

149 Lowe, C.R., Lowe, A.R., Gupta, G., New developments in affinity chromatography with potential application in the production of biopharmaceuticals. *J. Biochem. Biophys.* **2001**, *49*, 561–574.

150 Boschetti, E., The use of thiophilic chromatography for antibody purification: a review. *J. Biochem. Biophys. Methods* **2001**, *49*, 361–389.

151 Li, R., Dowd, V., Stewart, D.J., Burton, S.J., Lowe, C.R., Design, synthesis, and application of a Protein A mimetic. *Nature Biotechnol.* **1998**, *16*, 190–195.

152 Fish, B., Concepts in development of manufacturing strategies for monoclonal antibodies. In: *Antibodies, Volume 1, Production and Purification.* G. Subramanian (Ed.), Kluwer Academic New York, **2004**.

153 Hanna, L.S., Pine, P., Reuzinsky, G., Nigam, S., Omstead, D.R., Removing specific cell culture contaminants in a MAb purification process. *BioPharm* **1991**, *4*, 33–37.

154 Gosh, R., Protein separation using membrane chromatography, opportunities and challenges. *J. Chromatogr.* **2002**, *952*, 13–27.

155 Gottschalk, U., Fischer-Fruehholz, S., Reif, O., Membrane adsorbers. A cutting edge process technology at the threshold. *BioProcess Int.* **2004**, *2*, 56–65.

156 Warner, T.N., Nochumsnon, S., Rethinking the economics of chromatography. *BioPharm* **2003**, *16*, 58–60.

157 Food and Drug Administration, USA, Points to consider in the manufacture and testing of monoclonal antibody products for human use. **1997**.

158 Food and Drug Administration, USA, Guidance for industry, Bioanalytical method validation. **2001**.

159 Hoffmann, K., Strategies for host cell protein analysis. *BioPharm* **2000**, *13*, 38–45.

160 Richter, A., Jostameling, M., Müller, K., Herrmann, A., Pitschke, M., Quality control of antibodies for human use. In: *Antibodies – Production and Purification.* G. Subrama-

nian (Ed.), Kluwer Academic, New York, **2004**.

161 Committee for Proprietary Medicinal Products (CPMP), ICH Q5a. Note for Guidance on Quality of Biotechnology Products, Viral Safety Evaluation of Biotechnology Products Derived from Cell Lines of Human or Animal Origin (CPMP) ICH/295/95), **1997**.

162 Committee for Proprietary Medicinal Products (CPMP). Note for Guidance on Virus Validation Studies, The design, contribution and interpretation of studies validating the inactivation and removal of viruses. CPMP/BWP/268/95, **1996**.

163 Sofer, G., Virus Inactivation in the 1990s – Part 4, culture media, biotechnology products, and vaccines. *BioPharm* **2003**, *16*, 50–57.

164 Walter, J.K., Nothelfer, F., Werz, W., Validation of viral safety for pharmaceutical proteins. In: *Bioseparation and Bioprocessing*, Vol. 1. G. Subramanian (Ed.), Wiley-VCH Weinheim, **1998**, pp. 465–496.

165 Levy, R.V., Phillips, M.W., Lutz, H., Filtration and the removal of viruses from biopharmaceuticals. In: *Filtration in the biopharmaceutical industry.* T. Meltzer, M. Jornitz (Eds.), Marcel Dekker Inc., New York, Basel, Hong Kong, **1998**, pp. 619–646.

166 Sofer, G., Sofer, D.C., Boose, J.A., Inactivation methods grouped by virus. Virus inactivation in the 1990s – and into the 21st century. *BioPharm Int.* **2003**, Suppl. S37–S42.

167 Sofer, G., Validation: Ensuring the accuracy of scaled-down chromatography models. *BioPharm* **1996**, October, 51–54.

168 Plavsic, A.M., Bolin, S., Resistance of porcine circovirus to gamma irradiation. *BioPharm* **2001**, *14*, 32–36.

169 Wang, J., Mauser, A., Chao, S.-F., Remington, K., Treckmann, R., Kaiser, K., Pifat, D., Hotta, J., Virus inactivation and protein recovery in a novel ultraviolet-C reactor. *Vox Sanguinis* **2004**, *86*, 230–238.

170 Reif, O., Mora, J., Gottschalk, U., An integrated concept for reliable removal of viruses and process-related contaminants in biomanufacturing. IBC BioProcess International Conference October 4–7, Boston, USA.

171 Leibiger, H., Hansen, A., Schoenherr, G., Seifert, M., Wustner, D., Stigler, R., Marx, U., Glycosylation analysis of a polyreactive human monoclonal IgG antibody derived from a human-mouse heterohybridoma. *Mol. Immunol.* **1995**, *32*, 595–602.

172 Deng, X.K., Raju, T.S., Morrow, K.J., Achieving appropriate glycosylation during the scale up of antibody production. In: *Antibodies – Novel Technologies and Therapeutic Use.* G. Subramanian (Ed.), Kluwer Academic, New York, **2004**.

173 Wright, A., Morrison, S.L., Effect of glycosylation on antibody function. Implications for genetic engineering. *Trends Biotechnol.* **1997**, *15*, 26–32.

174 Jefferis, R., Lund, J., Glycosylation of antibody molecules. Structure and functional significance. *Chem. Immunol.* **1997**, *65*, 111–128.

175 Jenkins, N., Parekh, R.B., James, D.C., Getting the glycosylation right, implications for the biotechnology industry. *Nat. Biotechnol.* **1996**, *14*, 975–981.

176 Schenermann, M.A., Hope, J.N., Kletke, C., Singh, J.K., Kimura, R., Tsao, E.I., Folena-Wassermann, G., Comparability testing of a humanized monoclonal antibody, Synagis, to support cell line stability, process validation, and scale-up for manufacturing. *Biologicals* **1999**, *27*, 203–215.

177 Patel, T.P., Parekh, R.B., Moellering, B.J., Prior, C.P., Different culture methods lead to differences in glycosylation of a murine IgG monoclonal antibody. *Biochem. J.* **1992**, *285*, 839–845.

178 Siberil, S., Teillaud, J.L., Future prospects in antibody engineering and therapy. In: *Antibodies, Volume 2, Novel Technologies and Therapeutic Use.* G. Subramanian (Ed.), Kluwer Academic, New York, **2004**.

179 Yoo, E.M., Koteswara, R., Chintalacharuvu, M.L., Morrison, S.L., Myeloma expression systems. *J. Immunol. Methods* **2002**, *261*, 1–20.

180 Stanley, P., Glycosylation engineering. *Glycobiology* **1992**, *2*, 99–107.

181 Weikert, S., Papac, D., Briggs, J., Cowfer, D., Tom, S., Gawlitzek, M., Lowgren, J., Metha, S., Chisholm, V., Modi, N., Eppler, S., Carroll, K., Chamow, S., Peers, D., Berman, P., Engineering Chinese hamster ovary cells to maximize sialic acid content of recombinant glycoproteins. *Nat. Biotechnol.* **1993**, *17*, 1116–1121.

182 Sinclair, A., Biomanufacturing capacity, Current and future requirements. *J. Commercial Biotechnol.* **2001**, *8*, 43–50.

183 Gottschalk, U., Biotech manufacturing is coming of age. *BioProcess Int.* **2004**, *4*, 54–61.

184 Sadana, A., Beelaram, A., Efficiency and economics of bioseparation, some case studies. *Bioseparation* **1994**, *4*, 221–235.

185 Verma, R., Boleti, E., George, A.J., Antibody engineering, comparison of bacterial, yeast, insect and mammalian expression systems. *J. Immunol. Methods* **1993**, *216*, 165–181.

186 Plückthun, A., *Escherichia coli* producing recombinant antibodies. *Bioprocess Technol.* **1994**, *19*, 233–252.

187 Grant, S.D., Cupit, P.M., Learmonth, D., Byrne, F.R., Graham, B.M., Porter, A.J.M., Harris, W.J., Expression of monovalent and bivalent antibody fragments in *Escherichia coli. J. Hematotherapy* **1995**, *4*, 383–388.

188 Eldin, P., Pauza, M.E., Hieda, Y., Lin, G., Murtauch, M.P., Pentel, P.R., Pennell, C.A., High level secretion of two antibody single chain Fv fragments by *Pichia pastoris. J. Immunol. Methods* **1997**, *201*, 67–75.

189 Joosten, V., Lokman, C., Hondel, C., Punt, P.J., The production of antibody fragments and antibody fusion proteins by yeasts and filamentous fungi. *Microbial Cell Factories* **2003**, *2*, 1–15.

190 Young, M.W., Meade, H., Curling, J.M., Ziomek, C.A., Harvey, M., Production of recombinant antibodies in the milk of transgenic animals. *Res. Immunol.* **1998**, *149*, 609–610.

191 Fischer, R., Eman, N., Molecular Farming of pharmaceutical proteins. *Transgenic Res.* **2000**, *9*, 279–299.

192 Pollock, D.P., Kutzko, J.P., Brick-Wilson, E., Williams, J.L., Echelard, Y., Meade, H.M., Transgenic milk as a method for the production of recombinant antibodies. *J. Immunol. Methods* **1999**, *231*, 147–157.

193 Sensoli, S., Transgenic technology for monoclonal antibody production. In: *Antibodies – Novel Technologies and Therapeutic Use.* G. Subramanian (Ed.), Kluwer Academic, New York, **2004**.

194 Hiatt, G., Monoclonal antibody engineering in plants. *FEBS Lett.* **1992**, *307*, 71–75.

195 Knäblein, J., Biotech: A New Era In The New Millennium – Fermentation and Expression of Biopharmaceuticals in Plants. *SCREENING – Trends Drug Disc.* **2003**, *4*, 14–16.

196 Knäblein, J., Biopharmaceuticals expressed in plants – a new era in the new Millen-

nium. In: *Applications in Pharmaceutical Biotechnology.* R. Müller, O. Kayser (Eds.), Wiley-VCH, **2004**.

197 Valdes, R., Reyes, B., Alvarez, T., Garcia, J., Montero, J. A., Figueroa, A., Gomez, L., Padilla, S., Geada, D., Abrantes, M. C., Dorta, L., Fernandez, D., Mendoza, O., Ramirez, N., Rodriguez, M., Pujol, M., Borroto, C., Brito, J., Hepatitis B surface antigen immunopurification using a plant-derived specific antibody produced in large scale. *Biochem. Biophys. Res. Commun.* **2003**, *310*, 742–747.

198 Peeters, K., De Wilde, C., De Jaeger, G., Angenon, G., Depicker, A., Production of antibodies and antibody fragments in plants. *Vaccine* **2001**, *19*, 2756–2761.

199 Hellwig, S., Drossard, J., Twyman, R. M., Fischer, R., Plant cell cultures for the production of recombinant proteins. *Nature Biotechnol.* **2004**, *22*, 1393–1398.

200 Abiko, Y., Passive immunization against dental caries and periodontic disease, development of recombinant and human monoclonal antibodies. *Crit. Rev. Oral Biol. Med.* **2000**, *11*, 140–158.

201 Hodge, G., Disposable components enable a new approach to biopharmaceutical manufacturing. *BioPharm Int.* **2004**, *15*, 38–49.

202 Warner, T. N., Nochumsnon, S., Rethinking the economics of chromatography. *BioPharm* **2003**, *16*, 58–60.

203 Hawkins, R. E., Russell, S. J., Winter, G., Selection of phage antibodies by binding affinity, mimicking affinity maturation. *J. Mol. Biol.* **1992**, *226*, 889–896.

204 Hoogenboom, H. R., Designing and optimising library selection strategies for generating high-affinity antibodies. *Trends Biotechnol.* **1997**, *15*, 62–70.

205 Clynes, R. A., Towers, T. L., Presta, L. G., Ravetch, J. V., Inhibitory Fc receptors modulate in vivo cytotoxicity against tumor targets. *Nature Med.* **2000**, *6*, 443–446.

206 Shields, R. L., Namenuk, A. K., Hong, K., Meng, Y. G., Rae, J., Briggs, J., Xie, D., Lai, J., Stadlen, A., Li, B., Fox, J. A., Presta, L. G., High resolution mapping of the binding site on human IgG for Fcgamma RI, Fcgamma RII, Fcgamma RIII, and FcRn and design of IgG1 variants with improved binding to the Fcgamma R. *J. Biol. Chem.* **2001**, *276*, 6591–6604.

207 Radaev, S., Sun, P. D., Recognition of IgG by Fcgamma receptor. The role of Fc glycosylation and the binding of peptide inhibitors. *J. Biol. Chem.* **2001**, *276*, 16478–16483.

208 Umana, P., Jean-Mairet, J., Moudry, R., Amstutz, H., Bailey, J. E., Engineered glycoforms of an antineuroblastoma IgG1 with optimized antibody-dependent cellular cytotoxic activity. *Nat. Biotechnol.* **1999**, *17*, 176–180.

209 Boyd, P. N., Lines, A. C., Patel, A. K., The effect of the removal of sialic acid, galactose and total carbohydrate on the functional activity of Campath-1H. *Mol. Immunol.* **1995**, *32*, 1311–1318.

210 Cartron, G., Dacheux, L., Salles, G., Solal-Celigny, P., Bardos, P., Coloms, P., Watier, H., Therapeutic activity of humanized anti-CD20 monoclonal antibody and polymorphism in IgG Fc receptor FcγRIIIa gene. *Blood* **2002**, *99*, 754–758.

211 Joosten, V., Lokman, C., Hondel, C., Punt, P.J., The production of antibody fragments and antibody fusion proteins by yeasts and filamentous fungi. *Microbial Cell Factories* **2003**, *2*, 1–15.

2

Modern Antibody Technology: The Impact on Drug Development

Simon Moroney and Andreas Plückthun

Abstract

Antibodies are now the mainstream of bio-pharmaceuticals. By the end of 2003, 17 marketed therapeutic antibodies generated over $5 billion in combined annual sales, with market growth at 30%. Ten years earlier, this class of biopharmaceutical drugs was almost written off, based on disappointments experienced with the first generation of murine monoclonal antibodies. This chapter will look at how new technologies have provided solutions to problems that hampered early efforts to develop effective antibody therapeutics and transformed the market for antibody drugs. This includes the generation of fully human antibodies, their affinity maturation and the selection of antibodies to bind to particular epitopes on disease-relevant targets. The chapter will also highlight what distinguishes a therapeutic from a simple binding molecule – different modes of actions of antibodies in different molecular and cellular settings will be compared. Finally, some of the available formats of the antibody and their effect on molecular/pharmacological properties will be discussed.

Abbreviations

ADCC	antibody-dependent cellular cytotoxicity
CDC	complement-directed cytotoxicity
CDR	complementarity-determining region
CHO	Chinese hamster ovary
CTL	cytotoxic T lymphocyte
DOX	doxorubicin
EBV	Epstein–Barr virus
FcRn	neonatal Fc receptor
GlcNAc	*N*-acetylglucosamine
GM-CSF	granulocyte macrophage colony-stimulating factor
HIV	human immunodeficiency
HLA	human leukocyte antigen
IL	interleukin
mAb	monoclonal antibody
NK	natural killer
siglec	sialic acid-binding, immuno-globulin-like lectins
TNF	tumor necrosis factor
VEGF	vascular endothelial growth factor

2.1
Introduction

The initial promise of antibody-based biopharmaceuticals has taken a long time to

Modern Biopharmaceuticals. Edited by J. Knäblein
Copyright © 2005 WILEY-VCH Verlag GmbH & Co. KGaA, Weinheim
ISBN: 3-527-31184-X

be realized. The breakthrough that led to the routine generation of monoclonal antibodies (mAbs) [1] created expectations that antibodies would become a major class of drugs. That it has taken over 20 years for the potential of therapeutic antibodies to be translated into commercial success is attributable to the time needed to solve problems associated with the first generation of antibodies. While clinical efficacy and safety depend critically on the target against which the antibody is directed, as well as the exact binding epitope, the molecular properties of the therapeutic itself are equally important in a successful drug (see also Part IV, Chapter 16 and Part V, Chapter 1).

While some challenges have been largely solved, others remain. Over the next decade it is likely that additional improvements in the molecular properties of antibodies will be made, further increasing their importance as biopharmaceuticals.

The main factors that limit the clinical utility of antibodies are:

- Immunogenicity.
- Inability to reach a disease-relevant target in sufficient concentration.
- Inability to trigger a particular biological effect which translates into modification of the disease process.

Technological approaches to reduce problems in each of these areas have met with varying degrees of success and are discussed in detail in the following sections. It is useful to discuss in turn each of the molecular requirements for making a therapeutic antibody and we will begin with immunogenicity.

2.2
Immunogenicity

Early clinical applications of murine mAbs quickly encountered the problem of immunogenicity in humans. While the insight into cellular mechanisms that has been enabled by the use of mAbs in basic discovery research has been remarkable, the problems with early attempts to use antibodies in therapy cast a shadow of doubt over whether this class of molecule would ever be clinically useful. After the first clinical disappointments with mAbs in the early and mid 1980s, many gave up on their promise as therapeutic agents. The technological developments that have led to a reduction in the difficulties posed by immunogenic murine antibodies, i.e., the development of methods to make chimeric, then humanized and, finally, fully human antibodies (Fig. 2.1), count among the major success stories of the modern biotechnology era.

It is clear now that the immune reaction against the original murine format was a major factor limiting the therapeutic use of antibodies. A relatively small number of academic groups and biotechnology companies tackled these problems, and developed methods for solving them. Only within the last decade have the resulting technological solutions led to the creation of the successful class of drugs that antibodies today represent. As a result, enthusiasm for this class of drugs in the pharmaceutical industry is a very recent phenomenon.

Immunogenicity is undesirable because it can be the source of a number of safety concerns, such as hypersensitivity and allergic reactions, thrombocytopenia, anemia, etc. Very problematic with recombinant therapeutic proteins, but fortunately extremely unlikely with antibodies, are

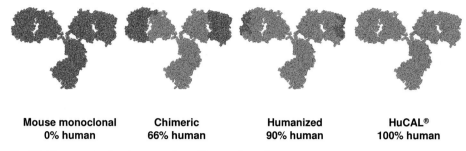

Mouse monoclonal	Chimeric	Humanized	HuCAL®
0% human	**66% human**	**90% human**	**100% human**

Fig. 2.1 Diagram showing the proportion of human (green) and murine (red) sequences in mouse, chimeric, humanized and human antibody structures, as exemplified by HuCAL.

cases in which the therapeutic protein elicits an immune response against one of the body's own proteins.

In addition to these safety concerns, there is still the problem of a loss of efficacy when the therapeutic molecule is removed by the immune system. This could be an issue in chronic indications, when the antibody has to be given repeatedly, as well as when the efficacy critically depends on half-life, as this may be reduced by an immune response against the therapeutic antibody. In certain other cases (for some examples, see below and Section 2.3.1), an immune response is elicited, but causes no major clinical effect.

Thus, the immunogenicity of antibodies (or any protein, for that matter) in humans is a very important parameter, but predicting it remains an inexact science. To date, no predictive scheme has emerged that can obviate the need for a clinical trial. The prediction of T cell epitopes from peptide sequence has been attempted and there are a number of websites available (*http://www.imtech.res.in/raghava/propred*, *http://mif.dfci.harvard.edu/Tools/rankpep.html* and *http://www.jenner.ac.uk/MHCPred*) [2–4] where this can be undertaken. Additionally, some antibody manufacturers test T cell epitopes experimentally in rather simple as-

says of T cell stimulation [5] by using a series of overlapping peptides covering the whole protein.

The rationale for these experiments is that the major human MHC alleles all require characteristic anchor residues in the peptides they bind. To elicit an immune reaction against a foreign protein, a T helper cell response is needed, which in turn requires that the protein is degraded to peptides and this also shows sequence specificity. A part of the protein is then presented in MHC class II on antigen-presenting cells. Peptides from the body's own proteins are also presented in MHC class II, but T cells that would recognize them do not normally exist, as they are eliminated in the thymus. In order to make a foreign protein "invisible" to T cells, none of its peptides must bind to MHC II, as T cells recognizing them will exist. Using available crystal structures and empirical data on peptide binding, profiles can be formulated for likely anchor residues [6]. While the prediction of the major antigenic epitope within a protein (or the absence of a clear hit) may be possible by exploiting the available structure and sequence information, it is still less clear whether such predictions can be extrapolated for engineering purposes. A sufficient number of clinical trials will be required to show whether proteins can be engineered by point mutations to completely evade any MHC binding and thus T cell surveillance without losing their folding and function.

Derivatization with polyethylene glycol, or "PEGylation" (see also Part VI, Chapter 3 and Part VI, Chapter 1) [7–9], while primarily regarded as a means of increasing serum half-life of small antibody fragments (see Section 2.4.3), can also be used to decrease the immunogenicity of foreign proteins. As antibodies of fully human composition can be now obtained, PEGylation, which introduces additional manufacturing problems, might be more appropriate for modifying potential nonhuman effector domains, such as toxins (see Section 2.5.5). Nevertheless, clinical data will be needed for each individual case.

Table 2.1 summarizes data on the immunogenicity of a number of therapeutic antibodies currently on the market. As is immediately apparent from this and a wealth of data on mouse monoclonals, antibodies of more human composition are, in general, less immunogenic than those of murine origin. However, drawing firm conclusions is difficult for a number of reasons. First, immunogenicity depends on a number of factors unrelated to the molecular composition of the drug, including dose, route of administration, type of formulation and immunocompetence of the patient. Second, the strict demarcation of antibody structures into the categories "chimeric", "humanized" or "human", terms reflective of the way the antibodies were made, diverts attention from the key issue of sequence homology at the amino acid level. As has been pointed out [10], since mouse and human antibodies are rather homologous, the closeness of a sequence to the nearest human germline gene is perhaps a more important

determinant of immunogenicity and it may thus be advantageous to create antibodies with this property. Similarly, any human protein that has been mutated, e.g., by diversifying a region, is potentially immunogenic.

That a number of the chimeric, humanized and human biopharmaceuticals in Table 2.1 are highly successful drugs proves that the movement away from murine monoclonals towards antibodies of more human composition has paid dividends in the clinic. The conclusion for drug development is that antibodies that are predominantly human in their composition are less likely to encounter problems of immunogenicity than murine antibodies.

This difference between murine antibodies and those comprising some human content is also evident in overall developmental success rates. Data from the Tufts Centre for the Study of Drug Development [11] show that the probabilities of chimeric, humanized or fully human antibodies progressing from entry into clinical trials to the market are 26, 18 and 14%, respectively, while for murine antibodies the corresponding probability is only 4.5%. Caution should be used in drawing any conclusions from the apparently higher success rates in developing chimeric over humanized and human antibodies since the sample size is limited to those 17 antibodies which had reached the market at the time the study was performed.

Although the advent of technologies that can provide fully human antibodies (see Section 2.3) would appear to have solved the problem of immunogenicity, it is to be expected that the solution to this problem is not complete. Some human antibodies are known to be immunogenic, typically through anti-idiotypic or anti-allotypic effects. For example, the fully human antibody adalimumab (Humira), which was

Table 2.1 Immunogenicity of a number of therapeutic antibodies currently on the market

Compound name	Trade name	Company	Antibody format	Antigen	Approved application	Year of first approval	Isotype	Immuno-genicity[a]
Antibodies affecting the immune system (inflammation, allergy, transplantation)								
Muromonab	Orthoclone OKT3	Johnson and Johnson	murine mAb	CD3	organ transplant rejection	1986	IgG2a	~80%
Daclizumab	Zenapax	Hoffmann La Roche	humanized mAb	CD25	kidney transplant rejection	1997	IgG1	14%
Basiliximab	Simulect	Novartis	chimeric mAb	CD25	kidney transplant rejection	1998	IgG1	<2%
Infliximab	Remicade	Johnson and Johnson	chimeric mAb	TNF-α	rheumatoid arthritis, Crohn's disease, ankylosing spondylitis	1998, 1999	IgG1	13%
Efalizumab	Raptiva	Genentech, Serono	humanized mAb	CD11a	psoriasis	2003	IgG1	6%
Omalizumab	Xolair	Genentech, Novartis	humanized mAb	IgE	allergic asthma	2003	IgG1	<0.1%
Adalimumab	Humira	Abbott	fully human mAb	TNF-α	rheumatoid arthritis	2003	IgG1	5%
Antibodies in oncology								
Rituximab	Rituxan	Genentech, Biogen, IDEC, Hoffmann La Roche	chimeric mAb	CD20	non-Hodgkin's lymphoma	1997	IgG1	<1%
Trastuzumab	Herceptin	Genentech, Hoffmann La Roche	humanized mAb	HER2/*neu*	breast cancer	1998	IgG1	<0.1%
Gemtuzumab ozogamicin	Mylotarg	Wyeth	humanized mAb/calicheamicin	CD33	acute myeloid leukemia	2000	IgG4	0%
Alemtuzumab	Campath-1H	Schering	humanized mAb	CD52	chronic lymphocytic leukemia	2001	IgG1	2%
Ibritumomab tiuxetan	Zevalin	Biogen, IDEC	murine mAb/^{90}Y	CD20	non-Hodgkin's lymphoma	2002	IgG1	4%
Tositumomab	Bexxar	GSK	murine mAb/^{131}I and murine mAb	CD20	non-Hodgkin's lymphoma	2003	IgG2a	>70%

Table 2.1 (continued)

Compound name	Trade name	Company	Antibody format	Antigen	Approved application	Year of first approval	Isotype	Immuno-genicity[a]
Cetuximab	Erbitux	ImClone, Merck KGaA	chimeric mAb	epidermal growth factor receptor	colorectal cancer	2003	IgG1	5% of evaluable patients
Bevacizumab	Avastin	Genentech, Hoffmann La Roche	humanized mAb	VEGF	colorectal cancer	2004	IgG1	0%
Antibodies in infectious diseases								
Palivizumab	Synagis	Medimmune, Abbott	humanized mAb	F protein of respiratory syncytial virus	lower respiratory tract disease in infants	1998	IgG1	<1%
Antibodies in perioperative care								
Abciximab	ReoPro	Centocor, Eli Lilly	chimeric Fab	GPIIb/IIIa	percutaneous coronary inter-vention	1995	Fab	6%
Antibodies in central nervous system diseases								
Natalizumab	Tysabri	Biogen Idec/Elan	humanized mAb	VLA4	multiple sclerosis	2004	IgG4	6–10%

a) Percentage of patients with antigen-antibody titer. From FDA-approved product label or other FDA submissions.

approved at the end of 2002 for the treatment of rheumatoid arthritis, is immunogenic in a significant number of patients. However, this has not hampered its successful use in the clinic.

As we learn more about the precise molecular features that contribute to immunogenicity, further improvements will be made in antibody composition to reduce this effect. However, for the foreseeable future, only clinical data in humans will allow determination of the severity of the effect and whether it has a negative impact on clinical utility. As a result of the reliance on human clinical data to determine immunogenicity, we may expect further progress in this area to be slow. It remains to be seen whether the molecular understanding of this complex process will advance to a point that immunogenicity can be engineered out at the amino acid level. It is also possible that a complete evasion of the immune system will simply not be possible in all cases.

2.3
Technology

The therapeutic antibodies currently on the market were developed at a time when molecular engineering had not progressed to its current level. Thus, because of the long development times typical of pharmaceutical development, all antibodies that are now on the market are derived in some way from mouse mAbs. Nevertheless, as will be clear from the following subsections, molecular engineering has now progressed to a point that therapeutic antibodies can be obtained without having an animal-derived mAb as a starting point. The great number of antibodies and derivatives in various phases of clinical trials that are derived from libraries bear witness to this point.

2.3.1
Chimeric and Humanized Antibodies

In the early and mid 1980s, as the first murine mAbs were being tested in the clinic, it was quickly recognized that in order to avoid the problems associated with their immunogenicity, antibodies of more human composition would be needed. While various methods have been investigated, such as immortalization of human B cells by Epstein-Barr virus (EBV transformation) (see Section 2.3.3), most success was achieved by starting with a murine mAb and engineering it to have a more human composition. We now summarize the different ways of achieving this goal.

Historically, the first generation of hybrid antibodies (part mouse/part human) comprised the entire murine variable domains of the original mAb, with the remainder of the IgG (constant C_L domain, usually κ, plus C_H1, hinge, C_H2 and C_H3 domains) coming from a human antibody (Fig. 2.1). Thus, in these so-called *chimeric* antibodies, four out of 12 domains in the IgG remain of murine origin (two V_L and two V_H). In total, approximately two-thirds of the sequence is of human origin, the remaining one-third being murine.

Currently, five chimeric antibodies are approved for human therapeutic use (Table 2.1). The immunogenicity of two chimeric antibodies, abciximab (ReoPro) and infliximab (Remicade) has recently been evaluated in some detail [12]. ReoPro is a chimeric Fab fragment that binds to the $a_{IIb}\beta_3$ integrin, also called platelet membrane glycoprotein GPIIb-IIIa [13]. GPIIb-IIIa is an adhesion receptor for fibrinogen and the von Willebrand factor, both of which carry multiple binding motifs for the integrin and thus mediate platelet aggregation. ReoPro is approved for use in percutaneous coronary intervention to prevent cardiac ischemic complications. In order to exert its effect by inhibiting thrombus formation, it must block a large fraction of its target integrins. At least 6% of patients develop an immune response to the antibody, but re-administration of the Fab fragment remains possible. Infliximab is a chimeric IgG1 specific for human tumor necrosis factor (TNF)-α, and is approved for the acute treatment of the signs and symptoms of Crohn's disease (an inflammation of the small intestine), and for the chronic treatment of rheumatoid arthritis. An immune response occurs in at least 10% of patients, although there does not seem to be a reduction in clinical efficacy.

The next improvement was *humanization* – the grafting of the complementarity-determining regions (CDRs) of a mouse antibody onto a human framework [14]. For this purpose, a human framework is chosen from the database of human genes for V_H, another for V_L (κ or λ), and an alignment of the murine and human sequences is made. In addition to the CDRs themselves, differences in the framework must be taken into account (Fig. 2.2). For example, the so-called outer loop (some-

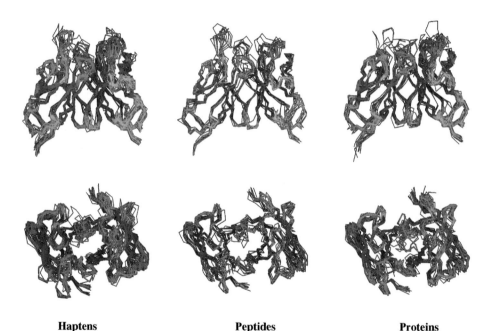

Haptens   **Peptides**   **Proteins**

Fig. 2.2 Antibodies binding haptens, oligopeptides and oligosaccharides or proteins. Superposed crystal structures of the variable regions were sorted into three classes according to the type of antigen. Structurally variable residues within in the CDRs of the antibodies are shown in green; those at the N-terminus, to the N-terminal side of CDR-1 and within the outer loop in cyan. Residues within the inner (dimer interface) β-sheet of V_L and V_H whose side-chains contribute both to the dimer interface and to antigen binding if it reaches deep into this pocket are shown in yellow, orange and red, depending on depth. Note that these residues formally belong to the framework. The structurally least-variable residues whose $C\alpha$ positions were used for the least-squares superposition of the antibody fragments are shaded gray.

times called CDR4) influences the conformation of CDR3 and a number of framework residues near the pseudo-2-fold axis (relating V_H and V_L by a rotation) are important for the binding of hydrophobic side-chains in a cavity in the binding site [15]. Such residues, even though formally from the murine donor, frequently need to be maintained. This involvement of framework residues is also the reason why the immune system uses different variable domain subtypes to bind antigens of many shapes and compositions (Fig. 2.3).

A structural model is usually built, exploiting the great number of three-dimensional structures of antibodies available today (*http://www.biochem.unizh.ch/antibody*) (over 300 structures of independent V_H, 220 independent V_κ and 40 independent V_λ sequences). Then a "loop-grafted" version of the antibody, carrying additional framework mutations as necessary, can be created with similar affinity to that of the original mouse monoclonal. While this goal can usually be achieved, albeit with significant effort in cloning, engineering, and assay development, the greatest limitation of this technology is that a mouse mAb with the desired specificity is needed as a starting point.

A related approach to humanization, i.e., antibody resurfacing [16–18], relies on

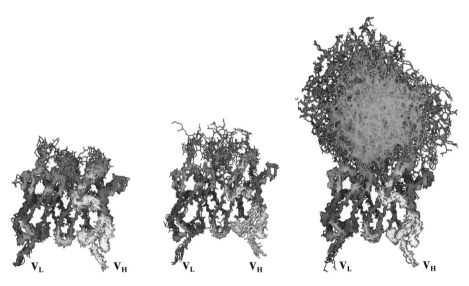

Haptens: 27 / 15 Structures   Peptides: 10 / 8 Structures   Proteins: 22 / 14 Structures

Fig. 2.3 Complexes of antibodies binding haptens, oligopeptides and oligosaccharides or proteins. The antigens are colored pink; parts of the antibody are colored as in Fig. 2.2. It can be seen that there is extensive binding to residues which formally belong to the framework, either in the dimer interface region for hapten binders or binders of peptides (which frequently use a side-chain in a hapten-like binding mode), and to the outer loop in protein binders. Hapten binders commonly form a deep, funnel-shaped binding pocket enlarged by a long CDR-L1 and open CDR-H3 conformation, while protein binders preferentially utilize a relatively flat antigen binding surface characterized by a short CDR-L1 and a closed CDR-H3 conformation. This is one of the main reasons why the immune system uses different frameworks, rather than different CDRs all on the same framework. It also highlights the points to consider in loop grafting. More details can be found elsewhere [15].

making point mutations in surface residues of the murine antibody, converting amino acids to those found in human frameworks. This process requires alignment of the sequences of the original murine antibody with various human congeners that fulfill requirements of sequence compatibility with the antibody being modified. One such antibody is now in a phase I clinical trial [19, 20].

The first humanized antibody to enter the clinic was alemtuzumab (CAMPATH-1H) which is directed against CD52, a glycosylphosphatidylinositol-anchored glycoprotein with unknown function, present on lymphocytes. CD52 is abundantly expressed on B and T cells, macrophages, monocytes, and eosinophils. Alemtuzumab is used for the treatment of non-Hodgkin's lymphoma [21] and was approved in 2001 as first-line treatment for chronic lymphocytic leukemia. It has been proposed [22] that alemtuzumab inhibits the growth of B and T cells by cross-linking of CD52. It has also been suggested that the antibody works entirely through its effector function, using both antibody-dependent cellular cytotoxicity (ADCC) and complement-directed cytotoxicity (CDC) [23]. Alemtuzumab has also been tested as an immunosuppressive reagent in transplantation and autoimmune diseases. When the humanized antibody was administered i.v., no immune response was elicited (but infusion site reactions were observed), while a s.c. injection did elicit an anti-idiotypic immune response in two of 32 patients [24].

Immunogenicity data for some other humanized antibodies are summarized in Table 2.1.

2.3.2
The Limitations Imposed by Immunization

The use of immunization to generate antibodies in animals is one of the oldest techniques in biology. The generation of mAbs [1] uses an elegant cellular cloning technique to obtain (usually murine) antibodies with single specificities, but still must start with an immunized animal. Chimerization and humanization are modern molecular engineering methods, but they also require immunization to provide a mAb as a starting point.

Even adalimumab (Humira), a fully human antibody directed against TNF-a, was obtained using a pre-existing mouse monoclonal as a starting point and a process termed "guided selection" [25]. In this approach, one chain of the murine antibody, in recombinant form, was used in phage display together with a library of human antibodies encoding the other chain, followed by the reciprocal experiment, thereby generating an "equivalent" human antibody in several steps. Adalimumab is on its way to becoming a highly successful biopharmaceutical drug for the treatment for rheumatoid arthritis and possibly also psoriasis.

Notwithstanding this spectacular success, the reliance on immunization is a major limitation in all antibody generation methods that use this step. For example, the possibility of directing the response to a particular binding epitope during immunization is very limited and usually restricted to screening hybridomas [26]. It may well be that the epitope desired for the biological effect is not particularly favored, and not one which leads to high-affinity antibodies. Furthermore, the animal repertoire is limited to those variations that the immune system introduces during somatic mutation, and these are only a small subset of the total theoretical repertoire. Therefore, the process of somatic mutation does not provide an exhaustive screen for the highest affinity or even for the optimal biological function. In contrast, modern library-based methods (see Section 2.3.3.1) provide the opportunity to select for antibodies with defined affinities, for particular epitope specificities, and even for pre-defined cross-reactivity patterns [27–29].

2.3.3
Fully Human Antibodies

Despite the outstanding achievements made with the techniques of chimerization and humanization, a means of routinely accessing fully human antibodies has always been a goal for developers of therapeutic antibodies. Historically, the first method for making human IgGs was the immortalization of human B cells with EBV. Since this method is rather inefficient, it has not been widely used. Recently, however, a new method has been introduced that dramatically increased the efficiency of transformation [30]. This offers the opportunity to immortalize memory B cells from patients after an infection, and potentially even from cancer patients, and thus complements cloning of such antibodies from patients and recovery of antibodies by display technologies [31]. Today, the most widely used technologies for making fully human antibodies are either library-based methods or transgenic mouse approaches. The fact that over 30 antibodies based on these technologies are currently in clinical trials indicates how well established they have become.

2.3.3.1 Library-based Methods
In the late 1980s, work was commenced on constructing antibody libraries. This development took antibody generation in a new direction, for the first time obviating the need for immunization of an animal. Library methods for antibody generation brought together several new technological developments, including construction of the library itself, but equally importantly, methods of screening the resulting library. This section summarizes both library construction and screening methods.

Library construction The first antibody libraries that had not been obtained from immunized animals were based on human genetic material isolated from natural sources [32, 33]. The most convenient sources of the appropriate genetic material are DNA or mRNA from human peripheral B cells, bone marrow B cells or tonsil B cells. The resulting library reflects the make-up of the human repertoire of the particular donor(s) and will to some degree reflect amplifications from recent infection events. Often, blood from many donors is pooled for this reason [33]. This bias can also be exploited – if peripheral blood lymphocytes from a patient with an infectious disease are used, such a library is an excellent source of antibodies against the infectious agent [31]. It is also possible to use non-rearranged genomic DNA as a starting point to make a library [34]. As the antibody genes in their germline configuration do not contain the V, D or J segments (and would thus be lacking CDR3 and the last β-strand of the domain), these elements need to be added in a subsequent PCR assembly.

An alternative approach, giving the highest level of control over the process, is to synthesize the antibody genes completely [35]. In this case the frameworks can be chosen to represent the optimal diversity of the antibody repertoire. In the case of HuCAL GOLD (unpublished), a fully synthetic human combinatorial library, seven frameworks for V_H and seven for V_L are used, giving 49 V_H–V_L combinations (see Fig. 2.4).

The system also comprises pre-synthesized libraries of cassettes for all six CDRs, each of which reflects the composition of the corresponding human CDRs.

To maximize diversity yet maintain structural integrity, the CDRs are not sim-

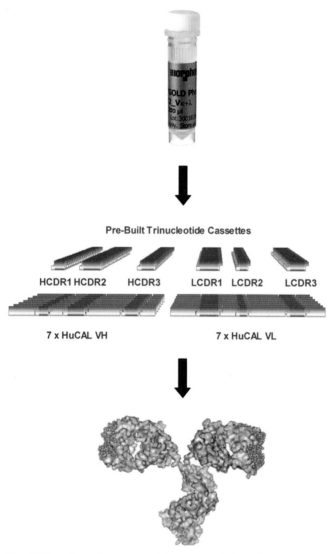

Fig. 2.4 The schematic structure of the genes underlying the HuCAL library, showing the modularity of the CDRs and the pre-assembled cassettes used during optimization.

ply randomized, but are diversified according to structural criteria, keeping a few key anchor residues constant and limiting the diversity of others to those observed in nature or compatible with the loop structure. Furthermore, the diversity of CDR1 and CDR2 is close to what is observed for a particular subtype, maximizing similarity to the germline. Conversely, the great length variation normally observed in CDR-H3 is also recaptured in HuCAL, allowing the resulting antibodies to bind to epitopes with a great variety of shapes and composition. Using this strategy, key structural residues are maintained and the variation observed in rearranged human antibody sequences, reflecting the process of affinity maturation and subsequent selection as documented in sequences of human antibodies, is well mimicked.

The HuCAL GOLD library also incorporates unique restriction sites bracketing the CDRs (Fig. 2.4), a feature made possible by the use of chemical synthesis for construction of the encoding genes. The ends of the CDR cassettes match the restriction sites bracketing their positions in the HuCAL library. A fully modular system is the result. The benefit of such a system is that antibody optimization can be rapidly and systematically carried out by replacing CDRs in turn to create new sublibraries based on one or more "hits" from a first screening. Multiple examples have shown that this is a reliable means of generating antibodies with predefined properties, while still retaining 100% "humanness" in the sequence. Such systematic optimization of antibodies builds on their inherent affinity and specificity to create substances with drug-like characteristics, including predefined cross-reactivity patterns as well as the ability to activate, deactivate and/or block certain biological processes.

The question sometimes arises as to why multiple frameworks are desirable in antibody libraries. There is a good reason why nature uses more than one framework in the antibody repertoire. As already mentioned above, antigens not only contact the CDRs, but also framework residues (Fig. 2.3). For example, large antigens make additional contact to the outer loop, or "CDR4" mentioned above, while small haptens, but also side chains of peptide antigens, often bind in a deep cavity near the pseudo-2-fold axis of the antibody [15]. These residues formally belong to the framework and thus vary between subgroups. Therefore, in order to capture the diversity of the human repertoire, a similar diversity of frameworks is necessary.

Single-framework libraries have also been made, and with a large enough diversity, high-affinity binders against many targets can be obtained [36, 37]. It must be remembered, however, that in the case of making therapeutic antibodies, usually only a very small subset of epitopes is of any utility. Therefore, obtaining high-affinity binders is a necessary, but not sufficient, condition for making a therapeutically active antibody. In general it will thus be necessary to have a technology available that can generate high-affinity binders to *any* epitope, so that cellular assays and preclinical experiments in animals can be used to identify the binders with highest potency. For example, a therapeutic antibody may need to block an interaction. In this case the antibody must bind to that part of the target engaged in the interaction, no matter whether it is protruding, recessed or flat. It is immediately obvious why a good antibody repertoire must therefore be able to bind any shape on any target molecule.

Libraries need to comprise diversity in the variable region if they are to effectively mimic the natural immune repertoire. The molecular arrangement of the library (scFv fragment or Fab fragments being the dominant formats) is, however, mostly dictated by the selection strategy used (discussed below). For example, in phage display both scFv and Fab libraries can be used, as they can be in yeast display. In contrast, in ribosome display only single polypeptides can be used, as in scFv fragments.

Screening technologies An essential technological adjunct to the creation of an antibody library is an effective means of screening it for binders having the desired characteristics. Over recent years, several technologies have been developed that allow the selection of proteins, including antibodies, from large repertoires. Some methods also allow directed molecular evolution [38]. All of them have in common that the fact that the phenotype of the protein (in the case of an antibody, its binding specificity) must be connected to the genotype (the sequence encoding the antibody in question). A schematic overview of these concepts is shown in Fig. 2.5. On affinity "enrichment" of the antibody, its DNA sequence is thereby enriched as well, allowing the identification of the precise molecular composition that gave the desired binding phenotype.

While selections in model systems and from complex antibody libraries have been described using a number of these technologies, we restrict the discussion here to a brief description of the three methods with which the greatest experience in the field of antibodies has been obtained to date.

Historically, the first technology for the selection of polypeptides from repertoires was display using filamentous phages (Fig. 2.5a), which was first demonstrated

for peptides [39] and subsequently adapted to antibodies [40, 41]. *Escherichia coli* are transformed with the library of interest in the form of an expression plasmid that encodes the antibody (either as an scFv fragment or as a Fab fragment) in fusion with a minor coat protein (usually g3p or its C-terminal domain) of the phage. Note that each *E. coli* cell receives, in general, only one plasmid molecule and will thus encode only one molecular species of antibody. When a helper phage is added to these bacteria, new phage particles are produced that incorporate the fusion protein into their coat and thus "display" the particular antibody, while packaging the plasmid, which contains the coding information of the antibody. The phage population is then added to an immobilized target and those phages that do not bind specifically to the target are removed by washing. As a result, phage-carrying antibodies with the desired binding properties can be enriched and, in principle, recovered by an appropriate elution step. Successful proof-of-principle experiments were initially reported for scFv and Fab fragments [40, 41], and many sources of libraries have been used with this selection technology since then. While space limitations do not permit a full account of the many permutations of this fundamental concept, we want to stress that it has found broad utility in selecting antibodies against purified targets, selected epitopes, but also antigens on whole cells [42]. Most antibody selections have been carried out with this technology.

A variant on this method uses a disulfide bond to link the antibody to the coat protein of the phage, in which a pendant cysteine residue is introduced [43]. The advantage of this method is that the elution step can be replaced by mild reduction of the disulfide bond, resulting in reliable re-

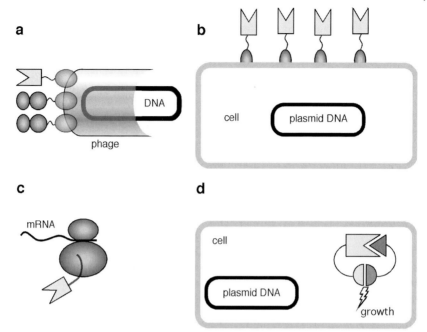

Fig. 2.5 The linking of phenotype and genotype in selection systems for antibodies. For more details on (a–c), see text. (a) In phage display, the antibody is fused to the minor coat protein g3p of filamentous phage, while the DNA is on the inside of the phage. (b) In yeast or bacterial cell surface display, the antibody is displayed on the outer surface of the cell, while the genetic information is encoded on a plasmid inside the cell. (c) In ribosome display, mRNA and protein product are linked by the ribosome, and the selection takes place in an *in vitro* translation system. Addition-ally, antibodies can be screened and selected intracellularly (d), and selection for growth puts high demands on the folding efficiency of the antibody, while the selection for affinity is not very stringent. Intracellular screening is achieved by fusing the antigen and antibody to two protein halves which, when brought together by the antibody–antigen interaction, allow cellular growth by allowing transcription of a selectable marker ("yeast two-hybrid system") [154] or reconstitution of a selectable enzyme activity ("protein-fragment complementation assay") [155].

covery of enriched phage. In contrast, elution that relies on a change of pH or a competing ligand may not necessarily lead to recovery of the most interesting antibodies, i.e., those that bind to the target with the highest affinity.

A second technology that has been used more widely recently is yeast (*Saccharomyces cerevisiae*) display [44] (Fig. 2.5 b). In this case, yeast cells are transformed with a plasmid library, which encodes a fusion with yeast *α*-agglutinin (Aga1p/Aga2p).

Again, this has been carried out both for scFv [45] and Fab [46] libraries. While transformation of yeast cells is somewhat more difficult than that of *E. coli*, methodologies do exist to achieve this. With antibodies displayed on the surface of yeast, not only a "mechanical" enrichment (e.g., with magnetic beads) is possible, but also the use of cell sorters together with a fluorescently labeled target. Therefore, thresholds for affinity can be defined and the affinity can be measured directly on the yeast cells by titration. It has been reported that the yeast libraries can be amplified without changing their composition, i.e., particular clones do not seem to be enriched or depleted upon copying the library [45].

A third display technology that has been used to screen antibody libraries is ribosome display [47–50] (Fig. 2.5 c). In contrast to the other methods, this works entirely *in vitro*, without using any cells at all. The library has to be in the scFv format, as only a single polypeptide chain can be displayed at a time. A library of PCR fragments is used, encoding a promoter and the open reading frame of the scFv fragment, fused to a "spacer", which runs to the physical end of the fragment and does not encode a stop codon. The function of the spacer is to allow the scFv fragment to exit from the ribosomal tunnel and fold into its correct three-dimensional structure. An *in vitro* translation is thus carried out with a quantity of ribosomes stoichiometric to mRNA. The mRNA is translated to the end and remains connected to the tRNA within the ribosome, the scFv protein thereby remaining connected to the ribosome, which is also still attached to the encoding mRNA. Thereby, the antibody and its encoding mRNA remain linked. The main advantage of this method is that very large libraries can be used (typically 10^{12} different variants), as unlike in the other technologies no diversity is lost in the transformation step of *E. coli* or yeast. Furthermore, by using polymerases without proof-reading capability or even, deliberately, error-prone polymerase chain reaction or other methods to increase diversity, combined with a stringent selection for affinity, binders with picomolar affinity can be selected, thereby mimicking somatic mutation *in vitro* [49, 50].

2.3.3.2 **Transgenic Mice**

A scientifically elegant development was the creation of transgenic mice, in which part of the human antibody repertoire is inserted into the mouse genome (see also Part III, Chapter 4) [51–58]. This has been achieved in a variety of ways, e.g., with yeast artificial chromosomes or pieces of human chromosomes and homologous recombination (see also Part III, Chapter 2). In order to get an efficient response of human antibodies, the mouse repertoire needs to be inactivated. This has been done by targeted deletion of the J_H and J_κ region (together with the constant κ region), to prevent V(D)J rearrangement of the murine antibody genes. This strategy permits well-established methods for the generation of mouse mAbs to be used to produce fully human antibodies. Historically, most therapeutic applications have used whole IgGs and the transgenic mouse approach produces them directly. In the case of the library approaches, which use antibody fragments during selection, a conversion to IgGs involves an additional (very straightforward) step. It can be expected, however, that a greater variety of formats for therapeutic antibodies will be employed in the future (see also Part V, Chapters 1 and 6), where the library technologies would have an additional advantage, since when transgenic mice are used, the antibody genes have first to be isolated by molecular cloning.

While a great scientific achievement, the disadvantages mentioned above in connection with immunization pertain here: lack of full control over the target epitope during the immunization process and inability to pre-determine cross-reactivity and affinity. In addition, and not unexpectedly, proteins that are highly conserved between man and mouse may not be immunogenic in this system.

2.4
Reaching the Target: The Importance of Specificity, Affinity and Format

An advantage of antibodies as potential therapeutics is their inherent affinity and specificity for their binding partner. These properties are a prerequisite for their therapeutic effectiveness. A third property is also crucial to the performance of a therapeutic antibody, i.e., its format. The importance of these molecular properties for the application of antibodies as biopharmaceuticals is considered here.

2.4.1
Epitope Specificity

Specificity for target is one of the main properties that distinguishes antibodies from other bioactive molecules. The first factor to consider is *where* exactly the antibody binds on the target. Some cases are easy to understand at the molecular level, such as in the case of blocking the action of a cytokine, such as TNF-*α*. This inflammatory cytokine is produced too abundantly in a number of diseases, such as inflammatory bowel disease (Crohn's disease and ulcerative colitis), rheumatoid arthritis and psoriasis. The therapeutic strategy thus consists of preventing the binding of this soluble, homo-trimeric molecule to its receptor. The antibody, obviously, must bind in a way that it overlaps with the binding interface to the receptor – it must bind to a "neutralizing" epitope.

In some cases, such as the one mentioned, such binders are quite straightforward to select, as the receptor contact surface is large. Indeed, TNF-*α* is a popular target for antibody and other protein-based therapeutic approaches, developers being encouraged by the success of etanercept (Enbrel, a soluble receptor–Fc fusion pro-

tein), infliximab (Remicade) and adalimumab (Humira) for the treatment of arthritis, psoriasis and Crohn's disease [59]. All three molecules are therapeutically active in rheumatoid arthritis. Interestingly however, the soluble receptor etanercept shows no activity in Crohn's disease, while infliximab does. A possible reason has been proposed [59]: to be effective, the transmembrane form of TNF-*α* must be targeted, which is present on T cells, where it may have a slightly different conformation, and this conformation is not recognized by the receptor, but by the antibody infliximab. Only the latter thus helps controlling inflammation by inducing apoptosis in T cells.

The antibody-binding site does not have to be identical to that of the receptor, it only has to overlap in such a way as to prevent simultaneous binding of the target to its receptor. In cases other than the ones described, this can be more difficult to achieve, e.g., if the binding site is small and not favored for binding. In such a case, using antibody library technologies it is often possible to "guide" the selection to the relevant epitope. This can be done in a variety of ways, such as using the "real" partner (e.g., the soluble receptor) as a competitor or using mutants of the soluble molecule to pre-bind all antibodies that are not directed to the desired epitope. The interested reader is directed to publications in which the technical approaches are discussed in more detail (see, e.g., [60]).

It is almost always necessary to test the binders so obtained in cell-based assays, in order to verify that they react with the antigen in its proper context on the cell. In some cases, on the other hand, it may not even be possible to obtain any soluble version of the protein of interest. In this case, selections can be carried out on whole cells and many of the same strategies applied. Again, the technical approaches and remaining challenges have been described [26–28], but space limitations do not permit us to discuss this in detail.

Blocking (or neutralizing) epitopes are easy to conceptualize. It should be noted, however, that in many cases the rationale why only binders to a particular epitope give a biological response is not at all clear. This is illustrated with two antibodies against HER2, a member of the epidermal growth factor receptor family overexpressed in about 25–30% of women with breast cancer and correlated with a poor prognosis (see also Part I, Chapter 5). The recently described crystal structure of 2C4 (Pertuzumab, Omnitarg) in complex with HER2 shows that this antibody inhibits the homodimerization of HER2 and also its heterodimerization with other members of the family, and thus signaling [61]. However, the antibody trastuzumab (Herceptin, 4D5) also inhibits signaling, yet without inhibiting dimerization [62, 63], and it binds close to the membrane, as shown in the crystal structure of its complex with HER2 [64]. The mechanistic reason for its inhibitory action, which is limited to tumors with high levels of homodimers of HER2, is still not entirely clear, and it is likely that internalization and/or proteolytic shedding are two of the factors influenced by trastuzumab binding, which then decrease signaling [62, 63]. Additionally, there is evidence [65] that Fc receptors on natural killer (NK) cells are recruited by the exposed Fc part of trastuzumab while bound to HER2 and this action is providing part of the therapeutic effect of this antibody. It is almost certain that the combined effect of all these factors is what gives Herceptin its efficacy.

With all the excitement surrounding this biopharmaceutical, it should not be forgotten that, for example, with Herceptin as

the sole treatment in metastatic breast cancer, only eight complete responses amongst 222 women were seen in a phase III trial [66]. Similarly, only 3–23% of all patients receiving Rituximab for relapsed or indolent refractory B cell non-Hodgkin's lymphoma showed a complete response [67]. These two examples, both concerning FDA-approved antibodies (see also Part VII, Chapter 4), underline the urgent need for further work in understanding the action – and subsequent improvement – of such antibodies, as well as the biological function of potential target antigens.

In cases where the structure and function of the target are known, presenting the "relevant" epitope in order to generate binders of the desired specificity is still the key challenge. A prime example is the difficulty in generating protective antibodies against HIV [68] (see also Part II, Chapter 7). The surface proteins of the virus, gp120 and gp41, are highly immunogenic, regardless of whether they are presented in the context of the virus particle (in infected patients) or as the soluble protein (as shed protein in infected patients). Antibodies can also be obtained readily, using immunization or display technologies. However, the great majority of antibodies that result are directed against non-protective surface epitopes. Indeed, the isolated protein does not even present the "protective" epitope, which is recessed between trimeric subunits when arranged as present on the virus. However, the virus particle normally does not elicit protective antibodies either, and thus it has been so far impossible to obtain an AIDS vaccine. Since HIV biology is well studied, a number of strategies are currently under way to overcome this challenge [69] (see also Part II, Chapter 8). Very few broadly neutralizing antibodies have been cloned from infected patients, but the mechanism of neutralization is not clear for all of them [70].

The difficulty of generating broadly neutralizing antibodies against HIV may have attracted much attention, but it is likely that inaccessible neutralizing epitopes, and epitopes that are only formed after an initial binding event, are much more common also in other biological systems and not restricted to infective agents.

In some cases, even the precise molecular definition of the desired epitope is unclear. Accordingly, the preclinical observation will frequently be that only very few antibodies will show a biological effect, even though many others bind to the same target with high affinity. It will usually be of great benefit to attempt to understand, at the molecular and structural level, the key features distinguishing the biologically active antibodies from the others. In the case of cellular targets, this may involve differences in receptor multimerization (the antibody either inducing or preventing it), in the ensuing receptor internalization, proteolytic shedding, blocking (or enhancing) the binding of an external ligand, or making the antibody Fc part accessible – while binding to the surface target – to macrophages, neutrophils, monocytes or NK cells carrying Fc receptors. It immediately follows that the type of Fc desired (or its desired absence) is another parameter important for engineering and it requires an understanding of the mode of action of the antibody.

A field of medicine where this lack of mechanistic understanding at the level of molecular structure is particularly notable and hampering progress is the treatment of cancer (see also Part II, Chapters 5 and 6). There are great challenges in targeting solid tumors, brought about by the enormous difficulty inherent in obtaining significant and selective enrichment of the antibody at the tumor site. As a consequence, the majority of antibodies approved today are directed against leukemia, myeloma and lymphoma (see also Part V, Chapters 5 and 6). In these cases, two factors favor clinically successful treatment with antibodies: (1) the target cells are easily accessible in the bone marrow, lymph nodes or blood, and (2) the tumors respond well to radiation and chemotherapy. Furthermore, they can be selectively targeted via several cell-lineage-specific markers. For example, CD20 is a marker specific for B cells (a more detailed description is given in Section 2.5.2). In this case, the antigen is not restricted to the diseased cell, but the redundancy and the self-regeneration of the immune system can sustain the temporary depletion of B cells. It is in general much more difficult to identify selective markers for solid tumors [71]. The number of such tumor markers that are suited for targeted therapy is small: despite massive attempts using a variety of techniques, including screening of healthy and diseased tissues with antibody libraries, only very few new tumor-associated surface proteins have been added to the list over the years [72–74].

2.4.2
Affinity

An important factor determining therapeutic efficacy is affinity. If the goal is to block the action of a soluble target such as a cytokine, then as little as possible of the cytokine should remain in an active form. The affinity directly determines the amount of cytokine that will be free at equilibrium. In general, the affinity should thus be as high as possible for such applications. It should be noted that in many cases the soluble protein to be inhibited is a monomer (trimeric TNF-α and its homologs being more unusual in this respect), so that the true monovalent thermodynamic affinity, i.e., the affinity of a single binding site, is the property of interest.

Another example where the importance of affinity has been clearly highlighted is the protective function of antibodies against toxic or infectious agents. For example, post-challenge protection against the anthrax toxin, a tripartite protein, correlated well with the dissociation constant of the antibody, all other properties being equal [75].

If the antibody is to be used for cellular targeting, however, more complicated relationships apply. A number of investigations in tumor targeting have uncovered some of these trends [76–78]. From these and other studies, it appears that tumor targeting in general improves with affinity, but seems to reach a plateau at affinities around 10^{-9} M. The steady-state concentration at the tumor does not get higher with higher affinity, as the dissociation rate from the tumor antigen is no longer limiting steady-state concentration once dissociation is very slow. Instead, cellular uptake and bulk flow become dominant parameters, and these are influenced by the format of the molecule. One should draw such conclusions on the importance of affinity only from comparisons of molecules which are point mutants of each other, but have otherwise exactly the same format – as different epitopes (some eliciting antigen internalization, others preventing dimerization), formats, molecular size, etc., will change targeting efficacy for different reasons (see Section 2.4.3).

Affinity is increased in the immune system by somatic mutation [79–81]. Space limitations do not allow us to discuss in detail strategies for the affinity maturation of antibodies using the various display technologies, and how the selection of high-affinity binders can be favored. This is, however, now possible in a variety of ways and the interested reader is referred to a number of articles [82–87].

2.4.3
Format and its Impact on Pharmacokinetics

An important factor to consider is also the format of the final therapeutic molecule. The availability of bivalent immunoglobulins as well as monovalent Fab and scFv fragments makes valency an important consideration in drug design. If the protein to be blocked is soluble and contains only one copy of the epitope, as is, for example, the case in many cytokines and protein hormones, no interaction strength is gained by having IgG molecules over Fab fragments or scFv fragments. This is also true for targets on the cell surface, if they are so far apart that the two arms of an IgG cannot reach two identical epitopes. In this case again, the binding affinity and thus the blocking affinity of a monovalent antibody fragment will be identical to that of a bivalent one.

Nevertheless, the longer half-life of IgG molecules may be advantageous as it guarantees a longer duration of the blocking function. The half-lives of therapeutic antibodies have been reviewed [88] and were mentioned in Section 2.2 in the context of immunogenicity (see also Part VI, Chapters 1 and 2). The longer half-life of an IgG is caused not only by the higher molecular weight of the IgG and thus the inability to be cleared through the kidney, but mostly because of a particular mechanism selectively protecting IgG from normal serum protein catabolism [89]. Over 5 days, the human vascular endothelium engulfs all serum by endocytosis. The content is processed through a complex network of endosomes and tubules with decreasing pH. The neonatal Fc receptor

(FcRn; also termed Brambell receptor after its discoverer) is expressed in hepatocytes, endothelial cells and phagocytic cells of the reticuloendothelial system, the main location of protein catabolism. The receptor binds the Fc part of IgGs using charged histidines and thus prevents antibodies from ending up in lysosomes. Instead, FcRn with the bound IgG recycles to the same cell surface, releasing the intact IgG at the serum pH of 7.4. By this mechanism, the half-life of IgG is increased by a factor of 10 compared to the absence of this receptor in transgenic animals [89, 90]. In short, the use of whole IgG guarantees a long blocking function through its long half-life, even if it binds only monovalently to its target.

Murine IgG does not bind to human FcRn, and this explains the shorter half-life of murine antibodies in human patients, typically 12 to 48 h [91]. The half-lives of endogenous human IgG isotypes have been well studied and they do differ – 3 weeks for IgG1, IgG2 and IgG4, while IgG3 has a half-life of 1 week [88, 92].

Abciximab (ReoPro), mentioned in Section 2.3.1 as an example of a chimeric antibody, is unique among therapeutic antibodies marketed thus far in being a Fab fragment. Its half-life in plasma is only 20–30 min [93], but when interacting with platelets, this rises to 4 h. It is now clear that the antibody is dynamically redistributed between individual target molecules and platelets in less than 1 h. Thus, while the half-life of dissociation from each individual integrin molecule is rather fast, the high local concentration of integrin molecules on the platelets provides the drug with a long platelet-bound half-life, with antibody detected on platelets as long as 2 weeks after therapy. This leads to a prolonged inhibition of platelet function in response to shear stress for 72 h to 1 week [94].

In the targeting of solid tumors, the situation is again far more complicated. Large IgG molecules, because of their long life time in serum, maintain a very high steady-state concentration. From this reservoir, levels at the tumor can reach very high percentages (20–30% of the injected dose per gram), but tumor to blood ratios are very small. The problem is that diffusion of large proteins such as IgG through a solid tumor is very slow and inefficient, because the antibody is in competition with removal by bulk flow. In addition, the tumor has a high hydrostatic interstitial pressure, is heterogeneous in composition and density of antigen expression, and has reduced vasculature. Furthermore, because of the slow accumulation and the long half-life, the antibody will at no time be truly selectively enriched at the tumor (when expressed as the percentage of the injected dose per gram of tissue or blood).

This essentially constitutes a limitation on the use of toxic molecules or radioactive isotopes (see Section 2.5.7) being conjugated to the antibody in many applications: at a dose approaching the toxicity limit, the tumoricidal effect is often not yet reached [95] (see also Part V, Chapters 1 and 6). The dose-limiting organ is usually determined by the action of the free drug, as some amount of drug can be cleaved non-specifically. Since IgGs are mainly degraded in the liver, there may be concerns for liver toxicity as well for antibody–toxin conjugates. In contrast, for radionuclides, bone marrow is usually the dose-limiting organ (see below and Section 2.5.7), as some radionuclides (e.g., yttrium) can be incorporated in bone marrow [96] (see also Part V, Chapters 4 and 5).

Radioimmunotherapy provides an example of a setting in which a shorter half-life can be advantageous. Currently, two antibody–radioisotope conjugates are on the market, both for the treatment of non-Hodgkin's lymphoma, i.e., ibritumomab tiuxetan (Zevalin) and tositumomab (Bexxar) (see also Part V, Chapter 7). Zevalin and Bexxar carry ^{90}Y and ^{131}I, respectively, but both are mouse antibodies. As detailed above, in the radioimmunotherapy setting, one of the main challenges is maximizing the dosage of radioactivity reaching the targeted tumor cells without delivering dangerous levels of non-specific radiation to organs, notably the bone marrow, the organ where hematopoietic stem cells are generated, the precursors of all blood cells. This balancing act requires that the antibody has a relatively short half-life, which is the reason why murine antibodies have been favored in this setting. The half-lives of the two marketed products illustrate this point: both ibritumomab tiuxetan [97, 98] and tositumomab [99] have half-lives of 1–3 days. To safely use rituximab in radioimmunotherapy, a chimeric antibody whose antigen (CD20) is the same as that of tositumomab would require dosimetry of individual patients [100].

In other words, the degree of humanness of an antibody, or in molecular terms, the lower affinity to the FcRn and thus the lack of selective prevention of degradation, can be used to achieve a half-life that is required for a particular therapeutic window. It is likely, however, that in the future fully human IgGs will be engineered with decreased FcRn receptor binding to obtain a particular half-life.

Smaller protein molecules (Fab fragments, scFv fragments) localize to a solid tumor much faster and also diffuse better through tumor tissue, but because of their faster excretion rate the steady-state level reached in serum is much lower. Particularly below about 25–50 kDa, clearance through the kidney becomes possible [101, 102]. It should be noted that this also

shifts safety concerns for antibodies conjugated with a toxic moiety from liver toxicity (for large proteins) to kidney toxicity (for smaller proteins), as a fraction of the recombinant molecules can be taken up by kidney parenchyma cells.

The increase in functional affinity to cell surfaces by having multiple binding sites – previously a hallmark of IgGs – has now been engineered into scFv and Fab fragments as well [103]. This "avidity effect" strongly increases the residence time on a surface-bound target molecule, provided that the antibody can reach two epitopes simultaneously. In combination with site-specific PEGylation remote from the antigen-binding site, the hydrodynamic properties and number of binding sites can now be engineered independently to achieve a compromise in the quest to optimize tumor targeting [7–9].

From these considerations it is clear that the optimal format of the antibody is dependent on the exact mode of action, its location, the effector mechanism, whether the antibody has been fused or conjugated with a toxin and what kind of toxin or toxic radioisotope is used. The use of an IgG is thus only one of many options. The advent of molecular engineering is thus pivotal to the further development of these therapeutic modalities.

2.5
Exerting an Effect at the Target

Ever since the first attempts to create antibodies for human therapy it has been recognized that, in most cases, mere binding to the target is necessary, but may not be sufficient. There may be additional prerequisites for an antibody to be a biopharmaceutical, including the blockade of a particular interaction and/or cell killing.

The largest number of therapeutic antibodies in development is in the area of cancer. This area also provides the most examples of the variety of ways in which antibodies can be used in medicine. Whereas in other diseases blockade of a particular interaction may be sufficient to have a therapeutic effect, the objective in cancer is to kill tumor cells, which usually requires some form of direct cytotoxicity. It is generally assumed [104, 105] that effective tumor killing by a naked antibody will use one or a combination of (1) blocking a growth signal, (2) delivering an inhibitory signal, (3) inducing apoptosis and (4) eliciting an immune response against the tumor. The relative importance of these factors depends on the tumor type and the targeted antigen. This section considers several such approaches in the context of anticancer drug development.

2.5.1
Blockade

Bevacizumab (Avastin) is a humanized antibody that is approved for the treatment of metastatic colorectal cancer [106] and is directed against vascular endothelial cell growth factor (VEGF), a molecule that stimulates angiogenesis. The antibody binds VEGF and thereby inhibits vascularization of tumors that overexpress the growth factor; preclinical studies show clear inhibition of tumor growth in a mouse xenograft model [107]. Bevacizumab is therefore a rare example of an anticancer antibody which exerts its effect by blocking a growth factor which is important for tumor cell proliferation, but without interacting directly with the cancer cell.

2.5.2
Naked Antibodies that Trigger Cell Killing

From a drug development perspective, an antibody that exerts its therapeutic effect without the requirement for further modification has a major advantage. Steps that require conjugation, chemical modification or new production methodology all add complexity, cost and risk to the development of a therapeutic antibody. Naked antibodies, which can be made using established, well-characterized cell lines and production/purification methods avoid these difficulties.

Several antibodies have been suggested to kill tumor cells by ADCC or CDC, or a combination of these effects. As noted above, CAMPATH is an example of an anticancer antibody that seems to work via such effects. Rituximab (Rituxan) is a chimeric antibody against CD20 that is approved for the treatment of non-Hodgkin's lymphoma [108]. The function of CD20, a well-known marker for B cell activation, is not precisely known, but it has been suggested to be involved in Ca^{2+} influx as a tetrameric molecule. Direct effects of Rituximab, including growth inhibition and apoptosis, have been shown *in vitro* and a cross-linking of lipid rafts by the antibody has been proposed, with a *trans*-activation of src kinases eventually leading to apoptosis [109]. However, it is unclear whether this contributes to the clinical benefit observed. In fact, the published data suggest that the predominant effector mechanism is ADCC, with a minor role played by complement [110]. This is the same target as that against which two radionuclide-conjugated antibodies (Zevalin and Bexxar; see Section 2.5.7) are directed, where the therapeutic mechanism is thought to be at least partially dependent on radiation damage.

A recent example of an anticancer antibody in development that functions via apoptosis is 1D09C3, a HuCAL-derived antibody specific for human leukocyte antigen (HLA)-DR [111, 112], which is highly expressed in B and T cell lymphomas. At the time of writing, this antibody is about to enter clinical trials for B cell lymphomas. The antibody causes cell killing selectively in activated tumor cells without the need for exogenous effector cells, via a mechanism that is caspase independent.

Safety concerns with drug- or radionuclide-conjugated antibodies are mostly due to the systemic effect of the toxic effector molecules. This problem is augmented by the high dose required, which in turn is due to the insufficient localization of the antibody, notably to solid tumors.

In contrast, naked antibodies are not systemically toxic per se. Of course, a particular antibody can be toxic by its biological effect on the chosen target – either at the intended tumor site or at another site where the target is expressed. For example, Herceptin shows incidence of cardiotoxicity, especially in combination with anthracyclines [113]. This is caused by HER2 being expressed in cardiac myoblasts, where it is involved in muscle spindle maintenance.

2.5.3
Modifying the Fc Portion to Enhance Effector Cell Recruitment

As antibodies are often able to trigger cell killing via Fc-mediated effector functions such as ADCC and CDC, much effort has been put into increasing this effect by modification of the Fc portion. A crucial experiment underlining the importance of the different receptors involved in mediating ADCC was reported recently [65].

While activator receptors, such as FcγRIIIA on NK cells, are responsible for arresting tumor growth by Herceptin in a transgenic mouse model, FcγRIIB receptors were found to be inhibitory to the killing action – mice deficient in this receptor showed much more pronounced ADCC. It is therefore tempting to modulate the relative binding to the activating and inhibitory receptor by engineering the Fc part.

Using a systematic series of point mutants in the Fc part, Presta and colleagues [114] identified residues which discriminate between binding to FcγRI, FcγRIIB and FcγRIIIA. A further complication is introduced by the presence of an allelic variant in human FcγRIIIA, which binds differentially to the mutants. Using a triple mutant IgG (S298A/E333A/K334A), improved binding to FcγRIIIA as well as more potent cytotoxicity in ADCC assays was observed [115]. Interestingly, in the recent crystal structure of the IgG1–Fc/FcγRIIIA complex, only one of these residues (Ser298) was found to make a direct contact to the receptor.

Using the crystal structure of the Fc/Fc receptor complex as a guide, other mutations in the protein sequence of the Fc part have been designed that were predicted to increase the interaction with the receptor (Dahiat et al. unpublished; Xencor).

The Fc part is glycosylated on an asparagine residue (Asn297), and while the recent crystal structure of a complex of an Fc with the FcγRIII and FcγRIIA [116–120] shows that the oligosaccharide makes only minimal contact to the receptor, it has been known for a long time that efficient binding to the receptors requires glycosylation [121]. Structurally, the oligosaccharides attached to the conserved Asn297 of IgG are a biantennary type with a core heptasaccharide that consists of four *N*-

acetylglucosamines (GlcNAc) and three mannoses, and variable fucose addition to the core at the first GlcNAc residue [121]. From the investigation of the crystal structure of the Fc/FcR complex and a series of glycosylation truncation mutants, it became clear that the sugars act both to increase the distance (with the wild-type being optimal) and to decrease mobility of the receptor-interacting segments of C_H2 domains [122].

One approach to improve FcγRIII interactions has been to add a bisecting GlcNAc sugar, by using engineered Chinese hamster ovary (CHO) cells expressing a glycosyl transferase [β(1,4)-N-acetylglucosaminyltransferase III] (see also Part IV, Chapters 2, 5 and 7). The resulting antibody killed neuroblastoma cells at 10- to 20-fold lower concentrations than when this bisecting GlcNAc was not present [123].

Recently, another strategy of "glyco-engineering" was reported to lead to an increased binding to FcγRIIIA and thereby to enhanced ADCC [124], i.e., the production of the antibody in host cells that do not add fucose. In a direct comparison, this strategy was more effective than adding the bisecting GlcNAc.

In an independent series of experiments using a CHO-derived cell line (lec 13) that is unable to add fucose, this lack of fucose was shown to have no effect on binding to FcγRI (CD64) and only a marginal effect on binding to FcγRIIA and FcγRIIB (CD32), but led to a significant increase in the binding to FcγRIIIA (CD16) [125] and even led to a further increase of binding of the engineered triple mutant IgG1 S298A/E333A/K334A, which indeed translated to improved ADCC *in vitro* [126]. Interestingly, the lack of fucose correlates with an increased receptor on-rate, suggesting Fc stabilization in the active con-

formation, while the point mutations lead to a decreased off-rate, suggesting higher interaction strength [127].

In contrast, the presence and absence of fucose had no effect on binding to FcRn (which controls half-life) or C1q (which controls complement activation). The binding of the Fc part to the neonatal receptor, FcRn, can also be influenced by mutations [128, 129]. These mutations must, however, be designed to differentially affect the binding to the receptor in a pH dependent manner, such that recycling of the IgG works properly. Encouraging mouse experiments have been reported (summarized in Presta [125]), but clinical trials in patients have not yet been carried out.

The role of carbohydrate in the clearance rate of IgG remains unclear. Binding of aglycosyl IgG to the FcRn appears identical to that of the glycosylated form. Nevertheless, there is some disagreement in the literature about the influence of glycosylation on the rate of clearance. While some studies detected a difference, others did not (summarized in Shields et al. [114]).

2.5.4
Low-molecular-weight Drug Conjugates

New effector functions may be needed that exceed the efficacy of the Fc part itself, since, in a particular tumor setting, ADCC and CDC may either not be elicited well or even not at all or not be effective enough by themselves. For many years, antibodies have been conjugated to cytotoxic agents with the intention that the antibody "targets" the desired payload to the tumor, thus providing the proverbial "magic bullet". This has been a long and winding road, with the therapeutic window frequently not opening wide enough between systemic toxicity and lack of efficacy. Nevertheless, improved molecular understanding, and consequently better molecular design, is likely to give these approaches a renewed chance. While totally selective targeting may never be achieved, engineering of the antibody with regard to target epitope, affinity, selectivity, molecular size and thus pharmacokinetics, with the consequences of degradation of the non-localized drug-loaded antibody firmly in mind, may increase tumor selectivity above the threshold of therapeutic utility, such that more toxic conjugates can be used as drugs. IgGs may not be the preferred molecules for such approaches and the advent of recombinant technologies has greatly increased the number of possibilities regarding the molecular format.

The conjugation of antibodies to low-molecular-weight cytotoxic agents has been investigated for many years. The target in these cases should be an internalizing surface protein, as most small molecule drugs act as inhibitors of cell replication and therefore need to reach the cytoplasm or nucleus to exert their effect [130]. A case in point is gemtuzumab ozogamicin (Mylotarg), the only antibody-based drug based on this approach to reach the market. Mylotarg is a chimeric anti-CD33 antibody conjugated to the highly potent enediyne drug calicheamicin and is approved for the treatment of acute myeloid leukemia [131]. CD33 belongs to a growing family of sialic acid-binding, immunoglobulin-like lectins (siglecs). It appears to be an inhibitory receptor in myeloid lineage development [132] and is highly expressed in myeloid leukemia. The conjugation of the drug to the antibody gives a more favorable therapeutic window compared to the drug alone, despite the fact that the marketed preparation is heterogeneous, comprising a significant fraction of unconjugated antibody.

The antibody BR96, which recognizes an extended form of the Lewis Y carbohydrate antigen present on many carcinomas [133, 134], has shown some promise as a conjugate with different small molecule drugs. Early attempts with doxorubicin (DOX), a drug of the anthracycline family, conjugated to BR96 provided excellent preclinical data. Nevertheless, phase II clinical trials of BR96–DOX for the treatment of metastatic breast cancer or gastric carcinoma showed limited efficacy, with elevated gastrointestinal toxicity, probably a consequence of the target being also expressed in healthy gastric epithelial cells [135, 136]. More recently, synergistic antitumor activity in animal models has been demonstrated for BR96–DOX in combination with the taxanes docetaxel and paclitaxel [137]. The conjugate is currently in clinical development for the treatment of non-small cell lung carcinoma in combination with docetaxel.

That factors beyond the antibody and cytotoxic agent are crucial in creating an effective biopharmaceutical is demonstrated by recent work with BR96. Conjugates comprising BR96 linked, via two different chemistries, to cytotoxic auristatin derivatives have shown promise in animal studies [138]. Clear superiority in this study was achieved by incorporating an enzyme-cleavable linker between antibody and cytotoxic agent, thereby increasing the efficiency and specificity of release of drug at the target.

In general, antibody–drug conjugates should be based on very highly potent small molecule drugs (see also Part V, Chapter 6). In addition to the cytotoxic agents mentioned above, examples of other small molecule drugs being investigated include tubulin polymerization inhibitors such as the maytansanoids, CC1065 and taxoids [130, 139]. As mentioned above, the linker chemistry, being either acid labile or enzymatically cleavable, possibly with matrix metalloproteinases as tumor-specific release mechanisms, are also factors being studied. While preclinical data are impressive, it remains to be seen whether a useful therapeutic window can be found for broad application of toxin immuno-conjugates in oncology.

2.5.5
Protein Toxin Conjugates

A conceptually similar approach is the conjugation of protein toxins to antibodies. Such toxins, typically from plants or bacteria, are enzymes that catalytically inactivate essential cellular processes such as translation. By covalently modifying a translation factor or the ribosome itself in an enzymatic process, a single enzyme molecule is sufficient to kill a cell [140–142]. The best clinically studied members of this group are *Pseudomonas* exotoxin A, a tripartite protein that enzymatically ADP-ribosylates elongation factor 2, and ricin, derived from the plant *Ricinus communis*, which modifies a critical nucleotide in eukaryotic ribosomal RNA. The natural toxins are produced with their own, unspecific uptake mechanism that allows them to infect any cell, exploiting receptor molecules ubiquitously expressed on mammalian cells. By deleting these cell-binding domains and replacing them by an internalizing antibody, typically in the single-chain format, tumor-selective killing can be achieved. The antibody thus mediates uptake of the enzyme by tumor cells.

Nevertheless, toxicity remains an issue for systemic applications of these immunotoxins, as does immunogenicity of the toxin part, which limits repeated dosing. In order to reach therapeutic levels in a solid tumor, high initial doses have to be used, but liver toxicity was observed in a trial against HER2 [143] (see also Part I,

Chapter 5). Therefore, more recent work has focused on applications in leukemia and lymphoma. Encouraging results were observed with an immunotoxin against CD22 in patients with hairy cell leukemia and chronic lymphocytic leukemia [144]. CD22 is a member of the siglec family, and serves as a receptor for sialic acid-bearing ligands expressed on erythrocytes and all leukocyte classes. CD22 appears to be primarily involved in the generation of mature B cells within the bone marrow, blood and marginal zones of lymphoid tissues [145].

A combined phase I/II trial for ricin conjugates with antibodies against the lymphocyte activation markers CD25 (the IL-2 receptor *a*-chain) or CD30 (a member of the TNF receptor superfamily, possibly involved among others in memory T cell development) for patients with Hodgkin's lymphoma showed some promise [146].

2.5.6
Cytokine Fusions

Since many tumors do elicit an immune response, albeit one which may be unable to eradicate the tumor, an attractive strategy would appear to be to enhance this response with immunostimulatory cytokines. In order to localize the cytokine to the tumor, fusion proteins with antibodies have been made. Constructs investigated include interleukin (IL)-2, IL-12, granulocyte macrophage colony-stimulating factor (GM-CSF) and members of the TNF superfamily [147, 148] (see also Part V, Chapters 1 and 6). In a recently reported phase I trial of a humanized mAb directed against the GD2 disialoganglioside, reversible clinical toxicities were reported together with the desired immune stimulation [149]. As is the case with bispecific antibodies (see below), the main challenge in these approaches will be to prevent systemic engagement of the cytokine receptor by the cytokine part of the conjugate in the absence of the antibody binding to the tumor, as this is the most likely source of side-reactions. The severity of the problem will depend on the complex interplay of pharmacokinetics of the antibody reaching the tumor or the cytokine receptor.

2.5.7
Antibody-Radioisotope Conjugates

Some aspects of radioimmunotherapy have already been discussed above in the context of half-life. The conjugation of an antibody to a radioactive element that should cause radiation damage at the tumor site is an idea that has been pursued for a number of years [95] (see also Part II, Chapter 5, and Part V, Chapters 5 and 7). Again, progress with solid tumors has been modest, while encouraging results are obtained in the treatment of hematopoietic neoplasms. As noted above, the fact that [^{131}I]tositumomab (Bexxar) and [^{90}Y]ibritumomab tiuxetan (Zevalin), the only FDA-approved radiolabeled therapeutic antibodies, are both of murine origin contributes significantly to a shorter half-life, which is desirable in this setting, but also creates a considerable immune response. This, however, is diminished in patients with hematopoietic disorders or prior chemotherapy. Interestingly, large quantities of the unlabeled antibody must be administered prior to or concomitantly with the radioconjugate to improve targeting. The relatively low dose that is sufficient for treating hematopoietic malignancies reduces adverse side-effects and may be the reason why a therapeutic window can be found in this case.

Improvements for solid tumors may potentially come from the use of pre-targeting strategies [95]. In this case the radionuclide is not directly coupled to the antibody. Instead, an antibody Fab fragment with a hapten binding function (e.g., a bispecific antitumor×antihapten construct) is injected first and allowed to concentrate at the tumor. Once the majority has left the circulation, the radionuclide, coupled to a monomeric or dimeric hapten [95] is injected. The hapten derivative's extremely short half-life, combined with the newly generated binding sites by the noninternalizing Fab fragment on the tumor, if present on a nonshedding surface antigen, allow excellent tumor selectivity. Nevertheless, it remains to be seen whether a useful therapeutic window can be obtained, with the concern of potentially new dose-limiting mechanisms for uptake of the radionuclide.

2.5.8
Bispecific Antibodies

Attempts have also been made to use bispecific antibodies to recruit effector cells. A number of challenges need to be overcome in this field. First, a robust method must be found by which such proteins can be produced. Initially, the co-expression of two antibodies in one cell and the separation of the one desired out of the 10 conceivable molecular forms did not seem an attractive proposition. However, a multitude of methods have been reported over the last few years [103, 150] to create bispecific formats of the antibody with a defined molecular composition. These include (1) the co-expression of heavy chains that have been engineered to allow only the desired pairing, (2) the direct chemical linkage of two different Fab fragments, (3) a number of different bispecific recombinant antibody constructs based on scFv fragments fused to heterodimerizing peptides and proteins or (4) the direct enforced pairing of the domains in so-called diabodies (see also Part IV, Chapter 16 and Part V, Chapter 1). It would be too early to favor one form over the others.

A second challenge is derived from the fact that binding with only one of the arms, e.g., the one engaging the effector cells, is a likely intermediate in the reaction: no systemic toxicity should result in such a case so as to avoid safety concerns. It is very likely that the binding to a solid tumor is much slower than binding to cells of the immune system found in the serum. The third challenge is the converse – the antibody may eventually bind to the tumor, but never reach an effector cell, because there is none there, or, for geometric reasons, its receptor cannot be reached or activated.

Factors such as these translate into practical limitations. Typically, excessively high concentrations of bispecific antibody are needed to see an effect *in vivo*. In addition, particularly in solid tumors, the effector cells are often ineffective in the absence of a local co-stimulatory signal, which usually requires the addition of an exogenous factor, a serious drawback for a viable therapeutic strategy.

A number of strategies have been developed to use bispecific antibodies to recruit different types of effector cells. Much of the earlier work sought to recruit effector cells via the IgA receptor FcαRI or the IgG receptors FcγRI or FcγRIII. In a phase I/II trial in 16 patients, a bispecific anti-CD30$\times$anti-FcγRIII construct led to one complete and three partial remissions plus four cases of stable disease [151]. Pretreatment with IL-2 cytokine resulted in augmented antitumor activity, possibly by an additional mechanism of activation of NK cells. In another trial, a bispecific anti-CD30$\times$anti-FcγRI construct was tested in 10 Hodgkin's lymphoma patients [149]: one complete and three partial remissions, plus four cases of stable disease were reported.

Quite in contrast to the situation with lymphoma, no responses were seen with solid tumors, using a bispecific anti-HER2$\times$anti-FcγRI antibody in combination with interferon-γ or GM-CSF [152] (see also Part VIII, Chapter 3).

Attention has also focused on recruiting cytotoxic T lymphocytes (CTLs) using CD3 as the trigger. Limitations of the type mentioned above have again hampered progress in this field: to date, no clinical efficacy has been observed on systemic administration of anti-CD3 based bispecific antibodies [150].

A newer technology, which seeks to overcome some of the disadvantages of previous bispecific CTL approaches, is unique in comprising two single-chain Fv fragments linked in tandem [150]. These types of molecule have been termed "bispecific T cell engager" or "BiTE" constructs. It also has an anti-CD3 recruitment arm and a second specificity directed against a tumor marker. The molecules tested seem to have two advantages over other CTL-recruiting bispecific antibodies described previously: (1) they do not appear to require co-stimulation of T cells and (2) they appear to catalyze killing of multiple target cells by a single T cell. These advantages, if they should translate into the clinic, might offer significant dosing and cost-of-goods benefits for therapeutic antibodies of this type.

2.6
Antibodies in their Natural Habitat: Infectious Diseases

There is one field of medicine where the antibody in its classical format may indeed constitute the optimal molecular design – in the defense against infectious agents, i.e., the normal function of an antibody. Palivizumab (Synagis) is a humanized mAb that prevents lower respiratory tract infection by respiratory syncytial virus, and it is used in pre-term infants and other young children at risk, such as with cardiopulmonary disease or immunosuppression [153].

In general, however, passive immunotherapy has not been the focus of many development projects, as small molecule antiviral and antibacterial agents would clearly be advantageous for ease of administration – provided they are available. However, the increasing spread of resistant strains of viruses and bacteria – not least caused by the indiscriminate use of antibiotics – may become one of the severe medical problems of the future. The logistics of passive immunotherapy, having available large doses of the required specificities in due time, are daunting, but perhaps not insurmountable with better production techniques, the use of well-designed cocktails of specificity and concentration on infections of great risk to global health. The multiple cases of outbreak of novel deadly diseases over the last few years, AIDS, SARS and Ebola fever, to name just a few, illustrate the need for rapid development of novel containment strategies. It is likely that recombinant antibodies will still have a role to play in this respect, as it may be faster in some cases to develop protective antibodies than a vaccine (see also Part II, Chapters 7 and 8).

2.7
Opportunities for New Therapeutic Applications Provided by Synthetic Antibodies

The advent of synthetic antibodies now allows a number of the previous key limitations in using antibodies for therapy to be addressed. In addition to solving the problem of immunogenicity, which may be a factor preventing therapeutic utility, as explained above, the new technologies for tailor-making the antibody molecule permit new approaches to improve efficacy.

First and foremost, the relevant epitope can be more easily targeted. If its location is known at the molecular level, selections for such binders are possible, as has been delineated above, and this will often be the decisive factor in determining whether a particular antibody has any *in vivo* potency at all. Secondly, affinity can be addressed independently. Again, as described above, high affinity is almost always a benefit. In the traditional immunizations of animals, even when a large number of different mAbs is obtained, there is no guarantee that high-affinity antibodies from the panel are directed against the relevant epitope. As a rule, other places on the large surface of a protein will give many opportunities for high-affinity binding, such that it would be rather unexpected that the highest affinity antibodies happen to be directed against the epitope of choice.

It is this particular situation, where a synthetic library technology for antibody generation such as HuCAL can play out to its full advantage: affinity maturation strategies can be applied to the antibodies targeting the relevant epitope. Therefore, having *any* lead compound will usually allow the generation of a molecule, which maintains the binding at the desired location but achieves an affinity commensurate with the desired mode of action.

Because of the modularity of the HuCAL design (see above), it is possible to tailor not only one, but several antibodies that have been identified as having the desired binding specificity. Since identical restriction sites in all members of HuCAL flank the CDR cassettes, they can be exchanged in several antibodies at once, and those with higher affinity can be selected. Thereby, it is possible to "walk" across the whole antigen-binding site and replace each CDR in turn. Since the cassettes are not random, but instead reflect the composition of the human repertoire, key structural properties of the antibody can be maintained during optimization. By separating the selection of efficacy (or binding epitope) from affinity, which can be subsequently improved for any hit, antibodies can be made to almost any specification.

The anti-HLA-DR antibody 1D09C3 mentioned above [111, 112] provides a good example of how antibody engineering can be applied to optimize antibody properties. The antibody was derived from a screen of a HuCAL library in the single-chain Fv format, which yielded a number of binders that, although of moderate affinity, efficiently killed tumor cells. The affinity of the initial lead antibodies was increased by sequential replacement of the L-CDR3 and L-CDR1, utilizing the modularity of the HuCAL gene library (see above). This process enables retention of the epitope specificity of the initial lead antibodies, which are vital for their cell-killing properties, while achieving the desired level of target affinity.

In practice, the ability to generate antibodies having particular properties from libraries is limited by the screen that is employed. Initial screens for affinity and/or specificity are used to reduce the number of potential hits from the initial library of 10^{10} to 10^3–10^5, which must then be screened for a particular biological function. As delineated above, the (usually few) binders with a biological effect can then be affinity matured, such that affinity does not have to be an initial selection criterion. HuCAL thus provides a systematic means of screening large parts of sequence and

structure space for antibodies with predefined properties. Importantly, this can be done while retaining the intrinsic human composition of antibodies emerging from primary screens of the library.

2.8
Future Directions and Concluding Statements

Antibody engineering has clearly helped to solve a number of problems that have hampered early attempts in successfully using mAbs as biopharmaceuticals. Because of the lengthy development times typical of drug development, many therapeutic antibodies entering the market recently have still been made with earlier technologies and did not even fully profit from the possibilities available today.

A pivotal development of the last few years is that it is now possible to make a human antibody to practically any specification – regarding epitope, specificity, mode of binding, affinity, format and any molecule that might be linked with it. While robust manufacturing of more complicated molecules still needs to be improved, the main processes for manufacturing recombinant IgGs and fragments thereof are established.

Encouraging progress has been made in the area of neoplastic diseases of the hematopoietic system (lymphomas and leukemias). These tumors are characterized by good accessibility to the drug and the body's immune defense, high antigen density and perhaps a more homogeneous tumor cell population. In contrast, progress in the area of solid tumors has been only incremental. This is a huge need for society, and thus a great opportunity for science and medicine and the industry. It is apparent that several therapeutic antibodies are directed not only against closely related diseases, but also against the same target (TNF-α, CD20 and CD25), while the medical need in many other areas is unmet.

The key challenge for the future is to back up today's molecular engineering capabilities with a much better molecular understanding of disease and the consequences of the application of particular molecules. A more detailed understanding of the exact molecular effect required for a particular target (blocking, dimerization, its prevention, exposure needed) will allow much better engineering of molecular properties. In particular, preclinical models must become more relevant and predictive. This is a difficult topic in complex diseases such as cancer, as not only the interaction between the therapeutic antibody and its target is of importance, but also the multitude of interactions with other cells and their surface proteins, the complex pharmacokinetics, and the detailed metabolism. For example, in many mouse models with tumors of human origin, the antigen is selectively expressed on the tumor, but not in the murine tissues. To better model the human situation, systematic approaches are dearly needed.

The use of antibodies as biopharmaceuticals to treat some of the most serious diseases affecting mankind today arises directly from impressive technological developments that have been made in this field over the last 20 years. This has been one of the most significant achievements in the field of modern biotechnology. The developments described here promise that this class of modern biopharmaceuticals will play an even more important role in the clinician's armamentarium for the foreseeable future. Molecular engineering holds the promise that the remaining problems, many of them due to incomplete molecular understanding of the most important dis-

eases, will be addressable in the future. Eventually, this will further lead to the ultimate biopharmaceutical, which, for example, "targets" the desired payload to the tumor, thus fully realizing Paul Ehrlich's vision of the proverbial "magic bullet".

Acknowledgments

We would like to thank Drs. Marlies Sproll, Uwe Zangemeister-Wittke and Michael Stumpp for critical reading of the manuscript, and Dr. Annemarie Honegger for Figs. 2.2 and 2.3.

References

1 Köhler, G., Milstein, C. Continuous cultures of fused cells secreting antibody of predefined specificity. *Nature* **1975**, *256*, 495–497.
2 Schirle, M., Weinschenk, T., Stevanovic, S. Combining computer algorithms with experimental approaches permits the rapid and accurate identification of T cell epitopes from defined antigens. *J. Immunol. Methods* **2001**, *257*, 1–16.
3 Flower, D.R. Towards *in silico* prediction of immunogenic epitopes. *Trends Immunol.* **2003**, *24*, 667–674.
4 Flower, D.R. Databases and data mining for computational vaccinology. *Curr. Opin. Drug Discov. Dev.* **2003**, *6*, 396–400.
5 Koren, E., Zuckerman, L.A., Mire-Sluis, A.R. Immune responses to therapeutic proteins in humans – clinical significance, assessment and prediction. *Curr. Pharm. Biotechnol.* **2002**, *3*, 349–360.
6 Sturniolo, T., Bono, E., Ding, J., Raddrizzani, L., Tuereci, O., Sahin, U., Braxenthaler, M., Gallazzi, F., Protti, M.P., Sinigaglia, F., Hammer, J. Generation of tissue-specific and promiscuous HLA ligand databases using DNA microarrays and virtual HLA class II matrices. *Nat. Biotechnol.* **1999**, *17*, 555–561.
7 Chapman, A.P., Antoniw, P., Spitali, M., West, S., Stephens, S., King, D.J. Therapeutic antibody fragments with prolonged *in vivo* half-lives. *Nat. Biotechnol.* **1999**, *17*, 780–783.

8 Greenwald, R. B., Choe, Y. H., McGuire, J., Conover, C. D. Effective drug delivery by PEGylated drug conjugates. *Adv. Drug Deliv. Rev.* **2003**, *55*, 217–250.

9 Yang, K., Basu, A., Wang, M., Chintala, R., Hsieh, M. C., Liu, S., Hua, J., Zhang, Z., Zhou, J., Li, M., Phyu, H., Petti, G., Mendez, M., Janjua, H., Peng, P., Longley, C., Borowski, V., Mehlig, M., Filpula, D. Tailoring structure–function and pharmacokinetic properties of single-chain Fv proteins by site-specific PEGylation. *Protein Eng.* **2003**, *16*, 761–770.

10 Clark, M. Antibody humanization: a case of the "Emperor's new clothes"? *Immunol. Today* **2000**, *21*, 397–402.

11 Reichert, J., Pavlou, A., Monoclonal antibodies market. *Nat. Rev. Drug Discov.* **2004**, *3*, 383–384.

12 Wagner, C. L., Schantz, A., Barnathan, E., Olson, A., Mascelli, M. A., Ford, J., Damaraju, L., Schaible, T., Maini, R. N., Tcheng, J. E. Consequences of immunogenicity to the therapeutic monoclonal antibodies ReoPro and Remicade. *Dev. Biol.* **2003**, *112*, 37–53.

13 Nurden, A. T., Nurden, P. GPIIb/IIIa antagonists and other anti-integrins. *Semin. Vascular Med.* **2003**, *3*, 123–130.

14 Jones, P. T., Dear, P. H., Foote, J., Neuberger, M. S., Winter, G. Replacing the complementarity-determining regions in a human antibody with those from a mouse. *Nature* **1986**, *321*, 522–525.

15 Ewert, S., Honegger, A., Plückthun, A. Stability improvement of antibodies for extracellular and intracellular applications: CDR grafting to stable frameworks and structure-based framework engineering. *Methods* **2004**, *34*, 184–199.

16 Padlan, E. A. A possible procedure for reducing the immunogenicity of antibody variable domains while preserving their ligand-binding properties. *Mol. Immunol.* **1991**, *28*, 489–498.

17 Roguska, M. A., Pedersen, J. T., Keddy, C. A., Henry, A. H., Searle, S. J., Lambert, J. M., Goldmacher, V. S., Blattler, W. A., Rees, A. R., Guild, B. C. Humanization of murine monoclonal antibodies through variable domain resurfacing. *Proc. Natl Acad. Sci. USA* **1994**, *91*, 969–973.

18 Roguska, M. A., Pedersen, J. T., Henry, A. H., Searle, S. M., Roja, C. M., Avery, B., Hoffee, M., Cook, S., Lambert, J. M., Blattler, W. A., Rees, A. R., Guild, B. C. A comparison of two murine monoclonal antibodies humanized by CDR-grafting and variable domain resurfacing. *Protein Eng.* **1996**, *9*, 895–904.

19 Helft, P. R., Schilsky, R. L., Hoke, F. J., Williams, D., Kindler, H. L., Sprague, E., DeWitte, M., Martino, H. K., Erickson, J., Pandite, L., Russo, M., Lambert, J. M., Howard, M., Ratain, M. J. A phase I study of cantuzumab mertansine administered as a single intravenous infusion once weekly in patients with advanced solid tumors. *Clin. Cancer Res.* **2004**, *10*, 4363–4368.

20 Tolcher, A. W., Ochoa, L., Hammond, L. A., Patnaik, A., Edwards, T., Takimoto, C., Smith, L., de Bono, J., Schwartz, G., Mays, T., Jonak, Z. L., Johnson, R., DeWitte, M., Martino, H., Audette, C., Maes, K., Chari, R. V., Lambert, J. M., Rowinsky, E. K. Cantuzumab mertansine, a maytansinoid immunoconjugate directed to the CanAg antigen: a phase I, pharmacokinetic, and biologic correlative study. *J. Clin. Oncol.* **2003**, *21*, 211–222.

21 Hale, G., Dyer, M. J., Clark, M. R., Phillips, J. M., Marcus, R., Riechmann, L., Winter, G., Waldmann, H. Remission induction in non-Hodgkin lymphoma with reshaped human monoclonal antibody CAMPATH-1H. *Lancet* **1988**, *2*, 1394–1399.

22 Rowan, W., Tite, J., Topley, P., Brett, S. J. Cross-linking of the CAMPATH-1 antigen (CD52) mediates growth inhibition in human B- and T-lymphoma cell lines, and subsequent emergence of CD52-deficient cells. *Immunology* **1998**, *95*, 427–436.

23 Dyer, M. J. The role of CAMPATH-1 antibodies in the treatment of lymphoid malignancies. *Semin. Oncol.* **1999**, *26*, 52–57.

24 Hale, G., Rebello, P., Brettman, L. R., Fegan, C., Kennedy, B., Kimby, E., Leach, M., Lundin, J., Mellstedt, H., Moreton, P., Rawstron, A. C., Waldmann, H., Osterborg, A., Hillmen, P. Blood concentrations of alemtuzumab and antiglobulin responses in patients with chronic lymphocytic leukemia following intravenous or subcutaneous routes of administration. *Blood* **2004**, *104*, 948–955.

25 Jespers, L. S., Roberts, A., Mahler, S. M., Winter, G., Hoogenboom, H. R. Guiding the selection of human antibodies from phage display repertoires to a single epitope of an antigen. *Biotechnology* **1994**, *12*, 899–903.

26 Jia, X. C., Raya, R., Zhang, L., Foord, O., Walker, W. L., Gallo, M. L., Haak-Frendscho, M., Green, L. L., Davis, C. G. A novel method

of multiplexed competitive antibody binning for the characterization of monoclonal antibodies. *J. Immunol. Methods* **2004**, *288*, 91–98.

27 Hoogenboom, H. R., Winter, G. By-passing immunisation. Human antibodies from synthetic repertoires of germline V_H gene segments rearranged *in vitro*. *J. Mol. Biol.* **1992**, *227*, 381–388.

28 Hoogenboom, H. R., Lutgerink, J. T., Pelsers, M. M., Rousch, M. J., Coote, J., Van Neer, N., De Bruine, A., Van Nieuwenhoven, F. A., Glatz, J. F., Arends, J. W. Selection-dominant and nonaccessible epitopes on cell-surface receptors revealed by cell-panning with a large phage antibody library. *Eur. J. Biochem.* **1999**, *260*, 774–784.

29 Mutuberria, R., Hoogenboom, H. R., van der Linden, E., de Bruine, A. P., Roovers, R. C. Model systems to study the parameters determining the success of phage antibody selections on complex antigens. *J. Immunol. Methods* **1999**, *231*, 65–81.

30 Traggiai, E., Becker, S., Subbarao, K., Kolesnikova, L., Uematsu, Y., Gismondo, M. R., Murphy, B. R., Rappuoli, R., Lanzavecchia, A. An efficient method to make human monoclonal antibodies from memory B cells: potent neutralization of SARS coronavirus. *Nat. Med.* **2004**, *10*, 871–875.

31 Burton, D. R., Barbas, C. F., 3rd, Persson, M. A., Koenig, S., Chanock, R. M., Lerner, R. A. A large array of human monoclonal antibodies to type 1 human immunodeficiency virus from combinatorial libraries of asymptomatic seropositive individuals. *Proc. Natl Acad. Sci. USA* **1991**, *88*, 10134–10137.

32 Marks, J. D., Hoogenboom, H. R., Bonnert, T. P., McCafferty, J., Griffiths, A. D., Winter, G. By-passing immunization. Human antibodies from V-gene libraries displayed on phage. *J. Mol. Biol.* **1991**, *222*, 581–597.

33 Vaughan, T. J., Williams, A. J., Pritchard, K., Osbourn, J. K., Pope, A. R., Earnshaw, J. C., McCafferty, J., Hodits, R. A., Wilton, J., Johnson, K. S. Human antibodies with sub-nanomolar affinities isolated from a large non-immunized phage display library. *Nat. Biotechnol.* **1996**, *14*, 309–314.

34 Hoogenboom, H. R. Overview of antibody phage-display technology and its applications. *Methods Mol. Biol.* **2002**, *178*, 1–37.

35 Knappik, A., Ge, L., Honegger, A., Pack, P., Fischer, M., Wellnhofer, G., Hoess, A., Wölle, J., Plückthun, A., Virnekäs, B. Fully synthetic human combinatorial antibody libraries (HuCAL) based on modular consensus frameworks and CDRs randomized with trinucleotides. *J. Mol. Biol.* **2000**, *296*, 57–86.

36 Söderlind, E., Strandberg, L., Jirholt, P., Kobayashi, N., Alexeiva, V., Aberg, A. M., Nilsson, A., Jansson, B., Ohlin, M., Wingren, C., Danielsson, L., Carlsson, R., Borrebaeck, C. A. Recombining germline-derived CDR sequences for creating diverse single-framework antibody libraries. *Nat. Biotechnol.* **2000**, *18*, 852–856.

37 Pini, A., Viti, F., Santucci, A., Carnemolla, B., Zardi, L., Neri, P., Neri, D. Design and use of a phage display library. Human antibodies with subnanomolar affinity against a marker of angiogenesis eluted from a two-dimensional gel. *J. Biol. Chem.* **1998**, *263*, 21769–21776.

38 Golemis, E. (ed.). *Protein–Protein Interactions: A Molecular Cloning Manual*. Cold Spring Harbor Laboratory Press, Cold Spring Harbor, NY, **2002**.

39 Smith, G. P. Filamentous fusion phage: novel expression vectors that display cloned antigens on the virion surface. *Science* **1985**, *228*, 1315–1317.

40 McCafferty, J., Griffiths, A. D., Winter, G., Chiswell, D. J. Phage antibodies: filamentous phage displaying antibody variable domains. *Nature* **1990**, *348*, 552–554.

41 Bass, S., Greene, R., Wells, J. A. Hormone phage: an enrichment method for variant proteins with altered binding properties. *Proteins* **1990**, *8*, 309–314.

42 Barbas, C. F., 3rd, Burton, D. R., Scott, J. K., Silvermann, G. J. *Phage Display: A Laboratory Manual*. Cold Spring Harbor Laboratory Press, Cold Spring Harbor, NY, **2001**.

43 Ostendorp, R., Frisch, C., Urban, M. Generation, engineering and production of human antibodies using HuCAL. In *Antibodies, Vol. 2: Novel Technologies and Therapeutic Use*, Subramanian, G. (ed.). Kluwer/Plenum, New York, **2004**.

44 Boder, E. T., Wittrup, K. D. Yeast surface display for screening combinatorial polypeptide libraries. *Nat. Biotechnol.* **1997**, *15*, 553–557.

45 Feldhaus, M. J., Siegel, R. W., Opresko, L. K., Coleman, J. R., Feldhaus, J. M., Yeung, Y. A., Cochran, J. R., Heinzelman, P., Colby, D., Swers, J., Graff, C., Wiley, H. S., Wittrup, K. D.

Flow-cytometric isolation of human antibodies from a nonimmune *Saccharomyces cerevisiae* surface display library. *Nat. Biotechnol.* **2003**, *21*, 163–170.

46 Weaver-Feldhaus, J. M., Lou, J., Coleman, J. R., Siegel, R. W., Marks, J. D., Feldhaus, M. J. Yeast mating for combinatorial Fab library generation and surface display. *FEBS Lett.* **2004**, *564*, 24–34.

47 Hanes, J., Plückthun, A. *In vitro* selection and evolution of functional proteins by using ribosome display. *Proc. Natl Acad. Sci. USA* **1997**, *94*, 4937–4942.

48 Hanes, J., Jermutus, L., Weber-Bornhauser, S., Bosshard, H. R., Plückthun, A. Ribosome display efficiently selects and evolves high-affinity antibodies *in vitro* from immune libraries. *Proc. Natl Acad. Sci. USA* **1998**, *95*, 14130–14135.

49 Jermutus, L., Honegger, A., Schwesinger, F., Hanes, J., Plückthun, A. Tailoring *in vitro* evolution for protein affinity or stability. *Proc. Natl Acad. Sci. USA* **2001**, *98*, 75–80.

50 Hanes, J., Schaffitzel, C., Knappik, A., Plückthun, A. Picomolar affinity antibodies from a fully synthetic naive library selected and evolved by ribosome display. *Nat. Biotechnol.* **2000**, *18*, 1287–1292.

51 Fishwild, D. M., O'Donnell, S. L., Bengoechea, T., Hudson, D. V., Harding, F., Bernhard, S. L., Jones, D., Kay, R. M., Higgins, K. M., Schramm, S. R., Lonberg, N. High-avidity human IgG kappa monoclonal antibodies from a novel strain of minilocus transgenic mice. *Nat. Biotechnol.* **1996**, *14*, 845–851.

52 Mendez, M. J., Green, L. L., Corvalan, J. R., Jia, X. C., Maynard-Currie, C. E., Yang, X. D., Gallo, M. L., Louie, D. M., Lee, D. V., Erickson, K. L., Luna, J., Roy, C. M., Abderrahim, H., Kirschenbaum, F., Noguchi, M., Smith, D. H., Fukushima, A., Hales, J. F., Klapholz, S., Finer, M. H., Davis, C. G., Zsebo, K. M., Jakobovits, A. Functional transplant of megabase human immunoglobulin loci recapitulates human antibody response in mice. *Nat. Genet.* **1997**, *15*, 146–156.

53 Tomizuka, K., Yoshida, H., Uejima, H., Kugoh, H., Sato, K., Ohguma, A., Hayasaka, M., Hanaoka, K., Oshimura, M., Ishida, I. Functional expression and germline transmission of a human chromosome fragment in chimaeric mice. *Nat. Genet.* **1997**, *16*, 133–143.

54 Green, L. L., Jakobovits, A. Regulation of B cell development by variable gene complexity in mice reconstituted with human immunoglobulin yeast artificial chromosomes. *J. Exp. Med.* **1998**, *188*, 483–495.

55 Nicholson, I. C., Zou, X., Popov, A. V., Cook, G. P., Corps, E. M., Humphries, S., Ayling, C., Goyenechea, B., Xian, J., Taussig, M. J., Neuberger, M. S., Brüggemann, M. Antibody repertoires of four- and five-feature translocus mice carrying human immunoglobulin heavy chain and kappa and lambda light chain yeast artificial chromosomes. *J. Immunol.* **1999**, *163*, 6898–6906.

56 Gallo, M. L., Ivanov, V. E., Jakobovits, A., Davis, C. G. The human immunoglobulin loci introduced into mice: V, (D) and J gene segment usage similar to that of adult humans. *Eur. J. Immunol.* **2000**, *30*, 534–540.

57 Magadan, S., Valladares, M., Suarez, E., Sanjuan, I., Molina, A., Ayling, C., Davies, S. L., Zou, X., Williams, G. T., Neuberger, M. S., Bruggemann, M., Gambon, F., Diaz-Espada, F., Gonzalez-Fernandez, A. Production of antigen-specific human monoclonal antibodies: comparison of mice carrying IgH/kappa or IgH/kappa/lambda transloci. *Biotechniques* **2002**, *33*, 680–684.

58 Ishida, I., Tomizuka, K., Yoshida, H., Tahara, T., Takahashi, N., Ohguma, A., Tanaka, S., Umehashi, M., Maeda, H., Nozaki, C., Halk, E., Lonberg, N. Production of human monoclonal and polyclonal antibodies in TransChromo animals. *Cloning Stem Cells* **2002**, *4*, 91–102.

59 Van den Brande, J. M., Braat, H., van den Brink, G. R., Versteeg, H. H., Bauer, C. A., Hoedemaeker, I., van Montfrans, C., Hommes, D. W., Peppelenbosch, M. P., van Deventer, S. J. Infliximab but not etanercept induces apoptosis in lamina propria T-lymphocytes from patients with Crohn's disease. *Gastroenterology* **2003**, *124*, 1774–1785.

60 Parsons, H. L., Earnshaw, J. C., Wilton, J., Johnson, K. S., Schueler, P. A., Mahoney, W., McCafferty, J. Directing phage selections towards specific epitopes. *Protein Eng.* **1996**, *9*, 1043–1049.

61 Franklin, M. C., Carey, K. D., Vajdos, F. F., Leahy, D. J., de Vos, A. M., Sliwkowski, M. X. Insights into ErbB signaling from the structure of the ErbB2–pertuzumab complex. *Cancer Cell* **2004**, *5*, 317–328.

62 Agus, D. B., Akita, R. W., Fox, W. D., Lewis, G. D., Higgins, B., Pisacane, P. I., Lofgren, J. A., Tindell, C., Evans, D. P., Maiese, K., Scher, H. I., Sliwkowski, M. X. Targeting ligand-activated ErbB2 signaling inhibits breast and prostate tumor growth. *Cancer Cell* **2002**, *2*, 127–137.

63 Badache, A., Hynes, N. E. A new therapeutic antibody masks ErbB2 to its partners. *Cancer Cell* **2004**, *5*, 299–301.

64 Cho, H. S., Mason, K., Ramyar, K. X., Stanley, A. M., Gabelli, S. B., Denney, D. W., Jr., Leahy, D. J. Structure of the extracellular region of HER2 alone and in complex with the Herceptin Fab. *Nature* **2003**, *421*, 756–760.

65 Clynes, R. A., Towers, T. L., Presta, L. G., Ravetch, J. V. Inhibitory Fc receptors modulate *in vivo* cytotoxicity against tumor targets. *Nat. Med.* **2000**, *6*, 443–446.

66 Cobleigh, M. A., Vogel, C. L., Tripathy, D., Robert, N. J., Scholl, S., Fehrenbacher, L., Wolter, J. M., Paton, V., Shak, S., Lieberman, G., Slamon, D. J. Multinational study of the efficacy and safety of humanized anti-HER2 monoclonal antibody in women who have HER2-overexpressing metastatic breast cancer that has progressed after chemotherapy for metastatic disease. *J. Clin. Oncol.* **1999**, *17*, 2639–2648.

67 Plosker, G. L., Figgitt, D. P. Rituximab: a review of its use in non-Hodgkin's lymphoma and chronic lymphocytic leukaemia. *Drugs* **2003**, *63*, 803–843.

68 Burton, D. R. Antibodies, viruses and vaccines. *Nat. Rev. Immunol.* **2002**, *2*, 706–713.

69 Pantophlet, R., Burton, D. R. Immunofocusing: antigen engineering to promote the induction of HIV-neutralizing antibodies. *Trends Mol. Med.* **2003**, *9*, 468–473.

70 Burton, D. R., Desrosiers, R. C., Doms, R. W., Koff, W. C., Kwong, P. D., Moore, J. P., Nabel, G. J., Sodroski, J., Wilson, I. A., Wyatt, R. T. HIV vaccine design and the neutralizing antibody problem. *Nat. Immunol.* **2004**, *5*, 233–236.

71 Urban, J. L., Schreiber, H. Tumor antigens. *Annu. Rev. Immunol.* **1992**, *10*, 617–644.

72 Agnantis, N. J., Goussia, A. C., Stefanou, D. Tumor markers. An update approach for their prognostic significance. Part I. *In Vivo* **2003**, *17*, 609–618.

73 Guillemard, V., Saragovi, H. U. Novel approaches for targeted cancer therapy. *Curr. Cancer Drug Targets* **2004**, *4*, 313–326.

74 Roselli, M., Guadagni, F., Buonomo, O., Belardi, A., Ferroni, P., Diodati, A., Anselmi, D., Cipriani, C., Casciani, C. U., Greiner, J., Schlom, J. Tumor markers as targets for selective diagnostic and therapeutic procedures. *Anticancer Res.* **1996**, *16*, 2187–2192.

75 Maynard, J. A., Maassen, C. B., Leppla, S. H., Brasky, K., Patterson, J. L., Iverson, B. L., Georgiou, G. Protection against anthrax toxin by recombinant antibody fragments correlates with antigen affinity. *Nat. Biotechnol.* **2002**, *20*, 597–601.

76 Adams, G. P., Schier, R., McCall, A. M., Simmons, H. H., Horak, E. M., Alpaugh, R. K., Marks, J. D., Weiner, L. M. High affinity restricts the localization and tumor penetration of single-chain Fv antibody molecules. *Cancer Res.* **2001**, *61*, 4750–4755.

77 Adams, G. P., Schier, R. Generating improved single-chain Fv molecules for tumor targeting. *J. Immunol. Methods* **1999**, *231*, 249–260.

78 Verel, I., Heider, K.-H., Siegmund, M., Ostermann, E., Patzelt, E., Sproll, M., Snow, G. B., Adolf, G. R., van Dongen, G. A. M. S. Tumor targeting properties of monoclonal antibodies with different affinity for target antigen CD44V6 in nude mice bearing head-and-neck cancer xenografts. *Int. J. Cancer* **2002**, *99*, 396–402.

79 Neuberger, M. S., Harris, R. S., Di Noia, J., Petersen-Mahrt, S. K. Immunity through DNA deamination. *Trends Biochem. Sci.* **2003**, *28*, 305–312.

80 Besmer, E., Gourzi, P., Papavasiliou, F. N. The regulation of somatic hypermutation. *Curr. Opin. Immunol.* **2004**, *16*, 241–245.

81 Sale, J. E., Bemark, M., Williams, G. T., Jolly, C. J., Ehrenstein, M. R., Rada, C., Milstein, C., Neuberger, M. S. *In vivo* and *in vitro* studies of immunoglobulin gene somatic hypermutation. *Philos. Trans. R. Soc. London B Biol. Sci.* **2001**, *356*, 21–28.

82 Gram, H., Marconi, L. A., Barbas, C. F., 3rd, Collet, T. A., Lerner, R. A., Kang, A. S. *In vitro* selection and affinity maturation of antibodies from a naive combinatorial immunoglobulin library. *Proc. Natl Acad. Sci. USA* **1992**, *89*, 3576–3580.

83 Hawkins, R. E., Russell, S. J., Winter, G. Selection of phage antibodies by binding affinity. Mimicking affinity maturation. *J. Mol. Biol.* **1992**, *226*, 889–896.

84 Schier, R., Bye, J., Apell, G., McCall, A., Adams, G. P., Malmqvist, M., Weiner, L. M., Marks, J. D. Isolation of high-affinity monomeric human anti-c-erbB-2 single chain Fv using affinity-driven selection. *J. Mol. Biol.* **1996**, *255*, 28–43.

85 Chen, Y., Wiesmann, C., Fuh, G., Li, B., Christinger, H. W., McKay, P., de Vos, A. M., Lowman, H. B. Selection and analysis of an optimized anti-VEGF antibody: crystal structure of an affinity-matured Fab in complex with antigen. *J. Mol. Biol.* **1999**, *293*, 865–881.

86 Plückthun, A., Schaffitzel, C., Hanes, J., Jermutus, L. *In vitro* selection and evolution of proteins. *Adv. Protein Chem.* **2000**, *55*, 367–403.

87 Zahnd, C., Spinelli, S., Luginbühl, B., Amstutz, P., Cambillau, C., Plückthun, A. Directed *in vitro* evolution and crystallographic analysis of a peptide-binding single chain antibody fragment (scFv) with low picomolar affinity. *J. Biol. Chem.* **2004**, *279*, 18870–18877.

88 Roskos, L. K., Davis, C. G., Schwab, G. M. The clinical pharmacology of therapeutic monoclonal antibodies. *Drug Dev. Res.* **2004**, *61*, 108–120.

89 Junghans, R. P. Finally! The Brambell receptor (FcRB). Mediator of transmission of immunity and protection from catabolism for IgG. *Immunol. Res.* **1997**, *16*, 29–57.

90 Telleman, P., Junghans, R. P. The role of the Brambell receptor (FcRB) in liver: protection of endocytosed immunoglobulin G (IgG) from catabolism in hepatocytes rather than transport of IgG to bile. *Immunology* **2000**, *100*, 245–251.

91 Trang, J. M. Pharmacokinetics and metabolism of therapeutic antibodies. In *Protein Pharmacokinetics and Metabolism*, Ferraiolo, B. L., Mohler, M. A., Gloff, C. A. (eds). Plenum, New York, **1992**.

92 Waldmann, T. A., Strober, W. Metabolism of immunoglobulins. *Prog. Allergy* **1969**, *13*, 1–110.

93 Schror, K., Weber, A. A. Comparative pharmacology of GP IIb/IIIa antagonists. *J. Thromb. Thrombolysis* **2003**, *15*, 71–80.

94 Tam, S. H., Sassoli, P. M., Jordan, R. E., Nakada, M. T. Abciximab (ReoPro, chimeric 7E3 Fab) demonstrates equivalent affinity and functional blockade of glycoprotein IIb/IIIa and $a_v\beta_3$ integrins. *Circulation* **1998**, *98*, 1085–1091.

95 Goldenberg, D. M. Advancing role of radiolabeled antibodies in the therapy of cancer. *Cancer Immunol. Immunother.* **2003**, *52*, 281–296.

96 Wilder, R. B., DeNardo, G. L., DeNardo, S. J. Radioimmunotherapy: recent results and future directions. *J. Clin. Oncol.* **1996**, *14*, 1383–1400.

97 Bischof Delaloye, A. The role of nuclear medicine in the treatment of non-Hodgkin's lymphoma (NHL). *Leuk. Lymphoma* **2003**, *44* (Suppl. 4), S29–S36.

98 Wiseman, G. A., Kornmehl, E., Leigh, B., Erwin, W. D., Podoloff, D. A., Spies, S., Sparks, R. B., Stabin, M. G., Witzig, T., White, C. A. Radiation dosimetry results and safety correlations from ^{90}Y-ibritumomab tiuxetan radioimmunotherapy for relapsed or refractory non-Hodgkin's lymphoma: combined data from 4 clinical trials. *J. Nucl. Med.* **2003**, *44*, 465–474.

99 Kaminski, M. S., Estes, J., Zasadny, K. R., Francis, I. R., Ross, C. W., Tuck, M., Regan, D., Fisher, S., Gutierrez, J., Kroll, S., Stagg, R., Tidmarsh, G., and Wahl, R. L. Radioimmunotherapy with iodine ^{131}I tositumomab for relapsed or refractory B-cell non-Hodgkin lymphoma: updated results and long-term follow-up of the University of Michigan experience. *Blood* **2000**, *96*, 1259–1266.

100 Scheidhauer, K., Wolf, I., Baumgartl, H. J., Von Schilling, C., Schmidt, B., Reidel, G., Peschel, C., Schwaiger, M. Biodistribution and kinetics of ^{131}I-labelled anti-CD20 MAB IDEC-C2B8 (rituximab) in relapsed non-Hodgkin's lymphoma. *Eur. J. Nucl. Med. Mol. Imaging* **2002**, *29*, 1276–1282.

101 Maack, T. Renal handling of proteins and polypeptides. In *Handbook of Physiology, Section 8: Renal Physiology*, Windhager, E. E. (ed.). APS, Washington, DC, **1992**.

102 Renkin, E. M., Gilmore, J. P. Glomerular filtration. In *Handbook of Physiology: Renal Physiology*, Orloff, J., Berliner, R. W. (eds.). APS, Washington, DC, **1973**.

103 Plückthun, A., Pack, P. New protein engineering approaches to multivalent and bispecific antibody fragments. *Immunotechnology* **1997**, *3*, 83–105.

104 Carter, P. Improving the efficacy of antibody-based cancer therapies. *Nat. Rev. Cancer* **2001**, *1*, 118–129.

105 Glennie, M. J., Johnson, P. W. Clinical trials of antibody therapy. *Immunol. Today* **2000**, *21*, 403–410.

106 Zondor, S. D., Medina, P. J. Bevacizumab: an angiogenesis inhibitor with efficacy in colorectal and other malignancies. *New Drug Dev.* **2004**, *38*, 1258–1264.

107 Kim, K. J., Li, B. Inhibition of vascular endothelial growth factor-induced angiogenesis suppresses tumour growth *in vivo*. *Nature* **1993**, *362*, 841–844.

108 Cartron, G., Watier, H., Golay, J., Solal-Celigny, P. From the bench to the bedside: ways to improve rituximab efficacy. *Blood* **2004**, *104*, 2635–2642.

109 Deans, J. P., Li, H., Polyak, M. J. CD20-mediated apoptosis: signalling through lipid rafts. *Immunology* **2002**, *107*, 176–182.

110 Maloney, D. G., Smith, B., Rose, A. Rituximab: mechanism of action and resistance. *Semin. Oncol.* **2002**, *29*, 2–9.

111 Nagy, Z. A., Mooney, N. A. A novel, alternative pathway of apoptosis triggered through class II major histocompatibility complex molecules. *J. Mol. Med.* **2003**, *81*, 757–765.

112 Nagy, Z. A., Hubner, B., Löhning, C., Rauchenberger, R., Reiffert, S., Thomassen-Wolf, E., Zahn, S., Leyer, S., Schier, E. M., Zahradnik, A., Brunner, C., Lobenwein, K., Rattel, B., Stanglmaier, M., Hallek, M., Wing, M., Anderson, S., Dunn, M., Kretzschmar, T., Tesar, M. Fully human, HLA-DR-specific monoclonal antibodies efficiently induce programmed death of malignant lymphoid cells. *Nat. Med.* **2002**, *8*, 801–807.

113 Cook-Bruns, N. Retrospective analysis of the safety of Herceptin immunotherapy in metastatic breast cancer. *Oncology* **2001**, *61* (Suppl. 2), 58–66.

114 Shields, R. L., Namenuk, A. K., Hong, K., Meng, Y. G., Rae, J., Briggs, J., Xie, D., Lai, J., Stadlen, A., Li, B., Fox, J. A., Presta, L. G. High resolution mapping of the binding site on human IgG1 for FcγRI, FcγRII, FcγRIII, and FcRn and design of IgG1 variants with improved binding to the FcγR. *J. Biol. Chem.* **2001**, *276*, 6591–6604.

115 Presta, L. G. Engineering antibodies for therapy. *Curr. Pharm. Biotechnol.* **2002**, *3*, 237–256.

116 Maxwell, K. F., Powell, M. S., Hulett, M. D., Barton, P. A., McKenzie, I. F., Garrett, T. P., Hogarth, P. M. Crystal structure of the human leukocyte Fc receptor, FcγRIIa. *Nat. Struct. Biol.* **1999**, *6*, 437–442.

117 Radaev, S., Motyka, S., Fridman, W. H., Sautes-Fridman, C., Sun, P. D. The structure of a human type III Fcγ receptor in complex with Fc. *J. Biol. Chem.* **2001**, *276*, 16469–16477.

118 Sondermann, P., Huber, R., Oosthuizen, V., Jacob, U. The 3.2-Å crystal structure of the human IgG1 Fc fragment–FcγRIII complex. *Nature* **2000**, *406*, 267–273.

119 Sondermann, P., Kaiser, J., Jacob, U. Molecular basis for immune complex recognition: a comparison of Fc-receptor structures. *J. Mol. Biol.* **2001**, *309*, 737–749.

120 Zhang, Y., Boesen, C. C., Radaev, S., Brooks, A. G., Fridman, W. H., Sautes-Fridman, C., Sun, P. D. Crystal structure of the extracellular domain of a human FcγRIII. *Immunity* **2000**, *13*, 387–395.

121 Jefferis, R., Lund, J., Pound, J. D. IgG-Fc-mediated effector functions: molecular definition of interaction sites for effector ligands and the role of glycosylation. *Immunol. Rev.* **1998**, *163*, 59–76.

122 Krapp, S., Mimura, Y., Jefferis, R., Huber, R., Sondermann, P. Structural analysis of human IgG-Fc glycoforms reveals a correlation between glycosylation and structural integrity. *J. Mol. Biol.* **2003**, *325*, 979–989.

123 Umana, P., Jean-Mairet, J., Moudry, R., Amstutz, H., Bailey, J. E. Engineered glycoforms of an antineuroblastoma IgG1 with optimized antibody-dependent cellular cytotoxic activity. *Nat. Biotechnol.* **1999**, *17*, 176–180.

124 Shinkawa, T., Nakamura, K., Yamane, N., Shoji-Hosaka, E., Kanda, Y., Sakurada, M., Uchida, K., Anazawa, H., Satoh, M., Yamasaki, M., Hanai, N., Shitara, K. The absence of fucose but not the presence of galactose or bisecting *N*-acetylglucosamine of human IgG1 complex-type oligosaccharides shows the critical role of enhancing antibody-dependent cellular cytotoxicity. *J. Biol. Chem.* **2003**, *278*, 3466–3473.

125 Presta, L. G. Antibody engineering for therapeutics. *Curr. Opin. Struct. Biol.* **2003**, *13*, 519–525.

126 Shields, R. L., Lai, J., Keck, R., O'Connell, L. Y., Hong, K., Meng, Y. G., Weikert, S. H., Presta, L. G. Lack of fucose on human IgG1 *N*-linked oligosaccharide improves binding to human Fcγ RIII and antibody-dependent cellular toxicity. *J. Biol. Chem.* **2002**, *277*, 26733–26740.

127 Okazaki, A., Shoji-Hosaka, E., Nakamura, K., Wakitani, M., Uchida, K., Kakita, S., Tsumoto, K., Kumagai, I., Shitara, K. Fucose depletion from human IgG1 oligosaccharide enhances binding enthalpy and association rate between IgG1 and FcγRIIIa. *J. Mol. Biol.* **2004**, *336*, 1239–1249.

128 Dall'Acqua, W. F., Woods, R. M., Ward, E. S., Palaszynski, S. R., Patel, N. K., Brewah, Y. A., Wu, H., Kiener, P. A., Langermann, S. Increasing the affinity of a human IgG1, for the neonatal Fc receptor: biological consequences. *J. Immunol.* **2002**, *169*, 5171–5180.

129 Hinton, P. R., Johlfs, M. G., Xiong, J. M., Hanestad, K., Ong, K. C., Bullock, C., Keller, S., Tang, M. T., Tso, J. Y., Vasquez, M., Tsurushita, N. Engineered human IgG antibodies with longer serum half-lives in primates. *J. Biol. Chem.* **2004**, *279*, 6213–6216.

130 Trail, P. A., King, H. D., Dubowchik, G. M. Monoclonal antibody drug immunoconjugates for targeted treatment of cancer. *Cancer Immunol. Immunother.* **2003**, *52*, 328–337.

131 Voutsadakis, I. A. Gemtuzumab ozogamicin (CMA-676, Mylotarg) for the treatment of CD33+ acute myeloid leukemia. *Anticancer Drugs* **2002**, *13*, 685–692.

132 Paul, S. P., Taylor, L. S., Stansbury, E. K., McVicar, D. W. Myeloid specific human CD33 is an inhibitory receptor with differential ITIM function in recruiting the phosphatases SHP-1 and SHP-2. *Blood* **2000**, *96*, 483–490.

133 Hellström, I., Garrigues, H. J., Garrigues, U., Hellström, K. E. Highly tumor-reactive, internalizing, mouse monoclonal antibodies to Le$_y$-related cell surface antigens. *Cancer Res.* **1990**, *50*, 2183–2190.

134 Saleh, M. N., Sugarman, S., Murray, J., Ostroff, J. B., Healey, D., Jones, D., Daniel, C. R., LeBherz, D., Brewer, H., Onetto, N., LoBuglio, A. F. Phase I trial of the anti-Lewis Y drug immunoconjugate BR96–doxorubicin in patients with lewis Y-expressing epithelial tumors. *J. Clin. Oncol.* **2000**, *18*, 2282–2292.

135 Ajani, J. A., Kelsen, D. P., Haller, D., Hargraves, K., Healey, D. A multi-institutional phase II study of BMS-182248-01 (BR96–doxorubicin conjugate) administered every 21 days in patients with advanced gastric adenocarcinoma. *Cancer J.* **2000**, *6*, 78–81.

136 Tolcher, A. W., Sugarman, S., Gelmon, K. A., Cohen, R., Saleh, M., Isaacs, C., Young, L., Healey, D., Onetto, N., Slichenmyer, W. Randomized phase II study of BR96-doxorubicin conjugate in patients with metastatic breast cancer. *J. Clin. Oncol.* **1999**, *17*, 478–484.

137 Wahl, A. F., Donaldson, K. L., Mixan, B. J., Trail P. A., Siegall, C. B. Selective tumor sensitization to taxanes with the mAb–drug conjugate cBR96–doxorubicin. *Int. J. Cancer* **2001**, *93*, 590–600.

138 Doronina, S. O., Toki, B. E., Torgov, M. Y., Mendelsohn, B. A., Cerveny, C. G., Chace, D. F., DeBlanc, R. L., Gearing, R. P., Bovee, T. D., Siegall, C. B., Francisco, J. A., Wahl, A. F., Meyer, D. L., Senter, P. D. Development of potent monoclonal antibody auristatin conjugates for cancer therapy. *Nat. Biotechnol.* **2003**, *7*, 778–784.

139 Prinz, H. Recent advances in the field of tubulin polymerization inhibitors. *Expert Rev. Anticancer Ther.* **2002**, *2*, 695–708.

140 Falnes, P. O., Sandvig, K. Penetration of protein toxins into cells. *Curr. Opin. Cell Biol.* **2000**, *12*, 407–413.

141 Perentesis, J. P., Miller, S. P., Bodley, J. W. Protein toxin inhibitors of protein synthesis. *Biofactors* **1992**, *3*, 173–184.

142 Stirpe, F. Ribosome-inactivating proteins. *Toxicon* **2004**, *44*, 371–383.

143 Pai-Scherf, L. H., Villa, J., Pearson, D., Watson, T., Liu, E., Willingham, M. C., Pastan, I. Hepatotoxicity in cancer patients receiving erb-38, a recombinant immunotoxin that targets the erbB2 receptor. *Clin. Cancer Res.* **1999**, *5*, 2311–2315.

144 Pastan, I. Immunotoxins containing Pseudomonas exotoxin A: a short history. *Cancer Immunol. Immunother.* **2003**, *52*, 338–341.

145 Poe, J. C., Fujimoto, Y., Hasegawa, M., Haas, K. M., Miller, A. S., Sanford, I. G., Bock, C. B., Fujimoto, M., Tedder, T. F. CD22 regulates B lymphocyte function *in vivo* through both ligand-dependent and ligand-independent mechanisms. *Nat. Immunol.* **2004**, *5*, 1078–1087.

146 Schnell, R., Borchmann, P., Staak, J. O., Schindler, J., Ghetie, V., Vitetta, E. S., Engert, A. Clinical evaluation of ricin A-chain immunotoxins in patients with Hodgkin's lymphoma. *Ann. Oncol.* **2003**, *14*, 729–736.

147 Helguera, G., Morrison, S. L., Penichet, M. L. Antibody–cytokine fusion proteins: harnessing the combined power of cytokines and antibodies for cancer therapy. *Clin. Immunol.* **2002**, *105*, 233–246.

148 Gillies, S.D., Lan, Y., Brunkhorst, B., Wong, W.K., Li, Y., Lo, K.M. Bi-functional cytokine fusion proteins for gene therapy and antibody-targeted treatment of cancer. *Cancer Immunol. Immunother.* **2002**, *51*, 449–460.

149 Borchmann, P., Schnell, R., Fuss, I., Manzke, O., Davis, T., Lewis, L.D., Behnke, D., Wickenhauser, C., Schiller, P., Diehl, V., Engert, A. Phase 1 trial of the novel bispecific molecule H22×Ki-4 in patients with refractory Hodgkin lymphoma. *Blood* **2002**, *100*, 3101–3107.

150 Kufer, P., Lutterbuse, R., Baeuerle, P.A. A revival of bispecific antibodies. *Trends Biotechnol.* **2004**, *22*, 238–244.

151 Hartmann, F., Renner, C., Jung, W., da Costa, L., Tembrink, S., Held, G., Sek, A., Konig, J., Bauer, S., Kloft, M., Pfreundschuh, M. Anti-CD16/CD30 bispecific antibody treatment for Hodgkin's disease: role of infu-sion schedule and costimulation with cytokines. *Clin. Cancer Res.* **2001**, *7*, 1873–1881.

152 James, N.D., Atherton, P.J., Jones, J., Howie, A.J., Tchekmedyian, S., Curnow, R.T. A phase II study of the bispecific antibody MDX-H210 (anti-HER2×CD64) with GM-CSF in HER2⁺ advanced prostate cancer. *Br. J. Cancer* **2001**, *85*, 152–156.

153 Pollack, P., Groothuis, J.R. Development and use of palivizumab (Synagis): a passive immunoprophylactic agent for RSV. *J. Infection Chemother.* **2002**, *8*, 201–206.

154 Visintin, M., Tse, E., Axelson, H., Rabbitts, T.H., Cattaneo, A. Selection of antibodies for intracellular function using a two-hybrid *in vivo* system. *Proc. Natl Acad. Sci. USA* **1999**, *96*, 11723–11728.

155 Mössner, E., Koch, H., Plückthun, A. Fast selection of antibodies without antigen purification: adaption of the protein fragment complementation assay to select antigen–antibody pairs. *J. Mol. Biol.* **2001**, *308*, 115–122.

3

Molecular Characterization of Autoantibody Responses in Autoimmune Diseases:
Implications for Diagnosis and Understanding of Autoimmunity

Constanze Breithaupt

Abstract

The production of antibodies directed at self-proteins is a hallmark of many auto-immune diseases, and the detection of specific autoantibody responses is used increasingly to aid diagnosis of various autoimmune diseases. Molecular characterization of autoantibody–antigen interaction sites may help to identify subsets of patients with certain clinical features or prognostic outcomes ("personalized medicine") (see Part I, Chapter 2 and 5). Moreover, it can facilitate the development of immunoassays that use recombinant or synthetic antigens as substrates for autoantibody detection. The vast majority of autoantibody epitopes is conformational – that is, comprised of amino acids from distant parts of the target protein sequence that cluster in the folded protein. When linear sequence stretches make dominant contributions to antibody binding, these regions may be identified by peptide mapping. In addition, many investigators are now making use of the growing number of known three-dimensional structures of autoantigens to guide mapping and mutagenesis studies. However, detailed information about strictly conformational epitopes can only be obtained by crystallographic studies which are so far confined to the structure of IgG_4 Fc complexed with the Fab of an IgM rheumatoid factor and the recently determined structure of the multiple sclerosis autoantigen myelin oligodendrocyte glycoprotein (MOG) complexed with the Fab of the pathogenic autoantibody 8-18C5. The MOG-(8-18C5) crystal structure identified a highly discontinuous epitope centered about MOG residues 101–108. These residues encompass a strained tight turn that is kept in its conformation by the protein environment; this explains the failure to detect this antigenic region by conventional peptide mapping. Interestingly, the immunodominant 8-18C5 epitope on MOG, sequestered behind the blood–brain barrier, is composed of residues that are least conserved between MOG and its homologues that are expressed outside the CNS and induce B cell tolerance; this point will be discussed in the chapter.

Abbreviations

ANAs	anti-nuclear antibodies
ANCAs	anti-neutrophil cytoplasmic antibodies
APF	antiperinuclear factor
BP	bullous pemphigoid
BTN	butyrophilin

CDRs	complementarity determining regions
CENP-C	centromer protein C
CNS	central nervous system
DDC	DOPA decarboxylase
EAE	experimental autoimmune encephalomyelitis
EIA	enzyme immunoassay
ERMAP	erythroid membrane-associated protein
GABA	gamma-aminobutyric acid
GBM	glomerular basement membrane
GP	Goodpasture disease
GST	glutathione-S-transferase
IIF	indirect immunofluorescence
LADA	latent autoimmune diabetes in adults
MBP	myelin basic protein
MOG	myelin oligodendrocyte glycoprotein
PLP	pyridoxal-5′-phosphate
snRNP	small nuclear ribonucleoprotein particles
TSH	thyroid-stimulating hormone
WG	Wegener's granulomatosis

3.1
Autoantibodies in Autoimmune Diseases

Autoimmune diseases form a heterogeneous group of chronic diseases in which the immune system erroneously attacks self-molecules (autoantigens) leading to the destruction of organs, tissue and cells. More than 80 clinically distinct autoimmune diseases have been identified [1], including systemic disorders such as rheumatoid arthritis and systemic lupus erythematosus (SLE), or organ-specific disorders such as multiple sclerosis (MS), type I diabetes mellitus, and autoimmune thyroid diseases. Though many autoimmune diseases are individually rare, they collectively affect an estimated 5–8% of the United States population [1] and, presumably, a similar percentage of the population elsewhere in the industrialized world. Affecting women disproportionately, autoimmune diseases are among the ten leading causes of death for young and middle-aged women [2].

3.1.1
Pathogenic Autoantibodies

The mechanisms that disrupt immunological self-tolerance and lead to autoimmune disease are largely unknown; environmental factors including infectious agents, chemicals and diet are all suspected to trigger or modulate autoimmune responses in genetically predisposed individuals. Cellular damage in autoimmune diseases is mediated by immune effector mechanisms triggered by autoaggressive T cells, high-affinity autoantibodies or a combination of both. A direct pathogenic effect mediated by autoantibodies is the hallmark of many of the best-characterized autoimmune diseases including myasthenia gravis and Graves' disease, where autoantibodies interfere with essential protein functions. Myasthenia gravis is caused by autoantibodies that bind the nicotinic acetylcholine receptor (AChR) expressed on skeletal muscle cells at the neuromuscular junction. Antibody binding leads to the reduction of functional AChR, and this results in fatigue and muscular weakness. The clinical signs of myasthenia gravis can be induced in experimental animals by the passive transfer of serum or purified antibodies from myasthenic patients, an experimental approach that defined the pathogenic role of autoantibodies in this disease. The majority of autoantibodies bind to one defined region exposed at the extracellular part of the a subunit of the pentameric transmembrane AChR. By cross-linking receptors they stimulate the physiological process of internalization and intracellular degradation, reducing the half-life of AChR to one-third. In addition, AChR-specific antibodies binding to the neuromuscular junction activate the complement cascade; this leads to focal destruction of the postsynaptic membrane and further internalization of AChR, probably due to disruption of the cytoskeleton [3, 4]. Graves' disease is the most common form of hyperthyroidism caused by autoantibodies directed against the thyroid-stimulating hormone (TSH) receptor. When activated by TSH, the TSH receptor stimulates the synthesis of thyroid hormones. In Graves' disease, autoantibody binding activates the TSH receptor that results in the production of overlarge amounts of thyroid hormones, as activation of the TSH receptor becomes independent of the physiological feedback regulation that operates by inhibiting the TSH synthesis. Another type of autoantibody-mediated dysfunction is due to the formation of large amounts of immune complexes as seen in SLE. Autoantibodies in SLE bind to various ubiquitous self-antigens such as DNA and nucleoprotein particles. The resulting immune complexes accumulate in small blood vessels of various organs causing severe inflammatory damage [5].

3.1.2
Autoantibodies as Markers for Autoimmune Disease

Irrespective of their role in disease pathogenesis, the detection of serum autoantibodies is becoming increasingly important for the diagnosis, prognosis and monitoring of autoimmune diseases. The use of autoantibody tests to identify individuals at risk of developing a specific autoimmune condition is based on the observation that serum autoantibodies often appear long before the onset of clinical disease [6]. Clinical manifestation of type I diabetes mellitus, for instance, is preceded by an asymptomatic prodromal period characterized by the appearance of autoantibodies directed at several antigens of the pancreatic islet cells. Numerous prediction studies have analyzed islet-associated antibodies in relatives of patients with autoimmune diabetes and genetically predisposed individuals, and the presence of antibodies against two or more specific islet autoantigens was found to be highly predictive for the development of clinical diabetes [7, 8]. Testing the serum of blood donors for IgM rheumatoid factor and anti-cyclic citrullinated peptide antibodies, both markers for rheumatoid arthritis (RA), showed that nearly half of the patients with RA were positive for one or both autoantibodies a medium of 4.5 years before onset of symptoms, compared to about 1% of false-positives in control samples [9]. Similarly, analysis of serum samples from members of the US Armed Forces with SLE, collected before onset of disease, identified autoantibodies characteristic for SLE that accumulated gradually before the development of the first clinical symptoms in 88% of the individuals [10].

The use of autoantibodies for diagnosis can be advantageous when diagnosis is complicated by the lack of specific symptoms in early autoimmune disease. Early SLE, for example, is associated with malaise, arthralgia, and fatigue [11]; MS patients can exhibit symptoms similar to those of infectious diseases, trauma or malignancy [1]. Vice versa, individuals with the same autoimmune disease can show different clinical phenotypes and distinct disease courses. Numerous autoantigens associated with autoimmune diseases have been identified so far, and tests for autoantibody responses directed against these antigens are currently applied to: 1) aid diagnosis; 2) monitor the degree of disease, the present immunologic activity or the response to therapy; or 3) provide prognostic information about the course and severity of the disease. An elevated level of anti-nuclear antibodies (ANAs), for example, constitutes one of the American College of Rheumatology criteria for the diagnosis of SLE. The production of ANAs is characteristic for various autoimmune connective tissue disorders including SLE, scleroderma, mixed connective tissue disease and Sjögren's syndrome. ANAs are detected by enzyme immunoassay tests (EIA) or by indirect immunofluorescence (IIF) on cultured cell lines that exhibit distinct fluorescence patterns depending on the particular disease or disease subset. Thus, nucleoli-positive IIF is associated with scleroderma, homogeneous fluorescence of cell nuclei with SLE. This diagnosis can be supported or further differentiated by enzyme-linked immunosorbent assays (ELISAs) detecting anti-Scl70 or anti-Sm antibodies that are highly specific for systemic sclerosis and SLE, respectively. Moreover, as the titer of SLE-specific autoantibodies directed against nuclear DNA correlates with disease activity, anti-nDNA ELISA can be used to monitor disease activity.

Eased by well-established methods of molecular biology and the accessibility of human genomic sequence data, autoantigens are currently being characterized in detail and new autoantigens are continuously identified. Recombinant protein expression and the insertion of protein tags allows for the production of large amounts of pure autoantigen for usage in immunoassays [12]. In concert with the shift from the microscopic cellular level to the molecular protein level, current efforts are directed at identifying particular B-cell epitopes on autoantigens. The determination of fine specificities of autoantibody binding is aimed at gaining a deeper understanding of the development and diversification of the autoantibody response, as well as at investigating a potential correlation of specific epitopes with a particular disease, a disease subtype, and the stage or the future course of the disease. In addition, epitope mapping studies could ideally result in the definition of synthetic peptidic antigens as substrates in immunoassays, presenting a lower cost, well reproducible alternative to the usage of whole proteins.

3.2
Autoantibody Epitopes

3.2.1
Structural Aspects of Antigen–Antibody Interaction

Extensive structural and biochemical studies of antigen–antibody complexes – a prime example of molecular recognition – have shown the substantial diversity of antigen binding but have also revealed several general features characteristic for antigen–antibody interfaces [13–15].

The antibody combining site (paratope) is mainly composed of residues that belong to the six complementarity determining regions (CDRs) formed by three loops at the tip of the variable domain of the light chain (L1–L3) and the three corresponding loops of the heavy chain (H1–H3). In large interfaces (as they occur in protein–antibody complexes), residues of the antibody framework sometimes contribute to binding. In most cases less than six CDRs are involved in binding the antigen, the heavy chain

CDRs (and of them especially H3) often dominate the interaction. Many paratopes are enriched in the aromatic residues tyrosine and tryptophan as well as the charged residue arginine that represent thermodynamic hotspots, probably due to their chemical composition that enables them to form multiple interactions via hydrophobic and polar contacts.

In general, complex formation buries 800 ± 200 Å^2 of the solvent-accessible surface of each the antibody and the antigenic protein which is in the typical range of protein–protein interactions. The chemical composition of the antigen's contact area (epitope) resembles that of the overall protein surface, as in other non-permanent protein complexes, with a slight preponderance of polar residues. The shape complementarity of the contact surfaces is good, but less perfect than for example that of protease inhibitor complexes; this is consistent with the lack of evolutionary optimization of the interaction between antibody and antigen. Slight packing imperfections are often compensated for by the introduction of water molecules that form hydrogen bonds to unpaired polar atoms and fill cavities of the interface.

In contrast to peptide or DNA–antibody complexes that exhibit a groove-like binding site, the combining site of antibodies that bind protein–antigens is generally quite planar. Structural characterization of protein–antibody complexes reveals that about 15–22 residues of each binding partner contribute to the interaction, corresponding to two to five separate segments of the polypeptide chain of the antigenic protein [16]. However, the functional epitope – defined as those residues which account for a large fraction of the overall binding energy – can be much smaller [17], as is also known for protein–protein interfaces in general [18, 19].

3.2.2
Classification of Epitopes

Epitopes ("epitope" stands for "B-cell epitope" throughout this text) are often divided into continuous (linear) and discontinuous epitopes, meaning that either residues contiguous in the polypeptide chain constitute the binding site or that the epitope is composed of residues separated in the sequence but located in close spatial proximity in the folded protein [16, 20, 21]. As in all protein–antibody complexes studied by X-ray crystallography so far, more than one segment of the antigen is involved in antibody binding, a continuous epitope more accurately denotes a sequence stretch that, when displayed by a peptide fragment, can bind antibodies raised against the whole antigenic protein. While continuous epitopes can still be recognized after denaturation of the protein, recognition of discontinuous or conformational epitopes requires the correctly folded native antigen.

A further classification of epitopes established in the 1960s to differentiate between assembled virions and separate coat protein subunits introduces cryptotopes and neotopes that are also found on autoantigens. Cryptotopes only become accessible after depolymerization, denaturation or fragmentation of the antigen; neotopes are newly created for instance by the assembly of protein subunits and depend, for example, on a particular conformation of a complex component adopted only after complex formation or on the contribution of residues from several subunits. Referring to autoimmunity, a cryptotope on the scleroderma autoantigen centromer protein C (CENP-C), revealed after cleavage by the apoptotic enzyme granzyme B, represents the preferred target for autoantibodies in a subset of patients with limited systemic sclerosis [22]. The 70-kDa subunit of the U1 small nuclear ribonucleoprotein particle recognized by autoantibodies of patients with SLE or SLE-overlap syndromes exhibits cryptotopes and neotopes after apoptotic cleavage and oxidative modification, respectively, that are associated with different clinical disease manifestations [23, 24].

Post-translationally modified epitopes are of growing interest in autoimmune research [25]. Enzymatic processes including phosphorylation, glycosylation, methylation, or citrullination can modify proteins, and (particularly in stressed and apoptotic cells) deamidation, isomerization or oxidative damage can occur spontaneously. Spontaneous modifications that create novel epitopes accumulate with age as well as in cases of defects in protein repair and degradation systems. That these modified autoantigens can become targets of autoimmune attack has been shown for instance in SLE, rheumatoid arthritis and type I diabetes mellitus. Recently, autoantigens that contain enzymatically modified arginine residues were found to be specifically targeted by autoantibodies [26]. Anti-Sm-antibodies, for instance, are anti-nuclear antibodies which are found in more than 30% of patients with SLE and that are highly specific for SLE. The seven homologous Sm proteins form the common core of the small nuclear ribonucleoprotein particles (snRNP) U1, U2, U4/U6 and U5 that are essential cofactors for pre-mRNA splicing in eukaryotes. One of the epitopes on the Sm proteins was mapped to the C-terminal glycine-arginine-rich region of the two Sm proteins D1 and D3 and was shown specifically to depend on the methylation of both aldimino groups of all arginines to form symmetrically dimethylated arginines [27].

Similarly, the antiperinuclear factor (APF; the first autoantibody discovered to be specific for rheumatoid arthritis [28]) and the related "antikeratin" antibody both bind to an epitope on the epithelial protein fillagrin that contains citrulline residues formed by the enzymatic deimination of arginines [29]. While the protein recognized by this antibody response in inflamed joints is still unknown, this epitope could be successfully mimicked by short citrulline-containing peptides. These results initiated the development of highly specific ELISAs for the diagnosis of rheumatoid arthritis that use cyclic citrullinated peptides as antigens [30].

3.2.3
Methods of Epitope Mapping

3.2.3.1 Continuous Epitopes

One common method for the determination of continuous epitopes is that of peptide mapping, overlapping synthetic or phage-displayed peptides (see Part IV, Chapter 16 and Part V, Chapters 1, 2, and 6) that cover the complete sequence of the antigenic protein are tested for binding antibodies specific for the antigen [21]. This procedure is often used to identify peptides that might induce antibodies cross-reactive with the antigenic protein, for instance aimed to develop a peptidic vaccine that is able to induce cross-reactive antibodies. For both vaccine and autoimmune research it is essential to verify that antibodies specific for certain peptides are really cross-reactive and thus also bind to the native protein [16]. It is important to realize that a linear epitope identified by peptide mapping may in reality be a mimotope present on a different protein (see Section 3.2.3.2). Immunization protocols, as used to induce experimental autoimmune diseases, can cause partial denaturation resulting in an antibody response that recognizes epitopes present on the highly immunogenic denatured antigen, but not on the correctly folded protein. In autoimmune disorders, antibody responses against linear epitopes not present on the native autoantigen are very common, and may represent responses to degradation products generated by proteolysis during necrosis and apoptosis [31]. Antibodies specific for linear epitopes of myelin oligodendrocyte glycoprotein (MOG) occur at high frequency in MS patients and were shown to be associated with myelin debris in actively demyelinating MS lesions [32]. However, as the antibody response against native MOG that induces severe demyelination in the experimental animal model of MS is focused on conformational epitopes [33–36], these MOG peptide-specific antibodies are probably binding to degraded MOG generated during myelin destruction and are incapable of triggering primary demyelination.

Epitopes determined by peptide mapping and associated with antibodies that recognize the folded antigen are often composed of polypeptide termini or surface-exposed, flexible loops and turns, as these can be most easily mimicked by linear peptides. Unlike helices and sheets, loops at the protein surface present a continuous stretch of accessible amino acid side chains that can be mirrored by a peptide. The same holds true for anti-peptide antibodies that cross-react with the whole protein [37]. In this case, the peptide conformation recognized by the antibody can be imposed on the flexible protein loop upon binding to the antibody. Furthermore, a few examples of continuous helical epitopes on autoantigens have been reported that could be mapped by the corresponding peptides. One such epitope comprises amino acids 231–245 of PM/Scl-100, one component of the multisubunit autoantigen of the polymyositis–scleroderma

overlap syndrome (PM/Scl) [38]. Mutational analysis of the 15mer peptide revealed that only the amino acids 234, 237, 240, and 241 are needed for autoantibody binding – residues that would be located at the same site of an α-helical structure. Moreover, secondary structure prediction and a protein structure containing a related sequence support the existence of an α-helix formed by amino acids 231–245 that can be mimicked by the corresponding peptide.

In general, the contact residues identified by peptide mapping form only part of a larger discontinuous epitope that contributes to the lower affinity of the antibody to the peptide compared to the corresponding antibody–protein interaction. Strategies to improve cross-reactivity between peptides and antigenic proteins include the introduction of conformational constraints by cyclization [39] as well as mutagenesis of specific amino acids [40], or the use of protein carriers aimed to shift the thermodynamic equilibrium of peptide conformations to the conformation most similar to the proteinaceous epitope. Peptides fused to glutathione-S-transferase (GST), for instance, were used to map the epitopes of the major bullous pemphigoid antigen 180 (BP180, type XVII collagen). Bullous pemphigoid (BP) is a subepidermal blistering disease characterized by pathogenic antibodies against the 155 kDa transmembrane protein BP180 that contributes to adherence of the epidermis to the basement membrane. The main target on BP180 is the extracellular 16th non-collagenous domain NC16a that is recognized by the large majority of sera of BP patients [41], and which is currently being tested as a substrate for diagnostic assays of BP [42, 43]. The analysis of sera immunoadsorbed with peptide fragments of NC16a coupled to GST revealed that the reactivity of almost all sera was restricted to the 40 N-terminal amino acids of NC16a [41, 44]. GST was shown to increase the level of ordered secondary structure for one of the peptides [45] that might contribute to the observed strong reactivity of the constructs.

3.2.3.2 Discontinuous Epitopes

In contrast to the cases presented above, the vast majority of clinically relevant autoantibody epitopes is supposed to be highly discontinuous. Epitopes *per se* are not preferentially composed of flexible loops or a particular helix, but can comprise any surface patch of the antigen, as shown for hen egg white lysozyme in biochemical and X-ray studies [46, 47]. The full characterization of a discontinuous epitope requires the structure determination of the antigen in complex with the antigen-binding domains (Fab or Fv) of the corresponding monoclonal antibody, ideally combined with mutagenesis studies of epitope residues to determine or to confirm the localization of hot spots in the binding interface. As this method is time-consuming and requires large amounts of sample, the three-dimensional structures of only two autoantigen Fab complexes have been determined so far [48, 49]. Some information about discontinuous epitopes, however, can also be obtained by testing if proteins homologous to the particular antigen maintain antibody binding. Similarly, if the antigen can be produced recombinantly, parts of the protein can be replaced by the corresponding fragments of a related protein or single residues can be changed by site-directed mutagenesis [50, 51]. Changes in the binding affinity of the resultant locally altered antigen point towards the participation of the particular region in antibody binding. It must be taken into account, however, that muta-

tions of epitope residues can be compensated for by the incorporation of water or the flexibility of the antibody-combining site that can even lead to increased binding affinity of the mutated antigen. Besides, mutations can provoke conformational changes distant from the site of mutagenesis, possibly within the epitope, that then result in the observed decreased binding affinity. Instead of abrogating antibody binding, Henriksson and colleagues restored it by introducing humanizing "gain of function" mutations into the homologous U1-70k autoantigen of *Drosophila* that led to the identification of a major conformational epitope [50].

Moreover, the ability of antibodies to protect the epitope of the bound antigen against chemical modifications [52, 53] and limited proteolytic degradation has been used to map discontinuous epitopes. As antibodies are highly resistant to enzymatic digestion [54], the epitope of the antigen bound to the immobilized antibody can be "excised" by endoproteolytic enzymes. After removal of the cleaved fragments of the antigen by a washing step, the individual segments of the epitope still bound to the antibody can be directly analyzed by laser desorption/ionization-based mass spectrometry [55–57]. Yet, unless the separate segments of the epitope exhibit a very high affinity to the antibody, limited proteolysis only results in the identification of a large protein fragment that cannot be cleaved further without dissociation.

Another approach to the elucidation of discontinuous epitopes is based on the identification of "mimotopes" [58], originally defined by Mario Geysen as "… molecule(s) able to bind to the antigen combining site of an antibody molecule, not necessarily identical with the epitope inducing the antibody, but an acceptable mimic of the essential features of the epitope" [59]. These putative mimics can be obtained from a library of random phage-displayed peptides after enrichment of binding peptides by several cycles of bio-panning. Though initially not expected [58], this method succeeded in identifying peptides that competed for antibody binding against discontinuous protein-epitopes and even DNA and carbohydrate antigens. Moreover, true mimotopes were shown to induce cross-reactive antibodies *in vivo*. The utilization of mimotopes in epitope mapping requires that: 1) mimics do not bind to an antibody subsite different from the site recognized by the antigen; and 2) mimicry is based on the side chains of amino acids and not confined to the arrangement of atoms. Supported by computer algorithms, consensus sequences of peptide subgroups can then be identified and mapped on the three-dimensional structure or possibly even on the primary structure to yield the discontinuous epitope [60–65].

3.3
Visualization of Epitopes

Knowledge of the molecular structure of autoantigens is of great value for the interpretation of existing mapping results (as shown for several examples below), as well as for the choice of peptides and mutations for future studies. The number of solved three-dimensional structures has increased tremendously within the past decade. In 2004, the coordinates of more than 27 000 proteins were available in the Protein Data Bank, among them a growing number of autoantigens.

3.3.1
Peptide Mapping of Epitopes on Proteinase 3

Antibodies specific for cytoplasmic antigens of neutrophils (anti-neutrophil cytoplasmic antibodies (ANCAs)) are typically found in patients with small-vessel vasculitis [66–68]. ANCAs can be differentiated by immunofluoresence assays where binding of ANCAs to ethanol-fixed neutrophils can produce a diffuse granular cytoplasmic (c-ANCAs) or a perinuclear (p-ANCAs) staining pattern. c-ANCAs that are primarily directed against proteinase 3 (PR3), found in azurophilic granules of neutrophils and granules of monocytes, are strongly associated with Wegener's granulomatosis (WG). A combination of IIF and ELISA to detect PR3-specific ANCAs that was tested in large multicenter studies showed high specificity and sensitivity for patients with active (99%, 91%) and inactive (99.5%, 63%) WG. In addition, some studies indicated that relapses in patients with PR3-ANCA-associated vasculitis were preceded by a rise of PR3-ANCA titers. Inhibition studies with PR3-specific ANCA sera and monoclonal antibodies revealed that PR3 presents a limited number of epitope areas that vary among patients and change during disease course. Putative linear epitopes on PR3 have been determined by three studies that used different detection methods and peptide constructs [69–71]. By mapping these epitopes onto the surface of the three-dimensional structure of PR3 [72], those parts of the peptide sequences that are surface-exposed in the structure of PR3 can be determined. Moreover, linear epitopes that combine in space to form a discontinuous epitope are detected. In addition, consistencies and discrepancies of the three studies can be visualized. Three regions are identified consis-

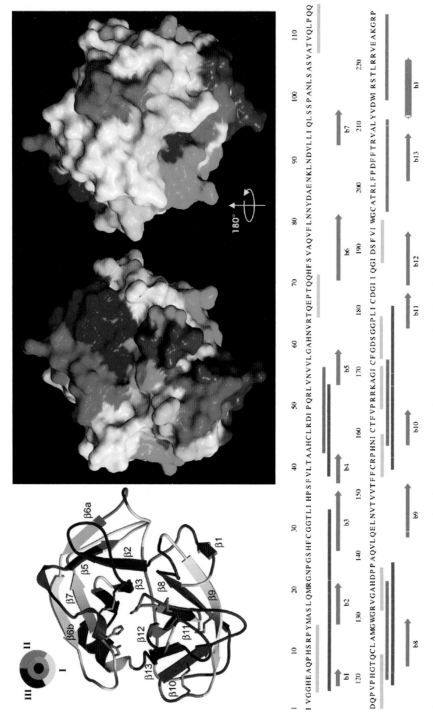

Fig. 3.1 Linear epitopes on proteinase 3. Colored regions correspond to epitope residues described by Williams et al. (yellow) [69], Griffith et al. (red) [70] and van der Geld et al. (blue) [71]; residues identified by more than one study are depicted in the respective combination colors.

tently in all three studies (Fig. 3.1). These regions all encompass surface-exposed protruding loops, two of them very flexible in the crystal structure, and are located in spatial proximity to each other (see Fig. 3.1), indicative of a possible immunodominant epitope region. The four main sequence stretches described by van der Geld et al. combine to form one surface patch adjacent to the active site. Altogether, the linear epitopes concentrate near the active site, this being consistent with reports that PR3-ANCA can interfere with the enzymatic activity of PR3 and with the binding of its physiological inhibitor-1 antitrypsin.

3.3.2
Cryptic Epitopes in Goodpasture Disease

Goodpasture disease (GP) is a prototype autoantibody-mediated, organ-specific autoimmune disease characterized by rapidly progressive glomerulonephritis with or without lung hemorrhage. The detection of pathogenic antibodies directed against the glomerular basement membrane (GBM) is used for the diagnosis of GP. Immunopathogenicity of anti-GBM antibodies can be inferred from the ability to induce the disease in primates by transfer of kidney-bound antibodies from patients with GP, and from the positive effect of plasma exchange in early disease. The major target of anti-GBM antibodies is the C-terminal non-collagenous (NC1) domain of the collagen $a3$(IV) chain, one of six homologous chains that constitute type IV collagen. $a3$ chains assemble with $a4$ and $a5$ chains to form triple helical protomers that generate collagenous networks by interaction of their N- and C-termini. Mapping of the purely discontinuous epitopes was performed using "gain of function" chimeras that consisted of the non- immunoreactive $a1$NC1,

in which sequence stretches most divergent among the a(IV) chains were replaced by the corresponding $a3$ sequence [73, 74]. When tested for reactivity with GP sera, the chimeras revealed a major (Ea, amino acids 17–31) and a minor (Eb, amino acids 127–141) epitope on $a3$NC1. Both epitopes were fully accessible only after dissociation of the hexamers, formed by two C-terminal NC1 trimers at the junction of two triple helices, into monomers. The recent elucidation of the crystal structure of the highly homologous $a1a1a2$ NC1 complex [75, 76] and investigation of the quaternary assembly of the $a3a4a5$ complex [77] allows localization of the cryptic epitopes on the homology model of $a3a4a5$ NC1. The major epitope forms a U-shaped structure composed of a β-strand that is partially hidden in the $a3a5$ NC1 interface – confirming the observed sequestered nature of the epitope – and of a loop followed by a coiled region near the interface. This loop is more easily accessible by antibodies after dissociation of the complex (Fig. 3.2a). In the trimer, pairs of neighboring monomers each form a shared β-sheet with one monomer contributing four β-strands, the other one contributing two. Disruption of this β-sheet upon dissociation of the complex probably causes conformational changes that may add to the higher accessibility of the Ea region in the monomer. Knowledge of the molecular structure of the complex will facilitate a detailed determination of the immunodominant region containing the Ea sequence that, regarding a recently reported correlation between unfavorable renal prognosis and antibodies binding to the Ea region [78], may be used to develop more specific diagnostic assays.

a

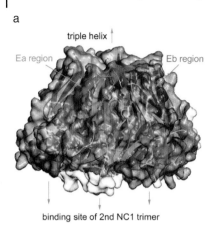

triple helix

Ea region   Eb region

binding site of 2nd NC1 trimer

Fig. 3.2 Visualization of conformational epitopes.
(a) The cryptic epitopes Ea and Eb of the trimeric NC1 are positioned at the interface of two monomers.
(b) Stereo view of the model of the dimeric glutamate decarboxylase 65, composed of a homology model of the middle domain, based on coordinates of the DOPA decarboxylase, and of the N- and C-terminal domain of DOPA decarboxylase. Selected epitope residues at the protein surface are marked.

b

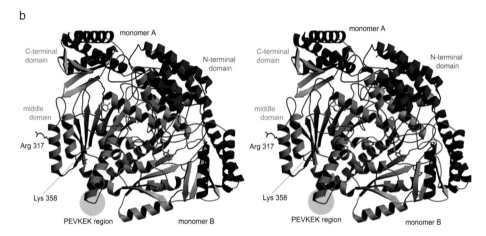

3.3.3
Modeling of Glutamate Decarboxylase 65

One of the islet cell autoantigens attacked in type I diabetes is the 65 kDa isoform of the pyridoxal-5′-phosphate (PLP)-dependent glutamate decarboxylase (GAD65) that decarboxylates glutamate, yielding the major inhibitory neurotransmitter γ-aminobutyric acid (GABA). Competition studies using recombinant Fab domains directed against different discontinuous epitopes of GAD65 revealed specific recognition patterns in patients with type I diabetes com-

pared to GAD65 antibody-positive patients with latent autoimmune diabetes in adults (LADA), first-degree relatives and healthy donors [79]. While precise structural mapping of the largely conformational epitopes is hampered by the lack of a three-dimensional structure of GAD65, some information can be inferred from the molecular structure of the related DOPA decarboxylase (DDC) [80]. Each monomer of the dimeric DDC consists of three domains, the large middle domain composed of eight helices surrounding a central seven-stranded β-sheet that harbors the PLP binding site, a C-terminal four-stranded β-sheet with three helices packed against it, both typical of the a-family of PLP-dependent enzymes, and in addition a small a-helical domain at the N-terminus. Based on the DDC structure, a tentative model of the middle domain of GAD65 that shows 28% identity (46% similarity) to DDC can be built (Fig. 3.2b) which allows the approximate positioning of amino acids found to be critical for antibody binding [81]. Evaluation of the accessibility of putative epitopes requires knowledge of the complete quaternary enzyme structure. Regarding the observed conservation of domain and dimerization structure in decarboxylases of the a-family, the overall structure of the family member GAD65 is likely to be similar to that of DDC. In this case, the PEVKEK region (amino acids 260–265) of GAD65, which is assumed to be a dominant epitope in type I diabetes [82, 83], is located in an accessible, surface-exposed loop at the dimer interface near the corresponding loop of the second monomer.

3.4
Structural Characterization of Autoantibody–Autoantigen Complexes

The most complete and reliable characterization of epitopes is provided by three-dimensional complex structures between antigen and Fab domain of the antibody. The only complex structures involving autoantigens that have been determined so far are that of IgG_4 Fc complexed with the Fab of an IgM rheumatoid factor [48, 84] and that of the MS autoantigen MOG in complex with the Fab of the pathogenic autoantibody 8-18C5 [49]. These structures allowed for a precise analysis of the epitopes, and both exhibited interesting char-

acteristics that were not expected from prior mapping studies, providing insight into formation and mechanism of the antibody response to these autoantigens. It must be remembered however that in each case a single antigen–antibody complex was analyzed, and additional studies are required to verify that the interactions are truly representative for the whole pool of antibodies specific for the autoantigen [84].

3.4.1
Antibody Responses against MOG

MOG is a 25 kDa, quantitatively minor myelin glycoprotein, which is highly conserved across species and expressed exclusively in the central nervous system (CNS), where it is sequestered from the immune system behind the blood–brain barrier [85]. It consists of an extracellular N-terminal immunoglobulin (Ig) -like domain, a transmembrane helix and a cytoplasmic domain that contains a second hydrophobic sequence, probably embedded in the intracellular half of the myelin membrane, and a short C-terminal tail. The physiological function of MOG is unknown; MOG-deficient "knock-out" mice develop normally and exhibit no detectable clinical, structural or pathological phenotype. Antibody binding to the Ig domain of MOG leads to the depolymerization of oligodendrocyte microtubuli *in vitro*, suggesting a role for MOG as a cell surface receptor that is involved in maintaining myelin stability. This concept is supported by the observation that MOG associates with lipid rafts, known as sites for the assembly of signaling complexes. Alternatively, MOG has been implicated in adhesion functions, in accord with the HNK1 glycan epitope, a marker for many cell adhesion proteins, which is also present on a subset of MOG molecules.

MOG was originally characterized two decades ago as the immunodominant target of demyelinating autoantibodies in the animal model of MS, experimental autoimmune encephalomyelitis (EAE) [86, 87]. EAE is an inflammatory autoimmune disease of the CNS, induced in susceptible laboratory animals by sensitization against various myelin proteins of the CNS that reproduces many of the clinical and pathological features of MS [88, 89]. While myelin-specific T cells initiate CNS inflammation in EAE, demyelination is autoantibody-mediated in rats and primates. In contrast to other encephalitogenic myelin autoantigens that are buried in the compact multilamellar myelin sheath, MOG is located on the outermost surface of the myelin sheath and is therefore easily accessible to antibodies present in the extracellular milieu. Immunization with the intracellular myelin antigen myelin basic protein (MBP) causes purely inflammatory EAE in rats and primates. The administration of the MOG-specific monoclonal antibody 8-18C5 at disease onset, however, leads to extensive demyelination and exacerbates the clinical disease dramatically, clearly demonstrating the pathogenic role of MOG-specific antibodies in EAE. The structural and immunopathological similarities between MOG-induced demyelinating lesions in EAE and the type of plaques seen in the majority of MS patients suggests that MOG also acts as autoantigen in MS.

MS is the most common chronic inflammatory demyelinating disease, affecting about one million people worldwide. Although MS is considered a primarily T-cell-dependent disorder, increasing evidence indicates that B cells and autoantibodies are also involved in immunopathogenesis, though so far no target autoantigen has been identified with certainty. The classification of MOG as candidate MS autoantigen is supported by its role in EAE, and by the identification of MOG-specific antibodies associated with myelin debris in actively demyelinating MS lesions. Additionally, several studies have shown that MOG-specific T- and B-cell levels are elevated in MS patients, although similar enhancements have also been observed in patients with other neurological disorders [90]. In a selected group of patients with a clinically isolated demyelinating syndrome, the presence of anti-MOG antibodies was prognostic for early conversion to clinically definite MS and rapid disease progression [91]. However, several groups report no significant differences in the frequency of MOG-specific antibody responses in MS patients compared to healthy donors. These controversial results most probably reflect differences in the selection of patients, the assay design or the antigen preparations used by various investigators.

Recent studies revealed that two sets of antibodies specific for the extracellular domain of MOG (MOG$_{ex}$) can be distinguished in EAE: antibodies specific for MOG-derived peptides and antibodies that recognize purely discontinuous, conformation-dependent epitopes on MOG [34, 36]. While MOG-peptide-specific antibodies were unable to bind native MOG on the cell surface and induced no or little demyelination in animals with EAE, injection of conformation-dependent MOG-specific antibodies caused extensive demyelination and reproduced the immunopathology and distribution of demyelinating lesions, typically seen in MS. The generation of large amounts of peptide-specific antibodies, that are not reactive with native MOG, following immunization with MOG$_{ex}$ produced recombinantly in *Escherichia coli*, is probably due to the fact that the antigen is denatured. This assumption is supported by the observation that animals immunized with refolded MOG_{ex} or vaccinated with MOG DNA all show a strong bias towards conformation-dependent MOG-specific antibodies [92]. A similar correlation between epitope specificity and demyelinating potential of MOG-specific antibodies might also exist in MS, as heterogeneous MOG-peptide specific antibody responses are observed in many MS patients as well as in healthy controls. Antibodies reactive to the MOG-peptide 21–40 were also identified in MS lesions [32], putatively binding to MOG epitopes newly exposed by protein degradation during myelin destruction. However, knowledge about the frequency of conformation-dependent MOG-specific antibodies in MS is very limited: Fabs directed at discontinuous MOG epitopes were shown to compete with antibodies of sera from MS patients [35], one study specifically identified conformation-dependent MOG-specific antibodies in the serum of one patient who benefited strikingly from plasmapheresis [34]. MS is a heterogeneous disease with an extremely variable course, possibly based on different pathogenetic mechanisms. The development of sensitive assays that differentiate between antibodies capable of binding to native MOG, and pathologically irrelevant antibodies recognizing linear peptide epitopes, is clearly essential to evaluate their importance in MS and to identify patients who may be responsive to therapies targeting pathogenic autoantibodies. The solution of the molecular structures of MOG_{ex} and MOG_{ex} bound to a Fab derived from the conformation-dependent antibody 8-18C5 is an important step towards achieving this goal. These data not only aid the interpretation of present epitope mapping results, but also provide the basis to design

mutant proteins to characterize the conformation-dependent MOG-specific antibody responses in EAE and MS.

3.4.2
The Three-dimensional Structure of MOG

The extracellular domains of rat and mouse MOG [49, 93] that have ~90% sequence identity to the human protein adopt an immunoglobulin (Ig) fold with features characteristic for IgV-like domains resulting in a sandwich of two antiparallel β-sheets comprised by strands A'GFCC'C'' and ABED (Fig. 3.3a). The Cα atoms of the two β-sheets align very well with those of other IgV-like proteins, such as various variable antibody domains, the $\gamma\delta$-T-cell receptor or their nearest structural relatives B7-2 and sialoadhesin (root mean square deviation <1.5 Å for ~100 aligned Cαs). Of interest, however, is the compactness of the MOG IgV fold that lacks long flexible loops but rather exhibits short connections between strands and a multitude of hydrogen bonds in the loop regions that tightly fix more than two-thirds of the loop residues. This may account for the lack of linear epitopes on the correctly folded MOG. One exception are peptides encompassing amino acids 63–87 that bind weakly to a subset of monoclonal mouse antibodies [33]. The corresponding region on MOG contains the protruding DE-loop (72–80) that is positioned at the top of MOG (is easily accessible to antibodies); this is consistent with these residues contributing to a partially linear epitope (Fig. 3.3b). The immunization of Lewis rats with MOG peptide 35–55 was shown to cause demyelination, possibly due to cross-reactivity of the peptide with whole-length native MOG [94]. In the MOG structure, amino acids 35–55 encompass the internal, buried C and C' strands and the exposed CC'-loop

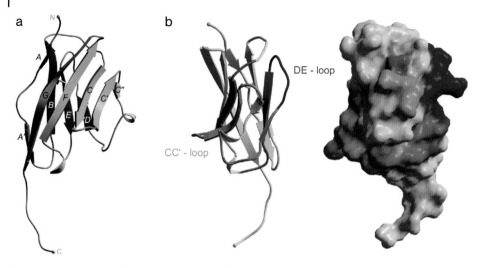

Fig. 3.3 Three-dimensional structure of myelin oligodendrocyte glycoprotein (MOG). (a) Overall structure. (b) Putative linear epitopes on MOG encompassing the CC'-loop and the DE-loop, respectively.

(41–46) that, devoid of any crystallographic contacts, is disordered in the rat MOG$_{ex}$ structure due to high flexibility, and thus is well-suited to provoke peptide–protein cross-reactivity.

3.4.3
The MOG$_{ex}$–(8-18C5)–Fab Complex

The monoclonal mouse antibody 8-18C5, used for this structural analysis, induces demyelination *in vitro* and *in vivo*, and probably recognizes an immunodominant epitope on MOG, as indicated by its ability to inhibit binding of 80% of a set of MOG-specific monoclonal mouse antibodies. The structure of MOG$_{ex}$ complexed with the Fab derived from 8-18C5 shows that 8-18C5 binds to the upper membrane-distal part of MOG, covering about 815 Å^2 of solvent-accessible MOG surface. As expected, the epitope is highly discontinuous, and consists of the N-terminus and the three upper loops of MOG, that correspond to

the CDRs in the related variable antibody domains, with additional contributions of three separate residues (Fig. 3.4). All epitope residues are totally conserved between human, rat and mouse MOG, consistent with the observed cross-reactivity of 8-18C5 between species. 8-18C5 was raised against native MOG and recognizes both the glycosylated and unglycosylated protein, indicating that the single glycosylation site Asn31 on MOG does not contribute to the epitope. This is supported by the structure of the complex in which Asn31, though directly adjoining the epitope, does not interact with the antibody.

The MOG$_{ex}$ (8-18C5)–Fab complex represents a rather typical antigen–antibody structure. MOG interacts centrally with the paratope of 8-18C5 formed by five of the six CDRs, with dominant contributions being made by the three heavy chain CDRs (Fig. 3.4). Several buried water molecules are visible in the interface. Aromatic and arginine residues – common

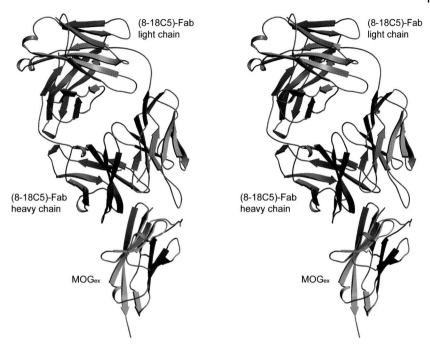

Fig. 3.4 Stereo view of the MOG (8-18C5)–Fab complex. MOG residues of the discontinuous epitope are colored in red.

paratope amino acids – form important interactions with MOG. The light chain, for instance, contributes two tyrosines to the paratope that fix the FG-loop of MOG at one side by strong hydrogen bonds. A tryptophan residue of the heavy chain CDR1 (H1-Trp33) forms a hydrogen bond to Gln106 and extensive stacking interactions with Arg101 of MOG; finally, two arginines of H2 that are the result of somatic mutations during maturation of the antibody response interact specifically with two MOG glutamate residues (Fig. 3.5). The light chain residues of the paratope correspond to their germline sequence; remarkable for CDR1, however, is its length of 17 amino acids – the maximum observed for Vκ-CDR1 sequences – that enables CDR1 to interact with MOG. Although the epitope of MOG is very discontinuous, amino acids 101–108 that correspond to the FG-loop and neighboring strand residues dominate the interaction by contributing 10 of the 12 hydrogen bonds and about 65% of the total contact area to the interaction. This raises the question of why this sequence has never been identified by epitope mapping using linear peptides (Fig. 3.5). The FG-loop (102–105) forms a tight turn, classified as a type II′ β-turn, that usually exhibits a glycine in the second position in order to avoid a sterically unfavorable conformation. In MOG, this position is occupied by His103, which results in a strained conformation that is stabilized by the protein environment but is highly unlikely to be adopted by linear peptides.

Based on the structure of the complex, mutagenesis studies were performed to

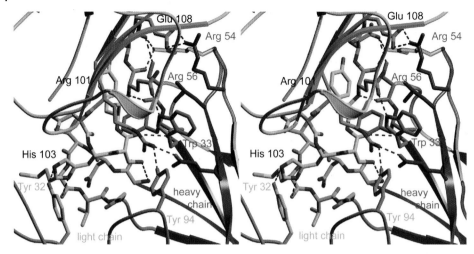

Fig. 3.5 Stereo view of MOG residues 101–108 (yellow, green) interacting with the 8-18C5 CDRs (heavy chain: blue, light chain: red).

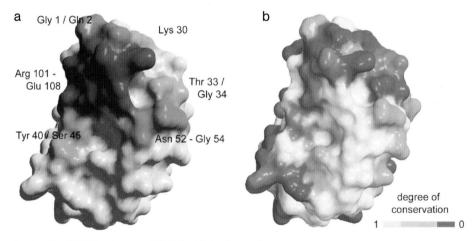

Fig. 3.6 The 8-18C5 epitope on MOG. (a) Molecular surface of MOG with epitope residues colored from orange to green according to their position in the MOG sequence. (b) Conservation between MOG and butyrophilin mapped onto the molecular surface of MOG.

evaluate the importance of the observed epitope. Interestingly, exchanging two amino acids of the FG-loop in the center of the epitope eliminated binding of nine out of 10 murine monoclonal antibodies, clearly demonstrating immunodominance of the 8-18C5 epitope in mice. Provided that a similarly biased reactivity exists in humans, assaying the differential binding of antibodies to mutant and wild-type

MOG may yield an easy and sensitive test for pathogenic MOG-specific antibodies in MS. Whereas MOG is a classical sequestered autoantigen proteins homologous to MOG, such as butyrophilin (BTN), BTN-like proteins or erythroid membrane associated protein (ERMAP) that are ~50% identical to MOG$_{ex}$ are expressed outside the CNS, and able to induce tolerance. Intriguingly, mapping the degree of conservation onto the molecular surface of MOG reveals a striking concordance of the epitope region and residues unique to MOG (Fig. 3.6). This correlation offers a simple rationale for the observed specificity of anti-MOG antibodies that might also be applicable to other sequestered autoantigens.

3.5
Conclusions

Autoantibody epitopes are currently being elucidated in a wide range of autoimmune diseases [95], with some of them having already become accepted tools in diagnosis [30, 96, 97]. The advent of miniaturized multiplex arrays in autoimmune research provides the opportunity to perform large-scale assays to detect specific antibody epitope profiles in autoimmune disorders [98, 99] (see also Part I, Chapter 3 and Part V, Chapter 8). The combined use of different classic and novel epitope mapping methods, supported by information derived from three-dimensional structures of autoantigens or their complexes with antibody domains, will further increase the quality and quantity of deciphered epitopes in order to support diagnosis and, in the long term, to guide the development of biopharmaceuticals aimed at removing or neutralizing pathogenic autoantibodies.

Acknowledgments

The author thanks Christopher Linington and Uwe Jacob for their critical reading of the manuscript.

References

1 Autoimmune diseases coordinating committee. (2003). Autoimmune Diseases Research Plan, National Institute of Allergy and Infectious Diseases, National Institutes of Health.
2 S. Walsh and L. Rau, (2000). Autoimmune diseases: a leading cause of death among young and middle-aged woman in the United States. *Am. J. Public Health* **90**, 1463–1466.
3 A. Vincent, O. Lily and J. Palace, (1999). Pathogenic autoantibodies to neuronal proteins in neurological disorders. *J. Neuroimmunol.* **100**, 169–180.
4 M. De Baets and M. H. W. Stassen, (2002). The role of antibodies in myasthenia gravis. *J. Neurol. Sci.* **202**, 5–11.
5 M. G. Robson and M. J. Walport, (2001). Pathogenesis of systemic lupus erythematosus (SLE). *Clin. Lab. Med.* **31**, 678–685.
6 R. H. Scofield, (2004). Autoantibodies as predictors of disease. Lancet **363**, 1544–1546.
7 D. Devendra, L. P. Yu and G. S. Eisenbarth, (2004). Endocrine autoantibodies. *Clin. Lab. Med.* **24**, 275.
8 M. Knip, (2002). Natural course of preclinical type 1 diabetes. *Hormone Res.* **57**, 6–11.
9 M. M. J. Nielen, et al., (2004). Specific autoantibodies precede the symptoms of rheumatoid arthritis – A study of serial measurements in blood donors. *Arthritis Rheum.* **50**, 380–386.
10 M. R. Arbuckle, et al., (2003). Development of autoantibodies before the clinical onset of systemic lupus erythematosus. *N. Engl. J. Med.* **349**, 1526–1533.
11 D. D'Cruz, (2002). Testing for autoimmunity in humans. *Toxicol. Lett.* **127**, 93–100.
12 J. Schmitt, (2003). Recombinant autoantigens for diagnosis and therapy of autoimmune diseases. *Biomed. Pharmacother.* **57**, 261–268.
13 S. Jones and J. M. Thornton, (1996). Principles of protein–protein interactions. *Proc. Natl. Acad. Sci. USA* **93**, 13–20.

14 S. J. Wodak and J. Janin, (2002). Structural basis of macromolecular recognition. *Adv. Protein Chem.* **61**, 9–73.

15 E. J. Sundberg and R. A. Mariuzza, (2002). Molecular recognition in antibody–antigen complexes. *Adv. Protein Chem.* **61**, 119–160.

16 M. H. V. Van Regenmortel, (1999). Molecular dissection or protein antigens and the prediction of epitopes. In: Synthetic Peptides as Antigens (S. Pillai and P. C. van der Vliet, Eds.), Vol. 28, pp. 1–79. Elsevier, Amsterdam.

17 L. Jin and J. A. Wells, (1994). Dissecting the Energetics of an Antibody–Antigen Interface by Alanine Shaving and Molecular Grafting. *Protein Sci.* **3**, 2351–2357.

18 G. A. Weiss, C. K. Watanabe, A. Zhong, A. Goddard and S. S. Sidhu, (2000). Rapid mapping of protein functional epitopes by combinatorial alanine scanning. *Proc. Natl. Acad. Sci. USA* **97**, 8950–8954.

19 T. Clackson and J. A. Wells, (1995). A Hot Spot of Binding Energy in a Hormone-Receptor Interface. *Science* **267**, 383–386.

20 M. Z. Atassi and J. A. Smith, (1978). A proposal for the nomenclature of antigenic sites in peptides and proteins. *Immunochemistry* **15**, 609–610.

21 S. Muller (1999). Peptides in diagnosis of autoimmune diseases. In: Synthetic Peptides as Antigens (S. Pillai and P. C. van der Vliet, Eds.), Vol. 28, pp. 247–280. Elsevier, Amsterdam.

22 L. Schachna, et al., (2002). Recognition of Granzyme B-generated autoantigen fragments in scleroderma patients with ischemic digital loss.[see comment]. *Arthritis Rheum.* **46**, 1873–1884.

23 E. L. Greidinger, L. Casciola-Rosen, S. M. Morris, R. W. Hoffman and A. Rosen, (2000). Autoantibody recognition of distinctly modified forms of the U1-70-kd antigen is associated with different clinical disease manifestations. *Arthritis Rheum.* **43**, 881–888.

24 K. C. R. Malmegrim, G. J. M. Pruijn and W. J. van Venrooij, (2002). The fate of the U1 snRNP autoantigen during apoptosis: Implications for systemic autoimmunity. *Israel Med. Assoc. J.* **4**, 706–712.

25 P. A. C. Cloos and S. Christgau, (2004). Posttranslational modifications of proteins: implications for aging, antigen recognition, and autoimmunity. *Biogerontology* **5**, 139–158.

26 M. A. van Boekel and W. J. van Venrooij, (2003). Modifications of arginines and their role in autoimmunity. *Autoimmunity Rev.* **2**, 57–62.

27 H. Brahms, et al., (2000). The C-terminal RG dipeptide repeats of the spliceosomal Sm proteins D1 and D3 contain symmetrical dimethylarginines, which form a major B-cell epitope for anti-Sm autoantibodies. *J. Biol. Chem.* **275**, 17122–17129.

28 R. L. F. Nienhuis and E. A. Mandema, (1964). A new serum factor in patients with rheumatoid arthritis: the antiperinuclear factor. *Ann. Rheum. Dis.* **23**, 302–305.

29 G. A. Schellekens, B. A. de Jong, F. H. van den Hoogen, L. B. van de Putte and W. J. van Venrooij, (1998). Citrulline is an essential constituent of antigenic determinants recognized by rheumatoid arthritis-specific autoantibodies. *J. Clin. Invest.* **101**, 273–281.

30 G. A. Schellekens, et al., (2000). The diagnostic properties of rheumatoid arthritis antibodies recognizing a cyclic citrullinated peptide. *Arthritis Rheum.* **43**, 155–163.

31 E. L. Greidinger, M. F. Foecking, S. Ranatunga and R. W. Hoffman, (2002). Apoptotic U1-70 kd is antigenically distinct from the intact form of the U1-70-kd molecule. *Arthritis Rheum.* **46**, 1264–1269.

32 C. P. Genain, B. Cannella, S. L. Hauser and C. S. Raine, (1999). Identification of autoantibodies associated with myelin damage in multiple sclerosis. *Nat. Med.* **5**, 170–175.

33 U. Brehm, S. J. Piddlesden, M. V. Gardinier and C. Linington, (1999). Epitope specificity of demyelinating monoclonal autoantibodies directed against the human myelin oligodendrocyte glycoprotein (MOG). *J. Neuroimmunol.* **97**, 9–15.

34 C. G. Haase, et al., (2001). The fine specificity of the myelin oligodendrocyte glycoprotein autoantibody response in patients with multiple sclerosis and normal healthy controls. *J. Neuroimmunol.* **114**, 220–225.

35 H. C. von Budingen, et al., (2002). Molecular characterization of antibody specificities against myelin/oligodendrocyte glycoprotein in autoimmune demyelination. *Proc. Natl. Acad. Sci. USA* **99**, 8207–8212.

36 H. C. von Budingen, et al., (2004). Frontline: Epitope recognition on the myelin/oligodendrocyte glycoprotein differentially influences disease phenotype and antibody effector func-

tions in autoimmune demyelination. *Eur. J. Immunol.* **34**, 2072–2083.

37 L. Craig, et al., (1998). The role of structure in antibody cross-reactivity between peptides and folded proteins. *J. Mol. Biol.* **281**, 183–201.

38 M. Bluthner, M. Mahler, D. B. Muller, H. Dunzl and F. A. Bautz, (2000). Identification of an alpha-helical epitope region on the PM/Scl-100 autoantigen with structural homology to a region on the heterochromatin p25beta autoantigen using immobilized overlapping synthetic peptides. *J. Molec. Med.* **78**, 47–54.

39 Y. Tian, et al., (2002). Structure–affinity relationships in the gp41 ELDKWA epitope for the HIV-(1) neutralizing monoclonal antibody F-2(5): effects of side-chain and backbone modifications and conformational constraints. *J. Peptide Res.* **59**, 264–276.

40 R. C. Landry, et al., (2001). Antibody recognition of a conformational epitope in a peptide antigen: Fv-peptide complex of an antibody fragment specific for the mutant EGF receptor, EGFRvIII. *J. Mol. Biol.* **308**, 883–893.

41 D. Zillikens, et al., (1997). Tight Clustering of Extracellular Bp180 Epitopes Recognized by Bullous Pemphigoid Autoantibodies. *J. Invest. Dermatol.* **109**, 573–579.

42 E. Schmidt, et al., (2002). A highly sensitive and simple assay for the detection of circulating autoantibodies against full-length bullous pemphigoid antigen 180. *J. Autoimmunity* **18**, 299–309.

43 D. Zillikens, et al., (1997). A highly sensitive enzyme-linked immunosorbent assay for the detection of circulating anti-BP180 autoantibodies in patients with bullous pemphigoid. *J. Invest. Dermatol.* **109**, 679–683.

44 A. Kromminga, et al., (2002). Cicatricial pemphigoid differs from bullous pemphigoid and pemphigoid gestations regarding the fine specificity of autoantibodies to the BP180NC16A domain. *J. Dermatol. Sci.* **28**, 68–75.

45 I. Laczkó, et al., (2000). Conformational consequences of coupling bullous pemphigoid antigenic peptides to glutathione-S-transferase and their diagnostic significance. *J. Peptide Sci.* **6**, 378–386.

46 D. R. Davies and G. H. Cohen, (1996). Interactions of protein antigens with antibodies. *Proc. Natl. Acad. Sci. USA* **93**, 7–12.

47 M. A. Newman, C. R. Mainhart, C. P. Mallett, T. B. Lavoie and S. J. Smith-Gill, (1992).

Patterns of antibody specificity during the BALB/c immune response to hen eggwhite lysozyme. *J. Immunol.* **149**, 3260–3272.

48 A. L. Corper, et al., (1997). Structure of human IgM rheumatoid factor Fab bound to its autoantigen IgG Fc reveals a novel topology of antibody–antigen interaction. *Nature Struct. Biol.* **4**, 374–381.

49 C. Breithaupt, et al., (2003). Structural insights into the antigenicity of myelin oligodendrocyte glycoprotein. *Proc. Natl. Acad. Sci. USA* **100**, 9446–9451.

50 E. W. Henriksson, M. Wahren-Herlenius, I. Lundberg, E. Mellquist and I. Pettersson, (1999). Key residues revealed in a major conformational epitope of the U1-70K protein. *Proc. Natl. Acad. Sci. USA* **96**, 14487–14492.

51 Y. Ma, et al., (2002). Key residues of a major cytochrome P4502D6 epitope are located on the surface of the molecule. *J. Immunol.* **169**, 277–285.

52 A. Burnens, S. Demotz, G. Corradin, H. Binz and H. R. Bosshard, (1987). Epitope mapping by chemical modification of free and antibody-bound protein antigen. *Science* **235**, 780–783.

53 W. Fiedler, C. Borchers, M. Macht, S. O. Deininger and M. Przybylski, (1998). Molecular characterization of a conformational epitope of hen egg white lysozyme by differential chemical modification of immune complexes and mass spectrometric peptide mapping. *Bioconj. Chem.* **9**, 236–241.

54 R. Jemmerson and Y. Paterson, (1986). Mapping epitopes on a protein antigen by the proteolysis of antigen–antibody complexes. *Science* **232**, 1001–1004.

55 D. Suckau, et al., (1990). Molecular epitope identification by limited proteolysis of an immobilized antigen–antibody complex and mass spectrometric peptide mapping. *Proc. Natl. Acad. Sci. USA* **87**, 9848–9852.

56 V. Legros, C. Jolivet-Reynaud, N. Battail-Poirot, C. Saint-Pierre and E. Forest, (2000). Characterization of an anti-*Borrelia burgdorferi* OspA conformational epitope by limited proteolysis of monoclonal antibody-bound antigen and mass spectrometric peptide mapping. *Protein Sci.* **9**, 1002–1010.

57 D. I. R. Spencer, et al., (2002). A strategy for mapping and neutralizing conformational immunogenic sites on protein therapeutics. *Proteomics* **2**, 271–279.

58 R.H. Meloen, W.C. Puijk and J.W. Slootstra, (2000). Mimotopes: realization of an unlikely concept. *J. Mol. Rec.* **13**, 352–359.

59 H.M. Geysen, S.J. Rodda and T.J. Mason, (1986). A priori delineation of a peptide which mimics a discontinuous antigenic determinant. *Mol. Immunol.* **23**, 709–715.

60 D. Enshell-Seijffers, et al., (2003). The mapping and reconstitution of a conformational discontinuous B-cell epitope of HIV-1. *J. Mol. Biol.* **334**, 87–101.

61 J.M. Davies, et al., (1999). Multiple alignment and sorting of peptides derived from phage-displayed random peptide libraries with polyclonal sera allows discrimination of relevant phagotopes. *Mol. Immunol.* **36**, 659–667.

62 M.J. Rowley, et al., (2000). Prediction of the immunodominant epitope of the pyruvate dehydrogenase complex E2 in primary biliary cirrhosis using phage display. *J. Immunol.* **164**, 3413–3419.

63 K. Beland, P. Lapierre, G. Marceau and F. Alvarez, (2004). Anti-LC1 autoantibodies in patients with chronic hepatitis C virus infection. *J. Autoimmunity* **22**, 159–166.

64 B.W. Bailey, et al., (2003). Constraints on the conformation of the cytoplasmic face of dark-adapted and light-excited rhodopsin inferred from antirhodopsin antibody imprints. *Protein Sci.* **12**, 2453–2475.

65 D. Bresson, et al., (2003). Localization of the discontinuous immunodominant region recognized by human anti-thyroperoxidase autoantibodies in autoimmune thyroid diseases. *J. Biol. Chem.* **278**, 9560–9569.

66 A. Wiik, (2003). Autoantibodies in vasculitis. *Arthritis Res. Ther.* **5**, 147–152.

67 J. Bartunkova, V. Tesar and A. Sediva, (2003). Diagnostic and pathogenetic role of antineutrophil cytoplasmic autoantibodies. *Clin. Immunol.* **106**, 73–82.

68 Y.M. Van der Geld, C.A. Stegeman and C.G. M. Kallenberg, (2004). B cell epitope specificity in ANCA-associated vasculitis: does it matter? *Clin. Exp. Immunol.* **137**, 451–459.

69 R.C. Williams, et al., (1994). Epitopes on Proteinase-3 Recognized by Antibodies from Patients with Wegeners Granulomatosis. *J. Immunol.* **152**, 4722–4737.

70 M.E. Griffith, A. Coulthart, S. Pemberton, A.J. George and C.D. Pusey, (2001). Anti-neutrophil cytoplasmic antibodies (ANCA) from patients with systemic vasculitis recognize restricted epitopes of proteinase 3 involving the catalytic site. *Clin. Exp. Immunol.* **123**, 170–177.

71 Y.M. van der Geld, et al., (2001). Antineutrophil cytoplasmic antibodies to proteinase 3 in Wegener's granulomatosis: Epitope analysis using synthetic peptides. *Kidney Int.* **59**, 147–159.

72 M. Fujinaga, M.M. Chernaia, R. Halenbeck, K. Koths and M.N. James, (1996). The crystal structure of PR3, a neutrophil serine proteinase antigen of Wegener's granulomatosis antibodies. *J. Mol. Biol.* **261**, 267–278.

73 K.O. Netzer, et al., (1999). The Goodpasture autoantigen – Mapping the major conformational epitope(s) of alpha 3(IV) collagen to residues 17-31 and 127-141 of the NC1 domain. *J. Biol. Chem.* **274**, 11267–11274.

74 T. Hellmark, H. Burkhardt and J. Wieslander, (1999). Goodpasture disease – Characterization of a single conformational epitope as the target of pathogenic autoantibodies. *J. Biol. Chem.* **274**, 25862–25868.

75 M.E. Than, et al., (2002). The 1.9-angstrom crystal structure of the noncollagenous (NC1) domain of human placenta collagen IV shows stabilization via a novel type of covalent Met-Lys cross-link. *Proc. Natl. Acad. Sci. USA* **99**, 6607–6612.

76 M. Sundaramoorthy, M. Meiyappan, P. Todd and B.G. Hudson, (2002). Crystal structure of NC1 domains – Structural basis for type IV collagen assembly in basement membranes. *J. Biol. Chem.* **277**, 31142–31153.

77 D.B. Borza, et al., (2002). Quaternary organization of the Goodpasture autoantigen, the alpha 3(IV) collagen chain – Sequestration of two cryptic autoepitopes by intraprotomer interactions with the alpha 4 and alpha 5 NC1 domains. *J. Biol. Chem.* **277**, 40075–40083.

78 M. Segelmark, T. Hellmark and J. Wieslander, (2003). The prognostic significance in Goodpasture's disease of specificity, titre and affinity of anti-glomerular-basement-membrane antibodies. *Nephron Clin. Pract.* **94**, C59–C68.

79 C.J. Padoa, et al., (2003). Recombinant Fabs of human monoclonal antibodies specific to the middle epitope of GAD65 inhibit type 1 diabetes-specific GAD65Abs. *Diabetes* **52**, 2689–2695.

80 P. Burkhard, P. Dominici, C. Borri-Voltattorni, J.N. Jansonius and V.N. Malashkevich, (2001). Structural insight into Parkinson's dis-

ease treatment from drug-inhibited DOPA decarboxylase. *Nature Struct. Biol.* **8**, 963–967.

81 H. L. Schwartz, et al., (1999). High-resolution autoreactive epitope mapping and structural modeling of the 65 kDa form of human glutamic acid decarboxylase. *J. Mol. Biol.* **287**, 983–999.

82 J. C. Tong, M. A. Myers, I. R. Mackay, P. Z. Zimmet and M. J. Rowley, (2002). The PEV-KEK region of the pyridoxal phosphate binding domain of GAD65 expresses a dominant B cell epitope for type 1 diabetes sera. *Ann. N. Y. Acad. Sci.* **958**, 182–189.

83 M. A. Myers, et al., (2000). Conformational epitopes on the diabetes autoantigen GAD65 identified by peptide phage display and molecular modeling. *J. Immunol.* **165**, 3830–3838.

84 B. Sutton, A. Corper, V. Bonagura and M. Taussig, (2000). The structure and origin of rheumatoid factors. *Immunol. Today* **21**, 177–183.

85 T. G. Johns and C. C. A. Bernard, (1999). The structure and function of myelin oligodendrocyte glycoprotein. *J. Neurochem.* **72**, 1–9.

86 R. Lebar, C. Lubetzki, C. Vincent, P. Lombrail and J. M. Boutry, (1986). The M2 autoantigen of central nervous system myelin, a glycoprotein present in oligodendrocyte membrane. *Clin. Exp. Immunol.* **66**, 423–434.

87 C. Linnington, M. Webb and P. L. Woodhams, (1984). A novel myelin-associated glycoprotein defined by a mouse monoclonal antibody. *J. Neuroimmunol.* **6**, 387–396.

88 A. Iglesias, J. Bauer, T. Litzenburger, A. Schubart and C. Linington, (2001). T- and B-cell responses to myelin oligodendrocyte glycoprotein in experimental autoimmune encephalomyelitis and multiple sclerosis. *Glia* **36**, 220–234.

89 H. C. Von Budingen, et al., (2001). Immune responses against the myelin/oligodendrocyte glycoprotein in experimental autoimmune demyelination. *J. Clin. Immunol.* **21**, 155–170.

90 E. Mathey, C. Breithaupt, A. S. Schubart and C. Linington, (2004). Commentary: Sorting the wheat from the chaff: identifying demyelinating components of the myelin oligodendrocyte glycoprotein (MOG)-specific autoantibody repertoire. *Eur. J. Immunol.* **34**, 2065–2071.

91 T. Berger, et al., (2003). Antimyelin antibodies as a predictor of clinically definite multiple sclerosis after a first demyelinating event. *N. Engl. J. Med.* **349**, 139–145.

92 C. Linington, personal communication.

93 C. S. Clements, et al., (2003). The crystal structure of myelin oligodendrocyte glycoprotein, a key autoantigen in multiple sclerosis. *Proc. Natl. Acad. Sci. USA* **100**, 11059–11064.

94 M. Ichikawa, T. G. Johns, J. L. Liu and C. C. A. Bernard, (1996). Analysis of the Fine B Cell Specificity During the Chronic/Relapsing Course of a Multiple Sclerosis-Like Disease in Lewis Rats Injected with the Encephalitogenic Myelin Oligodendrocyte Glycoprotein Peptide 35-55. *J. Immunol.* **157**, 919–926.

95 M. Mahler, M. Bluthner and K. M. Pollard, (2003). Advances in B-cell epitope analysis of autoantigens in connective tissue diseases. *Clin. Immunol.* **107**, 65–79.

96 B. G. Hudson, K. Tryggvason, M. Sundaramoorthy and E. G. Neilson, (2003). Alport's syndrome, Goodpasture's syndrome, and type IV collagen. *N. Engl. J. Med.* **348**, 2543–2556.

97 J. G. Routsias, A. G. Tzioufas and H. M. Moutsopoulos, (2004). The clinical value of intracellular autoantigens B-cell epitopes in systemic rheumatic diseases. *Clin. Chim. Acta* **340**, 1–25.

98 W. Hueber, P. J. Utz, L. Steinman and W. H. Robinson, (2002). Autoantibody profiling for the study and treatment of autoimmune disease. *Arthritis Res.* **4**, 290–295.

99 W. H. Robinson, et al., (2002). Autoantigen microarrays for multiplex characterization of autoantibody responses. *Nat. Med.* **8**, 295–301.

Find, Fight, and Follow –
Target-specific Troika from Mother Nature's Pharmacopoiea

4
Molecular Imaging and Applications for Pharmaceutical R&D

Joke G. Orsel and Tobias Schaeffter

Abstract

Classically, medical imaging provides structural information of the patient's body, and is used mainly for diagnostic purposes. The recent advances in the development of contrast agents that can highlight molecules or molecular structures and pathways allow researchers and physicians to obtain ever more detailed information. This field of molecular imaging promises to improve dramatically the future of healthcare, shifting the emphasis toward much earlier diagnosis and treatment. The combination of molecular imaging with the advent of devices for the imaging of small animals renders it increasingly interesting for use in drug discovery and drug development. This chapter provides an introduction to the different imaging techniques available, together with examples of contrast agents and applications for molecular imaging. Focus is then turned to the potential and implications of the technique for drug discovery and the development of modern biopharmaceuticals.

Abbreviations

CEA	carcinoembryonic antigen
CEST	chemical exchange-dependent saturation transfer
CT	computed tomography
DOT	diffuse optical tomography
EAE	experimental autoimmune encephalomyelitis
ED-B-FN	extra-domain B-fibronectin
FDG	fluorodeoxyglucose
FID	free induction decay
Gd-DTPA	gadolinium diethylenetri-aminepenta-acetic acid
ICAM-1	intercellular adhesion molecule-1
ICG	indocyanine green
MMP	matrix metalloproteinase
MP	microparticle
MRI	magnetic resonance imaging
NIH	National Institute of Health
NIR	near infrared
NMR	nuclear magnetic resonance
PET	positron emission tomography

Modern Biopharmaceuticals. Edited by J. Knäblein
Copyright © 2005 WILEY-VCH Verlag GmbH & Co. KGaA, Weinheim
ISBN: 3-527-31184-X

SPECT   single photon emission computed tomography

SPIO   superparamagnetic iron oxide

USPIO   ultra-small superparamagnetic iron oxide

VEGF   vascular endothelial growth factor

VEGFR2   vascular endothelial growth factor receptor-2

4.1
Introduction

In medical diagnostics, several techniques are available to obtain an image of a part or the whole of a patient's body. The use of imaging techniques, also called modalities, has grown exponentially during the past 30 years. Imaging techniques have now become essential tools, not only in clinical diagnostics and therapy monitoring, but also for pharmaceutical industries. The reason is clear: bringing a new drug to market is a time-consuming and expensive undertaking, requiring approximately 15 years and an investment of US$ 800–900 million. One costly factor is caused by the standard trial end-points of morbidity and mortality. Using imaging, other markers of drug efficacy and toxicity may be developed, potentially offering faster and more quantitative data. The advantages would be earlier decisions on whether to continue studies on a potential drug or to discard it, and earlier starts of clinical trials, thus reducing time to market. The United States Food and Drug Administration (FDA) has also recognized this and recently started an initiative to investigate the use of imaging results as biomarkers for biopharmaceutical development.

Traditionally, medical imaging provides information on the anatomical level, and diseases or effects of therapy are mainly detected when structural abnormalities occur. Additional contrast between different tissues may be gained through the use of contrast agents such as paramagnetic nanoparticles or radiolabeled molecules. About 10 years ago, a trend started towards the development of contrast agents that carry a recognition element for targeting to a certain molecule. Using such agents to visualize the concentration or activity of a specific molecule *in vivo* is called molecular imaging [1]. This technology has very important advantages for the diagnosis of cancer and other diseases, where subtle cellular changes occur well before an anatomical abnormality such as a tumor becomes detectable. Molecular imaging may also aid therapy decisions if it can be used to distinguish between structures with a similar morphology but different molecular malfunctions, such as benign and malignant tumors. Furthermore, imaging on a molecular level can provide information on therapy response before an anatomical effect on diseased tissue can be detected, enabling much faster treatment optimization. Molecular imaging not only has a high potential impact on medical diagnostics but, in combination with the advances in small animal imaging equipment and animal models for diseases, also on drug discovery and development. In this chapter, we will discuss these promising technologies and provide application examples. We would like to give a broad overview, and thus will by no means be comprehensive. Therefore, throughout the text we refer to outstanding reviews that deal more extensively with a certain subject. Before we dive into molecular imaging, we will first examine the different imaging techniques available.

4.2
Imaging Modalities and Contrast Agents

Several techniques have been developed that non-invasively provide images of the body *in vivo*. As each technique detects how another form of energy interacts with tissue, each has its own characteristics in terms of sensitivity, spatial resolution, available contrast agents, ease of use and costs, several of which are listed in Table 4.1 in Section 4.3.3. Together, the modalities offer a wide spectrum of possibilities and applications and can, in general, be considered complementary.

4.2.1
X-ray Imaging

X-ray imaging is a transmission-based technique, in which X-rays from a source pass through the patient and are detected on the opposite side. The contrast in the image arises from the different attenuation of X-rays in different tissues, and the amount of absorption depends on the tissue composition. For example, dense bone matter will absorb many more X-rays than soft tissues, such as muscle and fat. The contrast also depends on the energy of the X-rays – in soft tissue, low-energy X-rays result in better contrast. Thus, another important factor for the overall image quality of X-ray images is the X-ray tube. This tube consists of two electrodes, a negatively charged cathode and a positively charged anode. The heated cathode generates electrons, which are accelerated by applying a high voltage (20–150 kV). The accelerated electrons strike on the anode, generating X-rays. The resultant beam consists of X-rays of a broad range of energies, where the maximum energy depends on the applied voltage. In planar X-ray, the line integral of the spatially dependent attenuation is measured and the resulting intensity is displayed in a two-dimensional image. However, it is difficult to interpret overlapping layers of soft tissue and bone structures on such a projection image. In order to resolve such three-dimensional (3D) structures, X-ray computed tomography (CT) is used, which generates cross-sectional, two-dimensional (2D) images of the body. Due to the high spatial resolution of X-ray CT images, the technique is widely applied in structural imaging, that is it is used to depict morphology.

In CT, typically an X-ray tube and a detector are rapidly rotated 360° around the patient (see Fig. 4.1). The acquisition time of an image is determined by the rotation time of the detectors and the X-ray tube, which ranges from 0.3 to 1 second. Modern CT scanners use a number (16–64) of detector rows, which allows the simultaneous measurement of multiple slices. Further reduction of the scan time can be achieved by faster rotation times. There-

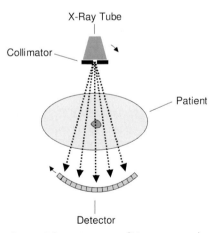

Fig. 4.1 Schematic set-up of X-ray computed tomography. An X-ray tube and a detector rotate around the patient. A collimator in front of the tube shapes the X-ray beam. Absorption of the transmitted X-rays is measured on the other side by a detector.

fore, the improvement of the mechanical design of the gantry, which carries the X-ray tube and the detector, is essential. The final image is reconstructed from the measurements by applying either a filtered backprojection algorithm or more advanced iterative reconstruction techniques. Each pixel contains the averaged attenuation values within the corresponding voxel (the smallest volume unit in the image). This number is compared to the attenuation value of water and displayed on a scale of Hounsfield units (HU), named after the inventor of CT, Sir Godfrey Hounsfield. The scale assigns an attenuation value of zero to water and regular attenuation values range from −1000 to 3000 HU. The attenuation value of soft tissue ranges between 40 and 100 HU.

The spatial resolution in X-ray CT depends on the focal spot of the X-ray tube and the size of the detector elements. The spatial resolution of a clinical CT scanner is less than 0.5 mm in the center of the CT scanner. For pre-clinical imaging, several micro-CT systems have been developed and are commercially available [2]. The spatial resolution of dedicated small-animal scanners is much higher than for clinical scanners, and is in the order of 20 μm. Fig. 4.2 shows high-resolution images obtained with an animal CT scanner. The major advantage of X-ray CT is its ease of use ("push-button-technology") for acquiring large 3D datasets with structural information at a very high spatial resolution. The disadvantage of X-ray CT is the use of ionizing radiation, which can lead to cell death or to cancer due to genetic mutations. For example, the effective dose of a clinical CT scan of the abdomen is about 10 mSv, which is about 400 times higher than the dose of a chest projection X-ray.

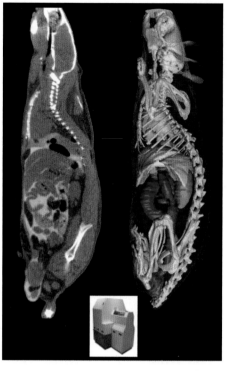

Fig. 4.2 Dedicated animal X-ray computed tomography system that allows high-resolution imaging (up to 18 μm) of rodents. The different attenuation of X-rays in various tissues and bones allows volume rendering of the different structures (right image) (Courtesy of ImTek, Inc., Knoxville, TN, USA).

4.2.1.1 X-ray Contrast Agents

X-ray contrast agents are chemicals that are introduced into the body to increase the image contrast. They contain substances with a high atomic number that increase the attenuation value in the regions where they accumulate. A typical X-ray agent used to image the gastrointestinal tract is barium sulfate, a solution of which the patient drinks hours prior to examination. For imaging of the colon, air is also used, which has a high negative Hounsfield value (−1000). Iodine-containing contrast agents are widely used for X-ray-based angiography, and are also applied in the detection of tumors. These agents are injected into the bloodstream, where they cause the attenuation of X-rays to be higher than in the surrounding tissue. The strength of the attenuation depends on the iodine load of the agent. However, a higher iodine load usually increases the osmolarity of the solution, which can change the cell volume and thus can lead to adverse reactions. Recently, a new generation of low- or iso-osmotic agents has been introduced.

4.2.2
Magnetic Resonance Imaging

Magnetic resonance imaging (MRI) is a non-ionizing imaging technique with superior soft-tissue contrast, high spatial resolution and good temporal resolution. MRI is capable of measuring a wide range of endogenous contrast mechanisms that include proton density, spin-lattice relaxation time (T_1), spin-spin relaxation time (T_2), chemical shift, temperature and different types of motion, such as blood flow, perfusion, or diffusion. MRI has become the modality of choice in many pre-clinical and clinical applications, because it can provide structural and functional information. With ongoing developments to improve the image quality, acquisition speed and quantitative accuracy, the range of applications for MRI continues to expand rapidly.

Given the short length of this chapter, the description of MRI must be superficial (for a detailed description, see Refs. [3, 4]). MRI is based on the nuclear magnetic resonance (NMR) effect that was observed by Bloch and Purcell in 1946. Three basic requirements must be satisfied to measure an MR-signal. First, the nucleus of interest must possess a non-zero magnetic moment. All nuclei that have an odd number

of protons and/or neutrons have this property and behave as small magnets. Typical nuclei are, for example, hydrogen (1-H), phosphorus (31-P), carbon (13-C), sodium (23-Na), or fluorine (19-F). In the absence of an external magnetic field, these individual magnetic moments are randomly oriented and there is no net magnetization. The second requirement is an external static magnetic field B_0. Typical field strengths for clinical MRI are between 0.23 and 3 Tesla, whereas they are much higher for pre-clinical systems (4.7–7 Tesla, and even >10 Tesla). Due to the Zeemann effect, magnetic moments align under specific angles along or opposed to the external field B_0, resulting in a precessional movement of the magnetic moments. The precessional frequency, also called Larmor frequency, is given by $f_0 = \gamma \, B_0$, where γ is the gyromagnetic ratio, a constant for a given nucleus. NMR of hydrogen is the most important for clinical applications, because hydrogen is highly present in biological tissues and its gyromagnetic ratio is the largest of all nuclei. Since an alignment parallel to the field is the lower energy state, it is preferred and slightly more nuclei will align along rather than opposed to the field. As a result, the tissue will exhibit a net magnetization, which is parallel to the external magnetic field and is called longitudinal magnetization. The amount of the net magnetization depends on the field strength and increases for higher magnetic fields. The third requirement is a time-varying magnetic field B_1 applied perpendicularly to the static B_0 field and at Larmor-frequency (i.e., at resonance condition). For this, an additional radiofrequency (RF-) coil produces a B_1-pulse of a certain amplitude and duration. Such a B_1-pulse flips the longitudinal magnetization to an arbitrary angle (also called flip-angle). The transverse component of the

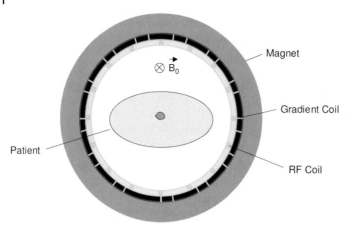

Fig. 4.3 Schematic set-up of an MRI system. The magnet causes a strong homogeneous static magnetic field, B_0. The gradient coil creates a linear variation of the magnetic field in all three dimensions. The RF-coil creates a time varying field that is perpendicular to the static magnetic field B_0.

flipped magnetization precesses around the static B_0 field at Larmor frequency and induces a time varying voltage signal in the RF-coil (Fig. 4.3).

The detected transverse magnetization does not remain forever, since two independent relaxation processes take place. First, the spin-lattice relaxation describes how fast the longitudinal magnetization recovers after applying the B_1-pulse. The rate of the recovery process is determined by the relaxation time T_1. Second, the spin-spin relaxation describes how fast the transversal magnetization loses its coherence and thus decays. The rate of dephasing is determined by the relaxation time T_2. In addition to spin-spin interactions, dephasing between the coherently precessing magnetic moments can also be caused by B_0-field inhomogeneities. As a result, an apparently stronger relaxation process is visible, which is called T_2^* relaxation. This T_2^* relaxation describes the decay of a time varying signal, which is called the free induction decay (FID).

Both the spin-lattice relaxation time T_1 and the spin-spin relaxation time T_2 vary among different types of tissue, and T_1 is always larger than T_2. In addition, the T_1 relaxation time depends on the field strength B_0, whereas the T_2 time is independent. The signal amplitude depends on the timing of the experiment and the relaxation times T_1 and $T_2(*)$. It can also be influenced by endogenous contrast mechanisms such as diffusion or blood-oxygenation.

In order to distinguish signals from different spatial locations, magnetic field gradients are applied by using gradient coils. The gradient coils create a linear variation in the z-component of the static magnetic field. Consequently, with the spatially varying field strength a spatially varying precessional frequency is connected. Usually, for 3D encoding, not all gradients are applied simultaneously and the image formation process can be separated into three phases: slice selection, phase encoding, and frequency encoding. After performing

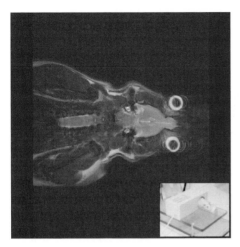

Fig. 4.4 Animal MR image with 100 μm resolution obtained on a clinical 3 Tesla MR scanner using a dedicated MR receive coil (Philips Research, Hamburg, Germany).

a number of experiments with different gradient values, an image can be reconstructed by using a Fourier-transform of the obtained signals. The spatial resolution strongly depends on the amplitude of the gradients and the acquisition bandwidth. Typical values of the spatial resolution of clinical scanners are in the order of 0.5–1 mm. However, high-resolution imaging on clinical MR-scanners is possible using dedicated RF-coils to increase the sensitivity. With this a spatial resolution of about 100 μm can be achieved (Fig. 4.4). A higher spatial resolution is possible in dedicated animal MR-scanners that operate at a higher magnetic field strength (e.g., 7 Tesla) and which apply strong gradients.

4.2.2.1 MRI Contrast Agents

In some clinical situations, the intrinsic contrast of the tissue is not sufficient to distinguish pathological from healthy tissue. Therefore, the use of contrast-enhancing agents has become an integral part of MR imaging. There are two basic classes of MRI contrast agents: paramagnetic and superparamagnetic agents. Paramagnetic agents primarily shorten the T_1 relaxation time of the tissue in which they accumulate. A more detailed description of MRI contrast agents can be found in Ref. [5].

Paramagnetic agents are based on metal ions with one or more unpaired electrons. These unpaired electrons result in a very large magnetic moment that interacts with the much smaller magnetic moments of the nucleus. Molecular motions result in random fluctuations of the dipolar magnetic interaction that reduces both the T_1 and the T_2 relaxation times. Gadolinium (Gd^{3+}) and manganese (Mn^{2+}) are examples of paramagnetic ions that are used in MR contrast agents. Since these metal ions are highly toxic, they must be contained in a chelate to prevent circulation of free ions in the body. Most clinically used agents base on gadolinium and differ only in the chelating agents; for example, the most commonly used clinical paramagnetic contrast agent is gadolinium diethylenetriaminepenta-acetic acid (Gd-DTPA; tradename Magnevist®, Schering AG, Berlin, Germany). The DTPA-chelate does not bind to blood proteins, ensuring rapid distribution through the bloodstream and fast clearance through the kidneys. Another class of paramagnetic agents is specifically designed to remain in the blood pool for a longer period of time. These blood pool agents either are of larger size, or they bind reversibly to albumin in blood plasma [6]. Recently, a new class of contrast agents was proposed that are based on chemical exchange-dependent saturation transfer (CEST) and which can be used to measure pH *in vivo* [7].

Superparamagnetic agents consist of small magnetic particles. These usually consist of a crystalline core comprising a mixture of iron oxides (Fe_2O_3 and Fe_3O_4) coated in a polymer matrix. The particles are divided into two classes according to their overall size: if the diameter is larger than 50 nm they are called superparamagnetic iron oxide (SPIO) particles, but if the diameter is smaller they are categorized as ultra-small superparamagnetic iron oxide (USPIO) particles. These contrast agents function by causing local field inhomogeneities that result in different relaxation regimes. Water molecules that diffuse through the local field inhomogeneities undergo T_2 and/or T_2^* decay. In addition, USPIO particles also have excellent T_1-enhancing properties.

4.2.3
Ultrasound Imaging

Ultrasound imaging is a non-invasive, portable and relatively inexpensive imaging modality, which is used extensively in the clinic. An ultrasound transducer (also called scanhead) sends short pulses of a high-frequency sound wave (1–10 MHz) into the body. At interfaces between two types of tissue, the wave will be refracted and part of the sound wave is reflected back due to Snell's law. How much is reflected depends on the densities of the respective tissues, and thus the speed of the sound wave within the different tissues. In addition, parts of the sound wave are also backscattered from small structures at tissue boundaries or within the tissue. High-frequency sound waves propagate well through soft tissue and fluids, but they are more or less stopped by air or bone. In clinical practice, this limitation is referred to as an "acoustic window". The transducer not only sends the wave into the body but also receives part of the reflected and/or backscattered wave, also named "echo". In clinical practice, ultrasound is used in a

wide variety of imaging situations including imaging of the heart (echocardiography), liver, kidney, ovaries, breast, peripheral vascular system, and even portions of the brain.

Diagnostic ultrasonic images can be generated using a variety of clinical imaging modes including A-lines, B-mode, and M-mode. The nomenclature is inherited from the world of radar. A-line mode is an older imaging mode where a one-dimensional amplitude line representing the propagation of sound along one line is shown. B-mode scanning is the most familiar clinical imaging mode and is represented by a 2D image. M-mode is the practice of rapidly firing one line of sight through a moving organ. This allows the tracking of motion of a structure such as a cardiac valve or cardiac wall with great time resolution.

Each line can be obtained in less than 100 µs. Thus, an image consisting of 100 lines can be obtained in less than 10 ms, which means that real-time imaging is possible. Therefore, in addition to imaging morphology, ultrasound is also capable of measuring the velocity of blood in circulation using the Doppler effect. The movement of red blood cells causes a shift in the frequency of returning ultrasound waves. The Doppler frequency shift is proportional to the blood flow velocity and thus allows quantification.

For the spatial resolution of an ultrasound system, three different dimensions must be considered: the axial, the lateral, and the elevation dimension. The axial resolution along the axis of the transducer is defined as the closest separation of two echoes that can be resolved and improves at higher frequencies. However, the penetration depth of ultrasound waves decreases with increasing frequencies. Therefore, lower frequencies (1–3 MHz) are used for studies of deep-lying structures, while higher frequencies (5–10 MHz) are used to image regions that are closer to the body surface. Typical values of the axial resolution are 1–2 mm at 1 MHz and 0.3 mm at 5 MHz. The lateral and elevation resolution of the ultrasound beam are determined by its thickness at the focal plane. For a single-crystal transducer, the lateral width of the ultrasound beam is determined by the diameter of the transducer and for clinical scanheads typically ranges from 0.5 to 2 mm at the focal plane. Nowadays, almost all commercially available transducers consist of an array of small piezoelectric crystals. Each of these array elements can be electronically controlled for transmission and reception of the ultrasound signal. In a phased array, a large number of array elements are controlled simultaneously to shape and to steer the location of focal plane of the ultrasound beam. Due to this dynamic focusing approach, the lateral resolution can be increased. The elevation dimension is determined by the length of the crystal elements and is in the order of 2–3 mm for clinical scanheads. However, with dedicated animal scanheads and/or systems operating at higher frequencies (up to 50 MHz) a much higher spatial resolution (below 100 µm) can be obtained on small animals. Fig. 4.5 shows an ultrasound image of a mouse heart taken on a clinical scanner using such a dedicated scanhead.

4.2.3.1 Ultrasound Contrast Agents

In several instances, the contrast in ultrasound is not high enough. For example, in 10–15% of patients normal echocardiography is not possible for anatomic or other reasons. In addition, Doppler-based methods to measure blood flow sometimes fail, due to masking of the signal by overlying tissue. For these applications, microbubbles can markedly enhance contrast. Microbubbles are gas-filled microspheres with a diameter of several micrometers, which are stabilized by a shell composed of, for example, albumin or lipids. Several versions of these diagnostic blood pool contrast agents are available commercially, and approved for better delineation of the chambers of the heart [8]. Microbubbles can be used to increase the echogenicity of blood via two different mechanisms. The first mechanism is resonance of microbubbles that expand and contract in an ultrasound field. At the resonant frequency, strong signals are generated at multiples of the transmitted frequency, also called harmonics. These can be very well detected with an ultrasound technique called harmonic imaging (Fig. 4.6), in which frequency filtering can be used to receive the harmonics. The second mechanism bases on differences in the acoustic impedance and thus increases the backscattering. In addition to resonance phenomena, microbubbles also show an increased scattering of the ultrasound wave [9], which is stronger than that for red blood cells. This property can be used to enhance

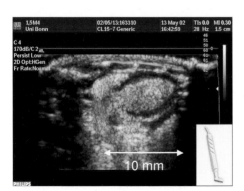

Fig. 4.5 Ultrasound image of a mouse heart using a dedicated animal scanhead on a clinical ultrasound scanner. A microbubble contrast agent provides a high contrast between blood (white) and myocardium. (Data courtesy K. Tiemann, University of Bonn, Germany).

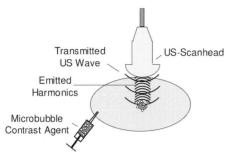

Fig. 4.6 Schematic set-up of harmonic ultrasound (US) imaging. Injected microbubbles resonate in a transmitted US field and emit US waves at higher harmonics that can be detected by the scanhead.

Doppler imaging of blood flow. Besides microbubbles, a targeted perfluorocarbon emulsion also showed an increase of backscattering upon accumulation and can thus be considered an "acoustic mirror" [10].

4.2.4
Nuclear Imaging

Nuclear imaging is based on the detection of gamma rays that are emitted by radionuclides. For this, small amounts (typically nanograms) of radionuclides are injected into the organisms. In contrast to other imaging modalities such as MRI, X-ray or ultrasound, nuclear imaging does not provide morphology information but images the spatial distribution of radionuclides in the organism. This distribution depends strongly on the biological behavior of a radiopharmaceutical. Therefore, the development and synthesis of radiopharmaceuticals is key in nuclear imaging to obtain physiological, metabolic, and molecular information. Two main nuclear imaging techniques can be distinguished due to the use of different types of radionuclides:

- Single photon emitters decay under the emission of gamma rays with energies between 100 and 360 keV.

- Positron emitters decay under the emission of positrons that result in a pair of high-energy gamma rays (511 keV) after annihilation with an electron.

The corresponding 3D imaging techniques are called single photon emission computed tomography (SPECT) and positron emission tomography (PET). As PET is discussed in detail elsewhere in this book (see Part V, Chapter 5), we will focus here on SPECT.

Fig. 4.7 shows the basic principles and instrumentation of SPECT. The injected radiopharmaceutical has accumulated in a suspicious region in the body. During decay of the radionuclides, gamma rays are emitted in all directions. Some of the gamma rays are attenuated and scattered in the body. In order to detect the gamma rays, a gamma camera is rotated around the body. The basic design of a gamma camera was described by Hal Anger in 1953 [11], which mainly consists of three parts: a collimator, a scintillation crystal, and a number of photomultipliers.

The collimator selects only those gamma rays that have a trajectory at an angle of 90° to the detector plane, and blocks all others. The collimator is generally a lead structure with a honeycomb array of holes, where the lead walls (septa) are designed to prevent penetration of gamma rays from one hole to the other. The parallel hole collimator is the most widely used collimator. Other types of collimators can be used to magnify the size of an object on the image. A pinhole collimator is an extreme form of a converging collimator and allows magnifying small objects placed very close to the pinhole [12]. The disadvantage of collimators is the low efficiency of gamma ray utilization, because they absorb most of the emitted gamma rays. Typically about one of 10000 emitted

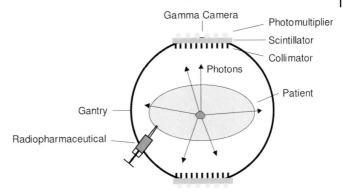

Fig. 4.7 Schematic set-up of a SPECT system. One or more gamma cameras slowly rotate around the patient. The gamma cameras detect gamma ray photons that are emitted by a radiopharmaceutical injected into the patient.

gamma rays is transmitted through a collimator, resulting in reduction of the signal-to-noise ratio.

The gamma rays that pass through the collimator are converted into a detectable signal. Usually, a single sodium iodide crystal doted with thallium is used. When a gamma ray strikes this scintillation crystal, it loses energy through photoelectric interactions. Consequently, light is emitted, the intensity of which is proportional to the energy of the gamma rays. Overall, approximately 15% of the absorbed energy is converted into visible light. Typically, about 100 photomultipliers are closely coupled to the scintillation crystal to convert the light signal into an electrical signal. The position of the scintillation point is determined from the relative signal outputs of the photomultipliers using an Anger position network. In addition, the sum of the signals is proportional to the energy of the absorbed gamma rays, which can be used to differentiate between non-scattered and scattered gamma rays. This energy window selection process can improve image quality, because scattered gamma rays produce a high background

noise. Furthermore, windows at different energies can also be used to discriminate between gamma rays of different energies, which are emitted from different radionuclides. Multi-energy windows thus allow for simultaneous imaging of various tracers.

In SPECT, one or more gamma cameras are rotated around the patient. The camera acquires a number of planar images from different view angles in a "stop-and-go" mode. Typically, 32 to 128 views are acquired to reconstruct a 3D image of the object using a filtered backprojection algorithm. Usually, images with a numerical resolution of 64×64 and 128×128 are reconstructed. In order to improve the image quality in SPECT, the tissue attenuation must be corrected for. The spatially dependent tissue attenuation can be determined from a pre-acquired CT-scan, which provides a 3D attenuation map. Alternatively, an approximation of an attenuation correction can be applied by performing a 360° rotation SPECT scan.

In first instance, the spatial resolution of SPECT is determined by the gamma camera. The resolution of the scintillator-photomultiplier combination is in the order of 3 mm, and depends on the thickness of the scintillation crystal and the diameter of the photomultiplier. However, the practical resolution of gamma cameras is less than 3 mm and is mainly limited by the collimator. The spatial resolution of a collimator depends on its geometry (length and the distance of the lead strips) and its distance to the gamma ray source. The overall spatial resolution of clinical SPECT ranges from less than 1 cm to about 2 cm, depending on the collimator type and its distance from the gamma ray source. In dedicated animal systems, special collimators are used, such as pinhole collimators and their resolution can be below 3 mm.

4.2.4.1 Nuclear Imaging Contrast Agents

Nuclear imaging can be considered as a pure molecular imaging technique, since it can directly detect the molecules in which the radionuclides have been embedded. These compounds are called "radiopharmaceuticals" or "radiotracers". Typical radionuclides for SPECT imaging are ^{99m}Tc, ^{111}In, ^{123}I, ^{201}Tl, and ^{67}Ga. Because of their toxicity, these isotopes are usually contained in chelates. The chemical structure of the radiopharmaceuticals determines their biodistribution and uptake in the body. Under pathological conditions, radiopharmaceuticals accumulate in certain regions and/or particular tissues, and thus can be used for early disease diagnosis. The major applications of SPECT imaging are assessment of cardiac function, measurement of blood perfusion in various organs (e.g., heart, brain or lung), detection of tumors and measurement of the renal function.

4.2.5
Optical Imaging

Optical imaging encompasses a large set of imaging technologies that use light from the ultraviolet to the infrared region to image tissue characteristics. These techniques rely on different contrast mechanisms such as transmission, absorption, reflectance, scattering, luminescence, and fluorescence. These mechanisms provide information on structure, physiology, biochemistry and molecular function. Optical imaging is a common tool for high-resolution imaging of surface structures. Microscopes are for instance used to characterize pathologies of the skin, whereas endoscopes are used to image structures inside the body. Naturally, these techniques currently are primarily limited to surface imaging or experimental imaging in small animals, because the penetration depth of light is very limited. However, light within a small spectral window of the near infrared (NIR) region (600–900 nm) can penetrate more than 10 cm into tissue due to the relatively low absorption rates at these wavelengths [13]. The lower boundary in this window is given by the high absorption rate of blood (hemoglobin), whereas the absorption above 900 nm rises due to the presence of water. For this reason, the NIR part of the spectrum is typically selected for non-surface optical imaging. The resolution of optical imaging is limited by the scattering of light: light photons propagating through tissue diffuse and follow random paths. Both, absorption and scattering are intrinsic contrast parameters of tissue that can be assessed by optical imaging. Measuring light absorption provides functional information on tissue, because oxyhemoglobin preferentially absorbs light at lower wavelengths than deoxyhemoglobin. This difference offers a non-invasive tool to quantify the vascularization and/or oxygenation status of tissue. On the other hand, scattering is associated with the structural properties of tissue. In general, two main optical imaging techniques can be differentiated to assess the absorption and scattering in tissue: transillumination and diffuse optical tomography.

In 1929, Cutler developed a technique called transillumination [14]: light was shined on one side of a breast, and the absorption behavior was examined on the other side. This approach is similar to projection X-ray imaging, but in transillumination the spatial resolution is significantly reduced due to the scattering of light photons. Because the absorption of hemoglobin depends on its oxygenation state, regions with increased vascularity could be detected with this technique. However, it did not provide sufficient specificity to distinguish between malignant and benign lesions. During the past 20 years, transillumination has been improved by employing advances in light sources (e.g., pulsed lasers) and detection techniques (e.g., charged coupled device detectors, time-of-flight techniques). Improvements in sensitivity and specificity for detection of breast tumors are currently under investigation.

During the past decade, diffuse optical tomography (DOT) was developed [15]. In this technique, light is applied from different angles, and scattered light is detected from all directions. In contrast to X-ray CT, in optical tomography proper modeling of the scattering process is essential. Typically, a numeric solution of the diffusion equation is used to describe the propagation of light photons in diffuse media and to predict the measurements of the experimental set-up. Due to the strong influence of scattering and since the reconstruction problem is ill-posed, the spatial resolution of optical tomography is rather poor and on the order of 5–10 mm. In comparison with transillumination, DOT allows better quantification of absorption, scattering, or fluorescence in three dimensions.

4.2.5.1 Optical Imaging Contrast Agents

Similar to other imaging modalities, the intrinsic contrast of optical imaging is not sufficient for certain applications, and imaging agents are necessary. In general, two different principles can be differentiated: fluorescence and bioluminescence. For fluorescence applications, fluorophore-labeled contrast agents are administered. The fluorophores are excited using light of an appropriate wavelength, which is generated either by a laser or by a white-light source using filters blocking light above the fluorophore absorption wavelengths. The emitted photons are detected using a high-sensitivity CCD camera. Two main set-ups are used [16]. In reflectance imaging, the light source and the CCD camera are both on the same side of the body. This provides relatively good images from probes that are no more than a few millimeters deep within the tissue. However, quantification of the signal is not possible, as it cannot be determined whether for example a lower signal is due to a deeper location or a lower concentration of the probes. This disadvantage is not present when capturing 3D images with tomographic set-ups such as DOT, which also permit the investigation of deeper-lying tissues, especially in animals. In these arrangements, fibers are used to guide the excitation light to different positions around the animal and to direct the emitted photons to the CCD camera. An algorithm designed especially for the reconstruction of fluorescence in media such as tissue improves and simplifies fluorescence imaging. The two types of set-up both allow multiplexing through the measurement of several different fluorophores and thus different targets in the same animal, either simultaneously or in fast sequence. In the NIR range, indocyanine green (ICG) is a widely used fluorescence agent, because it is safe and has been approved by the FDA. ICG is an intravascular agent that extravasates through vessels with a high permeability, such as those in tumors. Recently, more hydrophilic ICG derivatives were synthesized which showed a different biodistribution and provided a better contrast between healthy and tumor tissue [17].

In bioluminescence applications, an "internal signal source" is used – that is, photons are generated from an enzymatic reaction and no external light source is needed. This means, that bioluminescence imaging has practically no background. The enzymes used are luciferases, and their substrates are named luciferins. Luciferase:luciferin pairs occur in many different organisms that can glow or emit flashes of visible light, such as fireflies, bacteria, and many marine organisms. Luciferases and luciferins from different organisms are not necessarily structurally related. Each luciferase oxidizes its own specific substrate to form a product in an electronically excited state, which emits a photon upon decay. The emission spectra are relatively broad [18]. The luciferase:luciferin pair that is used most for optical imaging is that of the firefly, as a considerable part of its emission spectrum is above 600 nm and this light has better tissue-penetration properties. Bioluminescence is, in principle, detected with set-ups similar to those used in fluorescence imaging. Since bioluminescence imaging requires the stable expression of exogenous genes or modified endogenous genes in the organism under investigation, it is used only to investigate gene expression in transgenic or xenografted animals [19].

4.2.6
Multimodality Techniques

In general, the different imaging modalities provide different information and thus can be considered as being complementary rather than competitive. Therefore, the combination of techniques is of high interest. This can be done by image processing techniques (i.e., image registration) or by using integrated systems. In particular, the combination of structural imaging techniques (e.g., X-ray CT) and functional imaging (e.g., nuclear imaging) is of high interest, because it allows the co-registration of anatomy and molecular information. Nowadays, clinical PET-CT and SPECT-CT scanner combinations are commercially available, whereas other configurations such as PET-MR [20] and DOT-MR [21] are undergoing tests in academic research. In addition, micro-SPECT/CT systems are commercially available for animal imaging. Fig. 4.8 shows a micro-SPECT/CT (ImTek, Inc., Knoxville, TN, USA) [22]. This contains two 10×20 cm detector heads for whole-animal imaging [23], can accommodate different collimators, and provides high-resolution images of anatomy (20 μm) and function (2 mm).

In addition to the integration of different modalities into one system, a common table can be used to exchange the patient or animal rapidly and reproducibly. This approach can additionally be supported by a position tracking tool, which has already been used for animal imaging without the use of anesthetics [24]. Based on the concept of exchangeable tables, new scanner combinations are expected to become available in the near future.

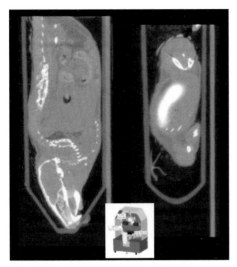

Fig. 4.8 Combined animal CT and SPECT system, which allows high-resolution SPECT imaging of mice (spatial resolution <2 mm). The image on the left shows the anatomy of a mouse, while the image on the right shows a SPECT overlay on the anatomy. (Data courtesy of ImTek, Inc., Knoxville, TN, USA).

4.3
Molecular Imaging

In the previous sections we have seen that substantial information can be gained from "traditional" imaging techniques. However, these provide mainly anatomical or functional information. Recently, an evolution in the development of contrast agents has resulted in an expansion of the number of probes that allow to image critical molecules and their interactions within the living body non-invasively. These techniques combine a regular contrast-conferring agent with a moiety that interacts specifically with a target molecule. Such moieties include receptor ligands, enzyme substrates and recognition elements such as an antibody or aptamer. They may be attached to the contrast agent via a linker molecule, for example to reduce steric hin-

drance for the interaction of the recognition moiety with its target. In the next sections, we will discuss different types of contrast agents, genetic technologies and suitable modalities for molecular imaging.

4.3.1
Contrast Agents for Molecular Imaging

After administration, traditional contrast agents will be distributed over and cleared from the body in patterns that depend on their physico-chemical properties. Characteristics such as molecular weight, hydrophobicity and charge may cause a contrast agent to accumulate preferentially in certain cells or tissues. This passive targeting is used, for example, in the imaging of tumors, as the increased permeability of tumor vasculature allows macromolecules to extravasate and build up to a larger degree in tumor tissue [25]. Active targeting of contrast agents will specifically increase their density at the target location. However, passive targeting is still needed to reach the site of interest. In general, both passive and active targeting contribute to the specificity of a contrast agent for molecular imaging.

Active targeting of contrast agents requires that molecular targets are known. With the advances in gene expression profiling and proteomic analyses of pathological human tissues, an increasing number of potential disease markers have been identified. Depending on their disease specificity and their accessibility – such as expression on the cell surface and in the vascular lumen – these markers may be used for molecular imaging.

Many potential disease markers and imaging targets are located inside the cell and cannot be reached by most types of contrast agents. Recently, several different peptides have been used for the intracellular delivery of imaging agents (for reviews, see Refs. [26, 27]). Generally, these peptides have no specificity for a certain cell type. Improving the specificity of delivery systems will reduce the amount of contrast agent that needs to be administered. Alternatively, more radionuclide-labeled small molecules with improved cellular permeation may be developed. When well-designed contrast agents interact with a target, the resulting image of probe localization and concentration should be directly related to the localization and concentration or activity of the target. In general, three main types of contrast agents for molecular imaging can be distinguished: indirect probes, which can be divided into accumulatable and activatable probes, and direct-binding probes [28].

Direct-binding or targeted probes bind their targets stoichiometrically and thus provide exact information on their localization and concentration. In principle, any protein can be targeted with labeled antibodies or aptamers. The imaging of cell surface-specific antigens with radiolabeled antibodies has developed over the past 30 years. The expression of receptors can also be monitored using labeled ligand analogs. As peptide receptors can be massively overexpressed in certain tumors, many studies have focused on the development of radiolabeled peptide derivatives for imaging (for reviews, see Refs. [29, 30]). A ^{111}In-labeled analog of somatostatin (Octreoscan®: ^{111}In-DTPA-[D-Phe1]-octreotide; Mallinckrodt, Inc., St. Louis, MO, USA) is one of the few FDA-approved peptides for imaging, and is used for the diagnosis of neuroendocrine cancer. More recently, annexin-V has been labeled with ^{99m}Tc for SPECT imaging of apoptosis [31, 32]. This protein binds to the phospholipid phosphatidylserine, which is present in higher concentrations in the outer leaflet of the cell membrane of apoptotic cells. In general, the imaging of receptors that are pathologically overexpressed, such as the HER2/neu receptor in breast cancer, with direct binding probes will allow the monitoring of global tumor burden as well as selection of patients for receptur tangeted therapy in a a "find, fight, follow" strategy (see Part I, Chapter 5).

For many disease processes, an increase in enzyme activation – not enzyme concentration – is an important marker. For example, in gastrointestinal stromal tumors, it is not the number but the kinase activity of c-Kit receptors that is increased [33]. This means that direct-binding probes cannot distinguish between healthy and diseased tissue. However, enzyme activity can be visualized using indirect probes, which do not bind their targets stoichiometrically but are changed upon interacting with them. These agents have a high potential for the imaging of very early therapy effects.

Accumulatable indirect probes become locally increased in concentration as a consequence of interaction with their target. The most well-known example is 18F-fluorodeoxyglucose (FDG), which becomes trapped within the cell after phosphorylation by the enzyme hexokinase [34]. Thus, a higher signal intensity visualized with PET indicates tissues with increased glucose utilization, and this is widely used to determine tumor malignancy, to detect metastases, and to follow therapy effects. In oncology in general, a major goal is the development of contrast agents that highlight the increased activity of critical kinases [35]. Such agents should remain inside the cell upon phosphorylation and be highly specific for the kinase under investigation. As cellular permeation is a prerequisite of accumulatable probes, only small labels can be incorporated, and consequentially PET and SPECT are used to visualize this type of contrast agent.

Activatable indirect probes are injected into the patient in a quenched state. The conversion of a probe molecule by its target enzyme increases its signal intensity, but has no effect on the concentration of the probes. Activatable probes are rather new and, until now, mainly fluorescent probes have been applied for the optical imaging of protease activity. For example, Bremer et al. used non-immunogenic copolymers to which fluorophores are attached via short peptides, which are substrates for matrix metalloproteinase 2 (MMP-2), a tumor marker. Due to their close proximity on the polymer, the fluorophores are quenched. Cleavage of the peptides releases the fluorophores, and their fluorescence signal increases. Using these probes, it was possible to visualize MMP-2 activity and its inhibition by the potential drug prinomastat in xenografted mice after 2 days of treatment [36]. Some principles of activatable probes for MRI have also been published [37, 38].

A fundamental difference between direct-binding and indirect probes is the intensity of the overall background signal. Direct-binding probes are visible throughout the entire body and require a waiting period until the probe is enriched at the target site and the non-bound probe has largely been cleared from the rest of the body. In contrast, many indirect probes can only be imaged after interaction with their target. In addition, one target enzyme can convert many probe molecules. This built-in amplification causes the background to be practically non-existent, even to the point of being a disadvantage, as it impedes the exact localization of the target-containing tissue.

To date, a variety of probes for a considerable number of targets have been developed [39]. However, the numbers of possible applications, targets and probes are daunting, and a specific imaging probe needs to be developed for each molecular target. Like drugs, contrast agents must be safe and specific, and they must also possess the right balance between clearance, biodegradability and stability to allow an optimal time window for imaging. In addition, they should preferably provide signal amplification to enable visualization of target molecules in physiological concentrations.

Different probes offer different levels of molecular information and have different application ranges. For example, a contrast agent targeted to a protein that is highly specific for a certain disease will provide very detailed information, but only for this one disease. On the other side of the spectrum is an agent such as FDG, which cannot reveal the biochemical reasons for a high glucose metabolism, but can be used in the detection of many different types of tumor. The sensitivity and specificity of each probe-target couple must be validated for its intended application. This can be a very time-consuming and costly process, and it may pose problems similar to those encountered in developing new drugs, while the criteria for imaging probes are often more stringent [40]. Therefore, new developments that will make the most headway into clinical practice will probably be those with a broad application range. This means that they should enable visualization of processes common to many diseases, such as apoptosis, angiogenesis, and inflammation. Another approach is to focus on platform technologies that can easily be adapted to various applications. Nevertheless, a general approval for such technologies remains a major obstacle.

4.3.2
Non-invasive Reporter Gene Assays

Important progress in pre-clinical studies is facilitated by so-called reporter gene assays (for excellent reviews, see Refs. [1, 40, 41]). For these assays, a measurable reporter gene is linked to a gene under investigation or only brought under control of its promotor. Consequentially, when the gene of interest is turned on, the reporter gene is transcribed and translated into a protein, usually an enzyme. The presence of this reporter gene product can then be assessed using a molecular imaging contrast agent, usually an indirect probe. Because a reporter gene can be linked to virtually any gene, a reporter gene assay is a general method to study non-invasively the expression of genes, eliminating the need to develop a specific contrast agent for each target.

For reasons of specificity, reporter genes can be chosen to be exogenous, stemming from a completely different type of organism than the animal under investigation, and the substrates for the enzymes they encode are selected to be not, or to a much lesser extent, convertible by endogenous proteins. Typically used exogenous enzymes are luciferases from the firefly or the sea pansy, which can be detected using bioluminescent imaging (see Section 4.2.5.1). Advantages of the firefly luciferase system are the very broad dynamic range and linearity of the reaction and the possibility of real-time measurements because of the high turnover rate of the enzyme. It is thus well suited for the monitoring of changes in gene expression on a relatively short timescale. Another much-used reporter gene system employs the thymidine kinase from type 1 herpes simplex virus (HSV1-TK). HSV1-TK activity can be assessed in a manner similar to hexokinase, namely by nuclear imaging of radiolabeled substrates that become intracellularly entrapped upon phosphorylation. The substrate label can be a positron emitter, for PET imaging, or a gamma-ray emitter, for SPECT. A few strategies were devised for reporter gene assays that can be visualized with MRI, such as EgadMe, a substrate for the enzyme beta-galactosidase. This method was demonstrated in vivo in Xenopus laevis embryos after injection of the substrate but cannot yet be used in mice until a version of EgadMe that can enter the cell has been developed [37]. Several reporter gene assays have been designed employing modified endogenous enzymes that only have a very narrow expression pattern under natural circumstances. Usually, these are receptors for which a radiolabeled ligand has already been developed. In case reporter gene assays will be applied in humans in the future, the lower or absent immunogenicity of endogenous enzymes would also be of advantage.

Reporter gene assays can be used for many different types of studies, such as the regulation of expression of genes of interest in xenografted and transgenic animals, as well as the tracking of migrating cells and even the assessment of gene therapy and the in vivo measurement of protein–protein interactions. Two examples of such applications will be provided at the end of this chapter. As the technology requires the stable expression of exogenous or modified endogenous genes in target tissues, it will be limited to animal studies in the near future.

4.3.3
Suitable Modalities for Molecular Imaging

Molecular imaging focuses on the visualization of molecules and molecular processes. Thus, especially in the case of direct-binding probes, the imaging techniques that are used should be sensitive enough to detect the molecule of interest in its physiological concentration. Since X-ray CT offers only millimolar sensitivity (see Table 4.1), it is not possible to detect sparse targets with this modality. However, it provides 3D anatomical information with a superior spatial resolution and is used in combination with PET or SPECT for a better localization of the radionuclide signal.

MRI also offers a good spatial resolution and superior soft tissue contrast, but it can only detect contrast agents in micromolar concentrations. Although MR imaging of receptors is possible considering the physics of MRI, there are biological limitations, such as delivery of the agent to the site in high enough quantities, which make this combination questionable [42]. In order to sufficiently amplify the signal, it requires a bulky reporter moiety, such as nanoparticles [43], dendrimers [44], buckeyballs [45], or polymers carrying a large number of lanthanide molecules or an iron oxide nanoparticle. Therefore, MRI contrast agents seem to be more practical for the visualization of intravascularly expressed targets.

Ultrasound imaging requires contrast agents that are even larger than those needed for MRI, but then a single micro-bubble can be visualized, which in principle can yield a very good sensitivity. Furthermore, its spatial resolution lies below 1 mm. In addition, ultrasound is a rather cheap and accessible imaging modality, and several targeted contrast agents have been developed.

Due to their high sensitivity, nuclear imaging techniques are well-suited to visualize targets present at low concentrations. PET and SPECT, with their picomolar sensitivity, allow for the imaging of most known targets using ligands that carry only one label each. As the radionuclide label is relatively small, the probes may even permeate into cells. However, the spatial resolution of nuclear techniques lies in the order of a few to tens of milli-

Table 4.1 Properties of imaging modalities

Modality	Sensitivity (concentration of contrast agent)	Spatial resolution	Acquisition time
X-ray-CT	Approx. 10^{-3} M	<500 µm	Sub-seconds
Animal CT		50 µm	
MRI	Approx. 10^{-6} M	250 µm–1 mm	Sub-minutes
Animal MRI		<100 µm	
Ultrasound	Single micro-bubble	100 µm–1 mm (depends on penetration depth)	Sub-seconds
Animal US		40 µm	
SPECT	Approx. 10^{-10} M	5–20 mm	Minutes
Animal SPECT		1–3 mm	
PET	Approx. 10^{-12} M	4–10 mm	Minutes
Animal PET		2–4 mm	
Optical (Fluorescence)	Approx. 10^{-10} M	100 µm–10 mm (depends on penetration depth)	Seconds

meters. In addition, the very low background in nuclear imaging often requires the use of an additional modality such as X-ray CT to provide structural information. Furthermore, nuclear imaging techniques (especially PET) require expensive equipment for probe synthesis, such as a nearby cyclotron to generate the short-lived isotopes, and rapid synthesis schemes to incorporate these isotopes into the final imaging probe.

Optical molecular imaging has a sensitivity similar to that of nuclear imaging, and its spatial resolution can be very high if surface structures are imaged. Optical molecular imaging is developing rapidly in the field of small-animal imaging for contrast agent and pharmaceutical R&D. The technology needed for optical imaging is relatively cheap and simple, and 3D optical imaging has been demonstrated [46]. The considerable attenuation of light by tissue does not pose a major problem in mice. However, translation of optical imaging to the clinic is still limited to fluorescence imaging applications that investigate accessible body surfaces. This may change in the future, as a penetration depth of 10 cm has been claimed for NIR fluorophores [13]. Fluorescence imaging also offers the advantage that no radionuclides are needed, and thus the synthesis of the probes is easier and cheaper. Bioluminescence imaging will probably remain exclusively a small-animal imaging modality for many years, because it requires the introduction of exogenous or modified endogenous genes into an organism.

4.4
Molecular Imaging for Drug Discovery and Development

Recent advances in genomics and molecular biology have raised new hopes for the prevention, treatment, and even cure of serious illnesses. However, many of the innovations of basic science are not transferred quickly enough into more effective and safe drugs. This is because the current medical product development path is becoming increasingly challenging and costly. Consequently, during the past few years, the number of new drugs submitted to the FDA has declined considerably, while the costs of product development have increased significantly. For example, in the year 2000, half of all potential drugs were discarded during clinical development because they lacked in safety and efficacy. The FDA identified two main reasons for this trend:

1. The profits from a decreasing number of successful products need to subsidize a growing number of expensive failures.
2. The path to market even for successful candidates is long, costly, and inefficient, due in large part to the current reliance on cumbersome assessment methods [47] (see Part VII, Chapter 4).

Fig. 4.9 depicts schematically the drug development process, which typically takes about 7 to 10 years to complete. Currently, an FDA-initiative [47] is trying to improve the predictability and efficiency along the critical path from laboratory concept to commercial product. In this context, molecular imaging promises many benefits, since it allows the measurement of drug absorption, distribution, and target binding. Even more importantly, molecular imaging has the potential to provide biomarkers. A biomarker is "... a characteris-

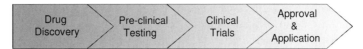

Fig. 4.9 Phases in drug development.

tic that is objectively measured and evaluated as an indicator of normal biological processes, pathogenic processes, or pharmacological responses to a therapeutic intervention" [48]. Several classes of biomarkers exist that can be used to determine drug safety and efficacy in a pre-clinical and clinical phase and thus can potentially reduce the costs of drug development [49]. A biomarker is called a "surrogate endpoint" when it is used to predict therapy effect without looking at patient well-being, functioning or survival. Well-known surrogate endpoints are the reduction of blood pressure or blood cholesterol level for the approval of drugs for cardiovascular diseases and stroke. The non-invasive nature of imaging technologies makes them very attractive for the use in surrogate endpoints. In oncology, the imaging of tumor size is a well-accepted marker of therapy response. With the advent of new medicines that do not necessarily reduce tumor size, molecular imaging has a high potential to provide rapid measures of therapy response. For example, recent progress in cancer therapies has resulted in new classes of drugs that can specifically target and inhibit a molecule, exerting more targeted cytostatic effects rather than overall cytotoxic effects. Examples are tyrosine kinase inhibitors such as Gleevec [50], angiogenic modulators [51] and inhibitors of proteases such as MMPs [52]. The usual measurements of drug plasma concentrations to determine dosage are inappropriate for such a drug, as these data do not necessarily reflect its concentration at the site of action, such as a tumor. Even more

important is that the standard clinical trial endpoints of morbidity and mortality – and newer ones such as tumor size – are insufficient to evaluate cytostatic therapies. In these cases, molecular imaging can help to answer the fundamental questions of drug discovery and development – namely, if and how those drugs work *in vivo*. Furthermore, if molecular imaging has been proven to show the effectiveness of a specific drug in drug development, the same technology could be applied in the clinical setting to monitor early drug response and to adapt the therapy to the individual patient ("personalized medicine") (see Part I, Chapter 2).

Obtaining molecular information *in vivo* can aid many steps in the drug discovery and development chain. In the following section, the role of molecular imaging will be discussed for the different phases of the drug development process.

4.4.1
Drug Discovery

The major task of the drug discovery phase is the identification and screening of new drug targets and lead compounds. For target identification, microarrays can be used that allow analysis of gene and protein expression patterns specific for diseased states. Recently, it was shown that molecular imaging can provide a cell-based, high-throughput method to screen thousands of samples against known target molecules and cells [53]. When a target is expressed, its functionality depends heavily on a number of factors such as expression level, turnover rate, post-transcriptional modifications and feedback regulations, which are affected by complex intra- and intercellular processes. Therefore, target expression and the effect of drugs should ideally be studied *in vivo*, and consequently molecular imaging has a high potential in drug discovery. It can provide methods to measure drug–target interactions *in vivo* and to determine whether a drug affects the expression and/or function of a specific target. In particular, molecular imaging techniques can help to identify a lead compound, by comparing the efficacy of different preselected molecules in small animals. For example, it can be assessed whether drug candidates find the target, and whether and how they interact with it. One possibility to determine the delivery and affinity of an unlabeled drug candidate to its target is to use nuclear imaging to determine how well it inhibits the specific binding of a well-characterized radiolabeled ligand. This precludes the need to design and synthesize a functional, radiolabeled analog of each potential drug. Alternatively, optical imaging techniques can be used to study target interaction *in vivo*. In addition, both SPECT and optical imaging allow multiplexing through the measurement of several different probes and thus different drug candidates in the same animal, either simultaneously or in fast sequence [21].

4.4.2
Pre-clinical Testing

Pre-clinical testing is used to study lead compounds with respect to their biodistribution (pharmacokinetics), dose, toxicity, and efficacy in small animals. Usually, time-consuming dissection of animals and histological analysis of the tissue is performed. In general, pre-clinical imaging allows the collection of data for different time points and doses for a single animal. Thus, the number of animals per study can be reduced, which is more ethical than sacrificing groups of animals for each data point. Furthermore, this procedure is more cost-effective and saves much time, as the dissection and histology procedures are slow and subject to sampling errors. Moreover, because each animal serves as its own control, the statistical relevance of a study increases as inter-animal variations become less important. Many technological developments have been made during the past 15 years for small animal imaging. For example, dedicated animal imaging equipment with a (much) higher resolution than for clinical scanners has been developed, and microsystems are now being marketed for almost all modalities. However, the findings based on structural and functional imaging are often too unspecific to replace histological techniques. Therefore, molecular imaging techniques have a great potential for the study of drug–target interactions as well as their functional consequences in living animals. In particular, nuclear imaging can be used to image the biodistribution of drugs. Notably, quantitative PET imaging is appropriate for this, since drugs can be labeled with ^{11}C or ^{18}F without changing the chemical properties of the compound to any degree [54, 55]. Another advantage of nuclear imaging techniques is that the uptake rate of the labeled drugs can be quantified more or less directly from the imaging data. As the time point of radionuclide measurement markedly affects the relative radiation intensity, maps of tracer kinetics are used. These are based on pharmacokinetic models, which describe the transport mechanisms of the tracer [56, 57].

In addition, molecular imaging can provide strategies to visualize the downstream consequences of administration of a lead compound on diseased and normal tissues, and to determine whether it has the desired disease-modifying effect [58], for example by visualizing cell proliferation, hypoxia, apoptosis, or (anti-)angiogenesis. Being able to use such markers of therapy success may drastically shorten the duration of pre-clinical studies. Most modalities can be used for these studies, depending on target localization and other properties, and examples will be provided in Section 4.4.5.

4.4.3
Clinical Trials

Molecular imaging is expected to have a dramatic impact on clinical trials, using strategies similar to those in pre-clinical studies. As in animal systems, pharmacokinetic data in patients can be obtained using labeled drug analogs. In fact, because the high sensitivity of nuclear imaging enables the detection of very low doses of radiolabeled molecules, it allows information to be gained on the biodistribution of a drug far below toxic levels, and even below therapeutic levels. This microdosing concept means that these data can be gathered at a very early stage in the drug development process [59]. Another possible application area of molecular imaging is the use of imaging biomarkers to stage patients and select those who will benefit from a drug (see Part I, Chapter 3). This will help in defining more stratified and relevant study groups.

Furthermore, molecular imaging methods that have been validated as a disease biomarker in the pre-clinical phase may also be applied to test drug efficacy in clinical trials. Although clinical trials – which represent the most expensive phase in drug development – would benefit heavily from molecular imaging techniques, there is currently a wide gap between their use in the pre-clinical and clinical phases. There are several reasons for this. First, only a few specific contrast agents are available that have been approved for patient studies. The development and approval of targeted contrast agents is the central challenge for molecular imaging, and this is a costly and time-consuming process (similar to that for drugs). Hence, the number of available targeted agents and probes is unlikely to increase in the near future. However, new PET agents may be an exception to this trend, if their approval process is changed due to the microdosing concept [59]. Another factor that limits the transfer from pre-clinical to clinical trials is the variation between pre-clinical and clinical imaging systems. Protocols that have been optimized on pre-clinical instruments cannot easily be transferred to clinical scanners. One way of circumventing these problems is to support pre-clinical imaging on clinical scanners, and this is possible by using dedicated animal handling, detectors, and software. Nevertheless, a strong interdisciplinary scientific effort is needed to move basic discoveries tested on animals into the clinic more efficiently ("from bench to bed side"). A number of initiatives are currently addressing the importance of this translational research for drug development, including the National Institutes of Health (NIH) roadmap [60] and the European organization for the treatment of cancer (EORTC) [61].

4.4.4
Clinical Applications

The main bottleneck of molecular imaging for clinical applications is the availability of approved targeted contrast agents. Currently, the greatest number of targeted agents is available for SPECT, and additional targeted agents are undergoing the approval process. In particular, numerous antibodies against different molecular targets and labeled with different gamma-emitting radionuclides have been developed over the past two decades. For example, arcitumomab-99m-technetium (CEA-Scan; Immunomedics, NJ, USA) is an antibody Fab fragment which is labeled with ^{99m}Tc and directed against carcinoembryonic antigen (CEA). The targeted contrast agent has been approved for the detection of colorectal cancer, and has also been found to provide imaging of both palpable and non-palpable breast lesions that appeared suspicious on screening mammograms. Currently, CEA-Scan is undergoing Phase III clinical trials for breast cancer. In addition to diagnostics, the radiolabeling of antibodies also has huge potential for cancer therapy. Zevalin® (Biogen Idec, Cambridge, MA, USA, and Schering AG, Berlin, Germany) is used for cancer radioimmunotherapy of non-Hodgkin's lymphoma, and consists of the monoclonal antibody ibritumomab and the attached chelator/linker tiuxetan. The chelator can bind indium-111 as well as the high-energy radioemitter yttrium-90. The biodistribution of the pharmaceutical can be visualized with SPECT using indium-111, before injecting the yttrium-90-labeled substance for therapy, and finally again with indium-111 to monitor the progress of treatment, using SPECT. This approach can be considered either as a "see and treat" approach or as a "find, fight, follow strategy". Gamma emitters are more suited for this approach than positron emitters due to their longer half-life.

Clinical applications can strongly benefit from molecular imaging in early diagnosis, disease staging, therapy monitoring, and follow-up studies. Molecular imaging

is also expected to be of major importance in the assessment of drug response. Consequently, the co-development of drugs and imaging-based biomarkers is likely to have a major impact on treatment schemes in the future. In particular, the measurement of drug response would allow the customization of treatment regimes and dose optimization for certain patient groups ("personalized medicine") (see Part I, Chapter 2). In addition, diagnosis and treatment could further be integrated into a "see and treat" approach, when imaging proves that a targeted probe does indeed interact with specific disease molecules, and the same targeted probe is subsequently loaded with a drug. The combination of such an approach with therapy monitoring then results in a "find, fight, follow strategy".

4.4.5
Molecular Imaging Examples

As described above, molecular imaging can be used in the different phases of drug development as well as in clinical applications. One important application of molecular imaging is the detection of receptors and their interactions with labeled ligands *in vivo*. In clinical oncology, the somatostatin/somatostatin receptor system is used most frequently for contrast-enhanced detection and radiotherapy of cancer [62]. Somatostatin receptors are overexpressed in many tumors, and are present in especially high concentrations in gastroenteropancreatic neuroendocrine tumors [29, 30]. Radiolabeled synthetic analogs of somatostatin have been applied successfully for routine molecular imaging of these tumors and their metastases for more than 15 years, and analog conjugates for radiation therapy have been under clinical trial for several years [63]. Recently, optical imaging of the somatostatin receptor has been demonstrated in a mouse model using a NIR fluorescent probe [64]. Fig. 4.10 illustrates results of this study. The indotricarbocyanine-conjugate of octreotate (a somatostatin analog) was specifically accumulated at the site of the tumor, and not taken up by surrounding fibroblasts, proving again the high specificity of this ligand–receptor interaction. Due to the active targeting of the agent, the tumor could be visualized with high contrast at only 1 hour after administration. Fluorophore-conjugated peptides show great pro-

mise for early tumor detection using fluorescence endoscopy, intraoperative imaging, and optical mammography.

Molecular imaging has also been used to measure general processes such as angiogenesis, inflammation, hypoxia. or arteriosclerosis. As many diseases result in abnormalities of these processes (also named "common denominators"), drug-based therapies aim at their modification. In other cases – such as the apoptosis of tumor cells – a change in this process is a desired therapeutic effect. For these reasons, quantitative measurement of the status of these processes may represent potential biomarkers of drug efficacy.

Angiogenesis occurs both in normal processes (e.g., wound repair) and in disease states, and is a common denominator in cancer and cardiovascular diseases. Usually, indirect imaging methods are used to characterize changes in angiogenesis, such as measurements of vessel density, perfusion, or vascular permeability. However, angiogenesis can also be detected at an early stage with targeted contrast agents, using a variety of modalities [65]. Angiogenesis is a complex process involving a large number of molecules, such as growth factors [e.g. vascular endothelial growth factor (VEGF), transforming growth factor, and fibroblast growth factor], MMPs, fibronectin, integrins, and endostatin. These molecules are potential targets for imaging and therapy. For example, a transgenic mouse model was recently developed to monitor the effect of potential (anti-) angiogenic therapeutics *in vivo* using bioluminescence imaging [66]. In these mice, firefly luciferase is brought under control of the promoter for murine VEGF receptor-2 (VEGFR2), a gene that is transcriptionally regulated during angiogenesis. The model was validated using skin wounding, which induced VEGFR2

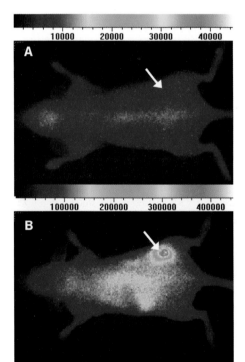

Fig. 4.10 Optical detection of a tumor in a mouse model using a fluorescently labeled somatostatin analog. The mouse is imaged before administration of the contrast agent (A) and 6 h after intravenous injection (B). Adapted from Ref. [64] with copyright permission from the Nature Publishing Group.

and thus luciferase expression. Treatment of the wounds with a glucocorticoid reduced VEGFR2 expression, which could be followed kinetically using bioluminescence imaging.

In contrast to many growth factors, extra-domain B-fibronectin (ED-B-FN) is a highly disease-specific target, meaning that it is present during angiogenesis in vessels of neoplastic tissues, but not in mature vessels [67]. Recently, NIR fluorescent fibronectin analogs have been applied for optical imaging of tumor angiogenesis in mice [68]. In addition, the $\alpha_v\beta_3$ integrin plays an important role during tumor-induced angiogenesis, and a ^{125}I-labeled antagonist analog was used in SPECT imaging of angiogenesis [69]. Since the $\alpha_v\beta_3$ epitope is highly expressed on activated neovascular endothelial cells, targeted ultrasound [70] and MR agents [71] can also be used. However, in order to detect low concentrations of such a target, the sensitivity of MRI must be improved. One possibility is to use a paramagnetic liquid perfluorocarbon nanoparticle with $\alpha_v\beta_3$ antag-

onist molecules and ca. 10^5 Gd-containing chelates on its surface. Due to the high number of Gd-chelates on these particles, the *in vivo* detection of $\alpha_v\beta_3$ receptors at nanomolar to picomolar concentrations in tumor neovasculature is possible [72]. Results of this study, in which early tumor-induced angiogenesis was measured on a clinical 1.5 Tesla MRI scanner are shown in Fig. 4.11. Hereto, targeted nanoparticles were injected systemically into rabbits carrying a Vx-2 tumor to fibronectin xenograft. At 2 hours after injection, an increased MR-signal could be seen on T_1-weighted images at the periphery of the tumor. In addition, enhanced MR contrast was observed within the walls of some vessels close to the tumor.

Angiogenesis is also a critical feature in the development of atherosclerotic plaques which, after rupture, can lead to myocardial infarction and stroke. Thus, the same $\alpha_v\beta_3$ integrin-targeted paramagnetic nanoparticles could be applied to detect atherosclerosis in rabbits [73]. Fibrin is another important target that is associated with

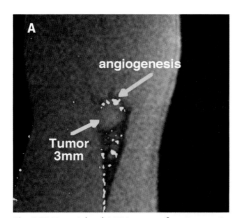
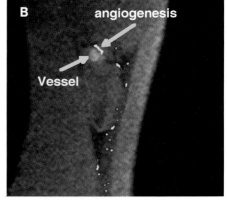

Fig. 4.11 T_1-weighted MR images of angiogenesis in a 3 mm Vx-2 tumor (rabbit model) obtained on a 1.5 Tesla clinical MR scanner. Perfluorocarbon nanoparticles studded with Gd-chelates and an $\alpha_v\beta_3$ epitope antagonist are used to confer tar-

geted contrast. Overlays are shown of regions of enhancement at 2 h post contrast (yellow pixels) of (A) the tumor and (B) a vein adjacent to the tumor. (Courtesy of P. Winter, G. Lanza, S. Wickline, Washington University, St. Louis, MO, USA) vulnerable plaque and thrombi (see Part II, Chapter 1). Therefore, anti-fibrin Fab fragments were conjugated to the nanoparticles, which were then capable of detecting vulnerable plaques *in vivo* [74]. Thus, the paramagnetic nanoparticles may be considered to be a contrast agent platform. Other studies with another fibrin-binding MR-agent (EP-2104R; EPIX Pharmaceuticals, Cambridge, MA, USA) and advances in coronary MRI techniques demonstrated the potential for direct imaging of coronary thrombosis [75]. In addition to specifically targeting molecules such as $a_v\beta_3$ integrin or fibrin, the detection of atherosclerotic plaques is also possible with gadofluorine-enhanced MRI [76] or with passive targeting using ultra-small iron oxide agents [77] that are non specifically phagocytosed by macrophages present in the plaques. These irone oxide particles also allow macrophage tracking associated with multiple sclerosis [78]. A more specific diagnosis and monitoring of treatment response in multiple sclerosis is possible when targeting the endothelial cell inflammatory marker intercellular adhesion molecule-1 (ICAM-1). Results of a quantitative ultrasound study in experimental autoimmune encephalomyelitis (EAE) rats using targeted ICAM-1 air-filled microparticles (MPs) [79] are shown in Fig. 4.12. In this study, targeted MPs were systemically injected into anesthetized EAE-rats (two groups, each n=4) and healthy control rats (n=4). A few minutes after compound administration (dose=5×10^8 MPs kg^{-1} body weight) the brains of one EAG group were examined *ex vivo* and of the other two groups *in vivo* with quantitative ultrasound

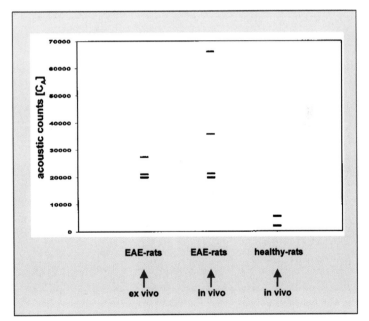

Fig. 4.12 Results of a quantitative ultrasound study in experimental autoimmune encephalomyelitis (EAE) rats. Application of ICAM-1-targeted microparticles results in a highly significant (P=0.001) increase of acoustic counts in EAE-rats in comparison to healthy rats. (Courtesy of P. Hauff, Schering AG, Berlin, Germany)

imaging. This imaging technique is based on the stimulated acoustic emission effect, which occurs after the destruction of air-filled MPs during Doppler ultrasound imaging, thereby allowing the quantification of MP-based acoustic counts [80]. The figure shows that a high number of acoustic counts was obtained at similar levels in both groups of EAE-rats due to the presence of targeted MPs, demonstrating a high correlation between *ex vivo* and *in vivo* examination. In contrast, in the healthy control rats and treated EAE rats only a few signals were obtained, mainly reflecting constitutive ICAM-1 expression at the blood–brain barrier.

Programmed cell death (apoptosis) plays a crucial role in the pathogenesis of autoimmune and neurodegenerative disease, cerebral and myocardial ischemia, and in the therapeutic response of cancer. One of the earliest events in apoptosis is the externalization of phosphatidylserine, a membrane phospholipid. Annexin-V has a high affinity for phosphatidylserine, and can thus be used for molecular imaging of apoptosis *in vivo* [81]. Nuclear imaging of apoptosis has also been demonstrated in patients with acute myocardial infarction [82]. In all patients of this study, reperfusion of myocardium was ensured by percutaneous transluminal coronary angioplasty. SPECT imaging of annexin-V-labeled ^{99m}Tc was performed several hours after the acute infarct. Fig. 4.13 (A) shows an increased uptake in specific regions of the heart, which was validated by a perfusion study performed several weeks later (Fig. 4.13 (B)).

Two fundamental processes which are also flawed in oncogenesis and may provide therapeutic targets are signal transduction and the regulation of gene transcription. At certain steps in these processes, protein dimerization is required, which can be triggered by small molecules. One naturally occurring example of such a molecule is rapamycin, of which

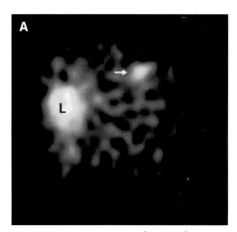 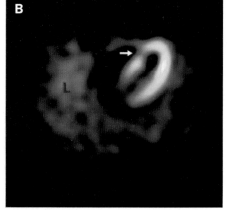

Fig. 4.13 Apoptosis imaging of acute infarction in a patient using transversal body slices. (A) The arrow shows increased Tc-99m-labeled annexin-V uptake 22 h after reperfusion. (B) Perfusion scintigraphy with sestamibi at 6–8 weeks after discharge shows an irreversible perfusion defect (arrow), which coincides with the previously measured area of increased Tc-99m-labeled annexin-V uptake. In both images, a high signal is obtained from the liver (L) due to clearance of annexin V. (Courtesy of L. Hofstra, Maastricht University, Maastricht, The Netherlands.)

several analogs have completed Phase I studies and have shown promising potential as anti-cancer drugs [35]. The efficacy of these drugs can now be imaged *in vivo* in mice using bioluminescence [83]. In this procedure, a luciferase gene is split and each half is linked to one of the proteins, the dimerization of which is under study. When a compound induces such dimerization, the luciferase halves can also bind each other and form a functional protein, and this results in light emission. The same trick can also be performed using split reporter genes, the expression of which can be imaged using either PET or fluorescence [84].

4.5
Concluding Remarks

Molecular imaging techniques promise many advantages for the field of drug discovery and development. They allow the non-invasive procurement of whole-body kinetic data on biodistribution and toxicity in a single animal, thus providing faster, cheaper, more ethical – and probably more relevant – information. Perhaps the greatest impact from molecular imaging will be seen in the measuring of surrogate endpoints: monitoring the target or a short-term therapy effect rather than morbidity and mortality will significantly reduce the time required to determine therapeutic success. The use of an imaging biomarker in patients that has already been validated and used in pre-clinical research allows the direct linking of results from different phases, and also facilitates pre-clinical to clinical transfer. Furthermore, real-time therapy monitoring and dose-optimization by testing different doses in one patient will help to reduce the time needed to establish therapy efficacy, and also accelerate therapy

development. Finally, by enabling the selection of those patients who will benefit from therapy, the size of a clinical trial can be markedly reduced, while increasing the chances of the success of a compound. Overall, molecular imaging promises a dramatic reduction in the time to market – and thus cost for new therapies.

In order to profit fully from these new advances in molecularly targeted drugs, and to improve attrition rates, the drug discovery and validation field will need to adopt techniques with which molecular information can be gained *in vivo*. Molecular imaging offers several such techniques. We have seen that ingenious compounds and strategies have been devised to obtain *in vivo* molecular information. In addition, imaging equipment with ever-improving sensitivity and resolution is available, and each of the modalities has shown its potential value in molecular imaging applications for therapeutic R&D. Although molecular imaging presents huge prospective benefits for drug discovery and development, it is still in its infancy. Increased development, validation and the approval of targeted contrast agents, biomarkers and clinical imaging endpoints are still required. As the field is highly interdisciplinary, close collaboration between pharmaceutical companies, developers of contrast agents and manufacturers of imaging equipment will accelerate progress and help to ensure that molecular imaging fulfils its promise in the development of modern biopharmaceuticals.

Acknowledgement

We are very grateful to Leo, the newborn son of Joke Orsel, for his cooperation, allowing us to finish this chapter in time.

References

1 R. Weissleder, U. Mahmood, *Radiology* 2001, 219, 316–333.

2 E. L. Ritman, *Annu. Rev. Biomed. Eng.* 2004, 6, 185–208.

3 E. M. Haacke, R. W. Brown, M. R. Thomson, et al., *Magnetic Resonance Imaging, Physical Principles and Sequence Design*, Wiley Liss, USA, 1999.

4 M. T. Vlaardingerbroek, J. A. den Boer, *Magnetic Resonance Imaging*, Theory and Practice, 3rd edn. Springer-Verlag, Germany, 2002.

5 A. E. Merbach, E. Toth, *The Chemistry of Contrast Agents in Medical Magnetic Resonance Imaging*, John Wiley and Sons, USA, 2001.

6 R. B. Lauffer, D. J. Parmelee, S. U. Dunham, *Radiology* 1998, 207, 529–538.

7 K. M. Ward, A. H. Aletras, R. S. Balaban, *J. Magn. Reson.* 2000, 143, 79–87.

8 J. R. Lindner, *Nat. Rev. Drug Discov.* 2004, 3, 527–532.

9 T. Fritzsch, M. Schartl, J. Siegert, *Invest. Radiol.* 1988, 23 (Suppl 1), S302–S305.

10 G. M. Lanza, K. D. Wallace, M. J. Scott, et al., *Circulation* 1996, 94, 3334–3340.

11 H. O. Anger, Radioisotope cameras. In: G. J. Hine (Ed.), *Instrumentation in nuclear medicine*, Vol. 1, pp. 485–552, Academic Press, USA, 1967.

12 F. J. Beekman, D. P. McElroy, F. Berger, et al., *Eur. J. Nucl. Med. Mol. Imaging* 2002, 29, 933–938.

13 W. F. Cheong, S. A. Prahl, A. J. Welch, *IEEE J. Quantum Electronics* 1990, 26, 2166–2185.

14 M. Cutler, *Surg. Gynecol. Obstet.* 1929, 48, 721–729.

15 S. R. Arridge, *Inverse Problems* 1999, 15, R41–R93.

16 V. Ntziachristos, C. Bremer, R. Weissleder, *Eur. Radiol.* 2003, 13, 195–208.

17 K. Licha, B. Riefke, V. Ntziachristos, et al., *Photochem Photobiol* 2000, 72, 392–398.

18 J. W. Hastings, *Gene* 1996, 173, 5–11.

19 P. R. Contag, I. N. Olomu, D. K. Stevenson et al., *Nat. Med.* 1998, 4, 245–247.

20 P. K. Marsden, D. Strul, S. F. Keevil, et al., *Br. J. Radiol.* 2002, 75 Spec No, S53–S59.

21 V. Ntziachristos, A. G. Yodh, M. Schnall, et al., *Proc. Natl. Acad. Sci. USA* 2000, 97, 2767–2772.

22 J. Gregor, S. Gleason, S. Kennel, et al., Small-Animal SPECT Workshop, Tucson, AZ, 2004.

23 A. G. Weisenberger, B. Kross, S. Majewski, et al., *Proc. IEEE Nuclear Science Symposium and Medical Imaging Conference*, Rome, Italy, October 2004.

24 S. S. Gleason, J. S. Goddard, M. J. Paulus, et al., *Proc. IEEE Nuclear Science Symposium and Medical Imaging Conference*, Rome, Italy, October 2004.

25 Y. Matsumura, H. Maeda, *Cancer Res.* 1986, 46, 6387–6392.

26 B. L. Franc, S. J. Mandl, Z. Siprashvili et al., *Mol. Imaging* 2003, 2, 313–323.

27 M. Zhao, R. Weissleder, *Med. Res. Rev.* 2004, 24, 1–12.

28 H. R. Herschman, *Science* 2003, 302, 605–608.

29 R. R. P. Warner, T. M. O'Dorisio, *Semin. Nucl. Med.* 2002, 32, 79–83.

30 J. C. Reubi, *Endocr. Rev.* 2003, 24, 389–427.

31 F. G. Blankenberg, P. D. Katsikis, J. F. Tait, et al. *Proc. Natl. Acad. Sci. USA* 1998, 95, 6349–6354.

32 L. Hofstra, I. H. Liem, E. A. Dumont, et al., *Lancet* 2000, 356, 209–212.

33 R. Capdeville, E. Buchdunger, J. Zimmermann, J. Matter, *Nat. Rev. Drug Discov.* 2002, 1, 493–502.

34 M. Reivich, D. Kuhl, A. Wolf, et al., *Acta Neurol. Scand. Suppl.* 1977, 64, 190–191.

35 Y. Lu, H. Wang, G. B. Mills, *Rev. Clin. Exp. Hematol.* 2003, 7, 205–228.

36 C. Bremer, C. H. Tung, R. Weissleder, *Nat. Med.* 2001, 7, 743–748.

37 A. Y. Louie, M. M. Hüber, E. T. Ahrens, et al., *Nat. Biotechnol.* 2000, 18, 321–325.

38 J. M. Perez, L. Josephson, T. O'Loughlin, et al., *Nat. Biotechnol.* 2002, 20, 816–820.

39 M. Rudin, R. Weissleder, *Nat. Rev. Drug Discov.* 2003, 2, 123–131.

40 R. G. Blasberg, J. G. Tjuvajev, *J. Clin. Invest.* 2003, 111, 1620–1629.

41 T. F. Massoud, S. S. Gambhir, *Genes Dev.* 2003, 17, 545–589.

42 A. D. Nunn, K. E. Linder, M. F. Tweedle, *Q. J. Nucl. Med.* 1997, 41, 155–162.

43 G. M. Lanza, C. H. Lorenz, S. E. Fischer, et al., *Acad. Radiol.* 1998, 5 (Suppl 1), S173–S176.

44 C. Wiener, M. W. Brechbiel, H. Brothers, et al., *Magn. Reson. Med.* 1994, 31, 1–8.

45 W. H. Noon, Y. Kong, J. Ma, *Proc. Natl. Acad. Sci. USA* 2002, 99, 6466–6470.

46 V. Ntziachristos, C. H. Tung, C. Bremer, et al., *Nat. Med.* 2002, 8, 757–760.

47 Food and Drug Administration. Innovation or Stagnation? Challenge and Opportunity on the critical path to new medical products, White Paper **2004**.

48 R. Frank, R. Hargreaves, *Nat. Rev. Drug Discov.* **2003**, 2, 566–580.

49 J. F. Niblack. In: G. J. Downing (Ed.), *Biomarkers and Surrogate Endpoints: Clinical Research and Applications.* Elsevier Science, **2000**.

50 L. K. Shawver, D. Slamon, A. Ullrich, *Cancer Cell* **2002**, 1, 117–123.

51 M. Cristofanillli, C. Charnsangajev, G. N. Hortobagyi, *Nature Rev. Drug Discov.* **2002**, 1, 415–426.

52 L. M. Coussens, B. Fingleton, L. M. Matrisian, *Science* **2002**, 295, 2387–2392.

53 D. Hogemann, V. Ntziachristos, L. Josephson, et al., *Bioconjug. Chem.* **2002**, 13, 116–121.

54 A. J. Fischman, N. M. Alpert, R. H. Rubin, *Clin. Pharmacokinet.* **2002**, 41, 581–602.

55 C. Halldin, B. Gulyas, L. Farde, *Ernst Schering Res. Found. Workshop* **2004**, 48, 95–109.

56 K. C. Schmidt, F. E. Turkheimer, *Q. J. Nucl. Med.* **2002**, 46, 70–85.

57 C. S. Patlak, R. G. Blasberg, *J. Cereb. Blood Flow Metab.* **1985**, 5, 584–590.

58 F. Brady, S. K. Luthra, G. D. Brown, *Curr. Pharm. Des.* **2001**, 7, 1863–1892.

59 G. Lappin, R. C. Garner, *Nat. Rev. Drug Discov.* **2003**, 2, 233–240.

60 See http://nihroadmap.nih.gov/overview.asp.

61 A. Eggermont, H. Newell, *Eur. J. Cancer* **2001**, 37, 1965.

62 C. Grotzinger, B. Wiedenmann, *Ann. N. Y. Acad. Sci.* **2004**, 1014, 258–264.

63 A. Heppeler, S. Froidevaux, A. N. Eberle, *Curr. Med. Chem.* **2000**, 7, 971–994.

64 A. Becker, C. Hessenius, K. Licha, et al., *Nat. Biotechnol.* **2001**, 19, 327–331.

65 M. Schirner, A. Menrad, A. Stephens, *Ann. N. Y. Acad. Sci.* **2004**, 1014, 67–75.

66 N. Zhang, Z. Fang, P. R. Contag, et al., *Blood* **2004**, 103, 617–626.

67 D. Neri, B. Carnemolla, A. Nissim, et al., *Nat. Biotechnol.* **1997**, 15, 1271–1275.

68 K. Licha, C. Perltz, P. Hauff, et al., *Proc. Soc. Molec. Imaging* **2004**, 264.

69 R. Haubner, H. J. Wester, U. Reuning, et al., *J. Nucl. Med.* **1999**, 40, 1061–1071.

70 H. Leong-Poi, J. Christiansen, A. L. Klibanov, et al., *Circulation* **2003**, 107, 455–460.

71 D. A. Sipkins, D. A. Cheresh, M. R. Kazemi, et al., *Nat. Med.* **1998**, 4, 623–626.

72 P. M. Winter, S. D. Caruthers, A. Kassner, et al., *Cancer Res.* **2003**, 63. 5838–5843.

73 P. M. Winter, A. M. Morawski, S. D. Caruthers, et al., *Circulation* **2003**, 108, 2270–2274.

74 S. Flacke, S. Fischer, M. J. Scott, *Circulation* **2001**, 104, 1280–1285.

75 R. M. Botnar, A. Buecker, A. J. Wiethoff, et al., *Circulation* **2004**, 110, 1463–1466.

76 J. Barkhausen, W. Ebert, C. Heyer, et al., *Circulation* **2003**, 108, 605–609.

77 S. G. Ruehm, C. Corot, P. Vogt, et al., *Circulation* **2001**, 103, 415–422.

78 M. Rausch, P. Hiestand, D. Baumann, et al., *Magn. Reson. Med.* **2003**, 50, 309–314.

79 P. Hauff, M. Reinhardt, M. Mäurer, et al., *Proc. Soc. Molec. Imaging* **2004**, 121.

80 M. Reinhardt, P. Hauff, A. Briel, et al., *Invest. Radiol.* **2004**, 39.

81 F. Blankenberg, C. Mari, H. W. Strauss, *Q. J. Nucl. Med.* **2003**, 47, 337–348.

82 L. Hofstra, I. H. Liem, E. A. Dumont, et al., *Lancet* **2000**, 356, 209–212.

83 R. Paulmurugan, T. F. Massoud, J. Huang, et al., *Cancer Res.* **2004**, 64, 2113–2119

84 G. D. Luker, V. Sharma, C. M. Pica, et al., *Proc. Natl. Acad. Sci. USA* **2002**, 99, 6961–6966.

5

Design and Development of Probes for *In vivo* Molecular and Functional Imaging of Cancer and Cancer Therapies by Positron Emission Tomography (PET)

Eric O. Aboagye

Abstract

Molecular imaging with positron emission tomography (PET) is now evolving as a unique non-invasive method for studying tumor and normal tissue biochemistry, physiology, and pharmacology. In oncology, a range of drugs can be radiolabeled for pharmacokinetic studies including "microdosing" of human subjects prior to Phase I trials. Gene delivery can be assessed by incorporating a reporter gene that is detectable by PET within the vector of interest. This chapter also reviews progress made in the PET imaging field in the design of pharmacodynamic markers including assays of cell surface receptor status, angiogenesis, apoptosis, proliferation, glucose metabolism, and hypoxia. Such probes are potentially useful in patient management and for drug development. PET is particularly attractive for the development of cancer-targeted therapies where assessment of plasma drug levels or assays of the target protein in peripheral blood cells are less specific than direct assessment of the tumor or tumor material. The recent introduction of dedicated small-animal scanners has helped bridge *in vitro* science with *in vivo* clinical studies for the efficient development of modern biopharmaceuticals.

Abbreviations

ATP	adenosine triphosphate
ATSM	diacetyl-bis(N4-methylthiosemicarbazone)
BrdU	bromodeoxy uridine
CDK	cyclin-dependent kinase
CEA	carcinoembryonic antigen
CT	computed tomography
D2R	dopamine D2 receptor
DACA	N-[2-(dimethylamino)ethyl]acridine-4-carboxamide
DLT	dose-limiting toxicity
DOTA	1,4,7,10-tetraazacyclododecane-N,N′,N″,N‴-tetraacetic acid
DPD	dihydropyrimidine dehydrogenase
DTPA	diethylenetriamine pentaacetic acid
ECM	extracellular matrix
EGFR	epidermal growth factor receptor
ELISA	enzyme-linked immunosorbent assay
ER	estrogen receptor
FDG	fluorodeoxyglucose
FES	16α-[^{18}F]fluoro-17β-estradiol
FESP	fluoroethylspiperone
FGF	fibroblast growth factor
FHBG	9-(4-[^{18}F]fluoro-3-hydroxy methyl butyl) guanine
FIAU	2′-fluoro-2′deoxy-1-β-D-arabinofuranosyl-5-[^{124}I]iodo-uracil
FLT	fluorothymidine

Modern Biopharmaceuticals. Edited by J. Knäblein
Copyright © 2005 WILEY-VCH Verlag GmbH & Co. KGaA, Weinheim
ISBN: 3-527-31184-X

FMAU 2-fluoro-5-methyldeoxyuracil-β-D-arabinofuranoside
FMISO fluoromisonidazole
FU fluorouracil
HDAC histone deacetylase
HIDAC high density avalanche chamber
HSP heat shock protein
HSV1 herpes simplex virus type I
HYNIC hydrazino nicotinamide
IUdR iododeoxyuridine
MAPK mitogen-activated protein kinase
MMP matrix metalloproteinase
MTD maximum tolerated dose
NIS nal symporter
PEG polyethylene glycol
PET positron emission tomography
PET-CT positron emission tomography – computed tomography
PK1 N-(2-hydroxypropyl)methacrylamide copolymer of doxorubicine
PS phosphatidylserine
PTAC Pharmakokinetic & Pharmacodynamic Technology Advisory Committee
RECIST response evaluation criteria in solid tumours
RGD arginine-glycine-aspartic acid
SIB N-succinimidyl-3-iodobenzoate
SPECT single photon emission computed tomography
TK1 thymidine kinase Typ 1
TMP thymidine monophosphate
TNF tumour necrosis factor
TRAIL tumor necrosis factor-related apoptosis-inducing ligand

5.1
What is Positron Emission Tomography?

Positron emission tomography (PET) imaging is one type of nuclear imaging that utilizes short-lived, positron-emitting isotopes to allow visualization and quantification of biological processes or drug kinetics [1]. Positron-emitting isotopes such as ^{15}O ($t_{1/2} = 2$ min), ^{11}C ($t_{1/2} = 20$ min), ^{18}F ($t_{1/2} = 109$ min), and ^{124}I ($t_{1/2} = 4.2$ days) can be incorporated into many compounds of biological interest to produce radiotracers such as [^{18}F]fluorodeoxyglucose (FDG) for PET studies. Some examples of positron-emitting isotopes are listed in Table 5.1.

A PET study begins with the injection or inhalation of a radiotracer, followed by scanning. When the radiotracer decays, it emits a positron that travels a short distance and annihilates with an electron (Fig. 5.1). Annihilation produces two 511-keV photons, which propagate at approximately 180° apart. These photons can be detected within a short time window called the coincidence time window (~10 ns). Many such events are summed to provide the distribution of the radiotracer. Radiotracer transport, washout and retention can be monitored by PET and, if calibrated, the PET images can yield quantitative estimates of the amount of radiotracer

Table 5.1 Positron-emitting radioisotopes

Radioisotope	Decay mode [% β^+]	Half-life [min]	Decay product
Carbon-11	99.8	20.38	Boron-11
Nitrogen-13	100	9.96	Carbon-13
Oxygen-15	99.9	2.03	Nitrogen-15
Fluorine-18	96.9	109.8	Oxygen-18
Iron-52	57	496	Manganese-52
Cobalt-55	77	1050	Iron-55
Copper-62	98	9.7	Nickel-62
Bromine-75	75.5	98	Selenium-75
Bromine-76	57	966	Selenium-76
Technetium-94m	72	52	Molybdenum-94
Iodine-124	25	6048	Tellurium-124
Gallium-68	90	68.3	Germanium-68

a

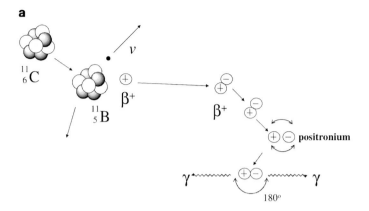

b

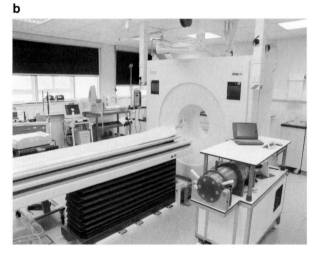

c

Fig. 5.1 (a) Positron emission. A positron emitter such as carbon-11 decays to boron-11 with the emission of a positron and a neutrino. The positron travels a finite distance, colliding with electrons to form a positronium. Annihilation of the positronium leads to the emission of two gamma photons at approximately 180° apart. These photons can be detected by coincident detectors within a scanner. (b) Example of a clinical PET scanner, Siemens HR+ (published with permission from Dr. Terry Spinks, Hammersmith Imanet, London, UK). (c) Example of a dedicated small-animal PET scanner, the Oxford Positron Systems nano-PET scanner, quad-HIDAC.

in specific parts of the body. Additional information of blood radioactivity levels or radioactivity in "reference regions" enables the calculation of exchange rate constants of transport, receptor binding, retention, and metabolic rate. Mathematical kinetic modeling is often employed to enhance data interpretation within a framework of important kinetic behaviors, and to obtain quantitative parameters of relevance and universal comprehension. As indicated below, PET is being used as a tool to understand focal pathophysiology, as well as to monitor response to drug treatment. Figure 5.1 also shows examples of clinical and pre-clinical PET scanners. Commercial clinical scanners have been in use for several years, and more recently the emergence of commercial PET-computed tomography (PET-CT) scanners implies that true fusion of anatomical and functional data can be achieved [1]. On the other hand, the recent development of commercial dedicated small-animal scanners has helped to bridge *in vitro* science with *in vivo* clinical studies [2].

Advancement in radiochemistry methods has made it possible to develop several probes to study biology. The development of PET radiopharmaceuticals for use in cancer and cancer therapeutics studies is presented in the next section.

5.2
Radiochemistry Considerations

Radiopharmaceuticals with very high specific activity ($5–15$ Ci mmol^{-1}) are used in PET studies so as to retain the pharmaceutical, biological, and biochemical properties of the compound being studied. This section reviews the common radiosynthetic methods, the selection of a suitable radiotracer, and limitations in the preparation of radiopharmaceuticals.

5.2.1
Radiosynthetic Methods

The labeling agents for PET radiopharmaceuticals are prepared from very simple chemicals such as [^{11}C]CO, [^{11}C]CO$_2$, [^{18}F]F$^-$ and [^{18}F]F$_2$ (Fig. 5.2). Of the various possibilities available, [^{11}C] methylation, nucleophilic substitution with [^{18}F]F$^-$ and electrophilic fluorination with [^{18}F]-labeled intermediates form the majority of labeling approaches used for biological

compounds [3]. The labeling agent is reacted with a precursor that is often prepared in a form such that a single-step (or a few steps) radiolabeling reaction is performed, followed by purification and formulation for intravenous injection. In some cases – for example, the preparation of 2-[^{11}C]thymidine [4] – single-step chemistry is not possible and a multi-step reaction is performed.

5.2.2
Selection of a Suitable Radiotracer

What is a suitable radiotracer for PET studies? The selection of any radiotracer for a biological application depends on the objectives of the study. Generic considerations include: the type of biological target; the affinity and specificity of the radiotracer for the target; the physico-chemical properties, pharmacokinetics and metabolism of the radiotracer; and the ability to synthesize the compound.

Small molecule-, peptide-, and antibody-based probes all tend to be suitable for targeting cell surface receptors. However, only small molecule-based probes and membrane-permeable peptides are suitable for targeting intracellular and nuclear enzymes or receptors. Consideration should be given to the existence of membrane carrier proteins such as nucleoside transporters or expression of efflux pumps such as the p-glycoprotein pump for poly-aromatic compounds in tissues. Additional considerations include barriers to radiotracer delivery (e.g., the blood–brain barrier), the concentration and location of the target, and the heterogeneity of the target within the tissue. The goal of most radiotracer studies is to achieve a high target-to-background ratio. This means that the affinity of the radiotracer for the receptor or enzyme system should be high, usually low

a

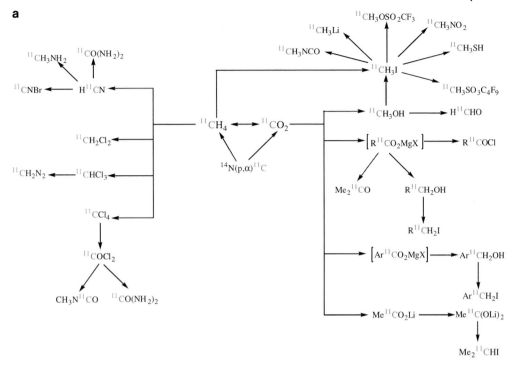

b

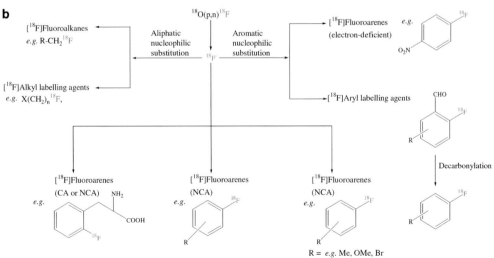

Fig. 5.2 Synthesis of labeling agents. (a) Small building blocks derived from $[^{11}C]CO_2$ and $[^{11}C]CH_4$. (b) Incorporation of $[^{18}F]F^-$ into small building blocks.

nanomolar binding affinity (often denoted as K_d, the concentration of tracer that is associated with half-maximal receptor binding, half-B_{max}); the concentration of target should be high (> 10 pmol mg^{-1} tissue); and non-specific accumulation in the tissue of interest should be low. For receptor-based studies, a general guiding principle is to select radioligands with high B_{max}/K_d of ≥ 4. Binding of a radioligand to a receptor, phosphorylation of a radiotracer by an intracellular enzyme or other mechanism that increases the charge of the compound, and covalent interactions with cellular macromolecules provide the specific signal. It is then desirable to have rapid clearance of the non-specific signal. Clearance is largely determined by the physicochemistry and metabolic fate of the compound. If a radiotracer has rapid clearance, then it is acceptable to radiolabel it with any isotope. For compounds with long half-lives (e.g., antibodies), long-lived isotopes such as ^{124}I are often used.

Non-specific accumulation of radiotracers, on the other hand, is more difficult to define for any particular radiotracer, and depends on specific activity (the presence of unlabeled, "cold" material increases non-specificity), specificity for the target versus other related targets, non-specific transport, and physico-chemical properties such as stereochemistry, inductive effect, lipophilicity, and ionization potential. The latter two characteristics can be altered fairly easily in the design of radiotracers to reduce non-specific binding. Whereas high lipophilicity (log p > 1.5 and < 3) is desirable in neuroscience applications to enable the radiotracer to traverse the blood–brain barrier, this is not necessarily the case in some oncology and cardiology applications, and it could lead to non-specific accumulation. Lipophilicity should be considered together with the ionization rate

constant, pK_a, of the compound, which could be designed to enhance or reduce membrane transport.

5.2.3
General Limitations in the Preparation of Radiopharmaceuticals for PET Studies

Due to the short physical half-lives of most positron-emitting radionuclides, radiotracer synthesis should be fast enough to allow the target drug to be isolated, purified, and formulated as a sterile, pyrogen-free, isotonic solution within two to three half-lives of the radionuclide. Consequently, large amounts of radioactivity must be handled, and the limitation of potential exposure to personnel is an extremely important consideration. For most compounds, these issues require the design and synthesis of precursors that can be radiolabeled in a single step, and in turn this has fostered the development of new methods for rapid remote-controlled and robotics-based chemistry. The presence of radiochemistry facilities and on-site expertise in radiochemistry are also required. The ability to radiolabel a compound depends on the availability of suitable functional groups. For example, compounds with N-, S- or O-methyl (or -ethyl) groups, and lysine and tyrosine functionalities, can be radiolabeled fairly easily. In some cases, the multi-step chemistry required to produce a radiotracer precludes radiolabeling and purification of molecules rapidly enough to avoid substantial decay of radioactivity. Constraints in the availability of suitable labeling reagents including precursors can also limit the ability to synthesize a radiotracer. The position of labeling should be robust towards metabolic degradation, which further limits the number of compounds that can be radiolabeled.

Table 5.2 Paradigm shift in oncology from DNA-targeting to targeting factors that initiate and drive tumorigenesis

- Classical targets of systemic chemotherapy
  - DNA: alkylators, platinators
  - Incorporation of nucleotides into DNA: antimetabolites
  - Topoisomerases: anthracyclines, camptothecins
  - Microtubules: vinca alkaloids, taxanes
- New targets for systemic therapy
  - Growth factors: erbB2, EGFR, VEGF, VEGFR
  - Angiogenesis/vascular: VEGF, FGF, PDGF, integrins
  - Signal transduction: protein kinases, cKit, ras, MAPK, TRAIL
  - Cell cycle regulation: cyclins, CDKs, p53
  - Invasion and metastasis: matrix metalloproteinases
  - Multiple targets: Hsp90, HDAC, Cox-2

EGFR, epidermal growth factor receptor; VEGF, vascular endothelial growth factor; VEGFR, VEGF receptor; MAPK, mitogen-activated protein kinase; TRAIL, TNF-related apoptosis-inducing ligand; CDKs, cyclin-dependent kinases; Hsp90, heat-shock protein 90; HDAC, histone deacetylase; Cox-2, cyclo-oxygenase type-2.

5.3
Pharmacological Objectives in Oncology Imaging Studies

In the past, most oncology therapeutics were designed to target DNA directly or indirectly via modulation of the enzymes involved in its synthesis, coiling/uncoiling, and segregation. In the past decade, our knowledge of the processes that initiate and drive tumorigenesis has increased, and so has the development of drugs that target these processes (Table 5.2). Such tumor-targeted drugs, however, bring along with them difficulties in drug development. In most cases, the use of systemic toxicity end-points (maximum tolerated

Table 5.3 Pharmacological objectives for testing targeted therapies

Objectives	Measurable end-points
Select patients expressing specific target	e.g., erbB2 status, hypoxia
Ensure adequate/optimal exposure in experimental animal model or patient	Pharmacokinetics in plasma or tissues
Demonstrate target modulation	e.g., kinase inhibition, demethylation
Demonstrate induction of desired biological effect	e.g., inhibition of proliferation, invasion, angiogenesis, or induction of differentiation, apoptosis
Resulting clinical response	e.g., disease-free survival, cytostasis, tumor shrinkage

dose, MTD; dose-limiting toxicity, DLT) are inappropriate. Furthermore, the drugs are largely cytostatic and do not cause overt changes in tumor size within several weeks to months of treatment. Methods which can show that the drug has reached its target, has modulated its biological target (or cognate biochemical events), and has caused a biological response are now required to provide proof of concept. The incorporation of such design into early-phase trials makes the trial an extension of the pre-clinical testing of the compound. A summary of generic pharmacological objectives in the development of novel targeted therapies is listed in Table 5.3. The end-points could be assessed directly on biopsy material from the patient. In the absence of biopsy material, investigators have used peripheral blood mononuclear cells and buccal scrapings to evaluate the mechanism of action. It should be noted that the target may not be expressed to the same extent in these tissues as in tumors, and where the target is expressed, target modulation could occur at a lower dose compared to that in tumors [5]. Molecular imaging methods (notably PET) are potential alternatives for clinical decision making, including dose and schedule selection for Phase II, and the selection of patient subpopulations enriched for response, as well as early detection of response. The imaging methods are attractive because:

- they are surgically non-invasive;
- they can be repeated in the same patient several times before and after treatment;
- they allow heterogeneity within tumors or between a primary tumor and metastases to be determined; and
- they provide quantitative information.

For a more detailed description of how to use imaging and non-imaging methods to develop novel targeted therapies, the reader should refer to the Cancer Research UK Pharmacokinetic & Pharmacodynamic Technologies Advisory Committee (PTAC) guidance document at the following URL (http://science.cancerresearchuk.org/reps/pdfs/PTACguidelines.pdf). The use of PET in studying the pharmacokinetics and pharmacodynamics of oncology drugs will be reviewed in the next section.

5.4
The Use of Radiolabeled Drugs to Image Tumor and Normal Tissue Pharmacokinetics

A number of drugs have been radiolabeled to enable their pharmacokinetics in tumor and normal tissues to be studied. In oncology, the large majority of these drugs are small aliphatic, aromatic or heterocyclic molecules identical (isotopic) or similar (non-isotopic; e.g., the replacement of a hydrogen or hydroxyl with fluorine) in structure to the compound of interest. For biopharmaceuticals such as antibodies and polymers, non-isotopic labeling methods are often employed. Examples of these radiopharmaceuticals will be given to illustrate the application of the technology in oncology. In most cases, the key objective of the study is to investigate whether the pharmacokinetic properties predicted *in silico* or seen *in vitro* are similar to those seen in animals and humans; this should enable the confirmation of drug-design objectives, and also enable dose versus tissue-exposure relationships to be assessed.

5.4.1
Pharmacokinetics of Small Molecules

In addition to toxicity, Phase I trials often involve assessment of the plasma pharmacokinetics of the drug to obtain parameters such as the elimination half-life (Ke), systemic clearance (sCL), area under the plasma drug concentration versus time curve ($pAUC$, the systemic exposure), and systemic volume of distribution (sV_d, the ratio of the amount of drug to the plasma drug concentration at steady state). The overall (systemic) extravascular distribution of drugs can be predicted from the sV_d; for example, if sV_d is higher than the plasma volume, then the drug shows extravascular distribution. sV_d does not indicate, however, the tissues to which the drug is distributing. The delivery, washout and retention of drugs in tumor and specific normal tissues are easily assessable by PET imaging of the radiolabeled compound in animal models and in patients. To this end, a number of anti-cancer drugs including 5-fluorouracil [6], cisplatin [7], temozolomide [8], and N-[2-(dimethylamino)ethyl]acridine-4-carboxamide (DACA) [9] have been radiolabeled for pharmacokinetic studies

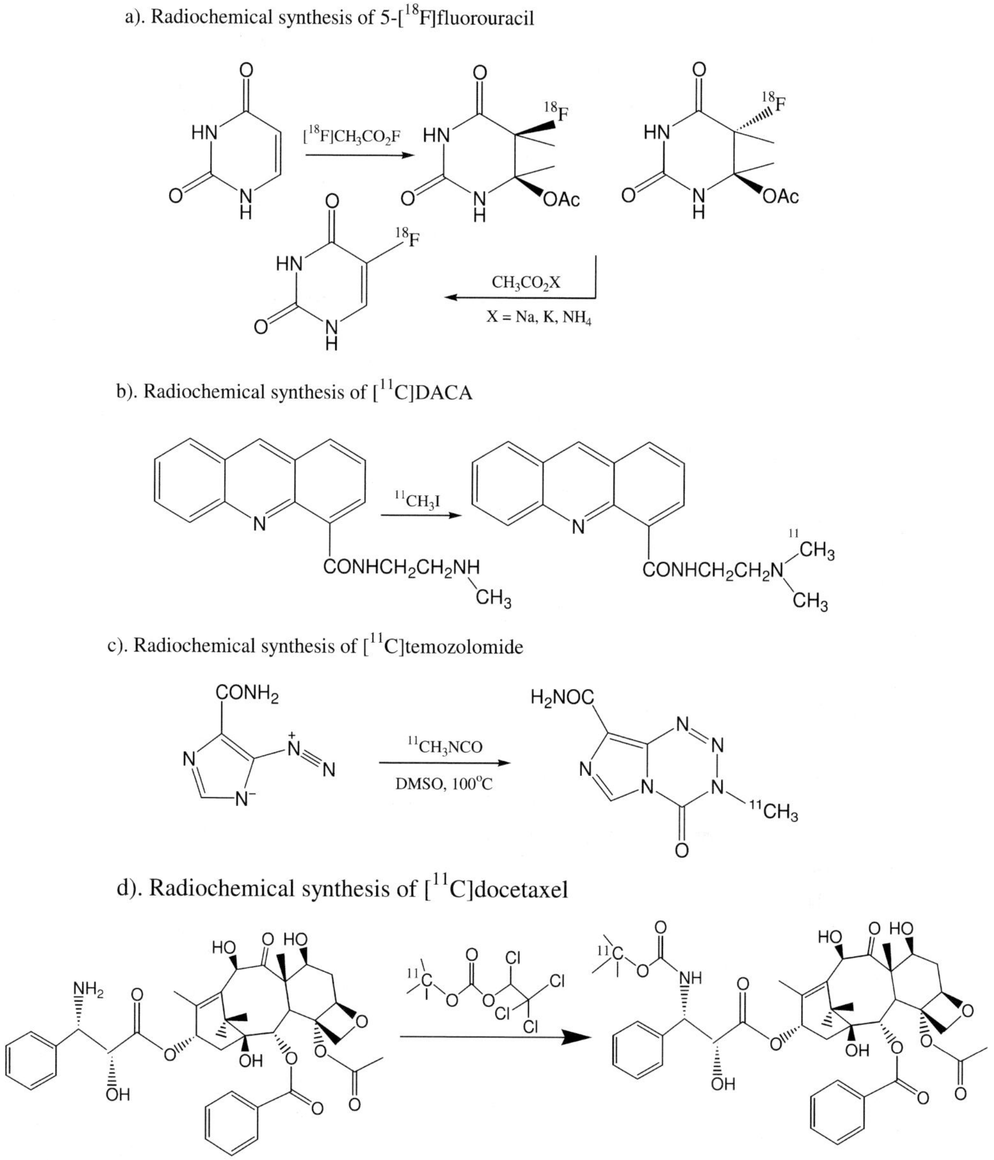
a). Radiochemical synthesis of 5-[18F]fluorouracil
[18F]CH3CO2F
CH3CO2X
X = Na, K, NH4
b). Radiochemical synthesis of [11C]DACA
11CH3I
CONHCH2CH2NH
CONHCH2CH2N
c). Radiochemical synthesis of [11C]temozolomide
11CH3NCO
DMSO, 100°C
d). Radiochemical synthesis of [11C]docetaxel

Fig. 5.3 Radiolabeling of anti-cancer drugs for pharmacokinetic studies. DACA, *N*-[2-(dimethyl-amino)]acridine-4-carboxamide.

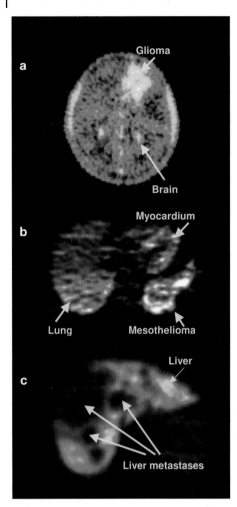

Fig. 5.4 Distribution of radiolabeled drugs in humans monitored by PET. (a) A brain image of [^{11}C]temozolomide, showing high localization of radioactivity in the tumor (glioma). (b) A thoracic image of [^{11}C]DACA scan, showing localization in the myocardium and tumor (mesothelioma). (c) An abdominal image of 5-[^{18}F]fluorouracil, showing a hyperintense normal liver and low uptake in the liver metastases.

beled, although clinical studies have not been performed with these materials [10–13]. Some examples of radiolabeling reactions are illustrated in Fig. 5.3.

The tissue distribution of a selection of radiolabeled drugs in cancer patients is shown in Fig. 5.4. From such studies, the time-course of drug distribution can be determined for any region of interest within the field of view. PET pharmacokinetic studies can be performed with the radiotracer alone (high specific activity), or mixed with a suitable dose of the unlabeled drug (low specific activity). Performing both studies in the same patient can give an indication of saturation effects. The studies can be carried out before Phase I trials (pre-Phase I, also called "micro-dosing") [9, 14], or as part of a Phase I trial [15]. An example of a pre-Phase I radiotracer study is that of [^{11}C]DACA, a DNA intercalating and topoisomerase I/II inhibitor. This study was conducted at one-thousandth of the Phase I starting dose at one year before the Phase I trial, and demonstrated that the drug distributed well to tumors [9]. Radioactivity localized to tissues in the order: vertebra < brain < tumor < kidney < lung < myocardium < spleen < liver. The low peak concentrations and overall exposure in brain and vertebra contrasted with the high distribution to brain in rodents, and suggested that neurotoxicity and myelotoxicity were less likely to be dose-limiting. On the other hand, the high localization of radioactivity in the myocardium (saturable at Phase I doses of the drug) [9, 15] warranted close monitoring of cardiovascular effects.

High distribution of [methyl-^{11}C]temozolomide, a DNA methylating agent, to tumors has also been demonstrated by PET. In patients with brain tumors, delivery and exposure of the radiotracer was found

in humans. More recently, the taxanes (docetaxel and paclitaxel), the oral fluoropyrimidine prodrug, capecitabine, and the epidermal growth factor receptor tyrosine kinase inhibitor, Iressa, have been radiolabeled to be higher in the tumor compared to normal brain tissue [16], and was suggestive of some selectivity for tumors. In another study, Saleem et al. showed that radiolabeled temozolomide (radiolabeled in the N-methyl or carbonyl position) undergoes ring opening selectively in tissues in comparison to plasma, but did not show selectivity for tumor versus brain [8]. The reason for the higher exposure of [methyl-^{11}C]-temozolomide seems to be a higher delivery (K_1) of the radiotracer to the tumor [16]. Studies with [methyl-^{11}C]temozolomide also illustrate one of the limitations of PET studies of drug pharmacokinetics. Often, the decay constant of the radioisotope (λ) is much higher than the plasma Ke, which means that only the initial delivery-phase of the drug is accurately measured.

Among anticancer drugs, 5-[^{18}F]fluorouracil ([^{18}F]FU) has been the most widely studied using PET, it having been shown that:

- retention of the drug in tumors is low [17];
- the drug is catabolized by the liver to a transiently trapped catabolite, [^{18}F]fluoro-β-alanine; a large proportion of the administered radioactivity is, therefore, localized in the liver [6, 17];
- eniluracil, a dihydropyrimidine dehydrogenase (DPD) inhibitor, can inhibit hepatic clearance of the drug and increase drug exposure in tumors; these effects occur in concert with an increase in plasma uracil levels (a systemic measure of DPD inhibition) [6, 17, 18];
- alpha interferon increases tumor exposure of 5-FU [19];
- folinic acid, a modulator of the thymidylate synthase activity, has no effect on [^{18}F]FU pharmacokinetic [19]; and
- intra-arterial administration gives rise to a higher tumor exposure of the radiotracer than the intravenous route [20].

These studies have demonstrated that PET imaging of small molecules radiolabeled with positron emitters can add value to classical studies of new drugs. The PET radiolabeling studies provide very important tissue pharmacokinetics information. It is worth noting, however, that the achievable pharmacokinetic parameters may be limited for three reasons:

- The use of radiotracers (high specific activity) at doses much below the point where metabolism and protein binding become important (K_m) can lead to altered systemic clearance than when the studies are carried out at relevant Phase I doses.
- The half-life of ^{11}C is often shorter than the half-life of the drug of interest; thus, delivery and partitioning of the radiotracer are more accurately estimated than retention parameters. The use of ^{18}F alleviates the problem of a short half-life in some cases.
- The metabolism of the radiotracer can complicate interpretation of the data.

Despite these difficulties, PET studies of radiolabeled drugs have provided unique pharmacokinetic information in patients.

5.4.2
Pharmacokinetics of Biopharmaceuticals

Although most radiolabeling and PET pharmacokinetic studies performed to date have utilized small-molecule drugs, there is potential to apply the technologies to macromolecular agents such as biopharmaceuticals. Applications to peptides and antibodies will be reviewed under pharmacodynamic studies, as the majority of such applications exist there. For polymer-based drugs, including *N*-(2-hydroxypropyl)methacrylamide copolymer doxorubicin (PK1; a doxorubicin-polymer conjugate), either the active agent or the polymer can be radio-labeled, depending on the clinical question asked. Long-lived isotopes such as ^{124}I are preferred for radiolabeling polymers as the pharmacokinetics can be measured over days, thereby allowing long-term preferential tumor localization (if any) to be assessed. In this case, the radiochemistry will require the synthesis of appropriate precursors containing for instance tyrosine or lysine residues to enable direct iodination (with Iodogen) or indirect iodination (with iodinated aromatic labeling agents). Previously, PK1 was radiolabeled with ^{131}I for gamma-camera imaging [21].

5.4.3
Pharmacokinetics of Gene Delivery Systems

An increasing number of PET studies are dedicated to detection of the efficiency of gene delivery. To date, utility with viral, liposome and stem cell delivery systems have been investigated. PET and other forms of molecular imaging modalities allow the location, magnitude and time-course of gene expression to be determined using a marker or reporter gene together with a reporter substrate – a substrate for the protein product of the reporter gene (Fig. 5.5). Gene expression is, thus, monitored indirectly by the detection of the activity of the reporter gene. Although reporter gene studies alone can be performed for characterization of vector systems, they are probably more useful when studied together with the therapeutic gene. This can be done by: 1) using a construct (with an internal ribosomal entry site) that allows both reporter and therapeutic gene to be transcribed as one mRNA and translated into two proteins; 2) using a bi-directional vector in which transcription of both reporter and therapeutic genes are initiated by a single event such

as doxycycline or tetracycline; and 3) using two separate vectors (for a review, see Ref. [22]). There are several reporter gene–reporter substrate pairs for use in gene expression studies, including: 1) herpes simplex virus type 1-thymidine kinase gene versus 2′-fluoro-2′-deoxy-1-β-D-arabinofuranosyl-5-[^{124}I]iodo-uracil (FIAU) or 9-(4-[^{18}F]-fluoro-3-hydroxy methyl butyl) guanine (FHBG); 2) dopamine type-2 receptor gene versus [^{18}F]fluoroethylspiperone (FESP); and 3) sodium iodide symporter versus [^{124}I]NaI [22]. The latter system is particularly useful for performing PET studies in institutions that do not have a cyclotron, since the half-life of iodine is 4.2 days and so can be transported to distant sites [23, 24]. More recently, PET has been used to assess the transcriptional activity of tissue/tumor-specific promoters, including the human telomerase RNA and protein promoters as a prelude to using such promoters for suicide gene therapy [25], and hypoxia response elements for reporting hypoxia-inducible factor 1 signal transduction [26]. As expected, viral therapy constitutes the bulk of these PET studies, although initial investigations involving liposome delivery vectors and stem cells are being pursued.

5.5
Pharmacodynamic Studies

As mentioned earlier, the early clinical development of tumor-targeted anticancer agents requires the use of non-traditional methods such as molecular drug end-points (Western blots, activity assays) in tumor or surrogate tissue, and functional imaging studies. A review of the literature [27] showed that such methods were not routinely incorporated into the study design for early trials of anticancer agents, and rarely formed the primary basis for

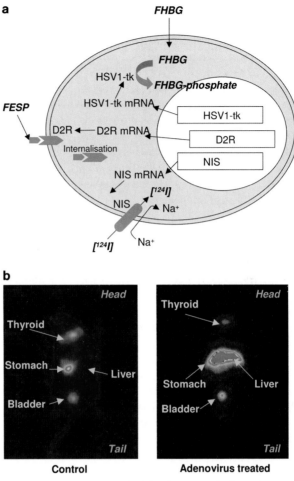

Fig. 5.5 PET imaging of gene expression. (a) Illustration of the three major methods for assessing gene expression *in vivo*: dopamine D₂ receptor (D2R), HSV1-thymidine kinase (HSV1-tk), and NaI symporter (NIS) methods. (b) [^{124}I]-PET images obtained from untreated and adenovirus-treated mice. The high localization in the liver after virus treatment is due to tropism of adenovirus for specific receptors expressed on liver cells. Uptake in the thyroid and stomach is due to physiological expression of NIS; the bladder signal is due to urinary excretion.

dose selection. This may be due to the lack of sufficiently validated functional imaging end-points for clinical studies. In this section, an overview is provided of new imaging assays that have the potential to be used in drug trials.

5.5.1
Assessment of Receptors, Antigens and Extracellular Matrix Proteins

Receptor imaging in cancer is less well developed than it is in neuroscience, in part due to the relative importance of post-receptor signal transduction mechanisms.

Receptor imaging is, however, particularly important for the early detection of cancer, staging and selecting patients for specific receptor-based therapies such as tamoxifen, herceptin and Iressa, which target the estrogen receptor (ER), c-erbB2 and epidermal growth factor receptor (EGFR), respectively. There is therefore active research in this area at present, albeit at an early stage, and these investigations will be discussed from the viewpoint of the type of radiotracer utilized, whether small molecules, peptides, or antibodies.

5.5.1.1 Small-molecule Radiotracers

A number of small-molecule radiotracers are currently under development for the imaging of EGFR. Overexpression of this transmembrane receptor has been found in breast, lung, ovarian, colon and prostate cancer, and is associated with poor prognosis in these cancers [28, 29]. In most specialist centers, the assay of receptor status by immunohistochemistry forms part of the patient work-up, before decisions are made on therapeutic management. Radiotracer methods are being developed to provide quantitative and non-invasive alternatives to biopsy-based methods. Most of the radiotracers are quinazoline derivatives, and similar to drug candidates for the receptors [29, 30]. Initial attempts to develop reversible ^{18}F- and ^{11}C-inhibitors of the intracellular ATP binding site of EGFR were unsuccessful, presumably because of competition for the radiotracers with the high levels of intracellular ATP in cells. A group at the Hebrew University, Israel has now developed irreversible EGFR inhibitors, including [^{11}C]ML03 (Fig. 5.6), with slower tumor washout kinetics because of covalent association through Michael addition, between the double bond of the acryl-amide group at

the 6-position of the quinazoline ring and Cys-773 at the receptor's tyrosine kinase domain [31]. It is hoped that these studies will yield clinical candidates for PET imaging of EGFR receptor status. Not all receptors are extracellular in nature, however. Nuclear receptors are also attractive candidates for imaging, although for these the radiotracers must traverse the plasma and nuclear membrane before binding to the receptor. The ER is an important nuclear receptor in breast cancer, and a target for drugs such as tamoxifen that bind to the receptor or aromatase inhibitors such as anastrozole that reduce circulating levels of the natural ligand, estrogen. The most widely used receptor ligand for ER is 16α-[^{18}F]fluoro-17-estradiol (FES). A good correlation between FES-PET imaging and biopsy-based measures of ER status has been reported [32]. Furthermore, ER occupancy as measured by FES-PET showed a greater decrease after tamoxifen in patients who responded to hormonal therapy than non-responders, thereby demonstrating promise of this radiotracer for predicting patients likely to respond to estrogen-based therapy [33, 34].

A number of PET radiotracers are being developed for imaging extracellular matrix (ECM) proteins. Of particular interest are the matrix metalloproteinases (MMPs); these are usually classified into five groups according to their domain structure – collagenases, gelatinases, stromelysins, membrane-type MMPs, and others – and are a family of zinc-dependent enzymes that degrade specific components of the ECM. The active forms of MMPs (after proteolytic cleavage of the inactive zymogen) are highly expressed during tumor growth and invasion compared to their expression in normal tissues. A suitable PET marker will thus find utility in disease prognosis, as well as for monitoring the pharmacody-

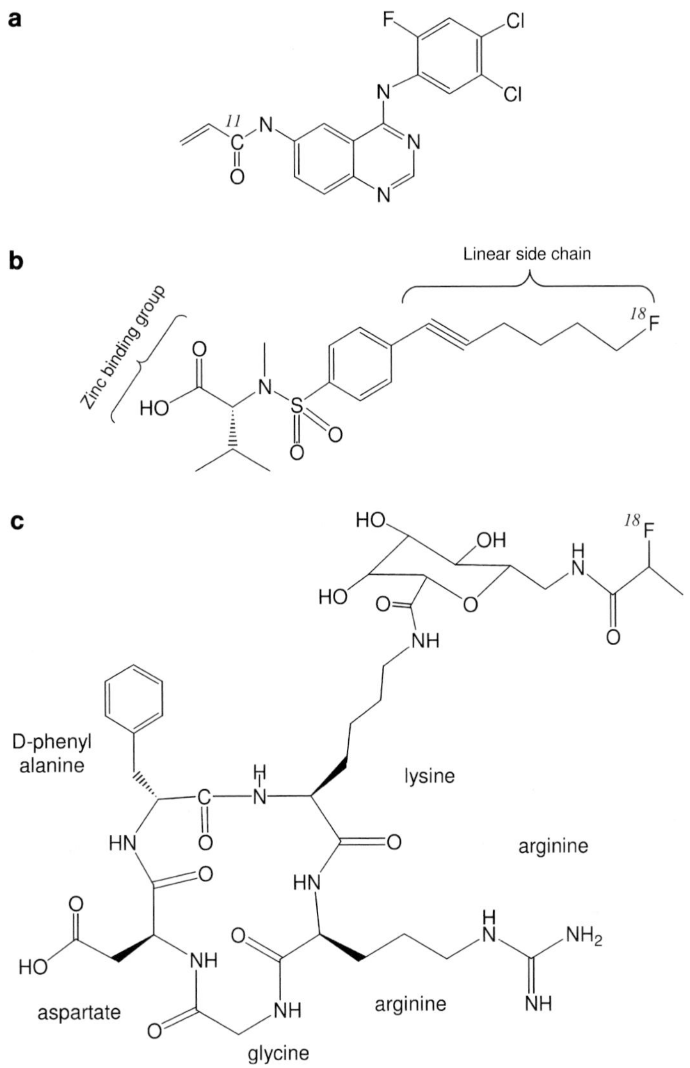

a

b

Linear side chain

Zinc binding group

c

D-phenyl alanine

lysine

arginine

HN

aspartate

glycine

arginine

Fig. 5.6 Examples of novel small molecule and peptide radiotracers for *in vivo* pharmacodynamic imaging. (a) [¹¹C]ML03, an irreversible inhibitor of EGFR tyrosine kinase and potential PET marker for the receptor (b) [¹⁸F]SAV03, an MMP-2 inhibi-tor developed for imaging the levels of this enzyme. (c) [¹⁸F]Galacto-RGD, an $\alpha_v\beta_3$ integrin receptor ligand developed for imaging the levels of this receptor, which is highly expressed on angiogenic vessels.

namics of MMP inhibitors [35]. Furumoto and co-workers have developed ¹⁸F probes ((2R)-2-[4-(6-[¹⁸F]fluorohex-1-ynyl)-benzene-sulfonylamino]-3-methylbutyric acid and its ester) for PET imaging of MMP type 2 (MMP-2) [36]. The key design features (Fig. 5.6) include a carboxylic acid group, which binds to active-site zinc ion, and a linear side chain that interacts with the channel-like S′-1 subsite of MMP [35]. The currently available radiotracers have a number of limitations, and new-generation probes are being developed to overcome this. Significant first-pass metabolism has been reported for the free carboxylic acid-containing radiotracer; ester analogues of these have reduced first-pass metabolism and enhanced tumor localization [36]. De-fluorination of the fluoroalkyl moiety leads to high non-specific uptake in the bone – a limitation that the new candidate radio-tracers will attempt to overcome. Some of the new candidates are based on pan-MMP-inhibitors. For example, Cheesman et al. have reported promising biodistribu-tion data with an [^{111}In]-DTPA-linked mac-rocyclic succinic acid hydroxamate conju-gate [37].

5.5.1.2 Antibody Radiotracers

Antibodies are promising alternatives to small molecules. Because of their size and charge, they do not cross the plasma membrane, and are therefore useful for imaging the extracellular domains of re-ceptors and antigens. Intact monoclonal antibodies (IgG, ~150 kDa) have extremely high selectivity for the target, but often present with poor pharmacokinetics and high immunogenicity. Interest in the use of antibody fragments, derived from pro-tein digestion or by recombinant methods for radiolabeling, stems from the potential gain in enhanced clearance of these mole-cules (see Part V, Chapters 1, 2 and 6).

The c-erbB2 receptor has also been traced using PET. Like EGFR, this receptor is also accessed routinely in breast cancer patients by immunohistochemistry for ex-pression, or by fluorescence *in situ* hybridi-zation for amplification, to inform patient management. Smith-Jones et al. [38] radi-olabeled a F(ab)$'_2$ fragment of the anti-c-erbB2 antibody herceptin with ^{68}Ga via a 1,4,7,10-tetra-azacyclododecane-N,N′,N″,N‴-tetra-acetic acid (DOTA) linker (see Part I, Chapter 5). The resultant radiotracer was used to image c-erbB2-expressing mouse BT-474 tumors; a reduction in tumor localization of the radiotracer was seen at 24 hours after treatment with geldana-mycin, an hsp90 inhibitor that degrades c-erbB2 [38]. Other investigators have explored single-chain Fv fragments (scFv: ~25–30 kDa) due to their potentially rapid clearance from the circulation and re-sultant high target-to-background ratio. Sundaresan et al. [39] radiolabeled anti-carcinoembryonic antigen (anti-CEA) scFv-C$_H$3 minibody and diabody with ^{124}I via the standard Iodogen method (which tar-gets the radiolabel predominantly to tyro-sine residues on the antibody). This allowed CEA-positive tumor xenografts (LS174T) which were <3 mm in diameter to be imaged *in vivo* by PET [39]. With the rapid development of recombinant strate-gies for preparing pharmacokinetically su-perior scFvs coupled with new ways of radiolabeling various antibodies, it is anti-cipated that most extracellular receptors could be imaged by PET in the very near fu-ture. These probes could be used to provide proof of the mechanism of action *in vivo*.

5.5.1.3 Peptide Radiotracers

Of the three classes of radiotracers used to image receptors, this is probably the most attractive. There are several reasons for this:

- Peptides are less-immunogenic mole-cules that show rapid distribution to the target tissue and which, unlike antibod-ies, have a rapid systemic clearance.
- There is rapid development in the field of peptide synthesis via solid phase and from phage libraries.
- There are new and attractive methods for metabolic stabilization and coupling

of peptides to chelators or prosthetic groups for indirect labeling, as well as methods for direct radiolabeling with radiohalogens to tyrosine residues.

Despite these advantages, there are a number of important considerations. First, the metabolic degradation of peptide radiotracers by endogenous peptidases or proteases can make their clearance too rapid, leading to low sensitivity for detection of target activity. The incorporation of linkers such as polyethylene glycol (PEG) can reduce the clearance (see Part VI, Chapter 2). Second, peptide labeling may lead to a loss of affinity and, like antibodies, affinity of the labeled peptide needs to be assessed (e.g., with an enzyme-linked immunosorbent assay; ELISA). Today, there are many diverse peptide radiopharmaceuticals under development, including radiotracers for the integrin $a_v\beta_3$ receptor expressed in neovascularization or angiogenesis, and radiopharmaceuticals for the somatostatin receptor expressed in neuroendocrine tumors/carcinoids, small cell lung cancer and lymphomas.

The integrin $a_v\beta_3$ receptor has attracted much interest in the field of PET, as it is known to be involved in angiogenesis. Molecules containing the tri-peptide sequence arginine-glycine-aspartate (RGD) have been shown to bind selectivity to the $a_v\beta_3$ receptor [40]. High *in vitro* affinity and *in vivo* tumor selectivity was demonstrated with the first candidate radiotracers {[^{124}I]cyclo(-Arg-Gly-Asp-D-Phe-Tyr-) and [^{124}I]cyclo(-Arg-Gly-Asp-D-Tyr-Val-)}. At the time, the fluorination of peptides was a rather challenging task, and was performed in multiple steps. The breakthrough in this field came with the development of [^{18}F]synthons and peptide precursors that permitted chemoselective, single-pot fluorination and simple purification of the radiolabeled product. For exam-

ple, an ^{18}F-labeling methodology based on the chemoselective oxide formation between an unprotected amino-functionalized RGD peptide and an ^{18}F-labeled aldehyde or ketone has been reported [40, 41]; Glaser and co-workers [42] also reported the use of [^{18}F]fluorothiols and methanesulfonyl precursors for labeling a model peptide that can potentially be used for radiolabeling RGD peptides. In addition to the use of cyclic and bicyclic peptides to improve systemic stability and receptor affinity and improved peptide labeling, a number of investigators have also attached PEG and sugar amino acids to improve the pharmacokinetics of candidate peptides [41, 43–45], the initial studies with which in tumor models have provided encouraging results.

5.5.2
PET Monitoring of End-points of Tumor Growth and Response to Treatment

Deranged proliferation and apoptosis are fundamental to the development of cancer [46]. "Tumor growth" is a term that describes the balance between cell division and cell death. This is arguably the most important biological end-point for determining the sensitivity of cancer cells to drugs, and is often estimated in patients using radiological methods (RECIST criteria [47]), by an analysis of biopsy material for visible mitosis, cell and cycle markers, and by autoradiography (for a review, see Ref. [48]). The limitations of the RECIST criteria and biopsy-based methods have led to an interest in alternative imaging methods for monitoring tumor growth. Thus, a review of PET methods for glucose metabolism, proliferation and apoptosis in monitoring the end-points of tumor growth and response to treatment will be presented in the next section.

a

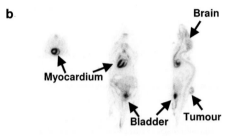

K_1		k_3	

$[^{18}F]FDG$ → $[^{18}F]FDG$ → $[^{18}F]FDG6P$
         k_2 k_4

vascular compartment   free space   metabolic compartment

b

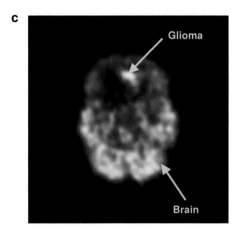

Brain

Myocardium

Bladder   Tumour

c

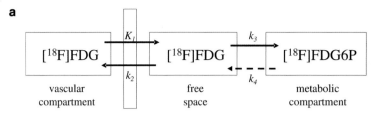

Glioma

Brain

Fig. 5.7 Imaging of glucose metabolism with FDG. (a) A model of FDG uptake. FDG is delivered and retained in tissues according to this model: k_1, k_2, k_3, and k_4 are rate constants for clearance into the cell from blood, clearance out of the cell, phosphorylation and dephosphorylation, respectively. (b) High-resolution FDG-PET images obtained using the nano-PET technology of a tumor-bearing mouse showing localization of the radiotracer in myocardium, brain, tumor, and bladder. Four-dimensional data (three-dimensional spatial + time) were obtained; 0.3-mm orthogonal slices are shown. (c) Brain image of a glioma patient, showing localization of the radiotracer in the tumor and forebrain.

5.5.2.1 PET Imaging of Glucose Metabolism

Fluorodeoxyglucose is the most commonly used radiotracer for imaging glucose utilization. The uptake of FDG into tissues is determined mainly by high glucose transporter and hexokinase activities and low glucose-6-phosphatase activity [49–51]. The model for FDG uptake is illustrated in Fig. 5.7. After transport into cells, FDG (like glucose) is phosphorylated by hexokinase to FDG-phosphate which, unlike glucose-6-phosphate, is not a substrate for further glycolytic metabolism. Trapping of the radiotracer is effected by the high charge of the phosphorylated product and its low rate of dephosphorylation in tumor tissues [52]. Tissues that have high glucose-6-phosphatase activity (e.g., liver) [51] show a low retention of FDG. Non-tumor tissues such as brain and myocardium (Fig. 5.7), as well as inflammatory cells [53] also take up FDG; consequently, care should be taken in interpreting FDG data. From the above account it is clear that FDG does not directly measure tumor proliferation. Rather, it is used as a surrogate for cell viability, which is related indirectly to cell number and proliferation. Other than applications in the diagnosis and staging of disease, FDG has found use in monitoring the response to drug therapy. Compared to cross-sectional imaging, FDG-PET is highly reproducible [54], differentiates between viable and fibrotic tissue, and changes in radiotracer uptake occur early after treatment [55, 56]. For cytotoxic therapies, the decrease in FDG uptake after therapy generally mimics the reduction in tumor cell viability. A number of response studies showing early reduction in FDG uptake after treatment have been published, including the treatment of brain tumors with temozolomide [57], of breast cancer patients with combined chemo- and hormonal therapy [58], and of non-Hodgkin's lymphoma with combination chemotherapy [59]. More recently, FDG-PET has been used to image the response of gastrointestinal stromal tumors to molecular therapeutics such as the c-Kit and bcr-abl inhibitor, Imatinib mesylate (Gleevec; Glivec; STI-571). In this setting, FDG uptake decreased dramatically as early as 24 hours after drug treatment [60, 61], and the early changes correlated with a clinical response at 1–3 months [60]. Due to a number of limitations with

FDG however, other tracers that more closely monitor cellular proliferation or cell death are being evaluated for imaging drug response.

5.5.2.2 PET Imaging of Cell Proliferation

A number of radiolabeled pyrimidine nucleosides have been synthesized for imaging proliferation. 2-[^{11}C]Thymidine and its analogues, 3'-deoxy-3'-[^{18}F]fluorothymidine (FLT), 2-fluoro-5-[^{11}C]methyldeoxyuracil-β-D-arabinofuranoside (FMAU), [^{76}Br]bromodeoxy uridine (BrdU), and [^{124}I]iododeoxyuridine (IUdR) (Fig. 5.8) have been evaluated pre-clinically and clinically for assessing proliferation (for a review, see Ref. [48]). These radiotracers are transported into the cell by diffusion, as well as by nucleoside transporters, and are phosphorylated by a thymidine kinase type-1 (TK1) enzyme to form the corresponding monophosphate (salvage pathway for DNA synthesis; Fig. 5.9). TK1 activity and the subsequent rates of phosphorylation of the monophosphate to di- and tri-phosphate vary for the different nucleosides [62]. Of importance to the clinical use of these radiotracers is the fact that the analogues with a halogen substitution in the sugar-ring have suitable *in vivo* stability [48]. Although more clinical studies have been performed with [^{11}C]thymidine, FLT is probably the most promising of all these radiotracers for monitoring proliferation due to its superior *in vivo* stability. Several *in vitro* and *in vivo* (rodent and clinical) studies have now been carried out demonstrating that FLT-PET uptake measures the TK-1-dependent phosphorylation [63, 64], which correlates with cellular proliferation as measured by S-phase fraction or Ki-67 immunohistochemistry [65–67]. This technology is currently being used pre-clinically in rodents to evaluate the efficacy of

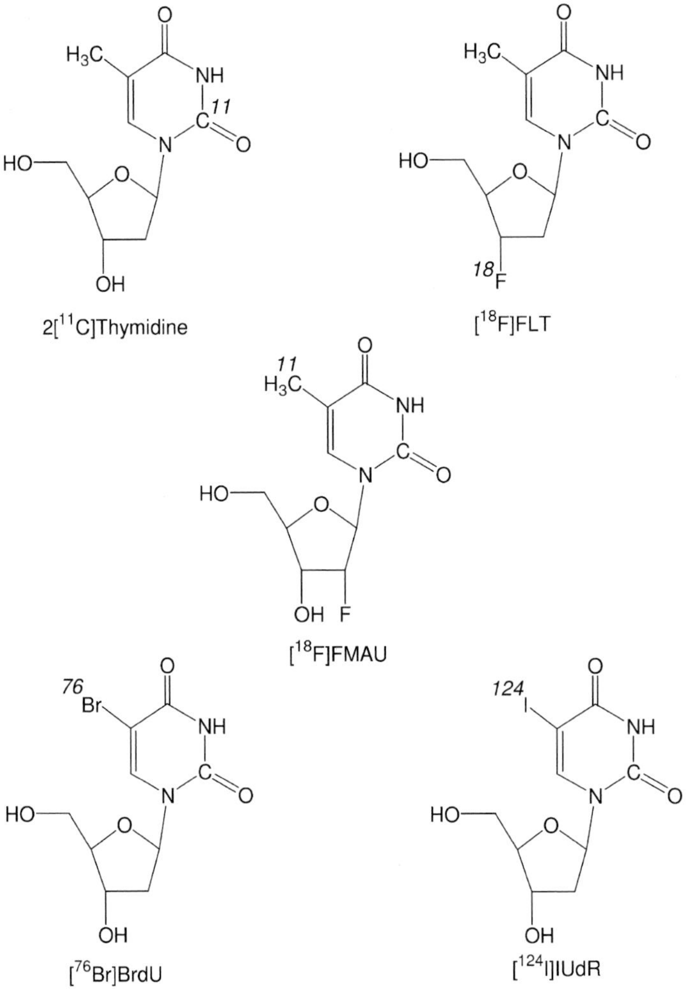

Fig. 5.8 Radiolabeled pyrimidine nucleosides for imaging cellular proliferation.

anti-cancer agents [68–70]. Prospective clinical studies of drug response are ongoing, but results have not been published for FLT. Such data do exist for [^{11}C]thymidine, however [71].

5.5.2.3 PET Imaging of Apoptosis
Programmed cell death, or apoptosis, is the major mechanism of cell death following anti-cancer drug therapy (see also video animation on supplement CD-ROM). Given its importance, several investigators are studying ways of imaging apoptosis *in vivo*. Among various possibilities, the use of radiolabeled Annexin V is currently the most promising. Annexin V is a 36-kDa calcium-dependent protein that binds to phosphatidylserine (PS) with very high affinity. When radiolabeled with positron emitters

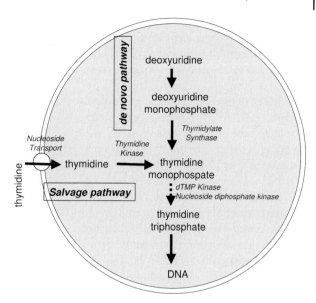

Fig. 5.9 Mode of action of radiolabeled thymidine analogues used for imaging cellular proliferation. The radiotracers are transported and phosphorylated by thymidine kinase (salvage pathway) to the corresponding monophosphate that is subsequently phosphorylated to the triphosphate and incorporated into DNA. For FLT, most of the label remains as the monophosphate being a poor substrate for deoxythymidine monophosphate kinase (dTPM). The *de-novo* pathway for DNA synthesis is also illustrated.

or gamma emitters, the resultant radiopharmaceuticals label apoptotic cells because, in apoptotic cells (unlike viable intact cells) PS translocates from the inner to the outer plasma membrane, making it accessible for binding. Annexin V has been indirectly radiolabeled with [124]I to provide [[124]I]-SIB-Annexin V for PET imaging studies [72, 73]. The most widely used radiopharmaceutical, however, is [^{99m}Tc]-HYNIC-annexin V for single photon emission computed tomography (SPECT) studies [74–76]. Clinical studies with this reagent have been performed [77, 78]. Given that apoptosis is a dynamic process, the optimal timing of these studies is crucial, and is likely to vary for different tumor types and for different therapeutic modalities. Indeed, this issue represents the biggest challenge for imaging apoptosis in tumors.

5.5.3
Assessment of Tumor Hypoxia

The final section of this chapter deals with an important predictor of therapeutic response in oncology – hypoxia. Imaging of this physiological property has relevance in selecting patients likely to benefit from therapies aimed at modulating hypoxia. The disorganized and inadequate vasculature and blood flow in tumors often leads to impaired oxygen delivery, and this leads to the creation of areas of low oxygen tension (hypoxic regions). Hypoxia is a powerful trigger of gene expression, and thus clonal selection of more aggressive phenotype, for example, diminished apoptotic potential [79]. Hypoxia also predicts for local tumor control by external beam radiotherapy, and predicts general treatment outcome, including metastatic potential and survival following radio-/chemotherapy and surgery in a number of human cancers [80–82]. Several radiotracers have been developed for imaging hypoxia by PET; these include 2-nitroimidazole-based probes such as [^{18}F]fluoromisonidazole (FMISO) and [^{18}F]fluoroetanidazole [83, 84], and copper bis-thiosemi-carbazones probes such as [^{60}Cu]ATSM [85]. In all cases, proof that the technique measures hypoxia involves comparisons with direct measurements of pO$_2$ by oxygen electrodes, assessment of uptake following modulation of hypoxia, and radiation sensitivity. Studies with FMISO have demonstrated the existence of hypoxic regions in a number of tumor types [83, 86]. With the development of more modern radiotracers, it is hoped that the measurement of hypoxia can be performed efficiently to enable patient selection for hypoxia-targeted therapy.

5.6
Conclusions

PET is a potentially powerful technology for monitoring drug pharmacokinetics, and for the prediction or assessment of response to anti-cancer treatment. The technique allows quantitative measurements to be made in animals and humans, in a non-invasive manner. It is hoped that developments during the next few years will provide the platform for incorporating this technology into the discovery and development of biopharmaceuticals and, in particular, for early drug trials in patients.

References

1 Valk, P. E., Bailey, D. L., Townsend, D. W., and Maisey, M. N. Positron emission tomography: basic science and clinical practice. Springer, London 2003.

2 Aboagye, E. O. Positron emission tomography of small animals in anticancer drug development. Mol. Imaging Biol., 7: 53–58, **2005**.

3 Elsinger, P. H. Radiopharmaceutical chemistry for positron emission tomography. Methods, 27: 208–217, **2002**.

4 Steel, C. J., Brady, F., Luthra, S. K., Brown, G., Khan, I., Poole, K. G., Sergis, A., Jones, T., and Price, P. M. An automated radiosynthesis of 2-[^{11}C]thymidine using anhydrous [^{11}C]urea derived from [^{11}C]phosgene. Appl. Radiat. Isot., 51: 377–388, **1999**.

5 Collins, J. M. Innovations in phase 1 trial design: where do we go next? Clin. Cancer Res., 6: 3801–3802, **2000**.

6 Saleem, A., Yap, J., Osman, S., Brady, F., Suttle, B., Lucas, S. V., Jones, T., Price, P. M., and Aboagye, E. O. Modulation of fluorouracil tissue pharmacokinetics by eniluracil: *in vivo* imaging of drug action. Lancet, 355: 2125–2131, **2000**.

7 Ginos, J. Z., Cooper, A. J., Dhawan, V., Lai, J. C., Strother, S. C., Alcock, N., and Rottenberg, D. A. [^{13}N]cisplatin PET to assess pharmacokinetics of intra-arterial versus intra-venous chemotherapy for malignant brain tumors. J. Nucl. Med., 28: 1844–1852, **1987**.

8 Saleem, A., Brown, G. D., Brady, F., Aboagye, E. O., Osman, S., Luthra, S. K., Ranicar, A. S. O., Brock, C. S., Stevens, M. F. G., Newlands, E., Jones, T., and Price, P. Metabolic activation of temozolomide measured in vivo using positron emission tomography. Cancer Res., 63: 2409–2415, **2003**.

9 Saleem, A., Harte, R. J., Matthews, J. C., Osman, S., Brown, G. D., Bleehen, N., Connors, T., Jones, T., Price, P. M., and Aboagye, E. O. Pharmacokinetic evaluation of N-[2-(Dimethylamino)ethyl]acridine-4-carboxamide (DACA; XR5000) in patients by positron emission tomography. J. Clin. Oncol., 19: 1421–1429, **2001**.

10 Fei, X., Wang, J. Q., Miller, K. D., Sledge, G. W., Hutchins, G. D., and Zheng, Q. H. Synthesis of [(^{18}F)]Xeloda as a novel potential PET radiotracer for imaging enzymes in cancers. Nucl. Med. Biol., 31: 1033–1041, **2004**.

11 Kurdziel, K. A., Kiesewetter, D. O., Carson, R. E., Eckelman, W. C., and Herscovitch, P. Biodistribution, radiation dose estimates, and in vivo Pgp modulation studies of ^{18}F-paclitaxel in nonhuman primates. J. Nucl. Med., 44: 1330–1339, **2003**.

12 De Jesus, O. T., Murali, D., Flores, L. G., Converse, A. K., Dick, D. W., Oakes, T. R., Roberts, A. D., and Nickles, R. J. Synthesis of [F-18]-ZD1839 as a PET imaging agent for epidermal growth factor receptors. J. Label. Compds. Radiopharm., 46: S1, **2003**.

13 van Tilburg, E. W., Franssen, E. J. F., van der Hoeven, J. J. M., Elshove, D., Lammertsma, A. A., and Windhorst, A. D. Radiosynthesis of [11C]docetaxel. J. Label. Compds. Radiopharm., 46: S3, **2003**.

14 Aboagye, E. O., Luthra, S. K., Brady, F., Poole, K., Anderson, H., Jones, T., Boobis, A., Burtles, S. S., and Price, P. Cancer Research UK procedures in manufacture and toxicology of radiotracers intended for pre-phase 1 positron emission tomography studies in cancer patients. Br. J. Cancer, 86: 1052–1056, **2002**.

15 Propper, D. J., de Bono, J., Saleem, A., Ellard, S., Flanagan, E., Paul, J., Ganesan, T. S., Talbot, D. C., Aboagye, E. O., Price, P., Harris, A. L., and Twelves, C. Use of positron emission tomography in pharmacokinetic studies to investigate therapeutic advantage in a phase 1 study of 120-hour intravenous infusion XR5000. J. Clin. Oncol., 21: 203–210, **2003**.

16 Meikle, S. R., Matthews, J. C., Brock, C. S., Wells, P., Harte, R. J., Cunningham, V. J., Jones, T., and Price, P. Pharmacokinetic assessment of novel anti-cancer drugs using spectral analysis and positron emission tomography: a feasibility study. Cancer Chemother. Pharmacol., 42: 183–193, **1998**.

17 Aboagye, E. O., Saleem, A., Cunningham, V., Osman, S., and Price, P. Extraction of 5-fluorouracil by tumor and liver: a non-invasive positron emission tomography study of patients with gastrointestinal cancer. Cancer Res., 61: 4937–4941, **2001**.

18 Bading, J. R., Yoo, P. B., Fissekis, J. D., Alauddin, M. M., D'Argenio, D. Z., and Conti, P. S. Kinetic modeling of 5-fluorouracil anabolism in colorectal adenocarcinoma: a positron emission tomography study in rats. Cancer Res., 63: 3667–3674, **2003**.

19 Harte, R. J., Matthews, J. C., O'Reilly, S. M., Tilsley, D. W., Osman, S., Brown, G., Luthra, S. J., Brady, F., Jones, T., and Price, P. M. Tumor, normal tissue, and plasma pharmacokinetic studies of fluorouracil biomodulation with N-phosphonacetyl-L-aspartate, folinic acid, and interferon alfa. J. Clin. Oncol., 17: 1580–1588, **1999**.

20 Dimitrakopoulou-Strauss, A., Strauss, L. G., Schlag, P., Hohenberger, P., Irngartinger, G., Oberdorfer, F., Doll, J., and van Kaick, G. Intravenous and intra-arterial oxygen-15-labeled water and fluorine-18-labeled fluorouracil in patients with liver metastases from colorectal carcinoma. J. Nucl. Med., 39: 465–473, **1998**.

21 Vasey, P. A., Kaye, S. B., Morrison, R., Twelves, C., Wilson, P., Duncan, R., Thomson, A. H., Murray, L. S., Hilditch, T. E., Murray, T., Burtles, S., Fraier, D., Frigerio, E., and Cassidy, J. Phase I clinical and pharmacokinetic study of PK1 [N-(2-Hydroxypropyl)methacrylamide copolymer doxorubicin]: first member of a new class of chemotherapeutic agents – drug-polymer conjugates. Clin. Cancer Res., 5: 83–94, **1999**.

22 Gambhir, S. S. Molecular imaging of cancer with positron emission tomography. Nature Med., 2: 683–693, **2002**.

23 Groot-Wassink, T., Aboagye, E. O., Glaser, M., Lemoine, N. R., and Vassaux, G. Adenovirus biodistribution and noninvasive imaging of gene expression in vivo by positron emission tomography using human sodium/iodide symporter as reporter gene. Hum. Gene Ther., 13: 1723–1735, **2002**.

24 Groot-Wassink, T., Aboagye, E. O., Wang, Y., Lemoine, N. R., Reader, A. J., and Vassaux, G. Quantitative imaging of NaI symporter transgene expression using positron emission tomography in the living animal. Mol. Ther., 9: 436–442, **2004**.

25 Groot-Wassink, T., Aboagye, E. O., Wang, Y., Lemoine, N. R., Keith, W. N., and Vassaux, G. Non-invasive imaging of the transcriptional activities of human telomerase promoter fragments in mice. Cancer Res., 64: 4906–4911, **2004**.

26 Serganova, I., Doubrovin, M., Vider, J., Ponomarev, V., Soghomonyan, S., Beresten, T., Ageyeva, L., Serganov, A., Cai, S., Balatoni, J., Blasberg, R., and Gelovani, J. Molecular imaging of temporal dynamics and spatial heterogeneity of hypoxia-inducible factor-1 signal transduction activity in tumors in living mice. Cancer Res., 64: 6101–6108, **2004**.

27 Parulekar, W. R. and Eisenhauer, E. A. Phase I trial design for solid tumor studies of targeted, non-cytotoxic agents: theory and practice. J. Natl. Cancer Inst., 96: 990–997, **2004**.

28 Herbst, R. S. Review of epidermal growth factor biology. Int. J. Radiat. Oncol. Biol. Phys., 59: 21–26, **2004**.

29 Silchenmyer, W. J., Elliott, W. L., and Fry, D. W. CI-1033, a pan-erb tyrosine kinase inhibitor. Semin. Oncol., 28: 80–85, **2001**.

30 Gaul, M. D., Guo, Y., Affleck, K., Cockerill, G. S., Gilmer, T. M., Griffin, R. J., Guntrip, S., Keith, B. R., Knight, W. B., Mullin, R. J., Murray, D. M., Rusnak, D. W., Smith, K., Tadepalli, S., Wood, E. R., and Lackey, K. Discovery and biological evaluation of potent dual erbB-2/EGFR tyrosine kinase inhibitors: 6-thiazolylquinazolines. Bioorg. Med. Chem. Lett., 13: 637–640, **2003**.

31 Mishani, E., Abourbeh, G., Rozen, Y., Jacobson, O., Laky, D., David, I. B., Levitzki, A., and Shaul, M. Novel carbon-11 labeled dimethylamino-but-2-enoic acid [4-(phenylamino)-quinazoline-6-yl]-amides: potential PET bioprobes for molecular imaging of EGFR-positive tumors. Nucl. Med. Biol., 31: 469–476, **2004**.

32 Dehdashti, F., Mortimer, J. E., Siegel, B. A., Griffith, L. K., Bonasera, T. J., Fusselman, M. J., Detert, D. D., Cutler, P. D., Katzenellenbogen, J. A., and Welch, M. J. Positron tomographic assessment of estrogen receptors in breast cancer: comparison with FDG-PET and in vitro receptor assays. J. Nucl. Med., 36: 1766–1774, **1995**.

33 Dehdashti, F., Flanagan, F. L., Mortimer, J. E., Katzenellenbogen, J. A., Welch, M. J., and Siegel, B. A. Positron emission tomographic assessment of "metabolic flare" to predict response of metastatic breast cancer to anti-estrogen therapy. Eur. J. Nucl. Med., 26: 51–56, **1999**.

34 McGuire, A. H., Dehdashti, F., Siegel, B. A., Lyss, A. P., Brodack, J. W., Mathias, C. J., Mintun, M. A., Katzenellenbogen, J. A., and Welch, M. J. Positron tomographic assessment of 16 alpha-[^{18}F] fluoro-17 beta-estradiol uptake in metastatic breast carcinoma. J. Nucl. Med., 32: 1526–1531, **1991**.

35 Zucker, S., Cao, J., and Molloy, C. J. Role of matrix metalloproteinases and plasminogen activators in cancer invasion and metastasis: therapeutic strategies. In: B. C. Baguley and D. J. Kerr (Eds.), Anticancer drug development, pp. 91–122. Academic Press, San Diego, 2002.

36 Furumoto, S., Takashima, K., Kubota, K., Ido, T., Iwata, R., and Fukuda, H. Tumor detection using ^{11}F-labeled matrix metalloproteinase-2 inhibitor. Nucl. Med. Biol., 30: 119–125, **2003**.

37 Cheesman, E. H., Rajopadhye, M., Tran, Y. S., Liu, S., Ellars, C., Onthank, D., Silva, P., Yalamanchili, P., Kavosi, M., Sachleben, R. A., Liu, R. Q., Robinson, S. P., and Edwards, D. S. Radiolabeled matrix metalloproteinase inhibitors with high uptake in mouse tumor models. J. Label. Compds. Radiopharm., 46: S5, **2003**.

38 Smith-Jones, P. M., Solit, D. B., Akhurst, T., Afroze, F., Rosen, N., and Larson, S. M. Imaging the pharmacodynamics of Her2 degradation in response to Hsp90 inhibitors. Nat. Biotechnol., 22: 701–706, **2004**.

39 Sundaresan, G., Yazaki, P. J., Shively, J. E., Finn, R. D., Larson, S. M., Raubitschek, A. A., Williams, L. E., Chatziioannou, A. F., Gambhir, S. S., and Wu, A. M. ^{124}I-labeled engineered anti-CEA minibodies and diabodies allow high-contrast, antigen-specific small-animal PET imaging of xenografts in athymic mice. J. Nucl. Med., 44: 1962–1969, **2003**.

40 Haubner, R., Wester, H.-J., Weber, W. A., and Schwaiger, M. Radiotracer-based strategies to image angiogenesis. Q. J. Nucl. Med., 47: 189–199, **2003**.

41 Poethko, T., Schottelius, M., Thumshirn, G., Hersel, U., Herz, M., Henriksen, G., Kessler, H., Schwaiger, M., and Wester, H.-J. Two-step methodology for high-yield routine radiohalogenation of peptides: ^{18}F-labeled RGD and octreotide analogs. J. Nucl. Med., 45: 892–902, **2004**.

42 Glaser, M., Karlsen, H., Solbakken, M., Arukwe, J., Brady, F., Luthra, S. K., and Cuthbertson, A. 18F-Fluorothiols: a new approach to label peptides chemoselectively as potential tracers for positron emission tomography. Bioconj. Chem., 15: 1447–1453, **2004**.

43 Chen, X., Liu, S., Hou, Y., Tohme, M., Park, R., Bading, J. R., and Conti, P. S. MicroPET imaging of breast cancer alphav-integrin expression with ^{64}Cu-labeled dimeric RGD peptides. Mol. Imaging Biol., 6: 350–359, **2004**.

44 Chen, X., Hou, Y., Tohme, M., Park, R., Khankaldyyn, V., Gonzales-Gomez, I., Bading, J. R., Laug, W. E., and Conti, P. S. Pegylated Arg-Gly-Asp peptide: ^{64}Cu labeling and PET imag-

ing of brain tumor alphavbeta3-integrin expression. J. Nucl. Med., 45: 1776–1783, **2004**.

45 Haubner, R., Kuhnast, B., Mang, C., Weber, W.A., Kessler, H., Wester, H.-J., and Schwaiger, M. [^{18}F]Galacto-RGD: synthesis, radiolabeling, metabolic stability, and radiation dose estimates. Bioconj. Chem., 15: 61–69, **2004**.

46 Hanahan, D. and Weinberg, R.A. The hallmarks of cancer. Cell, 100: 57–70, **2000**.

47 Therasse, P., Arbuck, S.G., Eisenhauer, E.A., Wanders, J., Kaplan, R.S., Rubinstein, L., Verweij, J., Van Glabbeke, M., van Oosterom, A.T., Christian, M.C., and Gwyther, S.G. New guidelines to evaluate the response to treatment in solid tumors. European Organization for Research and Treatment of Cancer, National Cancer Institute of the United States, National Cancer Institute of Canada. J. Natl. Cancer Inst., 92: 205–216, **2000**.

48 Kenny, L.M., Aboagye, E.O., and Price, P.M. Positron emission tomography imaging of cell proliferation in oncology. Clin. Oncol., 16: 176–185, **2004**.

49 Weber, G. Enzymology of cancer cells (second of two parts). N. Engl. J. Med., 296: 541–551, **1977**.

50 Weber, G. Enzymology of cancer cells (first of two parts). N. Engl. J. Med., 296: 486–492, **1977**.

51 Nelson, C.A., Wang, J.Q., Leav, I., and Crane, P.D. The interaction among glucose transport, hexokinase, and glucose-6-phosphatase with respect to 3H-2-deoxyglucose retention in murine tumor models. Nucl. Med. Biol., 23: 533–541, **1996**.

52 Gallagher, B.M., Fowler, J.S., Gutterson, N.I., MacGregor, R.R., Wan, C.N., and Wolf, A.P. Metabolic trapping as a principle of radiopharmaceutical design: some factors responsible for the biodistribution of [^{18}F] 2-deoxy-2-fluoro-D-glucose. J. Nucl. Med., 19: 1154–1161, **1978**.

53 Kubota, R., Yamada, S., Kubota, K., Ishiwata, K., Tamahashi, N., and Ido, T. Intratumoral distribution of ^{18}F-fluorodeoxyglucose in vivo: high accumulation in macrophages and granulation tissues studied by microautoradiography. J. Nucl. Med., 33: 1972–1980, **1992**.

54 Weber, W.A., Schwaiger, M., and Avril, N. Quantitative assessment of tumor metabolism using FDG-PET imaging. Nucl. Med. Biol., 27: 683–687, **2000**.

55 Bourguet, P., Blanc-Vincent, M.P., Boneu, A., Bosquet, L., Chauffert, B., Corone, C., Cour-

bon, F., Devillers, A., Foehrenbach, H., Lumbroso, J.D., Mazselin, P., Montravers, F., Moretti, J.L., and Talbot, J.N. Summary of the Standards, Options and Recommendations for the use of positron emission tomography with 2-[18F]fluoro-2-deoxy-D-glucose (FDP-PET scanning) in oncology (2002). Br. J. Cancer, 89: S84–91, **2003**.

56 Young, H., Baum, R., Cremerius, U., Herholz, K., Hoekstra, O., Lammertsma, A.A., Pruim, J., and Price, P. Measurement of clinical and subclinical tumour response using [^{18}F]-fluorodeoxyglucose and positron emission tomography: Review and 1999 EORTC recommendations. Eur. J. Cancer, 35: 1773–1782, **1999**.

57 Brock, C.S., Young, H., O'Reilly, S.M., Matthews, J., Osman, S., Evans, H., Newlands, E.S., and Price, P. Early evaluation of tumour metabolic response using [^{18}F]fluorodeoxyglucose and positron emission tomography: a pilot study following the phase II chemotherapy schedule for temozolomide in recurrent high-grade gliomas. Br. J. Cancer, 82: 608–615, **2000**.

58 Wahl, R.L., Zasadny, K., Helvie, M., Hutchins, G.D., Weber, B., and Cody, R. Metabolic monitoring of breast cancer chemohormonotherapy using positron emission tomography: initial evaluation. J. Clin. Oncol., 11: 2101–2111, **1993**.

59 Spaepen, K., Stroobants, S., Dupont, P., Van Steenweghen, S., Thomas, J., Vandenberghe, P., Vanuytsel, L., Bormans, G., Balzarini, J., De Wolf-Peeters, C., Mortelmans, L., and Verhoef, G. Prognostic value of positron emission tomography (PET) with fluorine-18 fluorodeoxyglucose ([^{18}F]FDG) after first-line chemotherapy in non-Hodgkin's lymphoma: is [^{18}F]FDG-PET a valid alternative to conventional diagnostic methods? J. Clin. Oncol., 19: 414–419, **2001**.

60 van Oosterom, A.T., Judson, I., Verweij, J., Stroobants, S., Donato di Paola, E., Dimitrijevic, S., Martens, M., Webb, A., Sciot, R., Van Glabbeke, M., Silberman, S., and Nielsen, O.S. Safety and efficacy of imatinib (STI571) in metastatic gastrointestinal stromal tumours: a phase I study. Lancet, 358: 1421–1423, **2001**.

61 Joensuu, H., Roberts, P.J., Sarlomo-Rikala, M., Andersson, L.C., Tervahartiala, P., Tuveson, D., Silberman, S., Capdeville, R., Dimitrijevic, S., Druker, B., and Demetri, G.D. Effect

of the tyrosine kinase inhibitor STI571 in a patient with a metastatic gastrointestinal stromal tumor. N. Engl. J. Med., 344: 1052–1056, **2001**.

62 Grierson, J. R., Schwartz, J. L., Muzi, M., Jordan, R., and Krohn, K. A. Metabolism of 3'-deoxy-3'[F-18]fluorothymidine in proliferating A549 cells: validation for positron emission tomography. Nucl. Med. Biol., 31: 829–837, **2004**.

63 Barthel, H., Perumal, M., Latigo, J., He, Q., Brady, F., Luthra, S. K., Price, P., and Aboagye, E. O. The uptake of 3'-deoxy-3'-[^{18}F]Fluorothymidine into L5178Y tumors in vivo is dependent on thymidine kinase 1 protein and ATP levels. Eur. J. Nucl. Med. Mol. Imaging, 32: 257–263, **2005**.

64 Rasay, J. S., Grierson, J. R., Wiens, L. W., Kolb, P. D., and Schwartz, J. L. Validation of FLT uptake as a measure of thymidine kinase-1 activity in A549 carcinoma cells. J. Nucl. Med., 43: 1210–1217, **2002**.

65 Buck, A. K., Halter, G., Schirrmeister, H., Kotzerke, J., Wurzinger, I., Glatting, G., Mattfeldt, T., Neumaier, B., Reske, S. N., and Hetzel, M. Imaging proliferation in lung tumors with PET: ^{18}F-FLT versus ^{18}F-FDG. J. Nucl. Med., 44: 1426–1431, **2003**.

66 Buck, A. K., Schirrmeister, H., Hetzel, M., von der Heide, M., Halter, G., Glatting, G., Mattfeldt, T., Liewald, F., Reske, S. N., and Neumaier, B. 3-deoxy-3-[(^{18})F]fluorothymidine-positron emission tomography for non-invasive assessment of proliferation in pulmonary nodules. Cancer Res., 62: 3331–3334, **2002**.

67 Toyohara, J., Waki, A., Takamatsu, S., Yonekura, Y., Magata, Y., and Fujibayashi, Y. Basis of FLT as a cell proliferation marker: comparative uptake studies with [^{3}H]arabinothymidine, and cell-analysis in 22 asynchronously growing tumor cell lines. Nucl. Med. Biol., 29: 281–287, **2002**.

68 Oyama, N., Ponde, D. E., Dence, C., Kim, J., Tai, Y. C., and Welch, M. J. Monitoring of therapy in androgen-dependent prostate tumor model by measuring tumor proliferation. J. Nucl. Med., 45: 519–525, **2004**.

69 Sugiyama, M., Sakahara, H., Sato, K., Harada, N., Fukumoto, D., Kakiuchi, T., Hirano, T., Kohno, E., and Tsukada, H. Evaluation of 3'-deoxy-3'-^{18}F-fluorothymidine for monitoring tumor response to radiotherapy and photodynamic therapy in mice. J. Nucl. Med., 45: 1754–1758, **2004**.

70 Barthel, H., Cleij, M. C., Collingridge, D. R., Hutchinson, O. C., Osman, S., He, Q., Luthra, S. K., Brady, F., Price, P. M., and Aboagye, E. O. 3'-deoxy-3'-[^{18}F]Fluorothymidine as a new marker for monitoring tumor response to anti-proliferative therapy in vivo with positron emission tomography. Cancer Res., 3: 3791–3798, **2003**.

71 Shields, A. F., Mankoff, D. A., Link, J. M., Graham, M. M., Eary, J. F., Kozawa, S. M., Zheng, M., Lewellen, B., Lewellen, T. K., Grierson, J. R., and Krohn, K. A. Carbon-11-thymidine and FDG to measure therapy response. J. Nucl. Med., 39: 1757–1762, **1998**.

72 Collingridge, D. R., Glaser, M., Osman, S., Barthel, H., Hutchinson, O. C., Luthra, S. K., Brady, F., Bouchier-Hayes, L., Martin, S. J., Workman, P., Price, P., and Aboagye, E. O. In vitro selectivity, in vivo biodistribution and tumour uptake of annexin V radiolabelled with a positron emitting radioisotope. Br. J. Cancer, 89: 1327–1333, **2003**.

73 Glaser, M., Collingridge, D. R., Aboagye, E. O., Bouchier-Hayes, L., Hutchinson, O. C., Martin, S., Price, P., Brady, F., and Luthra, S. K. Iodine-124 labelled annexin-V as a potential radiotracer to study apoptosis using positron emission tomography. Appl. Radiat. Isot., 58: 55–62, **2003**.

74 Blankenberg, F. G., Tait, J. F., and Strauss, H. W. Apoptotic cell death: its implications for imaging in the next millennium. Eur. J. Nucl. Med., 27: 359–367, **2000**.

75 Blankenberg, F. G. Recent advances in the imaging of programmed cell death. Curr. Pharm. Des., 10: 1457–1467, **2004**.

76 Blankenberg, F. G., Katsikis, P. D., Tait, J. F., Davis, R. E., Naumovski, L., Ohtsuki, K., Kopiwoda, S., Abrams, M. J., Darkes, M., Robbins, R. C., Maecker, H. T., and Strauss, H. W. In vivo detection and imaging of phosphatidylserine expression during programmed cell death. Proc. Natl. Acad. Sci. USA, 95: 6349–6354, **1998**.

77 van de Wiele, C., Lahorte, C., Vermeersch, H., Loose, D., Mervillie, K., Steinmetz, N. D., Vanderheyden, J. L., Cuvelier, C. A., Slegers, G., and Dierck, R. A. Quantitative tumor apoptosis imaging using technetium-99m-HYNIC annexin V single photon emission computed tomography. J. Clin. Oncol., 21: 3483–3487, **2003**.

78 Haas, R. L., de Jong, D., Valdes Olmos, R. A., Hoefnagel, C. A., van den Heuvel I., Zerp, S. F., Bartelink, H., and Verheij, M. In vivo imaging of radiation-induced apoptosis in follicular lymphoma patients. Int. J. Radiat. Oncol. Biol. Phys., 59: 782–787, **2004**.

79 Graeber, T. G., Osmanian, C., Jacks, T., Housman, D. E., Koch, C. J., Lowe, S. W., and Giaccia, A. J. Hypoxia-mediated selection of cells with diminished apoptotic potential in solid tumours. Nature, 379: 88–91, **1996**.

80 Horsman, M. R. and Overgaard, J. Overcoming tumour radiation resistance resulting from acute hypoxia. Eur. J. Cancer, 28A: 2084–2085, **1992**.

81 Hockel, M., Schlenger, K., Vorndran, B., Baussamann, E., Mitze, M., Knapstein, P. G., and Vaupel, P. Intratumoral pO_2 predicts survival in advanced cancer of the uterine cervix. Radiother. Oncol., 26: 45–50, **1993**.

82 Brizel, D. M., Scully, S. P., Harrelson, J. M., Layfield, L. J., Bean, J. M., Prosnitz, L. R., and Dewhirst, M. W. Tumor oxygenation predicts for the likelihood of distant metastases in human soft tissue sarcoma. Cancer Res., 56: 941–943, **1996**.

83 Rasey, J. S., Koh, W. J., Evans, M. L., Peterson, L. M., Lewellen, T. K., Graham, M. M., and Krohn, K. A. Quantifying regional hypoxia in human tumors with positron emission tomography of [18F]fluoromisonidazole: a pretherapy study of 37 patients. Int. J. Radiat. Oncol. Biol. Phys., 36: 417–428, **1996**.

84 Barthel, H., Wilson, H., Collingridge, D. R., Brown, G., Osman, S., Luthra, S. K., Brady, F., Workman, P., Price, P. M., and Aboagye, E. O. In vivo evaluation of [18F]fluoroetanidazole as a new marker for imaging tumor hypoxia with positron emission tomography. Br. J. Cancer, 90: 2232–2242, **2004**.

85 Dehdashti, F., Mintum, M. A., Lewis, J. S., Bradley, J., Govindan, R., Laforest, R., Welch, M. J., and Siegel, B. A. In vivo assessment of tumor hypoxia in lung cancer with 60Cu-ATSM. Eur. J. Nucl. Med. Mol. Imaging, 30: 844–850, **2003**.

86 Koh, W. J., Bergman, K. S., Rasey, J. S., Peterson, L. M., Evans, M. L., Graham, M. M., Grierson, J. R., Lindsley, K. L., Lewellen, T. K., Krohn, K. A., et al. Evaluation of oxygenation status during fractionated radiotherapy in human non-small cell lung cancers using [F-18]fluoromisonidazole positron emission tomography. Int. J. Radiat. Oncol. Biol. Phys., 33: 391–398, **1995**.

6
Ligand-based Targeting of Disease:
From Antibodies to Small Organic (Synthetic) Ligands

Michela Silacci and Dario Neri

Abstract

The targeted delivery of molecules to sites of disease *in vivo* promises to open new avenues for the imaging of pathologies, and for the development of more selective therapeutic agents. This chapter will review progress made in the identification of pathology-associated antigens and in the development of binding molecules (antibodies, peptides and small organic molecules). Furthermore, we will present the authors' views on molecular strategies for the conversion of binding molecules into novel imaging or therapeutic biopharmaceutical agents.

Abbreviations

ADCC	antibody-dependent cellular cytotoxicity
ADEPT	Antibody-Directed Enzyme Prodrug Therapy
aFGF	acid fibroblast growth factor
APβ42	amyloid-β-42 peptide
APP	amyloid precursor protein
ARMD	age-related macular degeneration
BBB	blood–brain barrier
bFGF	basic fibroblast growth factor
CEA	carcinoembryonic antigen
CLIO	cross-linked iron oxide
DTPA	diethylenetriamine penta-acetate
ECAM	endothelial cell adhesion molecules
ECM	extracellular matrix
ECs	endothelial cells
EDB	domain of the extracellular matrix protein fibronectin
ESACHEL	Encoded Self-Assembling Chemical Libraries
FMT	fluorescence-mediated tomography
FR-β	folate receptor β
ICAM	intercellular adhesion molecule
Gd-DTPA	gadolinium-DTPA
HIF-1	hypoxia-inducible factor
MMP	matrix metalloproteinases
MRI	magnetic resonance imaging
NIRF	near-infrared fluorescence
PD-ECGF	platelet-derived endothelial cell growth factor
PET	positron emission tomography
PSMA	prostate-specific membrane antigen
SAP	serum amyloid protein
SPECT	single-photon emission computed tomography
TGFα	transforming growth factor α
TGFβ	transforming growth factor β
tPA	tissue plasminogen activator

Modern Biopharmaceuticals. Edited by J. Knäblein
Copyright © 2005 WILEY-VCH Verlag GmbH & Co. KGaA, Weinheim
ISBN: 3-527-31184-X

uPA urokinase-plasminogen activator

VCAM vascular cell adhesion molecule

VEGF vascular endothelial growth factor

6.1
Introduction

Chemotherapy – that is, the administration of chemical compounds in order to confer a therapeutic benefit to the patient – is often limited by the doses of drug which can be reached, without observing limiting toxicities. For example, in oncology, many therapeutic strategies rely on the expectation that anti-cancer drugs will preferentially kill rapidly dividing tumor cells, rather than normal cells. Since a large proportion of tumor cells must be killed in order to obtain and maintain a complete remission, large doses of drugs are typically used, with significant toxicity towards proliferating non-malignant cells.

It is therefore not surprising that the search for improved potency and selectivity of therapeutic compounds is a common feature in most pharmaceutical development activities. In principle, several strategies could be considered in order to develop better, more selective therapeutic agents. In many cases, research is driven by the hope to identify macromolecular targets, which are not essential in normal physiology, but the inhibition of which may revert the pathological condition that one intends to fight. While such prerequisites may be met in certain therapeutic areas (e.g., the use of antibiotics inhibiting microbial protein targets which do not have a counterpart in the host), the discovery of selective targets remains a formidable challenge for many relevant pathologies.

The selective delivery of bioactive compounds to a site of disease (the "magic bullets" first envisioned by Paul Ehrlich at the end of the nineteenth century) appears to be a general strategy for the development of better, more selective therapeutic agents. In most cases, the selective accumulation of drugs at the site of disease will spare normal tissues and will increase the therapeutic index of the drug – that is, the relative activity towards the diseased tissue, compared to normal organs. In principle, targeted strategies based on the selective delivery of active compounds could be applicable both in diseases in which cell growth has to be limited (e.g., cancer) or promoted (i.e., tissues regeneration after infarction).

The words "targeting" and "targeted therapy" are often used for a variety of different pharmaceutical approaches, aimed at achieving better *in vivo* selectivities. In this chapter, however, we will concentrate solely on those targeting strategies, which rely on the ligand-based selective delivery of bioactive agents to sites of disease. Those readers interested in other targeting strategies, which achieve a selective biodistribution *in vivo* in the absence of a specific molecular recognition event and by means of other physical principles (e.g., the enhanced permeability and retention of polymers in tumors [1, 2]), are encouraged to consult other reviews which have been written on this topic [3–5].

The ligand-based targeting of diseases is a rational strategy for drug discovery. To some extent, in fact, the performance of a targeted drug can be predicted on the basis of how selective is its localization on the target tissue. Furthermore, a binding molecule capable of disease targeting may be useful not only for therapeutic applications, but also for imaging purposes, after modification with a suitable radionuclide or infrared fluorophore. A number of parameters are expected to influence the *in vivo* performance of a targeting agent. Molecular

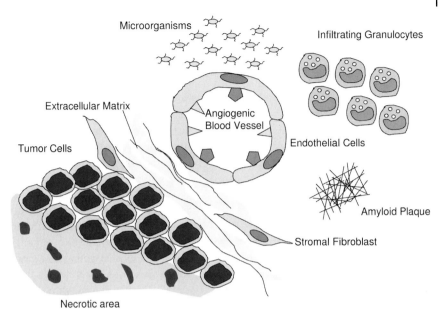

Fig. 6.1 Schematic representation of tissue and cellular components, which can be considered for targeting applications.

weight, binding affinity for the target, solubility, valence are some of the parameters which, in many cases, have been shown to contribute to the overall performance of the targeting process.

In this chapter, we will first review the classes of molecules that are currently used for targeting applications. We will then analyze classes of diseases which lend themselves to molecular targeting, and finally will discuss how a ligand can be converted into a diagnostic or therapeutic agent.

Different pathologies may require the ligand-based targeting of different antigens, located on different structures. A schematic representation of tissues and cellular components which can be considered for targeting applications is provided in Fig. 6.1. Whilst in oncology most targeting approaches focus on the tumor cells, other structures such as altered vascular structures, modified extracellular matrix, infiltrating leukocytes, areas of necrosis, pla-

ques and microbes can also be used as targets in oncology and in other diseases.

6.2
Ligands

6.2.1
Antibodies

At present, antibodies are the only general class of affinity reagents which can be generated rapidly against virtually any biomolecular target. Monoclonal antibodies represent an ideal alternative to hyperimmune sera for *in vivo* applications [6]. However, rodent antibodies are immunogenic in humans. Early studies showed that human monoclonal antibodies can be produced by immortalizing B cells with Epstein–Barr virus (EBV) [7, 8], or by fusing B cells with an appropriate partner to produce hybridomas [9, 10]. However, these methods have very low efficiency, and therefore alternative strategies have been developed. These include: 1) humanization of murine monoclonal antibodies through protein engineering [11] (see also Part V, Chapters 1 and 2) selection of antibodies from phage-display libraries of human antibody fragments [12, 13] origin (see also Part V, Chapters 2 and 3) immunization of transgenic mice carrying human immunoglobulin loci, followed by production of monoclonal antibodies using hybridoma technology [14].

Monoclonal antibodies exhibit a slow elimination from the blood, and accumulate in the liver. For these reasons, rapidly clearing antibody fragments are typically preferred for imaging applications in nuclear medicine. By contrast, intact immunoglobulins continue to represent the antibody format of choice for many therapeutic applications [15], which rely on the antibody's ability to interfere with signaling events, and to activate antibody-dependent cellular cytotoxicity (ADCC) mechanisms or complement.

The immunogenicity of rodent antibodies continues to be a concern for repeated administrations to humans, and the use of chimeric, humanized or fully human antibodies is generally preferred.

In our experience, antibody phage technology represents the most efficient avenue for producing good-quality human monoclonal antibodies, whenever sufficient quantities of pure antigen are available (1–2 mg). The display of antibody fragments on the surface of filamentous phage allows the facile construction of large ($>10^9$ antibodies) libraries of human antibodies, from which monoclonal antibodies can be isolated by panning the phage library onto an immobilized antigen [13, 16]. When required, antibody affinity can be "matured" using combinatorial mutagenesis of the antibody gene and stringent selection strategies [17, 18]. Recently, ribosome display has been pro-

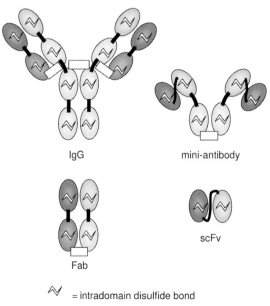

IgG

mini-antibody

Fab

scFv

∿ = intradomain disulfide bond

▭ = interdomain disulfide bond(s)

Fig. 6.2 Different antibody formats and antibody fragments.

posed as a fully *in vitro* avenue for the isolation and affinity maturation of human antibodies [19].

Antibody phage technology directly yields antibody fragments (typically in scFv or Fab format). However, other antibody formats (e.g., IgG) can easily be obtained by transplanting the genes coding for the variable antibody domains into suitable expression vectors (Fig. 6.2).

6.2.2
Peptides

A number of internalizing peptides, specific to receptors which are overexpressed in tumor cells, have been used for the imaging of tumors and for the selective delivery of therapeutic radionuclides to neoplastic lesions. The somatostatin analogue octreotide [20], for example, has been approved in Europe or the USA for the imaging of tumors. Several other agents are in development [21, 22], such as integrin binding peptides (RGD-peptides) [23] and bombesin peptide analogues [24]. Other areas in which naturally occurring peptides (or peptides derived from protein sequences) are used include the ligand-based targeting of thrombotic events, of microbial infections and of amyloidosis (see sections below).

In contrast to naturally occurring peptides, high-affinity peptidic ligands to protein targets are often difficult to isolate. Phage display libraries of linear and disulfide-constrained peptides are commercially available, and have been used for the isolation of binding specificities [25]. For example, peptides specific to human lung tumor cell lines have been selected from a phage library [26]. Novel technologies for the isolation of high-affinity binding peptides are available [27], but the *in vivo* stability of linear peptides remains a cause for concern. Peptide phage libraries have been used for

in vivo panning applications by the groups of Pasqualini and Ruoslahti [28, 29], but the real imaging and therapeutic potential of these phage-derived peptides remains to be investigated in advanced animal models, as well as in the clinic.

6.2.3
Small Organic Molecules

In contrast to antibody technology, the isolation of high-affinity small organic binders to protein antigens can be a difficult task, which often fails when the epitopes to be recognized do not contain hydrophobic pockets [30]. An increasing number of experimental evidences suggest that bidentate ligands, recognizing adjacent but not-overlapping surfaces of the target protein, may display high binding affinity and specificity, as a result of the chelate effect [31]. Methods for the identification of such bidentate ligands include SAR-by-NMR [32], dynamic combinatorial chemistry [33], and tethering approaches [34]. Our laboratory has recently developed a novel technology (termed "Encoded Self-Assembling Chemical Libraries", or ESACHEL), which allows the facile construction of very large libraries of chemical compounds by the DNA-mediated self-assembly of smaller sub-libraries [35]. Each pharmacophore in the library is covalently coupled to an oligonucleotide, which mediates the self-assembly of the library and provides the pharmacophore with a distinctive identification DNA tag. Similar to antibody phage display libraries, ESACHEL libraries can be panned in solution, thus enriching bidentate ligands which display a preferential binding to the target of interest.

After the capture of the desired binding specificities on the protein target, the "binding code" associated with the selected pharmacophores can be "decoded" by a number of experimental techniques (e.g., hybridization on DNA chips, by a modified PCR technique followed by sequencing). We have described the isolation of ESACHEL-derived bidentate molecules, with nanomolar affinity to carbonic anhydrase [35].

6.3
Classes of Diseases

6.3.1
Cancer

Cancer chemotherapy can be successful in certain specific indications, but suffers otherwise from major drawbacks. The lack of selectivity of anti-proliferative agents may give rise to severe side effects, thus limiting efficacy and facilitating the development of acquired drug resistance. The discovery of more selective anticancer drugs, with better discrimination between tumor and normal cells, is possibly the most important goal of modern anticancer research. The targeted delivery of bioactive moieties (drugs, cytokines, procoagulant factors, photosensitizers, radionuclides, etc.) by means of binding molecules (recombinant antibodies, peptides, etc.) specific to tumor-associated markers can improve the efficiency of tumor therapy and limit non-specific toxicity.

6.3.1.1 **Tumor-associated Markers**
The selection of a suitable molecular target is an essential step in the design of any ligand-based therapeutic. Tumor-associated markers are usually proteins or carbohydrates that are abnormally expressed or overexpressed in the tumor environment. Fundamental prerequisites of an ideal tumor-associated marker are specificity, abundance and stability, together with good accessibility for ligand molecules

transported by the blood stream. To date, only few good-quality tumor-associated markers are known (see also Part I, Chapter 5). Most existing candidate markers are also present in normal tissues, thus limiting their usefulness for *in vivo* targeting applications. Several methods, such as proteomic and transcriptomic techniques, biopanning of phage display libraries and serial analysis of gene expression, are now available and may help to identify new tumor-associated markers. However, the validation of the newly identified markers requires the generation of specific monoclonal antibodies, extensive immunohistochemical analysis and biodistribution experiments in tumor-bearing animals.

Tumor-associated markers (antigens) can be grouped into two main categories according to their localization in the tumor tissue: 1) antigens on the surface of tumor cells (tumor markers); or 2) stromal antigens, which can be located either around the tumor neovasculature or display a more diffuse staining pattern corresponding to the modified extracellular matrix (ECM) of solid tumors.

6.3.1.2 **Targeting Markers on Tumor Cells**
In this section we will discuss the most extensively studied tumor markers and the corresponding monoclonal antibodies that have been approved by the FDA (US Food and Drug Administration; www.fda.gov) and in Europe for the imaging and therapy of cancer, or that are currently being developed.

Tumor-associated glycoprotein 72 (TAG-72) This tumor marker [36] was identified by means of a murine monoclonal antibody, B72.3, raised against human, metastatic mammary carcinoma cells [37]. The expression pattern of TAG-72 was extensively analyzed in a number of different tumors, such as ovarian carcinoma, pancreatic adenocarcinoma, and colorectal adenocarcinoma [38–40]. Immunohistochemical studies showed that TAG-72 is expressed in more than 80% of colorectal carcinomas, but is rarely expressed in normal epithelium and benign diseases. TAG-72 can also be found in the body fluids of patients with adenocarcinomas, and its direct measurement can be used in conjunction with immunocytochemical analysis to help in discriminating benign from malignant effusions [38]. The murine monoclonal antibody satumomab pentedite indium-111 conjugate (OncoScint), specific to TAG-72, was the first monoclonal antibody approved by the FDA for tumor imaging (colorectal and ovarian cancer).

Carcinoembryonic antigen (CEA) This was first described in 1965 by Freedman and Gold [41]. CEA, which is a highly glycosylated membrane protein, has a restricted expression in normal tissues and is expressed at high levels in positive tumors (colon carcinoma). CEA became one of the most widely used tumor markers worldwide. Its main application is mostly in gastrointestinal cancer, especially in colorectal malignancy [42]. CEA-Scan, a murine monoclonal antibody fragment (Fab) linked to technetium-99m, was approved in both Europe and the USA in 1996 for the detection of recurrent/metastatic colorectal cancer. Further anti-CEA monoclonal antibody fragments were developed by the group of Begent. The scFv fragment MFE-23 [43], which shows high affinity to CEA, was tested in biodistribution experiments [44] and was genetically fused to several bioactive molecules, such has TNF-α [45] and carboxypeptidase G2 (CPG2) [46] used for Antibody-Directed Enzyme Prodrug Therapy (ADEPT).

Prostate-specific membrane antigen (PSMA) This is a type 2 membrane protein that represents an attractive target for cancer imaging and immunotherapy by virtue of its abundant and restricted expression on the surface of prostate carcinomas, and on the neovasculature of most other solid tumors. PSMA was originally discovered in the androgen-dependent LNCaP human prostatic adenocarcinoma cell line [47]. ProstaScint, approved by the FDA in 1996, is a murine monoclonal antibody imaging agent linked to indium-111 directed against PSMA [48,49] used for the detection, staging and follow-up of prostate adenocarcinoma.

HER2/neu oncogene This marker belongs to a family of human epidermal growth factor receptors (EGFRs) involved in the transmission of signals controlling normal cell growth and differentiation [50, 51]. HER2/neu is known to be overexpressed in many different types of human cancers, including breast, ovarian, lung, gastric, and oral cancers [52] (see also Part I, Chapter 5). The presence on their surface of high amounts of HER2 enhances the responsiveness to growth factors and malignant growth of breast tumor cells. The humanized monoclonal antibody Herceptin, specific for the protein product of HER2 (p185HER2) [53, 54], was approved by the FDA in 1998 and in Europe in 2000 for the treatment of metastatic breast cancer. Binding of Herceptin to the extracellular domain of the receptor results in down-regulation of HER2 by inducing receptor internalization, inhibition of cell-cycle progression and antibody-dependent cellular cytotoxicity by inducing an immune response [55]. Moreover, Herceptin is able to block cleavage of HER2, which would generate a membrane-bound truncated receptor that is constitutively active.

Two other HER2-specific monoclonal antibodies are currently in clinical trials, namely 2C4 – the activity of which (unlike that of Herceptin) is not dependent on HER2 amplification (Genentech) [56]; and Osidem, a bispecific antibody that was developed to target cytotoxic effector cells expressing Fc gamma receptor type I (Fc gammaRI, CD64) to HER2/neu-overexpressing tumor cells (Medarex) [57].

CD20 This is a signature B-cell differentiation antigen. The function of CD20 is unknown, although it is thought to be involved in B-cell activation, regulation of cellular growth, and transmembrane calcium flux [58]. There are two main classes of antibodies directed against the CD20 antigen that have been developed for therapeutic intent: unconjugated and radiolabeled antibodies. Rituxan, an unconjugated chimeric monoclonal antibody, was the first monoclonal antibody approved by the FDA in 1997 for the therapy of cancer, more precisely for the therapy of non-Hodgkin lymphoma (NHL). Zevalin and Bexxar are radiolabeled murine monoclonal antibodies that were approved for the therapy of NHL in 2002 and 2003, respectively [59, 60].

EpCAM This is a 40-kDa epithelial transmembrane glycoprotein expressed on the basolateral surface of simple, pseudostratified, and transitional epithelia. EpCam mediates epithelium-specific, Ca^{2+}-independent homotypic cell–cell adhesions. *In vivo* expression of EpCam is related to increased proliferation of epithelial cells, and correlates negatively with cell differentiation [61]. EpCam was found to be strongly expressed in carcinomas of various origins, including colon and rectum [62], prostate [63], liver [64], esophagus , lung, head and neck, pancreas, and breast [61]. Chimeric

and humanized antibodies have been generated, such as the chimeric antibody 17-1A (edrecolomab; Panorex, Glaxo Wellcome GmbH). Edrecolomab immunotherapy decreased the frequency of distant metastasis in patients with colorectal cancer and eliminated disseminated breast cancer tumor cells in the bone marrow [65]. A recently published study [66] could not show a significant effect of edrecolomab in stage III colon cancer therapy.

The instability and plasticity of tumor genomes represents a major drawback of the targeting approaches based on ligands (e.g., monoclonal antibodies) specific to antigens on the tumor cell membrane. Events such as partial or complete deletion of chromosomes, amplification of genes, translocations or rearrangements of chromosomes, and simple mutations ensure efficient selection and overgrowth of drug-resistant tumor cell during and after therapy. Furthermore, the accessibility of markers on tumor cells is not optimal for agents coming from the blood stream, as a high interstitial pressure and irregular tumor vasculature may hinder the antibody extravasation and tissue penetration. It is likely that the absolute amount of tumor-associated antigen in the neoplastic lesion influences the performance of ligand-based tumor targeting approaches.

6.3.2
Angiogenesis-related Diseases

6.3.2.1 **Angiogenesis and Tumor Angiogenesis**
Angiogenesis is the process through which new blood vessels form from pre-existing ones. It occurs primarily during embryogenesis as an essential process for the development of the vascular network of arteries, veins, arterioles, venules and capillary blood vessels that nourish and protect

the body's tissues [67]. Once the vascular network is in place in the adult, the endothelial cells (ECs) lining the blood vessels are quiescent, and angiogenesis is normally triggered only locally and transiently during some physiological processes such as the female reproductive cycle, hair growth, wound healing and inflammation [67]. Angiogenesis is a tightly controlled, multistep process in which pro-angiogenic and anti-angiogenic factors are in equilibrium to neutralize one another. Imbalance of this equilibrium, either by the up-regulation of pro-angiogenic or down-regulation of anti-angiogenic mediators, induces angiogenesis. Angiogenesis is an important feature of a range of different pathological conditions, cancer being one of the most prominent examples [68]. The growth of new capillaries is often triggered in conditions of cellular proliferation, ischemia or chronic inflammation, where an increase in blood supply may compensate for hypoxia and insufficient delivery of nutrient to the tissue [69, 70]. Unlike the situation in physiological conditions, blood vessels grow unabated in cancer and other pathologies, and tumor angiogenesis sustains the progression of the disease.

During angiogenesis, endothelial cells detach from the pre-existing destabilized vessel, migrate into the perivascular space, and proliferate to finally mature and form new vascular structures. A number of growth factors, proteases, adhesion molecules and other angiogenic mediators which enable endothelial cell migration or proliferation regulate this process. Vascular endothelial growth factor (VEGF) is considered to be one of the most important growth factors in angiogenesis [71]. It increases the permeability of existing blood vessels and acts as endothelial cell survival factor, as well as being a potent endothelial cell mitogen. The neutralizing humanized monoclonal anti-VEGF antibody Avastin has recently been approved for the treatment of colorectal cancer [72], but showed no survival benefit in patients with breast cancer [73].

Most of the current knowledge about angiogenesis stems from investigations on tumoral angiogenesis. A large number of molecules involved in angiogenesis have been first identified in tumors, and later confirmed in other pathological conditions. Many tumors in humans persist in situ without being accompanied by angiogenesis [74, 75]. At that stage they tend to be clinically undetectable and are rarely larger than 1–2 mm in diameter, because diffusion of oxygen and nutrients limit their size. The high rate of proliferation in these tumors is compensated by abundant internal apoptosis as a consequence of insufficient blood supply.

As the tumor adopts an angiogenic phenotype, the balance between pro- and anti-angiogenic factors is upset and angiogenesis is triggered. The tumor mass is allowed to overtake the apoptotic rate, and consequently expands. This process is referred to as "angiogenic switch" [68, 76]. Not only is angiogenesis required for tumors to grow beyond a certain size, but it also enables tumor cells to migrate into surrounding tissue and to colonize distant sites, forming metastases. Metastases again can only grow to threatening size if the metastatic cells are able to trigger angiogenesis [68].

Although the mechanisms eliciting the angiogenic switch are not entirely understood to date, it is believed that besides tumor-suppressor mutation and oncogene activation, hypoxia plays a pivotal role [77]. There are at least two hypoxia-dependent regulatory mechanisms which lead to VEGF expression. The first mechanism relies on the transcription factor hypoxia-inducible factor (HIF-1) which controls VEGF transcription [78]. The alpha subunit of HIF-1, HIF-1α, is degraded under normoxic conditions and stabilized under hypoxia [69, 79–81]. Second, VEGF mRNA becomes stabilized under hypoxic conditions [82]. VEGF concentrations stimulate the proliferation of endothelial cells, which in turn produce many unspecific angiogenic stimulators, including basic fibroblast growth factor (bFGF), acid fibroblast growth factor (aFGF), transforming growth factor α and β (TGFα and TGFβ) or platelet-derived endothelial cell growth factor (PD-ECGF). Additionally, tumor cells produce proteases, among which are matrix metalloproteinases (MMP) and serine proteases like urokinase-plasminogen activator (uPA) or tissue plasminogen activator (tPA). Endothelial cells display cell adhesion molecules such as integrins $\alpha_v\beta_3$ and $\alpha_v\beta_5$ which mediate interaction with the ECM. Laminin, type IV collagen and tenascin are synthesized to constitute the new basement membrane.

Reduced oxygen tension promotes angiogenesis not only by stimulating the production of inducers, but also by reducing the production of inhibitors.

Thrombospondin-1 was the first angiostatic protein for which anoxia-triggered down-regulation during tumorigenesis was demonstrated [83]. A number of endogenous angiogenesis inhibitors have since been identified.

The tumor vessels may be distinguished from their normal counterparts: architecturally, they are irregularly shaped, dilated, tortuous and even contain dead ends [84]. Extensive fenestration, an abnormal basement membrane and unusual wide gaps between adjacent endothelial cells make them leaky [85–87].

The treatment of cancer with an anti-angiogenic approach was first proposed more

than two decades ago [74]. Accordingly, various anti-angiogenic strategies have been investigated preclinically. This extensive research has culminated in the recent approval of Bevacizumab (Avastin, Genentech) as first-line treatment of metastatic colon carcinoma [72, 88]. Furthermore, several lines of evidence suggest that (at least in part) the action of chemotherapeutic agents against solid tumors may be related to the preferential killing of the tumor endothelium, rather than the endothelium of normal tissues [89].

6.3.2.2 **Non-tumor Angiogenesis**

A number of non-cancer disorders are strongly associated with the overexuberant proliferation of new blood vessels, and may benefit from anti-angiogenesis treatments. It is generally accepted that several potentially blinding ocular disorders (e.g., the exsudative form of age-related macular degeneration (ARMD), diabetic retinopathy, retinopathy of prematurity, rubeosis iridis, etc.) and chronic inflammatory conditions (e.g., rheumatoid arthritis, psoriasis) fall into the category of angiogenesis-related diseases [68, 70]. Consequently, the identification of markers for angiogenesis, and the validation of high-affinity ligands to such markers, is expected to lead to benefits both for the diagnosis and the therapy of these diseases. In a number of cases, anti-angiogenic treatments based on vascular targeting approaches (e.g., the use of laser irradiation and Visudine in ARMD) are already widely diffused in the clinical practice, even if their "targeting" component is limited. We expect to see more targeting approaches in the near future, for the treatment of non-cancer angiogenesis-related diseases.

6.3.2.3 Markers of Angiogenesis and Stromal Antigens

Several antigens have been proposed as putative markers of angiogenesis, but only a few have been extensively characterized by immunohistochemistry, by *in vivo* biodistribution analysis and by scintigraphic procedures in patients with cancer or other diseases.

The antigens which have been characterized more extensively are possibly the EDB domain of fibronectin [90, 91], the large isoforms of tenascin-C [92, 93], PSMA and the $a_v\beta_3$ integrin. Indeed, for all these antigens, extensive immunohistochemical studies have been reported in the literature, and monoclonal antibodies are currently undergoing clinical trials. We have described these antigens in detail in other reviews, and refer the interested readers to those articles [94–96]. Other markers of angiogenesis, which have displayed promising results but which are at later stages of development, include Endoglin (CD105) [97, 98], VEGF and VEGF-receptor complex [99, 100], CD44 [101], phosphatidyl serine phospholipids [102], magic roundabout (ROBO-4) [103], Aminopeptidase N [104] and Annexin A1 [100].

It may be worthwhile mentioning that recent technological breakthroughs may facilitate the discovery of markers of angiogenesis. For example, the group of Kinzler and Vogelstein have reported a transcriptomic analysis of endothelial cells purified from colorectal cancer and from normal tissues [105]. Recent experimental approaches based on terminal perfusion of tumor-bearing rodents have allowed, for the first time, a direct proteomic analysis of accessible antigens in vascular structures. The group of Schnitzer has reported the use of terminal perfusion protocols with silica beads for the identification of tumor endothelial markers. In their work,

Annexin A1 emerged as a promising antigen for the radiolabeled antibody-based imaging and therapy of cancer [100]. In our laboratory, we use terminal perfusion protocols featuring active esters of biotin for the selective chemical labeling of accessible proteins in vascular structures. Biotinylated proteins are then purified from different organs (collected separately) and are submitted to a comparative proteomic analysis [106].

6.3.3 Cardiovascular Diseases

Cardiovascular diseases are currently the leading cause of death and illness in developed countries.

6.3.3.1 Atherosclerosis

Complications of atherosclerosis are the leading cause of morbidity and mortality in developed countries. It can be considered as a chronic inflammation resulting from interaction between modified lipoproteins, monocyte-derived macrophages, T cells, and the normal cellular elements of the arterial wall [107–109]. The earliest lesion is a pure inflammatory lesion consisting of monocyte-derived macrophages and T cells. The presence of monocytes in every phase of atherosclerosis, and of hydrolytic enzymes secreted by these and other cells, play a central role in different stages of the disease, particularly in the resorption of the fibrous cap leading to plaque rupture.

None of the current imaging techniques (typically monitoring luminal diameter, volume, and thickness of the plaque, etc.) is capable of characterizing biological plaque activity to identify high-risk patients. Therefore, a considerable research effort concentrated in the development of techniques that allows the high-resolution detection of high-risk (or "active") atherosclerotic lesions.

Imaging of protease activity Proteolytic enzymes are produced at sites of atherosclerotic lesions by biologically active macrophages and endothelial cells. These enzymes, such as cathepsin B [110, 111] and MMPs [108, 112], seem to be involved in the degradation of the fibrous cap that can lead to the rupture of the atherosclerotic plaque. Recently, Chen et al. [113] described the imaging of cathepsin B activity *in vivo* by fluorescence-mediated tomography (FMT). This was achieved by using an autoquenched cathepsin B sensitive near-infrared fluorescence (NIRF) probe [114] that can generate a strong NIRF signal after enzyme activation (e.g., by cathepsin B activity). These activatable probes consist of NIR fluorochromes linked to a delivery vehicle via specific peptide sequences that serve as a substrate for the protease of interest [115]. The authors hypothesized that imaging of cathepsin B activity in atherosclerotic plaques may serve as a new method to measure plaque inflammation and vulnerability. Since MMP-2 has been suggested to be a specific mediator of fibrous cap destabilization [112], studies are currently being performed to image MMP-2 activity *in vivo* in an atherosclerosis animal model [109]. To this purpose, the MMP-2 peptide substrate was used to create an autoquenced NIRF probe. The feasibility of such an approach was proven by Bremer et al. in 2001, who demonstrated the *in vivo* imaging of MMP-2 expression using a MMP-2 sensitive NIRF probe [116].

Imaging of activated macrophages Given their ubiquitous presence in every stage of the atherosclerosis disease, activated macrophages are being recognized as an important target for atherosclerosis treatment and imaging. Detection of activated macrophages in atherosclerotic lesions was achieved by high-resolution MRI using superparamagnetic iron oxide nanoparticles [117]. Macrophages appear to phagocytose nanoparticles, and the resulting iron oxide accumulation generates strong T2 relaxation and MRI contrast.

Imaging of activated endothelial cells Endothelial cell adhesion molecules (ECAM) are expressed at high levels on the plaque surface, angiogenic vessels within the plaque, and adventitial vessels, with low expression levels in normal vessels [118]. ICAM-1, VCAM-1, P-selectin and integrins, such as $a_v\beta_3$, have been associated with advanced atherosclerosis. Several ultrasound contrast agents targeted to ECAMs were developed in the recent past. Acoustically active liposomes, conjugated with monoclonal antibodies specific to ICAM-1, were shown preferentially to accumulate in the endothelium overlying atherosclerotic lesions [119]. Lipid microbubbles conjugated to antibodies against P-selectin could be used to image early inflammatory responses [120]. Angiogenesis-targeted microbubbles were created by conjugating antibodies or peptides binding to a_v-integrins [121]. Recently, Joseph et al. described the creation of an air-filled microparticle conjugated to a L19 antibody derivative, specific to the EDB domain of fibronectin [122]. These microparticles are aimed at the ultrasound *in vivo* imaging of angiogenesis, but as yet no *in vivo* results have been reported.

6.3.3.2 Thrombosis

Thrombosis – that is, the formation of a solid mass of blood products in a vessel – is the pathological hallmark of a number of cardiovascular diseases (myocardial infarction, stroke, etc.). The imaging of molecules important for thrombogenesis could provide a highly specific diagnostic thrombosis imaging method.

Imaging of platelet deposits Activated platelets are usually found on the surface of the thrombus. Ligands binding to receptors found exclusively on the surface of activated platelets represent a highly specific approach to detect platelet deposits. The $a_{IIb}\beta_3$ integrin on platelets is the most commonly targeted receptor for the detection of platelet deposits. Several different peptide ligands to $a_{IIb}\beta_3$ have been developed and tested for the imaging of thrombosis [123]. A linear peptide based on the amino acid sequence of the binding domain of a monoclonal antibody directed against $a_{IIb}\beta_3$, PAC-1 [124] was synthesized and used as a base to develop peptides for imaging thrombi [125]. A second possible approach to image thrombi by means of peptides is to use cyclic peptides based on the simplest known integrin binding sequence, RGD. The cyclic peptide P280 (Apcitide) [126] is an approved thrombus-imaging radiopharmaceutical. The third class of thrombus-imaging peptides are natural polypeptides (disintegrins), with high affinity for the receptor. Disintegrins are usually composed of 48–84 amino acids, and are rich in cysteine residues. The formation of disulfide bridges confer to the polypeptides a defined structure where the RGD motif is exposed at the tip of a flexible loop. Radioactively labeled Bitistatin [127] produced images of intense uptake at the thrombus site which corresponded to the true dimension of the lesion.

Imaging of enzyme activities Thrombin, a serine protease, plays an important role in thrombogenesis, cleaving fibrinogen to form fibrin monomers, which subsequently polymerize to form fibrin, the scaffolding of thrombus [115] (see also Part II, Chapter 1). Jaffer et al. synthesized a NIRF probe that consisted of a human thrombin-cleavable peptide that contained an N-terminal NIR fluorochrome. The probe successfully detected thrombi in animal models [115]. Activated factor XIII is a tissue transglutaminase that cross-links fibrin chains and plasmin inhibitors to form mechanically and proteolytically stable thrombi [109]. Factor XIII activity in thrombi has been successfully imaged by both NIRF [115, 128] and MRI [129]. For MRI, the F13-CLIO agent consisted of a dextran-coated caged iron-oxide particle (CLIO) conjugated to an $a2$-antiplasmin peptide that can be cross-linked by factor XIII [129].

Imaging of fibrin As mentioned in the previous section, fibrin represents the scaffolding of the thrombi. It is a favorable thrombosis molecular imaging target, because it is usually present in all types of thrombi and its plasma concentration is low, thus minimizing the background signal [109]. Several fibrin-targeted molecular imaging agents have been developed for nuclear imaging [123], ultrasound imaging [119, 130] and for high-resolution MRI [131, 132]. Flacke et al. developed a novel fibrin-specific MR contrast agent consisting of a lipid-encapsulated liquid perfluorocarbon nanoparticle coated with an anti-fibrin Fab antibody fragment, which can carry high gadolinium-DTPA (Gd-DTPA) payloads for high detection sensitivity [131]. The authors demonstrated the selective accumulation of the nanoparticles in microthrombi overlying the atherosclerotic intima. Moreover, a radiolabeled 12 kDa fragment of fibronectin (Fibrin-binding domain) was shown by Taillefer selectively to bind to thrombi *in vivo* [133].

6.3.3.3 Heart Failure

Although heart failure is one of the most common human diseases in developed countries, a limited understanding of the underlying mechanisms leads to a lack of effective treatments. The origins of the disease can be very diverse, ranging from hypertension to viral infections to coronary occlusion, but end-stage heart failure shares many common pathologic features, such as loss in myocyte viability, interstitial remodeling, changes in gene expression and contractile dysfunction [134].

Imaging of myocardial apoptosis As mentioned above, cardiomyocyte apoptosis is a pathologic feature of heart failure. A characteristic feature of apoptosis is the externalization of phosphatidylserine phospholipids [135] – a lipid that, under normal conditions, is present in only the inner layer of the cell membrane. Radiolabeled annexin V, an intracellular phospholipid-binding protein, was used clinically in patients with acute myocardial infarction for the non-invasive imaging of apoptosis [136]. Moreover, the use of NIR fluorochrome-tagged annexin V has also been described [137].

6.3.4
Inflammation

6.3.4.1 Rheumatoid Arthritis

Rheumatoid arthritis is an autoimmune disease that affects multiple synovial joints and involves inflammation of the synovial membrane, often resulting in a loss of function due to the erosion of bone and cartilage [138].

Targeting activated macrophages Activated macrophages are known to constitute the key effector cells in rheumatoid arthritis [139]. A clear correlation between the levels of macrophage activity, joint inflammation, articular pain and bone erosion was defined. This correlation can be explained by the fact that activated macrophages secrete potent mediators of inflammation and tissue destruction. Moreover, macrophages participate in antigen presentation and therefore contribute to the activation and proliferation of antigen-specific T cells [140]. One possibility of reducing the destructive effects of rheumatoid arthritis might be the elimination of the cell population that is mainly responsible for the inflammation – that is, the activated macrophages. The folate receptor β (FR-β), a glycosylphosphatidylinositol-anchored protein that binds folic acid with high affinity, was shown to be expressed on monocytic and myelocytic lineages of hematopoietic cells in a functional inactive form unable to bind folic acid [141]. Interestingly, it was recently shown that activated synovial macrophages possess a functionally active FR-β [142]. EC20, a folate-conjugated radiopharmaceutical complex with ^{99m}Tc was shown to accumulate in arthritic extremities of diseased rat, but not in the extremities of healthy animals [143]. Fletcher et al. showed that a folate-targeted immunotherapy reduced the symptoms of rheumatoid arthritis in a similar way as methotrexate [144]. Animals that were previously immunized with haptens were injected with folate–hapten conjugates; in this way, the hapten-decorated active macrophages were killed by the mechanism of antibody-dependent cell cytotoxicity. A similar approach – the elimination of active macrophage – was undertaken by van Roon et al., who demonstrated the *in vitro* selective killing of activated macrophages, isolated from rheumatoid arthritis patients, by means of an Fcβ Receptor I-directed immunotoxin [145].

It is worth mentioning here that Wunder et al. [146] were able to image inflam-

mation in arthritic joints by means of a cathepsin B-specific autoquenced NIRF
probe. Cathepsin B activated NIRF probes
were shown to serve as reporters for the
imaging of treatment response to anti-
rheumatic drugs (e.g., methotrexate).

6.3.4.2 Other Inflammatory Diseases

The targeting of molecular markers associated with inflammation could, in principle, be useful for the imaging and therapy
of several other inflammatory diseases. Besides rheumatoid arthritis, these include
inflammatory bowel diseases (Crohn's disease and ulcerative colitis), psoriasis,
atherosclerosis, and diseases of the central
nervous system (Alzheimer's disease, multiple sclerosis, etc.).

The treatment of diseases such as psoriasis and inflammatory bowel could in particular take advantage of these targeting
strategies, as angiogenesis appears to play
an important role in the pathology of these
conditions [70].

Psoriasis, a chronic inflammatory skin
disease that affects approximately 1–3% of
the western population [109], is characterized by hyperproliferation of keratinocytes,
infiltration of inflammatory cells, and increased cytokine levels. Psoriasis is accompanied by an expansion of the superficial
dermal microvasculature and elongation of
capillary loops passing into dermal papil-
lae and the papillary tip [147].

Inflammatory bowel diseases (IBD) are
chronic inflammatory conditions which affect the gastrointestinal tract. They are characterized by a localized or diffuse granulo-
matous inflammatory process, accompanied by systemic manifestations. As ulcera-
tion and regeneration of the intestinal
epithelium occurs during the course of the
disease, angiogenesis is undoubtedly an integral part of the IBD pathology [148].

It is likely that the ligand-based delivery
of anti-inflammatory drugs or anti-inflammatory cytokines will improve the efficacy
of therapies of psoriasis and inflammatory
bowel diseases.

Today, several animal models for psoriasis and inflammatory bowel diseases are
available [149–151], and these should allow
investigations to be made of the benefits
of targeted therapy approaches in these
diseases.

6.3.5
Infection

Infections, which result from the invasion
of microorganisms, are usually diagnosed
on the basis of clinical history, physical examination, laboratory tests and the identification of pathogens in body fluids and
biopsy samples. The discrimination between infection and inflammation at an
early stage of the disease is considered to
be critical for a favorable outcome. Nuclear
medicine could contribute to the non-invasive detection of infections, provided that
specific tracers are available which can discriminate between infections and sterile
inflammations. To this purpose, two different groups of tracers are currently under
development. The first group consists of
radiolabeled antibiotics and antifungal
compounds, while the second group consists of radiolabeled peptides derived from
antimicrobial peptides/proteins.

Vinjamuri et al. showed that radiolabeled ciprofloxacin (Infecton), a fluoroqui-
nolone antimicrobial agent that binds to
the DNA gyrase in all dividing bacteria,
could successfully discriminate between
bacterial infections and sterile inflammations [152]. A further example of this class
of tracers is ^{99m}Tc-labeled fluconazole. Lupetti et al. showed that ^{99m}Tc-fluconazole
is a very good marker for *Candida albicans*;

in fact, this tracer can detect *C. albicans* infections but not bacterial infections in animal models [153]. The accumulation of ^{99m}Tc-fluconazole also correlated with the number of viable *C. albicans* microorganisms present at the infection site, making this agent suitable for monitoring therapy.

Antimicrobial peptides belong to the second group of tracers that are also under development. Antimicrobial peptides play a critical role in the defense system of multicellular organisms against bacteria, fungi, and viruses. They are produced by macrophages, epithelial and endothelial cells of all organisms, and their mechanism of action is based on the interaction between cationic residues of the peptides and the negatively charged bacterial surface. This feature determines the specificity of the antimicrobial peptides to bacteria, since in mammalian cells negatively charged lipids face the cytoplasm. Examples of this class of tracers are the ubiquicidin-derived peptides (UBI-peptides). Ubiquicidin is a 6.7-kDa linear peptide, which is a natural mammalian antimicrobial agent [154]. Different ^{99m}Tc-labeled UBI-derived peptides were tested for their ability to selectively detect infections *in vivo* [155]. ^{99m}Tc-labeled UBI 29-41 was able to detect bacterial infections in mice, with uptake at the infection site correlating with bacterial density. Interestingly, ^{99m}Tc-labeled UBI 29-41 was shown not to elicit any immune response in tested animals; this was in contrast to defensin-derived peptides (another class of antimicrobial peptides used for the imaging of infections), which have been reported to induce a potent immune response [156].

6.3.6
Amyloidosis

Amyloidosis is a condition which is characterized by the extracellular deposition of abnormal fibrillar proteins (amyloid), in amounts sufficient to impair correct organ function. Typical examples include Alzheimer's disease and prion disease.

6.3.6.1 Alzheimer's Disease
Alzheimer's disease (AD) is a complex neurodegenerative dementing illness. Extracellular amyloid plaques consisting predominantly of the amyloid-β-42 peptide (APβ42), a proteolytic derivative of the large transmembrane protein amyloid precursor protein (APP), are one of the pathological hallmarks of Alzheimer's disease (see also Part VIII, Chapter 4). Progress in an understanding of the mechanisms leading to the formation of extracellular amyloid plaques led to the development of new classes of drugs for the therapy of AD. β-Secretase is one of the proteases responsible for the proteolytic processing of APP that leads to the formation of APβ42, a peptide that is prone to aggregation. Several β-secretase inhibitors are currently being developed by companies such as Pfizer and Elan. In this section, we will focus on the emerging studies on imaging of amyloid plaques. Most efforts at *in vivo* neuroimaging of amyloid plaques have concentrated on developing radioactive ligands that can be detected by positron emission tomography (PET) or single-photon emission computed tomography (SPECT). In order to be used as an amyloid plaque-imaging agent, molecules must be able to cross the blood–brain barrier (BBB), and should therefore have a molecular weight of 400–600 kDa and form as few hydrogen bonds with water as

possible (see also Part VIII, Chapter 4). Congo Red (CR) and Thioflavin T are dyes that bind to amyloid plaques in brain sections but do not cross the BBB [157]. Several radiolabeled CR derivatives, such as Chrysamine G [158] and X-34 [159], have been created, but none of these gave satisfactory results *in vivo*, mainly due to their poor brain uptake. IBOX, a Thioflavin T derivative, was shown to cross the BBB, but no published *in vivo* studies are available [160]. Several peptide-based compounds have also been designed. 10H3 is a monoclonal antibody that binds specifically to Aβ1-28 [161], and radioactive labeled fragments of this antibody have been used for SPECT analysis. These studies failed due to unspecific binding and the inability of the peptide to cross the BBB. Further examples of peptide-based imaging agents are the serum amyloid protein (SAP) and β-amyloid peptide. SAP is a non-fibrillar glycoprotein, produced in the liver, which was shown to be effective in imaging systemic amyloidosis [162]. Although SAP is known to be able to cross the BBB, *in vivo* studies have not shown any difference between healthy and AD patients [163, 164]. As mentioned above, β-amyloid peptide can also be used as a plaque-imaging agent. Radiolabeled β-amyloid peptide 1-40 (Aβ1-40) was shown to bind with high affinity to amyloid plaques in brain sections, though *in vivo* studies have demonstrated only a limited brain uptake of Aβ1-40 [165].

To solve this problem, *in vivo* studies have been performed in rats using the Aβ1-40 conjugated with a monoclonal antibody specific to the rat transferrin receptor. The transferrin receptor allows receptor-mediated transcytosis through the rat's BBB. Conjugation of the peptide with this antibody increased its brain uptake and decreased its plasma clearance [166]. To our

knowledge, *in vivo* human studies have not yet been reported.

Highly lipophilic fluorescent molecules represent the last class of compounds that will be discussed in this section. The compound ^{18}FDDNP [167, 168], which binds with high affinity to Aβ1-40 fibrils *in vivo* was shown readily to cross the BBB due to its high lipophilicity. Human PET studies were performed, and ^{18}FDDNP was shown to localize in regions known to develop plaques in Alzheimer's disease [167].

6.3.6.2 Prion Diseases

Prion diseases in human and animals are fatal neurodegenerative diseases. The "protein-only" hypothesis proposes that the prion is a conformational isoform of the normal host prion protein PrP^C, which is found predominantly on the outer surface of neurons [169]. The abnormal PrP^{Sc} isoform, which is protease-resistant, is thought to cause the conversion of PrP^C into a likeness of itself. PrP^{Sc} has the tendency to aggregate and to accumulate mainly in the brain. The non-invasive diagnosis of prion diseases in humans is challenging due to the lack of specific and sensitive probes. However, during the past few years several PET ligands have been developed to image amyloid plaques in Alzheimer's disease (see Section 6.3.6.1). A CR derivative, methoxy-X04 [170], was tested for the ability of binding to prion deposits in 87V-infected mice. This compound was shown to localize at sites of prion deposition in both symptomatic and pre-symptomatic animals. The results of these studies suggest that similar compounds could be developed into useful PET imaging agents to improve the diagnosis of prion diseases in humans.

6.4

From a Ligand to a Product

6.4.1

Imaging

The identification of a pathology-associated antigen and the isolation of a specific binder (antibody, peptide, small organic molecule) does not automatically yield a novel imaging agent. Indeed, a number of conditions must be fulfilled in order to obtain a successful imaging result. In part, these conditions depend on the experimental modality chosen for the *in vivo* imaging application.

A general approach for converting a ligand into an imaging agent consists of radioactive labeling with a suitable radionuclide. Typically, short-lived gamma emitters must be used for gamma-camera-based imaging techniques (e.g., ^{99m}Tc, ^{123}I, ^{111}In) in order to minimize exposure of the patient to radiation.

Within the field of nuclear medicine, most developments are expected to originate in the field of PET. This can be thought of as a camera which can take pictures of a subject of interest, and requires an exposure time of a few seconds to several minutes. The camera does not image visible light, but rather high-energy gamma-rays that are emitted from inside the subject [171]. Natural biological molecules can be labeled with an isotope which produces two gamma rays by emitting a positron from its nucleus. Frequently used positron-emitting isotopes include ^{15}O, ^{13}N, ^{11}C and ^{18}F; the latter is often used as a substitute for hydrogen in the molecule of interest. ^{124}I can conveniently be used to radiolabel proteins. Labeled tracers are introduced into the subject, after which PET imaging is used to follow their distribution and tissue concentrations. PET is at least

tenfold more sensitive than SPECT, and positron-emitting isotopes can readily be substituted for naturally occurring atoms, producing less perturbation to the biochemical behaviour of the radiolabeled parent molecule. PET imaging might be used to couple high sensitivity with the possibility of obtaining very good spatial resolution and quantitative pharmacokinetic results. Recent advances in animal micro-PET will facilitate developments in this field [172, 173].

Ideal targeting agents for imaging applications will show a rapid and selective accumulation at the site of disease, and a rapid blood clearance. A number of antibody formats have been tested in animals and patients [174], and suggested that a mini-antibody format might be a good compromise between efficient localization at the site of disease and an acceptably rapid blood clearance [174, 175]. The small single-chain Fv fragments may yield better target:blood ratios at early time points, but their relatively low dose on the target may compromise immunoscintigraphic detection at acceptable doses of radioactivity.

A second approach to the ligand-based molecular imaging of markers of pathology relies on covalent modification of the ligand with NIR fluorophores. Light penetration of tissues is maximal at 800 nm, with a 10% transmittance through 1 cm of tissue [176]. A number of reactive ester derivatives of NIR fluorophores which absorb in this wavelength range are available (e.g., Cy7, Alexa750) [177, 178]. While imaging methodologies based on epi-illumination are limited to the detection of superficial lesions [179], and may be more appropriate for endoscopic applications [180], novel methodologies based on diffuse optical tomography may allow the detection of deeper lesions [181]. Compared to very sensitive radioactive imaging techniques, fluorescence-based detection methods may require the administration of a 100–1000 nanomoles of fluorophore derivative in order to obtain a sufficient signal on the site of interest.

As mentioned previously, the targeted delivery of suitable contrast agents for ultrasound or MRI imaging methodologies would be desirable for many applications. It remains to be seen whether microbubbles, magnetic nanoparticles or other active compounds can cross barriers *in vivo* (e.g., the endothelium) and can be delivered in sufficient amounts at sites of disease.

6.4.2
Therapy

Similar to imaging applications, most disease-targeting ligands must be modified in order to convert them into therapeutic agents.

In the special case of ligands capable of internalization (e.g., in tumor cells), a number of strategies may be considered, including modification with short-range-acting radionuclides, cytotoxic agents, and toxins [182–185]. However, in general, most ligands to sites of diseases will not be internalized. In oncology, the use of intact immunoglobulins which recognize antigens on tumor cells has led to the approval of several therapeutic antibodies in Europe and the USA. However, until now most of the anti-cancer antibodies approved have a signaling component which may contribute to their potency.

Small antibody fragments can be converted into therapeutic agents by modification with suitable radionuclides, drugs (using cleavable linkers or liposomal preparations), enzymes catalyzing prodrug conversion, cytokines, pro-coagulant factors, photosensitizers, and other classes of bioactive molecules (Fig. 6.3). Most of these experimental strategies have been tested in our laboratory using the L19 antibody, which is specific to the EDB domain of fibronectin, a marker of angiogenesis [186]. The relative advantages and disadvantages of the different methodologies are still, to a large extent, a matter of speculation, and this situation will only be resolved by an analysis of the outcome of clinical trials with different antibody derivatives.

The fact that both pro-inflammatory cytokines and anti-inflammatory cytokines can be fused to antibody fragments suggests a possible use of these fusion proteins not only for augmenting an immune response at suitable sites (e.g., against the tumor), but also to attenuate an immune response at sites of autoimmune diseases and in chronically inflamed areas.

6.5
Concluding Remarks

More than a hundred years after Paul Ehrlich's vision of "magic bullets", the targeted delivery of ligands to sites of diseases has been experimentally demonstrated for several pathological conditions. In some therapeutic areas (e.g., cancer), a large body of information is available, and the first conclusions can be drawn about the most promising strategies and about the parameters which are likely to influence targeting performance. For other diseases, ligand-based targeting approaches are still in their infancy. During the past decade, technological revolutions have greatly facilitated the isolation of specific binding molecules and the identification of disease-associated antigens. We believe that the next few years will continue to witness the comparative analysis of targeting agents in

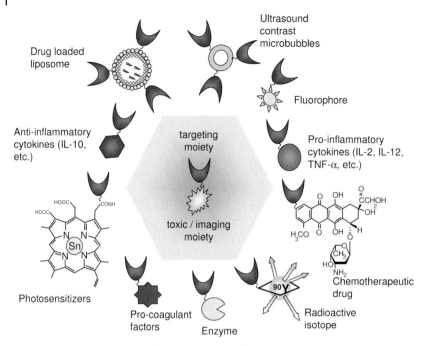

Fig. 6.3 Targeting moieties (ligands) can be converted into imaging or therapeutic agents by modification with suitable radionuclides, fluorophores, drugs, enzymes, cytokines or other bioactive molecules.

animal models of pathology, and translation of the most promising candidates from the bench, via the clinic, into the biopharmaceutical market.

References

1 Maeda, H., Fang, J., Inutsuka, T., and Kitamoto, Y. (2003) Vascular permeability enhancement in solid tumor: various factors, mechanisms involved and its implications, Int Immunopharmacol 3, 319–328.

2 Kopecek, J., Kopeckova, P., Minko, T., Lu, Z. R., and Peterson, C. M. (2001) Water soluble polymers in tumor targeted delivery, J Control Release 74, 147–158.

3 Duncan, R. (1999) Polymer conjugates for tumour targeting and intracytoplasmic delivery. The EPR effect as a common gateway, Pharm Sci Technol Today 2(11), 441–449.

4 Duncan, R. (1992) Drug–polymer conjugates: potential for improved chemotherapy, Anticancer Drugs 3, 175–210.

5 Maeda, H. (2001) The enhanced permeability and retention (EPR) effect in tumor vasculature: the key role of tumor-selective macromolecular drug targeting, Adv Enzyme Regul 41, 189–207.

6 Kohler, G., and Milstein, C. (1975) Continuous cultures of fused cells secreting antibody of predefined specificity, Nature 256, 495–497.

7 Steinitz, M., Klein, G., Koskimies, S., and Makel, O. (1977) EB virus-induced B lymphocyte cell lines producing specific antibody, Nature 269, 420–422.

8 Kozbor, D., and Roder, J.C. (1981) Requirements for the establishment of high-titered human monoclonal antibodies against tetanus toxoid using the Epstein-Barr virus technique, J Immunol 127, 1275–1280.

9 Kozbor, D., Roder, J.C., Chang, T.H., Steplewski, Z., and Koprowski, H. (1982) Human

anti-tetanus toxoid monoclonal antibody secreted by EBV-transformed human B cells fused with murine myeloma, Hybridoma 1, 323–328.

10 Karpas, A., Dremucheva, A., and Czepulkowski, B. H. (2001) A human myeloma cell line suitable for the generation of human monoclonal antibodies, Proc Natl Acad Sci USA 98, 1799–1804.

11 Jones, P. T., Dear, P. H., Foote, J., Neuberger, M. S., and Winter, G. (1986) Replacing the complementarity-determining regions in a human antibody with those from a mouse, Nature 321, 522–525.

12 McCafferty, J., Griffiths, A. D., Winter, G., and Chiswell, D. J. (1990) Phage antibodies: filamentous phage displaying antibody variable domains, Nature 348, 552–554.

13 Viti, F., Nilsson, F., Demartis, S., Huber, A., and Neri, D. (2000) Design and use of phage display libraries for the selection of antibodies and enzymes, Methods Enzymol 326, 480–505.

14 Green, L. L. (1999) Antibody engineering via genetic engineering of the mouse: Xeno-Mouse strains are a vehicle for the facile generation of therapeutic human monoclonal antibodies, J Immunol Methods 231, 11–23.

15 Walsh, G. (2003) Biopharmaceutical benchmarks – 2003, Nat Biotechnol 21, 865–870.

16 Winter, G., Griffiths, A. D., Hawkins, R. E., and Hoogenboom, H. R. (1994) Making antibodies by phage display technology, Annu Rev Immunol 12, 433–455.

17 Schier, R., Marks, J. D., Wolf, E. J., Apell, G., Wong, C., McCartney, J. E., Bookman, M. A., Huston, J. S., Houston, L. L., Weiner, L. M., and et al. (1995) *In vitro* and *in vivo* characterization of a human anti-c-erbB-2 single-chain Fv isolated from a filamentous phage antibody library, Immunotechnology 1, 73–81.

18 Pini, A., Viti, F., Santucci, A., Carnemolla, B., Zardi, L., Neri, P., and Neri, D. (1998) Design and use of a phage display library. Human antibodies with subnanomolar affinity against a marker of angiogenesis eluted from a two-dimensional gel, J Biol Chem 273, 21769–21776.

19 Schaffitzel, C., Hanes, J., Jermutus, L., and Pluckthun, A. (1999) Ribosome display: an *in vitro* method for selection and evolution of antibodies from libraries, J Immunol Methods 231, 119–135.

20 Kowalski, J., Henze, M., Schuhmacher, J., Macke, H. R., Hofmann, M., and Haberkorn, U. (2003) Evaluation of positron emission tomography imaging using [^{68}Ga]-DOTA-D Phe(1)-Tyr(3)-Octreotide in comparison to [^{111}In]-DTPAOC SPECT. First results in patients with neuroendocrine tumors, Mol Imaging Biol 5, 42–48.

21 Ginj, M., and Maecke, H. R. (2004) Radiometallo-labeled peptides in tumor diagnosis and therapy, Met Ions Biol Syst 42, 109–142.

22 Behr, T. M., Gotthardt, M., Barth, A., and Behe, M. (2001) Imaging tumors with peptide-based radioligands, Q J Nucl Med 45, 189–200.

23 Chen, X., Park, R., Hou, Y., Khankaldyyan, V., Gonzales-Gomez, I., Tohme, M., Bading, J. R., Laug, W. E., and Conti, P. S. (2004) MicroPET imaging of brain tumor angiogenesis with ^{18}F-labeled PEGylated RGD peptide, Eur J Nucl Med Mol Imaging 31, 1081–1089.

24 Okarvi, S. M., and al-Jammaz, I. (2003) Synthesis, radiolabelling and biological characteristics of a bombesin peptide analog as a tumor imaging agent, Anticancer Res 23, 2745–2750.

25 Felici, F., Luzzago, A., Monaci, P., Nicosia, A., Sollazzo, M., and Traboni, C. (1995) Peptide and protein display on the surface of filamentous bacteriophage, Biotechnol Annu Rev 1, 149–183.

26 Oyama, T., Sykes, K. F., Samli, K. N., Minna, J. D., Johnston, S. A., and Brown, K. C. (2003) Isolation of lung tumor specific peptides from a random peptide library: generation of diagnostic and cell-targeting reagents, Cancer Lett 202, 219–230.

27 Collins, J., Horn, N., Wadenback, J., and Szardenings, M. (2001) Cosmix-plexing: a novel recombinatorial approach for evolutionary selection from combinatorial libraries, J Biotechnol 74, 317–338.

28 Pasqualini, R., and Ruoslahti, E. (1996) Organ targeting *in vivo* using phage display peptide libraries, Nature 380, 364–366.

29 Trepel, M., Arap, W., and Pasqualini, R. (2002) *In vivo* phage display and vascular heterogeneity: implications for targeted medicine, Curr Opin Chem Biol 6, 399–404.

30 Arkin, M. R., and Wells, J. A. (2004) Small-molecule inhibitors of protein–protein interactions: progressing towards the dream, Nat Rev Drug Discov 3, 301–317.

31 Mammen, M., C.S.-K., Whitesides G.M. (1998) Polyvalent Interactions in Biological Systems: Implications for Design and Use of Multivalent Ligands and Inhibitors. Angewandte Chemie International Edition, 2754–2794.

32 Shuker, S.B., Hajduk, P.J., Meadows, R.P., and Fesik, S.W. (1996) Discovering high-affinity ligands for proteins: SAR by NMR, Science 274, 1531–1534.

33 Ramstrom, O., and Lehn, J.M. (2002) Drug discovery by dynamic combinatorial libraries, Nat Rev Drug Discov 1, 26–36.

34 Erlanson, D.A., Lam, J.W., Wiesmann, C., Luong, T.N., Simmons, R.L., DeLano, W.L., Choong, I.C., Burdett, M.T., Flanagan, W.M., Lee, D., Gordon, E.M., and O'Brien, T. (2003) In situ assembly of enzyme inhibitors using extended tethering, Nat Biotechnol 21, 308–314.

35 Melkko, S., Scheuermann, J., Dumelin, C.E., and Neri, D. (2004) Encoded self-assembling chemical libraries, Nat Biotechnol 22, 568–574.

36 Johnson, V.G., Schlom, J., Paterson, A.J., Bennett, J., Magnani, J.L., and Colcher, D. (1986) Analysis of a human tumor-associated glycoprotein (TAG-72) identified by monoclonal antibody B72.3, Cancer Res 46, 850–857.

37 Nuti, M., Teramoto, Y.A., Mariani-Costantini, R., Hand, P.H., Colcher, D., and Schlom, J. (1982) A monoclonal antibody (B72.3) defines patterns of distribution of a novel tumor-associated antigen in human mammary carcinoma cell populations, Int J Cancer 29, 539–545.

38 Guadagni, F., Roselli, M., Cosimelli, M., Ferroni, P., Spila, A., Cavaliere, F., Arcuri, R., Carlini, S., Mariotti, S., Gandolfo, G.M., Casciani, C.U., Greiner, J.W., and Schlom, J. (1996) TAG-72 expression and its role in the biological evaluation of human colorectal cancer, Anticancer Res 16, 2141–2148.

39 Guadagni, F., Roselli, M., Cosimelli, M., Mannella, E., Tedesco, M., Cavaliere, F., Grassi, A., Abbolito, M.R., Greiner, J.W., and Schlom, J. (1993) TAG-72 (CA 72-4 assay) as a complementary serum tumor antigen to carcinoembryonic antigen in monitoring patients with colorectal cancer, Cancer 72, 2098–2106.

40 Chieng, D.C., Rodriguez-Burford, C., Talley, L.I., Sviglin, H., Stockard, C.R., Kleinberg, M.J., Barnes, M.N., Partridge, E.E., Khazaeli, M.B., and Grizzle, W.E. (2003) Expression of CEA, Tag-72, and Lewis-Y antigen in primary and metastatic lesions of ovarian carcinoma, Hum Pathol 34, 1016–1021.

41 Gold, P., and Freedman, S.O. (1965) Demonstration of tumor-specific antigens in human colonic carcinomata by immunological tolerance and absorption techniques, J Exp Med 121, 439–462.

42 Duffy, M.J. (2001) Carcinoembryonic antigen as a marker for colorectal cancer: is it clinically useful?, Clin Chem 47, 624–630.

43 Verhaar, M.J., Chester, K.A., Keep, P.A., Robson, L., Pedley, R.B., Boden, J.A., Hawkins, R.E., and Begent, R.H. (1995) A single chain Fv derived from a filamentous phage library has distinct tumor targeting advantages over one derived from a hybridoma, Int J Cancer 61, 497–501.

44 Verhaar, M.J., Keep, P.A., Hawkins, R.E., Robson, L., Casey, J.L., Pedley, B., Boden, J.A., Begent, R.H., and Chester, K.A. (1996) Technetium-99m radiolabeling using a phage-derived single-chain Fv with a C-terminal cysteine, J Nucl Med 37, 868–872.

45 Cooke, S.P., Pedley, R.B., Boden, R., Begent, R.H., and Chester, K.A. (2002) In vivo tumor delivery of a recombinant single chain Fv: tumor necrosis factor-alpha fusion [correction of factor: a fusion] protein, Bioconjug Chem 13, 7–15.

46 Francis, R.J., Mather, S.J., Chester, K., Sharma, S.K., Bhatia, J., Pedley, R.B., Waibel, R., Green, A.J., and Begent, R.H. (2004) Radiolabelling of glycosylated MFE-23:CPG2 fusion protein (MFECP1) with ^{99m}Tc for quantitation of tumour antibody-enzyme localisation in antibody-directed enzyme pro-drug therapy (ADEPT), Eur J Nucl Med Mol Imaging 31, 1090–1096.

47 Horoszewicz, J.S., Kawinski, E., and Murphy, G.P. (1987) Monoclonal antibodies to a new antigenic marker in epithelial prostatic cells and serum of prostatic cancer patients, Anticancer Res 7, 927–935.

48 Gong, M.C., Chang, S.S., Sadelain, M., Bander, N.H., and Heston, W.D. (1999) Prostate-specific membrane antigen (PSMA)-specific monoclonal antibodies in the treatment of prostate and other cancers, Cancer Metastasis Rev 18, 483–490.

49 Freeman, L.M., Krynyckyi, B.R., Li, Y., Korupulu, G., Saleemi, K., Haseman, M.K., and Kahn, D. (2002) The role of ^{111}In Capromab

Pendetide (Prosta-ScintR) immunoscintigraphy in the management of prostate cancer, Q J Nucl Med 46, 131–137.

50 Rubin, I., and Yarden, Y. (2001) The basic biology of HER2, Ann Oncol 12 Suppl 1, S3–8.

51 Menard, S., Pupa, S. M., Campiglio, M., and Tagliabue, E. (2003) Biologic and therapeutic role of HER2 in cancer, Oncogene 22, 6570–6578.

52 Hung, M. C., and Lau, Y. K. (1999) Basic science of HER-2/neu: a review, Semin Oncol 26, 51–59.

53 Carter, P., Presta, L., Gorman, C. M., Ridgway, J. B., Henner, D., Wong, W. L., Rowland, A. M., Kotts, C., Carver, M. E., and Shepard, H. M. (1992) Humanization of an anti-p185HER2 antibody for human cancer therapy, Proc Natl Acad Sci USA 89, 4285–4289.

54 Park, J. W., Stagg, R., Lewis, G. D., Carter, P., Maneval, D., Slamon, D. J., Jaffe, H., and Shepard, H. M. (1992) Anti-p185HER2 monoclonal antibodies: biological properties and potential for immunotherapy, Cancer Treat Res 61, 193–211.

55 Fischer, O. M., Streit, S., Hart, S., and Ullrich, A. (2003) Beyond Herceptin and Gleevec, Curr Opin Chem Biol 7, 490–495.

56 Mass, R. D. (2004) The HER receptor family: a rich target for therapeutic development, Int J Radiat Oncol Biol Phys 58, 932–940.

57 Keler, T., Graziano, R. F., Mandal, A., Wallace, P. K., Fisher, J., Guyre, P. M., Fanger, M. W., and Deo, Y. M. (1997) Bispecific antibody-dependent cellular cytotoxicity of HER2/neu-overexpressing tumor cells by Fc gamma receptor type I-expressing effector cells, Cancer Res 57, 4008–4014.

58 Riley, J. K., and Sliwkowski, M. X. (2000) CD20: a gene in search of a function, Semin Oncol 27, 17–24.

59 Krasner, C., and Joyce, R. M. (2001) Zevalin: ^{90}yttrium labeled anti-CD20 (ibritumomab tiuxetan), a new treatment for non-Hodgkin's lymphoma, Curr Pharm Biotechnol 2, 341–349.

60 Vose, J. M. (2004) Bexxar: novel radioimmunotherapy for the treatment of low-grade and transformed low-grade non-Hodgkin's lymphoma, Oncologist 9, 160–172.

61 Went, P. T., Lugli, A., Meier, S., Bundi, M., Mirlacher, M., Sauter, G., and Dirnhofer, S. (2004) Frequent EpCam protein expression in human carcinomas, Hum Pathol 35, 122–128.

62 Herlyn, M., Steplewski, Z., Herlyn, D., and Koprowski, H. (1979) Colorectal carcinoma-specific antigen: detection by means of monoclonal antibodies, Proc Natl Acad Sci USA 76, 1438–1442.

63 Poczatek, R. B., Myers, R. B., Manne, U., Oelschlager, D. K., Weiss, H. L., Bostwick, D. G., and Grizzle, W. E. (1999) Ep-Cam levels in prostatic adenocarcinoma and prostatic intraepithelial neoplasia, J Urol 162, 1462–1466.

64 Ruck, P., Wichert, G., Handgretinger, R., and Kaiserling, E. (2000) Ep-CAM in malignant liver tumours, J Pathol 191, 102–103.

65 Braun, S., Hepp, F., Kentenich, C. R., Janni, W., Pantel, K., Riethmuller, G., Willgeroth, F., and Sommer, H. L. (1999) Monoclonal antibody therapy with edrecolomab in breast cancer patients: monitoring of elimination of disseminated cytokeratin-positive tumor cells in bone marrow, Clin Cancer Res 5, 3999–4004.

66 Punt, C. J., Nagy, A., Douillard, J. Y., Figer, A., Skovsgaard, T., Monson, J., Barone, C., Fountzilas, G., Riess, H., Moylan, E., Jones, D., Dethling, J., Colman, J., Coward, L., and MacGregor, S. (2002) Edrecolomab alone or in combination with fluorouracil and folinic acid in the adjuvant treatment of stage III colon cancer: a randomised study, Lancet 360, 671–677.

67 Bischoff, J. (1995) Approaches to studying cell adhesion molecules in angiogenesis, Trends Cell Biol 5, 69–74.

68 Folkman, J. (1995) Angiogenesis in cancer, vascular, rheumatoid and other disease, Nat Med 1, 27–31.

69 Pugh, C. W., and Ratcliffe, P. J. (2003) Regulation of angiogenesis by hypoxia: role of the HIF system, Nat Med 9, 677–684.

70 Carmeliet, P. (2003) Angiogenesis in health and disease, Nat Med 9, 653–660.

71 Gerber, H. P., and Ferrara, N. (2003) The role of VEGF in normal and neoplastic hematopoiesis, J Mol Med 81, 20–31.

72 Hurwitz, H., Fehrenbacher, L., Novotny, W., Cartwright, T., Hainsworth, J., Heim, W., Berlin, J., Baron, A., Griffing, S., Holmgren, E., Ferrara, N., Fyfe, G., Rogers, B., Ross, R., and Kabbinavar, F. (2004) Bevacizumab plus irinotecan, fluorouracil, and leucovorin for metastatic colorectal cancer, N Engl J Med 350, 2335–2342.

73 Cobleigh, M. A., Langmuir, V. K., Sledge, G. W., Miller, K. D., Haney, L., Novotny, W. F., Reimann, J. D., and Vassel, A. (2003) A phase I/II dose-escalation trial of bevacizumab in previously treated metastatic breast cancer, Semin Oncol 30, 117–124.

74 Folkman, J. (1971) Tumor angiogenesis: therapeutic implications, N Engl J Med 285, 1182–1186.

75 Folkman, J. (1972) Anti-angiogenesis: new concept for therapy of solid tumors, Ann Surg 175, 409–416.

76 Hanahan, D. (1998) A flanking attack on cancer, Nat Med 4, 13–14.

77 Bergers, G., and Benjamin, L. E. (2003) Tumorigenesis and the angiogenic switch, Nat Rev Cancer 3, 401–410.

78 Forsythe, J. A., Jiang, B. H., Iyer, N. V., Agani, F., Leung, S. W., Koos, R. D., and Semenza, G. L. (1996) Activation of vascular endothelial growth factor gene transcription by hypoxia-inducible factor 1, Mol Cell Biol 16, 4604–4613.

79 Ivan, M., Haberberger, T., Gervasi, D. C., Michelson, K. S., Gunzler, V., Kondo, K., Yang, H., Sorokina, I., Conaway, R. C., Conaway, J. W., and Kaelin, W. G., Jr. (2002) Biochemical purification and pharmacological inhibition of a mammalian prolyl hydroxylase acting on hypoxia-inducible factor, Proc Natl Acad Sci USA 99, 13459–13464.

80 Jaakkola, P., Mole, D. R., Tian, Y. M., Wilson, M. I., Gielbert, J., Gaskell, S. J., Kriegsheim, A., Hebestreit, H. F., Mukherji, M., Schofield, C. J., Maxwell, P. H., Pugh, C. W., and Ratcliffe, P. J. (2001) Targeting of HIF-alpha to the von Hippel-Lindau ubiquitylation complex by O$_2$-regulated prolyl hydroxylation, Science 292, 468–472.

81 Ivan, M., Kondo, K., Yang, H., Kim, W., Valiando, J., Ohh, M., Salic, A., Asara, J. M., Lane, W. S., and Kaelin, W. G., Jr. (2001) HIF-alpha targeted for VHL-mediated destruction by proline hydroxylation: implications for O$_2$ sensing, Science 292, 464–468.

82 Dibbens, J. A., Miller, D. L., Damert, A., Risau, W., Vadas, M. A., and Goodall, G. J. (1999) Hypoxic regulation of vascular endothelial growth factor mRNA stability requires the cooperation of multiple RNA elements, Mol Biol Cell 10, 907–919.

83 Tenan, M., Fulci, G., Albertoni, M., Diserens, A. C., Hamou, M. F., El Atifi-Borel, M., Feige,

J. J., Pepper, M. S., and Van Meir, E. G. (2000) Thrombospondin-1 is downregulated by anoxia and suppresses tumorigenicity of human glioblastoma cells, J Exp Med 191, 1789–1798.

84 Konerding, M. A., Fait, E., and Gaumann, A. (2001) 3D microvascular architecture of precancerous lesions and invasive carcinomas of the colon, Br J Cancer 84, 1354–1362.

85 Roberts, W. G., and Palade, G. E. (1997) Neovasculature induced by vascular endothelial growth factor is fenestrated, Cancer Res 57, 765–772.

86 Jain, R. K. (1987) Transport of molecules across tumor vasculature, Cancer Metastasis Rev 6, 559–593.

87 Hashizume, H., Baluk, P., Morikawa, S., McLean, J. W., Thurston, G., Roberge, S., Jain, R. K., and McDonald, D. M. (2000) Openings between defective endothelial cells explain tumor vessel leakiness, Am J Pathol 156, 1363–1380.

88 Ferrara, N., Hillan, K. J., Gerber, H. P., and Novotny, W. (2004) Discovery and development of bevacizumab, an anti-VEGF antibody for treating cancer, Nat Rev Drug Discov 3, 391–400.

89 Kerbel, R. S. (1991) Inhibition of tumor angiogenesis as a strategy to circumvent acquired resistance to anti-cancer therapeutic agents, BioEssays 13, 31–36.

90 Sekiguchi, K., Siri, A., Zardi, L., and Hakomori, S. (1985) Differences in domain structure between human fibronectins isolated from plasma and from culture supernatants of normal and transformed fibroblasts. Studies with domain-specific antibodies, J Biol Chem 260, 5105–5114.

91 Castellani, P., Viale, G., Dorcaratto, A., Nicolo, G., Kaczmarek, J., Querze, G., and Zardi, L. (1994) The fibronectin isoform containing the ED-B oncofetal domain: a marker of angiogenesis, Int J Cancer 59, 612–618.

92 Carnemolla, B., Borsi, L., Bannikov, G., Troyanovsky, S., and Zardi, L. (1992) Comparison of human tenascin expression in normal, simian-virus-40-transformed and tumor-derived cell lines, Eur J Biochem 205, 561–567.

93 Borsi, L., Carnemolla, B., Nicolo, G., Spina, B., Tanara, G., and Zardi, L. (1992) Expression of different tenascin isoforms in normal, hyperplastic and neoplastic human breast tissues, Int J Cancer 52, 688–692.

94 Brack, S. S., Dinkelborg, L. M., and Neri, D. (2004) Molecular targeting of angiogenesis for imaging and therapy, Eur J Nucl Med Mol Imaging 31, 1327–1341.

95 Ebbinghaus, C., Scheuermann, J., Neri, D., and Elia, G. (2004) Diagnostic and therapeutic applications of recombinant antibodies: targeting the extra-domain B of fibronectin, a marker of tumor angiogenesis, Curr Pharm Des 10, 1537–1549.

96 Alessi, P., Ebbinghaus, C., and Neri, D. (2004) Molecular targeting of angiogenesis, Biochim Biophys Acta 1654, 39–49.

97 Wang, J. M., Kumar, S., Pye, D., van Agtho-ven, A. J., Krupinski, J., and Hunter, R. D. (1993) A monoclonal antibody detects heterogeneity in vascular endothelium of tumours and normal tissues, Int J Cancer 54, 363–370.

98 Burrows, F. J., Derbyshire, E. J., Tazzari, P. L., Amlot, P., Gazdar, A. F., King, S. W., Letarte, M., Vitetta, E. S., and Thorpe, P. E. (1995) Up-regulation of endoglin on vascular endothelial cells in human solid tumors: implications for diagnosis and therapy, Clin Cancer Res 1, 1623–1634.

99 Brekken, R. A., Huang, X., King, S. W., and Thorpe, P. E. (1998) Vascular endothelial growth factor as a marker of tumor endothelium, Cancer Res 58, 1952–1959.

100 Oh, P., Li, Y., Yu, J., Durr, E., Krasinska, K. M., Carver, L. A., Testa, J. E., and Schnit-zer, J. E. (2004) Subtractive proteomic mapping of the endothelial surface in lung and solid tumours for tissue-specific therapy, Nature 429, 629–635.

101 Taniguchi, K., Harada, N., Ohizumi, I., Tsut-sumi, Y., Nakagawa, S., Kaiho, S., and Mayu-mi, T. (2000) Recognition of human activated CD44 by tumor vasculature-targeted antibody, Biochem Biophys Res Commun 269, 671–675.

102 Ran, S., Downes, A., and Thorpe, P. E. (2002) Increased exposure of anionic phospholipids on the surface of tumor blood vessels, Cancer Res 62, 6132–6140.

103 Huminiecki, L., Gorn, M., Suchting, S., Poulsom, R., and Bicknell, R. (2002) Magic roundabout is a new member of the roundabout receptor family that is endothelial specific and expressed at sites of active angiogenesis, Genomics 79, 547–552.

104 Pasqualini, R., Koivunen, E., Kain, R., Lah-denranta, J., Sakamoto, M., Stryhn, A., Ash-mun, R. A., Shapiro, L. H., Arap, W., and Ruoslahti, E. (2000) Aminopeptidase N is a receptor for tumor-homing peptides and a target for inhibiting angiogenesis, Cancer Res 60, 722–727.

105 St Croix, B., Rago, C., Velculescu, V., Traver-so, G., Romans, K. E., Montgomery, E., Lal, A., Riggins, G. J., Lengauer, C., Vogelstein, B., and Kinzler, K. W. (2000) Genes expressed in human tumor endothelium, Science 289, 1197–1202.

106 Rybak, J. N., Scheurer, S. B., Neri, D., and Elia, G. (2004) Purification of biotinylated proteins on streptavidin resin: a protocol for quantitative elution, Proteomics 4, 2296–2299.

107 Glass, C. K., and Witztum, J. L. (2001) Atherosclerosis. The road ahead, Cell 104, 503–516.

108 Libby, P. (2002) Inflammation in atherosclerosis, Nature 420, 868–874.

109 Jaffer, F. A., and Weissleder, R. (2004) Seeing within: molecular imaging of the cardiovascular system, Circ Res 94, 433–445.

110 Reddy, V. Y., Zhang, Q. Y., and Weiss, S. J. (1995) Pericellular mobilization of the tissue-destructive cysteine proteinases, cathepsins B, L, and S, by human monocyte-derived macrophages, Proc Natl Acad Sci USA 92, 3849–3853.

111 Leake, D. S., and Peters, T. J. (1981) Proteolytic degradation of low density lipoproteins by arterial smooth muscle cells: the role of individual cathepsins, Biochim Biophys Acta 664, 108–116.

112 Galis, Z. S., and Khatri, J. J. (2002) Matrix metalloproteinases in vascular remodeling and atherogenesis: the good, the bad, and the ugly, Circ Res 90, 251–262.

113 Chen, J., Tung, C. H., Mahmood, U., Ntzia-christos, V., Gyurko, R., Fishman, M. C., Huang, P. L., and Weissleder, R. (2002) *In vivo* imaging of proteolytic activity in atherosclerosis, Circulation 105, 2766–2771.

114 Weissleder, R., Tung, C. H., Mahmood, U., and Bogdanov, A., Jr. (1999) *In vivo* imaging of tumors with protease-activated near-infrared fluorescent probes, Nat Biotechnol 17, 375–378.

115 Jaffer, F. A., Tung, C. H., Gerszten, R. E., and Weissleder, R. (2002) *In vivo* imaging of

thrombin activity in experimental thrombi
with thrombin-sensitive near-infrared molecular probe, Arterioscler Thromb Vasc Biol
22, 1929–1935.

116 Bremer, C., Bredow, S., Mahmood, U.,
Weissleder, R., and Tung, C. H. (2001) Optical imaging of matrix metalloproteinase-2
activity in tumors: feasibility study in a
mouse model, Radiology 221, 523–529.

117 Kooi, M. E., Cappendijk, V. C., Cleutjens,
K. B., Kessels, A. G., Kitslaar, P. J., Borgers,
M., Frederik, P. M., Daemen, M. J., and van
Engelshoven, J. M. (2003) Accumulation of
ultrasmall superparamagnetic particles of
iron oxide in human atherosclerotic plaques
can be detected by *in vivo* magnetic resonance imaging, Circulation 107, 2453–2458.

118 Lindner, J. R. (2002) Detection of inflamed
plaques with contrast ultrasound, Am J Cardiol 90, 32L–35L.

119 Demos, S. M., Alkan-Onyuksel, H., Kane,
B. J., Ramani, K., Nagaraj, A., Greene, R.,
Klegerman, M., and McPherson, D. D.
(1999) *In vivo* targeting of acoustically reflective liposomes for intravascular and trans-
vascular ultrasonic enhancement, J Am Coll
Cardiol 33, 867–875.

120 Lindner, J. R., Song, J., Christiansen, J., Klibanov, A. L., Xu, F., and Ley, K. (2001) Ultrasound assessment of inflammation and renal tissue injury with microbubbles targeted
to P-selectin, Circulation 104, 2107–2112.

121 Leong-Poi, H., Christiansen, J., Klibanov, A.
L., Kaul, S., and Lindner, J. R. (2003) Noninvasive assessment of angiogenesis by ultrasound and microbubbles targeted to al-
pha(v)-integrins, Circulation 107, 455–460.

122 Joseph, S., Olbrich, C., Kirsch, J., Hasbach,
M., Briel, A., and Schirner, M. (2004) A real-
time *in vitro* assay for studying functional
characteristics of target-specific ultrasound
contrast agents, Pharm Res 21, 920–926.

123 Knight, L. C. (2001) Radiolabeled peptide ligands for imaging thrombi and emboli,
Nucl Med Biol 28, 515–526.

124 Shattil, S. J., Hoxie, J. A., Cunningham, M.,
and Brass, L. F. (1985) Changes in the platelet membrane glycoprotein IIb.IIIa complex
during platelet activation, J Biol Chem 260,
11107–11114.

125 Taub, R., Gould, R. J., Garsky, V. M., Cicca-
rone, T. M., Hoxie, J., Friedman, P. A., and
Shattil, S. J. (1989) A monoclonal antibody

against the platelet fibrinogen receptor contains a sequence that mimics a receptor
recognition domain in fibrinogen, J Biol
Chem 264, 259–265.

126 Lister-James, J., Knight, L. C., Maurer, A. H.,
Bush, L. R., Moyer, B. R., and Dean, R. T.
(1996) Thrombus imaging with a techne-
tium-99m-labeled activated platelet receptor-
binding peptide, J Nucl Med 37, 775–781.

127 Shebuski, R. J., Ramjit, D. R., Bencen, G. H.,
and Polokoff, M. A. (1989) Characterization
and platelet inhibitory activity of bitistatin, a
potent arginine-glycine-aspartic acid-containing peptide from the venom of the viper *Bi-
tis arietans*, J Biol Chem 264, 21550–21556.

128 Tung, C. H., Ho, N. H., Zeng, Q., Tang, Y.,
Jaffer, F. A., Reed, G. L., and Weissleder, R.
(2003) Novel factor XIII probes for blood coagulation imaging, Chembiochem 4, 897–899.

129 Jaffer, F. A., Tung, C. H., Houng, A. K.,
O'Loughin, T., Reed, G. L., Weissleder, R.
(2002) MRI of blood coagulation factor XIII
activity using a novel peptide derivatized
caged iron oxide nanoparticle (F13-CLIO),
Mol Imaging, 217–218.

130 Lanza, G. M., Wallace, K. D., Scott, M. J., Ca-
cheris, W. P., Abendschein, D. R., Christy,
D. H., Sharkey, A. M., Miller, J. G., Gaffney,
P. J., and Wickline, S. A. (1996) A novel site-
targeted ultrasonic contrast agent with broad
biomedical application, Circulation 94,
3334–3340.

131 Flacke, S., Fischer, S., Scott, M. J., Fuhrhop,
R. J., Allen, J. S., McLean, M., Winter, P., Si-
card, G. A., Gaffney, P. J., Wickline, S. A.,
and Lanza, G. M. (2001) Novel MRI contrast
agent for molecular imaging of fibrin: implications for detecting vulnerable plaques, Circulation 104, 1280–1285.

132 Barrett, J. A., Kolodziej, A. F., Caravan, P. D.,
Nair, S., Looby, R., Witte, S., Costello, C. R.,
Meslia, M. A., Drezwecki, L., Cesna, C.,
Pratt, C., McMurry, T. J., Lauffer, R. B., Yucel,
E. K., Zhao, L., Weisskoff, R. M., Carpenter,
A. P., Graham, P. B. (2002) EP-1873, a gado-
linium (Gd) labeled fibrin specific agent that
rapidly detects arterial and venous thrombi
with MRI, Circulation, II-120.

133 Taillefer, R. (2001) Radiolabeled peptides in
the detection of deep venous thrombosis,
Semin Nucl Med 31, 102–123.

134 Petrich, B. G., and Wang, Y. (2004) Stress-ac-
tivated MAP kinases in cardiac remodeling

and heart failure; new insights from transgenic studies, Trends Cardiovasc Med 14, 50–55.

135 Martin, S. J., Reutelingsperger, C. P., McGahon, A. J., Rader, J. A., van Schie, R. C., LaFace, D. M., and Green, D. R. (1995) Early redistribution of plasma membrane phosphatidylserine is a general feature of apoptosis regardless of the initiating stimulus: inhibition by overexpression of Bcl-2 and Abl, J Exp Med 182, 1545–1556.

136 Hofstra, L., Liem, I. H., Dumont, E. A., Boersma, H. H., van Heerde, W. L., Doevendans, P. A., De Muinck, E., Wellens, H. J., Kemerink, G. J., Reutelingsperger, C. P., and Heidendal, G. A. (2000) Visualisation of cell death *in vivo* in patients with acute myocardial infarction, Lancet 356, 209–212.

137 Schellenberger, E. A., Bogdanov, A., Jr., Petrovsky, A., Ntziachristos, V., Weissleder, R., and Josephson, L. (2003) Optical imaging of apoptosis as a biomarker of tumor response to chemotherapy, Neoplasia 5, 187–192.

138 Firestein, G. S. (2003) Evolving concepts of rheumatoid arthritis, Nature 423, 356–361.

139 Kinne, R. W., Brauer, R., Stuhlmuller, B., Palombo-Kinne, E., and Burmester, G. R. (2000) Macrophages in rheumatoid arthritis, Arthritis Res 2, 189–202.

140 Paulos, C. M., Turk, M. J., Breur, G. J., and Low, P. S. (2004) Folate receptor-mediated targeting of therapeutic and imaging agents to activated macrophages in rheumatoid arthritis, Adv Drug Deliv Rev 56, 1205–1217.

141 Reddy, J. A., Haneline, L. S., Srour, E. F., Antony, A. C., Clapp, D. W., and Low, P. S. (1999) Expression and functional characterization of the beta-isoform of the folate receptor on CD34(+) cells, Blood 93, 3940–3948.

142 Nakashima-Matsushita, N., Homma, T., Yu, S., Matsuda, T., Sunahara, N., Nakamura, T., Tsukano, M., Ratnam, M., and Matsuyama, T. (1999) Selective expression of folate receptor beta and its possible role in methotrexate transport in synovial macrophages from patients with rheumatoid arthritis, Arthritis Rheum 42, 1609–1616.

143 Turk, M. J., Breur, G. J., Widmer, W. R., Paulos, C. M., Xu, L. C., Grote, L. A., and Low, P. S. (2002) Folate-targeted imaging of activated macrophages in rats with adjuvant-induced arthritis, Arthritis Rheum 46, 1947–1955.

144 Fletcher, D. S., Widmer, W. R., Luell, S., Christen, A., Orevillo, C., Shah, S., and Visco, D. (1998) Therapeutic administration of a selective inhibitor of nitric oxide synthase does not ameliorate the chronic inflammation and tissue damage associated with adjuvant-induced arthritis in rats, J Pharmacol Exp Ther 284, 714–721.

145 van Roon, J. A., van Vuuren, A. J., Wijngaarden, S., Jacobs, K. M., Bijlsma, J. W., Lafeber, F. P., Thepen, T., and van de Winkel, J. G. (2003) Selective elimination of synovial inflammatory macrophages in rheumatoid arthritis by an Fcgamma receptor I-directed immunotoxin, Arthritis Rheum 48, 1229–1238.

146 Wunder, A., Tung, C. H., Muller-Ladner, U., Weissleder, R., and Mahmood, U. (2004) *In vivo* imaging of protease activity in arthritis: a novel approach for monitoring treatment response, Arthritis Rheum 50, 2459–2465.

147 Creamer, D., Sullivan, D., Bicknell, R., and Barker, J. (2002) Angiogenesis in psoriasis, Angiogenesis 5, 231–236.

148 Kapsoritakis, A., Sfiridaki, A., Maltezos, E., Simopoulos, K., Giatromanolaki, A., Sivridis, E., and Koukourakis, M. I. (2003) Vascular endothelial growth factor in inflammatory bowel disease, Int J Colorectal Dis 18, 418–422.

149 Xia, Y. P., Li, B., Hylton, D., Detmar, M., Yancopoulos, G. D., and Rudge, J. S. (2003) Transgenic delivery of VEGF to mouse skin leads to an inflammatory condition resembling human psoriasis, Blood 102, 161–168.

150 Boyman, O., Hefti, H. P., Conrad, C., Nickoloff, B. J., Suter, M., and Nestle, F. O. (2004) Spontaneous development of psoriasis in a new animal model shows an essential role for resident T cells and tumor necrosis factor-alpha, J Exp Med 199, 731–736.

151 Hoffmann, J. C., Pawlowski, N. N., Kuhl, A. A., Hohne, W., and Zeitz, M. (2002) Animal models of inflammatory bowel disease: an overview, Pathobiology 70, 121–130.

152 Vinjamuri, S., Hall, A. V., Solanki, K. K., Bomanji, J., Siraj, Q., O'Shaughnessy, E., Das, S. S., and Britton, K. E. (1996) Comparison of ^{99m}Tc infecton imaging with radiolabelled white-cell imaging in the evaluation of bacterial infection, Lancet 347, 233–235.

153 Lupetti, A., Welling, M.M., Mazzi, U., Nibbering, P.H., and Pauwels, E.K. (2002) Technetium-99m labelled fluconazole and antimicrobial peptides for imaging of *Candida albicans* and *Aspergillus fumigatus* infections, Eur J Nucl Med Mol Imaging 29, 674–679.

154 Knight, L.C. (2003) Non-oncologic applications of radiolabeled peptides in nuclear medicine, Q J Nucl Med 47, 279–291.

155 Welling, M.M., Paulusma-Annema, A., Balter, H.S., Pauwels, E.K., and Nibbering, P.H. (2000) Technetium-99m labelled antimicrobial peptides discriminate between bacterial infections and sterile inflammations, Eur J Nucl Med 27, 292–301.

156 Tani, K., Murphy, W.J., Chertov, O., Salcedo, R., Koh, C.Y., Utsunomiya, I., Funakoshi, S., Asai, O., Herrmann, S.H., Wang, J.M., Kwak, L.W., and Oppenheim, J.J. (2000) Defensins act as potent adjuvants that promote cellular and humoral immune responses in mice to a lymphoma idiotype and carrier antigens, Int Immunol 12, 691–700.

157 Sair, H.I., Doraiswamy, P.M., and Petrella, J.R. (2004) *In vivo* amyloid imaging in Alzheimer's disease, Neuroradiology 46, 93–104.

158 Klunk, W.E., Debnath, M.L., and Pettegrew, J.W. (1995) Chrysamine-G binding to Alzheimer and control brain: autopsy study of a new amyloid probe, Neurobiol Aging 16, 541–548.

159 Styren, S.D., Hamilton, R.L., Styren, G.C., and Klunk, W.E. (2000) X-34, a fluorescent derivative of Congo red: a novel histochemical stain for Alzheimer's disease pathology, J Histochem Cytochem 48, 1223–1232.

160 Zhuang, Z.P., Kung, M.P., Hou, C., Plossl, K., Skovronsky, D., Gur, T.L., Trojanowski, J.Q., Lee, V.M., and Kung, H.F. (2001) IBOX(2-(4'-dimethylaminophenyl)-6-iodobenzoxazole): a ligand for imaging amyloid plaques in the brain, Nucl Med Biol 28, 887–894.

161 Friedland, R.P., Majocha, R.E., Reno, J.M., Lyle, L.R., and Marotta, C.A. (1994) Development of an anti-A beta monoclonal antibody for *in vivo* imaging of amyloid angiopathy in Alzheimer's disease, Mol Neurobiol 9, 107–113.

162 Lovat, L.B., O'Brien, A.A., Armstrong, S.F., Madhoo, S., Bulpitt, C.J., Rossor, M.N., Pepys, M.B., and Hawkins, P.N. (1998) Scintigraphy with ^{123}I-serum amyloid P component in Alzheimer disease, Alzheimer Dis Assoc Disord 12, 208–210.

163 Hawkins, P.N., Rossor, M.N., Gallimore, J.R., Miller, B., Moore, E.G., and Pepys, M.B. (1994) Concentration of serum amyloid P component in the CSF as a possible marker of cerebral amyloid deposits in Alzheimer's disease, Biochem Biophys Res Commun 201, 722–726.

164 Hawkins, P.N., Tyrell, P., Jones, T., et al. (1991) Metabolic and scintigraphic studies with radiolabeled serum amyloid P component in amyloidosis: applications to cerebral deposits and Alzheimer disease with positron emission tomography, Bull Clin Neurosci, 178–190.

165 Maggio, J.E., Stimson, E.R., Ghilardi, J.R., Allen, C.J., Dahl, C.E., Whitcomb, D.C., Vigna, S.R., Vinters, H.V., Labenski, M.E., and Mantyh, P.W. (1992) Reversible *in vitro* growth of Alzheimer disease beta-amyloid plaques by deposition of labeled amyloid peptide, Proc Natl Acad Sci USA 89, 5462–5466.

166 Saito, Y., Buciak, J., Yang, J., and Pardridge, W.M. (1995) Vector-mediated delivery of ^{125}I-labeled beta-amyloid peptide A beta 1-40 through the blood-brain barrier and binding to Alzheimer disease amyloid of the A beta 1-40/vector complex, Proc Natl Acad Sci USA 92, 10227–10231.

167 Agdeppa, E.D., Kepe, V., Liu, J., Flores-Torres, S., Satyamurthy, N., Petric, A., Cole, G.M., Small, G.W., Huang, S.C., and Barrio, J.R. (2001) Binding characteristics of radiofluorinated 6-dialkylamino-2-naphthylethylidene derivatives as positron emission tomography imaging probes for beta-amyloid plaques in Alzheimer's disease, J Neurosci 21, RC189.

168 Barrio, J.R., Huang, S.C., Cole, G., et al. (1999) PET imaging of tangles and plaques in Alzheimer's disease with a highly hydrophobic probe, J Label Comp Pharm, S194–195.

169 Weissmann, C., Enari, M., Klohn, P.C., Rossi, D., and Flechsig, E. (2002) Transmission of prions, Proc Natl Acad Sci USA 99 Suppl 4, 16378–16383.

170 Klunk, W.E., Bacskai, B.J., Mathis, C.A., Kajdasz, S.T., McLellan, M.E., Frosch, M.P., Debnath, M.L., Holt, D.P., Wang, Y., and

Hyman, B. T. (2002) Imaging Abeta plaques in living transgenic mice with multiphoton microscopy and methoxy-X04, a systemically administered Congo red derivative, J Neuropathol Exp Neurol 61, 797–805.

171 Gambhir, S. S. (2002) Molecular imaging of cancer with positron emission tomography, Nat Rev Cancer 2, 683–693.

172 Ugur, O., Kothari, P. J., Finn, R. D., Zanzonico, P., Ruan, S., Guenther, I., Maecke, H. R., and Larson, S. M. (2002) Ga-66 labeled somatostatin analogue DOTA-DPhe1-Tyr3-octreotide as a potential agent for positron emission tomography imaging and receptor mediated internal radiotherapy of somatostatin receptor positive tumors, Nucl Med Biol 29, 147–157.

173 Burt, B. M., Humm, J. L., Kooby, D. A., Squire, O. D., Mastorides, S., Larson, S. M., and Fong, Y. (2001) Using positron emission tomography with [$^{(18)}$F]FDG to predict tumor behavior in experimental colorectal cancer, Neoplasia 3, 189–195.

174 Wu, A. M., Yazaki, P. J., Tsai, S., Nguyen, K., Anderson, A. L., McCarthy, D. W., Welch, M. J., Shively, J. E., Williams, L. E., Raubitschek, A. A., Wong, J. Y., Toyokuni, T., Phelps, M. E., and Gambhir, S. S. (2000) High-resolution microPET imaging of carcinoembryonic antigen-positive xenografts by using a copper-64-labeled engineered antibody fragment, Proc Natl Acad Sci USA 97, 8495–8500.

175 Borsi, L., Balza, E., Bestagno, M., Castellani, P., Carnemolla, B., Biro, A., Leprini, A., Sepulveda, J., Burrone, O., Neri, D., and Zardi, L. (2002) Selective targeting of tumoral vasculature: comparison of different formats of an antibody (L19) to the ED-B domain of fibronectin, Int J Cancer 102, 75–85.

176 Wan, S., Parrish, J. A., Anderson, R. R., and Madden, M. (1981) Transmittance of nonionizing radiation in human tissues, Photochem Photobiol 34, 679–681.

177 Licha, K., Riefke, B., Ebert, B., and Grotzinger, C. (2002) Cyanine dyes as contrast agents in biomedical optical imaging, Acad Radiol 9 Suppl 2, S320–322.

178 Licha, K., Riefke, B., Ntziachristos, V., Becker, A., Chance, B., and Semmler, W. (2000) Hydrophilic cyanine dyes as contrast agents for near-infrared tumor imaging: synthesis, photophysical properties and spectroscopic *in vivo* characterization, Photochem Photobiol 72, 392–398.

179 Birchler, M., Neri, G., Tarli, L., Halin, C., Viti, F., and Neri, D. (1999) Infrared photodetection for the *in vivo* localisation of phage-derived antibodies directed against angiogenic markers, J Immunol Methods 231, 239–248.

180 Funovics, M. A., Alencar, H., Su, H. S., Khazaie, K., Weissleder, R., and Mahmood, U. (2003) Miniaturized multichannel near infrared endoscope for mouse imaging, Mol Imaging 2, 350–357.

181 Ntziachristos, V., Yodh, A. G., Schnall, M., and Chance, B. (2000) Concurrent MRI and diffuse optical tomography of breast after indocyanine green enhancement, Proc Natl Acad Sci USA 97, 2767–2772.

182 Payne, G. (2003) Progress in immunoconjugate cancer therapeutics, Cancer Cell 3, 207–212.

183 Michel, R. B., Brechbiel, M. W., and Mattes, M. J. (2003) A comparison of 4 radionuclides conjugated to antibodies for single-cell kill, J Nucl Med 44, 632–640.

184 Mattes, M. J. (2002) Radionuclide-antibody conjugates for single-cell cytotoxicity, Cancer 94, 1215–1223.

185 Pastan, I., Beers, R., and Bera, T. K. (2004) Recombinant immunotoxins in the treatment of cancer, Methods Mol Biol 248, 503–518.

186 Neri, D. (2004) Tumor targeting, CHIMIA 58(10), 723–726.

7

Ultrasound Theranostics: Antibody-based Microbubble Conjugates as Targeted *In vivo* Contrast Agents and Advanced Drug Delivery Systems

Andreas Briel, Michael Reinhardt, Mathias Mäurer, and Peter Hauff

Abstract

One major challenge facing the pharmaceutical industry today is to develop contrast-enhancing agents for molecular imaging. Classic contrast agents primarily document the anatomy. For pathophysiological examinations using differential diagnostic techniques, i.e., characterizing the development of a disease, they are only suitable to a limited degree. Molecular imaging selectively tracks down molecules and cell structures to be able to establish proof of disease at a very early stage – and then to make decisions about highly individual treatment. The next straightforward vision of medical imaging quite clearly lies in the concept "Find, Fight and Follow!". In radiopharmaceuticals we are already pursuing the approach of a triad consisting of early diagnosis, therapy and therapy control. Utilizing the nanotechnological concepts of colloid and interface science, imaging on a molecular level can also be achieved via diagnostic ultrasound using tiny gas-filled polymer particles coupled to target-specific ligands. Additionally, nano-sized polymeric drug carriers for targeting and controlled release have been extensively studied in the past. Here, a nanoparticle or capsule acts like a container for a pharmacologically active agent. Passive and active targeting can be attained by carefully chosen size and surface modification of the carrier. Drug release can be controlled via desorption of surface-bound drugs, diffusion through the particle matrix or the capsule wall, or matrix erosion. Moreover, "smart" release can be achieved by using smart polymers (pH or temperature sensitive) or, more interestingly, by applying an external stress to the drug carrier. If the drug carrier is appropriately designed, release can be induced by diagnostic ultrasound. Building a bridge between therapy and diagnosis opens the field of "theranostics". With a "Find, Fight and Follow!" strategy, the tissue of interest can first be imaged via target-specific ultrasound contrast particles. In a second step, the same particles, now filled with a pharmacologically active agent, can be used for therapy. Finally, monitoring of treatment effects is possible by sequential imaging. This early approach demonstrates the success of a resolute implementation of nanobiotechnological concepts in a medical application, and will be presented with respect to polymer nanoparticle and microcapsule formation, the control of colloidal structure, surface modification, antibody-coupling strategies, and the resulting *in vitro* properties as an "ultrasound theranostic". *In vivo* results will be addressed with

Modern Biopharmaceuticals. Edited by J. Knäblein
Copyright © 2005 WILEY-VCH Verlag GmbH & Co. KGaA, Weinheim
ISBN: 3-527-31184-X

special emphasis on antibody-based targeting and gene delivery. Investigations with different drugs and targeting sites demonstrate that, in general, this approach can serve as a biopharmaceutical platform technology.

Abbreviations

3-D	three-dimensional
CFT	critical flooding temperature
CT	computerized tomography
EAE	experimental autoimmune encephalomyelitis
ED-B	extradomain B
EDC	1-ethyl-3-(3-dimethylaminopropyl) carbodiimide
FN	fibronectin
ICAM	intercellular adhesion molecule
LOC	loss of correlation
MS	multiple sclerosis
pDNA	plasmid DNA
PET	positron emission tomography
SAE	stimulated acoustic emission
SPAQ	sensitive particle acoustic quantification
SPECT	single photon emission computerized tomography
USCA	ultrasound contrast agent
VCAM	vascular cell adhesion molecule

7.1
Motivation: "Find, Fight and Follow!"

"D'ye see him?" cried Ahab after allowing a little space for the light to spread.
"See nothing, sir."
"Turn up all hands and make sail! he travels faster than I thought for; – the top-gallant sails! – aye, they should have been kept on her all night. But no matter – 'tis but resting for the rush." [...]

"There she blows – she blows! – she blows! – right ahead!" was now the mast-head cry.
"Aye, breach your last to the sun, Moby Dick!" cried Ahab, "thy hour and thy harpoon are at hand! – Down! down all of ye, but one man at the fore. The boats! – stand by!" [...] [1].

"Find, Fight and Follow!". It is not only Captain Ahab that relied on that concept to hunt whales with small harpoons by hand in former times; the same strategy was also used by the Indians in northern America for hunting buffalo – bows and arrows were not efficient enough to kill a strong animal like this immediately. In addition, the weapons of people in the stone age were obviously not suitable for knocking down a mammoth at first strike. Today, this is no longer a problem – people have developed specifically designed weapons to kill buffalos and whales (and even mammoths!) at the first shot.

However, what about our weapons to fight against cancer? Cancer is one of the most important "killers" today. Thousands of humans die every minute and no therapy can rescue them from their destiny – although researchers continue to make daily progress. We believe that we can learn from the hunters in former times – "Find, Fight and Follow!". With radiopharmaceuticals, we are already pursuing the approach of a triad consisting of early diagnosis, therapy and therapy control. We have built a bridge between diagnosis and therapy by coupling identical carriers with different active substances – diagnostic or therapeutic.

For example, the radionuclide indium-111 [^{111}In] (or Technetium-99m [^{99m}Tc] or Yttrium-86 [^{86}Y] as a positron emitter) can be linked via a chelator to a cancer-specific targeting moiety (a peptide or an antibody, etc.) (see also Part V, Chapter 2). Presuming superior specificity of the conjugate the contrast agent (or, even better, the tracer) will accumulate in the target tissue after administering to the patient. This leads to an increased signal-to-background ratio in the target and the position of a tumor can be imaged with a γ-camera [single photon emission computerized tomography (SPECT) or positron emission tomography (PET), respectively, or even better a fusion image with modern PET/CT settings] (see also Part V, Chapters 4–6).

Furthermore, because the imaging agent works by depositing radiation energy proportionally to its concentration within the tumor, imaging the target provides dosimetry about dose delivery and is therefore important for planning the treatment.

After "finding", "fighting" against the tumor can be realized by replacing the "weak" imaging radionuclide with a "strong" therapeutic radionuclide like yttrium-90 [^{90}Y], which shows chemically similar behavior with regard to coupling to the targeting moiety, but higher radiation energy [2, 3] (see also Part II, Chapter 5). Based on the dosimetry data, the therapeutic dose can be calculated and targeted locoregional radiotherapy can be started.

Finally, monitoring of treatment response is possible by sequential imaging using the same imaging agent. In sensitive tumors, there should be a marked reduction of contrast uptake [4, 5]. In the "follow" phase of the triad, these "reporters of efficacy" determine the endpoint of therapy or indicate further treatment. Differentiation of responder from nonresponder is a first step towards individualized medicine (see also Part I, Chapter 2).

As genomics is now coming to the point of indicating a predisposition for disease and is also able to identify markers of aggressive diseases, the combination with modern molecular imaging techniques provides exciting opportunities to exploit the strategy described above.

The "Find, Fight and Follow!" concept builds a bridge between therapy and diagnosis, and opens the field of "theranostics" – where genomics meets molecular imaging and advanced drug delivery. Several logistical challenges, e.g., short half-life of radiopharmaceuticals combined with the production in-time and on-site, and, furthermore, the radiation exposure to the patient (especially in the follow-up phase), make it important to evaluate alternative approaches to "theranostics".

In summary, a promising candidate needs:
- A high-resolution imaging modality which is furthermore able to quantify the amount of contrast conjugates in the target tissue in order to indicate responders and confirm therapeutic effects.
- Excellent sensitivity to be able to image small quantities of arresting contrast conjugates.
- High specificity and affinity to the target to yield a significant signal-to-background ratio shortly after administration.
- The possibility to deliver a drug and its localized release.
- Flexibility to act as a platform technology to couple different targeting molecules and deliver all kinds of relevant drug substances.

This chapter attempts to identify whether ultrasound in combination with smart ultrasound contrast agents (USCAs) can be a powerful technique in applying the "Find, Fight and Follow!" concept in future clinical practice.

7.2
Ultrasound: "Hear the Symptoms"

Ultrasound is a versatile imaging technique that provides real-time diagnostic information on the morphology of internal organs with high spatial resolution and sufficient penetration depth. Imaging is based on transmitting and receiving sound waves with frequency ranges from 1 to 50 MHz. The resolution is directly related to the frequency used, e.g., the higher the frequency, the better the resolution. On the other hand, high-frequency waves are more attenuated than low-frequency waves, which limits the penetration depth. Consequently, the frequency used for imaging is a trade-off between optimal resolution and desired penetration depth. Typical frequencies of 2–7.5 MHz are used to image large organs inside the body (with a submillimeter range resolution and penetration to about 30 cm) [6].

The acoustic energy used in diagnostic ultrasound (intensity below 0.1 W cm^{-2}) is very low compared to ultrasound for therapeutic applications [7] (usually much greater than 1 W cm^{-2} for clot lysis or transdermal drug delivery, which will not be addressed in this chapter). Noncontrast diagnostic ultrasound has been established as an important, rapid and cost-effective means in almost all medical fields. One major advantage of ultrasound is the fact that patients are not exposed to harmful radiation. There are no limits with regard to examination time and frequency, and even investigations of embryos in utero are now routine [an impressive video showing the three-dimensional (3-D) scanning of an embryo is shown on the supplementary CD-ROM]. Additionally, ultrasound energy can be focused on a small volume (down to cubic millimeters), which makes it a perfect tool to trigger actions inside the body (e.g., drug release from a "smart contrast agent").

To understand the following principles, only two of the more than 10 ultrasound-based imaging modalities are important. The so-called B-mode utilizes the specific ultrasound backscattering properties of different types of tissue and blood. Reconstruction of the backscattering signals leads to an image of the morphology. The second so-called Doppler mode detects and measures the blood flow by a frequency shift (Doppler shift) between transmitted and received ultrasound signals. Modern 3-D image processing enables a fused 3-D image of anatomical structure (gray scale) and functional blood flow (color coded) at the same time [8].

Additionally, many different color Doppler mode variations are implemented in high-end imaging devices (power Doppler, harmonic Doppler, harmonic power Doppler, etc.) and in principle these variations are suitable to highlight contrast agent-specific signals in a color-coded manner fused with a 3-D gray-scale image of the morphology.

In summary:
- Ultrasound imaging is safe (no harmful radiation used).
- The B-mode images the morphology at submillimeter resolution with sufficient penetration depth.
- Different kinds of Doppler modes are useful to highlight contrast-specific signals.
- Fused 3-D images are state of the art.
- Certain ultrasound pulses with focused energy can induce drug release from "smart contrast agents".

7.3
Ultrasound Contrast: "Tiny Bubbles"

The diagnostic capabilities have not only been improved by technological advances in the ultrasound systems, but also by the introduction of USCAs. Since 1980, extensive research has been performed in order to make contrast ultrasound an established diagnostic technique [9]. At the beginning of this period, investigators were forced to use home-made contrast agents and it was quickly realized that small bubbles increased the contrast dramatically.

This is due to the fact that the contrast factor in the underlying Rayleigh equation [10] depends on the density difference and compressibility difference of a dispersed matter in the surrounding media. In particular, if a liquid is dispersed in another liquid, the value of the contrast factor is about 0 (i.e., there is almost no density difference and, thus, no compressibility difference between liquids) and a solid dispersed in a liquid increases the value just slightly to about 1–2. However, in the case of air in water (or blood) this contrast factor raises to 10^{14}, which makes dispersed air bubbles a perfect contrast agent for ultrasound.

Another important parameter is the size of the bubbles. On the one hand, the scattering cross-section increases proportionally with the diameter of the bubble to the power of 6 (Rayleigh scattering). On the other hand, the bubbles must be small enough to pass through the smallest capillaries of the pulmonary system in order to prevent an embolus in the blood vessel. Considering safety and efficacy, a size range of 1–7 μm is most preferable for a gas-bubble-based contrast agent.

The first generation of commercialized USCAs comprises free air bubbles or surfactant-stabilized air bubbles (Echovist or Levovist). In order to increase their lifetime in blood, the second-generation USCAs (e.g., Echogen, Imagent, SonoVue or Optison) were made out of gas with low water solubility (like perfluorcarbon gases or sulfurhexafluoride) and a surface stabilizer to prevent aggregation.

A scientific breakthrough in order to design USCAs on demand can be seen in the third generation (Myomap, Quantison, BiSphere and Sonavist). Compared to the more or less free bubbles of the first and second generations, the novel type of USCAs consist of encapsulated microbubbles with a shell formed by a biopolymer (like human albumin) and/or a biocompatible synthetic polymer (like copolymers of polylactide and polyglycolide or derivatives of polycyanoacrylate). In addition to the prolongation of the lifetime in the blood stream, these polymer-stabilized microbubbles can be manufactured to fulfill certain needs, and to interact with diagnostic ultrasound in a defined and optimal manner.

The acoustic properties of encapsulated gas bubbles were intensively discussed by de Jong et al. [11, 12], Hoff [13] and Frinking and de Jong [14] in the 1990s. Based on their theoretical considerations, it can be shown that after choosing the nature of the gas and the type of polymer, the behavior of a gas-filled microsphere in an ultrasound field depends only on the geometry of the capsule; in particular, the size and shell thickness. This means that scattering and resonance properties can be tuned by the geometry of the USCA.

Moreover, based on stability relations in architecture it can be proven that the critical fragility of a hollow spherical body depends on the modulus of elasticity of the shell material (a physical constant determined by the nature and molecular weight of the polymer used) and the quotient of shell thickness over size. This means that we can fix the pressure threshold of a gas-filled microsphere by the geometry so that the sphere collapses above a certain, critical value of the applied acoustical pressure. This is extremely important due to the fact that the USCA should resist the blood pressure and a certain level of additional acoustic pressure that is just needed for ultrasound imaging. However, at higher acoustic pressure the shell of the gas-filled microcapsule should be destroyed.

It is now straightforward (see also Part V, Chapter 6) to fill the capsule not only with air, but a drug substance too, and to use an ultrasound pulse to trigger the release from outside the body at a defined place and time by bursting the bubble. Utilizing nanotechnological concepts of polymer, colloid and interface science we have established a novel process to manufacture gas-filled microcapsules. On demand, gas filling, elasticity, shell thickness and overall size can be tailored independently.

Using the so-called "two-step process" [15, 16], polymer nanoparticles are first synthesized via emulsion polymerization. The size of the resulting nanoparticles can be tuned by a simple process parameter and covers a range of about 30–400 nm. In a second step these nanoparticles are used to coat microbubbles in a controlled bubble formation process. The nanoparticles migrate to the surface of the bubbles (this is related to the interface activity of hydrophobic nanoparticles in general) and build a monolayer around the bubbles. Consequently, the size of the nanoparticles determines the shell thickness of the final microcapsules. Additionally, a carefully chosen nanoparticle concentration regime results in a certain microcapsule size distribution. In principle, particle sizes in the range of 0.5–10 μm can be adjusted and the microcapsule size distributions are ex-

tremely narrowly distributed. Last, but not least, the molecular weight of the polymer, which can be controlled in the first phase of the two-step process, determines the elasticity (within certain limits) of the shell-forming polymeric material without changing the chemical constitution.

An image of a gas-filled polymer-stabilized microcapsule obtained by electron microscopy is depicted in Fig. 7.1. The overall spherical shape and the substructure of nanoparticles are visible. The underlying "two-step process" is schematically described in Fig. 7.2. To highlight the flexibility of the manufacturing process, the tunable quality aspects like resonance behavior and pressure stability are summarized in Figs. 7.3 and 7.4, respectively.

In summary, the innovative "two-step process" is an enabling technology to manufacture gas-filled microparticles with defined colloidal architecture to fulfill all the

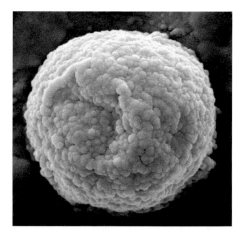

Fig. 7.1 An image of one selected gas-filled polymer-stabilized microcapsule obtained by electron microscopy. The "magic bullet" is spherical with a diameter of about 1.7 μm and the substructure made of nanoparticles is clearly visible. In order to indicate that the particles are in fact gas-filled, an imperfect capsule with a bump has been carefully selected.

1. Step: Emulsion-Polymerization

2. Step: Coating of microbubbles with nanoparticles

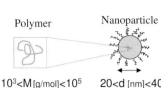

Polymer   Nanoparticle

$10^3 < M\,[\text{g/mol}] < 10^5$ $20 < d\,[\text{nm}] < 400$

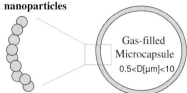

Gas-filled
Microcapsule
$0.5 < D[\mu m] < 10$

Shell thickness ~ Size of nanoparticles

Fig. 7.2 The principles of the "two-step process" suitable to control several parameters of an USCA independently. In the first step, polymer nanoparticles are synthesized via emulsion polymerization. The size of the resulting nanoparticles can be tuned by a simple process parameter and covers a range of about 30–400 nm. In the second step, these nanoparticles are used to coat microbubbles in a controlled bubble formation process. Consequently, the size of the nanoparticles determines the shell thickness of the final microcapsules. Additionally, a carefully chosen nanoparticle concentration regime yields a certain microcapsule size distribution. Particle sizes in the range of 0.5–10 µm can be adjusted and the size distributions are extremely narrowly distributed. The molecular weight of the polymer, which determines the elasticity (within certain limits) of the shell-forming polymeric material, can be controlled in the first phase of the two-step process.

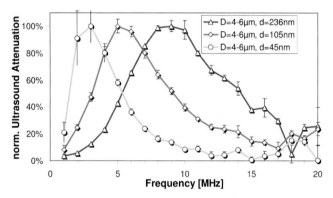

Fig. 7.3 The resonance behavior of three different microcapsule formulations obtained by ultrasonic spectroscopy. The ultrasound attenuation (y-axis, given relative to the maximum) is measured over a driving frequency from 1 to 20 MHz (x-axis). Each capsule population has the same mean diameter of about 5 µm, but the shell thickness differs significantly (45, 105 and 236 nm). Tailoring the geometry of the USCA tunes the resonance properties. For example, increasing the shell thickness at constant size of the gas-filled microcapsules yields an increase in resonance frequency (about 2, 5 and 9 MHz for the given examples). The sharply defined attenuation spectra indicate that the size distribution of the capsules is narrowly distributed. Using a mixture of bubbles with certain acoustic properties, combining them in one contrast agent could be the basis for ultrasound multiplexing (as in the case of optical spectroscopy, the basis for optical imaging).

physical requirements on an "ultrasound mode-specific contrast agent" (and advanced drug delivery system). Considering the strong sensitivity of ultrasound to small variations of the structure of USCAs, even multiplexing of microcapsules due to certain resonance behavior or critical fragility is possible.

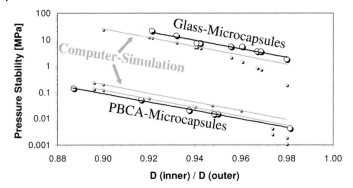

Fig. 7.4 To highlight a second tunable quality aspect of specially designed USCAs, the pressure stability against a geometry factor (inner diameter over the outer diameter) of capsules is plotted. This factor has been selected in order to compare the properties of small USCAs with commercially available glass capsules, which are usually larger (100–700 μm). The fact that both fitted lines are parallel proves that polymer and glass capsules follow the same physical principles. Just the modulus of elasticity of glass is about 100 times higher than in the case of the polymeric material used to manufacture the USCA. Computer-aided simulations have been performed to validate the findings. It is easy to understand that gas-filled polymer capsules with a quotient (inner to outer diameter) larger than about 0.94 cannot resist the blood pressure. Furthermore, the critical fragility of capsules can be again exploited for multiplexing and, most importantly, for drug delivery as well.

7.4
The Perfect Modality: "Sensitive Particle Acoustic Quantification (SPAQ)"

Depending on the sound pressure of an acoustic field, microbubbles can behave quite differently. Using low sound pressures, they oscillate and behave as linear scatterers improving the signal-to-noise ratio. Above a certain amplitude, gas-filled microcapsules disintegrate rapidly, thereby emitting a strong nonlinear signal, which is misinterpreted as a quick movement and mapped as a characteristic random color pattern in the color Doppler mode of the ultrasound device. This characteristic "bubble signature" firstly described by Reinhardt et al. 1992 [17] is called "stimulated acoustic emission" (SAE) or "loss of correlation" (LOC) [18–20], as depicted in Fig. 7.5. (Actually, a disintegrating bubble sends no acoustic signal, but the rapid shrinking process caused by the dissolution of gas changes the backscattering situation in such a way that the algorithms of common Doppler modes process an apparent movement).

However, the SAE effect enables the examiner to detect stationary microbubbles [21, 22] – a property that has generated a large field of interest; in particular, because the amplitude of SAE signals is even strong enough to detect individual microbubbles within a given tissue. Thus, the SAE effect is an extremely interesting tool and could be the key to highly sensitive molecular imaging.

Reliable quantification of microbubbles within the tissue, even in high concentrations, is needed to enable the efficient use of molecular imaging for temporal and spatial evaluation of molecular targets. A major constraint in the quantification of SAE signals is that SAE signals are displayed at millimeter-size in the image, be-

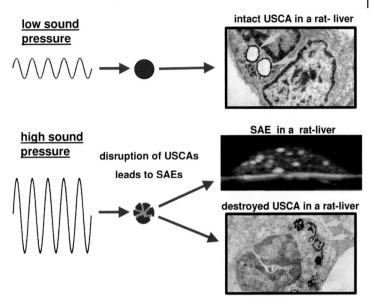

low sound pressure

intact USCA in a rat- liver

high sound pressure

disruption of USCAs

leads to SAEs

SAE in a rat-liver

destroyed USCA in a rat-liver

Fig. 7.5 Principle of SAE. Using high-amplitude sound pressure, the shell of the microcapsules breaks and the air bubble disintegrates rapidly. The sudden disappearance of the microbubble is misinterpreted by the ultrasound machine as a quick movement and color-coded in the image. This "pseudo-Doppler" effect is known as SAE.

Additionally, microcapsules coated with gold nanoparticles (to give increased contrast for identification in the electron micrograph) have been investigated in rat liver with electron microscopy before and after application of ultrasound (yellow arrows). Only fragments of shells can be identified in tissue post-ultrasound.

cause the device cannot locate the signal within the resolution cell (voxel). Due to the fact that a single SAE signal fills 1 voxel, more than a single microparticle within 1 voxel cannot be discriminated and any amount of contrast agent within 1 voxel would yield the same result in the image.

To overcome this inherent problem of SAE signal quantification, we developed a new quantification method called SPAQ, which is based on a defined overlap between consecutive ultrasound images. The SPAQ technology is schematically described in Fig. 7.6, and has been validated in agarose phantoms and ex-vivo in animal tissue [23, 24].

The experiments show that an accurate quantification of static microbubbles from a single bubble event up to 300 000 bub-

bles mL^{-1} and 3-D mapping of large volumes is possible with a resolution in a scan direction of down to 10 µm.

The great advantage of gas-filled microcapsules as reporters for molecular imaging is their outstanding detection sensitivity. With SPAQ, it will now become possible to quantify such targeted microbubbles with maximum precision in an automatic scan process that functions independently of the examiner and is highly reproducible.

7.5

Targeting and Molecular Imaging: "The Sound of an Antibody"

After discovering the suitable quantification technology, an USCA can be designed

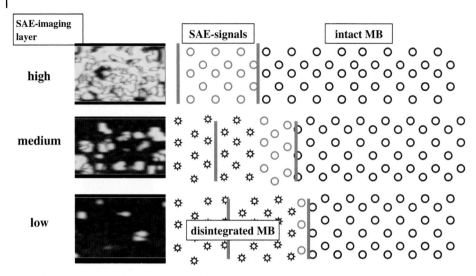

Fig. 7.6 SPAQ to enable the quantification of SAE signals of static USCAs with high accuracy. The principle of SPAQ is to create consecutive ultrasound images with a defined overlap. Thereby only microcapsules lying out of the overlapping region are subjects of SAE. The degree of the overlap determines the thickness of the SAE imaging layer. As a consequence of the higher spatial resolution of this method, single gas-filled microcapsules can be discriminated and therefore quantified even in high concentrations.

and optimized appropriately (i.e., with regard to gas filling, size, shell thickness, elasticity and pressure stability) utilizing the flexible "two-step process", which is perfectly suitable for SPAQ.

Now the question is: "How can we make the bubbles smart in a way that they can find targets inside a living organism?". The answer can be found again in the principles of colloid and interface science. It is well known that a surface modification of polymeric nanoparticles with biologically important antigens leads to particles which are useful for testing the presence of compliment antibodies in human blood. After an *ex vivo* mixing of blood together with the antigen–nanoparticle conjugates, large, visible aggregates are formed due to the interparticle connection via relevant antibodies.

Different coupling strategies have been established to manufacture these antigen–nanoparticle conjugates for the commonly called "aggregation assays" [25]. One of the most important is the [1-ethyl-3-(3-dimethylaminopropyl) carbodiimide] (EDC)-activated reaction which forms an amide bond between a primary amine of the biologically active substance and a carboxy group on the surface of particles. Thus, all we need to do is to introduce carboxy groups on the surface of the polymer-stabilized gas-filled microbubbles.

In the case of polybutylcyanoacrylate microbubbles a little caustic soda solution is needed to perform a hydrolysis of the ester groups just on the particle surface and the free carboxy groups can be used for the coupling of targeting ligands [15, 26]. To be more flexible in animal experiments we have preferred the coupling of streptavidin to the microcapsules. Biotinylation of antibodies is well established and the affinity of biotin to streptavidin is outstanding. To prepare the final formulation for *in vitro* and *in vivo* tests, incubation of the biotiny-lated antibodies together with streptavidin-coated microbubbles for about 10 min at room temperature is sufficient to produce the ready-to-use formulation.

Streptavidin and antibody loading can be quantified by a FACS assay. It could be shown that the streptavidin loading per microcapsule can be adjusted from several hundreds to about 1 million.

The first *in vitro* investigations using a flow-chamber assay proved the fact that target-specific microbubbles tag on the target even at high flow rates [27–29] and an antibody loading of not more than 10^5 per bubble is absolutely sufficient. Alternative coupling strategies to streptavidin–biotin have also been tested successfully [26, 30].

Different groups have shown that micro-bubbles can be coated with target-specific antibodies and could demonstrate the efficacy *in vivo* [31–36] with respect to a binary "yes or no" result. However, quantitative information about the amount of arrested USCA has only just recently been obtained.

Based on the microcapsules and coupling strategy described above, the *in vivo* feasibility was first demonstrated in mice and dogs. Using an i.v. administered L-selectin ligand-specific USCA, active lymph node targeting could be performed in both species [37]. Exemplarily, impressive results based on this investigation are presented in Fig. 7.7.

Furthermore, it could be shown that the induction of the microbubble-based SAE effect in color Doppler and the rapid blood clearance of nontargeted USCAs by cells of the reticuloendothelial system within 30–60 min, depending on the dose used, yields a sensitive detection method and a strong signal-to-noise ratio of targeted microcapsules. Moreover, it provides the op-portunity of repeated investigations within one session using USCAs targeted to the same or other endothelial cell receptors. This was the first time that active USCA targeting has been demonstrated in lymph nodes under normal conditions after i.v. administration.

L-selectin ligand-specific USCA could be a candidate for an indirect method of lym-phography for the safe and less-invasive ultrasonic identification of lymph nodes, e.g., when performing a biopsy. Lymph node-targeted microbubbles can be de-tected easily with any ultrasound device that has color Doppler capabilities.

After targeting healthy lymph nodes without further quantification of the re-sults, the SPAQ technique has been tested successfully in an animal model of inflam-mation [38–40]. By conjugating microbub-bles with antibodies to intercellular adhe-sion molecule (ICAM)-1 and vascular cell adhesion molecule (VCAM)-1, the authors were able to depict and quantify ICAM-1 and VCAM-1 in rat autoimmune encepha-lomyelitis (EAE), an established inflamma-tory disease model of human multiple sclerosis (MS) [41, 42]. Additionally, after treatment with methylprednisolone, the measured number of targeted anti-ICAM-1 and VCAM-1-microcapules (in spinal cord and brain) was significantly less ($P<0.01$) compared to untreated animals. This re-sult is extremely important, because this breakthrough highlights for the first time that a target-specific USCA can be suitable to act as a "reporter of efficacy" with re-spect to a therapeutic effect.

The workflow of a routine SPAQ proce-dure and *in vivo* images (through the skull in living animals) regarding ICAM-1 target-ing together with the quantitative result are presented in Figs. 7.8 and 7.9. Additionally, to validate the results regarding monitoring of therapy with the SPAQ technique, a first

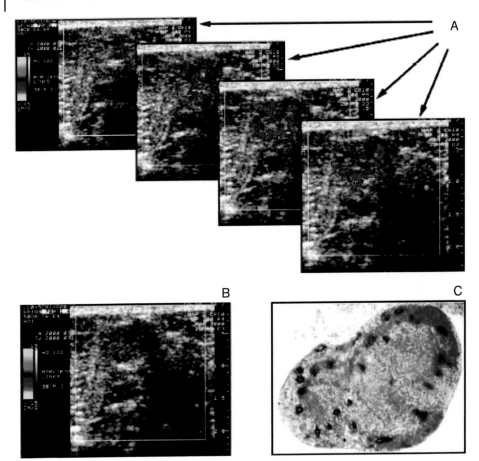

Fig. 7.7 Two-dimensional color Doppler ultrasound images show a dog lymph node treated
with administered L-selectin ligand-specific USCA.
(A) Four selected images from the first scan. (B)
Image from the second scan of the same lymph
node, which was performed immediately after the
first scan. The images in (A) show typical microbubble-based SAE signals in the paracortex,
whereas no signals are visible in the second scan
(B), which proves the destruction of microcapsules during the first scan. (C) The SAE signal
distribution in (A) corresponds immunohisto-
chemically to the high endothelial venule location.
(Reproduced with permission. © 2004 Radiology
Society of North America, Inc.)

quantitative *ex vivo/in vivo* correlation has
been performed [40, 43].

In order to round off the picture, the
"sounds of antibodies" were recently re-
corded in several tumor models. For example, the expression of extradomain B (ED-
B)-fibronectin (FN) in tumors could be
measured in appropriate animal models.

ED-B-FN is an angiogenesis-specific target exclusively found in the area of newly
formed blood vessels [44–47]. A target-specific cyanine dye with single-chain antibodies directed against ED-B-FN was reported
by Neri et al. [48] (see also Part V, Chapter
6). In addition, near IR fluorescent FN
analogs have recently been successfully ap-

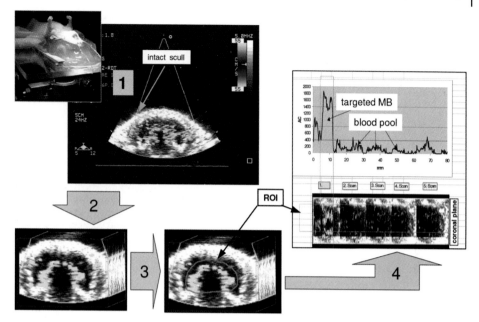

Fig. 7.8 The workflow of a routine SPAQ procedure. (Above left) A rat is placed in the SPAQ device. Four basic steps are illustrated. (1) Two-dimensional scans of the brain through the skull. (2) Generating a 3-D dataset. (3) Definition of a region of interest. (4) Data analysis. Subsequent investigation of the same region of interest identifies the amount of circulating USCAs (blood pool), which can be subtracted from the amount of microcapsules arrested in the target.

plied for optical imaging of tumor angiogenesis in mice [49, 50]. A new affinity-matured recombinant antibody single-chain fragment called AP39 has been used (see also Part V, Chapter 2).

The same antibody fragment with proven specificity *in vivo* based on optical imaging was used to manufacture AP39–microbubble conjugates. In a validated rat MTLN3 tumor model, a dose of just 5×10^8 microbubbles kg^{-1} body weight of AP39–microbubble conjugates has been administered. (Equal to about 8×10^{-16} mol kg^{-1}! Compared to about 5×10^{-8} mol kg^{-1} antibody–dye conjugates in the case of optical imaging or approximately the same 10^{-8} mol kg^{-1} dose for radiolabeled compounds.) Applying the SPAQ technique just 15 min post-injection [compared to an

imaging window of 4 (earliest point) to 24 h (maximum contrast-to-background ratio) in case of optical imaging [50]] the amount of 36461 target-specific microbubbles could be quantified within a tumor of 15 mm size. The control experiment under equal conditions, but using an isotype control antibody, leads to just 798 statistically arrested bubbles in the target tumor.

To give a final impression of the molecular imaging capabilities of the SPAQ technology, a typical result of the tumor-imaging study described above is given in Fig. 7.10.

Considering the results on targeting healthy, inflammatory and carcinogenic tissue, it is quite realistic to state that, depending on the antibody linked to the surface of the gas-filled microcapsules, the technique

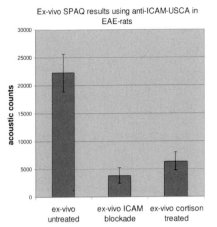

Ex-vivo SPAQ results using anti-ICAM-USCA in EAE-rats

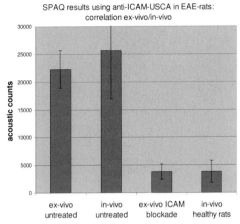

SPAQ results using anti-ICAM-USCA in EAE-rats: correlation ex-vivo/in-vivo

Fig. 7.9 Quantitative results of SPAQ investigations in accordance to the workflow described in Fig. 7.8. Bars show mean values and standard deviations (n=4) of SAE signals expressed as acoustic counts of ICAM-1 targeted ultrasound contrast agents at the blood brain barrier. The values are registered from the entire brain of healthy rats and rats with adoptive transfer experimental autoimmune encephalomyelitis (EAE-rats), respectively. For further experimental details please refer to Reinhardt et al [40]. Based on different investigations five important theses could be verified: (1) proof of principle – regarding molecular imaging, depiction and reproducible quantification of target-specific microcapsules; (2) proof of specificity – by blocking the target with high doses (200 µg) of anti-ICAM-1 antibodies prior to the administration of anti-ICAM-1 conjugated USCA; (3) proof of sensitivity – since glycocorticosteroids like methylprednisolone are known to cause a down regulation of adhesion molecules, the treatment of EAE-rats with Urbason (50 mg/kg b.w., daily injection for 3 days) results in a significant decrease in acoustic counts; (4) target-specific USCA as reporters of efficacy – both experiments, target blockade and cortison treatment, demonstrate the potential of targeted microbubbles together with SPAQ to monitor and quantify a therapeutic effect; (5) ex vivo/in vivo correlation – all validation experiments were performed ex vivo on isolated tissue (in a water tank) (graph left), while additional experiments were carried out on living anesthetized animals through the intact skull. The comparison (graph right) shows a first quantitative ex vivo/in vivo correlation as depicted based on the untreated experimental groups. Furthermore, the ICAM blockade experiment yields the a similar number of acoustic counts as the control group of healthy rats investigated with ICAM-1 targeted USCA as well.

can be used to quantify the expression of any accessible antigen expressed on the luminal surface of endothelial cells. Thus, the combination of target-specific USCAs with SPAQ is therefore a promising tool for the noninvasive and dynamic assessment of disease-related molecules.

In our view, the technique will certainly be first established as a molecular imaging-based drug discovery and development tool in preclinical practice. However, long-term possible clinical applications can be foreseen in safe and less-invasive ultrasound identification of certain tissue (e.g., lymph nodes), inflammatory regions of MS or the sensitive detection of small tumors. Also cardiovascular indications are conceivable due to the more vascular, blood-pool character of the USCAs. Finally, the great advantages of dynamic and quantitative assessment should lead to the use of target-specific USCAs as a disease-staging tool and as "reporters of efficacy" for evidence-based patient selection to reach

unspecific microparticles, 5e8 MPs/kg, 15´p.i.
(SPAQ-resolution: 50μm)

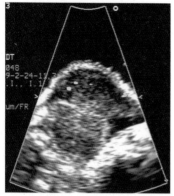

AP-39-bio/microparticle-conjugates, 5e8 MPs/kg, 15´p.i.
(SPAQ-resolution: 50μm)

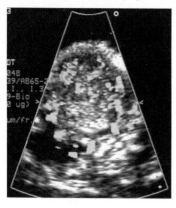

Fig. 7.10 *In-vivo* measurement of ED-B-FN in MTLN3 tumor of rats with AP39–microcapsule conjugates using SPAQ. (Above) Two-dimensional color Doppler image 15 min after administering unspecific USCAs. Only four SAE signals are visible in the target region. (Below) The same model, but 15 min after injection of target-specific microcapsules. Significantly more SAE (182 signals at a SPAQ resolution of 50 μm) are detectable. The investigation of the total tumor, which is about 15 mm in diameter, leads in this special case to an amount of 36 461 arrested microparticles. Assumed the tumor is spherical (and thus occupies a volume of about 1.8 mL), 20 633 contrast conjugates mL^{-1} tissue have been quantified. This shows a practical example of the outstanding sensitivity of the method.

the visionary 100% responder rate – a challenging, but worthwhile long-term aim.

7.6
Drug Delivery: "The Magic Bullet"

A perfect drug delivery system would be noninvasive, detectable from outside the body and should be able to target a diseased region (see also Part VI, Chapter 1). Furthermore, the release should be triggered and quantified by external means.

At the beginning of the last century Paul Ehrlich – the founder of chemotherapy – introduced the idea of "the magic bullet" ("Zauberkugel") [51] for medical applications. His vision of a receptor-specific drug, which acts in a highly specific manner for a disease without any side-effects, can be seen as the start of the search for advanced drug delivery systems.

For example, nanosized polymeric drug carriers for targeting and controlled release have been extensively studied in the past. In this case, a nanoparticle or capsule acts like a container for a pharmacologically active agent. Passive and active targeting can be attained by carefully choosing the size and surface modifications of the carrier (see also Part VI, Chapter 8). Conventionally, drug release can be controlled via desorption of surface-bound drugs, diffusion through the particle matrix or the capsule wall, or matrix erosion. In addition, "smart" release can be achieved by using smart polymers (pH or temperature sensitive) or, more interestingly, by applying an external stress to the drug carrier. As mentioned above, if the drug carrier is appropriately designed, release can be induced by diagnostic ultrasound.

The first patents on the combined diagnostic and therapeutic use of USCAs were filed by Stein et al. [52] in 1990 and Weit-schies et al. [53] in 1994, and early descriptions of an ultrasound contrast-based "micromachine" were published by Ishihara [54] in 1996. In the same year, Unger [55] presented a similar concept. Two years later Frinking et al. [56] published initial *in vitro* results on the forced release of an encapsulated compound from a microspherical carrier by high-intensity ultrasound. However, the applied spray-drying process showed several drawbacks (i.e., just 5% w/w loading in the shell, no accurate control of acoustical properties of the resulting microspheres, 20% immediate release of surface-bound drug, etc.), but it was the first published "proof-of-principle" *in vitro*.

Independently, at the same time we have developed a novel encapsulation procedure [15]. The starting material is the well-defined microsphere dispersion produced via the two-step process. These microcapsules are mixed with the drug substance of interest and heated above a certain temperature – the "critical flooding temperature" (CFT) – where the drug solution migrates through the now leaky shell into the capsules. (Several experiments have shown that the CFT is determined by the nature of polymer and its molecular weight.) Subsequent cooling stops the flooding process immediately and a carefully performed final freeze-drying procedure leads to solidification of drug inside the microcapsule.

Optionally, depending on the solubility of the drug in water, the filling CFT process can be repeated several times and it has been shown that the drug loading can be performed above 100% (w/w), which means that the shell thickness is more than double sized. (Of course, the filling influences the acoustical properties, but this can be considered in the planning of the "two-step process" to form the starting microcapsules with weaker shells).

Formulations which are appropriate for *in vivo* applications have been successfully tested *in vitro* with respect to drug release induced by diagnostic ultrasound. A typical release profile is shown in Fig. 7.11.

Unfortunately, the CFT process is limited to drugs with molecular weights smaller than 80 kDa. Alternatively, encapsulation of material with higher molecular weight, e.g., reporter plasmids (pDNA), in gas-filled microspheres can be performed by spray-drying or double-emulsion procedures [57]. Due to the energy impact in the case of double-emulsion procedures, for example, it is common to protect and stabilize the pDNA with polycations such as polylysine or polyethyleneimine. Fig. 7.12 schematically summarizes some of the encapsulation procedures.

Another strategy for drug delivery with gas-filled polymer-stabilized microbubbles is based on a surface binding of drugs, especially drugs with a high molecular weight. In addition to covalent and adsorptive coupling, linking can be easily performed by using electrostatic interactions.

A further look into the literature on colloid and interface science demonstrates that ionic interactions of oppositely charged polyions lead to extremely stable complexes [58]. This is also true for the interactions of charged surfaces and polyions. Considering that genetic material (naked DNA and plasmids, and also the surface of several types of virus) is in certain cases negatively charged, a complex can be formed by mixing with a positively charged polyion (e.g., polylysine). In addition to the protection and stabilization effect, such ionic complexes have also shown a better transfection rate than naked DNA (e.g., for synthetic nonviral vector systems) [59] (see also Part VI, Chap-

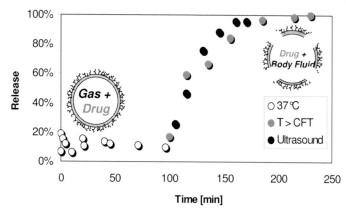

Fig. 7.11 Typical drug release profile of gas- and drug-filled microcapsule formulations. Percent release of an encapsulated model drug in physiological buffer media as a function of time. Storage under stirring for about 100 min at 37 °C shows that just a baseline concentration of free drug is measurable. By applying diagnostic ultrasound to a certain volume of the dissolution media, a strong release can be induced. Similar results can be obtained by increasing the temperature above the CFT of the capsules.

Approaches for USCAs as drug delivery systems

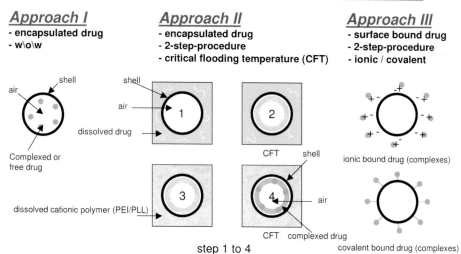

Fig. 7.12 Overview of some technical approaches for the preparation of drug-loaded USCAs. Approaches I and II describe encapsulation techniques based on solvent evaporation of a water/ oil/water emulsion and the CFT procedure, respectively. Additionally, surface coupling strategies are depicted in Approach III.

ters 6 and 7). Furthermore, complexes with a surplus of cationic polyelectrolytes lead to overall positively charged conjugates which can be attached again via electrostatic interactions to the surface of carboxy-modified gas-filled microcapsules [15, 26]. In an adaptation to the concept of "gene guns" based on gold nanoparticles, which are coated with DNA and can be injected using air pressure through the skin into surface cells [60], these pDNA-complex-coated microbubble conjugates may be called "noninvasive micro-gene guns".

7.7
Ultrasound, Microbubbles and Gene Delivery: "Noninvasive Micro-Gene Guns"

In gene therapy, efficient and target-site-specific *in vivo* gene delivery is a major challenge. Free DNA without any delivery system has been found to be highly susceptible to nuclease degradation [61] and is rapidly cleared from the plasma when injected i.v. [62]. Although viral-based delivery systems such as adenoviruses [63] and retroviruses [64] efficiently introduce genes, they suffer from drawbacks in terms immunogenicity and toxicity (see also Part I, Chapter 7 and Part II, Chapter 7). The immunogenicity of viral vectors limits repeated use of the delivery system. Therefore, nonviral gene delivery systems such as cationic lipids, liposomes and polymeric microspheres [59, 65–67] have been increasingly investigated as alternatives to viral vectors due to their potential advantages, such as ease of preparation and scale-up, as well as their relative safety and lack of immunogenicity (see also Part VI, Chapters 6, 7, and 8). Their disadvantages may be the lower transfection efficiency and the transient nature of transfection compared to that of viral vectors. Cur-

rently, one of the most important obstacles to viral and nonviral gene delivery systems is the lack of organ and cell specificity (see also Part VI, Chapter 5).

In order to overcome the drawbacks of viral vector systems and previously investigated nonviral systems, researchers focused their interest on microbubbles at the end of the last century. Unger et al. developed an USCA for gene delivery, where the gene material is entrapped in the center of the microbubbles [68]. These microbubbles are stabilized by a so-called soft shell consisting of a lipid bilayer. The first *in vivo* studies have shown that ultrasound treatment of the heart of rabbits led to gene expression after i.v. administration of encapsulated pDNA containing a marker gene [69]. Shohet et al. [70] demonstrated that albumin-stabilized microbubbles coated with adenovirus direct transgene expression to the myocardium of rats after their treatment with diagnostic ultrasound.

Additionally, diagnostic ultrasound can even promote and enhance released pDNA uptake by surrounding cells [71, 72] (e.g., Sonoporation™ does not even use microbubbles at all).

Overall, the major advantages seen in using USCAs as a gene delivery system are:
(1) Protection of the gene material in the blood stream against nuclease degradation.
(2) Site-specific and spatiotemporally controlled delivery of gene material.
(3) Monitoring of the release procedure (SAE).
(4) Use of the ultrasound-mediated sonoporation effect.

Most of the published reports on ultrasound-facilitated gene delivery experiments were focused on finding an effective method for the genetic treatment of ischemic diseases, mainly by intramuscular injection of DNA and/or microbubbles [73–78]. A review on the use of USCAs for gene delivery in cardiovascular medicine has been published recently by Bekeredjian et al. [79]. Only a very few researchers are investigating this approach for tumor gene therapy using marker genes that do not exert a therapeutic effect such as β-galactosidase and luciferase, mainly by using intratumoral substance administration [80, 81].

More recently, pDNA gas-filled microcapsule constructs have been investigated in combination with diagnostic ultrasound for ultrasound-mediated tumor gene delivery. Investigations have been performed in different rodent tumor models [82].

The authors used the model plasmid pUT651 containing the *Escherichia coli LacZ* gene for β-galactosidase to demonstrate the feasibility in CC531 liver tumors of rats. The pUT651-containing USCAs were administered in a first preliminary experiment intra-arterially and in a second experiment i.v. with simultaneous sonication of the tumors, both in a small number of animals.

Furthermore, the potential medical impact of this delivery system has been tested in Capan-1 (human pancreas adenocarcinoma)-bearing nude mice, using the plasmid pRC/CMV-p16 (containing the tumor suppressor gene p16). The tumor suppressor gene p16, which plays an important role in anoikis, is deleted in Capan-1 cells. The outcome of this study is summarized based on immunohistochemical evaluation and tumor volume doubling time in Figs. 7.13 and 7.14, respectively.

As a result, a clear expression of the pDNA was found in tumors of rats treated with a combination of pUT651containing USCA and ultrasound, while relevant controls showed a significantly lower expres-

sion of the marker gene. Additionally, the therapeutic effect of p16 was measured as an increase of the tumor volume doubling time. The controlled ultrasound-triggered release of the expression vector for the tumor suppressor gene p16 from USCAs leads to a strongly significant ($P < 0.002$) tumor growth inhibition in comparison with controls. This is the first time that a reliable tumor treatment effect was shown after i.v. injection of p16-containing USCA, and site- and time-controlled DNA release using a common diagnostic ultrasound device. In conclusion, it could be demonstrated that the combination of gene-material-carrying USCA and ultrasound provides a novel and effective approach for nonviral gene therapy-based cancer treatment.

A further improvement in this advanced drug delivery approach can be expected by the ability to target microbubbles via targeting moieties to different tissues or regions of disease (see also Part V, Chapters 2 and 6). This would allow site-specific delivery and release of therapeutic agents directly on the surface of the relevant cells. While a bubble increases and subsequently implodes within a very short time, the implosion results in so-called liquid "jet streams" at up to 400 km h^{-1}. The jets hit against the surface of the cell and make its membrane reversible leaky for a very short period of time. Utilizing these jet streams, the drug can be shot like a bullet into the cell, which again underlines the "gene gun" character of the microcapsules.

Furthermore, "magic bullet"-like properties can be derived from several attempts at manipulation. Dayton et al. [83] have shown that "primary" and "secondary" radiation force can be used to manipulate flowing contrast agents by ultrasound. Primary radiation forces may be used to direct bubbles to a specific region of the ves-

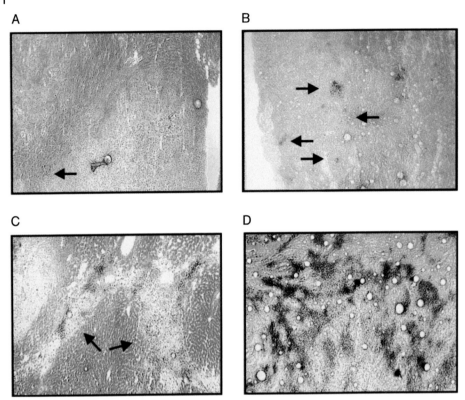

Fig. 7.13 β-Galactosidase staining in CC531 liver tumors to evaluate the gene transfection efficiency. (A–C) Beta-galactosidase activity after intra-arterial administration and (D) after i.v. substance administration. As indicated by the arrows, poor activity was seen with naked pUT651 and no ultrasound treatment (A), moderate activity was seen with naked pUT651 and ultrasound treatment (B), and strong activity was seen with encapsulated pUT651 and ultrasound treatment (C). (D) Strong activity of β-galactosidase 3 days after i.v. administration of encapsulated pUT651 and ultrasound treatment. (Reproduced with permission. © 2005 Radiology Society of North America, Inc.)

sel wall where the flow velocity is lower. Bubbles can even be stopped in a flow by primary radiation forces at certain acoustic pressures and high pulse repetition frequencies [84]. Additionally, secondary radiation force may be used to form aggregates of bubbles in the location to be treated. Finally, a SAE pulse induces the release of encapsulated and/or adsorbed drugs. In all the cases described, the amount of released drug can be monitored with SPAQ due to the fact that the concentration of the released drug is proportional to the encapsulation efficiency and number of ruptured microcapsules.

7.8
Summary: Ultrasound Theranostics "Building a Bridge between Therapy and Diagnosis"

Imagine you are a 30-year-old female. You take a blood test and receive a DVD that contains your genetic make-up. Using a

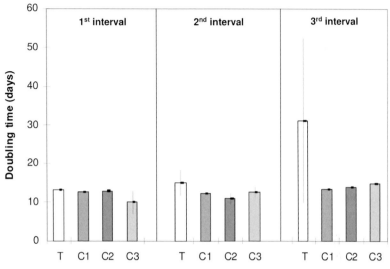

Fig. 7.14 Gene therapy effects confirmed by the tumor volume doubling times. Evaluation intervals of the tumor treatment study over three intervals. Tumor volume doubling times of each group are presented for each evaluation interval. T=treatment group (encapsulated pDNAp16 with ultrasound treatment), C1=control 1 (encapsulated pDNAp16 without ultrasound treatment), C2=control 2 (encapsulated pDNAneo with ultrasound treatment), C3=control 3 (0.9% NaCl with ultrasound treatment). (Reproduced with permission. © 2005 Radiology Society of North America, Inc.)

web service, you find that you are predisposed to an aggressive form of breast cancer. Your primary care physician refers you for an ultrasound exam that uses a new angiogenesis target-specific contrast agent. It shows signs that new microvasculature is being created in the left breast, characteristic for early tumor growth.

A follow-up exam using a mixture of new ultrasound contrast agents, which are suitable for multiplexing the evidence of several targets at the same time, confirms based on a quantitative finding that a couple of specific proteins are overexpressed, fueling aggressive tumor growth. Thus, the malignancy is proven by a quantitative evaluation on a concert of tumor-specific biomarkers.

A therapeutic drug encapsulated in gas-filled microcapsules targeted to one of the most overexpressed vascularly accessible protein is administered and released directly at the tumor, triggered from outside of the body via ultrasound. The amount of released drug can be monitored due to the visibility of each single disrupting microcapsule and the knowledge of its drug loading. After a sufficient concentration of therapeutic has reached the target, the treatment will be stopped.

Finally, a follow-up ultrasound exam with target-specific USCAs confirms that the overexpression is under control and the vasculature creation has returned to normal. [Scenario adapted from GE-Healthcare [85] 2005.]

Yes, this sounds like science fiction, but the use of SPECT or PET/CT (and some early magnetic resonance imaging applications) (see also Part V, Chapters 4 and 5) in combination with biochip-related diagnostic tools (see also Part I, Chapter 3 and Part V, Chapter 8) and highly specific antibody-based radiodiagnostics and radiotherapeutics (see also Part V, Chapters 2 and 6) makes this "Find, Fight and Follow!" scenario close to clinical practice. Today, interdisciplinary teams are professionally developing tomorrow's technology platforms for *ex vivo* diagnostic, molecular imaging and molecular medicine, and that is the key – exploiting the synergies between different disciplines of life sciences!

Although the largest number of targeting agents is available in the field of radiopharmaceuticals (and more targeted agents are in the approval process), the main bottleneck in an integrative "theranostics" approach for clinical application is the lack of available and approved targeting contrast agents and advanced drug delivery systems.

The role of microbubbles in the "Find, Fight and Follow!" triad is not near a clinical routine as yet. However, preclinical results reported so far are more than promising.

State-of-the-art ultrasound is a high-resolution imaging modality. Ultrasound is a safe and well-established tool to image morphology, and provides information about diseases at the anatomical level. In combination with contrast agents and the SEA-based sensitive particle acoustic quantification technique, findings can be confirmed based on actual figures. The concentration of microbubble contrast agent in the region of interest can be quantified with outstanding sensitivity (down to single contrast agent signals), and give maximum precision in an automatic scan process that works independently of the examiner and is highly reproducible. Recent progress in the manufacture of contrast agents enables us to design gas-filled microcapsules with

characteristic acoustic properties and a certain critical fragility, and enables the coupling of targeting moieties. Target-specific USCAs have proven their suitability for molecular imaging in many indications, and can selectively track down molecules and cell structures to be able to establish proof of disease at a very early stage. Furthermore, target molecules and therapeutic effects can be visualized and quantified to make decisions on highly individualized treatment (see also Part I, Chapter 2). Considering the strong sensitivity of ultrasound to small variations of the structure of USCAs, even multiplexing of microcapsules due to certain resonance behavior or critical fragility is possible.

Additionally, the USCA can act as a container to deliver drugs like small molecules and biopharmaceuticals, and even gene delivery is possible. Finally, controlled release regarding site, time and amount of drug can be realized and visualized.

We think that the described technique will be first established as a molecular imaging-based drug discovery and development tool in preclinical practice. However, possible clinical applications of the "technetium of ultrasound" can be foreseen in the safe and less-invasive ultrasound identification of certain tissue (e.g., lymph nodes), inflammatory regions or the sensitive detection of small lesions. Also, as stated before, cardiovascular indications are conceivable due to more vascular, blood-pool character of the USCAs.

The described advanced drug delivery qualifications are being investigated in preclinical settings, but considering that polymeric nano- and microparticle drug delivery systems are currently under investigation in several clinical trials, the step to ultrasound contrast-based drug delivery is just a small one.

The cover of *Modern Biopharmaceuticals* shows a sketch of a nanorobot working within the blood stream. The artist – inspired by Asimov – possibly thought about the diagnostic and therapeutic capabilities of Ehrlich's vision of the "magic bullet". We believe that the "ultrasound theranostics" concept is currently one of the most promising candidates to realize Paul Ehrlich's vision in the new millennium.

References

1 Adapted from: Moby Dick by Herman Melville; Ch. 134. Penguin Modern Classics, London, originally **1851**.
2 Kaminski MS, Zelenetz AD, Press OW, et al. *J Clin Oncol 19*, 3918–3928, *2001*.
3 Wiseman GA, White CA, Stabin M, et al. *Eur J Nucl Med 27*, 766–777, **2000**.
4 Vesselle H, Vesselle H, Grierson J, et al. *Clin. Cancer Res 8*, 3315–3323, **2002**.
5 Drevs J, Muller-Driver R, Wittig C, et al. *Cancer Res. 62*, 4015–4022, **2002**.
6 Wells PNT. Biomedical Ultrasonics. Academic Press, London, **1977**.
7 Tachibana K, Tachibana S. *Echocardiography 18*, 323–328, **2001**.
8 Nelson TR, Pretorius DH. *Ultrasound Med Biol 24*, 1243–1270, **1998**.
9 Ophir J, Parker KJ. *Ultrasound Med Biol 15*, 319–333, **1989**.
10 Lord Rayleigh (Strutt JW). *Philos Mag 34*, 94–98, **1917**.
11 de Jong N, Hoff L, Skotland T, Bom N, *Ultrasonics 30*, 95–103, **1992**.
12 de Jong N, Hoff L, *Ultrasonics 31*, 175–181, **1993**.
13 Hoff L. *Ultrasonics 34*, 591–593, **1996**.
14 Frinkink PJA, de Jong N. *Ultrasound Med Biol 24*, 523–533, **1998**.
15 Briel A. *J Pharm Pharmacol 56 (Suppl)*, 288, **2004**.
16 Budde U, Briel A, Rößling G, et al. Patent DE19925311.0, **1999** (and US 6652782 B1).
17 Reinhardt M, Fritzsch T, Heldmann D, Siegert J, Patent WO 93/25241, **1993**.
18 Uhlendorf V, Hoffmann C. *Proc IEEE Ultrasonics Symp 3*, Cannes, France, 1559–1562, **1994**.
19 Uhlendorf V, Scholle FD, Reinhardt M. *Ultrasonics 38*, 81–86, **2000**.
20 Tiemann K, Pohl C, Schlosser T, et al. *Ultrasound Med Biol 26*, 1161–1167, **2000**.
21 Blomley M, Albrecht T, Cosgrove D, et al. *Radiology 210*, 409–416, **1999**.
22 Hauff P, Fritzsch T, Reinhardt M, et al. *Invest Radiol 32*, 94–99, **1997**.
23 Reinhardt M, Hauff P, Briel A, Schirner M. *Proc 3th Annu Meet Soc Molecular Imaging*, St. Louis, MO, p. 148, **2004**.
24 Reinhardt M, Hauff P, Briel A, et al. *Invest Radiol 40*, 2–7, **2005**.
25 Hermanson GT. Bioconjugates. Academic Press, New York, **1996**.
26 Briel A, Rößling G, Hauff P, et al. Patent: DE 10013850.0, **2000** (and WO 01/68150).
27 Joseph S, Olbrich C, Kirsch J, Hasbach M, Briel A, Schirner M, *Pharm Res 21*, 920–926, **2004**.
28 Takalkara AM, Klibanova AL, Rychaka JJ, Lindner JR, Leya K, *J. Controlled Release 96*, 473–482, **2004**.
29 Joseph S, Schlehnsog M, Groenewold T, Quandt M, Olbrich C, Briel A, Schirner M, *Biosens Bioelectron 20*, 1829–1835, **2005**.
30 Olbrich C, Mossmayer D, Willuda J, Schmidt W, Schirner M. Presented at Controlled Release Society, Miami, FL, **2005**.
31 Lanza GM, Wallace KD, Scott MJ, et al. *Circulation 94*, 3334–3340, **1996**.
32 Klibanov AL, Hughes MS, Marsh JN, et al. *Acta Radiol. 412 (Suppl)*, 113–120, **1997**.
33 Unger E, Metzger P, Krupinski E, et al. *Invest Radiol 35*, 86–89, **2000**.
34 Lindner JR, Song J, Christiansen J, et al. *Circulation 104*, 2107–2112, **2001**.
35 Leong-Poi H, Christiansen J, Klibanov AL, et al. *Circulation 107*, 455–460, **2003**.
36 Ellegala DB, Leong-Poi H, Carpenter JE, et al. *Circulation 108*, 336–341, **2003**.
37 Hauff P, Reinhardt M, Briel A, et al. *Radiology 231*, 667–673, **2004**.
38 Mäurer M, Linker R, Hauff P, Reinhardt M, Briel A, et al. *J Neurol 249 (Suppl 1)*, I/57, **2002**.
39 Hauff P, Reinhardt M, Mäurer M, Linker R, Briel A, Schirner M. *Proc 3rd Annu Meet Soc for Molecular Imaging*, St. Louis, MO, p. 150, **2004**.
40 Reinhardt M, Hauff P, Briel A, Linker RA, Gold R, Rieckmann P, Becker G, Toyka K, Mäurer M, Schirner M. Neuroimage, in press **2005**.

41 Gold R, Hartung H-P, Toyka KV. *Mol Med Today 6*, 88–91, **2000**.

42 Archelos JJ, Hartung H-P. *Mol Med Today 3*, 310–321, **1997**.

43 Hauff P, Reinhardt M, Mäurer M, Linker R, Briel A, Toyka KV, Schirner M. Presented at 6th Int Symp on Ultrasound Contrast Imaging, Tokyo, **2004**.

44 Birchler M, Neri G, Tarli L, Halin C, Viti F, Neri D, *J Immunol Methods 231*, 239–248, **1999**.

45 Tarli L, Balza E, Viti F, et al. *Blood 94*, 192–198, **1999**.

46 Viti F, Tarli L, Giovannoni L, Zardi L, Neri D. *Cancer Res 59*, 347–352, **1999**.

47 Santimaria M, Moscatelli G, Viale GL, et al. *Clin Cancer Res 9*, 571–579, **2003**.

48 Neri D, Carnemolla B, Nissim A, et al. *Nat Biotechnol 15*, 1271–1275, **1997**.

49 Lich K, *J Pharm Pharmacol 56 (Suppl)*, 272, **2004**.

50 Licha K, Perlitz C, Hauff P, et al. *Proc Soc Mol Imaging 264*, **2004**.

51 Dieterle W (Director). Dr. Ehrlich's Magic Bullet [A movie about Paul Ehrlich's live.]. Warner Bros, Los Angeles, CA, **1940**.

52 Stein M, Heldmann D, Fritzsch T, Siegert J, Rößling G. Patent US 6264959 B1 based on application 07/536373, **1990**.

53 Weitschies W, Heldmann D, Hauff P, Fritzsch T, Stahl H. Patent US 6068857 based on application 08/605174, **1994**.

54 Ishihara K. *Proc 2nd Int Micromach Symp*, 69–75, **1996**.

55 Unger E, et al. Patent US 5542935, **1996**.

56 Frinking PJA, Bouakez A, de Jong N, ten Cate FJ, Keating S. *Ultrasonic 96*, 709–712, **1998**.

57 Seemann S, Hauff P, Schultze-Mosgau M, Lehmann C, Reszka R. *Pharm Res 19*, 250–257, **2002**.

58 Antonietti M, Conrad J, Thünemann AF, *Macromolecules 27*, 6007, **1994**.

59 Hart SL. *Exp Opin Ther Patents 10*, 199–208, **2000**.

60 Williams RS, Johnston SA, Riedy M, DeVit MJ, McElligott SG, Sanford JC. *Proc Natl Acad Sci USA 88*, 2726–2730, **1991**.

61 Kawabata K, Takkura Y, Hashida M, *Pharm Res 12*, 825–830, **1995**.

62 Yoshida M, Mahato RI, Kawabata K, Takakura Y, Hashida M. *Pharm Res 13*, 599–603, **1996**.

63 Matsuse T, Teramoto S. *Curr Ther Res Clin Exp 61*, 422–434, **2000**.

64 Kurian KM, Watson CJ, Wyllie AH. *J Clin Pathol, Mol Pathol 53*, 173–176, **2000**.

65 Deléfine P, Guillaume C, Floch V, et al. *J Pharm Sci 89*, 629–638, **2000**.

66 Simoes S, Slepuskin V, Pires P, Gaspar R, de Lima MCP, Duzgunes N. *Biochim Biophys Acta Biomembr 1463*, 459–469, **2000**.

67 Jones DH, Corris S, McDonald S, Clegg JCS, Farrar GA. *Vaccine 15*, 814–817, **1997**.

68 Unger EC, Hersh E, Vannan M, McCreery T. *Echocardiography 18*, 355–361, **2001**.

69 Unger EC, McCreery TP, Schweitzer R. Presented at Macromolecular Drug Delivery Conf, Breckenridge, CO, p. 20 (abstr), **1999**.

70 Shohet RV, Chen S, Zhou Y-T, et al. *Circulation 101*, 2554–2556, **2000**.

71 Anwer K, Kao G, Proctor B, et al. *Gene Ther 7*, 1833–1839, **2000**.

72 Bao S, Thrall BD, Miller DL. *Ultrasound Med Biol 23*, 953–959, **1997**.

73 Taniyama Y, Tachibana K, Hiraoka K, et al. *Gene Ther 9*, 372–380, **2002**.

74 Schratzberger P, Krainin JG, Schratzberger G, et al. *Mol Ther 6*, 576–583, **2002**.

75 Li T, Tachibana K, Kuroki M, Kuroki M. *Radiology 229*, 423–428, **2003**.

76 Lu QL, Liang HD, Partridge T, Blomley MJ. *Gene Ther 10*, 396–405, **2003**.

77 Yamashita Y, Shimada M, Tachibana K, et al. *Hum Gene Ther 13*, 2079–2084, **2002**.

78 Bekeredjian R, Chen S, Frenkel PA, Grayburn PA, Shohet RV. *Circulation 108*, 1022–1026, **2003**.

79 Bekeredjian R, Grayburn PA, Shohet RV. *J Am Coll Cardiol 45*, 329–335, **2005**.

80 Miller DL, Bao S, Gies RA, Thrall BD. *Ultrasound Med Biol 25*, 1425–1430, **1999**.

81 Manome Y, Nakamura M, Ohno T, Furuhata H. *Hum Gene Ther 11*, 1521–1528, **2000**.

82 Hauff P, Seemann S, Reszka R, et al. *Radiology*, in press **2005**.

83 Dayton PA, Morgan KA, Klibanov AL, Brandenburger GH, Nightingale K, Ferrara KW. *IEEE Trans Ultrason Ferr Freq Con 44*, 1264–1277, **1997**.

84 Dayton PA, Klibanov AL, Brandenburger GH, Ferrara KW. *Ultras Med Biol 25*, 1195–1201, **1999**.

85 Scenario adapted from GE-Healthcare (2005). http://egems.gehealthcare.com/geCommunity/nmpet/interest_groups/molecular_imaging/education/intro.jsp

Getting Insight – Sense the Urgency for Early Diagnostics

8
Development of Multi-marker-based Diagnostic Assays with the ProteinChip® System

Andreas Wiesner

Abstract

The earliest possible diagnosis of a disease is imperative for an efficient therapy. This simple fact is true for virtually any type of disease, and is best documented for tumorous conditions. At present, the single protein markers used exhibit high false-negative and/or false-positive rates, and the hope for more reliable DNA- and RNA-based screening tools has not been fulfilled. As a consequence, research activities have turned back to proteins as the real key players in pathological processes. Over the past few years, comparative protein profiling by surface-enhanced laser desorption/ionization time-of-flight mass spectrometry (SELDI TOF-MS) has become acknowledged as a promising means of detecting specific and predictive protein patterns that reflect certain stages of cancer, neurological disorders, and infectious diseases. The SELDI-based ProteinChip® System fulfils the needs of the new proteomic era by enabling comparative protein profiling by Expression Difference Mapping analysis of several hundreds of samples per day on a single, technology platform, using software support for the construction of multi-marker predictive models. The Interaction Discovery Mapping™ platform is introduced as the next methodical step for antibody-based assays and investigations into protein binding partners of possible importance in the diagnosis of conditions and the development of biopharmaceuticals.

Abbreviations

AFP	alpha-fetoprotein
AUC	area under the curve
CEA	carcinoembryonic antigen
EAM	energy-absorbing molecules
HCG	human chorionic gonadotropin
HPV	human papillomavirus
HR-PAC	host response protein amplification cascade

Modern Biopharmaceuticals. Edited by J. Knäblein
Copyright © 2005 WILEY-VCH Verlag GmbH & Co. KGaA, Weinheim
ISBN: 3-527-31184-X

ICAT	isotope-coded affinity tag
ID	identification
ITIH-IV	inter alpha trypsin inhibitor IV
LDH	lactate dehydrogenase
NSE	neuron-specific enolase
PAP	prostatic acid phosphatase
PCA	principal components analysis
POMC	pro-opiomelanocortin
PSA	prostate-specific antigen
ROC	receiver operating characteristics
RT-PCR	reverse transcription-polymerase chain reaction
SELDI TOF-MS	surface-enhanced laser desorption/ionization time-of-flight mass spectrometry
SEND	surface-enhanced neat desorption
TOF-MS	time-of-flight mass spectrometry
VEGF	vascular endothelial growth factor

8.1
The Urgency of Earlier Diagnosis

The earlier a disease is diagnosed, the more efficient a therapy can be. This is true for virtually any type of disease, including infectious conditions [1], Alzheimer's disease as well as other dementias [2, 3], and is best documented for tumorous diseases [4], where the current compilation of internationally acknowledged markers in human blood [5] contains not more than a handful of proteins: Prostate-specific antigen (PSA), prostatic acid phosphatase (PAP), CA 125, carcinoembryonic antigen (CEA), alpha-fetoprotein (AFP), human chorionic gonadotropin (HCG), CA 19-9, CA 15-3, CA 27-29, lac-

tate dehydrogenase (LDH), and neuron-specific enolase (NSE). Most of these markers were first described long ago – PSA in 1971, CA 125 in 1981, CEA in 1955, AFP in 1956, HCG in 1927, and CA 15-3 in 1984 [6] – and are mainly used for prognostic purposes. Applications in tumor diagnosis are very limited, especially for the early cancer stages [4], where the success rates are disappointingly low [7]. The survival rates of tumor patients diagnosed late in disease progression and suffering from regionally or distantly spread tumors have shown little improvement over the past 30 years. Despite huge research efforts, only a few targeted therapeutics, such as Herceptin® (trastuzumab), which is directed against the HER-2 receptor of breast cancer cells [8, 9] (see Part I, Chapter 5) are in use and in general, nonspecific cytotoxic agents are administered with limited success (see Part V, Chapter 6). It is clear that only an earlier diagnosis could improve the situation, and this will be certainly easier to achieve than the development of better targeted methods for the treatment of advanced-stage cancers.

The early detection of cervical cancer is an encouraging example for the beneficial result of early diagnosis by less invasive screening methods [4]. A cytological assay to detect the very early stages of cervical cancer was developed in the mid-twentieth century, and countries that included the "pap smear" assay in their national screening programs were able drastically to reduce the incidence of this disease. Later, when human papillomaviruses (HPVs) were recognized as causing the cancer by persistent infection, augmentation of the "pap smear" with molecular tests for HPV detection resulted in a powerful tool for early diagnosis. Many more such positive examples are urgently needed to save more lives, and to earn back the patients' trust in the benefits of diagnostics and healthcare [10].

8.2
Proteins are Best Choice Again

During the 1980s, the disappointments of single protein markers forced the greater part of the scientific community to turn to the field of genomics, with the aim being to develop more efficient diagnostic tools [11]. In most cases, reverse transcription-polymerase chain reaction (RT-PCR) was used to detect tumor-specific DNA alterations, cell-free and mitochondrial tumor DNA, tumor-associated viral DNA, and tumor RNA. Due to the high – but not always error-free – amplification of minute amounts of DNA or RNA, as well as the difficulty of predicting the stability of RNA, false-positive results were rather common. Furthermore, data reproducibility was difficult to achieve, and results from different laboratories were seldom comparable. In addition, the development of microsatellite analysis of tumor DNA could not be successfully established in the diagnosis of bladder, lung or colorectal cancers. The detection of tumor-associated DNA hypermethylation is another methodical PCR variation which has shown promising results, but has not been established in the clinical area, mainly because the levels of false-positive and/or false-negative results were too high [12]. Furthermore, interpretation of the data is often difficult. For example, DNA is released by not only malignant but also benign cells; hypermethylation is not only tumor-specific but is also age-related; consequently, all of the reported sensitivity and/or specificity values require major improvements to be made in order for the test to be useful [13–17]. Taken together, improved nucleic acid-based methods will certainly play some role in the future of tumor diagnostics, but they are far from fulfilling the promise shown at the beginning of the genomic era.

Without doubt, one of the main reasons for this lack of success by genomic methods is the fact that, on the DNA and RNA level, we are too far away from the actual physiologically active gene products, the proteins. The statement "...DNA makes RNA makes protein..." is simply wrong and misleading as it implies a carbon-copy-like process for the information pathway from DNA to messenger RNA and the final protein product. The statement even entitled a slide in the 1993 Nobel lecture of Phillip A. Sharp, who received the prize for the detection of split genes and RNA splicing. He described the process as similar to "...the work that a film editor performs: the unedited film is scrutinized, the superfluous parts are cut out and the remaining ones are joined to form the completed film." In such a way the fibronectin gene can code for up to 20 different fibronectin protein variants with different locations and functions [18].

However, even when the final sequence and quantity of mRNA is known, it is often not possible to predict the protein composition because of a poor correlation between mRNA and protein expression levels [19–21]. Furthermore, the exact knowledge of the finally processed m-RNA sequence will not help to predict the protein sequence. Many proteins are additionally spliced into smaller pieces before becoming biologically active [22]. For example, this phenomenon is well known for pro-opiomelanocortin (POMC)-derived peptides in the hypothalamus [23], where a 32 kDa precursor protein can be spliced into several different hormones, each with a different range of activity. Thus, the primary sequence of the protein does not depend on the DNA sequence or even the m-RNA sequence alone. Virtually any protein can be subject to a number of post-translational modifications such as phosphorylation, acylation, glycosylation and/or other types of covalent alterations. Many of these modifications are reversible, and determine the localization, structure and binding behavior of the protein, in turn resulting in other sophisticated regulatory activities of possible pathological importance [24].

Consequently, DNA tells what could happen, and RNA what might happen – but only proteins can tell what *actually* happens. Thus, real progress in the development of new diagnostic tools and therapeutic strategies will depend mainly on the new insights coming from investigations into proteins, their local concentration, structure, function, and interaction [25, 26].

8.3
Current Tools
for Protein Biomarker Detection

On the basis of mRNA processing and post-translational modifications, the estimated number of different human proteins is about 500 000 – more than 15 times higher than the estimated number of coding genes in humans. Although only a fraction of these proteins are present at any given time in any particular location, it will be a special challenge to catalogue their structures and functions, to describe their actual concentrations differing by orders of magnitude, and to detect their crosstalk-activities in complexes [27–29]. Identifying, cataloging and functionally describing all human proteins will be infinitely more challenging than it was to sequence the human genome [30].

On the other hand, realizing that the overall protein pattern of tissues or body fluids of a given individual is very similar to that of others from the same species (and is even more similar when indivi-

duals are of the same gender, age, and physiological state), the situation appears less complicated from the diagnostic point of view [31]. With regard to biomarker discovery, it is more promising to focus on the differences between individual protein patterns than to try to achieve a complete protein mapping of all the components. Learning more about which proteins are unique to a certain physiological state will allow new diagnostic assays to be developed, and to target the search for new biopharmaceuticals [32].

In analytical terms, ease of use and reliable instrumentation with quantitative capability and high sample throughput, combined with appropriate software tools for handling the data output, are needed. Only then can comparative profiling studies lead to the discovery and validation of new biomarkers. The system of two dimensional polyacrylamide gel electrophoresis (2D-PAGE) – where proteins are separated first by their isoelectric point and subsequently by their molecular weight – indeed contributed greatly to our understanding of the wide variety of proteins in a given sample. However, proteins in the peptide range as well as those of high hydrophobicity or of extreme isoelectric points are difficult to separate, are thus typically neglected, and result in a loss of potentially interesting proteins. Furthermore, the 2D-PAGE approach is time-consuming, difficult to standardize, and is in no way suited to the rapid screening of higher sample numbers, not even in the hands of experts in this method [33, 34]. Similarly, column chromatography – in all its variations – is well suited for purification purposes, but not for the efficient and reproducible comparative analysis of dozens or hundreds of samples per day. More recently, an isotope-coded affinity tag (ICAT) procedure was developed to overcome the difficulties in protein quantitation [35]. Here, two protein samples are differentially labeled on their cysteine residues, whereby the labeling molecules contain different isotopes. The proteins are then cleaved proteolytically, undergo HPLC-separation, and are finally analyzed using mass spectrometry, when the differently labeled peptides can be recognized by a specific shift in mass. Although this is a quite sophisticated approach, there are several practical drawbacks, caused mainly by incomplete labeling or artifact generation. Therefore, tagless extraction-retentate chromatography methods have been considered to be a more promising alternative [36].

Protein micro-arrays equipped with specific capture molecules such as antibodies, fragments thereof, peptides, or other bait molecules have been developed in recent years [37, 38]. Here, the challenges lay not only in the generation of libraries of cap-

ture molecules that are able to mirror the variety of proteins (and the slightly modified versions of each single type!) in a given sample, but also in the experimental design – where all problems known to be associated with protein interaction experiments and labeling procedures must be taken into account. This task is much more difficult to accomplish than it was with DNA microarrays, and it will take many more years before those types of micro-arrays are available as robust, affordable tools, and unpredicted screening will be very difficult to achieve [39].

8.4
The ProteinChip® System at a Glance

The ProteinChip System was developed by Ciphergen Biosystems, Inc. (www.ciphergen.com) to meet the demands of assay

Chemical Surfaces – Protein Expression Profiling (Expression Difference Mapping™)

Hydrophobic   Anionic   Cationic   Metal Ions   Hydrophilic

Biological Surfaces – Protein Interaction Assays (Interaction Discovery Mapping™)

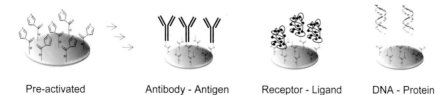

Pre-activated   Antibody - Antigen   Receptor - Ligand   DNA - Protein

Fig. 8.1 The different types of ProteinChip Arrays. The spots on the chromatographic ProteinChip Arrays possess hydrophobic, cationic, anionic, metal ion affinity, or hydrophilic properties. These "chemical surfaces" are best suited for protein expression profiling studies. Another series of Pro-

teinChip Arrays have pre-activated "biological surfaces" designed for coupling of biomolecules onto the spots with applications in antibody–antigen assays, receptor–ligand interaction studies and DNA-protein binding experiments.

development in the post-genomic era. Based on the patented SELDI process [40–42], it combines two formerly well-established methods – solid-phase chromatography and time-of-flight mass spectrometry (TOF-MS) – into one easy-to-use system for carrying out Expression Difference Mapping applications on a single integrated platform [43–46].

The basic procedure for Expression Difference Mapping studies with the ProteinChip System is quite simple. Virtually any type of protein-containing solution can be applied to the spots of the ProteinChip Arrays. These spots present either chromatographic surfaces with certain physicochemical characteristics (hydrophobic, cationic, anionic, metal ion-presenting or hy-

drophilic), or are pre-activated for the coupling of capture molecules (protein, DNA, or RNA) prior to sample loading (Fig. 8.1).

Typically, for Expression Difference Mapping experiments, chromatographic surfaces are used. The sample requirement is low (1–10 µg total protein per spot), and the sample volume can be freely chosen from 0.5 µL up to around 300 µL. After a short incubation period, unbound proteins and any contaminants on the spot surface are washed away. A solution of energy-absorbing molecules is applied to each of the spots, and the ProteinChip Array is ready for the analysis in the ProteinChip Reader, a highly sensitive laser desorption/ionization TOF-MS instrument. The results are initially visualized in a graph, with the mass

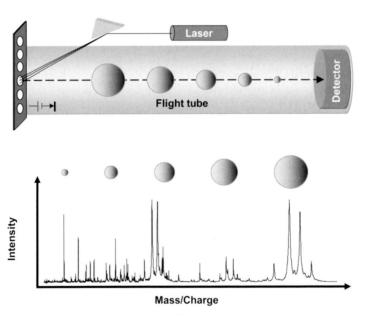

Fig. 8.2 Sample analysis in the ProteinChip Reader. After loading the ProteinChip Array into the ProteinChip Reader, a laser beam is directed onto the sample on the spot. Thereby protons are transferred onto the peptides and proteins that are subsequently accelerated by electromagnetic fields through a flight tube. The time-of-flight is inversely proportional to the molecular mass. This results in a spectrum being generated where the molecules in the sample are represented in a graph with the mass to charge ratio of the native compounds on the x-axis and the corresponding signal intensity on the y-axis.

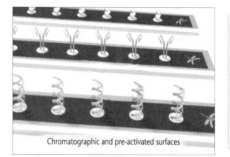

Step 1: Choosing an array
ProteinChip® Arrays are available with different chromatographic properties, including hydrophobic, hydrophilic, anion exchange, cation exchange, and immobilized-metal affinity surfaces. Other ProteinChip Arrays with pre-activated surfaces are available for covalently coupling protein, DNA, RNA or other "bait" molecules by the user.

Chromatographic and pre-activated surfaces

Step 2: Sample application
Crude biological samples can be applied directly to the ProteinChip Arrays. Application can be done manually by pipetting or by employing Ciphergen's customized configuration of the Beckman Coulter Biomek® 2000 laboratory automation station.

Step 3: Washing
After a short incubation period, unbound proteins are washed off the surface of the ProteinChip Array. Only proteins interacting with the chemistry of the array surface are retained for analysis. After washing, energy absorbing molecules are applied to the array as a final step.

SELDI

Step 4: Analysis
ProteinChip Arrays are then analyzed in the ProteinChip Reader, a time-of-flight mass spectrometer. The mass values and signal intensities for the detected proteins and peptides can be viewed in several formats and then transferred to Ciphergen's software suites for further in-depth analysis.

Fig. 8.3 The four steps of sample analysis. Despite covering a wide range of application areas, the basic methodical procedure of sample analysis with the ProteinChip System is very similar in all the different types of experiments.

to charge ratio of the sample components plotted on the x-axis, and the corresponding signal intensities plotted on the y-axis (Fig. 8.2). ProteinChip Software enables the user to control and influence the automated detection process, and also incorporates a wide range of univariate and multivariate software tools for the comparative analysis of higher numbers of samples. Thus, the entire Expression Difference Mapping technique can be achieved in a fast, four-step procedure (Fig. 8.3).

The ProteinChip System Series 4000 Enterprise AutoBiomarker Edition, comprised a ProteinChip Reader with autoloader capacity to automatically run – simultaneously – up to 168 bar-code-labeled ProteinChip Arrays with, in total, 1344 samples, a customized Biomek® 2000 liquid handling robot and CiphergenExpress™ Data Manager Software for automated sample tracking and advanced data analysis together with the Biomarker Patterns™ Software for the direct develop-

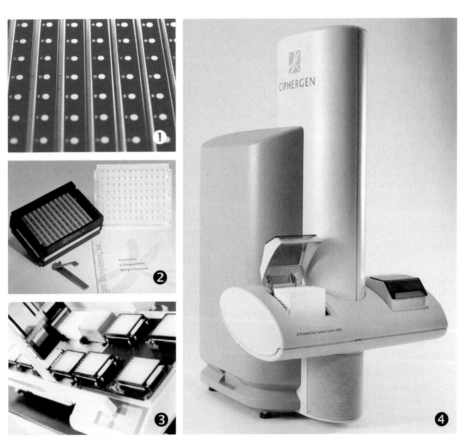

Fig. 8.4 Hardware components of the ProteinChip System. The ProteinChip Arrays (1) are arranged in bioprocessors to match the 96-well microtiter plate format (2). Samples can be applied using a customized Biomek® 2000 liquid-handling robot (3). The ProteinChip System Series 4000 Enterprise Edition (4) automatically runs up to 168 bar-code labeled ProteinChip Arrays with, in sum, 1344 samples at once.

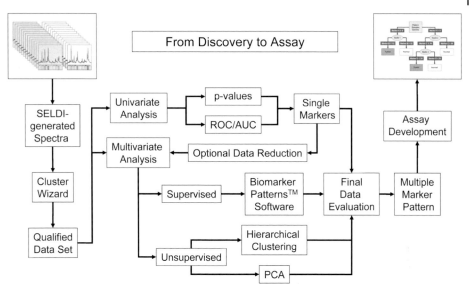

Fig. 8.5 Software-enabled assay development. The CiphergenExpress™ Software with Data Manager and Biomarker Analysis Modules is a relational database system with a client-server architecture designed for automated sample tracking and advanced data analysis. After identification and selection of clusters for meaningful data reduction, univariate and multivariate data analysis tools are used for the detection of single and multiple markers. ROC (Receiver Operating Characteristics), AUC (Area Under the Curve), and PCA (Principal Components Analysis) calculations are implemented together with appropriate data visualization formats such as cluster diagrams, ROC plots, heat maps and 3-D projection of PCA data. The Biomarker Patterns™ Software is the key component for the direct development of multi-marker-based diagnostic assays.

ment of diagnostic assays on the same platform (Figs. 8.4 and 8.5).

8.5
Distinctions of the SELDI Process

As described in Section 8.4, only those proteins which interact actively with the spot surfaces are analyzed in the ProteinChip Reader, since all other components are washed off in advance. Such washing provides one of the most obvious advantages of the SELDI process – that certain components (e.g., salts or detergents) which commonly cause problems with other analytical tools are removed.

Additionally, as each analysis is also linked to an on-spot fractionation step, the complexity of the samples is reduced, thus providing much higher chances for the detection of lower concentrated markers. Normally, a number of different array types are combined with washing steps of varying stringencies to retain different protein subsets from the original samples. Equivalent array–buffer combinations of the different sample groups can then be analyzed comparatively to decipher the formerly unknown markers. With complex samples such as serum, pre-fractionation is preferable as it enhances the total number of signals resolved. Subsequent biomarker purification is easier to achieve because, in addition to the exact molecular weight, the basic physico-chemical characteristics of the protein of interest can be elucidated from its binding behavior under certain washing conditions. Last – but not least – active interaction of the analyte with the spot surface ensures another important feature of the ProteinChip technology, namely its ability to quantify individual protein levels in a given sample; this is a prerequisite for the successful use of any technology in comparative profiling. As the proteins are equally distributed on the spot surfaces and the laser positioning is conducted very exactly, the resulting signal intensities correspond very well to the concentration of the proteins (Fig. 8.6).

8.6
The Pattern Track™ Process: From Biomarker Discovery to Assay Development

Many recent publications have provided proof of the suitability of the ProteinChip System in the discovery and validation of new biomarkers for a wide range of diseases. This is true for ovarian [47], prostatic [48], pancreatic [49] and head and neck [50] cancer, as well as for a large number of other tumorous diseases [7]. Similarly, successful studies have been reported for Alzheimer's disease [51], viral [52], bacterial [53], and parasitic [54] infections. The same applies to investigations about possible new markers to monitor the effect of drug treatment [55–57], and also to predict transplant rejections [58, 59]. The ProteinChip System is currently used by more

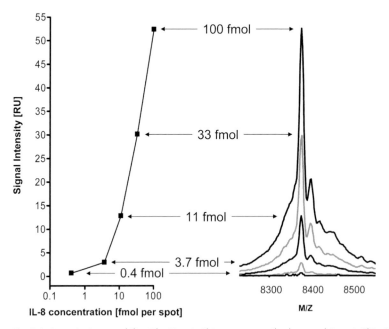

Fig. 8.6 Quantitation capability. The ProteinChip System allows the relative and absolute quantitation of proteins. In this example, interleukin-8 (IL-8) was spiked into human serum and captured on an antibody-coated ProteinChip Array. After analysis, the resulting signal intensities correspond very well to the concentration of the proteins.

than 1000 customers worldwide in pharmaceutical and biotechnology laboratories, as well as in academic and governmental institutes. As diverse as the research areas might be, one result is common to most of them: Multimarker panels have been discovered and were found to be superior over single marker-based traditional assays. For the elucidation of such patterns, a specialized tool – the Biomarker Patterns Software – has been developed. This supervised learning program performs multivariate analyses and is able to classify the processed data by the use of specially adapted algorithms [60]. It creates tree-like structured decision diagrams by splitting the original data set (parent node) in two sub-groups (child nodes) of highest possible purity. Each child node then becomes a parent node at the time of creation, and can be the origin of a new split. The program calculates which signals (i.e., proteins) are best suited to act as splitters, whereby the splitting rules define what intensity (i.e., protein abundance) a signal must have to be sorted into one of the groups (Fig. 8.7). After the initial learning period where the tree building takes place, the prediction success of the resulting model can be tested with new sets of data. The Biomarker Patterns Software is a true prediction tool, enabling the user to conduct multivariate analysis on a professional level without the need to be a specialist in statistical evaluations. In contrast to other classifiers, such as neural networks or nearest-neighbor calculations, the models are easier to understand and each sample can be tracked down the tree [61].

Even with all these developments in technology, the importance of well-designed marker discovery studies should not be forgotten. For example, in the past the success of tumor marker studies has often been hampered by uncertainties in tumor grading [62] and a lack in generally accepted guidelines for reporting the results [63]. Furthermore, questions about the numbers needed to screen [64] and how to define surrogate end points in cancer research [65] are still under discussion. In other words, there exists an urgent need for generally accepted rules in marker discovery to assure the best possible outcome from the corresponding projects. A detailed review addressing this topic can be found elsewhere [66].

In the Pattern Track Process, only marker proteins of proven importance will become identified. In most cases, the protein of interest needs to be purified or at least enriched for subsequent identification experiments [e.g., 67–70]. Mini-spin columns, microplate-formatted chromatographic devices or traditional preparative scale columns are suitable for separation. One of the advantages of the SELDI process is that the best-suited chromatographic material and the correct elution conditions can be predicted from the binding behavior of the protein on the ProteinChip Arrays [71–73]. Optimized separation materials have been developed by BioSepra, the Process Division of Ciphergen Biosystems, Inc. For identification, the purified or enriched protein is cleaved proteolytically, and the masses of the resulting peptide fragments are determined using the ProteinChip Reader. Afterwards, searches in public databases are used for the identification by peptide mass finger printing. For unequivocal identifications and partial sequencing of the protein, the ProteinChip Arrays can be read in high-resolution tandem-MS instruments [74]. For this purpose, an exchangeable UV laser desorption ion source, the ProteinChip Interface PCI 1000, is directly linked to a tandem-MS system (Tandem MS QSTAR®; from Sciex-MDS/Applied Biosystems).

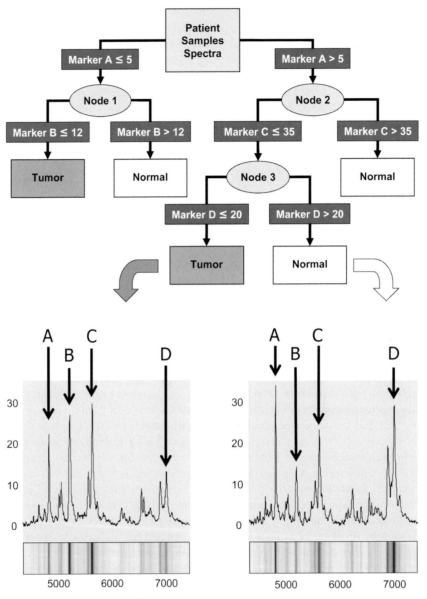

Fig. 8.7 The Biomarker Patterns™ Software. This supervised learning program performs multivariate analyses and creates tree-like structured decision diagrams by splitting the original data set into sub-groups of highest possible purity. The program calculates which signals (i.e., proteins) are best suited to act as splitters, and the splitting rules define what intensity (i.e., protein abundance) a signal must have to be sorted into one of the groups. The resulting model can be tested with new data sets from blinded samples to verify the reliability of the model under realistic conditions. In the example given, four markers are part of the model. According to the rules, the spectrum from the left-hand side is recognized as derived from a tumor patient, whereas the one on the right-hand side represents the normal state.

Recently, a new type of ProteinChip Array was commercialized to further facilitate the analysis of peptide mass fingerprints by Surface-Enhanced Neat Desorption (SEND) technology for protein identification (ID). In SEND ID, the energy-absorbing molecules (EAM) are incorporated in the surface chemistry of the array. Thus, separate EAM addition is not needed and the chemical noise introduced by the application of EAM is significantly reduced, enabling low molecular-weight species to be detected with greatly improved signal-to-noise ratio, whilst also minimizing the variability associated with matrix addition. The spot surface contains a hydrophobic interaction functional group to enable the on-spot clean-up of samples. Most other approaches to this application require clean-up of the sample in a small chromatography column in addition to pre-mixing with matrix. By avoiding these steps with SEND, the work flow is simplified and the losses of "sticky" peptides are minimized. In addition, the suppression of interfering matrix signal results in superior sensitivity and improved sequence coverage.

The knowledge of the marker identities will not only enable to transfer the assay on antibody-coupled ProteinChip Arrays [66] but also will support a better understanding of the underlying biology. Furthermore, by coupling the marker protein and capturing physiological binding partners, secondary markers might be found. Those interaction experiments can be performed either with pre-activated affinity beads and elution on chromatographic arrays, or directly on pre-activated ProteinChip Arrays. The Interaction Discovery Mapping platform complements the earlier established Expression Difference Mapping application. As with the entire ProteinChip System, both methods profit from the successful combination of chromatographic principles with mass spectrometric detection methods.

8.7
Protein Variants as Disease Markers

Many of the studies cited above revealed a very interesting phenomenon – namely, that a number of the markers belong to quite common proteins known to be of higher abundance in biological fluids. At first glance it seemed difficult to recognize them as specific indicators for a certain disease. However, from the exact molecular weights and identities delivered by the SELDI process, it became clear that these common proteins exhibit truncations or other types of covalent modifications which make them different from the respective corresponding proteins in the other patient groups. The ability to detect and to quantify those protein variants is another advantage of the SELDI-process, and one which could not be achieved by any traditional approach.

For example, such variants are reported for serum amyloid A in nasopharyngeal cancer [75] and renal cancer [76], as well as the vascular endothelial growth factor (VEGF) in bronchial cancer and for apolipoprotein A1, transthyretin and inter alpha trypsin inhibitor IV (ITIH-IV) in ovarian cancer [47]. In contrast, in other areas of research (e.g., for Alzheimer's disease), the importance of the different truncated versions of the amyloid-beta peptides has already been recognized and used to monitor the pathology of the disease [77]. The diagnostic value of protein variants represents a new insight of promising perspectives.

According to the traditional hypothesis, marker proteins are released into the bloodstream by tumor cells, and must reach a certain concentration to become detectable. It follows that the tumor must have reached a reasonable size to produce sufficient marker proteins above a certain threshold. However, tumor cells release not only passively floating proteins but also active enzymes which modify and cleave host proteins. Inflammation and cancer progression are very closely related [78], with cytokines and proteases acting in concert [79, 80]. Proteases released by tumors attack the surrounding host proteins [81], and enzyme activity seems to be correlated to the aggressiveness of the lesions [82]. Furthermore – and this is of special interest for diagnostic purposes – the protease composition exhibits a very high specificity when compared between different types of tumor cells and normal cells [83]. Thus, tumor-released enzymes could act as generators of specifically modified

marker proteins. In this way, an amplification process could be postulated with a few enzymes that produce a large amount of modified host proteins as putative biomarkers, and these markers could signal the presence of tumors at a very early stage (Fig. 8.8). Ciphergen has applied for patents on this concept, which it terms the Host Response Protein Amplification Cascade (HR-PAC), and on the use of the approach in diagnostic testing.

8.8
Conclusion and Outlook

Over the past few years, the ProteinChip System has proven to be a valuable tool for the discovery and validation of newly detected protein biomarker patterns. It is used by more than 1000 customers worldwide, including academic and governmen-

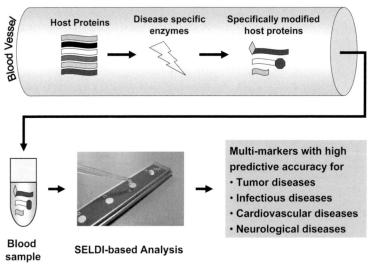

Fig. 8.8 The Host Response Protein Amplification Cascade (HR-PAC). According to this new theory, disease-specific enzymes act as generators of specifically modified marker proteins which can serve as valuable biomarkers by signaling a disease in a very early stage. The covalently modified marker proteins appear in biological fluids, become detected by SELDI-based ProteinChip technology, and can be used in multimarker assays to predict clinical conditions.

tal institutes as well as pharmaceutical companies. The SELDI process – with its unique combination of chromatographic principles and mass spectrometric detection – meets the challenges of the new proteomic era in diagnostic assay development. The Pattern Track process includes all steps from biomarker discovery to assay development on one single experimental platform. All of the assays developed thus far are still of research grade, and have not been released for commercial use. Currently, multi-institutional studies are proceeding to evaluate the clinical robustness of SELDI-based multi-marker detection with the necessary scientific diligence before making it available for the public. These studies will help in the establishment of diagnostic assays, which in turn will foster the development of modern biopharmaceuticals. For up-to-date information regarding these ongoing activities, please refer to http://www.ciphergen.com.

References

1 Pfaller, M.A. *Emerging Infectious Diseases* **2001**, *7(2)*, 312–318.
2 Blennow, K., Hampel, H. *The Lancet Neurology* **2003**, *2*, 605–613.
3 Blennow, K. *J. Int. Med.* **2004**, *256*, 224–234.
4 Etzioni, R., Urban, N., Ramsey, S., McIntosh, M., Schwartz, S., Reid, B., Radich, J., Anderson, G., Hartwell, L. *Nat. Rev. Cancer* **2003**, *3(4)*, 243–252.
5 National Cancer Institute *Cancer facts – Tumor markers*, http://cis.nci.nih.gov/fact/5_18.htm, **1998**.
6 Bidart, J.M., Thuillier, F., Augereau, C., Chalas, J., Daver, A., Jacob, N., Labrousse, F., Voitot, H. *Clin. Chem.* **1999**, *45(10)*, 1695–1707.
7 Wiesner, A. *Curr. Pharm. Biotechnol.* **2004**, *5*, 45–67.
8 Jarvinen, T.A., Liu, E.T. *Breast Cancer Res. Treat.* **2003**, *78(3)*, 299–311.
9 Jarvinen, T.A., Liu, E.T. *Comb. Chem. High Throughput Screen* **2003**, *6(5)*, 455–470.
10 Moynihan, R., Smith, R. *Br Med J* **2002**, *324(7342)*, 859–860.
11 Goessl, C. *Curr. Med. Chem.* **2003**, *10(8)*, 691–706.
12 Simon, R. *Br. J. Cancer* **2003**, *89*, 1599–1604.
13 Johnson, P.J., Lo, Y.M. *Clin. Chem.* **2002**, *48(8)*, 1186–1193.
14 Bernard, P.S., Wittwer, C.T. *Clin. Chem.* **2002**, *48(8)*, 1178–1185.
15 Ransohoff, D.F. *Science* **2003**, *299(5613)*, 1679–1680.
16 Laird, P.W. *Nat. Rev. Cancer* **2003**, *3(4)*, 253–266.
17 Sidransky, D. *Nat. Rev. Cancer* **2002**, *2(3)*, 210–219.
18 Lim, L.P., Sharp, P.A. *Mol. Cell. Biol.* **1998**, *18(7)*, 3900–3906.
19 Gygi, S.P., Rochon, Y., Franza, B.R., Aebersold, R. *Mol. Cell. Biol.* **1999**, *19(3)*, 1720–1730.
20 Chen, G., Gharib, T.G., Huang, C.C., Taylor, J.M., Misek, D.E., Kardia, S.L., Giordano, T.J., Iannettoni, M.D., Orringer, M.B., Hanash, S.M., Beer, D.G. *Mol. Cell. Proteomics* **2002** *1(4)*, 304–313.
21 Anderson, L., Seilhamer, J.A. *Electrophoresis* **1997**, *18(3-4)*, 533–537.
22 Wallace, C.J. *Protein Sci.* **1993**, *2(5)*, 697–705.
23 Pritchard, L.E., Turnbull, A.V., White, A. *J. Endocrinol.* **2002**, *172*(3), 411–421.
24 Hanahan, D., Weinberg, R.A. *Cell* **2000**, *100*(1), 57–70.
25 Pandey, A., Mann, M. *Nature* **2000**, *405(6788)*, 837–846.
26 Service, R.F. *Science* **2001**, *294(5549)*, 2074–2077.
27 Gavin, A.C., Bosche, M., Krause, R., et al. *Nature* **2002**, *415(6868)*, 141–147.
28 Kumar, A., Snyder, M. *Nature* **2002**, *415(6868)*, 123–124.
29 Gavin, A.C., Superti-Furga, G. *Curr. Opin. Chem. Biol.* **2003**, *7*(1), 21–27.
30 Tyers, M., Mann, M. *Nature* **2003**, *422(6928)*, 193–197.
31 Yarmush, M.L., Jayaraman, A. *Annu. Rev. Biomed. Eng.* **2002**, *4*, 349–373.
32 Srinivas, P.R., Srivastava, S., Hanash, S., Wright Jr., G.L. *Clin. Chem.* **2001**, *47(10)*, 1901–1911.
33 Gygi, S.P., Corthals, G.L., Zhang, Y., Rochon, Y., Aebersold, R. *Proc. Natl. Acad. Sci. USA* **2000**, *97*(17), 9390–9395.
34 Rabilloud, T. *Proteomics* **2002**, *2*(1), 3–10.

35 Gygi, S.P., Rist, B., Gerber, S.A., Turecek, F., Gelb, M.H., Aebersold, R. *Nat. Biotechnol.* **1999**, *17*(10), 994–999.

36 Weinberger, S.R., Viner, R.J., Ho, P. *Electrophoresis* **2002**, *23*(18), 3182–3192.

37 Weinberger, S.R., Morris, T.S., Pawlak, M. *Pharmacogenomics* **2000**, *1*(4), 395–416.

38 Jenkins, R.E., Pennington, S.R. *Proteomics* **2002**, *1*(1), 13–29.

39 Kodadek, T. *Trends Biochem. Sci.* **2002**, *27*(6), 295–300.

40 Weinberger, S.R., Davis, S., Makarov, A., Thompson, S., Purves, R., Whittal, R.M., in: *Encyclopedia of Analytical Chemistry* (Meyers, R.A., Ed.), John Wiley & Sons Ltd, Chichester, **2000**, pp. 11915–11984.

41 Weinberger, S.R., Dalmasso, E.A., Fung, E.T. *Curr. Opin. Chem. Biol.* **2001**, *6*(1), 86–91.

42 Tang, N., Tornatore, P., Weinberger, S.R. *Mass Spec. Rev.* **2004**, *23*, 34–44.

43 Yip, T.T., Lomas, L. *Technol. Cancer Res. Treat.* **2002**, *1*(4), 273–280.

44 Issaq, H.J., Conrads, T.P., Prieto, D.A., Tirumalai, R., Veenstra T.D. *Anal. Chem.* **2003**, *75*(7), 148A–155A.

45 Issaq, H.J., Veenstra, T.D., Conrads, T.P., Felschow, D. *Biochem. Biophys. Res. Commun.* **2002**, *292*(3), 587–592.

46 Chapman, K. *Biochem. Soc. Trans.* **2002**, *30*(2), 82–87.

47 Zhang, Z., Bast, R.C. Jr., Yu, Y., et al. *Cancer Res.* **2004**, *64*(16), 5882–5890.

48 Adam, B.L., Qu, Y., Davis, J.W., et al. *Cancer Res.* **2002**, *62*(13), 3609–3614.

49 Koopmann, J., Zhang, Z., White, N., et al. *J. Clin. Cancer Res.* **2004**, *10*, 860–868.

50 Wadsworth, J.T., Somers, K.D., Cazares, L.H., et al. *Cancer Res.* **2004**, *10*, 1625–1632.

51 Carrette, O., Demalte, I., Scherl, A., et al. *Proteomics* **2003**, *3*, 1486–1494.

52 Poon, T.C.W., Chan, K.C.A., Ng, P.C., et al. *Clin. Chem.* **2004**, *50* (8), 1452–1455.

53 Gravett, M.G., Novy, M.J., Rosenfeld, R.G., et al. *JAMA* **2004**, *292*, 462–469.

54 Papadopoulos, M.C., Abel, P.M., Agranoff, D., et al. *The Lancet* **2004**, *363*, 1358–1363.

55 Xiao, Z., Luke, B.T., Izmirlian, G., et al. *Cancer Res.* **2004**, *64*, 2904–2909.

56 Boot, R.G., Verhoek, M., de Frost, M., et al. *Blood* **2004**, *103*, 33–39.

57 Ginestse, C., Ho, L., Pompl, P., Bianchi, M., Pasinetti, G.M. (2003) *Drugs* **2003**, *63* (Suppl.1), 23–29.

58 Schaub, S., Rush, D., Wilkins, J., et al. *J. Am. Soc. Nephrol.* **2004**, *15*, 219–227.

59 Clarke, W., Silverman, B.C., Zhang, Z., Chan, D.W., Klein, A.S., Molmenti, E.P. *Ann. Surg.* **2003**, *237*, 660–665.

60 Fung, E.T., Enderwick, C. *Biotechniques, Suppl. Computational Proteomics* **2002**, *32*, S34–S41.

61 Adam, B.L., Qu, Y., Davis, J.W., et al. *Cancer Res.* **2002**, *62*(13), 3609–3614.

62 Hayes, D.F. *Recent Results Cancer Res.* **1998**, *152*, 71–85.

63 Riley, R.D., Abrams, K.R., Sutton, A.J., Lambert, P.C., Jones, D.R., Heney, D., Burchill, S.A. *Br. J. Cancer* **2003**, *88*(8), 1191–1198.

64 Rembold, C.M. *Br. Med. J.* **1998**, *317*(7154), 307–312.

65 Schatzkin, A., Gail, M. *Nat. Rev. Cancer* **2002**, *2*(1), 19–27.

66 Fung, E. *Preclinica* **2004**, *2*(4), 253–258.

67 Zhang, L., Yu, W., He, T., et al. *Science* **2002**, *298*(5595), 995–1000.

68 Uchida, T., Fukawa, A., Uchida, M., Fujita, K., Saito, K. *J. Proteome Res.* **2002**, *1*(6), 495–499.

69 Diamond, D.L., Zhang, Y., Gaiger, A., Smithgall, M., Vedvick, T.S., Carter, D. (2003) *J. Am. Soc. Mass. Spectrom.* **2003**, *14*(7), 760–765.

70 Shiwa, M., Nishimura, Y., Wakatabe, R., Fukawa, A., Arikuni, H., Ota, H., Kato, Y., Yamori, T. *Biochem. Biophys. Res. Commun.* **2003**, *309*(1), 18–25.

71 Santambien, P., Brenac, V., Schwartz, W., Boschetti, E., Spencer, J. (2002) *Genet. Eng.* **2002**, *22*, 13.

72 Weinberger, S.R., Boschetti, E., Santambien, P., Brenac, V. *J. Chromatogr. B Analyt. Technol. Biomed. Life Sci.* **2002**, *782*(1-2), 307–316.

73 Shiloach, J., Santambien, P., Trinh, L., Schapman, A., Boschetti, E. *J. Chromatogr. B Analyt. Technol. Biomed. Life Sci.* **2003**, *790*(1-2), 327–336.

74 Reid, G., Gan, B.S., She, Y.M., Ens, W., Weinberger, S., Howard, J.C. *Appl. Environ. Microbiol.* **2002**, *68*(2), 977–980.

75 Cho, W.C.S., Yip, T.T.C., Yip, C., et al. *Clin. Cancer Res.* **2004**, *10*, 43–52.

76 Tolson, J., Bogumil, R., Brunst, E., et al. *Lab. Invest.* **2004**, *84*, 845–856.

77 Lewczuk, P., Esselmann, H., Groemer, T.W., et al. *Biol. Psychiatry* **2004**, *55*, 524–530.

78 Coussens, L.M., Werb, Z. *Nature* **2002**, *420*(6917), 860–867.

79 Van Damme, J., Struyf, S., Opdenakker, G. *Semin. Cancer Biol.* **2004**, *14*(3), 201–208.

80 Fowlkes, J. L., Winkler, M. K. *Cytokine Growth Factor Rev.* **2002**, *13*(3), 277–287.

81 DeClerck, Y. A., Mercurio, A. M., Stack, M. S., et al. *Am. J. Pathol.* **2004**, *164*(4), 1131–1139.

82 Mahmood, U., Weissleder, R. *Mol. Cancer Ther.* **2003**, *2*(5), 489–496.

83 Nuttall, R. K., Pennington, C. J., Taplin, J., Wheal, A., Yong, V. W., Forsyth, P. A., Edwards, D. R. *Mol. Cancer Res.* **2003**, *1*(5):333–345.

9

Early Detection of Lung Cancer: Metabolic Profiling of Human Breath with Ion Mobility Spectrometers

Jörg Ingo Baumbach, Wolfgang Vautz, Vera Ruzsanyi, and Lutz Freitag

Abstract

Early detection of lung cancer by metabolic profiling of human breath with ion mobility spectrometry (IMS) – dream or reality? Volatile metabolites occurring in exhaled air are correlated directly to different kinds of diseases. Some metabolites are biomarkers, i.e., acetone is related to diabetes, nitric acid to asthma and ammonia to hepatitis, whereas other arise from bacteria. In the present chapter, IMS coupled to a multi-capillary column as a pre-separation unit is used to identify and quantify volatile metabolites occurring in human breath down to the nanogram and picogram per liter range of analytes. The spectra obtained from patients suffering from chronic obstructive pulmonary disease and pneumonia are discussed in detail. Furthermore, IMS chromatograms of metabolites of *Serratia marcescens*, *Enterobacter aerogenes* and *Escherichia coli* are compared. In addition, the effect of drug delivery on a patient showing angina lateralis is presented as an example to show the potential of the method developed in the field of detection of pathways, effective dosage and decisions of effective time intervals to deliver biopharmaceuticals. The aim of the studies is to introduce the investigation of metabolites in human breath as a method for the early recognition of selected diseases and monitoring of the effectiveness of drug delivery based on IMS data.

Abbreviations

COPD   chronic obstructive pulmonary disease
GC   gas chromatography
IMS   ion mobility spectrometry
MCC   multi-capillary column
MS   mass spectrometry
RIP   reactant ion peak
VOC   volatile organic compound

9.1
Introduction

The general aim of this chapter is to contribute to the establishment of a fast and low-cost device for human breath analysis in addition to investigations of blood and urine as a non-invasive standard method in hospitals and point-of-care centers for different medical applications. On the basis of miniaturized ion mobility spectrometry (IMS), the full procedure, including sampling, pre-separation and identification of metabolites in human exhaled air, will be developed and implemented with re-

Modern Biopharmaceuticals. Edited by J. Knäblein
Copyright © 2005 WILEY-VCH Verlag GmbH & Co. KGaA, Weinheim
ISBN: 3-527-31184-X

spect to future application in hospitals. Metabolic profiling of human breath of healthy individuals and those suffering from different diseases, in particular chronic obstructive pulmonary disease (COPD) and pneumonia, will be considered.

In the medical community it is well known that humans exhale volatile metabolites which potentially carry important information about their health status. Thus, successful and fast detection of potential products of different metabolic processes becomes attractive, especially if the detection limits of the spectrometric methods used are low enough and the instruments become available at moderate price levels to be used as standard methods in hospitals or point-of-care centers. The vision of the authors is to contribute to the use of human breath as a carrier of information of the health status of the body in addition to human blood and urine.

Human exhaled breath contains numerous volatile metabolites derived from both endogenous metabolism and external exposure to vapors/gases. Approximately 200 different analytes have been detected in human breath; some are correlated to various common disorders like diabetes, heart disease and lung cancer [1–14]. The composition of different constituents in respired air is mostly representative for blood-borne concentrations. Such analytes were detected through gas exchange at the blood–breath interface in the lungs [15]. Thus, the presence and also the quantitative variations of specific metabolites in exhaled air are directly linked to volatile organic compounds (VOCs) occurring in the blood. However, as the blood contacts all parts of the human body, including diseased tissues or organs, the point of origin is not fixed in all cases. Furthermore, metabolites derived from local bacterial infections in the airway system can also be de-

tected directly in human breath. Currently, a number of different techniques are used in the field of breath analysis. The most popular sampling method seems to be the use of Teflon bags to collect human breath. Also sorbent- or cryo-trapping of exhaled air followed by desorption into an analytical instrument is widely used. In particular, mass spectrometric methods are used for the identification of the metabolites. Mostly, chromatographic pre-separation is also applied. Gas chromatography/mass spectrometry (GC/MS) is used in the majority of cases reported [16] (see also Part V, Chapter 8). Overall, this is a rather time-consuming process with numerous different steps. All may lead to a significant loss of analytes [2, 17] and, sometimes, analytes may adsorb on the surface of the bag [18].

In the literature, the major VOCs found in the breath of healthy persons are isoprene (10–600 ppb$_v$), acetone (1–2000 ppb$_v$), ethanol (10–1000 ppb$_v$) and methanol (150–200 ppb$_v$). All are products of the standard metabolic processes in the human body [7].

Considering the analytical methods mentioned above, the high moisture content of breath samples is a major problem. In addition, mass spectrometers, currently undergoing a process of miniaturization, have the status of laboratory instruments, which are comparatively large and expensive, and offer an analysis time of 1–2 h, depending on the sample preparation steps necessary. Therefore, there exists a need for instruments applicable for direct, on-line analysis of the breath from a patient and delivering results nearly "just in time". In particular, in the case where the number of steps for sample handling could be minimized, no additional laboratory steps become necessary and no additional carrier gases of high purity (like nitrogen or helium as used in GC/MS) are required, on-line methods for effective breath analysis procedures become attractive for point-of-care applications. In such cases, investigations on humans could be handled in hospitals by standard personal.

Recently, IMS devices have become comparatively small and effective devices to determine traces of quantities of VOCs down to the low ppb$_v$ range, especially in air [19]. The major advantages of IMS are that no vacuum system is required for operation and ambient air can be used as a carrier gas. Worldwide, more than 70 000 units are in service, especially to detect chemical warfare agents, narcotics and explosives, and many different instruments are available on the market [20, 21]. It was shown that VOCs are detectable at the nanogram, and sometimes picogram, per liter levels in air [22–37].

9.2
Material and Methods: IMS

For the measurements described below, a custom-designed IMS equipped with a ^{63}Ni β-ionization source was used [22, 25]. The operational principles and details of IMS are summarized in [21, 28], and therefore only a brief description will follow. The term IMS refers to the method of characterizing analytes in gases by their gas-phase ion mobility. Normally, the drift time of ion swarms formed using suitable ionization sources and electrical shutters is measured. Ion mobilities are characteristic for analytes, and can provide a means for detecting and identifying vapors. The drift velocity is related to the electric field strength by the mobility. Therefore, the mobility is proportional to the inverse drift time, which will be measured at a fixed drift length.

IMS combines both high sensitivity and relatively low technical expenditure with high-speed data acquisition. The time to acquire a single spectrum is in the range of 10–100 ms. Thus, IMS is suitable for process control, but due to the occurrence of ion–molecule reactions and relatively poor resolution of the species formed, generally not for the identification of unknown compounds. Compared with MS, the mean free path of the ions is smaller than the dimensions of the instrument. Therefore, an ion swarm drifting under such conditions experiences a separation process that is based on different drift velocities of ions with different masses or geometrical structures. Collection of these ions on a Faraday plate delivers a time-dependent signal corresponding to the mobility of the arriving ions. Such an ion mobility spectrum contains information on the nature of the different trace compounds present in the sample gas.

The most important parts are the ionization/reaction region and the drift region of the instrument (see Fig. 9.1). The external and homogenous electric field will be established within the drift tube using several drift rings for stabilization. The carrier gas will take sample molecules within the ionization region. The ionization of the analytes occurs by chemical ionization on collisions of the analyte with ionized carrier gas molecules by application of radioactive ionization sources. A so-called drift gas will flow from the Faraday plate towards the ionization region. Normally, if the shutter is closed, no ions can reach the drift region. The drift gas will protect the drift region and no neutral analyte molecules should enter the drift region. If the shutter is held closed, all analyte molecules, neutrals and ions will be washed out of the gas outlet. During the shutter opening time, some ions will enter the

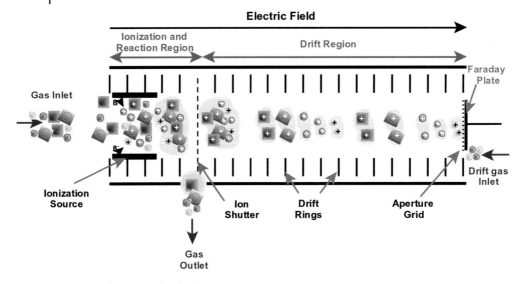

Fig. 9.1 Working principle of IMS.

Table 9.1 Main parameters of ^{63}Ni-IMS.

Ionization source	^{63}Ni β-radiation source, 510 MBq
Length of the drift region	12 cm
Electrical field strength	326 V cm^{-1}
Drift voltage	4 kV
Shutter opening time	300 μs
Drift and sample gas	synthetic air
Drift gas flow	100 mL min^{-1}
Sample gas flow	150 mL min^{-1} (optimized for breath analysis)
Temperature	25 °C (ambient)
Pressure	101 kPa (ambient)

drift region. During several collisions with the surrounding gas molecules, a steady drift velocity will be reached. If no chemical reactions occur, in the ideal case total separation will be reached at the Faraday plate. The time-dependent voltage or current during a time interval measured from the half of the shutter opening pulse is called the ion mobility spectrum. Using air as carrier gas, the carrier gas molecules are normally ionized by the β-particles directly. In the present case, positive ions are under consideration. These primary positive ions of the carrier gas (called reaction or reactant ions) will undergo different chemical reactions with the analyte ions to form so-called product ions by proton transfer, nucleophilic attachment, hydride abstraction, etc. Charge transfer and proton abstraction could also occur.

All of the parts of the IMS which are in contact with the analytes are formed from inert materials. The relevant parameters of the IMS are summarized in Tab. 9.1.

To realize an effective pre-separation of the rather complex mixtures occurring in exhaled air, a 17-cm long polar multi-capillary column (MCC; OV-5; Sibertech, Novosibirsk, Russia) made by combining approximately 1000 capillaries with an inner diameter of 40 μm and a film thickness of 0.2 μm was coupled to the [63]Ni-IMS [21, 22, 25].

The total column diameter of 3 mm allows operation with a carrier gas flow up to 150 mL min[-1], which is the optimum flow rate for IMS. In addition, the effective separation of humidity is one major advantage of the MCC used [38].

The heating of the column is indispensable for the reproducibility of the chromatographic results. To achieve comparable retention times, the MCC was held at 30 °C during breath analysis. To realize isothermal separation, a simple heating construction is needed, which means a concise size decline of the instrument.

In the sampling process, a subject blows through a mouthpiece coupled with a brass adapter designed at ISAS to a Teflon tube (1/4 in.; Bohlender, Lauda, Germany),

which is connected to a 10-mL stainless steel sample loop of an electric six-port valve (Nalco; Macherey-Nagel, Düren, Germany). By switching the six-port valve, breath is transported by the carrier gas from the sample loop into the MCC. The separated substances can be directly analyzed by IMS. Therefore, the results can be obtained within 10 min depending on the separation time of the compounds. This construction enables direct and rapid sampling at a known breath volume. A schematic drawing showing sampling and detection using MCC-[63]Ni-IMS is shown in Fig. 9.2.

9.3
Results and Discussion

Two typical spectra of acetone in air at concentrations of 100 and 500 ng L[-1] are shown in Fig. 9.3. The two peaks observed arise from air [reactant ion peak (RIP)] at 17.69 ms and the analyte acetone at 19.59 ms. The ionization of acetone is realized by charge transfer from the air to the

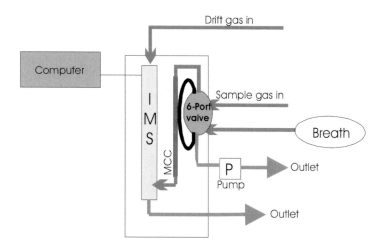

Fig. 9.2 Diagram of the breath sampling unit and adaptation to MCC-[63]Ni-IMS.

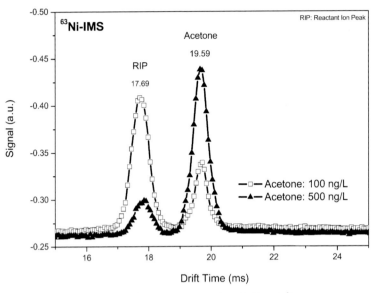

Fig. 9.3 Ion mobility spectra of acetone in air (100 and 500 ng L^{-1}).

acetone molecule. Thus, it becomes comes clear that with higher concentration of the analyte in air, the RIP will decrease and the acetone-related peak will increase. In general, the peak position is relevant for the identification procedure and the peak area is related to the concentration of the analyte.

On the other hand, it is shown in Fig. 9.3 that much higher concentrations than 500 ng L^{-1} of the analyte acetone in air will not be ionized effectively. In such a case not enough further reactant ions of the analyte, which are needed for effective ionization of the molecule, are available. Here, it should be noted that IMS is suitable for trace gas analysis.

The combination of a chromatographic column with the IMS enables multidimensional data analysis, and allows peak identification by the use of chromatographic data (retention times) and also by use of the specific ion mobility data (arrival time at the Faraday plate) of the ions formed from the analytes. Therefore, retention times of the compounds separated by the MCC and drift time values of the analytes are shown in so-called IMS chromatograms. The MCC will reduce the number of interactions in the ionization region of the IMS as the formation of clusters and avoids concurrent charge discharge transfer by time-delayed entrance of the analytes. Thus, the effective ionization of nearly all analytes in a given sample is organized by selection of proper GC and effective temperature programming of the chromatographic column or keeping the temperature constant, as in the present case.

Fig. 9.4 shows the results of investigations of exhaled air of a healthy human containing ammonia, acetone and ethanol peaks, and showing also the RIP as major signals. Smaller portions will not be considered here. In addition, the mobilities of the ions calculated from the drift time of the ion peak K_0 values (for details, see

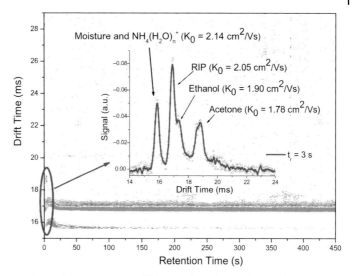

Fig. 9.4 IMS chromatogram of human exhaled air – inlet: single spectrum at a fixed retention time of 3 s showing signals of the major constitutions ammonia, ethanol and acetone.

[21]) are included in the inlet, showing a single spectrum at a single, fixed retention time.

It is also clearly visible in Fig. 9.4 that the humidity, which can reach really high values, particularly when considering exhaled air, is separated effectively by using the MCC. During the first 20 s or so high values of humidity enter the ionization region of the IMS. In general, the MCC will reduce the influence of humidity as acquired. On the other hand, the main interest will focus on an interval for higher drift times of about 17 ms at any retention time, as shown in the following section.

9.4
Clinical Study

In cooperation with the Lung Hospital in Hemer, on-site measurements were carried out with the exhaled air of 40 subjects, including 22 patients suffering from differ-

ent pulmonary lung infections. Breath samples of 18 different healthy persons were also analyzed on-line using MCC-^{63}Ni-IMS. The full analysis was always performed in the same room, where room air was determined before each of the breath measurements. To reduce the risk of cross-contamination from other sources, subjects had not drunk, eaten or smoked for at least 2 h before the breath measurements.

An IMS chromatogram from the exhaled air of a patient suffering bacterial lung infection is shown in Fig. 9.5. The room air values are subtracted in all cases, as discussed above. The colors refer to different peak height levels. The peaks identified are labeled with the mobility values K_0 calculated from the drift time of the ions of the analytes, and also taking into account the values of the electric field strength in the drift tube, the pressure and the temperature conditions during the experiment (for further details, see [21]).

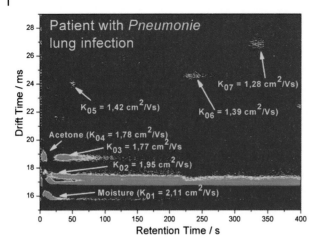

Fig. 9.5 IMS chromatogram of the breath of a patient suffering from a lung infection (K_0 refers to the mobility values calculated for each major peak).

As mentioned above, the formation of so-called reactant (arising from the carrier gas air) and product ions (formed by the analytes) in IMS can be altered by moisture. Therefore, separation of water molecules is of great importance when working with breath samples. Figure 9.5 shows that the high breath moisture content (mobility values $K_{01} = 2.11$ cm^2 Vs^{-1}) could be separated effectively using the MCC. The signals correlated to moisture are always localized at the beginning of the IMS chromatogram. Thus, moisture has no disturbing effect for peaks showing longer retention times. In the topographic plot of the exhaled air of the healthy person, acetone was identified based on its reduced mobility value ($K_{04} = 1.78$ cm^2 Vs^{-1}). Acetone has a similar retention time as water, but possesses a stronger proton affinity. This is enough for acetone to be ionized via ion–molecule reactions.

Comparing the two IMS chromatograms of Figs. 9.4 and 9.5, the differences are conspicuous. In Fig. 9.5, the two-dimensional plot exhibits additional peaks. Two major peaks occur at the retention times of 28 s ($K_{02} = 1.95$ cm^2 Vs^{-1}) and 36 s ($K_{03} = 1.77$ cm^2 Vs^{-1}). Both are probably de-

gradation products of antibiotics or other drugs, because they were also found in the breath of several persons who had taken similar antibiotics. The peak at the position of 50 s and $K_{05} = 1.42$ cm^2 Vs^{-1} was also analyzed from the exhaled air of other patients suffering from pneumonia infection. Two other larger analytes with lower mobility values ($K_{06} = 1.39$ cm^2 Vs^{-1} and $K_{07} = 1.28$ cm^2 Vs^{-1}) and longer retention times ($t_{ret6} = 244$ s and $t_{ret7} = 330$ s) were often detected in patients with bacterial infection and airway inflammation.

The exhaled air of a patient with COPD is shown in Fig. 9.6. Two major peaks occur. The peak at lower retention times (P-Co01-01) is identically to peaks found in *Staphylococcus aureus* cell cultures (B-Sa-01). The second peak at higher retention and drift times, indicating a higher-molecular-weight substance, is not correlated with *S. aureus*. No other bacteria were identified in additional microbiological investigations. However, it becomes obvious that metabolites of bacteria occurring in the lung should be taken into consideration. Therefore, investigations of emissions of bacteria were carried out in addition. As a reference, Fig. 9.7 shows the IMS chro-

Fig. 9.6 IMS chromatogram (right) of a patient suffering from COPD and two typical single spectra (below).

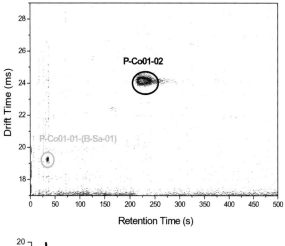

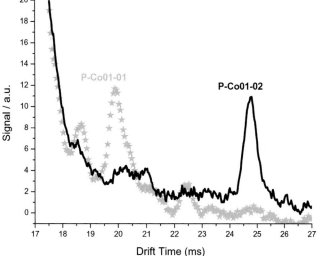

matogram of volatile emissions obtained from *S. aureus*. A single major peak (B-Sa-01) with an ion mobility of 1.86 cm² Vs⁻¹ and a retention time of 33 s (B-Sa-01) is clearly visible.

To demonstrate the rather complex situation between different diseases and bacteria involved, Fig. 9.8 shows the IMS chromatogram of a COPD patient with different infections. There are five, clearly separated, major peaks. One among them, called P-Co02-04,

is identical to the peak B-Ec-05 arising from *Escherichia coli* (see Fig. 9.9). With regard to *E. coli*, five major peaks could also be recognized. In addition to the rather small volatiles (B-Ec-01, B-Ec-02 and B-Ec-03), two peaks [$K_0 = 1.33$ cm² Vs⁻¹ (B-Ec-04) and $K_0 = 128$ cm² Vs⁻¹ (B-Ec-05)] occur at larger retention times ($t_{\text{retB-Ec04}} = 129$ s and $t_{\text{retB-Ec05}} = 327$ s). As mentioned, peaks B-Ec-05 and P-Co02-04 are identical if the mobility and not the drift time scale is considered, be-

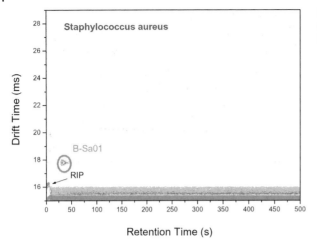

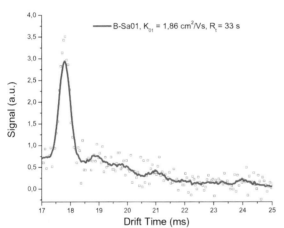

Fig. 9.7 IMS chromatogram (left) of *S. aureus* and the corresponding single spectrum (below).

cause of small changes in experimental conditions affecting the mobility values.

To demonstrate the analytical potential of IMS considering the direct detection of the effectiveness of drug delivery, Fig. 9.10 shows two IMS chromatograms of a patient suffering on angina lateralis. The main peak at $K_0 = 1.51$ cm^2 Vs^{-1} and retention time $t_{ret} = 48$ s is considered as being related to angina lateralis. On the other hand, the peak was also identified as related to *Enterobacter aerogenes* (B-Eb04). Note, the three additional peaks are correlated with *E. coli* (B-Ec-03, B-Ec-04 and B-EC-05). The first IMS chromatogram shows the situation before drug delivery (Amoxibeta T 1000 containing amoxicillin trihydrate). The other one was obtained after 72 h of continuous amoxicillin trihydrate delivery once per day. It becomes obvious that the major peak related to angina lateralis and *E. aerogenes* decreases. In addition, in the special case of the antibiotic delivered to the patient, *E. coli* peaks are also reduced. Two peaks at higher retention times disappeared completely. Only one peak remains after 3 days of application of amoxicillin. The related single

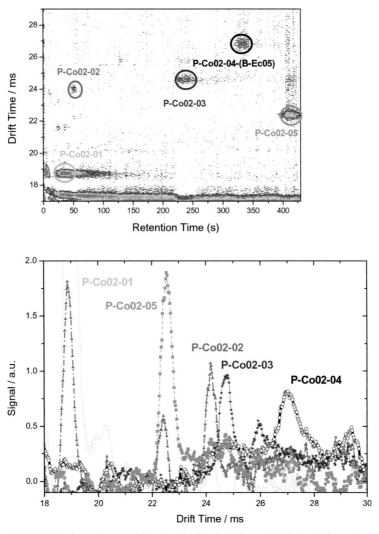

Fig. 9.8 IMS chromatogram (above) and related single spectra (below) of a patient suffering infections and COPD.

spectra for the main peak related to *E. aerogenes* are shown in Fig. 9.11 together in a single plot. The effect of the amoxilin causes a peak height reduction at drift times of 22.37 ms and retention time of 46 s during the first 3 days of application of amoxicillin, as summarized in Fig. 9.12. As expected, the signal intensity reduction follows an exponential decrease. Thus, an effective way of quickly determining the effectiveness of drug delivery and/or of actual use of drugs within IMS measurements should be considered in more detail in the future for different cases and specific metabolites in exhaled air.

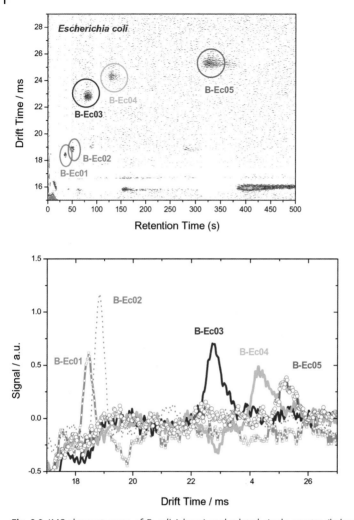

Fig. 9.9 IMS chromatogram of *E. coli* (above) and related single spectra (below).

To verify the characteristic peaks of volatile metabolites detected, further measurements will be carried out including a larger number of patients and healthy persons. Data will be processed using statistical methods to clarify the assignment between the diseases and the characteristic pattern of the IMS topographic plots. In the near future, the results should be confirmed by comparative analysis using mass spectroscopy to identify the nature of the metabolites detected by IMS.

9.5 Conclusions

By coupling IMS and a MCC for pre-separation, investigations were carried out to directly detect volatile metabolites in human exhaled air. The total analysis was re-

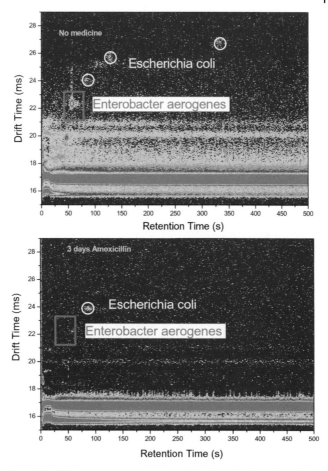

Fig. 9.10 IMS chromatograms of the exhaled breath of a patient with angina lateralis – signals of *E. coli* and *E. aerogenes* were identified (surrounded in white and red, respectively) before and 3 days after the start of amoxicillin therapy.

alized within time intervals of less than 500 s. The system constructed shows sufficient sensitivity to detect the metabolites directly by showing characteristic IMS chromatograms (peak height diagrams). The exhaled air is introduced directly into the MCC using a sample loop. Results of investigations on patients suffering from pneumonia show different metabolites in comparison to healthy persons.

Further investigations using MS methods should allow the identification of the metabolites found by IMS. The possibilities for on-site and short-time analysis using air as a carrier gas and working at ambient pressure are the most important benefits of the technique developed.

Using MCC-IMS directly in the diagnostic clinic allows results to be obtained within minutes, delivers additional infor-

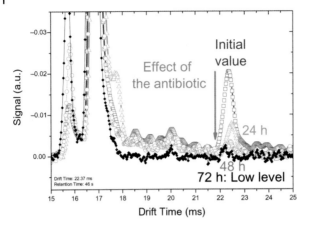

Fig. 9.11 Effect on the antibiotic on the peak height in an ion mobility spectrum.

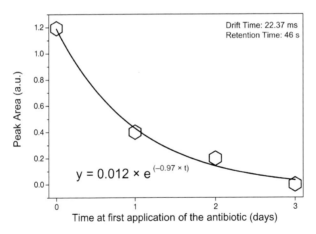

Fig. 9.12 Decrease of the signal (peak area) of the specific peak considered in the ion mobility spectrum of a patient suffering angina lateralis during drug delivery (amoxicillin) for 3 days.

mation for the therapeutic strategy and facilitates the building of databases for several diseases. The general aim of the studies could introduce the investigation of volatile metabolites in human breath as a marker for the early recognition of selected diseases, e.g. lung cancer, on the basis of IMS data. Furthermore, especially when considering biopharmaceuticals, metabolic profiling using different kinds of IMS opens the possibility to also directly investigate metabolic processes (including single reactions and interactions of biopharmaceuticals at the cell level up to the entire human body), using human breath as the carrier of information.

Acknowledgments

Financial support of the Bundesminister-ium für Bildung und Forschung and the Ministerium für Wissenschaft und For-schung des Landes Nordrhein-Westfalen, and the funding of the common project by G. A. S. Gesellschaft für Analytische Senso-rensysteme mbH is gratefully acknowl-edged. The possibility for the analysis in the Lung Hospital in Hemer and the help of the assistants and doctors should be mentioned with thanks. The authors wish to express their special thanks to Mrs. Oberdrifter for realizing laboratory sup-port, Dr. Litterst and Dr. Westhoff for help-ful discussions and organizing the clinical measurements (all at the Lung Hospital Hemer), and Mrs. Güssgen and Mrs. Sei-fert (both at ISAS, Dortmund) for support in the operation of the ion mobility spec-trometer and preparation of the data han-dling.

References

1 Phillips, M. Method for the collection and as-say of volatile organic compounds in breath. *Anal. Biochem. 247*, 272–278, **1997**.

2 Phillips, M., Gleeson, K., Hughes, J. M., Greenberg, J., Cataneo, R. N., Baker, L., McVay, W. P. Volatile organic compounds in breath as markers of lung cancer: a cross-sec-tional study. *Lancet 353*, 1930–1933, **1999**.

3 Phillips, M., Greenberg, J. Ion trap detection of volatile organic compounds in alveolar breath. *Clin. Chem. 38*, 60–65, **1992**.

4 Rooth, G., Lund, M. D., Östenson, S. Acetone in alveolar air, and the control of diabetes. *Lancet* 1102–1104, **1966**.

5 Benolt, M. F., Davidson, W. R., Lovett, A. M., Nacson, S., Ngo, A. Breath analysis by atmo-spheric pressure ionization mass spectrome-try. *Anal. Chem. 55*, 805–807, **1983**.

6 Cheng, W.-H., Lee, W.-J. Technology develop-ment in breath microanalysis for clinical diag-nosis. *J. Lab. Clin. Med. 133*, 218–228, **1999**.

7 Fenske, J. D., Paulson, S. E. Human breath emissions of VOCs. *J. Air Waste Manag. Ass. 49*, 594–598, **1999**.

8 Grote, C., Pawliszyn, J. Solid-phase microex-traction for the analysis of human breath. *Anal. Chem. 69*, 587–596, **1997**.

9 Hansel, A., Jordan, A., Holzinger, R. Proton transfer reaction mass spectrometry: on-line trace gas analysis at the ppb level. *Int. J. Mass Spectrom. Ion Proc. 149*, 609–619, **1995**.

10 Jansson, B. O., Larsson, B. T. Analysis of or-ganic compounds in human breath by gas chromatography mass spectrometry. *J. Lab. Clin. Med. 74*, 961–966, **1969**.

11 Lindinger, W., Hansel, A., Jordan, A. On-line monitoring of volatile organic compounds at pptv levels by means of proton-transfer-reac-tion mass spectrometry (PTR-MS) medical ap-plications, food control and environmental re-search. *Int. J. Mass Spectrom. Ion Proc. 173*, 191–241, **1998**.

12 Manolis, A. The diagnostic potential of breath analysis. *Clin. Chem. 29*, 5–15, **1983**.

13 Ruzsanyi, V., Sielemann, S., Baumbach, J. I. Determination of VOCs in human breath using IMS. *Int. J. Ion Mobility Spectrom. 5*, 45–48, **2002**.

14 Stein, V. B., Narang, R. S., Wilson, L., Aldous, K. M. A simple, reliable method for the deter-mination of chlorinated volatile organics in human breath and air using glass sampling tubes. *J. Anal. Toxicol. 20*, 145–150, **1996**.

15 Pleil, J. D., Lindstrom, A. B. Exhaled human breath measurement method for assessing ex-posure to halogenated volatile organic com-pounds. *Clin. Chem. 43*, 723–730, **1997**.

16 Gordon, S. M., Szidon, J. P., Krotoszynski, B. K., Gibbons, R. D., O'Neill, H. J. Volatile or-ganic compounds in exhaled air from patients with lung cancer. *Clin. Chem. 31*, 1278–1282, **1985**.

17 Pleil, J. D., Lindstrom, A. B. Collection of a single alveolar exhaled breath for volatile or-ganic compounds analysis. *Am. J. Ind. Med. 28*, 109–121, **1995**.

18 Wilson, H. R. Breath analysis: physiological basis and sampling techniques. *Scand. J. Work Environ. Health 12*, 174–192, **1986**.

19 Li, F., Xie, Z., Schmidt, H., Sielemann, S., Baumbach, J. I. Ion mobility spectrometer for online monitoring of trace compounds. *Spec-trochim. Acta B 57*, 1563–1574, **2002**.

20 Spangler, G. E., Carrico, J. P., Campbell, D. N. Recent advances in ion mobility spectrometry for explosives vapor detection. *J. Test. Eval. 13*, 234–240, **1985**.

21 Baumbach, J. I., Eiceman, G. A. Ion mobility spectrometry: arriving on site and moving beyond a low profile. *Appl. Spectrosc. 53*, 338–354, **1999**.

22 Sielemann, S. Detektion flüchtiger organischer Verbindungen mittels Ionenmobilitätsspektrometrie und deren Kopplung mit Multi-Kapillar-Gas-Chromatographie. *Dissertation*, University of Dortmund, **1999**.

23 Sielemann, S., Baumbach, J. I., Pilzecker, P., Walendzik, G. Detection of *trans*-1,2-dichloroethene, trichloroethene and tetrachloroethene using multi-capillary columns coupled to ion mobility spectrometers with UV-ionization sources. *Int. J. Ion Mobility Spectrom. 2*, 15–21, **1999**.

24 Sielemann, S., Xie, Z., Schmidt, H., Baumbach, J. I. Determination of MTBE next to benzene, toluene and xylene within 90 s using GC/IMS with multi-capillary column. *Int. J. Ion Mobility Spectrom. 4*, 69–73, **2001**.

25 Xie, Z., Ruzsanyi, V., Sielemann, S., Schmidt, H., Baumbach, J. I. Determination of pentane, isoprene and acetone using HSCC-UV-IMS. *Int. J. Ion Mobility Spectrom. 4*, 88–91, **2001**.

26 Xie, Z., Sielemann, S., Schmidt, H., Li, F., Baumbach, J. I. Determination of acetone, 2-butanone, diethyl ketone and BTX using HSCC-UV-IMS. *Anal. Bioanal. Chem. 372*, 606–610, **2002**.

27 Baumbach, J. I., Sielemann, S., Xie, Z., Schmidt, H. Detection of the gasoline components methyl tert-butyl ether, benzene, toluene, and *m*-xylene using ion mobility spectrometers with a radioactive and UV ionization source. *Anal. Chem. 75*, 1483–1490, **2003**.

28 Eiceman, G. A., Karpas, Z. *Ion Mobility Spectrometry*. CRC Press, Boca Raton, FL, **1994**.

29 Eiceman, G. A., Leasure, C. S., Vandiver, V. J. Negative ion mobility spectrometry for selected inorganic pollutant gases and gas mixtures in air. *Anal. Chem. 58*, 76–80, **1986**.

30 Eiceman, G. A., Nazarov, E., Miller, R. A. A micro-machined ion mobility spectrometer–mass spectrometer. *Int. J. Ion Mobility Spectrom. 3*, 15–27, **2000**.

31 Eiceman, G. A., Nazarov, E., Miller, R. A., Krylov, E. V., Zapata, A. M. Micro-machined planar asymmetric ion mobility spectrometer as a gas chromatographic detector. *Analyst 127*, 466–471, **2002**.

32 Eiceman, G. A., Shoff, D. B., Harden, C. S., Snyder, A. P., Martinez, P. M., Fleischer, M. E., Watkins, M. L. Ion mobility spectrometry of halothane, enflurane, and isoflurane anesthetics in air and respired gases. *Anal. Chem. 61*, 1093–1099, **1989**.

33 Hill, C. A., Thomas, C. L. P. A pulsed corona discharge switchable high resolution ion mobility spectrometer–mass spectrometer. *Analyst 128*, 55–60, **2003**.

34 Hill, H. H., Eatherton, R. L. Ion mobility spectrometry after chromatography – accomplishments, goals, challenges. *J. Res. Natl Bureau Stand. 93*, 425–426, **1988**.

35 Hill, H. H., Jr., Siems, W. F., St. Louis, R. H., McMinn, D. G. Ion mobility spectrometry. *Anal. Chem. 62*, 1201A–1209A, **1990**.

36 St. Louis, R. H., Siems, W. F., Hill, H. H., Jr. Detection limits of an ion mobility detector after capillary gas chromatography. *J. Microcolumn Sep. 2*, 138–145, **1990**.

37 St. Louis, R. H., Hill, H. H., Jr. Ion mobility spectrometry in analytical chemistry. *Crit. Rev. Anal. Chem. 21*, 321–355, **1990**.

38 Vautz, W., Sielemann, S., Baumbach, J. I. Determination of terpenes in humid ambient air using ultraviolet ion mobility spectrometry. *Anal. Chim. Acta 513*, 393–399, **2004**.